# MASS SPECTROMETRY OF INORGANIC AND ORGANOMETALLIC COMPOUNDS

# Physical Inorganic Chemistry

*A collection of monographs edited by*

## M. F. LAPPERT

*The Chemical Laboratory, University of Sussex,*
*Falmer, Brighton, Sussex*

MONOGRAPH 2

535·336·2 : 543·51 : 661·8

## Rocket Propulsion Establishment
## Library

Please return this publication, or request renewal, before the last date stamped below.

| Name | Date |
|------|------|
| C. Place | 19/3/81 |
| | |
| | |
| | |
| | |
| | |
| | |
| | |

RPE Form 243 (revised 6/71)                739490

# Mass Spectrometry of Inorganic and Organometallic Compounds

M. R. LITZOW

*Union Carbide Corporation,*
*Cincinnati, Ohio, U.S.A.*

and

T. R. SPALDING

*Chemistry Department, The City University,*
*London, England*

ELSEVIER SCIENTIFIC PUBLISHING COMPANY

*Amsterdam - London - New York*

1973

ELSEVIER SCIENTIFIC PUBLISHING COMPANY
335 Jan van Galenstraat
P.O. Box 1270, Amsterdam, The Netherlands


AMERICAN ELSEVIER PUBLISHING COMPANY, INC.
52 Vanderbilt Avenue
New York, New York 10017

Library of Congress Card Number: 79-190678

ISBN 0-444-41047-3

With 77 illustrations and 161 tables

Printed in The Netherlands

# Preface

The mass spectral study of organometallic compounds and a large number of inorganic compounds has increased at an accelerating rate over the last decade. The usefulness of mass spectrometry in providing accurate molecular weights and formulae, and otherwise unobtainable data about gas phase properties of molecules and ions has come to be recognised by most inorganic and organometallic chemists. The extent of the recognition has proceeded to the point that by now most newly prepared, volatile, compounds are subjected to mass spectrometric investigation as part of the accepted routine of characterisation.

It is hoped that the first part of this book will enable chemists with little or no working knowledge of mass spectrometry to understand and appreciate the technique and its limitations. The second part of the book may be used as a guide to help the reader interpret his results in the context of what has already been reported up to the end of 1970 and beginning of 1971. Although a number of review articles have appeared (see, *e.g.*, "General Bibliography", p. 77), the authors feel that none are as comprehensive as the coverage given in the second half of this book.

M. R. LITZOW

T. R. SPALDING

# CONTENTS

*Chapter 7* (T. R. SPALDING)

*Chapter 12* (M. R. Lɪᴛᴢᴏᴡ)

PART 1

# Introduction

The first part of this book is concerned with the observation of mass spectra, the energetics of ionisation processes and the decomposition of ions, and the general treatment of mass spectral data. Although the basic instrumentation required for the recording of mass spectral data is described, no attempt has been made to discuss it in great detail. This is partly because the authors anticipate that most readers will be interested mainly in the interpretation of mass spectra after having their samples run for them, partly because we acknowledge that each mass spectrometer has its own design peculiarities, and partly because there are already several more specialised books on the subject. The features of the types of ions observed in mass spectra, namely parent molecular ions, fragment ions, multiply charged ions, rearrangement ions, ions formed by ion–molecule reactions, metastable ions and negative ions, are discussed. Since nearly all the reported spectra concern only positively charged ions our discussion of the types of ions observed will be predominantly in terms of positive ions.

The information obtainable from mass spectral data, which includes the identification of compounds from the determination of their molecular weight, structural information and bond dissociation energy data is discussed. A general bibliography referring to previous books dealing with instrumentation and mass spectra of organic compounds and reviews of mass spectra of organic compounds and reviews of mass spectra of inorganic and organometallic compounds is included at the end of Part 1.

*Chapter 1*

# Basic Instrumentation

M. R. LITZOW

The forerunner of the modern mass spectrometer is generally considered to be the instrument used by J. J. Thomson to study positive rays in the period 1910–1920. Many types have been developed and refined since, and a wide range of commercially produced instruments is now available. Mass spectrometry involves the production and analysis of ionic species which may be either mono- or polyatomic in nature and either positively or negatively charged. Positive ions are much more numerous however, and the vast majority of investigations is concerned with these.

In general the instruments consist of three major sections, the ion source, the mass analyser, and the detector. Vacuum conditions are maintained throughout. The gaseous ions are produced in the ion source; after acceleration from this source they are analysed according to their mass-to-charge ($m/e$) ratio, and finally they are collected and recorded. The intensity of the recorded signal is proportional to the abundance of the particular ionic species.

In the past, instruments were constructed primarily for either precise mass or ion abundance measurements. The former employed photographic plates for detection of the ions and were called mass spectrographs, while in the latter detection was accomplished electrically and these instruments were referred to as mass spectrometers. The term mass spectroscope may be used to refer to either type of instrument. These distinctions are now somewhat artificial, especially since both types of detection system are sometimes incorporated into the same instrument.

## 1. ION SOURCES

A large variety of ion sources has been employed. Some of these, such as the gaseous discharge source, are now mainly of historical interest only. When a choice may be made between various types of ion sources, this choice is dictated by the particular problem to be investigated. Some have very limited use. The nature of the sample and the kind of information sought must be considered. By far the most versatile and widely used is the electron-impact ion source. Others are generally employed only in particular investigations; the photoionisation source, for example, is especially suited to the determination of ionisation and appearance potentials. Some of the latest instruments incorporate a combined ion source with a capability of, for example, both electron-impact ionisation and field ionisation.

## 1.1 Electron-impact ion source

Briefly, a gaseous stream of sample molecules at a reduced pressure, generally of the order of $10^{-5}$–$10^{-6}$ torr, is directed so that collisions occur between molecules and a collimated beam of electrons. The electrons are emitted from an electrically heated filament and are coerced into a narrow beam by the combined efforts of an accelerating potential and a narrow slit. The filament in commercial instruments is usually of tungsten or of rhenium. Rhenium filaments have the distinct advantage that they operate at a somewhat lower temperature although their cost is higher and they tend to have a shorter life than do those of tungsten. Still lower temperatures are obtained with certain coated filaments. These lower operating temperatures are especially important in studies involving appearance potential measurements since the unwanted thermal energy of the electrons is lowered, and the Maxwellian spread of their energy is reduced (Chapter 3, Section 3). In addition, undesirable pyrolysis effects within the ion source may be eliminated, or at least reduced. Normal operating temperatures of an electron-impact source are 50–250 °C. The electron beam travels from the filament to a trap which is held at a positive potential of about 20 V relative to the ionisation chamber in order to prevent the escape of secondary electrons. A small magnetic field may be maintained along the filament–electron trap path as a further aid in preventing divergence of the collimated beam. Collisions occur between the electrons and a small fraction of the sample molecules resulting in the formation of ionic species if the energy of the electrons is sufficiently high. The ionisation processes which occur and the effect of the magnitude of the electron energy is discussed in Chapter 2, Section 2 and Chapter 3, Section 3.

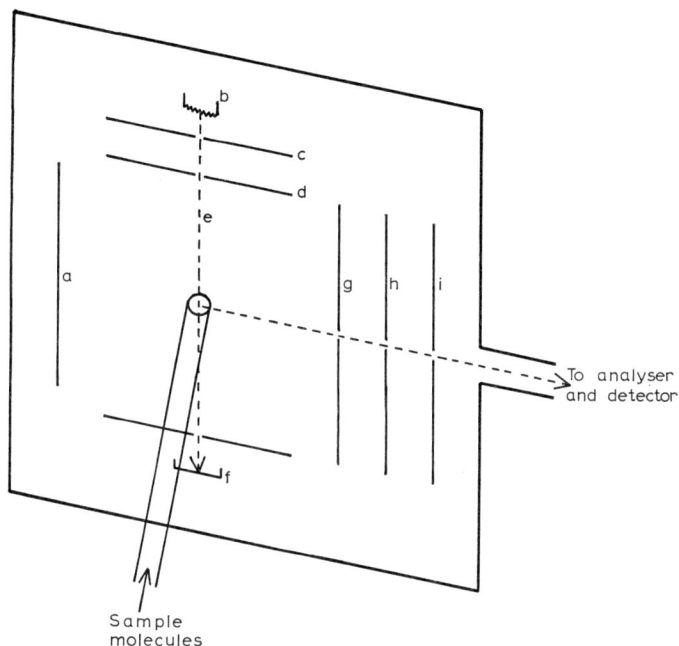

Fig. 1.1. Schematic representation of an electron-impact ion source.

A schematic diagram of a simple electron-impact source is shown in Fig. 1.1. The stream of gaseous sample molecules is directed to meet the beam of electrons, e, emitted from the heated filament, b at an angle of 90°. That part of the source where ionisation takes place (between a and g in Fig. 1.1) is usually referred to as the ionisation chamber.

Plates c and d serve two purposes. (*i*) Plate c is maintained at a small positive potential with respect to b and therefore attracts electrons emitted by the filament. The electrons are then accelerated by a potential drop between d and b usually of from 0 to 100 V and their actual energy is controlled by the magnitude of this potential difference which is very accurately maintained at the desired value. (*ii*) The slits in c and d are very small, resulting in a finely collimated beam of electrons; as mentioned previously, divergence of the electron beam may be considerably reduced by a small magnetic field along the direction bf. The electron trap f is maintained at a low positive potential with respect to the ionisation chamber in order to prevent the escape of secondary electrons. It is essential that the ionising beam current be held constant if reproducible ion intensities are to be obtained and this is accomplished by an electronic feedback system. Thus the electron trap current, which is dependent upon the number of electrons arriving, is maintained constant at a predetermined value by an automatic adjustment to the filament current as soon as any variation begins to take place.

Collision of the electrons with sample molecules results in the formation of positive and, to a lesser extent, negative ions. Plate a, usually referred to as the ion repeller, is maintained at a positive potential so that positive ions are repelled in the direction of the analyser tube whereas negative ions are attracted to it and dissipated. A very small potential difference between the repeller plate and g assists in drawing out the positive ions which are then acelerated by a large potential difference, usually of a few thousand volts, between h and i.

The probability of ionisation varies for different substances at any particular electron energy, but is normally a maximum in the range 50–100 electron volts (eV). The most common value employed in routine mass spectral analysis is 70 eV. This is discussed more fully in later sections.

## 1.2 Field-ionisation or field-emission ion source

Positive ions are produced from molecules and atoms by very high electric fields of the order of $10^8$ V/cm. Early workers used fine metal points to produce the field, but fine wires, sharp metal edges, and multiple thin layer films have also been used as field anodes.

Little fragmentation occurs with the field-ionisation source so that mass spectra obtained are much simpler than those obtained with an electron-impact source. This has obvious advantages with compounds which do not display a parent molecular ion under electron-impact conditions, and also in analysis of mixtures.

Commercial instruments are now available with combined electron-impact–field-ionisation sources.

## 1.3 Photoionisation source

Ionisation of the sample molecules at reduced pressure ($10^{-4}$–$10^{-6}$ torr) is accomplished in the photoionisation source by a beam of UV light. The fragmentation is again much less than with the electron-impact source, resulting in a simpler spectrum.

This source is particularly suited to the determination of ionisation and appearance potentials and in the study of the ionisation process itself.

Details of other, less common, ion sources used in the study of inorganic and organometallic compounds may be found in refs. 1–3.

## 2. MASS ANALYSERS

Once the ions have been formed in the ion source it is necessary to separate them into the individual species present. The various mass analysers which accomplish this are sometimes classified as either static or dynamic analysers. In the former, ions of a particular $m/e$ value are brought to a focus by steady fields whereas in dynamic mass analysers the time dependence of one or more parameters of the system, such as electric field strength, magnetic field strength, or ion movement, is fundamental to the mass analysis[4]. A number of seemingly unrelated analysers are thus included in this second category.

Mass spectrometers are generally classified according to the analyser system employed. The most common are briefly described here.

### 2.1 Single-focusing instruments

Suppose the positive ions formed in the ion source and initially at rest are accelerated by a potential drop **V**. For an ion of charge $e$ and mass $m$

$$eV = \tfrac{1}{2} mv^2 \tag{1.1}$$

where $v$ is the velocity of the ion, since the kinetic energy is equal to the work done on the ion. If the ion is then directed into a magnetic field perpendicular to the direction of motion of the ion, the ion experiences a force at right angles to both its direction and the direction of the magnetic field, and is therefore deflected. This principle was employed by Classen in 1907 in his study of the mass-to-charge ratio of electrons, and later by Dempster in his mass spectrometer which is represented schematically in Fig. 1.2. In this instrument, ions travel through a 180° arc under the influence of a magnetic field. The force on the ion, $Hev$, due to the magnetic field of strength $H$, must be balanced by the centrifugal force of the ion, so that

$$Hev = \frac{mv^2}{r} \tag{1.2}$$

where $r$ is the radius of the semicircular path travelled by an ion of mass $m$

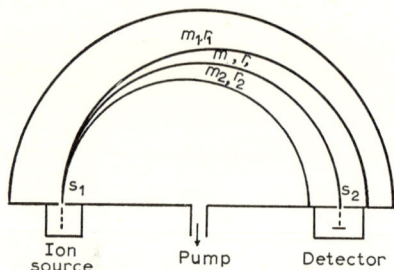

Fig. 1.2. Schematic diagram of Dempster's mass spectrometer.

which passes through the collector slit $S_2$ and arrives at the detector. From eqn. (1.2)

$$v = \frac{Her}{m}$$

Substituting for $v$ in eqn. (1.1)

$$eV = \tfrac{1}{2}m\left(\frac{Her}{m}\right)^2 \qquad\qquad (1.3)$$

$$m/e = \frac{H^2r^2}{2V} \qquad\qquad (1.4)$$

Since $r$ is fixed for a particular instrument, either $H$ or $V$ is adjusted experimentally, enabling the ratio $m/e$ to be determined*. Ions of different mass, for example $m_1$ and $m_2$, will travel arcs of different radii, $r_1$ and $r_2$, at fixed values of $H$ and $V$ (Fig. 1.2), but if one of these parameters is varied, both $r_1$ and $r_2$ may separately be made to equal the fixed radius of the instrument and focusing of each is in turn accomplished at the detector.

As illustrated in Fig. 1.3, the paths of ions of a particular $m/e$ value and the same velocity travelling in diverging directions at the entrance slit ($S_1$) to the magnetic field will come to a reasonable focus at the exit slit $S_2$. Ions of other $m/e$ values will come to a focus at other points, but can be swept across the exit slit by variation of either the magnetic field or the accelerating voltage. This makes use of the 'lens' action of the magnetic field; the instrument is said

---

* It should be noted that only the ratio $m/e$ can be determined; however, the majority of ions will carry a single positive charge. This is discussed in greater detail in later sections. In the following discussion it will be assumed that ions are singly charged.

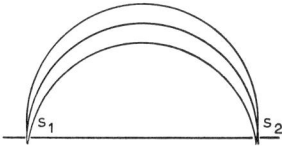

Fig. 1.3. Refocusing of divergent ions in a magnetic field.

to be direction focusing. For highest resolution, an essentially monoenergetic beam of electrons is required since the resolving power of the instrument will be lowered as the spread of the velocity, and hence energy, of the resultant ions is increased. Figure 1.3 also shows that focusing achieved cannot be perfect.

It was later shown that the focusing action of a 180° magnetic field just discussed was merely a special case of that of any wedge-shaped magnetic field which is uniform within the wedge's boundaries. If the wedge is considered to be a section of the semicircular field, then its apex must be the centre of curvature of the ion beam. This condition is met if an incident beam homogeneous in mass and energy meets the boundary of the magnetic field at right angles; it then leaves the other face of the field at right angles and is brought to a focus at a point which lies on an extension of the line joining the point of origin and the apex of the wedge (Fig. 1.4). A mechanical ad-

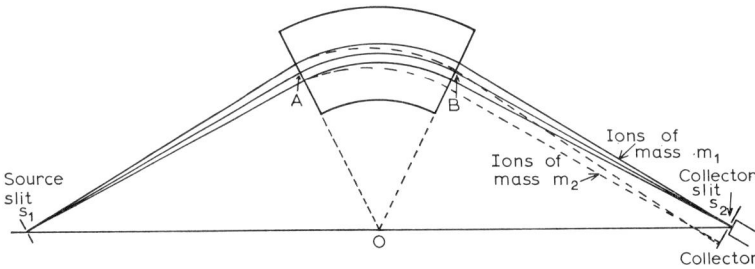

Fig. 1.4. Diagram of a sector magnetic field illustrating the trajectories and the focusing action of the field. By varying the strength of the magnetic field or the accelerating potential, ions of mass $m_2$ may be brought to a focus at the collector slit. Angles $S_1AO$ and $OBS_2$ are right angles.

justment of the magnet is usually made until optimum response of the detector is attained for ions of a given $m/e$. If perfect refocusing were obtained, the resolving power would be given by $m/\delta m = OA/(S_1 + S_2)$ where $S_1$ and $S_2$ are the widths of the two slits. It is found in practice, however, that the resolving power is lowered by unavoidable aberrations due to edge effects of the magnetic field. These are not encountered in the 180° instrument since the flight path is completely within the field. Instruments employing this type of magnetic field are referred to as sector-type mass spectrometers, and sector angles of 15°, 60°, and 90° have all been employed. The magnitude of this angle has no effect on the resolving power of the instrument.

### 2.2 Double-focusing instruments

The magnetic analyser is very dependent on the velocity spread, and hence the energy spread, of the ions of any particular mass entering the field. The maximum resolution that can be obtained is therefore limited by the velocity spread of the electrons leaving the source. Since an electrostatic field possesses velocity focusing properties, resolution can be considerably improved by incorporating an electrostatic analyser prior to the magnetic analyser.

It was stated previously that the kinetic energy of an ion leaving the source is equal to the work done on the ion

$$eV = \tfrac{1}{2} mv^2 \tag{1.1}$$

The force acting on the ion and due now to the electrostatic field must again be balanced by the centrifugal force of the ion

$$\frac{mv^2}{r_e} = eE \tag{1.5}$$

where $r_e$ is the average radius of curvature of the ions (Fig. 1.5) and $E$ is the electrostatic field existing at this mean radius. From eqns. (1.1) and (1.5) it can be seen that

$$r_e = \frac{2V}{E} \tag{1.6}$$

The radius of curvature of the ion is therefore dependent on its energy but not on its mass. By means of the electrostatic analyser one can select ions of the same mass which are homogeneous in energy. In addition, a lens effect occurs which is similar to that observed in a magnetic field. Ions with the same energy but with slightly diverging directions are brought to a focus by the electrostatic field (Fig. 1.5). If these ions are then passed through a

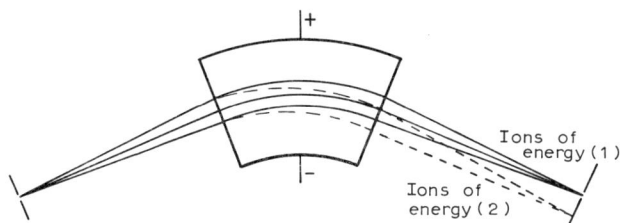

Fig. 1.5. Deflection of ions in an electrostatic field and focusing of the divergent ion beam.

magnetic analyser, the resolution obtained will be much greater than that obtained with the single focusing instrument.

Schematic diagrams of the analyser systems of the two main variants of double-focusing instruments are shown in Figs. 1.6 and 1.7. In the first, the Nier–Johnson type (Fig. 1.6), the resolved beams are brought to a focus at a slit (the exit slit) behind which is situated the detector. Thus one must vary either the accelerating voltage or the field of the magnet in order to record

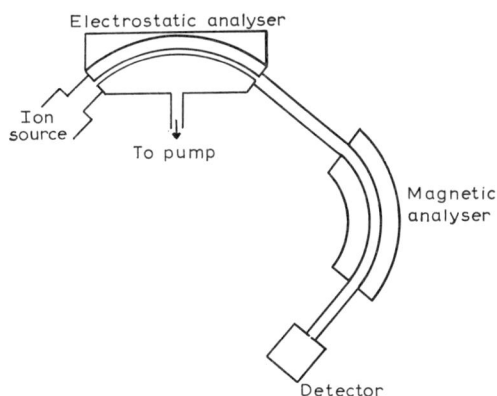

Fig. 1.6. Double-focusing mass spectrometer of the Nier–Johnson type.

successively the different ion beams. In the instrument of Mattauch–Herzog geometry (Fig. 1.7), the resolved ion beams come to a focus on a plane, so the detector is usually a photographic plate and all the ion beams are recorded simultaneously. However, the spectrum can be scanned across an exit slit. In most commercially available instruments, electronic detection is now available as an alternative to the photographic plate.

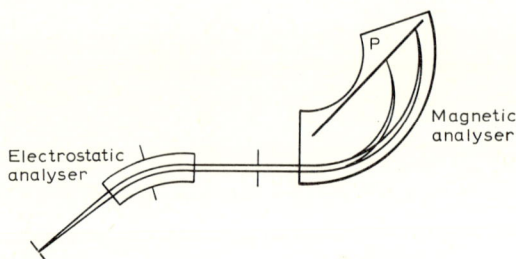

Fig. 1.7. Double-focusing mass spectrometer of the Mattauch–Herzog type. The photographic plate, P, is situated within the magnetic field.

## 2.3 The time-of-flight mass spectrometer

Analysis in this instrument is based on an entirely different principle. If all ions with the same charge are accelerated from the source with the same energy, then ions of different masses must leave the source with different velocities. The time taken to travel a linear path to the detector will therefore depend on the mass of the particular ion. The spectrum is a plot of the intensity *versus* the time of flight of the particle, this time being shorter for ions of lower mass. The differences in the times of arrival of ions of slightly different mass will be extremely small and so very fast electronics are required in the detector system. In addition, it is necessary to produce ions in bunches, the production of a subsequent bunch being delayed until analysis of the first is completed. This requires electronic pulsing techniques.

A typical linear, non-magnetic time-of-flight mass spectrometer is shown in Fig. 1.8. The electron beam, e, arises from the filament, a. At the start of each cycle the electron beam is allowed to pass into the ionisation region by

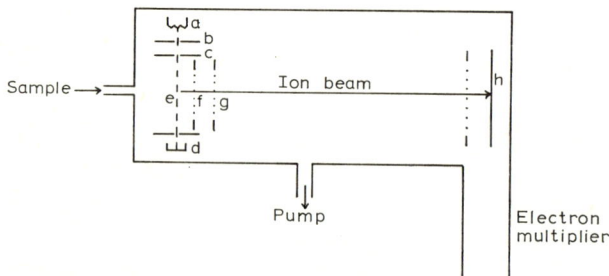

Fig. 1.8. Schematic diagram of a simple time-of-flight mass spectrometer.

the control grid, b, and the electron energy is controlled by the potential difference between b and c. The electron trap is represented by d. The ions are accelerated from the source into the field-free drift space between source and detector by grids f and g. The ions finally arrive at the detector h and the signals are amplified by an electron multiplier.

The mass range of such instruments is much lower than that of previously discussed mass spectrometers. Similar instruments employ either a magnetic or a radio-frequency field in separating ions of different mass.

## 2.4 Quadrupole mass analysers

The quadrupole analyser was developed in the early 1950's and has recently gained considerable popularity. Four cylindrical rods are arranged

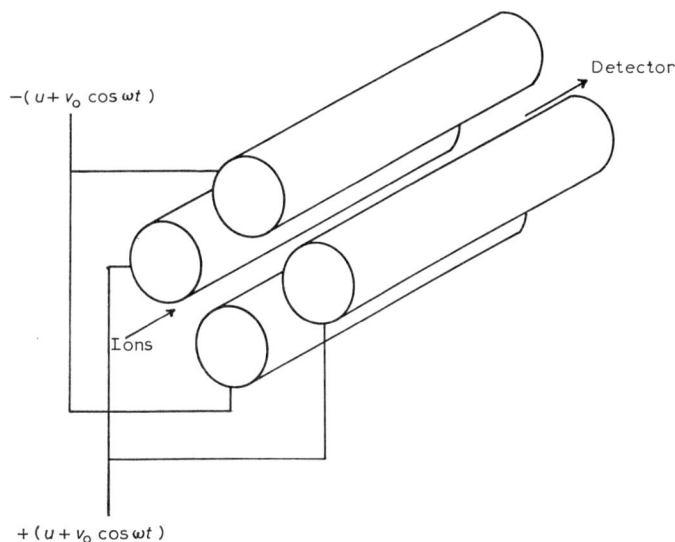

$-(u + v_0 \cos \omega t)$

Detector

Ions

$+(u + v_0 \cos \omega t)$

Fig. 1.9. Schematic representation of a quadrupole mass analyser.

so that their longitudinal axes define a parallelepiped (Fig. 1.9). Opposite rods are electrically connected and both d.c. and rf voltages applied so that the voltage on the positive electrodes is $+(U + V_0 \cos \omega t)$ and that on the negative electrodes is $-(U + V_0 \cos \omega t)$. The ions are directed down the central axis and mass scanning is accomplished by either varying the rf frequency, $\omega$, at constant voltages, or more frequently by varying both the rf and d.c. voltages while keeping both their ratio, $U/V_0$, and the frequency constant. For given values of these parameters, only ions with a specific $m/e$ value are able to move along the axis defined by the four rods and be collected. Ions of other masses are affected in a complicated manner and eventually lose their charges by coming into contact with one of the rods.

## 3. ION DETECTORS

The principal ion detectors used are (*i*) the photographic plate, (*ii*) the Faraday cup, combined with a d.c. amplifier or vibrating-reed electrometer, and (*iii*) the electron multiplier.

The photographic plate was the first ion detector used in mass spectrometry and its use in now confined principally to precise mass measurements. It is less convenient than electrical detectors for ion abundance measurements and their sensitivity is lower. However, distinct advantages are their ability to record, simultaneously, complete spectra at high resolution and, of course, the absence of complex electronics.

The electron multiplier is noted for extreme sensitivity and fast response. The bombardment of the ion cathode (conversion dynode) by the highly energetic ions results in secondary electron emission. The resultant electron beam is amplified in a number of steps involving bombardment of successive dynodes. Many commercial instruments now feature 'combination' detectors systems incorporating the Faraday cup and the electron multiplier detectors.

Detailed descriptions of these various detector systems are beyond the scope of this text.

## 4. SAMPLE INLET SYSTEMS

Sample introduction is dependent upon the type of ion source employed. In the spark source, for example, the solid material to be analysed acts as the electrodes, while in thermal ionisation sources, vaporisation and ionisation of the sample take place in a single process. As discussed previously, the electron-impact source requires a beam of gaseous sample molecule and the sample inlet system must be capable of providing this. If relative abundances of ions are to be obtained with any degree of accurracy, the sample pressure in the ion source must be kept constant during the time taken to record the complete mass spectrum. This does not present any problem with gaseous samples or with volatile liquid samples unless the amount of sample available is very small.

*Gases.* A storage reservoir of 1–5 l is connected to the ion source by way of a slow controlled leak which usually consists of either a porous material or of pinholes in thin gold foil. Thus it is possible to provide the low sample pressure required in the ion source and to maintain this sample pressure constant during the time necessary to record the mass spectrum. The pressure in the reservoir is of the order of $10^{-2}$ torr, with the source pressure $10^{-5}$–$10^{-6}$ torr.

*Liquids.* With many compounds whose vapour pressure is too low to allow diffusion through a leak and into the ion source at room temperature, a heated reservoir capable of being maintained at any temperature up to about 300 °C is often used. Introduction of the liquid sample into the reservoir may be achieved in a number of ways such as (*i*) injection through a septum, (*ii*) evaporation through a sinter separated from the atmosphere by warm gallium (the end of a micro pipette or capillary tubing containing the sample is touched to the sintered surface below the warm gallium), (*iii*) by means of a sample holder which can be frozen for evacuation of air and then heated. The inlets are represented diagrammatically in Fig. 1.10. The tube connecting the reservoir to the source, as well as the source itself, is maintained at the same temperature as the reservoir to prevent condensation. A controlled leak is again incorporated between reservoir and source. The reservoir may be of glass, but more usually consists of metal which is lined with glass or enamel; a heated metal surface may cause catalytic decomposition. Care should be taken that the chemical composition of all or part of the sample is not altered as a result of being maintained at the temperature of the heated inlet system (see discussion in Chapter 3, Section 1).

*Solids.* Highly volatile solids may be introduced by means of the heated inlet system but most of them, and also many high boiling liquids, have vapour pressures too low to allow this. The spectra of the majority of these can be recorded, however, by means of some form of direct insertion probe. The sample is admitted directly into the ionisation chamber on the end of the probe which may be either heated electrically, or heated indirectly by the ion

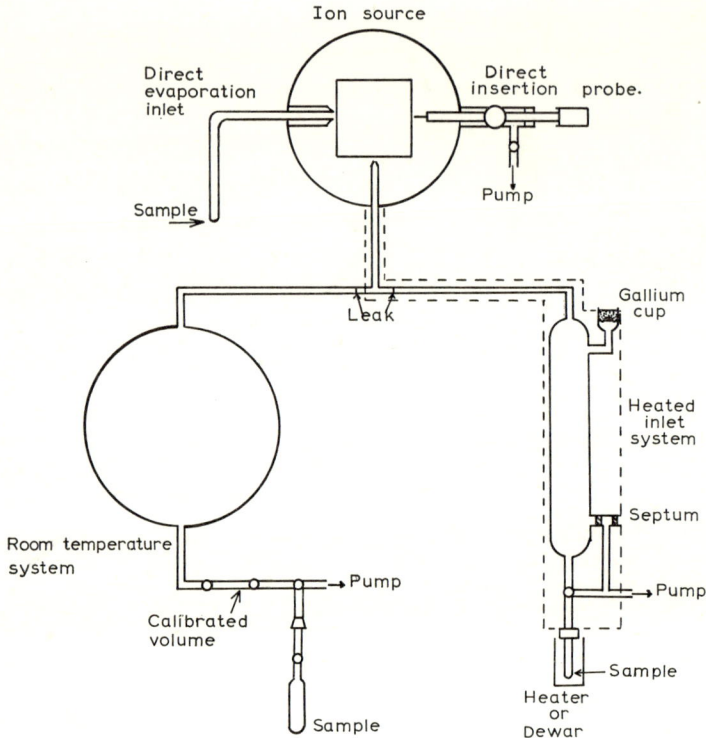

Fig. 1.10. Representation of various inlet systems.

source. The probe is introduced into the source by means of a vacuum lock.

Since no reservoir is involved and there is no pressure decrease across a leak, the amount of sample required for the direct insertion probe may be much lower than that for the previously discussed inlet systems. Low-temperature probes have been developed, allowing direct insertion of compounds which are highly volatile at normal operating temperatures.

A system allowing direct evaporation into the ion source may also be employed. Such a system may be especially favourable for air-sensitive compounds of low volatility or of low thermal stability. In the latter case the rate of evaporation into the source may be controlled by a constant temperature bath.

*Knudsen cell.* Mass spectrometry has proved to be extremely convenient for high-temperature studies. Thus it has been possible to identify many previously unknown molecular species and to measure relevant thermo-dynamic data such as heats of formation, heats of dimerisation, dissociation energies etc. A Knudsen cell, usually constructed of molydenum, tantalum or tungsten, is attached directly to the ion source and the sample under investigation is heated within it. Molecules of the vapour in thermodynamic equilibrium with the condensed vapour phase diffuse from a cell through a small aperture into the ionisation chamber. Studies up to temperatures of 2300 °C have been carried out.

Greater detail of the basic components of the mass spectrometer may be obtained from refs. 1–3.

REFERENCES

1 J. H. Beynon, *Mass Spectrometry and its Application to Organic Chemistry*, Elsevier, Amsterdam, 1960.
2 R. W. Kiser, *Introduction to Mass Spectrometry and its Applications*, Prentice-Hall, Englewood Cliffs, N.J., 1965.
3 J. Roboz, *Introduction to Mass Spectrometry: Instruments and Techniques*, Interscience, New York, 1968.
4 E. W. Blauth, *Dynamic Mass Spectrometers*, Elsevier, Amsterdam, 1966.

*Chapter 2*

# Mass Spectra and Types of Ions Observed

M. R. LITZOW

## 1. THE MASS SPECTRUM

Mass spectrometry is becoming increasingly important in the characterisation of new compounds. Even workers who have no intention of carrying out detailed studies with such an instrument need to be familiar with the information which may be readily obtained from a mass spectrum recorded at low to medium resolution. Mass spectrometers are becoming standard equipment in major research laboratories and with the improvements in instrumentation allowing compounds of low volatility to be studied, the gap between the large volume of available information on mass spectral behaviour of organic compounds and that available on inorganic and organometallic compounds is closing rapidly.

In Sections 2 and 3 of this chapter, ionisation of molecules and subsequent decomposition to produce fragment ions are discussed. First of all, however, the general format of a mass spectrum should be considered.

It was shown in Chapter 1, Section 2.1 that the ratio $m/e$, can be determined for a particular ion observed in an instrument with a magnetic deflection analyser if both the magnetic field strength $H$ and the ion accelerating voltage $V$ are known (eqn. (1.4)). In practice, a continuous scanning process is employed so that ions of either increasing or decreasing $m/e$ values are successively brought to focus at the detector. For each species a signal is produced due to a flow of electrons and is usually referred to as the ion current. The intensity of this signal is directly proportional to the number of ions impinging on the detector. The mass spectrum then consists of a plot of ion current *versus* $m/e$ (Fig. 2.1). The mass-to-charge ratios for the various ions in a spectrum could be determined by accurately monitoring $H$ or $V$, whichever is being varied, during the complete scan. Several commercial

*References pp. 48–50*

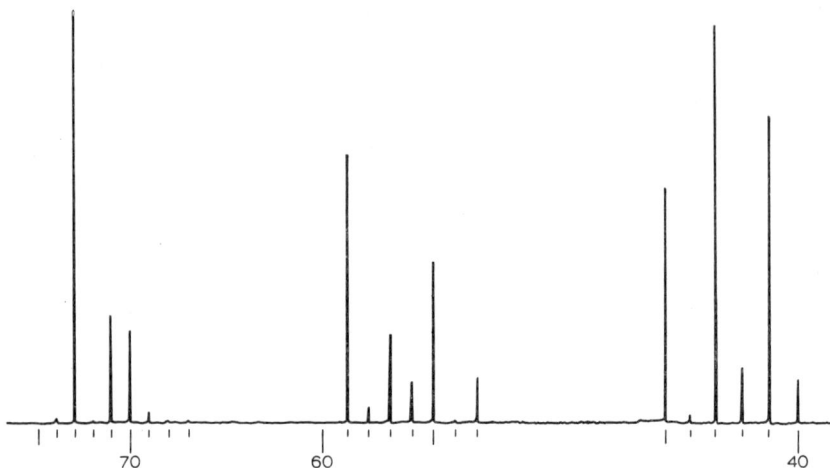

Fig. 2.1. Small portion of a mass spectrum; plot of ion current *versus* mass-to-charge ratio ($m/e$).

mass spectrometers do feature "mass markers" such as, for example, those based on a high precision measurement of the magnetic field by making use of the Hall effect. However, $m/e$ values can usually be determined without the assistance of such mass markers, but their convenience and time saving are strong arguments in their favour to counterbalance the extra cost involved.

In most cases mass markers are not available. Ions formed from residual amounts of water and air normally present in the instrument serve as calibrants at the lower end of the mass scale, and peaks at $m/e$ 18, 28 and 32 due to $H_2O^+$, $N_2^+$ and $O_2^+$ respectively can be readily identified *. Counting from these, it is possible to determine the $m/e$ values for the remaining peaks in the spectrum. Where peaks occur at every integral $m/e$ value, assignment is exceedingly simple. Unambiguous assignment is still relatively easy where small "gaps" in the spectrum occur; by taking into account the distances between adjacent peaks on each side of the "gap", the number of $m/e$ values not represented by peaks can be estimated. When large peak-free regions occur, calibration can be effected by recording the spectrum of the sample

---

* These and other peaks due mainly to compounds desorbing from the walls of the instrument are referred to as the "background" in the instrument. Such a "background spectrum" is usually recorded, therefore, before the sample is introduced.

compound alone and in the presence of a reference material whose spectrum has been established. Perfluorohydrocarbons are usually employed as calibrants because of their relatively simple spectra and suitable volatility, even for high molecular weight members. It is most important that masses are correctly assigned in the spectrum.

Because the ions produced are analysed according to their mass-to-charge ratio, a doubly charged ion of mass $m$ gives rise to the same peak in the mass spectrum as a singly charged ion of mass $m/2$. No distinction between ions of $m/e$ ratio 20/1 and 40/2, for example, would therefore be observed in the spectrum. Where $m$ is an odd number, however, the doubly charged ion would occur at a non-integral $m/e$ value, or "half mass number", and would be easily detected. This often occurs since many elements have more than one isotope; the existence of one such isotope with an odd mass number would give rise to this situation. An example where doubly charged ions were easily identified is shown in Fig. 2.2. Under normal operating conditions most of

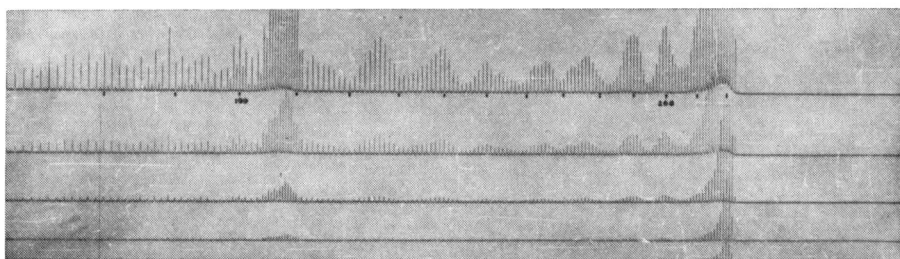

Fig. 2.2. Portion of the mass spectrum of 1-phenyl-1,2-dicarba-*closo*-dodecarborane(12). The five galvanometer tracings differ in sensitivity thus allowing widely varying ion intensities to be determined.

the ions produced in a mass spectrometer are singly charged species. Doubly charged ions are much less frequently observed, while ions carrying higher charges are not encountered in significant concentration. Therefore since $e$ has a value of one in the vast majority of cases, the ion beam from the source is separated according to the respective masses of the ions.

Ions formed in low abundance may be just as interesting or important as the very prominent ones in a spectrum, and an accurate determination of their abundance is therefore desirable. Various attenuation procedures have been employed to allow this. With the development of the recording oscillograph the accurate determination of these abundances has been greatly facilitated. A number of channels, usually ranging from 2 to 6, are employed so that, in effect, several spectra of differing sensitivity are recorded simultaneously. An example is shown in Fig. 2.2.

Spectra obtained from instruments which do not employ magnetic analysers are similar. With quadrupole mass spectrometers, for example, the continuous scan may be realised by variation of both the rf and d.c. voltages.

The spectrum as obtained from the mass spectrometer is invariably quite lengthy. For convenience, therefore, it is normally converted to a bar graph representation. In such a representation the intensity is usually expressed as the relative abundance or in terms of total ionisation units. The former is the percent intensity relative to that of the most intense ion in the spectrum. The peak corresponding to this ion is often called the base peak. Total ionisation is defined as the sum of all of the ion intensities in the spectrum of a compound multiplied by the sensitivity (ion current per unit of pressure) of the most intense peak in the spectrum. To express the intensity of an ion in total ionisation units, therefore, its ion current is divided by the total ionisation, which is characteristic of the particular instrument. A compound such as $n$-butane is often taken as a standard and then the total ionisation of any compound relative to that of the standard can be obtained. This relative value should be independent of the instrument employed. In spite of this, however, most mass spectra are reported in terms of relative abundance since this is much simpler and less tedious. Figure 2.3 illustrates a typical bar graph.

The intensity of a peak is sometimes also expressed as a percentage of the sum of the intensities of all peaks from a particular mass number, $m$, to the parent molecular ion. This is written as $x\%\Sigma_m$.

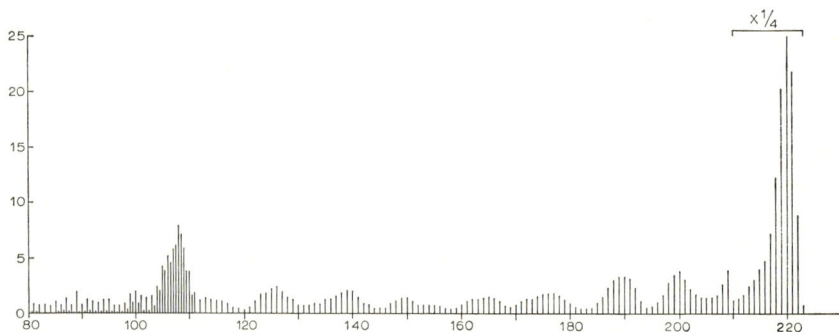

Fig. 2.3. Bar graph representation of the mass spectrum shown in Fig. 2.2.

## 1.1 Resolving power or resolution

It may be said that two adjacent peaks are "just resolved" when it becomes possible to determine that there are two components present. Ions with very similar masses, for example $^{10}BH^+$ and $^{11}B^+$, will be observed as a single peak in a low-resolution spectrum. In high-resolution instruments, no difficulty would be experienced in resolving the peaks due to these two ions with masses of 11.0209 and 11.0093 respectively. The relative abundances of the two could then be determined. Even with medium-resolution instruments it may be possible to show that both of these ions are present. Norman et al.[1] were able to resolve a number of such multiplets in the spectrum of tetraborane-10. Similarly, Baylis et al.[2] were able to obtain an accurate value for the relative abundance of the low intensity $^{10}B_3H_6^+$ ion in spite of the presence of background peaks due to $H^{35}Cl^+$ and $^{12}C_3^+$ all at nominal mass 36.

To obtain a meaningful description of the resolution obtained, a more quantitative basis must be employed. Most commonly used is the "% valley" definition. Thus two peaks of masses $M$ and $M + \delta M$ and of equal intensity can be considered to be resolved when $\Delta H/H$ (Fig. 2.4) is less than or

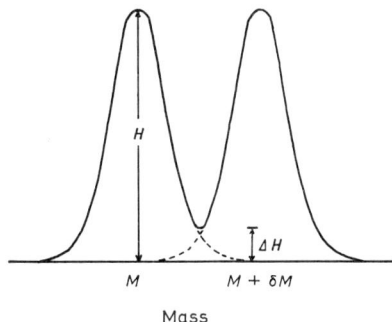

Fig. 2.4. Measurements necessary to calculate resolution from the "% valley" definition.

equal to some arbitrary value. This is normally 0.1 (10%) although other values such as 0.5, 0.05, 0.01, and even 0.001 have been used. When $\Delta H/H$ just equals this value, the resolving power is given by the corresponding $M/\delta M$. It can be seen therefore, that a whole range of doublets with the two components having equal intensities would be needed for direct measurement of resolving power using the above definition. It becomes necessary to determine this characteristic of an instrument in terms of the shape of a

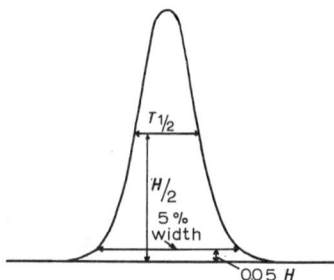

Fig. 2.5. Measurements necessary to calculate resolution from the "5% peak width" definition.

single peak. The width of a peak at a distance of 0.05 of its height above the base line (Fig. 2.5) is equivalent to $\Delta H/H = 0.1$ for two equal peaks, assuming all curves are Gaussian, and is also equal to 2.08 times the peak width at half height of the peak (2.08 $T_{\frac{1}{2}}$). Either the "5% peak width" or the $2T_{\frac{1}{2}}$ value (rather than 2.08 $T_{\frac{1}{2}}$) is often adopted, therefore, as a definition of resolving power.

## 2. THE PARENT MOLECULAR ION

It is interesting and informative to observe the changes which occur as a sample is bombarded with an electron beam of gradually increasing energy. With the electron energy below the ionisation potential of the compound, no positive ions whatever will be produced. When the electron energy equals the ionisation potential, usually in the region 7–15 eV, a bombarding electron may result in the formation of a parent molecular ion according to the equation

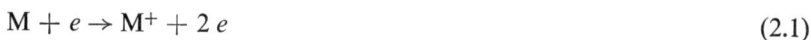

$$M + e \rightarrow M^+ + 2e \qquad (2.1)$$

Since all of the electron's energy must be transmitted to the molecule to cause this ionisation, however, the probability that this process will occur is very small and the intensity of the parent molecular ion will be exceedingly low. As the energy of the electron beam is increased further, the probability that ionisation will occur must also increase, with a resultant increase in the intensity of the parent molecular ion.

At still higher electron energies, sufficient energy may be imparted to the molecule to cause bond dissociation to occur with the resultant formation of the fragment ion. Its probability of formation will now rise with increasing electron energy, similar to the parent molecular ion case. Of course the number of fragmentation processes will increase with further increases of electron energy, and doubly charged ions will also begin to appear. The abundances of these fragment ions will increase relative to the abundance of the parent molecular ion. In some cases the intensity of the parent molecular ion may actually fall as fragmentation increases, and may not be observed at all at higher electron energies. Once the energy of the bombarding electrons has reached 40–50 eV, the relative intensities of ions are more or less unaffected by small changes in this energy. Most spectra, therefore, are recorded at either 50 or 70 eV.

Figure 2.6 serves to illustrate the previous discussion; the spectrum of chromium hexacarbonyl is shown at various energies of the bombarding electrons. In (k)–(t) the plots of ion current $vs.$ $m/e$ emphasize the increase in the number of ions formed as the electron energy increases, while it can be seen in (a)–(j) that when the spectrum is represented in the normal way, plotting relative intensity against $m/e$, little change occurs at energies higher than 40–50 eV. Kiser and co-workers have utilised clastograms (see, for example, ref. 3) which serve to illustrate this point more clearly; the intensity of each ion is represented as a proportion of the total ion current and is plotted against the electron energy. Those corresponding to the spectra of Fig. 2.6 are shown in Fig. 2.7. From a similar plot in conjunction with appearance potential measurements, Winters and Kiser[3] postulated that the principal ions formed from the transition metal Group VI hexacarbonyls resulted from successive losses of CO groups $viz.$

$$M(CO)_6{}^+ \rightarrow M(CO)_5{}^+ \rightarrow \ldots \rightarrow MCO^+ \rightarrow M^+ \qquad (2.2)$$

The ion formed in the process represented by reaction (2.1) is referred to here as the parent molecular ion although it is often called either the parent ion or the molecular ion. Both of these terms have sometimes tended to have other connotations associated with them, however, and confusion may result. Some workers have used the term parent ion to described any ion which fragments to produce an ion of lower mass; such parent ions may have themselves been formed by a fragmentation processes. During fragmentation, ions corresponding to the removal of an electron from a neutral

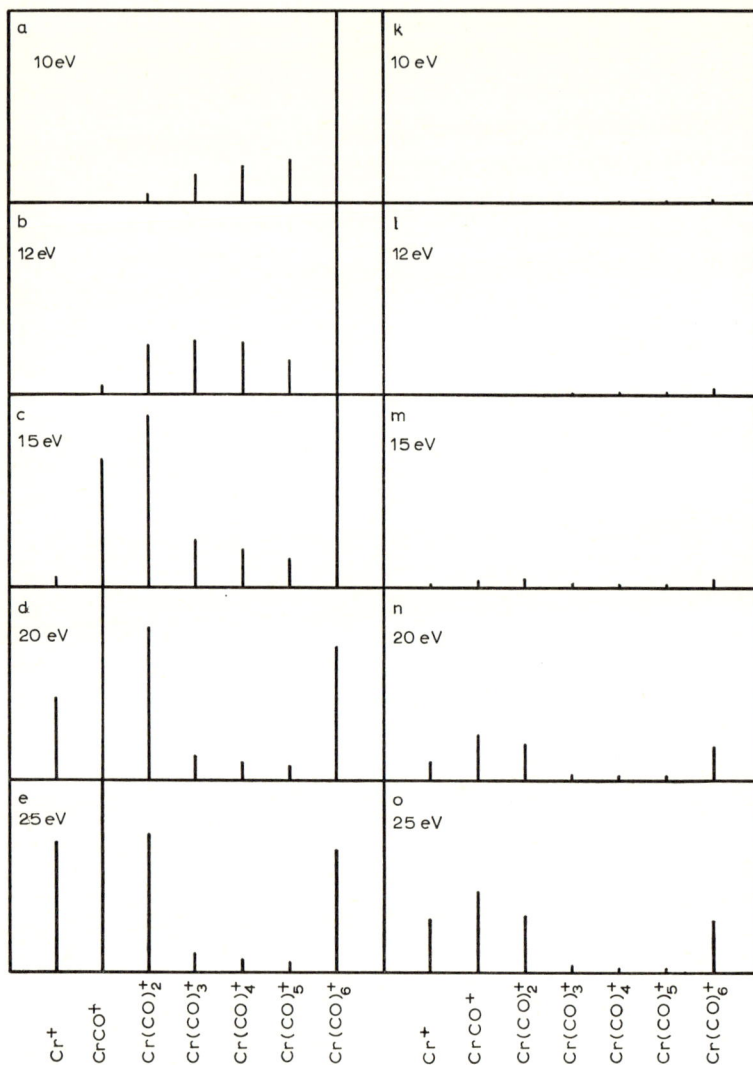

Fig. 2.6. Monoisotopic mass spectrum of $Cr(CO)_6$ at various energies of the bombarding electrons, $Cr(CO)_x^+$ ions only, where $x = 0$–6. In (a)–(j), absolute ion intensities are used (arbitrary units) while in (k)–(t), relative intensities are employed.

molecule may be formed, often by a process involving rearrangement. Such ions are called molecular or molecule ions (reactions (2.3)–(2.5)). Reed[4] has suggested that such ions be called "daughter molecular ions". To avoid

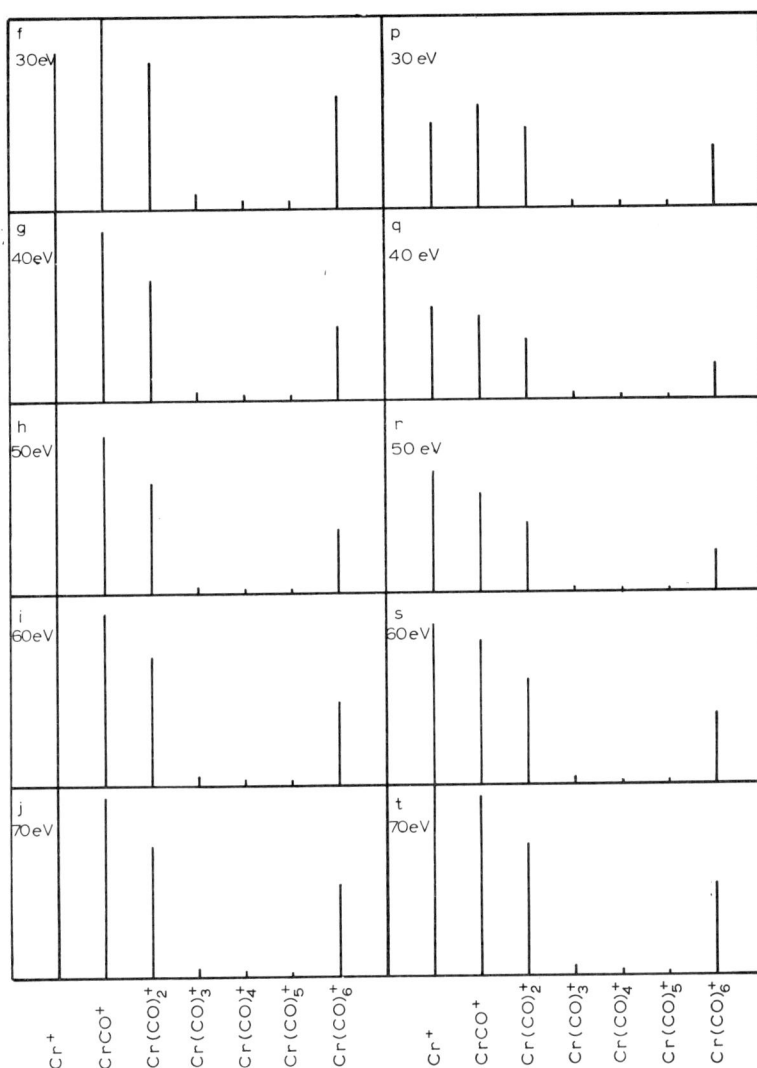

ambiguity, there fore, the term parent molecular ion as used by Reed will be used here to describe the ion formed by the removal of a single electron from a molecule of the compound admitted to the ion source.

References pp. 48–50

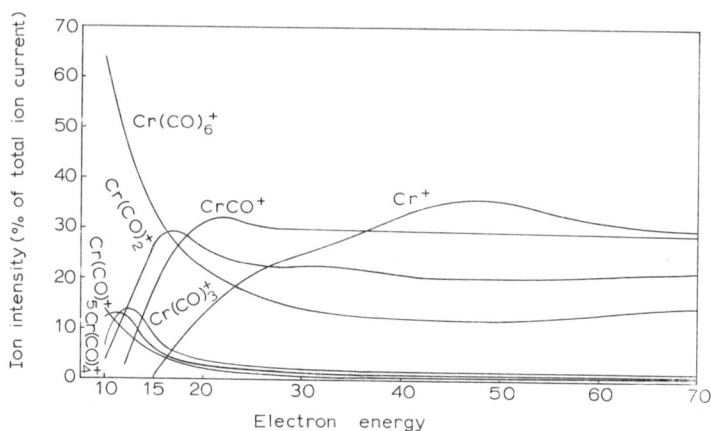

Fig. 2.7. Clastograms of ions of type $Cr(CO)_x^+$ formed from $Cr(CO)_6$.

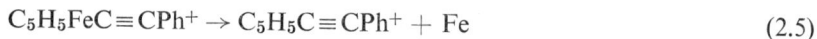

$$B(OR)_3^+ \rightarrow B(OH)(OR)_2^+ + R\text{-}H \qquad\qquad (2.3)$$

$$C_6H_6Cr^+ \rightarrow C_6H_6^+ + Cr \qquad\qquad (2.4)$$

$$C_5H_5FeC \equiv CPh^+ \rightarrow C_5H_5C \equiv CPh^+ + Fe \qquad\qquad (2.5)$$

In general, a direct relationship exists between the stability of a parent molecular ion and its relative abundance compared with the relative abundance of fragment ions formed from it. In some cases, such as the interhalogen compounds $IF_7$ (ref. 5) and $BrF_5$ (ref. 6), the intensity of the parent molecular ion may be too small to be detected. This occurs when the rate of decomposition is so high that all of these ions have decomposed in the time elapsed between formation and collection of ions—of the order of $10^{-5}$ sec in most instruments. Fragmentation of an ion occurs when it possesses sufficient vibrational and excitational energy, transmitted to it from the bombarding electrons, to cause the rupture of bonds. It may be possible to observe unstable parent molecular ions, therefore, by decreasing the energy of the electron beam, perhaps to a value only slightly above the ionisation potential of the molecule. The spectrum will become simpler with such a decrease, with fewer fragment ions being formed. It must be remembered, however, that an overall decrease in sensitivity will accompany this decrease in electron energy. Although fewer parent molecular ions will undergo fragmentation, fewer will actually be formed, but they will account for a larger percentage of the total ionisation (Fig. 2.6 (a)–(d), (k)–(n)). Thus the abundance of all ions decreases markedly as the electron energy falls, but the relative intensity of the parent molecular ion is considerably enhanced.

The abundance of the parent molecular ion may be critically dependent on the temperature of the ionisation chamber. The number of molecules in the source will decrease as the temperature rises, resulting in a slight decrease in the intensities of all ions in the spectrum[7]. Of greater importance, however, is that higher temperatures often result in an increased rate of decomposition of the parent molecular ions, so that the number reaching the collector is decreased. The relative abundance of fragment ions will therefore increase. Pyrolysis of a sample may also occur in the inlet system. Thus a parent molecular ion of tetraborane-10 has not been observed in a conventional mass spectrometer. However, this ion ($B_4H_{10}^+$) has been detected by Norman et al.[8] who employed an inlet system with variable temperature conduits (total contact time of sample ca. 0.5 sec) and a "zero source contact" mass spectrometer, in which molecules coming in contact with the ion source walls were rejected. Pignataro and Lossing[9] found that iron pentacarbonyl decomposed in the mass spectrometer at low temperatures giving mainly $Fe(CO)_4$ and CO; as the temperature was raised, the ratio $Fe(CO)_4^+/Fe(CO)_5^+$ increased. Pyrolysis of many inorganic and organometallic compounds has occurred in mass spectral investigations and is discussed in greater detail in Chapter 3, Section 1.

The absence of a parent molecular ion may therefore be attributed either to (i) thermal decomposition prior to electron impact, or (ii) fast unimolecular decomposition after electron impact so that all parent molecular ions decompose in a time less than that required to reach the detector.

## 3. FRAGMENT IONS

The fragmentation produced as the energy of the electron beam is increased above the ionisation potential of a compound has been discussed in the previous section. It is important, however, to realise the wide range of fragment ions which may be observed. Dissociation of the parent molecular ion may occur by several different routes. In addition, fragment ions so produced may themselves undergo further cleavage; the resultant spectrum recorded at 50–70 eV may therefore contain an extremely large number of peaks. The processes discussed above may be represented by the following, depicting the possible fragmentation paths of a hypothetical molecule ABCD.

(I)    $ABCD^+ \longrightarrow A^+ + BCD$

(II)   $ABCD^+ \longrightarrow A^. + BCD^+$

$\qquad\qquad\qquad B^+ + CD^.\qquad B^. + CD^+$                               (2.6)

$\qquad\qquad\qquad\qquad\quad C^. + D^+\qquad C^+ + D^.$

(III)  $ABCD^+ \longrightarrow ABC^. + D^+$

(IV)   $ABCD^+ \longrightarrow ABC^+ + D^.$  etc.

(V)    $ABCD^+ \longrightarrow AB^. + CD^+$

$\qquad\qquad\qquad\qquad\quad C^. + D^+\qquad C^+ + D^.$

(VI)   $ABCD^+ \longrightarrow AB^+ + CD^.$  etc.

In each case fragmentation results in the formation of (*i*) a positively charged species, and (*ii*) a neutral species. The latter may be either a radical or a neutral molecule.

By straightforward cleavage processes, therefore, it is possible for the parent molecular ion $ABCD^+$ to yield nine fragment ions. In addition, somewhat less probable multi-charged (Section 4) and rearrangment ions (Section 5) may appear.

The number of ions in a spectrum, their relative abundance, and their composition are characteristic of the molecule from which they arise. In general, they reflect both the relative stability of the various bonds (*i.e.* bond dissociation energies) and the stabilities of the neutral fragments. An approximate correlation can usually be drawn between the intensity of various ions formed by dissociation of a single bond and the strength of that bond, in that a higher intensity suggests a lower bond dissociation energy. However, care must always be taken when making predictions concerning bond dissociation energies from ion intensities. The relative intensities of the various ions in the spectra of the three compounds $Mn_2(CO)_{10}$, $MnRe(CO)_{10}$, and $Re_2(CO)_{10}$, for example, indicated that the order of stability of the metal–metal bonds was Mn–Mn $<$ Mn–Re $<$ Re–Re[10–12]. Appearance potential measurements, however, indicated that the bond dissociation energies of these bonds were in the order Mn–Mn $<$ Re–Re $<$ Mn–Re[12].

Even a relatively simple molecule, therefore, may produce a spectrum depicting a very large number of ions (Fig. 2.8) and so it is highly characteristic of that compound. It may appear that the complexity of the spectrum only serves to complicate the assignment of structure. If one also considers

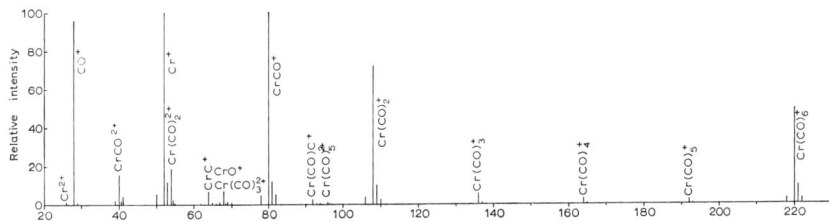

Fig. 2.8. Mass spectrum of $Cr(CO)_6$ recorded at 70eV.

the relative intensity of these fragment ions, however, and makes comparisons with the spectra of similar compounds of known structure, a much clearer picture emerges.

## 4. MULTIPLY CHARGED IONS

The processes described previously have depicted the formation of ions carrying a single positive charge. Although these account for the vast majority of ions occurring in the mass spectrometer, multiply charged ions may also be formed and observed.

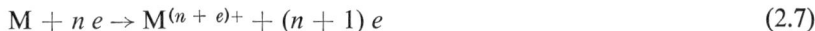

$$M + n e \rightarrow M^{(n + e)+} + (n + 1) e \qquad (2.7)$$

Multiply charged monatomic ions such as those of mercury and the rare gases were recognized very early[13] and these have been closely studied. Subsequent observation of multiply charged polyatomic species followed. Despite low abundances in most cases, detection and recognition is facilitated in many instances by the appearance of peaks in the spectrum at non-integral mass numbers, since the spectrum is a plot of ion current against the mass-to-charge ratio, $m/e$. Consider doubly charged ions. Because of the prevalence of elements with more than one isotope, most inorganic compounds will have some molecules whose isotopic composition is responsible for those molecules having an odd mass number. Such species carrying a double positive charge will be observed at half mass numbers. Examples are show in Figs. 2.2 and 2.8.

Numerous doubly charged parent moecular ions have been observed and in addition the triply charged ions $COS^{3+}$, $CS_2^{3+}$, and $C_2N_2^{3+}$ have been detected[14] in very low abundance. Triply charged ions have been observed in the mass spectra of phosphonitriles[15]. The intensities of peaks due to doubly charged ions in the mass spectra of organic compounds rarely exceed 5% of the base peak intensity[16,17]. Aromatic or heteroaromatic molecules which do not contain bonds that are easily ruptured, however, have been found to possess an increased ability to sustain two positive charges[17]. This also appears to be true for certain aminoboranes. Dewar and Rona[18] reported that in the mass spectra of the three compounds tris(dimethylamino)borane (I), bis(dimethylamino)phenylborane (II), and 1,3-dimethyl-2-phenyl-1,3,2-diazoborolidine (III), the relative abundances of the doubly charged parent molecular ions were quite high. Since these ions

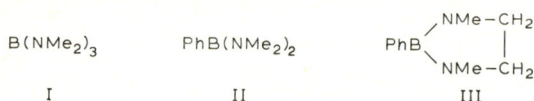

$$B(NMe_2)_3 \qquad PhB(NMe_2)_2 \qquad PhB\overset{NMe-CH_2}{\underset{NMe-CH_2}{\big\backslash}}$$

$$\text{I} \qquad\qquad\qquad \text{II} \qquad\qquad\qquad \text{III}$$

were also formed at relatively low ionisation potentials, it is clear that they are unusually stable. Doubly charged ions observed in the spectra of boron hydrides, on the other hand, have been present in very low abundance.

The doubly charged parent molecular ion formed from bis($\pi$-indenyl) iron was reported to have a relative intensity of 6%[19,20].

Recent reports [21-23] have noted some exceptional cases in which the doubly charged parent molecular ion was more abundant than the corresponding singly charged species. This can only occur where the doubly charged ion is much more stable than its singly charged counterpart and therefore decomposes much more slowly.

Doubly charged parent molecular ions may fragment to lower mass species in the same way that their singly charged analogs do. However, the two charges may either remain with one of the resulting fragments or else be shared by two separating fragments.

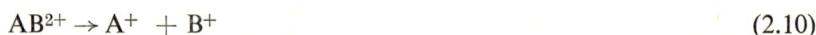

$$AB^{2+} \rightarrow A^{2+} + B \qquad\qquad\qquad\qquad (2.8)$$

$$AB^{2+} \rightarrow A \;\; + B^{2+} \qquad\qquad\qquad\qquad (2.9)$$

$$AB^{2+} \rightarrow A^+ \;\; + B^+ \qquad\qquad\qquad\qquad (2.10)$$

The following decompositions have been shown to occur in the fragmentation of iron pentacarbonyl[24].

$$Fe(CO)_4^{2+} \xrightarrow{-CO} Fe(CO)_3^{2+} \xrightarrow{-CO} Fe(CO)_2^{2+} \xrightarrow{-CO} FeCO^{2+} \qquad (2.11)$$

A large number of doubly charged ions of high abundance have also been observed in the mass spectrum of tungsten hexacarbonyl; the ion $W(CO)_2{}^{2+}$, for example, was present to the extent of 23 % of the base peak[25]. The separate spectra of the singly and doubly charged ions were very similar, providing a sharp contrast to observations made on the spectra of hydrocarbons[17]. This similarity suggested that the doubly charged species were formed by an analogous process to that by which the singly charged ions were formed, namely elimination of neutral CO groups.

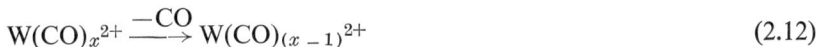

$$W(CO)_x{}^{2+} \xrightarrow{\;-CO\;} W(CO)_{(x-1)}{}^{2+} \qquad\qquad (2.12)$$

Observed fragmentations of doubly charged boron hydride ions have also been of this type; for example, in the spectrum of pentaborane-9 the following decomposition was reported[26].

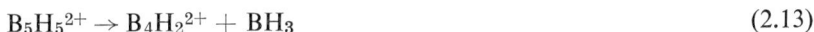

$$B_5H_5{}^{2+} \rightarrow B_4H_2{}^{2+} + BH_3 \qquad\qquad (2.13)$$

The number of decompositions of type (2.10) which have been reported are much fewer. This does not necessarily mean that their occurrence is rarer than those of types (2.8) and (2.9). The product ions $A^+$ and $B^+$ are identical with those formed in the fragmentation of the singly charged ion $AB^+$ and a decomposition of this type is established or perhaps even suspected only if the corresponding "metastable peak" is detected. Reactions of this type which have been shown to occur include[27-29]

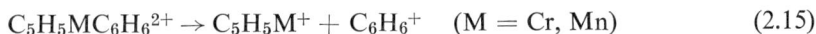

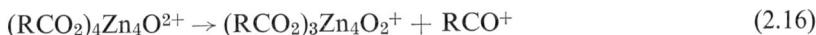

$$CO_2{}^{2+} \rightarrow CO^+ + O^+ \qquad\qquad (2.14)$$

$$C_5H_5MC_6H_6{}^{2+} \rightarrow C_5H_5M^+ + C_6H_6{}^+ \quad (M = Cr, Mn) \qquad (2.15)$$

$$(RCO_2)_4Zn_4O^{2+} \rightarrow (RCO_2)_3Zn_4O_2{}^+ + RCO^+ \qquad\qquad (2.16)$$

## 5. REARRANGEMENT IONS

The spectra of many compounds indicate the presence of ions containing bonds between atoms which were not directly bonded in the original molecule. The ion $A-B-D^+$, for example, cannot be formed from the molecule $A-B-C-D$ by simple bond cleavage. Such ions must be formed by rearrangement processes and if care is not taken the observation of them may hinder a correct interpretation of the molecular structure. The interpretation of the spectra of organic compounds has reached the stage, however, where the identification of such ions can often be of considerable assistance in structure determination.

These rearrangements are intramolecular in nature and so must occur in the very short time between formation of the parent molecular ion and acceleration of it from the ion source. Migration of hydrogen atoms is extremely common, but rearrangements involving bulky groups have also been observed on numerous occasions. Indeed, one of the features of the mass spectra of organometallic compounds is the prominence of such ions. Transfer of atoms or groups to a metal atom appears to be favoured in many cases both by the formation of new strong bonds to the metal and by the elimination of comparatively stable neutral molecules. Typical of ions formed by hydrogen atom transfer are $HZnCH_3^+$ and $HZn^+$ observed by Winters and Kiser[30] in the mass spectrum of dimethylzinc. Numerous hydrocarbon–metal compounds give rise to ions of this type. A similar class of rearrangement is that undergone by certain alkoxy derivatives, for example[31]

$$B(OR)_3^+ \rightarrow B(OH)(OR)_2^+ \rightarrow BO(OR)_2^+ \rightarrow BO(OH)(OR)^+ \qquad (2.17)$$

where $R = C_2H_5, \textit{n-}C_3H_7, \textit{n-}C_4H_9$.

Rearrangement ions have been reported[32] in the fragmentation of substituted ferrocene complexes (2.18), and in the spectra of numerous metal carbonyl complexes (see, for example, refs. 33 and 34).

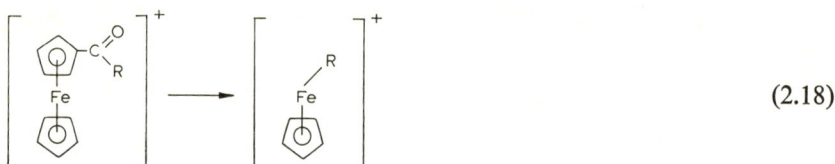

$$(2.18)$$

Mays and Simpson[34] have pointed out that reactions involving group or atom transfer to a metal atom are well known in organometallic chemistry and so these rearrangements are not surprising. In the spectrum of the compound $CH_3COMn(CO)_4NH_3$, for example, the ions $CH_3MnNH_3^+$, $CH_3Mn^+$, and $HMn^+$ were observed[34]. $\sigma$-Bonded to $\pi$-bonded rearrangements have been invoked[35,36] to explain the high intensity of certain ions. These will be discussed in more detail later.

One other example of rearrangement ions will suffice here. In the spectrum of the compound $Ph_3GeSnMe_3$, ions due to $(C_6H_5)_2GeCH_3^+$ and $C_6H_5Sn^+$ are found, viz.

$$(C_6H_5)_3GeSn(CH_3)_2^+ \xrightarrow{\phantom{xxxxx}} \begin{array}{l} (C_6H_5)_2GeCH_3^+ \\ \\ C_6H_5Sn^+ \end{array} \qquad (2.19)$$

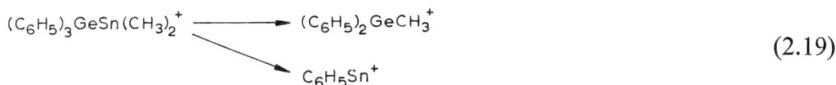

Analogous rearrangements to those mentioned here, except for hydride transfer, are not often encountered in the mass spectra of pure organic molecules (see, for example, refs. 37 and 38).

Decomposition processes such as[39]

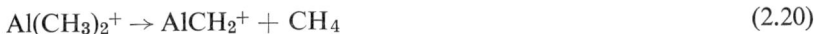

$$Al(CH_3)_2^+ \rightarrow AlCH_2^+ + CH_4 \qquad (2.20)$$

are not generally considered when referring to rearrangements in the mass spectrometer since $AlCH_2^+$ is not a rearrangement ion. However, formation of $CH_4$ obviously must involve a rearrangement.

### 6. METASTABLE IONS

Many of the ions formed in the ionisation chamber of a mass spectrometer are stable enough to traverse the length of the analyser tube and arrive at the collector without undergoing decomposition, while others possess sufficient energy from the bombarding electrons to dissociate before leaving the ion source. Some, however, are of intermediate stability so that they dissociate during flight; these are referred to as metastable ions*. The metastable decompositions which are detected in an instrument with a magnetic analyser occur in flight before this analyser is reached. Thus the precursor or parent ion is accelerated from the ion source, the velocity it attains being determined by its mass $m_1$. After this initial acceleration it

---

* The term metastable is often used loosely and may be employed to refer to any of the following: the precursor ion which decomposes, the decomposition, the resultant ion, and the observed peak corresponding to detection of the resultant ion.

decomposes so that the magnetic field acts on the resultant daughter ion of mass $m_2$. Since the velocity of this daughter ion is the same as its precursor and therefore less than if it were formed in the source, before acceleration it will appear at an $m/e$ value in the spectrum corresponding neither to the mass $m_1$ nor to mass $m_2$, but, except for a certain type discussed later, at a lower mass than either. Hipple and Condon[40] considered the general case where an ion of mass $m_1$ and charge $e_1$ decomposes to produce two particles of masses $m_2$ and $(m_1\text{-}m_2)$ and of charges $e_2$ and $(e_1\text{-}e_2)$ respectively. They showed that the ion of mass $m_2$ appeared at a position on the mass scale ("metastable peak") designated as $m^*$, the "apparent mass",

$$m^* = \frac{(m_2/e_2)^2}{m_1/e_1} \left[ 1 + \frac{m_1 - m_2}{m_2} \cdot \frac{\mathbf{V} - V^1}{\mathbf{V}} \right] \qquad (2.21)$$

where $\mathbf{V}$ is the accelerating voltage applied to the ion of mass $m_1$ and $V^1$ the potential difference through which this ion falls before decomposition occurs. This approximates very closely to

$$m^* = \frac{(m_2/e_2)^2}{m_1/e_1} \qquad (2.22)$$

and in the most common case where the precursor ion carries a single charge

$$m^* = \frac{m_2^2}{m_1} \qquad (2.23)$$

This equation applies to both sector and 180° instruments, and to those which are double-focusing. Only a small proportion of daughter ions of mass $m_2$ which are formed in metastable decompositions reach the collector. In the single-focusing instrument they have the greatest probability of detection if the decomposition takes place at the magnet entrance slit, or just prior to it. In the double-focusing mass spectrometer discrimination occurs in the electrostatic analyser against ions with less than the full accelerating voltage energy. Thus, in general, the only ions produced in metastable decompositions which are observed are those formed in the field-free region between the electrostatic and magnetic analysers; this is referred to as the second field-free region (Fig. 2.9).

Usually it is found that some of the parent ions are collected before decomposition occurs, while others fragment before leaving the source,

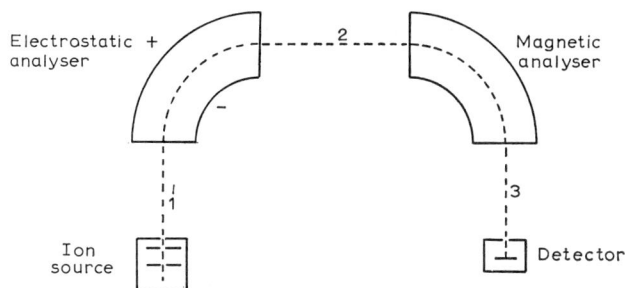

Fig. 2.9. Field-free regions in a double-focusing mass spectrometer.

with the result that both precursor and daughter ions will also be detected at their true $m/e$ values. These are, in fact, usually collected in high abundance, although it should be remembered that some of the daughter ions observed in the spectrum may also be formed by an entirely different process.

It can be seen from eqn. (2.23) that the "metastable peaks" need not, and in general do not, occur at an integral mass number. Their intensity is very low, usually less than 1% of the base peak in the spectrum. Their intensities relative to those of other ions in the spectrum may be increased somewhat by increasing the repeller voltage and by opening the source exit slit. The characteristic feature which initially serves to identify them, however, is their broad, diffuse nature, roughly Gaussian in shape and sometimes extending over a number of mass units (Fig. 2.10). A trial and error procedure may be adopted to identify the process resulting in the observed "metastable peak". Various combinations of the major ions in the spectrum are inserted into eqn. (2.23) until that combination is found which results in an apparent mass value corresponding to that observed. Of considerable assistance in such a procedure is the nomogram[41]. These methods are slow, however, and

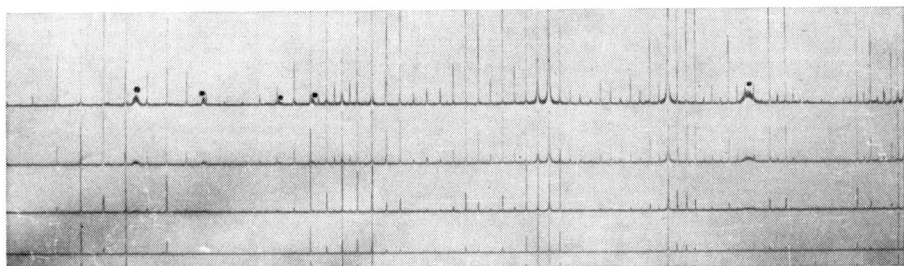

Fig. 2.10. Metastable peaks are indicated in the above spectrum by m*.

require considerable effort. Recently the computer has been employed to facilitate metastable identification. Beynon *et al.*[42] have published a computer-calculated set of tables for values of $m_1 \leq 500$ and of $m^*$ in the range $500 \geq m^* \geq 1$. Other computer programs have also been written (see, for example, ref. 43).

The observation of metastable transitions is extremely valuable in establishing fragmentation pathways*. According to the "Statistical Theory of Mass Spectra"[44,45] the observation of these small diffuse peaks at non-integral mass numbers indicates an energetically favoured path of decomposition and thus provides a means of examining the formation of the fragment ions from the parent molecular ion. They provide direct evidence of particular decompositions. It had been proposed by two groups of workers[46,47] for example, that transition metal carbonyl ions decompose by a series of consecutive unimolecular reactions involving successive loss of neutral carbon monoxide groups, *viz.*

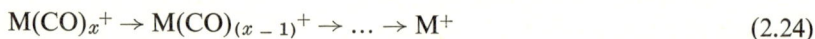

$$M(CO)_x^+ \rightarrow M(CO)_{(x-1)}^+ \rightarrow \ldots \rightarrow M^+ \tag{2.24}$$

This proposed decomposition scheme was later substantiated, at least for $Fe(CO)_5$, by the observation[24] of a number of metastable decompositions in its spectrum (Fig. 2.11).

Another excellent illustration of their usefulness concerns $\pi$-cyclopentadienylmetal carbonyls. It is well known that $\pi$-cyclopentadienyliron compounds containing bridging carbonyl groups thermally decompose to ferrocenes, and it was therefore assumed that the ion $(C_5H_5)_2Fe^+$ present in the spectra of many such compounds arose from ferrocene produced by pyrolytic decomposition within the mass spectrometer, and played no part in the fragmentation of the compounds. However, studies of $[C_5H_5Fe(CO)_2]_2$ showed that, while pyrolysis may have been partially responsible for the

---

* Throughout this book, the $m/e$ value at which the metastable peak occurs will be designated by $m^*$ and a decomposition for which there is 'metastable evidence' will be written as,

$$m_1^+ \underset{*}{\rightarrow} m_2^+ + (m_1 - m_2)$$

or

$$m_1^+ \xrightarrow[*]{-(m_1 - m_2)} m_2^+$$

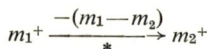

The asterisk indicates the reaction to be metastable supported.

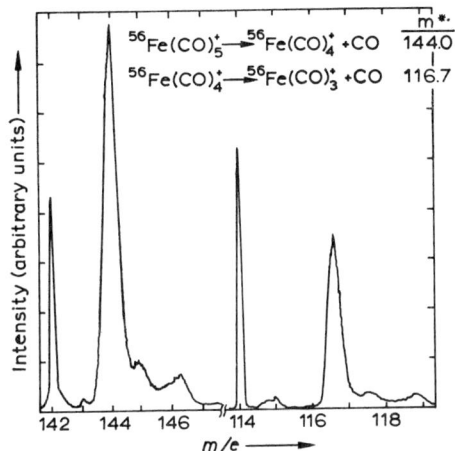

Fig. 2.11. Magnetic scan of mass regions 144 and 116 for Fe(CO)$_5$. The diffuse peaks correspond to the metastable transitions noted. (Reproduced with permission from ref. 24.)

highly abundant ion $(C_5H_5)_2Fe^+$, a metastable transition was observed ($m^*$ at $m/e = 143.5$) corresponding to its formation from $(C_5H_5)_2Fe_2^+$ by migration of a cyclopentadienyl ring from one iron atom to the other[48,49].

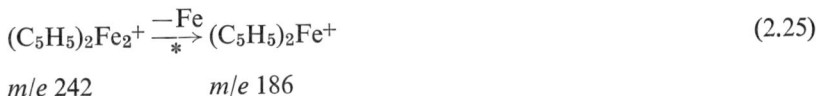

$$(C_5H_5)_2Fe_2^+ \xrightarrow[*]{-Fe} (C_5H_5)_2Fe^+ \qquad (2.25)$$

$m/e$ 242 $\qquad m/e$ 186

Similar rearrangement processes have since been observed in a large number of compounds, for example from $[C_5H_5CoNO]_2$ and $[C_5H_5NiCO]_2$,

$m^*$ at $m/e = 116$

$$[C_5H_5CoNO]_2 \text{ (ref. 50) } (C_5H_5)_2Co_2(NO)_2^+ \xrightarrow[*]{-(Co + 2NO)} (C_5H_5)_2Co^+ \qquad (2.26)$$

$m/e$ 308 $\qquad\qquad m/e$ 189

$m^*$ at $m/e = 144$

$$[C_5H_5NiCO]_2 \text{ (ref. 49) } (C_5H_5)_2Ni_2^+ \xrightarrow[*]{-Ni} (C_5H_5)_2Ni^+ \qquad (2.27)$$

$m/e$ 246 $\qquad m/e$ 188

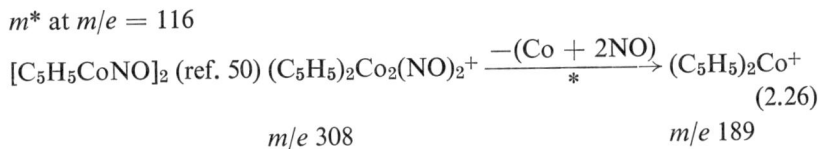

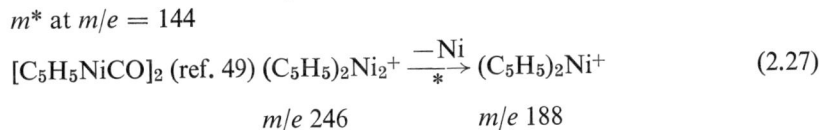

Because of the information gained from the detection of metastable transitions, therefore, considerable effort has been expended in recent years on the study of them and on the improvement of techniques and instrumentation to enable more such transitions to be observed. Consider the metastable decomposition of an ion of mass $m_1$ to one of mass $m_2$ in a double-focusing mass spectrometer, and suppose it decomposes in the first field-free region, that is between the source and the electrostatic analyser. The energy of the ion $m_1^+$ at an accelerating voltage $\mathbf{V}$ is $e\mathbf{V}$, and after decomposition the energy of the ion $m_2^+$ and of the neutral particle of mass $(m_1 - m_2)$ is given by

$$e\mathbf{V} = \frac{m_2}{m_1} e\mathbf{V} + \frac{m_1 - m_2}{m_1} e\mathbf{V} \qquad (2.28)$$

The ion $m_2^+$ cannot pass the electric field because it has an energy of only $(m_2/m_1)e\mathbf{V}$. If, therefore, the accelerating voltage is increased to $V''$, where $V'' = (m_1/m_2)\mathbf{V}$, (while the electrostatic analyser voltage is kept constant) the energy of the ion $m_2^+$ produced in the first field-free region will be

$$\frac{m_2}{m_1} eV'' = \frac{m_2}{m_1} e\left[\frac{m_1}{m_2} \mathbf{V}\right] = e\mathbf{V} \qquad (2.29)$$

By varying the accelerating voltage from $\mathbf{V}$ to $V''$, the ion $m_2^+$ formed in the metastable decomposition can therefore be detected at the same place as a similar $m_2^+$ ion formed instead in the ionisation chamber when the accelerating voltage is $\mathbf{V}$. In practice, normal operating conditions are employed to bring $m_2^+$ ions into focus, then the accelerating voltage is varied while the electrostatic analyser voltage and magnetic field are kept constant[51,52]. In

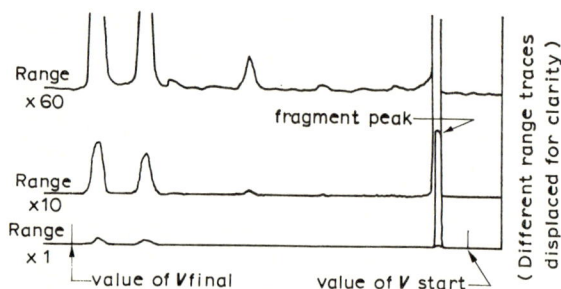

Fig. 2.12. Metastable ion scan for the fragment ion of $m/e$ 178 in a steroid. (Reproduced with permission from ref. 52.)

this way a spectrum of metastable transitions is obtained (Fig. 2.12) and since $m_2$ is known and the voltage ratio for each can be measured, $m_1$ is uniquely determined. An added advantage is the greatly increased sensitivity, so that many more metastable decompositions may be observed and identified than is possible from normal low-resolution spectra depicting those occurring in the second field-free region. This technique has been utilised in instruments of both Nier–Johnson[51,52] and Mattauch–Herzog[53] geometry. Sensitivity is increased by a factor of 50 over conventional spectra so that there is a corresponding increase in the number of metastable transitions observed. However, a separate scan of the accelerating voltage is required for each fragment ion.

Similar results may be obtained by holding the accelerating potential $V$ constant while varying the electrostatic sector voltage of a double-focusing mass spectrometer[54]. Tajima and Seibl[55] used this technique to focus fragment ions formed in a metastable transition in the first field-free region, and then by varying the magnetic field they were able to observe further fragmentation of these ions in the second field-free region. These appeared in the spectrum as ordinary "metastable peaks" in normal spectra and could be evaluated in the same manner. In this way, two- and three-step degradation sequences were established.

Metastable decompositions of doubly-charged ions are of two types, *viz.*

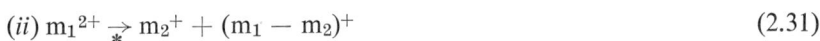

$$(i)\ m_1^{2+} \xrightarrow{\ *\ } m_2^{2+} + (m_1 - m_2) \tag{2.30}$$

$$(ii)\ m_1^{2+} \xrightarrow{\ *\ } m_2^{+} + (m_1 - m_2)^{+} \tag{2.31}$$

From eqn. (2.22)

$$m^* = \frac{(m_2/2)^2}{m_1/2} = \frac{m_2^2}{2m_1} \tag{2.32}$$

Under normal operating conditions, type $(i)$ transitions therefore give rise to Gaussian peaks at $m/e = (m_2^2/2m_1)$. Winters and Collins[24] showed that fragmentation of doubly charged ions formed from iron pentacarbonyl was of this type, and so neutral CO molecules were expelled as in the decomposition of the corresponding singly charged species, *viz.*

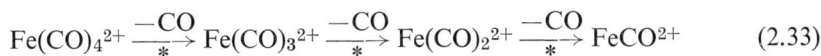

$$\mathrm{Fe(CO)_4^{2+}} \xrightarrow[\ *\ ]{-CO} \mathrm{Fe(CO)_3^{2+}} \xrightarrow[\ *\ ]{-CO} \mathrm{Fe(CO)_2^{2+}} \xrightarrow[\ *\ ]{-CO} \mathrm{FeCO^{2+}} \tag{2.33}$$

The fragmentation of tetraborane was shown[26] to include the following decompositions.

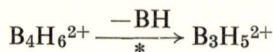

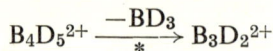

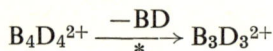

$$B_4H_6^{2+} \xrightarrow[*]{-BH} B_3H_5^{2+}$$

$$B_4D_5^{2+} \xrightarrow[*]{-BD_3} B_3D_2^{2+}$$

$$B_4D_4^{2+} \xrightarrow[*]{-BD} B_3D_3^{2+}$$

Certain tris($\pi$-cyclopentadienyl) lanthanide ions carrying a double positive charge were also observed[56] to undergo a hydrogen rearrangement process.

$$(C_5H_5)_3M^{2+} \xrightarrow[*]{} C_5H_5MC_5H_4^{2+} + C_5H_6 \tag{2.34}$$

where M = Pr, Ho or Lu.

For each decompostion of type ($ii$), however, two metastable peaks may be observed, corresponding to the detection of ions $m_2^+$ and $(m_1 - m_2)^+$ respectively. Substituting in eqn. (2.22) again

$$m_2^* = \frac{(m_2/1)^2}{m_1/2} = \frac{2m_2^2}{m_1} \tag{2.35}$$

$$(m_1-m_2)^* = [(m_1-m_2)^2/1]/m_1/2 = \frac{2(m_1-m_2)^2}{m_1} \tag{2.36}$$

where $m_2^*$ and $(m_1 - m_2)^*$ represent the apparent masses of fragment ions $m_2^+$ and $(m_1 - m_2)^+$ respectively. The simplest molecule observed[57] to undergo this type of decomposition is $CO_2$.

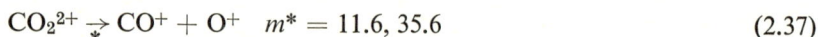

$$CO_2^{2+} \xrightarrow[*]{} CO^+ + O^+ \quad m^* = 11.6, 35.6 \tag{2.37}$$

In an investigation of the mass spectra of the $\pi$-benzene complexes $C_5H_5Cr$-$C_6H_6$ and $C_5H_5MnC_6H_6$, Müller and Göser[58] observed three such decompositions.

($i$) $m^* = 140.0$ (calc. 140.4)  $C_5H_5CrC_6H_6^{2+} \xrightarrow[*]{} C_5H_5Cr^+ + C_6H_6^+$

$m^* = 62.4$ (calc. 62.4)

(ii) $m^* = 146.0$ (calc. 145.5) $C_5H_5MnC_6H_6^{2+} \xrightarrow{*} C_5H_5Mn^+ + C_6H_6^+$

$m^* = 62.0$ (calc. 61.5)

(iii) $m^* = 178.5$ (calc. 178.7) $C_5H_5MnC_6H_6^+ \xrightarrow{*} C_5H_5^+ + MnC_6H_6^+$

$m^* = 42.6$ (calc. 42.7)

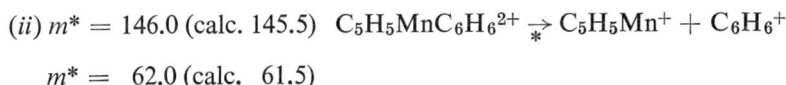

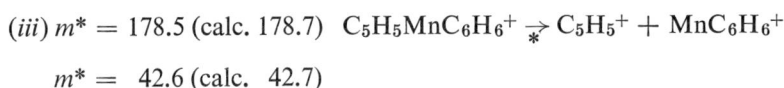

A number of similar metastable decompositions, for example

$m^*$ at $m/e = 1229.4$

$[(C_7H_{15}CO_2)_4Zn_4O]^{2+} \xrightarrow{*} [(C_7H_{15}CO_2)_3Zn_4O_2]^+ + C_7H_{15}CO^+$

$m/e$ 849.7 $\qquad\qquad\qquad$ $m/e$ 722.7

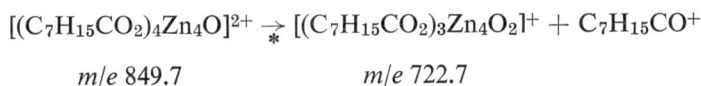

were observed[59] in the mass spectra of basic zinc carboxylates. It is immediately obvious then that in decompositions of this type, the "metastable peak" may occur at an $m/e$ value higher than that of the precursor ion.

Jennings and Whiting[60] have reported a decomposition of the type

$m_1^{3+} \xrightarrow{*} m_2^{2+} + m_3^+$

in the fragmentation of 9,10-diphenylanthracene.

In most metastable decompositions a neutral fragment is expelled, and so studies of these decompositions prove to be a convenient method of determining some of the neutral fragments formed in the mass spectrometer. However, it should be noted that the neutral fragment lost need not have been present in the molecule as a structural entity. Metastable transitions have been reported[61] for borazine corresponding to the decompositions

$$B_3N_3H_x^+ \xrightarrow{*} B_3N_2H_{(x-3)}^+ + NH_3 \qquad x = 3,4,5 \qquad (2.38)$$

in which the $NH_3$ must have been formed by a rearrangement process. The structure of the neutral fragment is not known, although in many cases such as that above, one particular arrangement is much more likely than any other.

The cycloidal mass spectrometer appears to be specially suited to studying metastable transitions although relatively few such investigations have been carried out. The apparent mass values in this particular instrument may be greater than the mass of the parent ion[62,63]. Modification of a time-of-flight mass spectrometer to incorporate the use of a retarding potential technique[64] provides a promising means of observing the neutral fragments produced.


## 7. IONS FORMED FROM ION–MOLECULE REACTIONS

The number of ions present in the source is only very small compared with the number of sample molecules which have not been ionised. The vast majority of these ions are formed by unimolecular decomposition mechanisms, but occasionally a collision may occur between an ion and a neutral molecule, resulting in the formation of a secondary ion whose mass is usually greater than the molecular weight of the compound. Such a bimolecular process will obviously be dependent on sample pressure. The intensity of the ions formed will vary with the second power of the pressure, and at those employed in normal operation (about $10^{-5}$–$10^{-6}$ torr in most instruments) this intensity will usually be too low for the ions to be detected. There is a very low probability that an ion formed in the ion source will collide with a molecule, and an even lower probability, because of increased electrostatic repulsion, that two ions will collide. In general, the only secondary ions of importance at normal operating pressures result from abstraction of a hydrogen atom from a neutral molecule by the parent molecular ion. While this reaction is often encountered in mass spectral studies of organic compounds, it appears to be fairly rare in organometallic chemistry. Identification of this (parent molecule + 1)[+] ion may be somewhat difficult when dealing with polyisotopic molecules.

The presence in a spectrum of an ion produced by an ion–molecule interaction may be recognised by three characteristics which they possess.

(*i*) Their relative intensity is proportional to the square of the pressure.

(*ii*) Their relative intensity will often vary with the ion repeller voltage.

(*iii*) Most of them occur at mass-to-charge ratios higher than that of the parent molecular ion.

The intensities of peaks at high $m/e$ values due to impurities in the sample will increase linearly with pressure and so their relative intensity will remain unchanged as the pressure is varied.

Schumacher and Taubenest[65] investigated the mass spectral behaviour of both ferrocene and a mixture of ferrocene and nickelocene, and observed low intensity, high mass ions which they attributed to products of ion–

molecule reactions. Ions such as $(C_5H_5)_2Fe_2^+$, $(C_5H_5)_2Fe_2C_3H_3^+$, $C_5H_5$-$Fe_2C_3H_3^+$, $(C_5H_5)_3Fe_2^+$, $(C_5H_5)_3Ni_2^+$ and $(C_5H_5)_3FeNi^+$ were detected with a source pressure of $10^{-5}$ torr. The mass spectra of a number of carboxylato derivatives of bis(cyclopentadienyl)titanium(III) also contained low intensity ions which were apparently formed by ion–molecule reactions[66]. These were of the type $(P + C_3H_3)^+$ and $(P + xCH_2)^+$ where P represents the parent molecule and $x = 1, 2,$ or 3.

The reactions between $He^+$, $Ne^+$, $Ar^+$, $Kr^+$, and $Xe^+$ ions and pentaborane-9 have been studied using a two-stage mass spectrometer and a reaction chamber pressure of $10^{-3}$–$10^{-4}$ torr[67]. As might be expected, the fragmentation of the boron hydride varied markedly as a function of the ionisation potential of the rare gas. Xenon has the lowest ionisation potential, and with $Xe^+$ ions only $B_5H_7^+$ and $B_5H_5^+$ were produced in significant amounts, while $He^+$ ion bombardment resulted in a variety of $B_5H_n^+$, $B_4H_n^+$, and $B_3H_n^+$ ions. By plotting the relative abundances of the products as a function of the ionisation potentials of the rare gases, breakdown curves were constructed; integration of these curves resulted in a mass spectrum in reasonably good agreement with the 70 eV electron-impact spectrum.

In recent years many studies of ion–molecule reactions at "high pressures" have been reported. Giardini-Guidoni and Volpi[68] observed the low-intensity ion $PH_4^+$ in the mass spectrum of phosphine, and Halmann and Platzner[69] later employed pressures of 0.2–5.0 torr to facilitate a study of the formation of this ion. They showed by appearance potential measurements that it was produced in the ion–molecule reaction

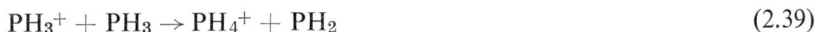

$$PH_3^+ + PH_3 \rightarrow PH_4^+ + PH_2 \tag{2.39}$$

The rare gas molecular ions have been the subject of a number of investigations[70–72]. Kebarle et al.[73] studied the formation of positive ions from alpha-particle irradiation of the rare gases in the pressure range 5–40 torr. $Xe_2^+$ was found to be a major ion in krypton containing 10 ppm Xenon, and it was concluded that it was formed by the reactions

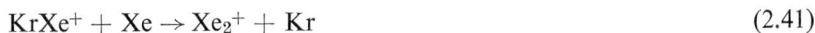

$$Kr_2^+ + Xe \rightarrow KrXe^+ + Kr \tag{2.40}$$

$$KrXe^+ + Xe \rightarrow Xe_2^+ + Kr \tag{2.41}$$

Other high pressure studies have involved $N_3^+$ and $N_4^+$ in nitrogen[74,75],

$H_3^+$ and $H_5^+$ in hydrogen[76], $O_3^+$ and $O_4^+$ in oxygen[77], and $H_3O^+$, $H_5O_2^+$, $H_7O_3^+$, and $H_9O_4^+$ in water[78]. In mixtures of hydrogen both with rare gases and with mercury, Norton[79] observed the formation of rare gas and mercury hydride ions.

Negative ion–molecule reactions in $O_2$, $CO_2$, $H_2O$ and CO, and in mixtures of these gases at high pressures have also been studied[80].

## 8. NEGATIVE IONS

Negative ions may be formed in the mass spectrometer by three different mechanisms.

(*a*) Resonance capture

$$AB + e \rightarrow AB^- \tag{2.42}$$

(*b*) Dissociative resonance capture

$$AB + e \rightarrow A^- + B \tag{2.43}$$

(*c*) Ion-pair production

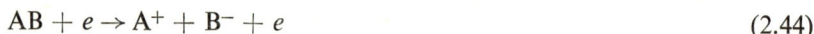

$$AB + e \rightarrow A^+ + B^- + e \tag{2.44}$$

Most instruments may be adjusted to observe negative ions. However, the number of negatively charged species formed is usually much less than the number of positive ions and so a high detection sensitivity is a necessary requirement. Enrione and Rosen[81] reasoned that the electron deficiency of the boron hydrides, for example, may result in the ready formation of negative ions. They found, however, that the total intensities of positive ions formed from decaborane-14, pentaborane-9, and diborane-6 were respectively $10^2$, $10^3$, and $10^6$ times the total intensities of the negative ions. This is an unfavourable feature of negative ion spectra, although in the mass spectrum of perchloryl fluoride[82], $ClO_3F$, the most abundant ion, positive or negative, was $ClO_3^-$.

The resonance capture process depicted above usually occurs with low-energy electrons and differs significantly from all other processes of ion formation in that there is no electron to carry away the excess kinetic energy. The energy of the bombarding electron must therefore lie within a very

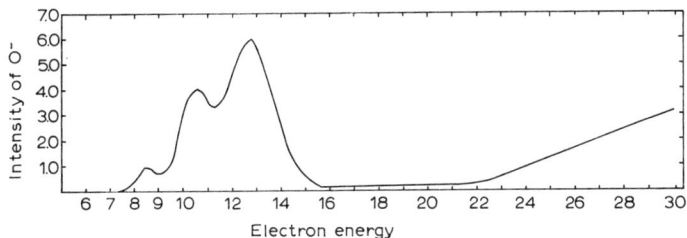

Fig. 2.13. Appearance potential curve for $O^-$ from $H_2O$ illustrating how the ion intensity is critically dependent on the electron energy. (Reproduced with permission from ref. 83.)

narrow range for capture to occur. As a result, few negatively charged parent molecular ions such as $AB^-$ are produced under normal operating conditions of the mass spectrometer. Those that are formed usually dissociate very quickly. Negative ion mass spectra are critically dependent on the energy of the ionising electrons (Fig. 2.13; for a detailed discussion the reader is referred to the review by Melton[83].)

The negative ion mass spectra of the metal carbonyls $Ni(CO)_4$, $Fe(CO)_5$, $Cr(CO)_6$, $Mo(CO)_6$, and $W(CO)_6$ have been recorded[84]. None of the parent molecular ions could be detected using low-energy electrons, indicating that these ions were unstable and quickly dissociated. Kiser[85] suggested that one would not expect to observe these ions since the addition of an electron would require it to enter an antibonding molecular orbital. He pointed out, however, that the $V(CO)_6$ molecule would differ in that the added electron could enter a non-bonding orbital. He and his coworkers[86] later detected the ion $V(CO)_6^-$ in the mass spectrum of this compound. From their observations of the ions formed by ion pair production processes, Winters and Kiser[84] proposed the following mechanism for the origin of the 70 eV negative ion mass spectra of the above carbonyls.

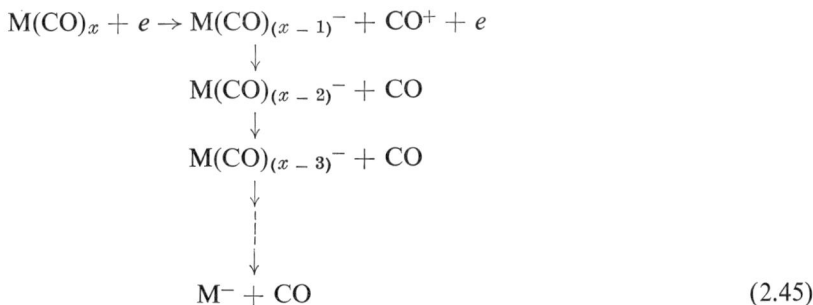

$$M(CO)_x + e \rightarrow M(CO)_{(x-1)}^- + CO^+ + e$$
$$\downarrow$$
$$M(CO)_{(x-2)}^- + CO$$
$$\downarrow$$
$$M(CO)_{(x-3)}^- + CO$$
$$\downarrow$$
$$\vdots$$
$$\downarrow$$
$$M^- + CO \qquad (2.45)$$

A negative parent molecular ion was observed[87] in the spectrum of bis-(pentafluorophenyl)mercury at electron energies ranging from 20 to 70 eV. Its formation must have occurred by secondary electron capture, and it is one of the few examples of such ions derived from organometallic compounds.

## REFERENCES

1  A. D. Norman, R. Schaeffer, G. A. Pressley, Jr. and F. E. Stafford, *J. Am. Chem. Soc.*, 88 (1966) 2151.
2  A. B. Baylis, G. A. Pressley, Jr., E. J. Sinke and F. E. Stafford, *J. Am. Chem. Soc.*, 86 (1964) 5358.
3  R. E. Winters and R. W. Kiser, *Inorg. Chem.*, 4 (1965) 157.
4  R. I. Reed, *Applications of Mass Spectrometry to Organic Chemistry*, Academic Press, New York, 1966, p. 13.
5  C. J. Schack, D. Pilipovich, S. N. Cohz and D. F. Sheehan, *J. Phys. Chem.*, 72 (1968) 4697.
6  A. P. Irsa and L. Friedman, *J. Inorg. Nucl. Chem.*, 6 (1958) 77.
7  G. D. Tantsyrev, L. L. Dekabrun and V. L. Tal'roze, *Zh. Tekhn. Fiz.*, 25 (1955) 1983.
8  A. D. Norman, R. Schaeffer, A. B. Baylis, G. A. Pressley, Jr. and F. E. Stafford, *J. Am. Chem. Soc.*, 88 (1966) 2151.
9  S. Pignataro and F. P. Lossing, *J. Organometal. Chem.*, 11 (1968) 571.
10  J. Lewis, A. R. Manning, J. R. Miller and J. M. Wilson, *J. Chem. Soc. A*, (1966) 1663.
11  B. F. G. Johnson, J. Lewis, I. G. Williams and J. M. Wilson, *J. Chem. Soc. A*, (1967) 341.
12  H. J. Svec and G. A. Junk, *J. Am. Chem. Soc.*, 89 (1967) 2836.
13  J. J. Thomson, *Phil. Mag.*, 24 (1912) 668.
14  A. S. Newton, *J. Chem. Phys.*, 40 (1964) 607.
15  C. E. Brion and N. L. Paddock, *J. Chem. Soc. A*, (1968) 388.
16  J. H. Beynon, *Mass Spectrometry and Its Applications to Organic Chemistry*, Elsevier, Amsterdam, 1960, pp. 282, 344.
17  K. Biemann, *Mass Spectrometry: Organic Chemical Applications*, McGraw-Hill, New York, 1962, p. 158.
18  M. J. S. Dewar and P. Rona, *J. Am. Chem. Soc.*, 87 (1965) 5510.
19  R. B. King, *Can. J. Chem.*, 47 (1969) 559.
20  R. B. King, unpublished tables deposited with the Depository of Unpublished Data, National Science Library, National Research Council of Canada, Ottawa, Canada.
21  E. S. Waight, in R. Bonnett and J. G. Davis (Eds.), *Some Newer Physical Methods in Structural Chemistry*, United Trade Press, London, 1967, p. 67; *Chem. Commun.*, (1969) 1258.
22  M. Solomon and A. Mendelbaum, *Chem. Commun.*, (1969) 890.
23  D. Hellwinkel and C. Wünsche, *Chem. Commun.*, (1969) 1412.
24  R. E. Winters and J. H. Collins, *J. Phys. Chem.*, 70 (1966) 2057.
25  R. E. Winters and R. W. Kiser, *J. Phys. Chem.*, 70 (1966) 1680.

26 T. P. Fehlner and W. S. Koski, *J. Am. Chem. Soc.*, 86 (1964) 581.
27 A. S. Newton and A. F. Sciamanna, *J. Chem. Phys.*, 40 (1964) 718.
28 J. Müller and P. Göser, *J. Organometal. Chem.*, 12 (1968) 163.
29 W. L. Mead, W. K. Reid and H. B. Silver, *Chem. Commun.*, (1968) 573.
30 R. E. Winters and R. W. Kiser, *J. Organometal. Chem.*, 10 (1967) 7.
31 P. J. Fallon, P. Kelly and J. C. Lockhart, *Intern. J. Mass Spectrom. Ion Phys.*, 1 (1968) 133.
32 A. Mandelbaum and M. Cais, *Tetrahedron Letters*, (1964) 3847.
33 B. F. G. Johnson, J. Lewis, I. G. Williams and J. M. Wilson, *J. Chem. Soc. A*, (1967) 338.
34 M. J. Mays and R. N. F. Simpson, *J. Chem. Soc. A*, (1967) 1936.
35 M. I. Bruce, *Inorg. Nucl. Chem. Letters*, 3 (1967) 157; *J. Organometal. Chem.*, 10 (1967) 495.
36 J. D. Hawthorne, M. J. Mays and R. N. F. Simpson, *J. Organometal. Chem.*, 12 (1968) 407.
37 F. Komitsky, J. E. Gurst and C. Djerassi, *J. Am. Chem. Soc.*, 87 (1965) 1399.
38 J. H. Bowie, R. Grigg, D. H. Williams, S. O. Lawesson and G. Schroll, *Chem. Commun.* (1965) 403.
39 D. B. Chambers, G. E. Coates, F. Glockling and M. Weston, *J. Chem. Soc. A*, (1969) 1712.
40 J. A. Hipple and E. V. Condon, *Phys. Rev.*, 69 (1946) 347.
41 R. W. Kiser, *Introduction to Mass Spectrometry and Its Applications*, Prentice-Hall, Englewood Cliffs, New Jersey, 1965, p. 298.
42 J. H. Beynon, R. A. Saunders and A. E. Williams, *Table of Meta-Stable Transitions for Use in Mass Spectrometry*, Elsevier, Amsterdam, 1965.
43 A. Mendelbaum, *Israel J. Chem.*, 4 (1966) 161.
44 H. M. Rosenstock, M. B. Wallenstein, A. L. Wahrhafig and H. Eyring, *Proc. Natl. Acad. Sci. U.S.*, 38 (1952) 667.
45 H. M. Rosenstock, A. L. Wahrhafig and H. Eyring, *J. Chem. Phys.*, 23 (1955) 2200.
46 R. E. Winters and R. W. Kiser, *Inorg. Chem.*, 3 (1964) 699; 4 (1965) 157; *J. Phys. Chem.*, 69 (1965) 1618.
47 A. Foffani, S. Pignataro, B. Cantone and F. Grasso, *Z. Physik. Chem. (Frankfurt)*, 45 (1965) 79.
48 J. Lewis, A. R. Manning, J. R. Miller and J. M. Wilson, *J. Chem. Soc. A*, (1966) 1663.
49 E. Schumacher and R. Taubenest, *Helv. Chim. Acta*, 49 (1966) 1447.
50 H. Brünner, *J. Organometal. Chem.*, 12 (1968) 517.
51 K. R. Jennings, in *Some Newer Physical Methods in Structural Chemistry*, R. Bonnet and J. G. Davis (Eds.), United Trade Press Ltd., London, 1967, p. 105.
52 M. Barber, W. A. Wolstenholme and K. R. Jennings, *Nature*, 214 (1967) 664.
53 E. Watanabe, Y. Itagaki, T. Aoyama and E. Yamauchi, *Anal. Chem.*, 40 (1968) 1000.
54 J. H. Beynon, W. E. Baitinger and J. W. Amy, *Intern. J. Mass Spectrom. Ion Phys.*, 3 (1969) 55.
55 E. Tajima and J. Seibl, *Intern. J. Mass Spectrom. Ion Phys.*, 3 (1969) 245.
56 J. Müller, *Chem. Ber.*, 102 (1969) 152.
57 A. S. Newton and A. F. Sciamanna, *J. Chem. Phys.*, 40 (1964) 718.
58 J. Müller and P. Göser, *J. Organometal. Chem.*, 12 (1968) 163.

59 W. L. Mead, W. K. Reid and H. B. Silver, *Chem. Commun.*, (1968) 573.

60 K. R. Jennings and A. F. Whiting, *Chem. Commun.*, (1967), 820.

61 *Mass Spectral Data*, American Petroleum Institute Research Project 44, Institute of Technology, Pittsburg, Pennsylvania, Serial No. 1346.

62 H. Benz and H. W. Brown, *J. Chem. Phys.*, 48 (1968) 4308.

63 J. H. Beynon, *Mass Spectrometry and its Applications to Organic Chemistry*, Elsevier, Amsterdam, 1960, p. 257.

64 R. E. Ferguson, K. E. McCulloh and H. M. Rosenstock, *J. Chem. Phys.*, 42 (1965) 100.

65 E. Schumacher and R. Taubenest, *Helv. Chim. Acta*, 47 (1964) 1525.

66 R. S. P. Coutts and P. C. Wailes, *Australian J. Chem.*, 20 (1967) 1529.

67 G. R. Hertel and W. S. Koski, *J. Am. Chem. Soc.*, 87 (1965) 404.

68 A. Giardini-Guidoni and E. G. Volpi, *Nuovo Cimento*, 17 (1960) 919.

69 M. Halmann and I. Platzner, *J. Phys. Chem.*, 71 (1967) 4522.

70 J. A. Hornbeck and J. P. Molnar, *Phys. Rev.*, 84 (1951) 621.

71 D. M. Morris, *Proc. Phys. Soc.* (*London*), A68 (1955) 11.

72 W. Kaul and R. Taubert, *Z. Naturforsch.*, 17a (1962) 88.

73 P. Kebarle, R. M. Haynes and S. K. Searles, *J. Chem. Phys.*, 47 (1967) 1684.

74 R. K. Curran, *J. Chem. Phys.*, 38 (1963) 2974.

75 R. K. Asundi, G. J. Schulz and P. J. Chantry, *J. Chem. Phys.*, 47 (1967) 1584.

76 M. Saporoschenko, *J. Chem. Phys.*, 42 (1965) 2760.

77 J. L. Franklin, F. H. Field and F. W. Lampe, *Advan. Mass Spectrometry, Proc. Conf. Univ. London, 1958*, Pergamon Press Ltd., London, 1959, p. 308.

78 M. S. B. Munson, *J. Am. Chem. Soc.*, 87 (1965) 5313.

79. F. J. Norton, *Natl. Bur. Standards* (*U.S.*) *Circ. No. 522*, 1953, p. 201; *Nature*, 169 (1952) 542.

80 J. L. Moruzzi and A. V. Phelps, *J. Chem. Phys.*, 45 (1966) 4617.

81 R. E. Enrione and R. Rosen, *Inorg. Chim. Acta*, 1 (1967) 169.

82 V. H. Dibeler, R. M. Reese and D. E. Mann, *J. Chem. Phys.*, 27 (1957) 176.

83 C. E. Melton, in *Mass Spectrometry of Organic Ions*, F. W. McLafferty (Ed.), Academic Press, New York, 1963, p. 163.

84 R. E. Winters and R. W. Kiser, *J. Chem. Phys.*, 44 (1966) 1964.

85 R. W. Kiser, in *Characterization of Organometallic Compounds, Part 1*, M. Tsutsui (Ed.), Interscience, New York, 1969, p. 181.

86 R. E. Sullivan, M. S. Lupin and R. W. Kiser, *Chem. Commun.*, (1969) 655.

87 S. C. Cohen and E. C. Tifft, *Chem. Commun.*, (1970) 226.

*Chapter 3*

# Information Obtainable Using Mass Spectrometry

M. R. LITZOW

## 1. MASS SPECTRA

Although the mass spectrometer may be put to a variety of uses in studies of inorganic chemistry, the most readily available information is that which can be obtained from the low-resolution spectrum of a particular compound. In the majority of cases the molecular weight of the compound is readily determined by noting the mass of the parent molecular ion. Each compound gives rise to a characteristic mass spectrum which can serve as a ready means of identification although it should be remembered that the spectrum may vary somewhat from one instrument to another. Correlations can frequently be made between the spectrum and molecular structure. Although the total number of hypothetical fragmentation paths that a polyatomic molecule can follow is very large (Chapter 2, Section 3) some of these are much more likely to occur than others. The metal carbonyls, for example, readily undergo sequential loss of carbon monoxide groups, but the abundance of ions formed by C–O bond rupture is low and this only occurs after a number of CO ligands are lost.

The initial problem is the assignment of fragment ions to the various peaks in the spectrum. This may be straightforward from a consideration of the chemistry of the compound being studied. The characteristic isotope abundance patterns of polyisotopic elements can be extremely helpful in this regard. Any uncertainty, or an inability to choose between two possible ion compositions can usually be resolved by a precise mass measurement. Once the fragment ions have been identified, the decomposition reactions responsible for their formation are of interest. However, the complete fragmentation scheme can only be established by the observation of a large number of metastable decompositions. This is only possible if a thorough search is made (Chapter 2, Section 6). Frequently, some of these can be detected in low-resolution spectra and this allows the corresponding fragmentations to be established with certainty. Isotope labelling experiments may also indicate a particular decomposition path. Detailed fragmentation schemes are often postulated from the low-resolution spectrum. It must be remembered, however, that a large portion of such schemes are purely speculative although a consideration of all available data increases the probability that the postulated reactions are indeed responsible for the observed species.

General fragmentation patterns can often be established for various classes of compounds and these are then of considerable assistance in determining the structure of some new compound. Whereas metal carbonyl halides with the halogen in a terminal position undergo competitive loss of CO groups and halogen atoms for example, those with bridging halogens tend to lose all CO groups before loss of the halogen occurs. Similarly, tri-B-substituted and tri-N-substituted borazines can usually be distinguished since B–H bond rupture in these compounds occurs very readily under electron-impact conditions whereas N–H bond rupture does not.

Rearrangement ions are a potential source of confusion. As long as they are recognised and taken into consideration, however, they too can be of assistance in structure determination.

It is often tempting to assign orders of bond dissociation energies in compounds from the relative intensities of ions, and in many cases correct conclusions can result. In general, however, it must be considered dangerous to predict ground-state molecular properties from mass spectral fragmentation patterns (see Chapter 11, Section 4.2 and ref. 1).

*1.1 Molecular weight*

In most cases the parent molecular ion is readily identified and so allows for an extremely simple molecular weight determination.

It must be remembered that where atoms of more than one isotope are involved, the parent molecular ion will extend over more than one $m/e$ value. The molecular weight is sometimes expressed as the value incorporating the isotopes of highest natural abundance, so that there may be peaks of lower intensity at higher mass due to molecules containing less abundant isotopes. This is an entirely different situation from that involving peaks of high $m/e$ due to the occurrence of ion–molecule reactions, discussed in Chapter 2, Section 7. In some cases the peaks at highest $m/e$ do not correspond to the parent molecular ion of the compound. There are several possible reasons for this including decomposition of the compound in the mass spectrometer by thermal or hydrolytic reactions and these are discussed later in this chapter. Furthermore, caution must be exercised that the peak of highest $m/e$ in the spectrum is not caused by either an impurity in the sample or by spectrometer background.

Where the parent molecular ion is not observed, the molecular weight of the compound cannot be so easily determined. A method has recently been described[2] whereby the weight of the unobserved parent molecular ion may be calculated to within $\pm 1$ atomic mass unit in some such cases. It involves detection of metastable ions corresponding to the decomposition of the parent molecular ion in a high-resolution mass spectrometer, and appears to be useful for many types of compound, although it has not proved successful in all cases. Thus, the parent molecular ion need only live long enough to escape the ion source (approximately $10^{-6}$ sec) for its mass to be determined. It is suggested that this method may also be useful for compounds which display parent molecular ions that are obscured by background or by impurities.

Since spectra of most compounds can be obtained on a quantity of sample of the order of 1 $\mu$g, the analytical importance of mass spectrometry is at once obvious. It is essential, however, that care is always taken in assigning the molecular weight and in making deductions pertaining to the composition and structure of the compound from this value. Chemical, spectral, and magnetic measurements, for example, indicate that the Schiff base complex of nickel(II), *viz.*

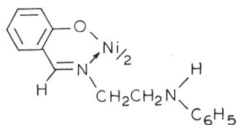

is a square planar, four-coordinate nickel compound with a ligand(L): Ni ratio of 2:1 (ref. 3). Dudek *et al.*[4] found that the mass spectrum exhibited a peak corresponding to the normal parent molecular ion $L_2Ni^+$, but in addition, a multiplet whose intensity was twenty times that of this normal parent molecular ion appeared at higher mass and was attributed to the ion $[(L-H)_2Ni_2]^+$, *viz.*

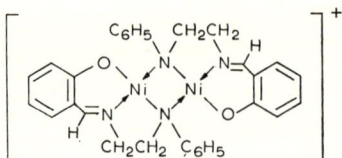

It was postulated that a molecule–molecule reaction at the sample surface was responsible for formation of the dimeric species. Other similar Schiff base complexes of nickel(II) behaved analogously in the mass spectrometer[4].

It may be possible to decide whether the peak of highest $m/e$ is or is not the parent molecular ion from a consideration of the spectrum. In the previous example, let us assume that the ion $(L–H)_2Ni_2^+$ was a normal parent molecular ion. The ion $L_2Ni^+$ which was present could only have been formed by some process involving abstraction of hydrogen, such as

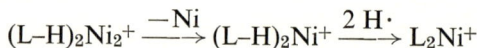

$$(L–H)_2Ni_2^+ \xrightarrow{-Ni} (L–H)_2Ni^+ \xrightarrow{2\,H\cdot} L_2Ni^+$$

or

$$(L–H)_2Ni_2^+ \xrightarrow{2\,H\cdot} L_2Ni_2^+ \xrightarrow{-Ni} L_2Ni^+$$

One would expect, therefore, to see either the ion $L_2Ni_2^+$ or $(L–H)_2Ni^+$ in the spectrum, but their complete absence throws suspicion on the original assumption that the compound had the composition $(L–H)_2Ni_2$.

*1.2 Isotopic abundances*

A feature of inorganic and organometallic compounds is the wide occurrence of polyisotopic elements. The characteristic patterns can make for quick and easy identification of ions. Where more than one atom of a polyisotopic nature is present, the expected pattern must be calculated from the respective isotopic abundances.

From elementary probability theory, the number of ways of selecting $x$ items from a total of $N$ such items is given by

$$^N C_x = {}^N C_{N-x} = \frac{N!}{x!\,(N\text{-}x)!} = \frac{N(N-1)\,(N-2)...(N-x+1)}{x!} \qquad (3.1)$$

If the total $N$ is actually a mixture consisting of $P_1$ items of type $A_1$ and $P_2$ items of type $A_2$, the probability of selecting one of $A_1$ at random is given by

$P_1/(P_1 + P_2)$ $(=p_1)$. Similarly, the probability of selecting one of type $A_2$ at random is $P_2/(P_1 + P_2)$ $(=p_2)$.

If the total number $N$ is very large, the probabilities $p_1$ and $p_2$ will remain essentially constant as a small sample is selected. In such a case, the probabilities of selecting two of type $A_1$, two of type $A_2$, and one each of $A_1$ and $A_2$ are $p_1^2$, $p_2^2$, and $2p_1p_2$ respectively. In the general case where the total number selected is $n$ (which is small compared with $N$) the compositions of the samples and their corresponding probabilities are given by

| No. of $A_1$ | $n$ | $n-1$ | $n-2$ | ... | $n-x$ | ... | 2 | 1 | 0 |
|---|---|---|---|---|---|---|---|---|---|
| No. of $A_2$ | 0 | 1 | 2 | ... | $x$ | ... | $n-2$ | $n-1$ | $n$ |
| Probability | $p_1^n$ | $np_1^{n-1}p_2$ | $^nC_2p_1^{n-2}p_2^2$ | ... | $^nC_xp_1^{n-x}p_2^x$ | ... | $^nC_2p_1^2p_2^{n-2}$ | $np_1p_2^{n-1}$ | $p_2^n$ |

These probabilities are the terms of the expansion of $(p_1 + p_2)^n$; the coefficients $^nC_x$ are the well known binomial coefficients.

Similarly, if an element has two naturally occurring isotopes $A_1$ and $A_2$ of fractional abundances $p_1$ and $p_2$, the probability of having $[(A_1)_a(A_2)_{n-a}]$ or $[(A_1)_a(A_2)_b]$ where $a + b = n$ is given by the appropriate term in the expansion of $(p_1 + p_2)^n$, namely $^nC_ap_1^ap_2^b$ or $(n!/a!b!)p_1^a p_2^b$.

Consider now the molecule $A_nB_m$ where $A$ has two isotopes $A_1$ and $A_2$ of fractional abundances $p_1$ and $p_2$, and $B$ likewise has two isotopes $B_1$ and $B_2$ of fractional abundances $q_1$ and $q_2$. The general molecule is represented by $(A_1)_a(A_2)_b(B_1)_u(B_2)_v$, where $a + b = n$ and $u + v = m$, and its fractional abundance is given by

$$^nC_ap_1^ap_2^b \cdot {}^mC_uq_1^uq_2^v \quad \text{or} \quad \frac{n!}{a!b!} p_1^ap_2^b \cdot \frac{m!}{u!v!} q_1^uq_2^v$$

Finally, if element $A$ consists of isotopes $A_1$, $A_2$, $A_3$, $A_4$, $A_5$, and $A_6$ with fractional abundances $p_1$, $p_2$...$p_6$, and element $B$ consists of isotopes $B_1$...$B_6$ with fractional abundances $q_1$...$q_6$ respectively, then for the molecule $A_nB_m$ the fractional abundance of $(A_1)_a(A_2)_b(A_3)_c(A_4)_d(A_5)_e(A_6)_f(B_1)_u(B_2)_v$ $(B_3)_w(B_4)_x(B_5)_y(B_6)_z$ is given by

$$\frac{n!}{a!b!c!d!e!f!} p_1^ap_2^bp_3^cp_4^dp_5^ep_6^f \cdot \frac{m!}{u!v!w!x!y!z!} q_1^uq_2^vq_3^wq_4^xq_5^yq_6^z$$

The expected relative abundances of the various compositions of the ion $BCl_2Br^+$ are calculated as shown in Table 3.1.

Table 3.1

Calculation of Peak Intensities for the $BCl_2Br^+$ ion using the Isotopic Abundances $^{10}B$ = 0.2, $^{11}B$ = 0.8, $^{35}Cl$ = 0.755, $^{37}Cl$ = 0.245, $^{79}Br$ = 0.505, and $^{81}Br$ = 0.495

| Ion composition | $m/e$ | Fractional abundance | | Rel. abundance |
|---|---|---|---|---|
| $^{10}B^{35}Cl_2{}^{79}Br$ | 159 | $(0.2)(0.755)^2(0.505)$ | = 0.058 | 0.153 |
| $^{11}B^{35}Cl_2{}^{79}Br$ | 160 | $(0.8)(0.755)^2(0.505)$ | = 0.230 | 0.613 |
| $^{10}B^{35}Cl^{37}Cl^{79}Br$ | 161 | $(0.2)2(0.755)(0.245)(0.505)$ | = 0.037 ⎫ | 0.250 |
| $^{10}B^{35}Cl_2{}^{81}Br$ | 161 | $(0.2)(0.755)^2(0.495)$ | = 0.056 ⎭ | |
| $^{11}B^{35}Cl^{37}Cl^{79}Br$ | 162 | $(0.8)2(0.755)(0.245)(0.505)$ | = 0.150 ⎫ | 1.000 |
| $^{11}B^{35}Cl_2{}^{81}Br$ | 162 | $(0.8)(0.755)^2(0.495)$ | = 0.226 ⎭ | |
| $^{10}B^{37}Cl_2{}^{79}Br$ | 163 | $(0.2)(0.245)^2(0.505)$ | = 0.006 ⎫ | 0.113 |
| $^{10}B^{35}Cl^{37}Cl^{81}Br$ | 163 | $(0.2)2(0.745)(0.245)(0.495)$ | = 0.036 ⎭ | |
| $^{11}B^{37}Cl_2{}^{79}Br$ | 164 | $(0.8)(0.245)^2(0.505)$ | = 0.024 ⎫ | 0.450 |
| $^{11}B^{35}Cl^{37}Cl^{81}Br$ | 164 | $(0.8)2(0.755)(0.245)(0.495)$ | = 0.145 ⎭ | |
| $^{10}B^{37}Cl_2{}^{81}Br$ | 165 | $(0.2)(0.245)^2(0.495)$ | = 0.006 | 0.016 |
| $^{11}B^{37}Cl_2{}^{81}Br$ | 166 | $(0.8)(0.245)^2(0.495)$ | = 0.024 | 0.063 |

Deviation of the isotope pattern from that calculated indicates that either the expected ion is not present, or that overlap is occurring between it and another ion. This situation often arises with the parent molecular ion and (parent molecule − 1)⁺ ions. Such an example is illustrated in Fig. 3.1 where the intensity of the $(BCl_2NMe_2-H)^+$ ion is fairly high so that the isotopic pattern observed differs considerably from that expected. Consideration of isotopic patterns often provides valuable assistance, therefore, in the interpretation of mass spectra.

## 1.3 Thermal reactions

Thermal reactions within either the inlet system or the ion source of the mass spectrometer have been largely ignored or else not recognised, but many inorganic and organometallic compounds are extremely susceptible.

The boron hydrides readily undergo pyrolysis and the mass spectra of many of them display ions due to decomposition products when recorded on conventional instruments (see Fig. 6.1 and 6.4). The metal carbonyls $Fe(CO)_5$, $Cr(CO)_6$, $Mo(CO)_6$, and $W(CO)_6$ have been shown[5,6] to decompose in the mass spectrometer at fairly low temperatures. An important observation was that $Cr(CO)_6$ and $Mo(CO)_6$ began to decompose at an appreciably

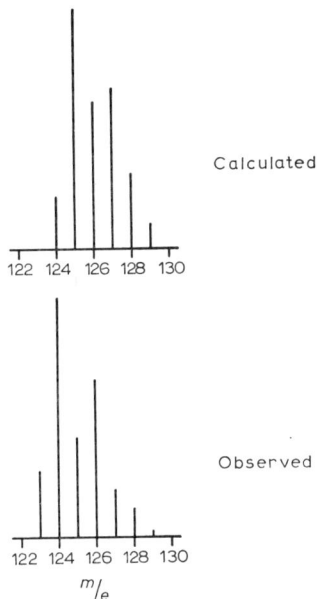

Fig. 3.1. Calculated relative intensities of the various isotopic $BCl_2NMe_2^+$ ions ($m/e$ 124–129) compared with the intensities actually observed in the mass spectrum of $BCl_2NMe_2$; the difference is due to the presence of the $(BCl_2NMe_2–H)^+$ ion.

lower temperature during a second heating of the pyrolysis furnace. These results suggested that the compounds can generate catalysts by depositing active metals and/or free radicals on heated wall surfaces inside the mass spectrometer. There is always the possibility, therefore, that an organo-metallic compound may deposit such a catalyst which could enhance thermal decompositions of subsequent samples.

Svec and Junk[6] studied the behaviour of manganese pentacarbonyl chloride in the mass spectrometer. On the basis of known solution chemistry and established bond energies the thermal reactions

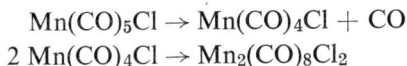

$$Mn(CO)_5Cl \rightarrow Mn(CO)_4Cl + CO$$
$$2\ Mn(CO)_4Cl \rightarrow Mn_2(CO)_8Cl_2$$

were expected to occur. Instead, the reactions

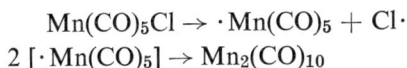

$$Mn(CO)_5Cl \nrightarrow \cdot Mn(CO)_5 + Cl\cdot$$
$$2\ [\cdot Mn(CO)_5] \nrightarrow Mn_2(CO)_{10}$$

were considered to take place on the wall surface inside the ion chamber. A range of 8.2–11.4 eV was observed for the appearance potential of the $Mn(CO)_5^+$ ion.

Detailed investigations of most of these thermal reactions require complex equipment[7,8]. It is not necessary to carry out such investigations but the possibility of thermal decomposition should be checked and this can be done fairly readily. If the direct insertion probe is being used, suspected pyrolysis may be checked by recording the mass spectrum at varying ion source temperatures. Svec and Junk[6] have shown that for compounds introduced through the reservoir inlet system, on the other hand, temperature effects can be readily checked by installing a small electric furnace around the inlet line between the molecular leak and the ion source. Monitoring of the parent molecular ion as the temperature of the furnace is varied yields a temperature at which thermal reactions begin. The ion source temperature must be lower than that of the furnace.

As an elementary precaution, all temperatures, whether of the inlet system or of the ion source itself, should be kept as low as possible. Where suitable, the use of low-temperature probes and also water-cooled sources should be of great assistance in minimising these unwanted thermal reactions.

*1.4 Hydrolysis*

Many inorganic and organometallic compounds are moisture sensitive to some degree. Although the amount of moisture present in the inlet system and ion source is small, reactions between it and highly sensitive compounds do occur. A check should be made to see if hydrolysis products are present in the mass spectrum. When powerful water scavengers such as the boron trihalides are first introduced, the parent molecular ion is often not observed, but the peaks due to $HX^+$ ions are very strong. It may therefore be necessary to "condition" the ion source by allowing the compound to diffuse into it for some time.

*1.5 Memory effects*

Another potential source of confusion is provided by so-called "memory effects". Normal operating procedure involves pumping out the previous compound thoroughly and checking the background before introducing any subsequent sample. Some compounds are troublesome in that they be-

come adsorbed on the walls of the ion source and pumping must be con-
tined for some time to remove them. Such an occurrence is of nuisance value
only since one can observe the effect readily and be aware of it merely by
checking the background. A more insidious problem is generated by reactive
species which are adsorbed on the walls of the ion source and are not
detected in the background spectrum, but which can react with a subsequent
sample. King[9] has reported certain metal carbonyl iodide ions in the mass
spectrum of halogen free compounds, while exchange of adsorbed iodine
and bromine with metal chloride compounds to produce the corresponding
bromide and iodide species may occur.


2. PRECISE MASS MEASUREMENTS

In low-resolution spectra discussed previously, masses of ions are measured
to the nearest half mass unit by peak counting. However, atomic numbers
differ slightly from whole numbers (except for $^{12}C$ which is taken as the
standard) due to the binding energy involved. Accurate mass measurements
take advantage of this to determine the elemental composition of ions. Masses
can be measured with a precision of at least 10 p.p.m. A compound can
therefore be identified from a precise mass measurement of its parent
molecular ion. The same is true for fragment ions, thus resolving any
ambiguity remaining after a consideration of other information such as
isotope abundance patterns. When ions exhibit a spread of $m/e$ values due to

Table 3.2

Some of the Ions Observed[10] in the Spectrum of $(CH_3)_3SnOS(O)C_6H_5$

| $m/e$ obs. | Assignment | $m/e$ calc. |
|---|---|---|
| 305.9740 | $Me_3{}^{120}SnSO_2C_6H_5$ | 305.97365 |
| 274.9856 | $Me_3{}^{120}SnSC_6H_5$ (one $^{13}C$) | 274.98718 |
| 274.9566 | $Me_3{}^{120}SnSOC_6H_5$ | 274.95530 |
| 274.9222 | $Me^{119}SnSO_2C_6H_5$ | 274.92785 |
| 260.0085 | $Me_3{}^{122}SnOC_6H_5$ | 260.00791 |
| 259.9613 | $Me_2{}^{120}SnSC_6H_5$ (one $^{13}C$) | 259.96471 |
| 259.9316 | $Me^{120}SnSOC_6H_5$ | 259.93179 |
| 259.9086 | $^{119}SnSO_2C_6H_5$ | 259.90432 |

the presence of isotopes, difficulties may arise due to the overlapping of peaks from different ions. This occurs to a large extent in the spectra of the boron hydrides and of compounds such as $C_5H_5(CO)_3Mo–SnMe_3$ (ref. 35) where considerable overlap occurs with, for example, the ions (parent molecule–CO)$^+$ and (parent molecule–2Me·)$^+$. In this case the presence of both ions was determined from a consideration of isotope patterns, but this is not always possible. Fragmentation of the $SO_2$ insertion products $(CH_3)_3$-SnOS(O)R under electron impact resulted in the formation of a very large number of ions, many of them in rearrangement processes[10]. In situations such as this, accurate mass measurements enable the complete range of ions formed to be established. Some of the results obtained for the compound $(CH_3)_3SnOS(O)C_6H_5$ are shown in Table 3.2.

### 3. ENERGETIC STUDIES BY ELECTRON-IMPACT METHODS

Mass spectrometry offers a means of obtaining fundamental information concerning bonding fairly readily from the study of the energetics of systems by means of ionisation and appearance potential measurements. Unfortunately, the techniques employed in these measurements are still somewhat unreliable and very little is known about the structures of the gaseous ions formed. As long as the limitations are recognised, however, valuable information may be obtained.

### 3.1. Ionisation efficiency curves

An ionisation efficiency curve* is a graphical representation illustrating the variation in the ion current of a particular ion with increasing energy of the bombarding electrons. Instead of scanning a wide mass range, therefore, the instrument is adjusted so that it collects only the ion of interest. The energy of the electron beam is then varied and a plot made of this energy against the resultant ion current. As mentioned in Chapter 2, Section 2, until this energy reaches a certain value for a particular ion, the ion will not appear. This threshold value is characteristic of both the ion and its origin.

---

* It has become standard practice to apply this name to such curves obtained for all ions, both parent molecular ions and fragment ions, although, as Kiser[11a] has pointed out, this is improper terminology for the latter.

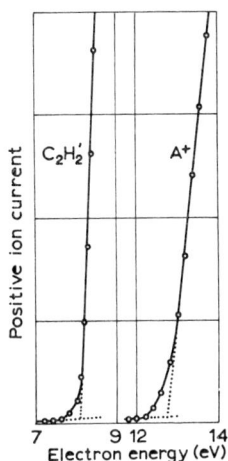

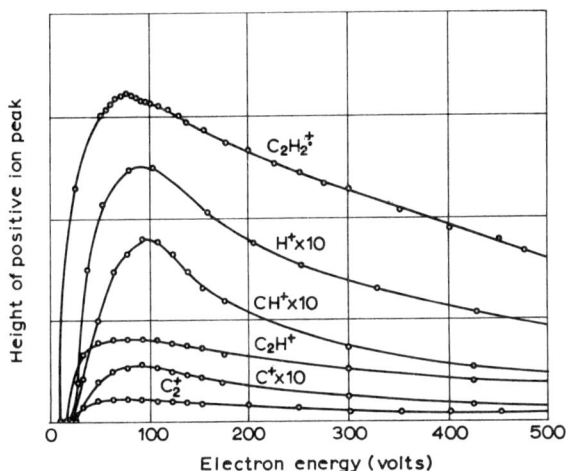

Fig. 3.2. Ionisation efficiency curves for several ions formed from acetylene. (Reproduced with permission from ref. 34.)

The whole of the ionisation efficiency curve is also characteristic of a particular ion although the shapes of all curves are basically similar.

From Fig. 3.2 it can be seen that the curves consist of four sections: (*i*) an initial curved portion for 1–2 V above the ionisation threshold; (*ii*) a steep linear portion for the next 10–20 V; (*iii*) a broad maximum which is better defined in some cases than in others and ranging from 50–100 V in width;

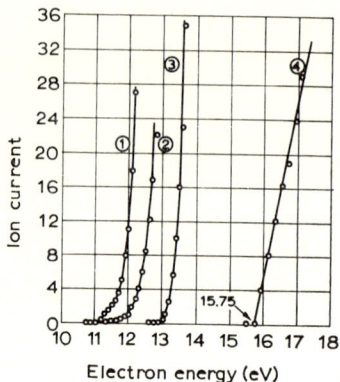

Fig. 3.3. Ionisation efficiency curves for argon. Runs (1) and (2) were taken by conventional methods whereas in runs (3) and (4) a retarding potential was applied to the electrons to reduce their energy spread to 0.1 eV. (Reproduced with permission from ref. 13.)

and (iv) a gradual decrease in ion current as the voltage is further increased.

The largest part of the initial curved portion or tail of the curve is the result of the beam of electrons being energetically inhomogeneous (Fig. 3.3). The electron beam is obtained from a heated filament, and the emitted electrons have appreciable thermal energies, the distribution of which is essentially Maxwellian in character (Fig. 3.4). Other factors which may contribute to the tail are discussed by Kiser[11b]. For molecular ions, and also fragment ions formed by a single bond-rupture process, the tail of the curve is normally shorter and the slope of the linear portion greater than for ions formed by more complicated mechanisms.

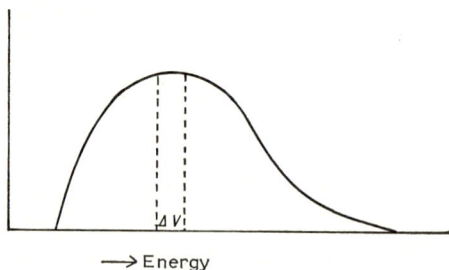

Fig. 3.4. Energy spread of the bombarding electrons in a normal electron-impact source; the distribution is essentially Maxwellian in character. Electrons with the small energy spread $\Delta V$ are those selected by the retarding potential difference source.

In the vast majority of experiments, the ionisation efficiency curves are recorded on instruments whose electron beams possess a considerable energy spread which has the effect of smearing out any fine structure. Fine structure is predicted, for example, corresponding to the production of an ionic species in its various electronic states, but where the energy separation of these states is less than the half-width of the electron energy distribution they cannot be resolved by inspection of the curves. A good deal of effort has been expended in attempts to reduce the energy spread[12-16] and it has been found possible to confine it to 0.1 eV or less. Photoionisation techniques are also useful for the observation of fine structure in ionisation efficiency curves.

The fine structure can arise from a number of factors such as (*i*) formation of ionic species in vibrational or electronic excitation levels, (*ii*) autoionisation, or (*iii*) onset of a second process resulting in the formation of the particular ion. Examples of (*i*) have been observed by a number of workers [17-19] for the $O_2^+$ ion, and from the breaks in the ionisation efficiency curve the separation of the $^2P_{3/2}$ and $^2P_{1/2}$ levels can be obtained. The occurrence of autoionisation with oxygen has also been reported[19,20]. McDowell and Warren[21] observed a break in the ionisation efficiency curve for the fragment ion $CH_2^+$ formed from methane at a point about 4.5 eV above the threshold value. This is about the bond dissociation energy of the $H_2$ molecule. Prior to this point on the curve, the $CH_2^+$ ions were assumed to be formed solely by the process

$$CH_4 + e \rightarrow CH_2^+ + H_2 + 2e$$

Above the break, the ion current increases sharply and this was interpreted as being due to onset of the process

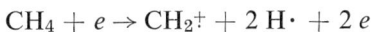

$$CH_4 + e \rightarrow CH_2^+ + 2H\cdot + 2e$$

On the other hand, Cantone *et al.*[22] recorded the ionisation efficiency curves (using a non-monoenergetic electron beam) for $W(CO)_6^+$ and the fragment ions $W(CO)_x$ ($x = 0-5$) and found that above an initial linear increase of the ion current of $W(CO)_6^+$, for example, with electron energy, the slope suddenly decreased at a point corresponding to the appearance potential of the $W(CO)_5^+$ ion. This was interpreted as being due to fragmentation of the parent molecular ion to form the daughter ion $W(CO)_5^+$. Indeed, addition of the two ionisation efficiency curves produced a curve with a wider linear portion. Thus the fragmentation pattern was determined without the assistance of metastable transitions, which were not observed.

With instruments producing an inhomogeneous electron beam, however, it is necessary that the energy spread be significantly less than the difference in energy required for the two processes if this break is to be observed. Otherwise the onset of the second process is ill-defined. The only result then is that the initial curvature in the ionisation efficiency curve will be relatively long. Whenever such a long tail is observed, therefore, the possibility of the ion under investigation being formed by more than one process must be considered. This is especially important if an attempt is being made to determine the appearance potential of the ion.

Of the various attempts which have been made to reduce the electron energy spread, the most successful have been those in which a narrow section is taken out of the initial energy distribution and this is then used to produce the ionic species. This reduction of the energy spread is experimentally difficult, but, in addition, the electron current obtainable decreases as the section is made narrower, and the signal-to-noise ratio in the recorded data increases. Various attempts have therefore been made to remove the energy spread by analytical methods. One of the most successful has been the deconvolution method developed by Morrison[23] which reduces the effective electron energy spread by a factor of four or five times. Fourier transforms are applied to experimental ionisation efficiency curves recorded on a normal mass spectrometer with a large electron energy spread. A second analytical method is described by Winters et al.[24]. This is discussed in the following section.

*3.2 Ionisation and appearance potentials*

The first ionisation potential of an atom is defined as the minimum energy required to completely remove an electron from the atom in its ground state to form the corresponding atomic ion, also in its ground state. The minimum energy implies that no kinetic energy is imparted to the ion. With molecules however, one must consider vibrations which result in changes in the internuclear distances. The first ionisation potential of a molecule may therefore be defined as the minimum energy required to remove an electron from the highest occupied molecular orbital, both the original molecule and the resultant molecular ion being in their ground vibrational states, and, of

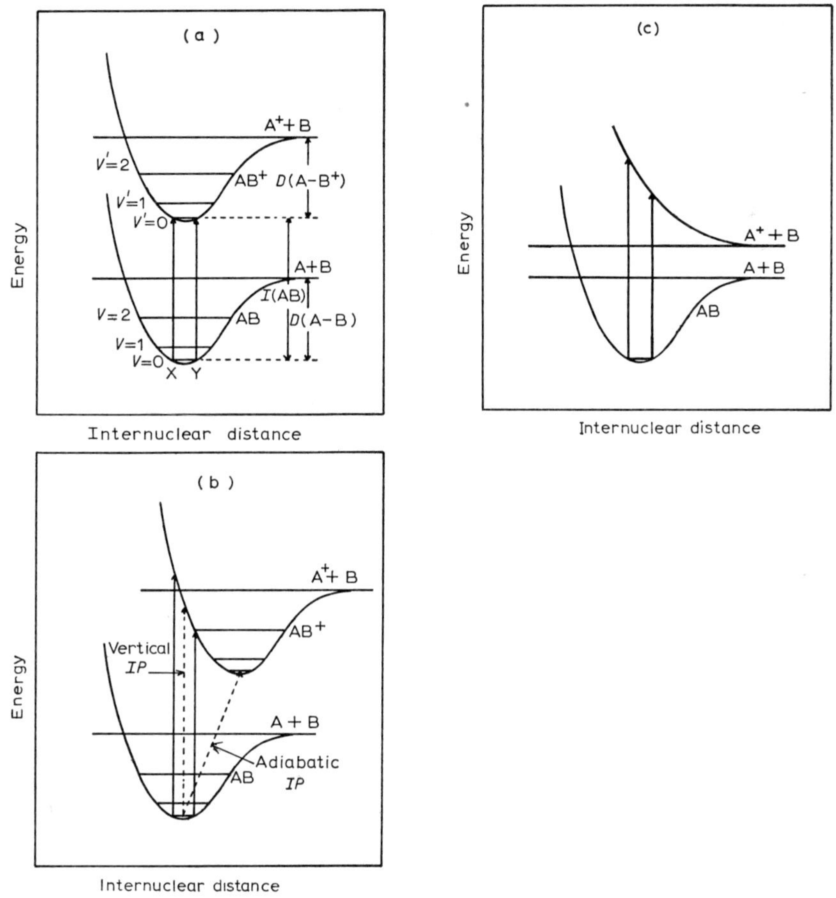

Fig. 3.5. Potential energy curves of diatomic molecules illustrating the transitions expected according to the Franck–Condon principle.

course, their lowest electronic levels. This energy, $I(AB)$, is the difference between the $v = 0$ and $v' = 0$ vibrational levels in Fig. 3.5(a) and is termed the adiabatic ionisation potential.

For simplicity, consider the diatomic molecule AB. According to the Franck–Condon principle, the passage of an electron from one molecular level to another occurs in too short a time for the much more massive atoms to have suffered appreciable displacement. The transitions therefore occur along a vertical line in Fig. 3.5. Every electronic state has its own potential energy curve, of particular shape and position relative to others, and if the minima in these curves correspond to sufficiently different internuclear distances, as is often the case, a change in the vibrational levels accompanies the electronic excitation. In addition, the Franck–Condon principle states that the transitions from one state to another are most probable when the nuclei are in their extreme vibrational positions, X and Y in Fig. 3.5(a), for it is there, when the vibrational kinetic energy is zero, that they spend the longest time. When the equilibrium distances are very much the same for the neutral molecule and the ion, the probability that the transition will occur to the lowest vibrational level of the ion is high, and the resultant ionisation potential is expected to be very close to the adiabatic value (Fig. 3.5(a)). Usually, however, the ionic species $AB^+$ is of lower stability than the neutral molecule AB, and the equilibrium distance in $AB^+$ tends to be increased. It is possible that some transitions from the ground vibrational state of the molecule, such as that shown in Fig. 3.5(b), will be accompanied by dissociation if the internuclear equilibrium distances are appreciably different.

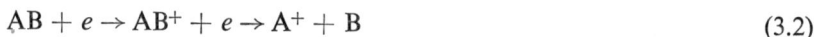

$$AB + e \rightarrow AB^+ + e \rightarrow A^+ + B \tag{3.2}$$

For some molecules the upper electronic levels possess potential energy curves of the type shown in Fig. 3.5(c). Every transition from the lower to the upper state results in dissociation of the molecule and the parent molecular ion is not observed.

The energy change involved in the formation of a molecular ion, as measured by electron-impact methods, is represented by a vertical transition and so rarely corresponds to the adiabatic ionisation potential which may be determined spectroscopically, at least in favourable cases. This measured value should therefore always be taken as an upper limit. In addition, the actual determination of the energy involved in such a vertical transition by electron-impact has often been unsatisfactory, but in recent years improvements in instruments, techniques, and interpretation of the collected data have enabled more accurate values to be obtained.

If the electron energy is high enough, ionisation of a molecule followed by dissociation may occur. The minimum energy required to produce the fragment ion in its ground vibrational state from the original molecule in its ground vibrational state is referred to as the appearance potential of that ion.

A rough estimate of an ionisation potential may be quickly obtained by gradually increasing the energy of the bombarding electrons and noting the value of this energy when the ion is first observed. The energy scale must be calibrated by use of an element or compound whose ionisation potential has been accurately determined by some other technique such as UV spectroscopy. The rare gases are commonly used for this purpose. Because of the very low values and gradual increase of the ion current in the threshold region, as shown by ionisation efficiency curves, it is expected that the electron energy at which a particular ion is first observed would provide a highly unreliable value for the ionisation potential. Such a value is found, in fact, to be subject to influence by a number of factors. All procedures which have been adopted, therefore, to determine an ionisation potential or appearance potential involve recording the ionisation efficiency curves of both the unknown and the calibrant, followed by evaluation of these curves to yield the desired value. This has been achieved by a number of essentially empirical methods, some definitely better than others, but none entirely satisfactory.

The determination of the absolute energy of the electron beam is extremely difficult and is dependent on a number of factors such as contact and surface potentials of the electrodes, the undesirable thermal energy of the electrons (of unknown magnitude) from the heated filament in addition to the energy resulting from the known applied potential to achieve acceleration, and the presence of an electric field and potential gradients within the ionisation chamber. In addition, the thermal energy of the electrons originating from the heated filament is not constant but as stated previously the energy spread is essentially Maxwellian in character. For accurate determinations of ionisation potentials, a monoenergetic beam of electrons is essential. The vast majority of values reported, however, has been calculated from ionisation efficiency curves obtained using an inhomogeneous electron beam.

It is necessary, therefore, that the calibrating gas is admitted simultaneously with the sample gas to the ion source and the two ionisation efficiency curves are recorded with as little time lapse as possible so as to reduce errors caused by an alteration in the magnitude of the above effects to a minimum. In addition, it is preferable that the pressure of the calibrating gas be adjusted so that the ion currents of it and the unknown are of similar intensity, because of the sensitivity of the detecting device to pressure. It is also advantageous to employ a reference gas whose ionisation potential is similar to that of the unknown. A number of determinations should be made (say 6 or 7) preferably during two different periods of use of the instrument, and then averaged. The instrument is thus used as a comparator.

## 3.3  Methods of determining ionisation and appearance potentials

### 3.3.1  Vanishing current method

The earliest determinations of ionisation potentials were made by noting the electron energy at the onset of ionisation, or alternatively by recording the electron energy at which the ion current vanishes as this energy is decreased[25]. This method obviously depends on the sensitivity of the instrument employed.

### 3.3.2  Initial break method

This method is a logical extension of the previous one and resulted from the difficulty experienced in determining the energy of the electron beam at

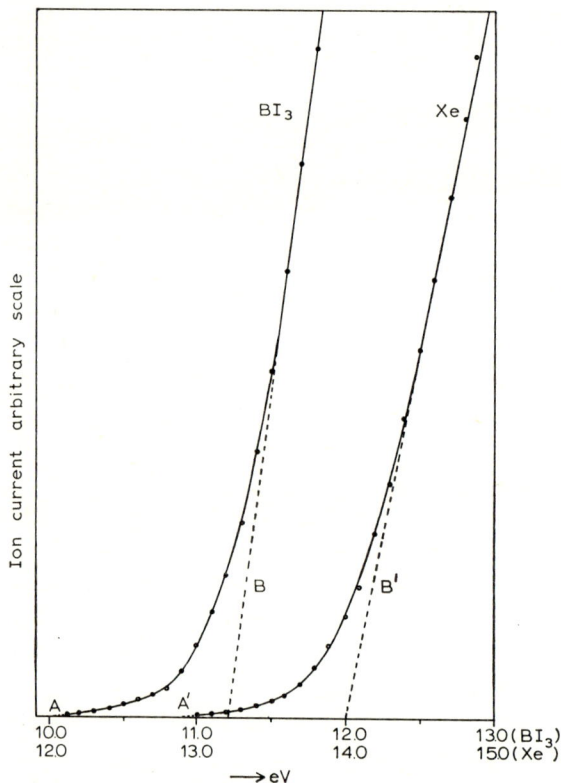

Fig. 3.6. Lower portions of the ionisation efficiency curves for $BI_3^+$ and the reference standard $Xe^+$, illustrating initial break (AA′) and linear extrapolation (BB′) methods of obtaining the ionisation potential.

which ionisation first begins. It involves extrapolation of the initial portion of the ionisation efficiency curve back to intersect with the electron axis, and is illustrated in Fig. 3.6. Because of the low sensitivity in this region and the very small slope of the foot of the ionisation efficiency curve, such a method provides unsatisfactory results. It has also been shown that the values obtained are dependent on the sample pressure[26]. Either of the terms "vanishing current" or "initial break" is often used to include both methods.

### 3.3.3 Linear extrapolation method

In an attempt to overcome the shortcomings of the initial break method, Vought[27] suggested extrapolation of the linear portion of the ionisation efficiency curve back to the electron energy axis (Fig. 3.6). Evaluation is clearly easier, but the results obtained, especially for appearance potentials, are often high.

### 3.3.4 The critical slope method

This was the next attempt to determine reliable values of ionisation and appearance potentials. Based on the shape of the initial portion of the ionisation efficiency curve, it takes into account the energy spread of the electrons. It was proposed by Honig[28] who assumed that the probability of ionisation was proportional to a power (usually the second) of the excess electron energy above the ionisation potential. A tangent of a certain critical slope was applied to the semi-log plot of ion current *versus* electron energy. This method is usually considered to provide results which are more satisfactory than those of the first two methods but is now rarely used.

### 3.3.5 Extrapolated voltage differences method

This method, proposed by Warren[29] in 1950, has proved to be fairly reliable and has been widely used. The first requirement is to record the ionisation efficiency curves to at least 4 eV above the onset of ionisation so that the linear portions are clearly apparent and their slopes may be determined (Fig. 3.7(a)). The curve of either the sample ion or the calibrant ion is then plotted on a suitably adjusted ion current scale so that the linear portions of the curves are parallel (Fig. 3.7(b)). This is conveniently accomplished by multiplying the ion current values of one of the curves by the appropriate factor and then plotting on the original scale. The voltage difference $\Delta V$ between the sample and calibrant curves at a particular value of the ion current is determined (Fig. 3.7(b)) and repeated for a succession of points on

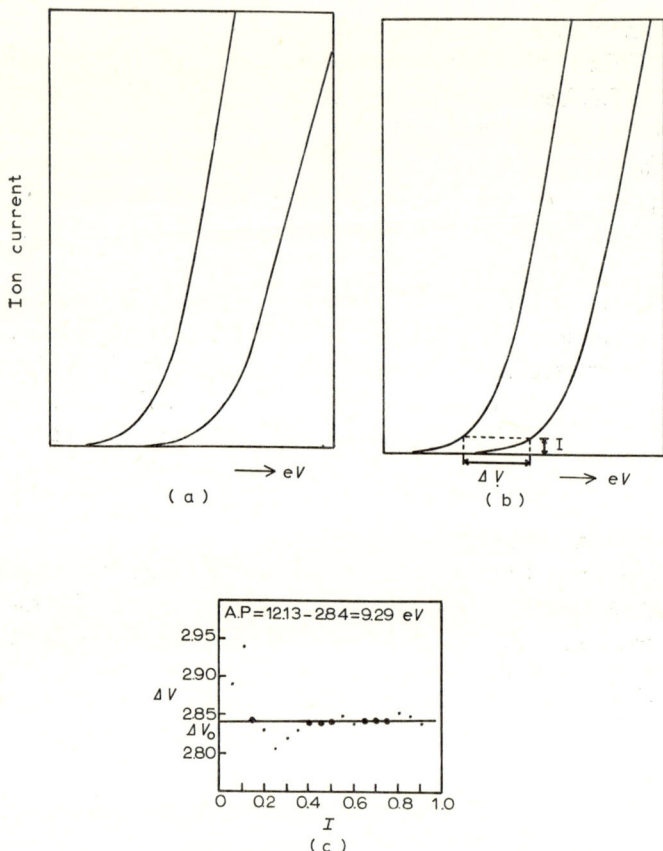

Fig. 3.7. Determination of the ionisation potential of $BI_3$ using the extrapolated voltage differences method, with xenon as the reference standard.

the initial sections of the curves. The values of $\Delta V$ so obtained are plotted against the ion current (Fig. 3.7(c)), and a linear extrapolation back to zero ion current made to obtain a value $\Delta V_0$, which is taken as the difference between the ionisation potentials of the calibrant and sample.

This method has proved fairly reliable for a wide range of compounds, and is probably the most convenient for ionisation efficiency curves with long tails.

### 3.3.6 Semi-logarithmic plot method

Lossing et al.[30] examined the initial portions of the ionisation efficiency curves for a number of gases. They plotted the ion current expressed as a percent of its value at 50 eV on a log scale against the electron energy, after the amounts of the gases had been adjusted to give approximately the same ion current at 50 eV. It was found that the curves so obtained were very nearly parallel over a range of 0–10%. They were parallel within the error of measurement for values of 1% or less. Good results have been obtained for ionisation potential and sometimes appearance potential measurements and

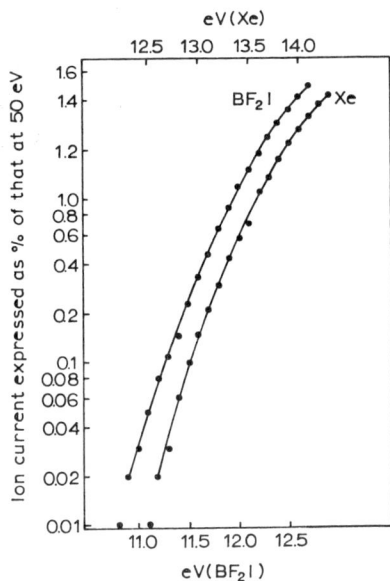

Fig. 3.8. Determination of the ionisation potential of BF$_2$I using the semi-logarithmic plot method, with xenon as reference standard. (Reproduced with permission from ref. 33.)

the method is widely used. The curve obtained[33] for BF$_2$I is shown in Fig. 3.8. Often, however, such plots for a fragment ion and an inert gas ion differ considerably from parallelism in this region. Use of the extrapolated voltage differences method is advantageous in such cases.

In practice, the mixture of the calibrating gas and the sample of unknown ionisation potential is adjusted until the respective ion currents at 50 eV are approximately equal. The electron energy is plotted against the ion current (expressed as a percent of that at 50 eV), and $\Delta V$ at, say, 1% noted. This is taken as the difference in the ionisation potentials of the two molecules.

### 3.3.7 Energy-distribution difference method

Winters et al.[24] have described an energy-distribution difference technique by means of which they were able to eliminate many of the difficulties arising from the thermal energy spread of the ionising electrons, while using a conventional ion source. Use of the equation

$$\Delta I(V) = I(V) - b\,I(V + \Delta V) \tag{3.3}$$

where $I(V)$ and $I(V + \Delta V)$ are the ion currents observed at the electron energies $V$ and $V + \Delta V$, respectively, and $b$ is a constant, effectively reduces this energy spread and eliminates the foot of the ionisation efficiency curve obtained by plotting $\Delta I$ vs. $V$ (Fig. 3.9). The calculation of the ionisation potential is not subject, therefore, to as much uncertainty as in previous methods. The method has also enabled the observation of fine

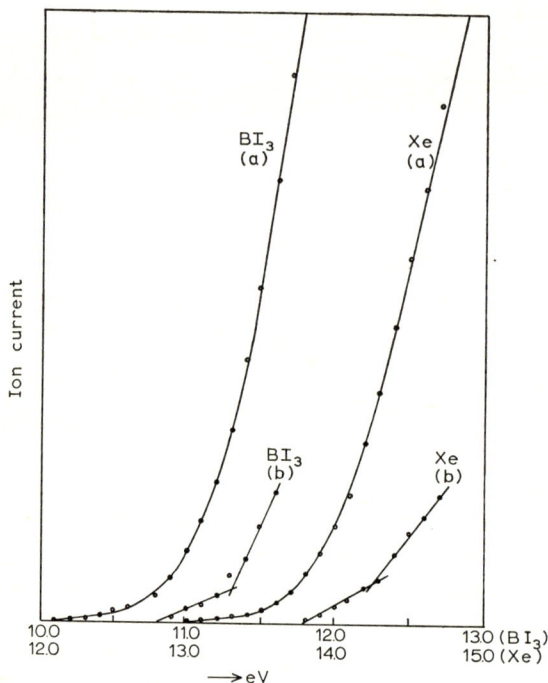

Fig. 3.9. (a) Ionisation efficiency curves for $BI_3^+$ and $Xe^+$. (b) Curves obtained by applying the energy-distribution difference technique.

structure or breaks in the ionisation efficiency curves. This is illustrated in Fig. 3.9.

The reduction in the effective electron energy spread is provided by a suitable choice of the parameter $b$, which serves to eliminate the high energy tail of the thermal energy distribution of the ionising electron beam. The ionisation potential is calculated from the intersections of the adjusted ionisation efficiency curves of unknown and calibrant with the electron energy axis. Compared with those of the other analytical method of reducing the electron energy spread discussed previously, Morrison's deconvolution method, the calculations involved in the energy-distribution difference technique are less involved and so this latter method is favoured for ionisation potential measurement. Allinson and Sedgwick[31] have recently reported a computer programme designed to select the best value of the parameter $b$, and to remove manual line fitting to the points obtained.

### 3.3.8 Monoenergetic electron-impact methods

A different approach involves the use of apparatus which produces an essentially monoenergetic electron beam, effectively removing the tail of the ionisation efficiency curve discussed previously (Fig. 3.3). The instrumentation is complex and beyond the scope of this text.

A number of relatively simple adjustments may be made to the normal electron-impact source during ionisation and appearance potential measurements so that the energy spread of the bombarding electrons is kept to a minimum. These include

(*i*) Maintenance of the filament current, and also the temperature, as low as possible. A higher filament temperature produces an increased thermal energy spread of the electrons, which in turn results in an extended tail of the ionisation efficiency curve.

(*ii*) Shorting of the ion repeller to the source. The normal positive potential of the repeller plate with respect to the ionisation chamber also tends to result in a wider energy spread of the bombarding electrons.

(*iii*) Reduction of the trap current to 5–10 $\mu$A. This will help to prevent space–charge effects and subsequent broadening of the electron energy distribution.

Although factors affecting the accuracy and reliability of ionisation potential measurements by electron-impact methods have been given only cursory mention, it should be remembered that the number of such factors is large and the magnitude of their effect uncertain. Appearance potentials are, in general, more prone to erroneous determination than are ionisation potentials, although it is felt that with recent improvements in instrumentation and technique, more accurate and reliable values are now obtained.

Although this book deals with mass spectrometric methods of determining ionisation potentials, the introduction of photoelectron spectrometers[36] as commercially available instruments provides an alternative means of determining first and lower orbital energy levels. Usually the results from photoelectron spectrometry are more accurate (to $\pm 0.05$ eV or better) than from mass spectrometry, and for most compounds mass spectrometry may now be regarded as a secondary method for the measurement of ionisation potentials. This is no way detracts from the overwhelming importance of mass spectrometry as a method of measuring appearance potentials.

Indirect determinations of ionisation potentials of molecules can be made from the heat of formation of the molecule and the heat of formation of the ion. This is discussed in the following section.

### 4. HEATS OF FORMATION OF IONS AND BOND DISSOCIATION ENERGIES

For the reaction

$$AB + e \rightarrow A^+ + B + 2e$$

the appearance potential of the ion $A^+$ is related to the heats of formation of the various species by the equation

$$AP(A^+) = \Delta H_{\text{reaction}} = \Delta H_f(A^+) + \Delta H_f(B) - \Delta H_f(AB) + E_k + E_e \quad (3.4)$$

where $E_k$ is the kinetic energy of the particles produced and $E_e$ is their excitation energy (electronic, vibrational, and rotational). It is normally assumed that $E_k$ and $E_e$ are quite small and are therefore neglected, although this may not be the case with $E_k$. Equation (3.4) then reduces to

$$AP(A^+) = \Delta H_{\text{reaction}} = \Delta H_f(A^+) + \Delta H_f(B) - \Delta H_f(AB) \quad (3.5)$$

These thermodynamic functions may also be related to the bond dissociation energy $D(A-B)$ by the equation

$$AP(A^+) = D(A-B) + I(A) \quad (3.6)$$

where $I(A)$ is the ionisation potential of the atom or radical A. The ionisation potentials of many radicals have been measured directly by pyrolyzing a suitable compound in a furnace adjacent to the ion source. The path to the ionisation chamber must be as short as possible to prevent radical recombination. Many other such values have been determined indirectly by means of eqn. (3.6). If the bond dissociation energy for a molecule A–B has been determined by some other method, measurement of the appearance potential of the ion $A^+$ from A–B will provide a means of calculating the ionisation potential of A. It should always be remembered that the determination of a bond dissociation energy by mass spectrometric measurements provides an upper limit to its value because of the terms $E_k$ and $E_e$.

It has been pointed out by Stevenson[32] that a necessary condition for eqn. (3.6) to hold is that $I(A) > I(B)$. If, instead, $I(A) < I(B)$, then

$$AP(A^+) > D(A–B) + I(A) \qquad (3.7)$$

and the second product of the reaction, B, is either electronically excited or dissociated into smaller fragments, and the products may possess kinetic energy.

In many instances, the values of bond dissociation energies obtained mass spectrometrically are in good agreement with values obtained independently by, say, kinetic methods, whereas considerable discrepancies sometimes exist. In spite of the limitations of the mass spectrometric method a great deal of valuable information has been obtained with it, especially in making comparisons within series of similar compounds.

REFERENCES

1  M. F. Lappert, J. B. Pedley, J. Simpson and T. R. Spalding, *J. Organometal. Chem.*, 29 (1971) 195.
2  L. A. Shadoff, *Anal. Chem.*, 39 (1967) 1902.
3  L. Sacconi, P. Nannelli and U. Campigli, *Inorg. Chem.*, 4 (1965) 818.
4  E. P. Dudek, E. Chaffee and G. Dudek, *Inorg. Chem.*, 7 (1968) 1257.
5  S. Pignataro and F. P. Lossing, *J. Organometal. Chem.*, 11 (1968) 571.
6  H. J. Svec and G. A. Junk, *Inorg. Chem.*, 7 (1968) 1688.
7  A. B. Baylis, G. A. Pressley, Jr. and F. E. Stafford, *J. Am. Chem. Soc.*, 88 (1966) 2428.
8  A. D. Norman, R. Schaeffer, A. B. Baylis, G. A. Pressley, Jr. and F. E. Stafford, *J. Am. Chem. Soc.*, 88 (1966) 2151.
9  R. B. King, *Fortschr. Chem. Forsch.*, 14 (1970) 92.
10  C. W. Fong, W. Kitching and M. R. Litzow, unpublished results, 1969.

11 R. W. Kiser, *Introduction to Mass Spectrometry and Its Applications*, Prentice-Hall, Englewood Cliffs, N.J., 1965, (a) p. 165; (b) p. 196.

12 W. B. Nottingham, *Phys. Rev.*, 55 (1939) 203.

13 R. E. Fox, W. M. Hickam, T. Kjeldaas and D. J. Grove, *Phys. Rev.*, 84 (1951) 859; *Rev. Sci. Instr.*, 26 (1955) 1101.

14 E. M. Clark, *Can. J. Phys.*, 32 (1954) 764.

15 P. Marmet and L. Kerwin, *Can. J. Phys.*, 38 (1960) 787, 972.

16 P. Marmet and J. D. Morrison, *J. Chem. Phys.*, 35 (1961) 746.

17 D. C. Frost and C. A. McDowell, *Proc. Roy. Soc. (London)*, 232 (1965) 277; *J. Am. Chem. Soc.*, 80 (1958) 6183.

18 F. H. Dorman, J. D. Morrison and A. J. C. Nicholson, *J. Chem. Phys.*, 32 (1960) 378.

19 C. E. Melton and W. H. Hamill, *J. Chem. Phys.*, 41 (1964) 546.

20 A. J. C. Nicholson, *J. Chem. Phys.*, 39 (1963) 954.

21 C. A. McDowell and J. W. Warren, *Discussions Faraday Soc.*, 10 (1951) 53.

22 B. Cantone, F. Grasso and S. Pignataro, *J. Chem. Phys.*, 44 (1966) 3115.

23 J. D. Morrison, *J. Chem. Phys.*, 39 (1963) 200.

24 R. E. Winters, J. H. Collins and W. L. Courchene, *J. Chem. Phys.*, 45 (1966) 1931.

25 See, for example, H. D. Smyth, *Proc. Roy. Soc. (London)*, *Ser. A*, 102 (1922) 283.

26 J. D. Waldron and K. Wood, *Mass Spectrometry Conference*, Institute of Petroleum, London, 1952, p. 16.

27 R. H. Vought, *Phys. Rev.*, 71 (1947) 93.

28 R. E. Honig, *J. Chem. Phys.*, 16 (1948) 105.

29 J. W. Warren, *Nature*, 165 (1950) 810.

30 F. P. Lossing, A. W. Tickner and W. A. Bryce, *J. Chem. Phys.*, 19 (1951) 1254.

31 I. I. O. Allinson and R. D. Sedgwick, in E. Kendrick (Ed.), *Advances in Mass Spectrometry*, Vol. 4, Inst. of Petroleum, London, 1968, p. 99.

32 D. P. Stevenson, *Discussions Faraday Soc.*, 10 (1951) 35.

33 M. F. Lappert, M. R. Litzow, J. B. Pedley, P. N. K. Riley and A. Tweedale, *J. Chem. Soc. A*, (1968) 3105.

34 J. T. Tate, P. T. Smith and A. L. Vaughan, *Phys. Rev.*, 48 (1935) 525.

35 D. J. Cardin, S. A. Keppie, M. F. Lappert, M. R. Litzow and T. R. Spalding, *J. Chem. Soc. A*, (1971) 2262.

36 D. W. Turner, C. Baker, A. D. Baker and C. R. Brundle, *Molecular Photoelectron Spectroscopy*, Wiley-Interscience, London, 1970.

GENERAL BIBLIOGRAPHY

*Instrumentation and Mass Spectra of Organic Compounds*

1 J. H. Beynon, R. A. Saunders and A. E. Williams, *The Mass Spectra of Organic Molecules*, Elsevier, Amsterdam, 1968.

2 J. H. Beynon, *Mass Spectrometry and its Application to Organic Chemistry*, Elsevier, Amsterdam, 1960.

3 K. Biemann, *Mass Spectrometry*, McGraw-Hill, New York, 1962.

4 E. W. Blauth, *Dynamic Mass Spectrometers*, Elsevier, Amsterdam, 1966.

5 H. Budzikiewicz, C. Djerassi and D. H. Williams, *Interpretation of Mass Spectra of Organic Compounds*, Holden-Day, San Francisco, 1964.
6 H. Budzikiewicz, C. Djerassi and D. H. Williams, *Structure Elucidation of Natural Products by Mass Spectrometry*, Vol. 1. *Alkaloids*; Vol. 2. *Steroids, Terpenoids, Sugars and Miscellaneous Classes*, Holden-Day, San Francisco, 1964.
7 R. M. Elliot (Ed.), *Advances in Mass Spectrometry*, Vol. 2, Pergamon Press, London, 1963.
8 F. H. Field and J. L. Franklin, *Electron Impact Phenomena*, Academic Press, New York 1957.
9 R. W. Kiser, *Introduction to Mass Spectrometry and its Applications*, Prentice-Hall, Englewood Cliffs, N.J., 1965.
10 C. A. McDowell (Ed.), *Mass Spectrometry*, McGraw-Hill, New York, 1963.
11 F. W. McLafferty (Ed.), *Mass Spectrometry of Organic Ions*, Academic Press, New York 1963.
12 C. E. Melton, *Principles of Mass Spectrometry of Negative Ions*, Marcel Dekker, New York, 1970.
13 R. I. Reed, *Ion Production by Electron Impact*, Academic Press, New York, 1962.
14 R. I. Reed (Ed.), *Mass Spectrometry*, Academic Press, New York, 1965.
15 R. I. Reed, *Applications of Mass Spectrometry to Organic Chemistry*, Academic Press, New York, 1966.
16 R. I. Reed, *Modern Aspects of Mass Spectrometry*, Plenum Press, New York, 1968.
17 J. Roboz, *Introduction to Mass Spectrometry: Instrumentation and Techniques*, Interscience, New York, 1968.
18 J. D. Waldron (Ed.), *Advances in Mass Spectrometry*, Vol. 1, Pergamon Press, London, 1959.

*Mass Spectra of Inorganic and Organometallic Compounds*

1 M. I. Bruce, *Advan. Organometal. Chem.*, 6 (1968) 273.
2 M. Cais and M. S. Lupin, *Advan. Organometal. Chem.*, 8 (1970) 211.
3 D. B. Chambers, F. Glockling and J. R. C. Light, *Quart. Rev.*, 22 (1968) 317.
4 F. Glockling, *The Chemistry of Germanium*, Academic Press, New York, 1969.
5 R. B. King, *Fortschr. Chem. Forsch.*, 14 (1970) 92.
6 R. W. Kiser, *Characterization of Organometallic Compounds, Part 1*, M. Tsutsui (Ed.), Interscience, New York, 1969, p. 137.
7 J. Lewis and B. F. G. Johnson, *Accounts Chem. Res.*, 1 (1968) 245.
8 M. Lesbre, P. Mazerolles and J. Satgé, *The Organometallic Compounds of Germanium*, Interscience, New York, 1971.
9 J. L. Margrave (Ed.), *Mass Spectrometry in Inorganic Chemistry*, *Advan. Chem. Ser.*, Vol. 72, American Chemical Society, Washington, D.C., 1968.
10 T. R. Spalding, *Spectroscopic Methods in Organometallic Chemistry*, W. O. George (Ed.), Butterworths, London, 1970, p. 95.

# PART 2

# Introduction

T. R. SPALDING

It is convenient to discuss the mass spectra of inorganic and organometallic compounds of main group elements and rare gases in terms of the central (usually metallic) *element* with other groups attached. Transition metal compounds, on the other hand, are more conveniently treated in terms of the *ligands* which are attached to the central metal atom. The reason for this distinction is that the influence of the main group elements on the mass spectra of their compounds is, on the whole, very marked whereas transition metals influence mass spectra much more subtly and the major features are usually ligand-oriented. Main group compounds are therefore discussed before transition metal compounds since most ligands which are not organic are derived from compounds of the main group elements. In all the chapters in Part 2 the elements and their inorganic compounds, that is their oxides, sulphides, etc., and halides are discussed before the compounds which contain metal-to-carbon bonds.

The most common use of mass spectrometry is the determination of molecular weight and the types of groups present in a compound. In inorganic and organometallic chemistry as in organic chemistry, mass spectrometry has proved to be enormously useful for this purpose. One example here will suffice. The structure of a compound of the formula $C_{16}H_{16}Ge$ was originally suggested[1] to be I. The mass spectrum showed that the compound was, in fact, $C_{32}H_{32}Ge_2$ (II) with a parent molecular ion in the range $m/e$ 556–568, and had the expected fragments corresponding to the ions (parent molecule–$GeMe_2$)$^+$, and $Me_4Ge_2^+$ (ref. 2).

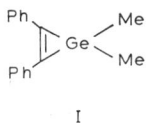

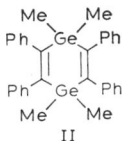

In other cases, high-resolution mass spectrometry has been used to establish the exact formula of a compound, *e.g.* $Me_3SnSO_2Ph$ (ref. 3) and $[(Cl_2Ge)-Fe(CO)_4]_2$ (ref. 4). The determination of molecular weight by mass spectrometry should always be done bearing in mind the known characteristics of the compound being investigated. Chambers *et al.*[5] in a study of some organoberyllium compounds have shown that the spectrum of diethylberyllium at 45–50° is quite different from that at 196 °C. At the lower temperature ions derived from the trimeric $(Et_2Be)_3$ species were observed, but at the higher temperature only ions due to the dimeric species were found. Therefore, in compounds where association is known to occur, a careful investigation of the effect of experimental conditions on the mass spectrum should be undertaken if the highest molecular weight species is to be observed. Apart from pyrolysis effects, other effects which have commonly to be taken into account are the stability of a compound towards hydrolysis, the possibility of catalytic decomposition and "memory" effects (see Chapter 3, Section 1).

A convenient theory for the quantitative prediction of the mass spectra of inorganic and organometallic compounds has not yet been developed. The "quasi-equilibrium" theory proposed by Rosenstock *et al.*[6], which has been applied with some success to a few simple organic molecules such as paraffin hydrocarbons up to $C_5$, ethyl chloride, and propanol, has not been extended to inorganic and organometallic compounds. It seems doubtful if it would have greater success in this field. Because of this the interpretation of mass spectra relies heavily on empiricism. This is a fact that few mass spectroscopists can be happy about, but which most have come to accept. However, several generalisations about ion fragmentation and the factors affecting ion abundance can be formulated. The abundance of any ion is related to both its rate of formation and its rate of decomposition. In terms of the "quasi-equilibrium" theory the mass spectrum is a record of the position of the quasi-equilibrium of these rates for the ions observed. Factors which will influence these rates will include the ionisation potentials of the species from which the ions are derived, the electron configuration of the ions, bond dissociation energies in the ions and the possibility of ion stabilisation by conjugation effects or rearrangements and the possibility of decomposition by facile elimination reactions. With regard to the first factor, the effect of the ionisation potential of the species from which the ions are derived can be seen in the following way. Suppose the ion $ML_n^+$ where M is the main group element, rare gas or transition metal and L is some attached atom or group, may fragment by two paths (1) or (2) giving an ion and a radical\*.

$$ML_n^{\ddagger} \rightarrow ML_{(n-1)}^+ + L\cdot \tag{1}$$

$$\rightarrow L^+ + \cdot ML_{(n-1)} \tag{2}$$

It is generally found that the path which is favoured is the one which produces the ion whose neutral species has the lower ionisation potential, *i.e.* if the ionisation potential of $\cdot ML_{(n-1)}$ is lower than $L\cdot$ then path (1) will be favoured and if the reverse then path (2) will be favoured. Now, since M-containing species generally have lower ionisation potentials than the L (organic or inorganic) ligands, nearly all of the ions in the spectra of inorganic and organometallic compounds are M-containing ions. In succeeding chapters, the reader will come to realise the importance of this generalisation. Two examples from the vast number of cases are given here, (3) and (4).

$$BCl_3^+ \rightarrow BCl_2^+ + Cl\cdot \tag{3}$$

I.P. $BCl_2\cdot = 7.52$ eV (ref. 7)

I.P. $Cl\cdot \quad = 13.01$ eV (ref. 8)

$$Me_4Sn^{\ddagger} \rightarrow Me_3Sn^+ + Me\cdot \tag{4}$$

I.P. $Me_3Sn\cdot = 6.83$ eV (ref. 9)

I.P. $Me\cdot \quad = 9.95$ eV (ref. 8)

This generalisation is not, however, an exclusive rule since quite often sufficient energy is available to produce both $ML_{(n-1)}^+$ and $L^+$ ions, particularly if the ionisation potentials of $\cdot ML_{(n-1)}$ and $L\cdot$ are close. For example both $Me_3Sn^+$ and $Me_3C^+$ were observed from $Me_3Sn\text{–}CMe_3$ (I.P. $Me_3C\cdot = 7.45$ eV)[9]. Relatively few cases where the L radical has a lower ionisation potential than $\cdot ML_{(n-1)}$ and $L^+$ ions are favoured are known. Some examples which are metastable supported are (5) and (6).

---

\* The notation adopted in this book for even and odd electron ions, where it is possible to distinguish them, is to write even electron ions as $ML_m^+$ and odd electron ions as $ML_p^{\ddagger}$ (the dot signifying the odd electron).

From tetra o-(or m-)tolylgermanium

$$C_7H_7Ge^+ \rightarrow C_7H_7^+ + Ge \tag{5}$$

I.P. $C_7H_7\cdot$ (tropylium radical) = 6.60 eV (ref. 8)
     (benzyl radical) = 7.72 eV (ref. 8)
I.P. Ge                    = 7.88 eV (ref. 8)

From tricyclopentadienyl arsenic

$$C_5H_5As\stackrel{+}{\cdot} \rightarrow C_5H_5^+ + As\cdot \tag{6}$$

I.P. $C_5H_5\cdot$           = 8.69 eV (ref. 8)
I.P. As$\cdot$              = 9.81 eV (ref. 8)

Several authors have observed that in the mass spectra of organic compounds, ions which have an even rather than odd electron configuration are favoured[10]. This also appears to be true for main group compounds. For transition metal compounds the situation is not so clear because the underlying 'd' shell may or may not be able to interact with the electrons in the valency orbitals in such a way as to invalidate the significance of an even or odd electron configuration. Therefore discussion of the effects of even and odd electron configurations cannot have the same meaning for main group and transition metal compounds. As far as the main group compounds are concerned, some even electron ions may be expected to be particularly stable. For instance, the ion $ML_{(n-1)}^+$ produced by the loss of an L radical from a $ML_n\stackrel{+}{\cdot}$ parent molecular (odd electron) ion may be expected to be reasonably stable since it would have the electron configuration of the corresponding stable compound $M'L_{(n-1)}$ of the element $M'$ in the preceding group. An example of this is the occurrence of $Me_3M^+$ ions as the base peaks in the spectra of $Me_4M$ compounds (M = Si, Ge, Sn, Pb)[9]. These ions will have the electron configuration of the corresponding stable $Me_3M'$ compounds ($M'$ = Al, Ga, In, Tl, respectively). On this basis, one may also expect the $ML_{(n-3)}^+$ ions to be stabilised to some extent by an "inert pair" effect and that this effect will be more pronounced in the compounds of the heavier elements in each group. The spectra of $Me_4M$ compounds (M = Ge, Sn, Pb) lend support to this view, the corresponding $CH_3M^+$ ions being the second most abundant ions. The situation observed with main group compounds has no obvious equivalent in compounds of transition metals. However, the negative ion spectrum of $V(CO)_6$ shows the

$V(CO)_6^-$ ion[11], and the completion of the valency shell (the "18 electron rule") may contribute to the stability of this ion. Mass spectra are sometimes discussed with reference to bond energies or bond dissociation energies. Most of the discussions ignore the fact that nearly all of the rather scarce bond energy or bond dissociation energy data available concern neutral species and not ions. However, the use of what are, for the most part, intuitive ideas of chemical bond strengths cannot be denied particularly when discussing, say, a series of compounds $L_nM-X$ which fragment by breakage of a single bond to produce $L_nM^+$ and $L_{(n-1)}MX^+$ ions. For instance, $Ph_3GeX$ compounds (X = Cl, Br, or I) produce $Ph_3Ge^+$ and $Ph_2GeX^+$ ions in abundances which suggest the Ge–X bond strengths lie in the order Ge–Cl > Ge–Br > Ge–I (Table 7.20, page 263, in accord with chemical intuition. However, one must remember that the process considered above represents only the formation of the ions and ignores their relative stability to further decomposition. That the decomposition processes are as important as those of formation can be seen easily by consideration of the corresponding ions from the tin compounds $Ph_3SnX$ (X = F, Cl, Br or I; Table 7.20). The abundances of the $Ph_3Sn^+$ and $Ph_2SnX^+$ ions suggest the Sn-X bond strengths in the order Sn–F $\approx$ Sn–Cl $\sim$ Sn–Br $\gg$ Sn–I. The suggestion that the Sn–F bond is about as strong as the Sn–Br bond in $Ph_3SnX$ compounds is highly suspect. Hence one must be extremely careful about interpreting ion abundances in terms of bond strengths as was pointed out in Chapter 2, Section 3, although quite often there appears to be some correlation. Other factors affecting stability of ions include the possibility of the delocalisation of charge in ions which contain aromatic groups or conjugated groupings. The importance of such effects is difficult to assess. In some cases, for example the ion formed by the elimination of hydrogen from $Ph_2P^+$, III, they may be very important.

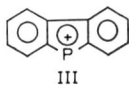

III

Factors such as the possibilities of rearrangements and eliminations are generally more obviously important. Rearrangement ions in the mass spectra of inorganic and organometallic compounds are often numerous and very abundant *. All the possible rearranged ions of the types $^+MPh_nMe_{(3-n)}$

---

* The common convention adopted by most mass spectroscopists that a rearrangement involving the shift of one electron from one part of the ion to another is shown by an

$(n = 1–3)$ and $^+M'Ph_mMe_{(3 - m)}$ $(m = 0–2)$ have been observed[12] from $Me_3Ge–SnPh_3$. Rearrangements of fluorine atoms from fluorocarbon groups to give M–F bonds are very common in the spectra of fluorocarbon derivatives. The decomposition of ions by the loss of a group or atom as a radical is extremely common. So, too, is the elimination of a neutral group (usually an entity which is chemically stable). Groups which are commonly seen to be eliminated are hydrogen, olefins (from alkyl groups) other hydrocarbons (including $CH_4$, $:CH_2$, benzene, biphenyl, and acetylene especially from aromatic ligands), CO, $N_2$, $NH_3$, $H_2O$, $COCl_2$, HX (X = halogen or pseudohalogen), RX (R may be aliphatic or aromatic) and complete organometallic entities such as $SiMe_4$ (e.g. from some siloxanes) and $BeMe_2$ (e.g. from some polymeric ions in the spectrum of dimethylberyllium). The elimination of neutral groups takes place more often from even electron ions to produce other even electron ions than from odd electron ions which prefer to decompose by the loss of a radical. Although the reader will come across numerous examples of eliminations in succeeding chapters, one example is given here since it illustrates several of the points discussed above. The spectrum of dimethylaluminium hydride[5], a trimer in inert solvents, shows ions due to trimeric species at low source temperatures, and high abundance of dimeric ions which include $Me_4HAl_2^+$ and $Me_3H_2Al_2^+$ both even electron ions. The metastable supported decompositions (7) and (8) are observed from these ions, eliminating the even electron neutral fragments $Me_2AlH$ and $MeAlH_2$ respectively to produce the even electron ion $Me_2Al^+$, which has the electron configuration of $MgMe_2$.

$$Me_4HAl_2^+ \quad\xrightarrow[-Me_2HAl]{*}\quad Me_2Al^+ \quad (7)$$

$$Me_3H_2Al_2^+ \quad\xrightarrow[\phantom{-MeH_2Al}]{-MeH_2Al \quad *}\quad Me_2Al^+ \quad (8)$$

It should be noted that in (7) and (8) the species eliminated are those with the most Al–H bonds which also have higher ionisation potentials than the $Me_2Al$ radical. (The alternatives—loss of $Me_3Al$ and $Me_2AlH$ or $Me_3Al$ to produce $MeAlH^+$ or $H_2Al^+$—are less favourable on ionisation

arrow ⤵ whereas a rearrangement involving the shift of two electrons is shown by ⤵ is used in this book provided the stereochemical nature of the rearrangement can be reasonably assessed.

potential grounds, I.P. $H_2Al\cdot$ > I.P. $MeHAl\cdot$ > I.P. $Me_2Al\cdot$). The ion $Me_2Al^+$ further fragments by loss of $CH_4$ or $C_2H_4$ or the successive loss of two methyl radicals.

It has been the general policy in Part 2 to discuss in detail the spectra of those compounds whose spectra are fully reported. Compounds for which relatively little data have been reported are listed at the end of the relevant section, or references are given to them in appendices at the end of the relevant chapter. Comparatively few studies of the energetics of ion formation have been reported for inorganic and organometallic compounds. The results are presented and discussed in the relevant sections except for the work on main Group IV organometallic compounds which is discussed in a separate section (Chapter 7, Section 8).

REFERENCES

1 M. E. Vol'pin and D. N. Kursanov, *Izv. Akad. Nauk SSSR*, (1960) 1903; M. E. Vol'pin, Yu. D. Koreshkov, V. G. Dulova and D. N. Kursanov, *Tetrahedron*, 19 (1962) 107.
2 F. Johnson, R. S. Golke and W. A. Nasutavicus, *J. Organometal. Chem.*, 3 (1965) 233.
3 C. W. Fong, W. Kitching and M. R. Litzow, unpublished results, 1969.
4 R. Kummer and W. A. G. Graham, *Inorg. Chem.*, 7 (1968) 1208.
5 D. B. Chambers, G. E. Coates and F. Glockling, *Discussions Faraday Soc.*, 47 (1969) 157.
6 H. M. Rosenstock, M. B. Wallenstein, A. L. Wahrhartig and H. Eyring, *Proc. Natl. Acad. Sci. U.S.*, 38 (1952) 667; H. M. Rosenstock and M. Krauss, *Mass Spectrometry of Organic Ions*, F. W. McLafferty (Ed.), Academic Press, New York, 1963.
7 V. H. Dibeler and J. A. Walker, *Inorg. Chem.*, 8 (1969) 50.
8 V. I. Vedeneyev, L. V. Gurvich, V. N. Kondrat'yev, V. A. Medvedev and Ye. L. Frankevich, *Bond Energies, Ionisation Potentials and Electron Affinities*, Arnold, London, 1966.
9 M. F. Lappert, J. B. Pedley, J. Simpson and T. R. Spalding, *J. Organometal. Chem.*, 29 (1971) 195.
10 F. W. McLafferty (Ed.), *Mass Spectrometry of Organic Ions*, Academic Press, New York, 1963.
11 R. E. Sullivan, M. S. Lupin and R. W. Kiser, *Chem. Commun.*, (1969), 655.
12 D. B. Chambers and F. Glockling, *J. Chem. Soc. A*, (1968), 735.

*Chapter 4*

# The Group I Elements

M. R. LITZOW

1. INTRODUCTION

The majority of the mass spectrometric investigations of Group I compounds has involved high-temperature studies of the metal halides and other inorganic derivatives. The mass spectra of only a few organo-derivatives have been reported; no parent molecular ions were observed in these spectra, even at low ionising voltages.

2. THE ELEMENTS

Both sodium and caesium have only one naturally occurring isotope, while lithium, potassium, and rubidium have two, three, and two respectively (Table 4.1., Fig. 4.1).

Table 4.1

Natural Abundance of Main Group I Isotopes[1]

| Element | Mass no. | Abundance (%) |
|---------|----------|---------------|
| Li      | 6        | 7.42          |
|         | 7        | 92.58         |
| Na      | 23       | 100           |
| K       | 39       | 93.10         |
|         | 40       | 0.01          |
|         | 41       | 6.88          |
| Rb      | 85       | 72.15         |
|         | 87       | 27.85         |
| Cs      | 133      | 100           |

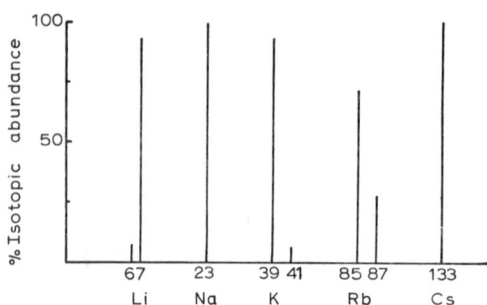

Fig. 4.1. Natural isotopic abundances of main Group I elements (ref. 1).


3. HIGH-TEMPERATURE STUDIES

Homonuclear diatomic molecules of all of the alkali metals and also the intermetallic molecules NaK, NaRb, KHg, and CsHg have been detected in low concentrations and their dissociation energies measured[2] (Table 4.2); those of the former decrease as the group is descended.

The alkali metal halides have been extensively studied since the combination of high-temperature Knudsen cell and mass spectrometric techniques was introduced. Berkowitz and Chupka[3] investigated the molecular species present in the alkali halide vapours in equilibrium with their condensed phases by observing mass spectrometrically the ions produced by electron impact. They established that both dimers and trimers were present and, in addition, tetramers of LiF, LiCl, LiBr, and NaF were observed. Gorokhov[4] reported monomers, dimers, trimers, and tetramers in LiI vapour with

Table 4.2

Bond dissociation energies of diatomic molecules $Li_2$, $Na_2$, $K_2$, $Rb_2$, $Cs_2$ and NaK, NaRb, KHg, and CsHg

| Molecule | $D$ (M–M) (kcal·mole$^{-1}$) | Molecule | $D$ (M–M') (kcal·mole$^{-1}$) |
|----------|------------------------------|----------|-------------------------------|
| $Li_2$   | $25.0 \pm 1.2$               | NaK      | $14.3 \pm 0.7$                |
| $Na_2$   | $17.3 \pm 0.7$               | NaRb     | $13.1 \pm 1.5$                |
| $K_2$    | $11.8 \pm 1.2$               | KHg      | $1.08 \pm 0.05$               |
| $Rb_2$   | $10.8 \pm 1.2$               | CsHg     | $1.10 \pm 0.05$               |
| $Cs_2$   | $10.4 \pm 1.0$               |          |                               |

predominance of the $Li_2I_2$ species. A large number of independent electron-impact studies of these compounds has been made but the sensitivity of detection was often not as high as that attained by Berkowitz and Chupka. No other reports of tetramers have been made, but monomers, dimers, and trimers were observed for the compounds MX where M = Li, X = F (refs. 5, 6), Cl (refs. 7–9), Br (ref. 9), I (ref. 10); M = Na, X = F (refs. 5, 6, 9), Cl (ref. 7); M = K, Rb, Cs, X = F (ref. 5); M = Cs, X = I (ref. 11). Miller and Kusch[9] observed dimers of all the alkali halides and, in addition, other workers reported dimers, but no higher polymeric species, of the following: M = Na, X = Cl (refs. 8, 9), I (ref. 9); M = K, X = Cl (refs. 7–9), I (ref. 9); M = Rb, X = Cl (refs. 7–9); M = Cs, X = Cl (refs. 7, 12). Berkowitz and Chupka[13] have since employed photoionisation rather than the electron-impact technique in a study of sodium iodide in which they detected the monomer and dimer. The same authors noted[3] that the lithium halides exhibited more dimeric than monomeric species, and in their investigation of lithium and sodium fluorides, Porter and Schoonmaker[6] observed that the lithium compound had the higher concentration of dimers and trimers. In another photoionisation study, Berkowitz[14] observed the monomer, dimer and trimer of CsF, and monomers and dimers of CsCl, CsBr, and CsI. The alkali halide dimer molecules have been calculated[15] to be 40–60 kcal.mole$^{-1}$ more stable energetically than their separated monomers. Negative ion spectra of the lithium, sodium, and potassium chlorides have also been reported[8].

Binary mixtures of the compounds have been studied[5,6,12,16] and formation of species such as $MM'X_2$ observed. The mixtures $MCl_2$ + CsCl (M = Cd, Pb) produced results[12] which were interpreted to indicate the existence of complex molecules of the type $CsMCl_3$.

No metastable transitions have been detected. Thermodynamic properties such as heats of sublimation[10,17], heats of dimerisation[5,9,10,17,18], and dissociation energies D(M–X)[10,13] determined in these high-temperature investigations, have also been reported.

Of considerable concern in this field is the lack of agreement among reported appearance potential measurements. The very high values obtained in some instances suggest that the ions under scrutiny may be formed with excess kinetic energy[12,19-21]. Although it has been observed[22] that this is indeed so for the $Cs^+$ ion derived from CsCl, it has been shown[8] that such is not the case with ions derived from LiCl, NaCl, and KCl. Bloom et al.[23] believed that the disagreement in the reported appearance potential values was primarily due to the methods employed to derive them from the ionisation efficiency curves. They showed that application of the linear extrapolation method to curves with a large "tail" gave values 2–3 eV higher than if the energy of first appearance was compared with that of the standard. Other possible reasons for the discrepancies were also discussed. Values which have been reported are summarised in Table 4.3. Bond dissociation energies, derived from these appearance potential measurements, are, of course, also discordant.

Table 4.3

Appearance Potential Data for Alkali Halide Vapours

| Salt | Ion | Exptl. A.P. (eV) | Method | Calcd. A.P. (eV) | Ref. |
|------|-----|------------------|--------|------------------|------|
| LiF | $Li^+$ | 11.5 $\pm$ 0.3 | a | $11.35^d$ | 18 |
|  | $LiF^+$ | 11.3 $\pm$ 0.3 |  |  |  |
|  | $Li_2F^+$ | 11.5 $\pm$ 0.3 |  |  |  |
| LiCl | $Li^+$ | 10.6 $\pm$ 0.3 | a | $10.4^d$ | 18 |
|  | $LiCl^+$ | 10.1 $\pm$ 0.3 |  |  |  |
|  | $Li_2Cl^+$ | 10.6 $\pm$ 0.3 |  |  |  |
| LiBr | $Li^+$ | 9.9 $\pm$ 0.3 | a | $9.8^d$ | 18 |
|  | $LiBr^+$ | 9.4 $\pm$ 0.3 |  |  |  |
|  | $Li_2Br^+$ | 9.9 $\pm$ 0.3 |  |  |  |
| LiI | $Li^+$ | 9.2 $\pm$ 0.3 | a | $8.9^d$ | 18 |
|  | $LiI^+$ | 8.6 $\pm$ 0.3 |  |  |  |
|  | $Li_2I^+$ | 8.8 $\pm$ 0.3 |  |  |  |
| NaCl | $Na^+$ | 9.5 $\pm$ 0.2 | a | $9.4^d$ | 23 |
|  |  | 11.9 $\pm$ 0.3 | b |  | 23 |
| NaI | $Na^+$ | 8.7 $\pm$ 0.3 | a | $8.2^d$ | 20 |
|  |  | 9.6 $\pm$ 0.3 | ? |  | 21 |
|  | $NaI^+$ | 8.7 $\pm$ 0.3 | a |  | 20 |
|  |  | 8.8 $\pm$ 0.3 | ? |  | 21 |
|  |  | 7.64 | c |  | 13 |
|  |  | 8.0 $\pm$ 0.3 | ? |  | 25 |
|  | $Na_2I^+$ | 7.80 $\pm$ 0.05 | c |  | 13 |
| KF | $K^+$ | 9.5 $\pm$ 0.3 | a | $9.4^d$ | 24 |
| KCl | $K^+$ | 9.1 $\pm$ 0.3 | a | $8.7^d$ | 23 |
|  |  | 11.8 $\pm$ 0.3 | b |  | 23 |
|  | $KCl^+$ | 8.0 $\pm$ 0.3 | a |  | 23 |
| KI | $K^+$ | 8.6 $\pm$ 0.3 | a | $7.7^d$ | 20 |
|  | $KI^+$ | 8.2 $\pm$ 0.3 | a |  | 20 |

Table 4.3 (continued)

| | | | | | | |
|---|---|---|---|---|---|---|
| RbCl | Rb⁺ | 8.8 | ± 0.3 | b | 8.6 [d] | 3 |
| | | 6 | | | 5.0 [e] | 3 |
| | | 8.9 | ± 0.5 | a | 8.6 [d] | 23 |
| | | 11.8 | ± 0.3 | b | | 23 |
| RbI | Rb⁺ | 8.1 | ± 0.3 | a | 7.6 [d] | 20 |
| CsF | Cs⁺ | 9.2 | ± 0.3 | a | 9.2 [d] | 23 |
| | CsF⁺ | 8.75 | ± 0.1 | c | | 14 |
| CsCl | Cs⁺ | 12.0 | ± 0.2 | ? | | 21 |
| | | 11.9 | ± 0.2 | b | | 23 |
| | | 8.4 | ± 0.3 | a | 8.3 [d] | 23 |
| | CsCl⁺ | 8.3 | ± 0.3 | a | | 23 |
| | | 8.47 | ± 0.07 | c | | 14 |
| CsBr | CsBr⁺ | 8.06 | ± 0.05 | c | | 14 |
| | Cs₂Br⁺ | 8.14 | ± 0.06 | c | | 14 |
| CsI | Cs⁺ | 7.3 | ± 1.0 | ? | 7.2 [d] | 11 |
| | | 8.3 | ± 0.3 | a | | 20 |
| | CsI⁺ | 7.3 | ± 1.0 | ? | | 11 |
| | | 7.46 | ± 0.05 | c | | 14 |
| | Cs₂I⁺ | 10.7 | ± 1.0 | ? | | 11 |

[a] The method of first appearance, or initial break method.
[b] The linear extrapolation method.
[c] Photoionisation.
[d] The process considered was $MX + e \rightarrow M^+ + X\cdot + 2e$.
[e] The process considered was $MX + e \rightarrow M^+ + X^- + e$.

A mass spectral examination of the vapour above heated lithium oxide[26] revealed the presence of the species $Li_2O$, $LiO$, and $Li_2O_2$ but the latter species was not observed in three similar studies[27–29]. The work by Berkowitz et al.[27] demonstrated that lithium oxide sublimed mainly by decomposition to the elements at 1100 °C but an appreciable partial pressure of $Li_2O$ (g) in equilibrium with the solid was indicated. The presence of sodium as an impurity in lithium oxide led to the identification of LiONa which appeared to have considerable stability[27]. The reaction of water vapour with lithium oxide was followed by mass spectrometry and the vapour phase was shown to consist mainly of LiOH but, in addition, the dimeric and trimeric species $Li_2(OH)_2$ and $Li_3(OH)_3$ were present[30]. Sodium hydroxide, on the other hand, was shown[31] to vaporise mainly as gaseous dimers in the temperature range 300–400 °C. Other compounds reported to exist in dimeric as well as monomeric forms were the nitrates and nitrites of both lithium and sodium[32], sodium and potassium perrhenates[33], and sodium cyanide[34].

Thermodynamic properties such as heats of sublimation[26,27,33,35], heats of dimerisation[31,32,36], dissociation energies[2,26,27], and heats of formation[26,30] were reported.

The electric deflection of mass spectrometrically detected beams was used by Buchler et al.[29] to study the molecular geometry of high-temperature species by determining whether or not they possess permanent dipoles. The results were in agreement with the generally assumed planar structure for the lithium halide dimers and with a linear structure for $Li_2O$, which is in accord with an infrared study of the molecule[37].

### 4. ORGANOMETALLIC COMPOUNDS OF GROUP I

Only a few reports of mass spectrometric investigations of organo-derivatives of Group I elements have appeared. It was shown[38,39] that in the vapour state ethyl-lithium, I, was present as the hexamer (Table 4.4), and on the basis of appearance potential measurements (Table 4.5) it was concluded that tetramer units were also present. In addition, it was also postulated from appearance potential data that both polymeric forms consist of an inner lithium core surrounded by ethyl groups. This was confirmed by subsequent infrared and nuclear magnetic resonance studies[40], and X-ray analysis[41,42]. The compounds lithiomethyltrimethylsilane, II, and lithium tert.-butoxide, III, have been similarly investigated[39,43]; II was present in the vapour state solely as a tetramer while III was obviously present as the hexamer and appearance potential data showed clearly that this was the only species present to any significant extent in the vapour. These observations are basically in agreement with association behavior of these compounds in solution[39,43].

The fragmentation patterns of all three compounds were very similar (Table 4.4). Parent ion species $Li_nR_n^+$ were not present in sufficient concentration for unambiguous identification, and the most intense peak observed in each spectrum was due to the $Li_2R^+$ ion. The initial reaction occurring with ethyl-lithium[38] was therefore assumed to be

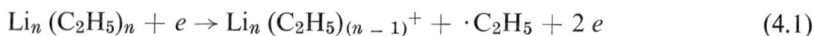

$$Li_n (C_2H_5)_n + e \rightarrow Li_n (C_2H_5)_{(n-1)}^+ + \cdot C_2H_5 + 2e \qquad (4.1)$$

Table 4.4

Relative Intensities of Ions in the Mass Spectra of Organolithium and Organopotassium Compounds

| Ion | $LiC_2H_5$ S[a] (refs. 38, 39) | $LiC_2H_5$ US[a] | $LiCH_2Si(CH_3)_3$ (refs. 39, 43) | | $LiOC(CH_3)_3$ (refs. 39, 43) | $KOC(CH_3)_3$ (ref. 45) |
|---|---|---|---|---|---|---|
| $M_6R_6-\cdot CH_3^+$ | | | | | 4 | |
| $M_6R_5^+$ | 24 | 15 | 16 | | | 87 |
| $M_5R_4^+$ | 1.3 | 1.3 | 1 | | | 5 |
| $M_4R_4-\cdot CH_3^+$ | | | 2 | | | |
| $M_4R_3^+$ | 47 | 46 | 67 | 24 | 40 | 55 |
| $M_4R_3-CH_4^+$ | | | | | | 7 |
| $M_3R_3-\cdot CH_3^+$ | | | | | | 2 |
| $M_3R_2^+$ | 15 | 14 | 15 | 2 | 11 | 10 |
| $M_3R_2-CH_4^+$ | | | | | | 1 |
| $M_2R_2-\cdot CH_3^+$ | | | | | | 1 |
| $M_2R^+$ | 100 | 100 | 100 | 100 | 100 | 100 |
| $M_2R-CH_4^+$ | | | | | | 8 |
| $MR-\cdot CH_3^+$ | | | | | | 1 |
| $M^+$ | 14 | 17 | 3 | 38 | 0.5 | 13 |

[a] Relative intensities of the peaks are given for when the vapour in the effusion cell was in equilibrium with solid (S), and for when it was undersaturated (US).

Table 4.5

Appearance Potentials of Organolithium Fragment Ions (eV)

| Ion | $LiC_2H_5$ (refs. 38, 39) | $LiCH_2Si(CH_3)_3$ (refs. 38, 43) | | $LiOC(CH_3)_3$ (refs. 38, 43) |
|---|---|---|---|---|
| $Li_6R_6-\cdot CH_3^+$ | | | | 8.8 |
| $Li_6R_5^+$ | 7.7 ± 0.5 | 8.2 | | 9.6 |
| $Li_5R_4^+$ | 12.5 ± 0.5 | | | 25.7 |
| $Li_4R_4-\cdot CH_3^+$ | | 9.7 | | |
| $Li_4R_3^+$ | 8.0 ± 0.5 | 7.6 | 8.4 | 17.4 |
| $Li_3R_2^+$ | 11.7 ± 0.5 | 14.0 | 15.2 | 25.8 |
| $Li_2R_2-\cdot CH_3^+$ | | 11.9 | | >20 (est.) |
| $Li_2R^+$ | 11.6 ± 0.5 | 10.8 | 11.8 | 25.5 |
| $Li^+$ | 14 ± 2 | 30 | 24.9 | 52 |

followed by the fragmentation processes

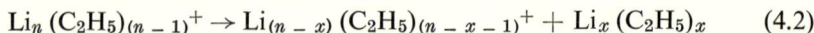

$$Li_n (C_2H_5)_{(n-1)}^+ \rightarrow Li_{(n-x)} (C_2H_5)_{(n-x-1)}^+ + Li_x (C_2H_5)_x \qquad (4.2)$$

However, the appearance potential values[38] (Table 4.5) for the various fragment ions did not rule out a stepwise fragmentation process for the tetramer, *viz.* (4.3)

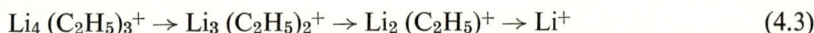

$$Li_4 (C_2H_5)_3^+ \rightarrow Li_3 (C_2H_5)_2^+ \rightarrow Li_2 (C_2H_5)^+ \rightarrow Li^+ \qquad (4.3)$$

Indeed the fact that $Li_3 (C_2H_5)_2^+$ and $Li_2 (C_2H_5)^+$ had identical appearance potentials (within experimental error), together with the low relative intensity of the former, suggested that $Li_4 (C_2H_5)_3^+$ may have been decomposing to $Li_3 (C_2H_5)_2^+$ which was then readily decomposing further to $Li_2 (C_2H_5)^+$. Subsequent appearance potential measurements[39] differed considerably and ruled out the stepwise decomposition mechanism. By incorporating appropriate thermochemical values with appearance potentials for the $Li_n$ $(C_2H_5)_{(n-1)}^+$ ions, the ionisation potentials of the radicals $Li_n (C_2H_5)_{(n-1)}$ were calculated[44] to be approximately 4.4 eV.

A similar fragmentation mechanism appears to be operating with compounds II and III.

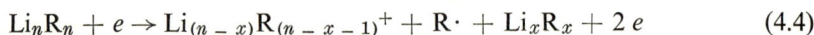

$$Li_nR_n + e \rightarrow Li_{(n-x)}R_{(n-x-1)}^+ + R\cdot + Li_xR_x + 2\,e \qquad (4.4)$$

The process is energetically more favoured when $x$ is an even integer.

Alkene eliminations to produce metal hydride ions, so common in the mass spectra of many alkyl metal compounds, were not observed. No legitimate comparison can be made, however, since these lithium alkyls and substituted alkyls have an entirely different structure from the alkyls of Groups II, III, and IV.

Of additional interest was the observation of the ions $(Li_nR_n - \cdot CH_3)^+$ in the spectra of compounds II and III. In the case of II the appearance potential of this ion was 1.3 eV higher than the appearance potential of $Li_nR_{n-1}^+$, whereas for III the appearance potential of the corresponding ion was 0.8 eV lower than that of $Li_nR_{n-1}^+$. As Hartwell and Brown[43] point out, the Li–C and Li–O bonds are expected to be quite polar, the latter more so than the former. They argued that dissociation of a methyl group

THE GROUP I ELEMENTS

from each of these species led to the formation of a species in which some double bond character was possible.

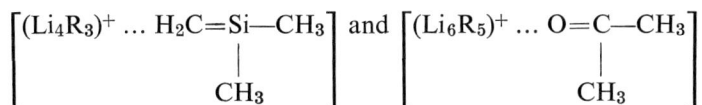

$$\left[(Li_4R_3)^+ \ldots H_2C{=}Si\underset{\underset{CH_3}{|}}{-}CH_3\right] \text{ and } \left[(Li_6R_5)^+ \ldots O{=}C\underset{\underset{CH_3}{|}}{-}CH_3\right]$$

The observed trends in the appearance potentials, therefore, are not unexpected.

Because of the higher polarity of the Li–O bond as compared with that of the Li–C bond, one might also expect that the appearance potential of $Li_6[OC(CH_3)_3]_5{}^+$ would be less than that of $Li_4[CH_2Si(CH_3)_3]_3{}^+$. Presumably the ionisation potential of the parent molecule must be much higher for the oxygen compound, indicating that the electron removed originated from an orbital localized largely on the organic group which is surprising.

The mass spectrum of potassium *tert.*-butoxide has also been recorded[45] and indicates that the compound exists as tetrameric units in the gas phase, in contrast to the hexamers of the corresponding lithium compound. The spectrum is very similar to those of the lithium compounds discussed previously: (*i*) a parent molecular ion was not observed even at low electron energies, (*ii*) the base peak was $K_2R^+$, and (*iii*) all possible ions of general formula $K_nR_{n-1}{}^+$ were observed. Two other types of fragment ions deserve mention, namely $K_nR_n{-}\cdot CH_3{}^+$ and $K_nR_{(n-1)}{-}CH_4{}^+$ which were present in low intensities. The latter type had not been previously reported in other Group I compounds. The ions observed and their intensities are recorded in Table 4.4.

REFERENCES

1 R. C. Weast (Ed.), *Handbook of Chemistry and Physics*, The Chemical Rubber Publishing Co., Cleveland, Ohio, U.S.A., 49th edn., 1968–69.
2 J. Drowart and P. Goldfinger, *Angew. Chem. Intern. Ed. Engl.*, 6 (1967) 581.
3 J. Berkowitz and W. A. Chupka, *J. Chem. Phys.*, 29 (1958) 653.
4 L. N. Gorokhov, *Dokl. Akad. Nauk SSSR*, 142 (1962) 113.
5 R. C. Schoonmaker and R. F. Porter, *J. Chem. Phys.*, 30 (1959) 283.
6 R. F. Porter and R. C. Schoonmaker, *J. Chem. Phys.*, 29 (1958) 1070.
7 T. A. Milne, H. M. Klein and D. Cubicciotti, *J. Chem. Phys.*, 28 (1958) 718.
8 R. M. Hobson, *J. Chem. Phys.*, 23 (1955) 2463.
9 R. C. Miller and P. Kusch, *J. Chem. Phys.*, 25 (1956) 860.

10 L. Friedman, *J. Chem. Phys.*, 23 (1955) 477.
11 P. Winchell, *Nature*, 206 (1965) 1252.
12 H. Bloom and J. W. Hastie, *Australian J. Chem.*, 19 (1966) 1003.
13 J. Berkowitz and W. A. Chupka, *J. Chem. Phys.*, 45 (1966) 1287.
14 J. Berkowitz, *J. Chem. Phys.*, 50 (1969) 3503.
15 T. A. Milne and D. Cubicciotti, *J. Chem. Phys.*, 29 (1958) 846.
16 T. A. Milne, *J. Chem. Phys.*, 32 (1960) 1275.
17 T. A. Milne and H. M. Klein, *J. Chem. Phys.*, 33 (1960) 1628.
18 J. Berkowitz, H. A. Tasman and W. A. Chupka, *J. Chem. Phys.*, 36 (1962) 2170.
19 R. C. Schoonmaker, *Ph.D. Thesis*, Cornell University, Ithaca, N.Y., 1960.
20 G. Platel, *J. Chim. Phys.*, 62 (1965) 1176.
21 N. I. Ionov, *Dokl. Akad. Nauk SSSR*, 59 (1948) 467.
22 P. A. Akashin and L. N. Gorokhov, *Vestn. Mosk. Univ.*, *Ser. II*, 15 (1960) 3.
23 H. Bloom, J. W. Hastie and J. D. Morrison, *J. Phys. Chem.*, 72 (1968) 3041. ·
24 J. W. Hastie, K. F. Zmbov and J. L. Margrave, *J. Inorg. Nucl. Chem.*, 30 (1968) 729.
25 J. Berkowitz and W. A. Chupka, unpublished result quoted in ref. 12.
26 D. White, K. S. Seshadri, D. F. Dever, D. E. Mann and M. J. Linevsky, *J. Chem. Phys.*, 39 (1963) 2463.
27 J. Berkowitz, W. A. Chupka, G. D. Blue and J. L. Margrave, *J. Phys. Chem.*, 63 (1959) 644.
28 A. Büchler, J. L. Stauffer, W. Klemperer and L. Wharton, *J. Chem. Phys.*, 39 (1963) 2299.
29 A. Büchler, J. L. Stauffer and W. Klemperer, *J. Am. Chem. Soc.*, 86 (1964) 4544.
30 J. Berkowitz, D. J. Meschi and W. A. Chupka, *J. Chem. Phys.*, 33 (1960) 533.
31 R. C. Schoonmaker and R. F. Porter, *J. Chem. Phys.*, 28 (1958) 454.
32 A. Büchler and J. L. Stauffer, *J. Phys. Chem.*, 70 (1966) 4092.
33 K. Skudlarski, J. Drowart, G. Exsteen and A. Van der Auwera-Mahieu, *Trans. Faraday Soc.*, 63 (1967) 1146.
34 R. F. Porter, *J. Chem. Phys.*, 35 (1961) 318.
35 A. Büchler, J. B. Berkowitz-Mattuck, *J. Chem. Phys.*, 39 (1963) 286.
36 R. F. Porter and R. C. Schoonmaker, *J. Phys. Chem.*, 63 (1959) 626; 64 (1960) 457.
37 D. White, K. S. Seshadri, D. F. Dever, M. J. Linevsky and D. F. Mann, private communication cited in ref. 32.
38 J. Berkowitz, D. A. Bafus and T. L. Brown, *J. Phys. Chem.*, 65 (1961) 1380.
39 G. E. Hartwell, *Ph.D. Thesis*, University of Illinois, Urbana, Ill., 1966.
40 T. L. Brown, D. W. Dickerhoof and D. A. Bafus, *J. Am. Chem. Soc.*, 84 (1962) 1371.
41 H. Dietrich, *Acta Cryst.*, 16 (1963) 681.
42 E. Weiss and E. A. C. Lucken, *J. Organometal. Chem.*, 2 (1964) 197.
43 G. E. Hartwell and T. L. Brown, *Inorg. Chem.*, 5 (1966) 1257.
44 T. L. Brown, *Ann. N.Y. Acad. Sci.*, 136 (1966) 98.
45 E. Weiss, H. Alsdorf, H. Kühr and H. F. Grutzmacher, *Chem. Ber.*, 101 (1968) 3777.

*Chapter 5*

# The Group II Elements

M. R. LITZOW

## 1. INTRODUCTION

As with the alkali metals, inorganic compounds of Group II elements have stimulated considerable activity in the field of high-temperature mass spectrometry. These studies, however, are not of major interest here and are discussed only briefly.

Beryllium provides the only examples of dialkyl compounds of Group II elements other than Zn, Cd, and Hg, which have been investigated mass spectrometrically, and their behaviour under electron impact shows some significant differences from that of dialkyls of Group IIB and tetraalkyls of main Group IV. Very few organo-derivatives of the Group II metals have been investigated, however, and considerable scope remains for work in this area.

## 2. THE ELEMENTS

All of the elements in Group II, with the exception of beryllium, are polyisotopic; the naturally occurring isotopes are shown in Table 5.1 and Fig. 5.1.

A number of Aston's early mass spectrometric studies[2] of isotope abundances of Zn, Cd, and Hg utilised the dimethyl derivatives, and led to reports of new isotopes. It was later demonstrated[3], however, that the formation of the hydride ions $MH^+$, and the failure to recognise them as such, had often led Aston to erroneous conclusions. Dibeler and Mohler[4,5] have investigated the mass spectra of dimethyl-, diethyl-, and di-$n$-butylmercury for isotopic analysis of mercury.

Table 5.1

Natural Abundances of Group II Isotopes[1]

| Element | At.wt. | Natural abundance (%) | Element | At.wt. | Natural abundance (%) |
|---------|--------|------------------------|---------|--------|------------------------|
| Be      | 9      | 100                    | Zn      | 64     | 48.89                  |
|         |        |                        |         | 66     | 27.81                  |
|         |        |                        |         | 67     | 4.11                   |
| Mg      | 24     | 78.70                  |         | 68     | 18.57                  |
|         | 25     | 10.13                  |         | 70     | 0.62                   |
|         | 26     | 11.17                  |         |        |                        |
|         |        |                        | Cd      | 106    | 1.22                   |
| Ca      | 40     | 96.97                  |         | 108    | 0.88                   |
|         | 42     | 0.64                   |         | 110    | 12.39                  |
|         | 43     | 0.145                  |         | 111    | 12.75                  |
|         | 44     | 2.06                   |         | 112    | 24.07                  |
|         | 46     | 0.0033                 |         | 113    | 12.26                  |
|         | 48     | 0.18                   |         | 114    | 28.86                  |
|         |        |                        |         | 116    | 7.58                   |
| Sr      | 84     | 0.56                   |         |        |                        |
|         | 86     | 9.86                   |         |        |                        |
|         | 87     | 7.02                   | Hg      | 196    | 0.146                  |
|         | 88     | 82.56                  |         | 198    | 10.02                  |
|         |        |                        |         | 199    | 16.84                  |
|         |        |                        |         | 200    | 23.13                  |
| Ba      | 130    | 0.101                  |         | 201    | 13.22                  |
|         | 132    | 0.097                  |         | 202    | 29.80                  |
|         | 134    | 2.42                   |         | 204    | 6.85                   |
|         | 135    | 6.59                   |         |        |                        |
|         | 136    | 7.81                   |         |        |                        |
|         | 137    | 11.32                  |         |        |                        |
|         | 138    | 71.66                  |         |        |                        |

## 3. HIGH-TEMPERATURE STUDIES

### 3.1 The elements

As with the alkali metals, homonuclear diatomic molecules of beryllium, magnesium, calcium, zinc, cadmium, and mercury have been detected and their dissociation energies measured[6]. Again they decrease smoothly as the

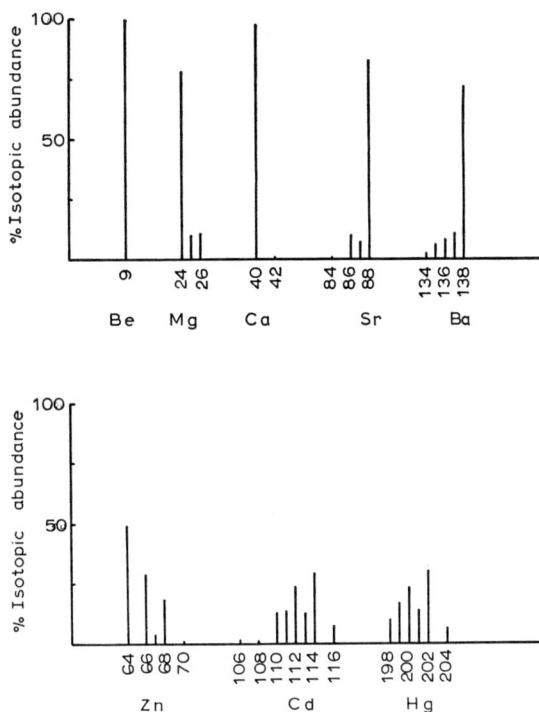

Fig. 5.1. Natural isotopic abundances of Group II elements (ref. 1).

group is descended (Table 5.2). The intermetallic molecules MAu (M = Be, Mg, Ca, Sr, Ba), MIn (M = Zn, Cd), and MHg (M = K, Cs, Tl) have likewise been identified and their dissociation energies measured[6] (Table 5.2).

Table 5.2

Bond dissociation energies of diatomic molecules, $M_2$, and MAu, MIn, and MHg

| Molecule | $D$ (M–M) (kcal·mole$^{-1}$) | Molecule | $D$ (M–M′) (kcal·mole$^{-1}$) | Molecule | $D$ (M–M′) (kcal·mole$^{-1}$) |
|---|---|---|---|---|---|
| $Be_2$ | (14) | BeAu | 73 | ZnIn | 7.7 |
| $Mg_2$ | ( 7) | MgAu | 61 | CdIn | 32 |
| $Ca_2$ | ( 5) | CaAu | 57 | HgK | $1.08 \pm 0.05$ |
| $Zn_2$ | ( 6) | SrAu | 88 | HgCs | $1.10 \pm 0.05$ |
| $Cd_2$ | $2.1 \pm 0.5$ | BaAu | 66 | HgTl | 0.7 |
| $Hg_2$ | $1.4 \pm 0.07$ | | | | |

The heats of sublimation of zinc and cadmium were determined mass spectrometrically by Tickner and coworkers[7,8] but no dimeric or polymeric molecules could be detected in these studies.

## 3.2 Halides

The halides, particularly the fluorides, have again been investigated in some detail but dimeric and polymeric species were not nearly as common as for the alkali halides. The magnesium halides exhibited both dimeric (approx. 1 mole %) and trimeric (approx. 0.01 mole %) species[9,10] but evidence for similar association of strontium and barium fluorides was lacking[10]. Dimers of $BeF_2$(ref. 11), $BeCl_2$ (ref. 12), $ZnCl_2$ (ref. 13), and $ZnBr_2$ (ref. 13) were also detected. Evidence has been presented for the existence of all of the mono-fluorides (MF) of Be, Mg, and the Group IIA elements. Their dissociation energies have been measured in each case (Table 5.3) and indicate consider-

Table 5.3

Mass Spectrometrically Determined Dissociation Energies of Main Group II Monohalides

| Compound | Dissociation energy (eV) | Ref. |
|---|---|---|
| BeF | <6.4 | 11 |
|  | 5.85 ± 0.10 | 17 |
| MgF | 4.62 ± 0.1 | 19 |
|  | 4.79 ± 0.05 | 14 |
| CaF | 5.4 ± 0.2 | 20 |
| SrF | 5.43 ± 0.1 | 19 |
|  | 5.61 ± 0.07 | 14 |
| BaF | 5.83 ± 0.1 | 19 |
|  | 6.07 ± 0.07 | 14 |
| BeCl | 3.98 ± 0.1 | 12 |

able stability for the molecules with bond strengths approaching the average bond energies in the corresponding difluorides. Hildenbrand[14] has presented by far the most accurate study and discussed his results in terms of the bonding in the alkaline earth fluorides. This work has recently been extended to beryllium monochloride[12]. Ionisation and appearance potentials of ions formed from $MCl_2$ (M = Mg, Ca, Sr, and Ba) have been measured and bond dissociation energies determined[15]; these values were in excellent agreement with those obtained from thermodynamic equilibrium measurements. Ionisation potentials of a number of MX molecules have been compared with calculated values[16]. The heat of formation of BeF has also been measured, and ion intensities used to calculate equilibrium constants of several isomolecular reactions involving this molecule[17]. An attempt was made[18] to detect the species Mg I but it could not be positively identified in experiments carried out with $MgI_2$. The negative ions $ZnCl_2^-$, $ZnBr_2^-$, and $ZnBr^-$ have been observed, as well as a number of doubly charged positive ions in the mercury halide spectra[21].

Appearance potential measurements have figured prominently in a number of investigations[9,11,12,14,15,17,19,21-24]. The thermodynamic quantities, heats of sublimation[9,10,12,20] heats of formation[11,17,21,24], and heats of vaporisation[13] have also been determined for various metal halides of the Group II elements. The zinc compounds $ZnCl_2$ and $ZnBr_2$ were found to be unique among salts that have polymeric gaseous species in that the heat of vaporisation of the monomer was greater than the heat of vaporisation of the dimer whereas the monomer is the predominant species in the temperature range investigated[13].

Results of electric deflection in mass spectrometrically detected beams have indicated that beryllium, magnesium, zinc, and mercury dihalides are linear while calcium and barium dihalides have a bent structure[25].

### 3.3 Compounds with main Group VI elements

Two separate studies of the vapour above beryllium oxide in tungsten Knudsen cells have resulted in the identification of several polymeric molecules. Chupka et al.[26] reported that the vapour consisted primarily of Be and O atoms and $(BeO)_3$ and $(BeO)_4$ molecules, with smaller amounts of $O_2$, BeO, $(BeO)_2$, $(BeO)_5$, and $(BeO)_6$. Heats of vaporisation of these polymeric molecules were interpreted as being consistent with ring structures. Theard and Hildenbrand[27] repeated the experimental work and obtained similar results. In addition to the previously reported ions $Be^+$, $O^+$, $O_2^+$, and $(BeO)_n^+$ ($n = 1-6$), low-intensity ions $Be_nO_{(n-1)}^+$ ($n = 2-5$) were observed. Appearance potentials of the lower mass ions were measured. Evidence of the existence of the suboxides $Be_2O$ (ref. 27), $Sr_2O$ (ref. 28), and $Zn_2O$ (ref. 29) was also obtained.

The reaction of barium oxide with water was studied mass spectrometrically[30]; the $Ba(OH)_2^+$ ion was not observed but $BaOH^+$ was inferred to have arisen from both $Ba(OH)_2$ and BaOH. Goldfinger and Jeunehomme[31] examined the vapour above the nine MX compounds (M = Zn, Cd, Hg; X = S, Se, Te) and found no gaseous molecular compounds to the detection limit of one part in $10^3$–$10^5$. Decomposition occurred into atomic Zn, Cd, and Hg and diatomic S, Se, and Te, except in the case of HgS and HgSe where polymeric S and Se species were observed. De Maria et al.[32] investigated ZnS, ZnSe, and ZnTe in the temperature range 1000–1200 °C and found that the vapours contained about one gaseous molecule in $10^4$–$10^5$ of decomposition products. This permitted measurements of the dissociation energies which are shown in Table 5.4.

Berkowitz and Marquart[33] examined the vapours above the sulphides of Ca, Sr, Zn, Cd, and Hg and found them to consist mainly of sulphur species, $S_2$ to $S_{10}$; the molecular species ZnS, CdS, and HgS were again detected in extremely small quantities. Molecular species MS are much more prevalent

Table 5.4

Dissociation Energies of Simple Compounds of Group II Determined Mass Spectrometrically (kcal.mole$^{-1}$)

| Element | MO | (ref.) | MS | (ref.) | MX | (ref.) |
|---|---|---|---|---|---|---|
| Be | 106 ± 2 | (26) | | | | |
| Mg | 86 ± 5[a] | (28) | | | | |
| | < 90[a] | (39) | | | | |
| Ca | 84.4 ± 6[a] | (35) | 73.7 ± 4.5[a] | (35) | | |
| | 93 ± 5[a] | (28) | 79.5[a] | (41) | | |
| Sr | 92.2 ± 6[a] | (35) | 74.1 ± 4.5[a] | (35) | 59.1 ± 3 | (42) |
| | 102 ± 5[a] | (28) | 80[a] | (41) | (X = Se) | |
| | 83[a] | (39) | | | | |
| Ba | 130.4 ± 6[a] | (35) | 94.7 ± 4.5[a] | (35) | 107 | (30) |
| | 131.8 ± 1 | (40) | | | (X = OH) | |
| Zn | ≤ 66[b] | (29) | 48.0 ± 3.0[a] | (32) | 31.7 ± 3.0[a] | (32) |
| | | | ≤ 50[a] | (41) | (X = Se) | |
| | | | | | 10–20[a] | (32) |
| | | | | | (X = Te) | |
| Cd | | | ≤ 47[a] | (41) | | |
| Hg | | | ≤ 50[a] | (41) | | |

[a] $D_0$ (M–X) i.e. D at 0° K.
[b] $D_{298}$ (M–X) i.e. D at 298 °K.

in the vapour above the Group IIA sulphides and much higher dissociation energies are characteristic of these compounds (Table 5.4). Similarly, the vapour above $Cd_3As_2$ gave rise[34] to the ions $Cd^+$, $As^+$, $As_2^+$, $As_3^+$, and $As_4^+$. Appearance potential measurements demonstrated that the ions were formed from Cd and $As_4$ rather than from molecules of $Cd_3As_2$, which must therefore dissociate under the conditions employed.

A large number of dissociation energies has been measured and the results are summarised in Table 5.4. Other thermodynamic values deduced were heats of vaporisation[26,32,35], heats of atomisation of beryllium[27], strontium[28], and barium[36] suboxides, and heats of sublimation[37].

A mass spectrometric study[38] of the vapour above both magnesium and calcium hydrides in a Knudsen effusion cell failed to find any evidence of either MH or $MH_2$ species. The results indicated that the bond dissociation energies of MgH and CaH must be less than 2.6 and 2.1 eV respectively.


4. VOLATILE HYDRIDES AND ORGANOMETALLIC COMPOUNDS


*4.1 $BeB_2H_8$*


The structure of beryllium borohydride has long been considered to be that represented by I, but as a result of a recent electron diffraction study structure II was proposed[43] for the compound. The validity of II was questioned by Cook and Morgan[44] who interpreted their infrared and mass spectral results in favour of structure III. The mass spectrum was recorded at both 70 and 15 eV and the ions observed are listed in Table 5.5. No relative intensity details were given except to say that ($i$) in the 15 eV spectrum


Table 5.5

Ions observed[44] in Mass Spectrum of Beryllium Borohydride at an Ionising Voltage of 70 eV

| | | |
|---|---|---|
| $BeB_2H_8^+$ | $B_2H_5^+$ | $B^{+a}$ |
| $BeB_2H_6^+$ | $B_2H_2^+$ | $BeH^{+a}$ |
| $BeB_2H_4^+$ | $BeBH_2^+$ | $Be^{+a}$ |
| $BeB_2H_2^+$ | $BH_2^{+a}$ | |

a Not observed at an ionising voltage of 15 eV.

*References pp. 112–113*

the relative intensities of the larger fragments increase relative to the smaller fragments (as one would expect) and (*ii*) the ion $BeB_2H_4^+$ was the most abundant in the 70 eV spectrum. The predominance of this ion indicated an especial stability for this grouping and it was claimed that this was the "core" of the postulated structure III. On this argument, however, one would be tempted to predict either I or II where the "core" $BeB_2H_4$ contains only bridging hydrogens and expect the predominant ion to be $BeB_2H_3^+$ if III were the correct structure.

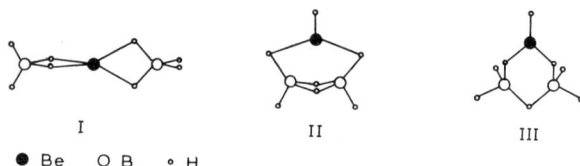

I                                  II                              III

● Be      O B      ○ H

A number of metastable transitions corresponding to loss of $H_2$ and $D_2$ from $BeB_2H_x^+$ and $BeB_2D_x^+$ ions, respectively, were observed.

*4.2 Organoderivatives of beryllium and magnesium*

Most of the dialkylberyllium compounds for which data are available are polymers with di-*tert.*-butylberyllium being a known exception since it exists solely as a monomer[45]. Dimethylberyllium exhibits various degrees of association; it possesses a long-chain polymeric structure in the solid, whereas the vapour consists of a mixture of monomer, dimer, and trimer, with appreciable amounts of higher polymers under near saturation conditions[45]. A mass spectrometric investigation[46] of the unsaturated vapour detected the monomer only; the source temperature was not reported. Its spectrum was basically similar to those of the dimethyl derivatives of Group IIB (see Table 5.8, p. 108). However, it was reported[47] that mass spectral analysis of dimethylberyllium vapours allowed to equilibrate in a chamber showed, in addition to the peaks due to ions arising from the monomer, very low-intensity peaks which were assigned to octameric, heptameric, hexameric and pentameric dimethylberyllium.

The most extensive mass spectral investigations of organoberyllium compounds have been summarised by Chambers *et al.*[48,49]. They have pointed out that, because of the low energy of the Be–C–Be bridging bonds, one can expect to observe dimers and trimers only under carefully controlled conditions. The abundances of all beryllium-containing ions decreased with increasing source temperatures because of thermal decomposition before ionisation. The dialkyls $R_2Be$ ($R = Et$, $Pr^n$, $Pr^i$) have all been shown mass spectrometrically to be associated in the vapour phase. For example, the ions $Et_3Be_2^+$ and $Et_5Be_3^+$, resulting from the dimer and trimer respectively, have been observed. Fragmentation paths such as

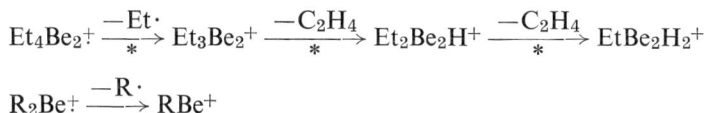

$$Et_4Be_2^+ \xrightarrow[*]{-Et\cdot} Et_3Be_2^+ \xrightarrow[*]{-C_2H_4} Et_2Be_2H^+ \xrightarrow[*]{-C_2H_4} EtBe_2H_2^+$$

$$R_2Be^+ \xrightarrow{-R\cdot} RBe^+$$

reflect the greater stability of even-electron ions. In addition, cleavage of C–C and C–H bonds occurs in the higher alkyl derivatives. Alkane elimination from the parent molecular ion is reported to be a favoured process.

$$(C_nH_{2n+1})_2Be^+ \rightarrow C_nH_{2n}Be^+ + C_nH_{2n+2}$$

Although corresponding $C_nH_{2n}M^+$ ions are observed (in much lower abundances) in the mass spectra of Group IIB dialkyls (Table 5.8, p. 108), no evidence has been presented to suggest their mode of formation. The behaviour of the beryllium dialkyls, however, is different from that of the main Group IV metal alkyls.

The most abundant ion in the spectrum of diisopropylberyllium was $C_3H_6Be^+$; appearance potential measurements on this odd-electron ion have indicated that its formation was accompanied by considerable rearrangement, possibly to an allylic hydride.

$$Pr^i_2Be^+ \rightarrow C_3H_8 + C_3H_6Be^+ (CH_2{=}CH{-}CH_2{-}BeH^+)$$

It was further reported that alkene elimination from the monomer was always observed, although the hydride ion produced was usually of low abundance. Ionisation and appearance potential measurements were made[49], (Table 5.6) but it was feared that "tailing" of the ionisation efficiency curves of fragment ions resulted in values for the bond dissociation energies which must be regarded as upper limits.

During early investigations of the novel bis-cyclopentadienyl and cyclopentadienide compounds of metals, a mass spectrometric study[50] was undertaken with the objective of establishing a correlation between spectra and molecular structure. This will be further discussed in conjunction with

Table 5.6

Ionisation and Appearance Potentials of Ions Derived from Beryllium Dialkyls, and Calculated Bond Dissociation Energies[49]

| Compound | Ion | I.P. or A.P. (eV) | Bond | Dissociation energy (kcal.mole$^{-1}$) |
|---|---|---|---|---|
| $Me_2Be$ | $(CH_3)_2Be^{+\cdot}$ | $10.67 \pm 0.07$ | $(CH_3Be–CH_3)^+$ | $46.1 \pm 3.2$ |
| | $CH_3Be^+$ | $12.67 \pm 0.02$ | | |
| | $CH_2Be^{+\cdot}$ | $11.92 \pm 0.05$ | $(BeCH_2–H)^+$ | $86.7 \pm 3.2$ |
| $Et_2Be$ | $(C_2H_5)_2Be^{+\cdot}$ | $9.46 \pm 0.05$ | $(C_2H_5Be–C_2H_5)^+$ | $47.3 \pm 3.2$ |
| | $C_2H_5Be^+$ | $11.51 \pm 0.05$ | | |
| | $C_2H_4Be^{+\cdot}$ | $10.35 \pm 0.03$ | $(BeC_2H_4–H)^+$ | $71.5 \pm 3.2$ |
| $Pr^n_2Be$ | $(C_3H_7)_2Be^{+\cdot}$ | $8.71 \pm 0.06$ | $(C_3H_7Be–C_3H_7)^+$ | $42.7 \pm 3.2$ |
| | $C_3H_7Be^+$ | $10.81 \pm 0.05$ | | |
| | $C_3H_6Be^{+\cdot}$ | $9.86 \pm 0.05$ | $(BeC_3H_6–H)^+$ | $70.7 \pm 3.2$ |
| $Pr^i_2Be$ | $(C_3H_7)_2Be^{+\cdot}$ | $8.80 \pm 0.02$ | $(C_3H_7Be–C_3H_7)^+$ | $48.7 \pm 3.2$ |
| | $C_3H_7Be^+$ | $10.65 \pm 0.01$ | | |
| | $C_3H_6Be^{+\cdot}$ | $9.60 \pm 0.01$ | $(BeC_3H_6–H)^+$ | $76.0 \pm 3.2$ |
| $Bu^i_2Be$ | $(C_4H_9)_2Be^{+\cdot}$ | $8.74 \pm 0.05$ | $(C_4H_9Be–C_4H_9)^+$ | $29.1 \pm 3.2$ |
| | $C_4H_9Be^+$ | $10.00 \pm 0.05$ | | |
| | $C_4H_8Be^{+\cdot}$ | $9.14 \pm 0.03$ | $(BeC_4H_8–H)^{+\cdot}$ | $77.7 \pm 3.2$ |

Table 5.7

Mass Spectrum of Magnesium Cyclopentadienide and Appearance Potentials of Principal Ions[50]

| Ion | Relative intensity[a] | A.P.[b] (eV) |
|---|---|---|
| $C_{10}H_{10}Mg^+$ | 58.5 | $7.76 \pm 0.1$ |
| $C_8H_8Mg^+$ | 3.8 | |
| $C_5H_5Mg^+$ | 100 | $10.98 \pm 0.1$ |
| $CH_3Mg^+$ or $C_3H_3^+$ | 8.6 | |
| $Mg^+$ | 80.7 | $14.36 \pm 0.2$ |
| $C_5H_5^+$ | 3.5 | |

[a] Relative intensity at ionising voltage of 50 eV.
[b] Vanishing current method.

transition metal compounds but some pertinent comments about magnesium cyclopentadienide will be made here. The major ions observed were the parent molecular ion (even though extensive ionic character is a feature of the $Mg-C_5H_5$ bond) and products of the rupture of these metal-to-ring bonds (Table 5.7). Only minor amounts of dissociation involving loss of C or H, or hydrocarbon aggregates smaller than $C_5H_5$ from the $C_{10}H_{10}Mg$ unit were observed. This is a significant difference from the behaviour of other Group II organometallic compounds such as dimethylberyllium, dimethylzinc, and dialkylmercury which undergo extensive fragmentation involving rupture of C–H and C–C bonds.

The relative intensity of the parent molecular ion (and also that of manganese cyclopentadienide) was somewhat less than those of $\pi$-cyclopentadienyl transition metal compounds suggesting that the magnesium-to-ring bond was of quite a different nature from the corresponding bonds in the latter compounds. This, of course, has been supported by additional chemical and physical evidence. Appearance potential measurements allowed the bond dissociation energies $D(M-C_5H_5)$ to be calculated; these supported the concept of weaker bonding in $Mg(C_5H_5)_2$ than in most transition metal cyclopentadienyl compounds.

*4.3 Organoderivatives of Group IIB*

The relative intensities of the various ions observed in the reported mass spectra of zinc and mercury dialkyls are shown in Table 5.8. Dimethylzinc[51] and dimethylmercury[4,52] give rise to similar spectra; the most abundant ion in each case, considering only those containing the metal, was $MCH_3^+$. Ions corresponding to loss of hydrogen atoms from $MCH_3^+$ were also observed, as were the rearrangement ions $HMCH_3^+$ and $MH^+$ (refs. 4, 51). (These ions were not detected in the mass spectrum of dimethylmercury by Hobrock and Kiser[52,53].) Dibeler and Mohler[4] noted that when the pressure of dimethylmercury was increased by a factor of three, the $HHg^+/Hg^+$ ratio was unchanged suggesting that the hydride ion was formed by an intramolecular rearrangement rather than by an ion–molecule reaction. The $HHg^+$ ion was also observed in the spectra of diethyl- and di-*n*-butylmercury and, in addition, the dihydride $H_2Hg^+$ was detected in small quantities in the spectrum of the diethyl derivative. As with the main Group IV metal alkyls, the elimination of (R–H) from $R_2M$ to from metal hydride species appears to be an energetically favoured process.

Table 5.8

Relative Intensities of Ions Observed in the Electron-Impact Mass Spectra of Organo-
Derivatives of Group II Elements

| Ion (X = H, F) | Relative intensity | | | | | |
|---|---|---|---|---|---|---|
| | $Be(CH_3)_2$ (ref. 46) | $Zn(CH_3)_2$ (ref. 51) | $Hg(CH_3)_2$ [a] (ref. 4) | $Hg(CF_3)_2$ (ref. 56) | $Hg(C_2H_5)_2$ (ref. 4) | $Hg(C_4H_9)_2$ (ref. 4) |
| $MC_8X_{18}^+$ | | | | | | 51 |
| $MC_8X_{17}^+$ | | | | | | 2.4 |
| $MC_6X_{14}^+$ | | | | | | 3.4 |
| $MC_6X_{13}^+$ | | | | | | 1.8 |
| $MC_4X_{10}^+$ | | | | | 83 | 1.7 |
| $MC_4X_9^+$ | | | | | 1.5 | 9.7 |
| $MC_2X_6^+$ | 15 | 36.6 | 40.2 | 3.12 | 12 | |
| $MC_2X_5^+$ | | | 1.36 | 19.6 | 100 | 13 |
| $MC_2X_4^+$ | | | | | 13 | 18 |
| $MCX_4^+$ | | 7.6 | 0.44 | | | |
| $MCX_3^+$ | 100 | 100 | 100 | 84.5 | 2.8 | 0.8 |
| $MCX_2^+$ | 60 | 9.0 | 13.7 | 7.43 | 6.2 | 3.4 |
| $MCX^+$ | | 2.8 | 5.9 | 0.79 | 0.5 | 1.5 |
| $MC^+$ | | 1.3 | 2.8 | 1.10 | | |
| $MX_2^+$ | | | | | 0.7 | |
| $MX^+$ | 5 | 2.3 | 4.8 | 3.53 | 14 | 5.1 |
| $M^+$ | 15 | 16.6 | 41.8 | 100 | 68 | 100 |
| $CX_3^+$ | | 21.4 | 110 | 125 | | |

[a] Other hydrocarbon ions were also reported.

A comparison of the spectra of $(CH_3)_2Zn$, $(CH_3)_2Hg$, and $(C_2H_5)_2Hg$ reveals that the abundance of the $M^+$ ion increases in this order, paralleling the decrease in the bond dissociation energy of the M–C bond. Bis(trifluoro-methyl)mercury, however, appears to behave anomolously. The bond dissociation energy of the Hg–C bond in this compound has not been determined but, in general, $M–CF_3$ bonds are more stable than $M–CH_3$ bonds. However, a large increase in the abundance of $Hg^+$ and $CX_3^+$ ions (X = H, F) is observed on going from dimethylmercury to the bis(trifluoromethyl) derivative, illustrating the danger involved in making conclusions concerning trends in bond dissociation energies from relative intensities in mass spectra. Hydrocarbon ions contribute a large proportion of the total ion current in all the mercury alkyls metnioned above, but this may be due in part to thermal decomposition before ionisation. A photoionisation study of the dimethyl derivatives of zinc, cadmium, and mercury has recently been reported[54]. Hydride ions and ions formed by loss of one hydrogen atom were not detected.

Vinyl and perfluorovinyl derivatives of mercury were reported[55] to display weakness of the Hg–C bond under electron-impact conditions, $(CH_2{=}CH)_2Hg$ apparently being the only compound investigated which yielded some ions with the Hg–C bond intact.

No peaks due to metastable transitions have been observed. Winters and Kiser[51] plotted clastograms for the main fragment ions from dimethylzinc and these indicated that the ion $ZnCH_3^+$ participated as an intermediate in the consecutive unimolecular decompositions

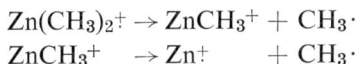

$$Zn(CH_3)_2^+ \rightarrow ZnCH_3^+ + CH_3 \cdot$$
$$ZnCH_3^+ \rightarrow Zn^+ + CH_3 \cdot$$

This was supported by appearance potential measurements which were used in conjunction with additional thermochemical data to ascertain the most probable processes accounting for the principal positive ions observed. (Obviously competitive processes must also be in operation to account for ions such as $HZnCH_3^+$ and $HZn^+$.) Heats of formation of the ions were also deduced and are listed in Table 5.8. The appearance potentials and heats of formation of the ions $Hg(CH_3)_2^+$ and $HgCH_3^+$ have similarly been determined (Tables 5.9 and 5.10). The appearance potential of the latter ion and the characteristics of the associated clastogram led to the process of simple bond cleavage being proposed[52,53]. Ionisation and appearance potentials of ions derived from the dimethyl derivatives of Zn, Cd, and Hg have since been determined by a photoionisation technique, and are compared with the electron-impact values in Table 5.9, while heats of formation of the various ions are compared in Table 5.10. Calculated values of the ionisation potential of the $ZnCH_3$ radical[51,54] are in poor agreement.

Exchange reactions between $(CD_3)_2Hg$ and $(CH_3)_2M$ (where M = Mg Zn, Cd, and Hg) have been followed by mass spectrometry[57] and species such as $CH_3HgCD_3^+$ observed.

The mass spectra of methylzinc methoxide, ethoxide, isopropoxide, and *tert.*-butoxide all show peaks attributable to $(CH_3)_3Zn_4(OR)_4^+$ indicating the presence of tetramers in the vapour phase[58]. It was stated that fragmen-

Table 5.9

Ionisation and Appearance Potentials of Ions Derived from Organo Derivatives of Zinc, Cadmium, and Mercury (eV) References are given in brackets.

| Ion | Compound | | | | | | |
|---|---|---|---|---|---|---|---|
| | $Zn(CH_3)_2$ | $Zn(C_2H_5)_2$ | $Cd(CH_3)_2$ | $Hg(CH_3)_2$ | $ClHgCH_3$ | $Hg(C_2H_5)_2$ | $Hg[CH(CH_3)_2]_2$ |
| $MR_2^{+\cdot}$ | 8.86±0.15(51) 9.00±0.02(54)[a] | ~12(63) | 8.48±0.02(54)[a] 8.56±0.02(54)[a] | 8.90±0.2 (52) 9.02±0.2 (64) 9.1 ±0.1 (65) 9.02±0.02(54)[a] 9.10±0.05(54)[a] | 11.5±0.2 (65) | 8.5 ±0.1(65) | 7.6±0.1(65) |
| $MR^+$ | 11.2 ±0.2 (51) 10.22±0.02(54)[a] | | 9.69±0.02(54)[a] | 10.4 ±0.2 (52) 10.5 ±0.1 (65) 10.10±0.02(54)[a] | 12.35±0.2(65)[b] | 9.65±0.1(65) | 9.1±0.1(65) |
| $MH^+$ | 13.9 ±0.4 (51) | | | | | | |
| $M^{+\cdot}$ | 13.4 ±0.3 (51) 13.13±0.02(54)[a] | | 12.05±0.02(54)[a] | 13.05±0.02(54)[a] | | | |
| $R^+$ | 15.1 ±0.5 (51) | | | | | | |

[a] Photoionisation value.
[b] Appearance potential of the $HgCH_3^+$ ion.

Table 5.10

Heats of Formation of the Ions of Dimethyl Compounds of Zinc, Cadmium, and Mercury (kcal.mole$^{-1}$)

References are given in brackets.

| Ion | M | | |
|---|---|---|---|
| | Zn | Cd | Hg |
| M(CH$_3$)$_2^+$ | 218 (51) | | 158 (52) |
| | 220.2 ± 0.5 (54) | 221.7 ± 0.5 (54) | 232.4 ± 1 (54) |
| MCH$_3^+$ | 241 (51) | | 160 (52) |
| | 215.2 ± 0.5 (54) | 214.5 ± 0.5 (54) | 222.3 ± 0.5 (54) |
| MH$^+$ | 233 (51) | | |
| M$^+$ | 258 (51) | | |
| | 249.1 ± 0.5 (54) | 235.65 ± 0.5 (54) | 257.1 ± 0.5 (54) |

tation to Me$_2$Zn$_3$(OR)$_3^+$ occurred more readily for the *tert.*-butoxide than for the other alkoxides, but no further details were reported.

The mass spectrum of bis(pentafluorophenyl)mercury has been reported [59,60]; metastable peaks observed indicate that the main processes are

$$(C_6F_5)_2Hg^+ \rightarrow C_6F_5Hg^+ \rightarrow C_6F_5^+$$

Subsequent breakdown of the C$_6$F$_5^+$ ion differs from that observed for transition metal complexes containing the C$_6$F$_5$ group. Bruce[59] has commented that there appears to be little tendency to form ions containing a Hg–F bond, in contrast to the fluorine abstraction by elements of Groups IV and V during fragmentation of their pentafluorophenyl derivatives[61]. The ion C$_6$F$_5$HgF$^+$ has a relative intensity of only 0.2%, while HgF$^+$ and HgF$_2^+$ were not observed. However, the ion HgF$^+$ was formed in considerable quantities[56] in the electron-impact induced fragmentation of (CF$_3$)$_2$Hg. The doubly charged ions Hg$^{2+}$, C$_6$F$_5$Hg$^{2+}$, and (C$_6$F$_5$)$_2$Hg$^{2+}$ appeared with a relatively high intensity. The corresponding cadmium compound was reported[62] to give rise to (C$_6$F$_5$)$_2$Cd$^+$, C$_6$F$_5$Cd$^+$, and fragment ions of C$_6$F$_5$; no further details were given. Spectra of the compounds C$_6$F$_5$HgX (X = Cl, Br) have also been published[60]. The rearrangement ions C$_6$F$_5$X$^+$ were observed, while the ions C$_6$F$_5$X$^-$ were observed in the negative ion mass spectra of these compounds. Of greater interest was the negative parent molecular ion $(C_6F_5)_2Hg^-$ in the spectrum of the bis(pentafluorophenyl) derivative, since it represents the first such ion of its type in organometallic compounds.

REFERENCES

1  R. C. Weast (Ed.), *Handbook of Chemistry and Physics*, The Chemical Rubber Publishing Co., Cleveland, Ohio, U.S.A., 49th edn., 1968–69.
2  F. W. Aston, *Phil. Mag.*, 45 (1923) 934; *Proc. Roy. Soc. (London), Ser. A*, 130 (1931) 302; 149 (1935) 396; *Nature*, 130 (1932) 847.
3  K. T. Bainbridge and E. B. Jordan, *Phys. Rev.*, 50 (1936) 282; K. T. Bainbridge, *Phys Rev.*, 39 (1932) 847.
4  V. H. Dibeler and F. L. Mohler, *J. Res. Natl. Bur. Std.*, 47 (1951) 337.
5  V. H. Dibeler, *Anal. Chem.*, 27 (1955) 1958.
6  J. Drowart and P. Goldfinger, *Angew. Chem. Intern. Ed. Engl.*, 6 (1967) 581.
7  K. H. Mann and A. W. Tickner, *J. Phys. Chem.*, 64 (1960) 251.
8  J. B. Westmore, H. Fujisaki and A. W. Tickner, *Advan. Chem. Ser.*, 72 (1968) 231.
9  J. Berkowitz and J. R. Marquart, *J. Chem. Phys.*, 37 (1962) 1853.
10  J. W. Green, G. D. Blue, T. C. Ehlert and J. L. Margrave, *J. Chem. Phys.*, 41 (1964) 2245.
11  D. L. Hildenbrand and L. P. Theard, *J. Chem. Phys.*, 42 (1965) 3230.
12  D. L. Hildenbrand and L. P. Theard, *J. Chem. Phys.*, 50 (1969) 5350.
13  F. J. Keneshea and D. Cubicciotti, *J. Chem. Phys.*, 40 (1964) 191.
14  D. L. Hildenbrand, *J. Chem. Phys.*, 48 (1968) 3657.
15  D. L. Hildenbrand, *Intern. J. Mass Spectrom. Ion Phys.*, 4 (1970) 75.
16  J. W. Hastie and J. L. Margrave, *J. Phys. Chem.*, 73 (1969) 1105.
17  D. L. Hildenbrand and E. Murad, *J. Chem. Phys.*, 44 (1966) 1524.
18  J. Berkowitz and W. A. Chupka, *J. Chem. Phys.*, 45 (1966) 1287.
19  T. C. Ehlert, G. D. Blue, J. W. Green and J. L. Margrave, *J. Chem. Phys.*, 41 (1964) 2250.
20  G. D. Blue, J. W. Green, R. G. Bautista and J. L. Margrave, *J. Phys. Chem.*, 67 (1963) 877.
21  R. W. Kiser, J. G. Dillard and D. L. Dugger, *Advan. Chem. Ser.*, 72 (1968) 153.
22  T. C. Ehlert and J. L. Margrave, *J. Chem. Phys.*, 41 (1964) 1066.
23  P. D. Foote and F. L. Mohler, *Phys. Rev.*, 17 (1921) 394.
24  D. L. Hildenbrand and E. Murad, *J. Chem. Phys.*, 43 (1965) 1400.
25  A. Büchler, J. L. Stauffer and W. L. Klemperer, *J. Chem. Phys.*, 40 (1964) 3471; *J. Am. Chem. Soc.*, 86 (1964) 4544.
26  W. A. Chupka, J. Berkowitz and C. F. Giese, *J. Chem. Phys.*, 30 (1959) 827.
27  L. P. Theard and D. L. Hildenbrand, *J. Chem. Phys.*, 41 (1964) 3416.
28  J. Drowart, G. Exsteen and G. Verhaegen, *Trans. Faraday Soc.*, 60 (1964) 1920.
29  D. F. Anthrop and A. W. Searcy, *J. Phys. Chem.*, 68 (1964) 2335.
30  F. E. Stafford and J. Berkowitz, *J. Chem. Phys.*, 40 (1964) 2963.
31  P. Goldfinger and M. Jeunehomme, *Trans. Faraday Soc.*, 59 (1963) 2851.

32 C. De Maria, P. Goldfinger, L. Malaspina and V. Piacente, *Trans. Faraday Soc.*, 61 (1965) 2146.
33 J. Berkowitz and J. Marquart, *J. Chem. Phys.*, 39 (1963) 275.
34 J. B. Westmore, K. H. Mann and A. W. Tickner, *J. Phys. Chem.*, 68 (1964) 606.
35 R. Colin, P. Goldfinger and M. Jeunehomme, *Trans. Faraday Soc.*, 60 (1964) 306.
36 M. G. Inghram, W. A. Chupka and R. F. Porter, *J. Chem. Phys.*, 23 (1955) 2159.
37 T. P. J. H. Babeliowsky, *J. Chem. Phys.*, 38 (1963) 2035.
38 T. C. Ehlert, R. M. Hilmer and E. A. Beauchamp, *J. Inorg. Nucl. Chem.*, 30 (1968) 3113.
39 R. F. Porter, W. A. Chupka and M. G. Inghram, *J. Chem. Phys.*, 23 (1955) 1347.
40 R. J. Ackermann and R. J. Thorn, *Progress in Ceramic Sciences*, Vol. 1, Pergamon Press, London, 1961, p. 39.
41 J. R. Marquart and J. Berkowitz, *J. Chem. Phys.*, 39 (1963) 283.
42 J. Berkowitz and W. A. Chupka, *J. Chem. Phys.*, 45 (1966) 4289.
43 A. Almenningen, G. Gundersen and A. Haaland, *Acta Chem. Scand.*, 22 (1968) 859.
44 T. H. Cook and G. L. Morgan, *J. Am. Chem. Soc.*, 91 (1969) 774.
45 G. E. Coates, M. L. H. Green and K. Wade, *Organometallic Compounds*, Vol. 1,Methuen, London, 1967, p. 103.
46 R. A. Kovar and G. L. Morgan, *Inorg. Chem.*, 8 (1969) 1099.
47 G. E. Coates, unpublished results cited in ref. 46.
48 D. B. Chambers, F. Glockling and J. R. C. Light, *Quart. Rev.*, 22 (1968) 317.
49 D. B. Chambers, G. E. Coates and F. Glockling, *Discussions Faraday Soc.*, 47 (1969) 157.
50 L. Friedman, A. P. Irsa and G. Wilkinson, *J. Am. Chem. Soc.*, 77 (1955) 3689.
51 R. E. Winters and R. W. Kiser, *J. Organometal. Chem.*, 10 (1967) 7.
52 B. G. Hobrock and R. W. Kiser, *J. Phys. Chem.*, 66 (1962) 155.
53 B. G. Hobrock, *M. S. Thesis*, Kansas State University, 1961.
54 G. Distefano and V. H. Dibeler, *Intern. J. Mass Spectrom. Ion Phys.*, 4 (1970) 59.
55 S. S. Dubov, F. N. Chelobav and R. N. Sterlin, *Zh. Vses. Khim. Obshchestva im. D. I. Mendeleeva*, 7 (1962) 585.
56 *Mass Spectral Data*, American Petroleum Institute Research Project 44, Carnegie Institute of Technology, Pittsburg, Pennsylvania; Serial No. 702.
57 R. E. Dessy, F. Kaplan, G. R. Coe and R. M. Salinger, *J. Am. Chem. Soc.*, 85 (1963) 1191.
58 J. M. Bruce, B. C. Cutsforth, D. W. Farren, F. G. Hutchinson, F. M. Rabagliati and D. R. Reed, *J. Chem. Soc. B*, (1966) 1020; G. Allen, J. M. Bruce, D. W. Farren and F. G. Hutchinson, *J. Chem. Soc. B*, (1966) 799.
59 M. I. Bruce, *J. Organometal. Chem.*, 14 (1968) 461.
60 S. C. Cohen and E. C. Tifft, *Chem. Commun.*, (1970) 226.
61 J. M. Miller, *J. Chem. Soc. A*, (1967) 828; *Can. J. Chem.*, 47 (1969) 1613.
62 M. Schmeiber and M. Weidenbruck, *Chem. Ber.*, 100 (1967) 2306.
63 P. D. Foote and F. L. Mohler, *Phys. Rev.*, 17 (1921) 394.
64 R. M. Reese and V. H. Dibeler, private communication cited in ref. 52.
65 B. G. Gowenlock, R. M. Haynes and J. R. Majer, *Trans. Faraday Soc.*, 58 (1962) 1905.

*Chapter 6*

# The Main Group III Elements

M. R. LITZOW

## 1. INTRODUCTION

The number of mass spectral studies on compounds of these elements decreases sharply as the group is descended. The hydrides and the halides of boron have been the subject of numerous investigations while, in contrast to the elements of main Group IV, the organo-derivatives of the elements have received scant attention. Other than the hydrides and the halides, the mass spectra of only random compounds have been reported so that it is practically impossible to discern general trends in a particular class. The boron–nitrogen compounds are a case in point—the borazines have received some attention but the mass spectra of only a few examples of simpler boron–nitrogen derivatives have been reported. In addition, results of energetic studies of any of these compounds are practically non-existent. It is possible with main Group IV compounds to observe trends as the group is descended, but similar comparisons cannot be made in main Group III because of the lack of information on the behaviour of indium, gallium, and thallium compounds under electron impact.

Whereas the application of mass spectrometry to inorganic chemistry has, in general, been very recent, the boron hydrides have been the subject of considerable mass spectral activity during the last twenty years. Several reasons have been responsible for this: (*i*) most were readily volatile and lent themselves to easy manipulation in the instruments available (volatility is no longer such a limiting factor with contemporary instruments), (*ii*) mass spectrometry has proved a ready means of characterising these (often) air-sensitive compounds, (*iii*) comparisons were sought with hydrocarbons which were the subject of numerous mass spectral investigations, (*iv*) fragmentation patterns were investigated during initial attempts to elucidate the structure of certain members, (*v*) mass spectrometry provided a method of establishing the $^{10}B/^{11}B$ isotopic ratio, and (*vi*) studies of the energetics of boron hydrides were made in attempts to learn more about the bonding involved. Appearance potential measurements provide access to bond dissociation energies and heats of formation if suitable thermodynamic data are available. Unfortunately, the lack of these data, especially in comparison with those available for hydrocarbons, has limited the applications of these studies.

The mass spectrometric work carried out on these compounds has paralleled the development of instruments and techniques. At first sight there appears to be considerable repetition of work, especially for $B_2H_6$, $B_4H_{10}$, and $B_5H_9$, but each successive study has been able to contribute more knowledge and accuracy, largely because of these improved instruments and techniques.

Another difficulty not always recognised has been the pyrolysis of samples within the inlet system and the ionisation chamber. The boron hydrides readily undergo pyrolysis and even when the problem has been fully appreciated, the precautions which could be taken to lessen this undesirable effect have been severely limited. Norton[1] was one of the first to recognise the problem and therefore introduced a low-temperature inlet system during his investigation of diborane, but still found that large quantities of hydrogen, a product of the pyrolysis of diborane[2], were produced. Overcoming this problem of pyrolysis in the ion source is much more difficult. Use of a rhenium filament rather than a tungsten one will result in a lower source temperature, but the temperature of the filament itself is still very high and molecules coming in contact with it will be pyrolysed. This whole problem has been overcome in the zero source-contact mass spectrometer[3,4] in which molecules that have come in contact with the spectrometer walls or the hot ion source assembly are rejected, and only the very small percentage which do not are utilised. Unfortunately, the instrumental problems are somewhat formidable and only a few investigations have been carried out with this instrument.

There is no doubt then that during the recording of many of the mass spectra of the boron hydrides the compounds have been decomposing, with the result that a spectrum of a "pure" compound has rarely been obtained.

Table 6.1

Natural Abundances of Isotopes of main Group III Elements[5]

| Element | Mass no. | Abundance (%) |
|---------|----------|---------------|
| B       | 10       | 19.6          |
|         | 11       | 80.4          |
| Al      | 27       | 100           |
| Ga      | 69       | 60.4          |
|         | 71       | 39.6          |
| In      | 113      | 4.28          |
|         | 115      | 95.72         |
| Tl      | 203      | 29.50         |
|         | 205      | 70.50         |

## 2. THE ELEMENTS

All of the elements of main Group III have two naturally occurring isotopes with the exception of aluminium, which has only one (Table 6.1, Fig. 6.1). Except for aluminium, therefore, interpretation of mass spectra of compounds of these elements is in general facilitated, but yet the difficulties and the more tedious processes associated with the assignment of ion intensities when polyisotopic species are present, are not introduced.

A search of the chemical literature shows considerable variation among the cited values of the natural isotopic abundance ratio of $^{10}B$ and $^{11}B$ *. Inghram[7]

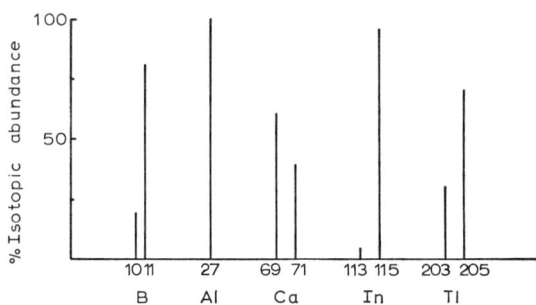

Fig. 6.1. Natural isotopic abundances of main Group III elements (ref. 5).

* A summary of these values is presented in ref. 6.

Table 6.2

Calculated Intensities of Ions with Varying Isotopic Composition Based on $^{11}B/^{10}B$ Ratios of 4.00 and, in Brackets, 4.10

| $B_x$ | $^{11}B_x$ $(m/e\ x)$ | $^{11}B_{(x-1)}{}^{10}B$ $(m/e\ x-1)$ | $^{11}B_{(x-2)}{}^{10}B_2$ $(m/e\ x-2)$ | $^{11}B_{(x-3)}{}^{10}B_3$ $(m/e\ x-3)$ | $^{11}B_{(x-4)}{}^{10}B_4$ $(m/e\ x-4)$ |
|---|---|---|---|---|---|
| $B_1$ | 1.0000 | 0.2500 | | | |
| | (1.0000) | (0.2439) | | | |
| $B_2$ | 1.0000 | 0.5000 | 0.0625 | | |
| | (1.0000) | (0.4878) | (0.0595) | | |
| $B_3$ | 1.0000 | 0.7500 | 0.1875 | 0.0156 | |
| | (1.0000) | (0.7317) | (0.1784) | (0.0145) | |
| $B_4$ | 1.0000 | 1.0000 | 0.3750 | 0.0625 | 0.0039 |
| | (1.0000) | (0.9756) | (0.3569) | (0.0580) | (0.0035) |
| $B_5$ | 0.8000 | 1.0000 | 0.5000 | 0.1250 | 0.0156 |
| | (0.8200) | (1.0000) | (0.4878) | (0.1190) | (0.0145) |
| $B_6$ | 0.6667 | 1.0000 | 0.6250 | 0.2083 | 0.0391 |
| | (0.6833) | (1.0000) | (0.6097) | (0.1982) | (0.0363) |
| $B_7$ | 0.5714 | 1.0000 | 0.7500 | 0.3125 | 0.0781 |
| | (0.5857) | (1.0000) | (0.7317) | (0.2974) | (0.0725) |
| $B_8$ | 0.5000 | 1.0000 | 0.8750 | 0.4375 | 0.1367 |
| | (0.5125) | (1.0000) | (0.8536) | (0.4164) | (0.1269) |
| $B_9$ | 0.4444 | 1.0000 | 1.0000 | 0.5833 | 0.2188 |
| | (0.4555) | (1.0000) | (0.9756) | (0.5552) | (0.2031) |
| $B_{10}$ | 0.3556 | 0.8889 | 1.0000 | 0.7500 | 0.3281 |
| | (0.3736) | (0.9111) | (1.0000) | (0.6504) | (0.2776) |

determined a $^{11}B/^{10}B$ ratio of 4.31 from the mass spectrum of $BF_3$ and this became the accepted value, but subsequently a number of workers[*] found that 4.0 was the optimum ratio in the calculation of mono-isotopic spectra for the boron hydrides. Further investigations[9,10] indicated that impurities and factors such as selective desorption of $^{11}BF_3$ could have accounted for the high value obtained from the study of this compound. To cloud the issue further, it was found that at times different ratios had to be employed for high- and low-mass fragments in boron hydrides to obtain minimum residues in the "stripping" procedures.

Shapiro and coworkers have pursued this problem in some detail. A thorough examination of the mass spectrum of trimethylborane[8] found that

---

[*] See, for example, publications listed in ref. 8.

| $^{11}B_{(x-5)}{}^{10}B_5$ $m/e\ x-5$ | $^{11}B_{(x-6)}{}^{10}B_6$ $(m/e\ x-6)$ | $^{11}B_{(x-7)}{}^{10}B_7$ $(m/e\ x-7)$ | $^{11}B_{(x-8)}{}^{10}B_8$ $(m/e\ x-8)$ | $^{11}B_{(x-9)}{}^{10}B_9$ $(m/e\ x-9)$ | $^{10}B_{10}$ $(m/e\ x-10)$ |
|---|---|---|---|---|---|
| 0.0008 | | | | | |
| (0.0007) | | | | | |
| 0.0039 | 0.0002 | | | | |
| (0.0035) | (0.0001) | | | | |
| 0.0117 | 0.0010 | 0.00003 | | | |
| (0.0106) | (0.0009) | (0.00003) | | | |
| 0.0273 | 0.0034 | 0.0002 | 0.000008 | | |
| (0.0248) | (0.0030) | (0.0002) | (0.000007) | | |
| 0.0547 | 0.0091 | 0.0010 | 0.00006 | 0.000002 | |
| (0.0495) | (0.0081) | (0.0008) | (0.00005) | (0.000001) | |
| 0.0984 | 0.0205 | 0.0029 | 0.0003 | 0.00002 | 0.0000004 |
| (0.0813) | (0.0165) | (0.0023) | (0.0002) | (0.00001) | (0.0000003) |

the only differences in the spectral patterns of $Me_3{}^{11}B$ and $Me_3{}^{10}B$ were in the intensities of their rearrangement ions. The $Me_3B$-$d_9$ pattern, however, differed considerably from the other two. This is not surprising since it has been established that the fragmentation probabilities for C–H and C–D bonds are quite different[11–13], and similar trends have been observed for B–H and B–D bonds[14,15]. Possible isotope effects have also been reported [16–18] for $^{12}C$ and $^{13}C$ but these are much smaller and any similar effects for $^{10}B$ and $^{11}B$ would also be expected to be small[19]. A normal abundance ratio of 4.13 ($^{10}B/^{11}B$ 0.242) was calculated[8] for the $m/e$ 34–45 range of $Me_3B$, which incorporates the $Me_2B^+$ fragment ion, while ratios of 3.775 and 3.86 were required for the low-mass ranges of $Me_3B$ and $Me_3B$-$d_9$ respectively. The authors suggested that none of these values represented the true boron isotopic abundance but were instead only apparent abundance ratios which included adjustment factors that took into account the differences in fragmentation for the various isotopic species.

Further insight into the problem was sought with the compound trimethylboroxine[20]; in this instance the $^{11}B/^{10}B$ isotopic ratio determined from the $m/e$ region 107–113, which again includes the (parent molecule $-\cdot Me)^+$ fragment ion, was 3.985 ($^{10}B/^{11}B$ 0.251). By introducing theoretical considerations involving molecular vibrations, the authors showed that there were minimum isotope effects on fragmentation, and consequently on the calculated abundance ratio, when bond rupture was produced by vibration of light atoms or groups against heavy boron-containing groups. One would therefore expect the different ratios obtained from ions formed by $H_3C–B$ bond rupture and $H_3C–(B_3O_3$ ring) bond rupture respectively.

The value[5] of the $^{11}B/^{10}B$ ratio now generally accepted is 4.10 ($^{10}B/^{11}B$ 0.244); in the calculation of monoisotopic mass spectra a value of 4.00 has most often been utilised. De Bievre and Debus[21] found significant variations between various natural boron stocks, and suggested that this was probably due to their geological origin. They accurately determined an absolute value for the isotope ratio of a particular natural boron stock by thermal ionisation mass spectrometry of borax: $^{10}B/^{11}B = 0.24726 \pm 0.00032$; $^{11}B/^{10}B = 4.0443$. The range 3.95–4.10 represents the natural variation of the boron isotope ratio.

One of the difficulties immediately encountered in the interpretation of mass spectra of boron hydrides is the mass range covered by particular ions because of the two isotopes of boron, $^{10}B$ and $^{11}B$. In any monoboron species such as $B(CH_3)_3^+$, each ion contributes to peaks at two $m/e$ values. This spread increases, however, as the number of borons increases and, in general, for a $B_n^+$ ion the spread covers $n + 1$ mass numbers (Table 6.2). It can be readily seen, therefore, that since fragmentation of these compounds in the mass spectrometer will involve loss of hydrogen, ions such as $^{11}B_nH_x^+$, $^{11}B_{(n-1)}{}^{10}BH_{(x+1)}^+$, $^{11}B_{(n-2)}{}^{10}B_2H_{(x+2)}^+$, etc. will possess almost identical masses and therefore be superimposed in the low-resolution spectra. "Stripping" procedures to obtain monoisotopic spectra of the compounds have been described in many papers (see, for example, refs. 14, 22, and 23). The spread of peaks due to ions differing only in isotopic content may become useful, however, in separating the spectrum of a particular hydride from a mixture, since one can usually select a peak that is unique for the compound of interest. Another characteristic of the spectra of boron hydrides is the occurrence of peaks at practically every mass number from that of highest $m/e$ value down to 10. The ions at $m/e$ 10 and 11 are unique to compounds of boron so that the presence of this element is readily established.

## 3. HIGH-TEMPERATURE STUDIES

The dissociation energies of the homonuclear diatomic molecules of the elements have been determined (Table 6.3)[6,24,25]. The gaseous phase of boron has been shown[24,26] to be essentially monatomic, with only very small amounts of dimer present. A study of the B–C and B–C–Si systems[27] produced atomisation energies of the molecules $BC_2$, $B_2C$, $BCSi$, and $BSi$ which indicated that these species were very strongly bonded and, in addition, they

Table 6.3

Bond dissociation energies of diatomic molecules

| Molecule | $D$ (M–M) (kcal. mole$^{-1}$) |
|---|---|
| $B_2$ | $65.5 \pm 5.5$ |
| $Al_2$ | $46 \pm 5$ |
| $Ga_2$ | $32 \pm 5$ |
| $In_2$ | $19.5 \pm 2.5$ |
| $Tl_2$ | (14) |

were shown to be more stable than predicted by assuming bond dissociation energies ascribed to the individual bonds. The molecules $Al_2C_2$ and $Ga_2C_2$ have likewise been identified and their heats of formation calculated[28]. Other investigations report the heat of sublimation[26,29] and the evaporation coefficient[26] of boron, and the heat of sublimation and heat of formation of $ZrB_2$ (ref. 29).

The boron trihalides and aluminium trichloride and tribromide all have high vapour pressures and have been studied mass spectrometrically at normal operating temperatures. They will therefore be discussed fully in Section 4.2 of this chapter. High-temperature studies of other main Group III halides are few. $AlF_3$ exhibits a spectrum[30] similar to that of $BF_3$ except for the presence of the ion $Al_2F_5^+$, the only polymeric ion of significant intensity. A careful examination[31] of the species effusing from a Knudsen cell containing $MgF_2$ and Al revealed that the ion $AlF_2^+$ appeared at an ionising voltage below that necessary for the ion to be formed from $AlF_3$, thus providing evidence for the existence of $AlF_2$. The stepwise dissociation energies for $AlF_3$ were calculated from the results obtained and showed a decreased bond strength in $AlF_2$. It was suggested that this was due to lack of spin correlation in this "odd-electron" molecule. The main Group III monofluorides and TlI have all been studied mass spectrometrically and values for their dissociation energies reported (Table 6.4). These values are

Table 6.4

Mass Spectrometrically Determined Dissociation Energies of Main Group III Compounds (kcal. mole$^{-1}$)

References are given in brackets.

| Com-pound | M | | | | |
|---|---|---|---|---|---|
| | B | Al | Ga | In | Tl |
| MF | 180±3 (32) | 159±3 (35) | 138±3.5 (32) | 121±3.5 (32) | 101±3.5 (32) |
| | 181±3.5 (35) | 159 (31) | | | |
| MI | | | | | 64 (53) |
| $M_2O$ | | 245±7 (54) | 210±4 (56) | 178 ± 4 (40) | |
| | | 256±7 (55) | 206.8±7.0 (39) | 179.2±4.0 (39) | |
| | | 241.6±7.0 (39) | | | |
| MO | 191.0±3.5 (6) | 115±5 (54) | 90.2±3.5 (39) | < 76 (40) | |
| | | 116±3.0 (39) | | | |
| $M_2O_2$ | | 365 (55) | | | |
| $MO_2$ | 322±10 (6) | | | | |
| MP | | 50.8±3.0 (57) | | | |

in good agreement with those obtained spectroscopically*. No evidence was found[33] for the association of AlF to polymeric species, but it was reported[32,34] that the vapour of thallium monofluoride consisted primarily of monomer and dimer with small amounts of trimer and tetramer also being present[34].

Heats of formation of the monofluorides have also been reported[32,35], as have heats of sublimation of $AlF_3$ and $Al_2F_6$ (ref. 36). Ionisation potentials

---

* Summarised and compared with spectroscopic values in ref. 32.

of a number of MX and $MX_2$ molecules have been compared with calculated values[37].

Mass spectral studies on the vapour above the oxides $M_2O_3$ (M = B, Al, Ga, In) have been carried out. Whereas the ion $B_2O_3^+$ was observed[38], the only ions produced by electron bombardment of the species evaporating from the solid and liquid surfaces of the remaining three oxides were $Al^+$, $AlO^+$, $Al_2O^+$, $O^+$ (ref. 39); $Ga^+$, $GaO^+$, $Ga_2O^+$, $O^+$, $O_2^+$ (ref. 39); and $In^+$, $In_2O^+$, $O_2^+$ (refs. 39, 40), respectively. The low appearance potential of each of the ions indicated that they resulted from simple ionisation of a molecule without fragmentation. Dissociation energies of the oxides MO and $M_2O$ (M = Al, Ga, In) were calculated[39] from the data obtained (Table 6.4). The vapour phase in equilibrium with a condensed mixture of B and $B_2O_3$ was shown to consist of B, BO, $B_2O_2$, and $B_2O_3$, with $B_2O_2$ predominating[41]. The polymerisation energy of $(BO)_2$ was calculated[41] and values for the atomisation energy of $BO_2$ (ref. 6) and the heat of vaporisation of $B_2O_3$ (ref. 42) have been reported. Electron impact of the vapour above a mixture of $B_2O_3$ and $Al_2O_3$ produced the ions $Al^+$, $Al_2O^+$, $B^+$, $BO^+$, $B_2O_3^+$, $AlBO^+$, and $AlBO_2^+$. Appearance potentials and the temperature dependence of ion intensities showed that, with the exception of $B^+$, all were parent molecular ions[43].

The equilibrium $(BOF)_3(g) \rightleftharpoons 3\ BOF(g)$ in the vapour above $MgF_2 + B_2O_3$ at high temperatures has been studied[44]; the heat of the reaction and the heat of formation of BOF were determined. The molecular species $H_4B_4O_4$ and $H_4B_6O_7$ were identified mass spectrometrically[45] as products of high-temperature reactions in the H–B–O system. Heats of formation were calculated and structures postulated. The reaction of $H_2$ with B–$B_2O_3$ mixtures and of water with elemental boron at high temperatures produced boroxine, $H_3B_3O_3$, and a smaller yield of hydroxyboroxine, $H_3B_3O_4$ (ref. 46). As a result of the reaction between water vapour and boric oxide in the temperature range 790–1200 °C, the ions $H_2O^+$, $HBO_2^+$, $H_3BO_3^+$, and $(HBO_2)_3^+$ were observed, the latter to the extent of <1% of that of its monomer[47]. The reaction of $BF_3$ with $B_2O_3$ produced[48] $(BOF)_3$ as the primary product in the range 400–1000 °C. The heat of atomisation of GaOF has been reported[49].

At temperatures above 250 °C, boron sulphide has been shown[50,51] to form a number of polymeric species with molecular weights up to 800.

The gaseous molecules BH, $BH_3$, $HBC_2$, and $H_2BC_xH_{(2x+1)}$ (x = 1,2) were identified[52] as products of a high-temperature reaction of hydrogen with boron carbide. Values obtained for $D(BH-H)$ and $D(BH_2-H)$ were critically compared with previously reported figures and $D(H-BC_2)$ was also determined.

## 4. MAIN GROUP III HYDRIDES

### 4.1 Boron hydrides

Throughout the discussion on the boron hydrides it has been assumed that ions containing the same number of boron atoms as the parent compound possess the intact boron skeleton present in that compound. It is emphasized, however, that conclusive proof substantiating this assumption is normally lacking, but appearance potential measurements in certain cases indicate that high energy processes involving decomposition of the compact boron arrangement have not occurred.

#### 4.1.1 Borane, $BH_3$
Identification of this unstable species, postulated as an intermediate in many reactions of the boron hydrides, has been achieved mass spectrometrically by pyrolysis of diborane[3,58,59] and borane carbonyl[60,61] and is discussed more fully in conjunction with diborane. Its mass spectrum is similar to those of the boron trihalides, and is compared with them in Table 6.22 (p. 162).

The dissociation of borane carbonyl, $BH_3CO$, into $BH_3$ and CO has been the subject of a detailed kinetic study[60] using a special mass spectrometer. The bond dissociation energy $D(H_3B-CO)$ was estimated from the kinetic measurements to be $23.1 \pm 2$ kcal mole$^{-1}$, while a rough measurement of the appearance potential of $BH_3^+$ from $BH_3CO$ resulted in a value of $21 \pm 9$ kcal.mole$^{-1}$ for this parameter. A recent study[61] of the molecular energetics of borane carbonyl resulted in a value of 33.7 kcal.mole$^{-1}$. The mass spectrum of this compound recorded on a conventional instrument displayed peaks arising from pyrolysis products, namely $B_2H_6$ and higher boron hydrides; a zero source contact mass spectrometer allowed an authentic spectrum to be obtained[62] (Fig. 6.2).

#### 4.1.2 Diborane-6
In 1940, Hipple[63] compared the mass spectra of diborane and the hy-

Conventional : inlet,30°; source,270°

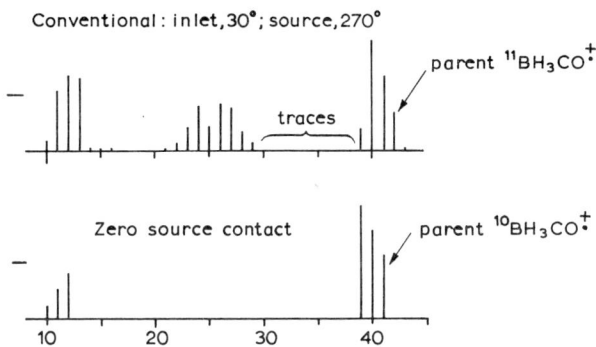

parent $^{11}BH_3CO^+$

traces

Zero source contact

parent $^{10}BH_3CO^+$

10 20 30 40

Fig. 6.2. Comparison of the borane carbonyl, BH$_3$CO, mass spectrum recorded under zero ion source contact conditions (bottom) and conventional conditions where the material may be pyrolysed by contact with the hot ion source (top). (Reproduced with permission from ref. 62.)

drocarbon ethane since he assumed that these compounds had similar structures. Dibeler and Mohler[64] reported that the spectrum of diborane resembled that of ethylene more than that of ethane and suggested that BH$_2$ and not BH$_3$ was the unit structure in the compound with two of the hydrogen atoms not bound directly to the boron. They suggested, in effect, that the hydrogens in the molecule were not all equivalent. The mass spectrum of essentially monoisotopic diborane containing 96% of the $^{10}$B isotope as well as that of a normal isotopic sample were recorded by Norton[1] and fairly good agreement was obtained with the previously reported results.

Dibeler and associates[14] continued their mass spectrometric investigation of diborane by examining the deuterium-labeled compounds diborane-$d_6$ (containing approximately 7% B$_2$D$_5$H) and ethane-$d_6$. The spectra of diborane and diborane-$d_6$ proved to be quite similar. The sensitivities of the base peaks of both compounds agreed within experimental error, demonstrating the nearly equal probability of dissociating a deuterium atom from B$_2$D$_6$ and a hydrogen atom from B$_2$H$_6$. In the B$_2$ group, the relative probabilities of dissociating successive deuterium atoms from B$_2$D$_6$ and successive hydrogen atoms from B$_2$H$_6$ were similar but always somewhat less for the deuterated compound. In the B$_1$ group, the probabilities were roughly equal. This compares reasonably well with the similarities reported[65] for methane and methane-$d_4$, and for ethane and ethane-$d_6$ (ref. 14). It was found that purely statistical considerations failed to account for the B$_2$D$_5$H spectrum. This is not surprising, however, since analysis of the mass spectra of simple hydrocarbons containing one deuterium atom[65,66] indicates that removal of the deuterium and hydrogen atoms is not equally probable but in a ratio of approximately 0.5 to 1 for these molecules. In addition, the case of diborane is more complicated since not all six hydrogen atoms are equivalent.

The spectra of both [10]B- and D-labeled diboranes were also recorded by Koski et al.,[15] and that of normal isotopic diborane by Quayle[67]. Good agreement was observed with previously reported spectra. It was concluded[15] that the ionisation probabilities for $^{10}B_2$, $^{10}B^{11}B$, and $^{11}B_2$ molecules must be very nearly the same. A comparison of the values obtained for the monoisotopic spectra of the diboranes and the deuterated diboranes confirmed the previous observation that there is an isotope effect in the fragmentation of partially deuterated diborane.

The spectrum of diborane recorded on a zero source contact mass spectrometer[3] (Fig. 6.3) was essentially similar to those recorded on conventional

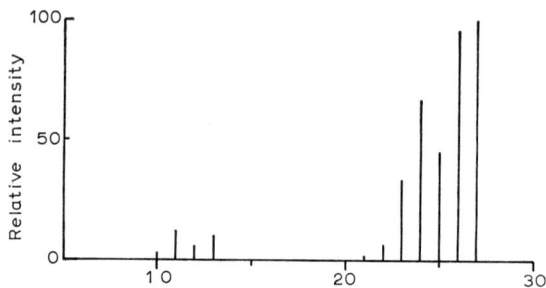

Fig. 6.3. Zero source contact mass spectrum of normal isotopic $B_2H_6$ (ref. 3).

instruments, but a definite decrease in the intensities of the monoborane species was obvious indicating that pyrolysis had probably occurred in the conventional mass spectrometers. The small variations in the intensities of the $B_2$ ions appear to have been due to instrument characteristics, however, since Baylis et al.[3] found that the intensities of the $B_2H_x^+$ peaks relative to one another remained constant regardless of the amount of diborane decomposition.

Metastable transitions which correspond to loss of mass 2 are by far the most frequent type occurring in hydrocarbons, and this was found to be true for diborane also[14,64,69,70].

The major peaks in the spectrum of $\mu$-diethylaminodiborane, $\mu$-$Et_2NB_2H_5$, were reported by Jennings and Wade[68] but there is little resemblance to the

Appearance Potentials of Ions from Diborane-6 (eV)
References are given in brackets.

| Ion[a] $B_nX_x^+$ | Isotopic composition of ion | | | | | |
|---|---|---|---|---|---|---|
| | $^{10}B_nH_x$ | $^{11}B_nH_x$ | $^{10}B_nD_x$ | $^{11}B_nD_x$ | $^{10}B_nD_{(x-1)}H$ | Normal isotopic $B_nH_x$ |
| $B_2X_6^+$: | 12.1 ± 0.2 (71) | 12.0 ± 0.1[b] (69) | 12.0 ± 0.1 (15) | 12.0 ± 0.1[b] (69) | | |
| | 11.9 ± 0.1 (15) | | | | | |
| $B_2X_5^+$ | 12.0 ± 0.2 (71) | 12.0 ± 0.1 (69) | 12.1 ± 0.1 (15) | 12.0 ± 0.1 (69) | 12.0 ± 0.1 (15) | 11.3 ± 0.5 (71) |
| | 11.9 ± 0.1 (15) | 11.84 ± 0.1 (74) | | | | |
| $B_2X_4^+$: | 12.3 ± 0.5 (71) | 12.3 ± 0.2 (69) | 12.3 ± 0.1 (15) | 12.3 ± 0.2 (69) | 12.3 ± 0.1 (15) | 12.6 ± 0.5 (71) |
| | 12.3 ± 0.1 (15) | | | | | |
| $B_2X_3^+$: | 14.8 ± 0.5 (71) | 14.3 ± 0.2 (69) | 14.3 ± 0.1 (15) | 14.5 ± 0.2 (69) | 14.2 ± 0.1 (15) | 12.0 ± 1.5 (71) |
| | 14.2 ± 0.1 (15) | | | | | |
| $B_2X_2^+$: | 13.8 ± 0.2 (71) | 14.1 ± 0.2 (69) | 14.0 ± 0.1 (15) | | 14.0 ± 0.1 (15) | 13.9 ± 0.5 (71) |
| | 13.8 ± 0.1 (15) | | | | | |
| $B_2X^+$ | 21.4 ± 0.5 (71) | | 18.3 ± 0.1 (15) | | 20.1 ± 0.1 (15) | |
| | 20.1 ± 0.1 (15) | | | | | |
| $B_2^+$: | 26.3 ± 0.5 (71) | | 21.8 ± 0.2 (15) | | | |
| | 21.1 ± 0.2 (15) | | | | | |
| $BX_3^+$: | 12.1 ± 0.2 (71) | 14.88 ± 0.05 (74) | 12.7 ± 0.2 (15) | | | |
| | 13.1 ± 0.2 (15) | | | | | |
| $BX_2^+$ | 13.5 ± 0.2 (71) | | 13.6 ± 0.1 (15) | | | 12.0 ± 1.0 (71) |
| | 13.4 ± 0.1 (15) | | | | | |
| | 15.5 ± 0.05 (74) | | | | | |
| $BX^+$: | 16.6 ± 0.2 (71) | | 14.8 ± 0.1 (15) | | | |
| | 14.9 ± 0.1 (15) | | | | | |
| | 16.39 ± 0.3 (74) | | | | | |
| $B^+$ | 19.2 ± 0.2 (71) | | 18.6 ± 0.1 (15) | | | 20.1 ± 0.5 (71) |
| | 18.7 ± 0.1 (15) | | | | | |
| | 18.39 ± 0.02 (74) | | | | | |

[a] $n - 1, 2; x - 0-6; X - H$ or $D$.
[b] Estimated value.

Table 6.6

Proposed Processes and Configurations, Calculated Ionisation Potentials, and Measured Appearance Potentials of Fragments from Diborane and Deuterated Diborane[15]

| Ion | Process | Configuration | A.P. from $^{10}B_2X_6$ (X=H or D) (eV) | I.P. (eV) |
|---|---|---|---|---|
| $B_2H_6^{\dagger}$ | $B_2H_6 + e \rightarrow$ $B_2H_6^{\dagger} + 2e$ | | $11.9 \pm 0.1$ | $11.9$ |
| $B_2H_5^+$ | $B_2H_6 + e \rightarrow$ $B_2H_5^+ + H\cdot + 2e$ | | $11.9 \pm 0.1$ | $7.8_6$ |
| $B_2H_4^{\dagger}$ | $B_2H_6 + e \rightarrow$ $B_2H_4^{\dagger} + H_2 + 2e$ | | $12.3 \pm 0.1$ | $10.9_3$ |
| $B_2H_6^+$ | $B_2H_6 + e \rightarrow$ $B_2H_3^+ + H_2 + H\cdot + 2e$ | | $14.2 \pm 0.1$ | $8.7_9$ |
| $B_2H_2^{\dagger}$ | $B_2H_6 + e \rightarrow$ $B_2H_2^+ + 2H_2 + 2e$ | $H-B \equiv B-H^{\dagger}$ | $13.8 \pm 0.1$ | $11.3_6$ |
| $B_2H^+$ | $B_2H_6 + e \rightarrow$ $B_2H^+ + 2H_2 + H\cdot + 2e$ | $H-B-B^+$ | $20.1 \pm 0.1$ | $10.6_2$ |
| $B_2^{\dagger}$ | $B_2H_6 + e \rightarrow$ $B^{\dagger} + 3H_2 + 2e$ | $B-B^{\dagger}$ | $21.1 \pm 0.1$ | $12.0_6$ |
| $B_2D_6^{\dagger}$ | | | $12.0 \pm 0.1$ | $12.0$ |
| $B_2D_5^+$ | | | $12.1 \pm 0.1$ | $8.0_1$ |
| $B_2D_4H^+$ | | | $12.0 \pm 0.1$ | $7.9_1$ |
| $B_2D_4^{\dagger}$ | | | $12.3 \pm 0.1$ | $10.9_0$ |
| $B_2D_3H^{\dagger}$ | | | $12.3 \pm 0.1$ | $10.9_0$ |
| $B_2D_3^+$ | | | $14.3 \pm 0.1$ | $8.8_1$ |
| $B_2D_2H^+$ | | | $14.2 \pm 0.1$ | $8.7_1$ |
| $B_2D_2^{\dagger}$ | | | $14.0 \pm 0.1$ | $11.5_0$ |
| $B_2DH^{\dagger}$ | | | $14.0 \pm 0.1$ | $11.5_0$ |
| $B_2D^+$ | | | $18.3 \pm 0.1$ | $11.7_1$ |
| $B_2^{\dagger}$ | | | $21.8 \pm 0.2$ | $12.6_7$ |

spectrum of diborane. The strength of the B–N bond is the dominating influence and this bond is present in all the major fragments.

A number of appearance potential studies of diborane have been reported. Margrave[71] examined the fragment ions of both normal isotopic diborane and also $^{10}B_2H_6$ but considerable disagreement exists between the values he obtained for many of the corresponding ions in these two samples (Table 6.5). Koski et al.[15] determined the appearance potentials of the fragment ions containing two boron atoms derived from diborane and combined these measurements with bond dissociation energy data to calculate the ionisation potential of the fragment after a consideration of its most likely structure (Table 6.6). Their appearance potential values of corresponding ions from $^{10}B_2H_6$ and $^{10}B_2D_6$ agreed fairly closely (Table 6.5) although the potentials of the deuterated ions tended to be slightly higher. When one considers the reported limits of accuracy, however, there is some doubt that the difference of approx. 0.1 eV is significant. Replacement of a hydrogen atom bonded to another element by a deuterium would be expected to raise the bond disso- ciation energy slightly. This in turn would lead to a corresponding increase in the appearance potential of a fragment ion formed by dissociation of the bond to deuterium rather than to hydrogen. This factor would not influence the ionisation potential of a deuterated species since no bond dissociation is involved. It has been reported[72] that a difference of 0.12–0.18 eV exists between the ionisation potentials of methane and methane-$d_4$ but such a large difference is difficult to explain. The ionisation potentials of $C_2H_2$ and $C_2D_2$, and of $C_2H_4$ and $C_2D_4$, were identical within experimental error[73].

Calculation of the ionisation potentials of the $BH_x^+$ fragments proved to be much more difficult, but in general agreement was obtained between the appearance potentials of the $BH_x^+$ and corresponding $BD_x^+$ ions. It was suggested, however, that there may be a fundamental difference in the nature of the bridging in the completely protonated and completely deuterated boron hydrides. This appears unlikely.

Improvements in instrumentation and technique were accompanied by a number of more detailed investigations of the energetics of diborane. These served to check the earlier figures obtained, to establish more accurate values than was previously possible, and to extend knowledge of the system con- siderably.

Prior to 1964 no direct experimental observation of the borane molecule $BH_3$ had been reported although it had been postulated[75,76] as an inter- mediate in many reactions. The energy required to dissociate diborane into two boranes, $D(BH_3-BH_3)$, had been calculated from electronegativities[77], kinetic data for the decomposition of $BH_3CO$ (ref. 78), and thermochemical measurements[79], but the values reported varied from 24 to 35 kcal.mole$^{-1}$. The direct experimental observation of $BH_3$ was reported almost simultaneously by Sinke et al.[58] and by Fehlner and Koski[59], and was accomplished mass spectrometrically in each case. Unfortunately, however, a large discrepancy existed between the values obtained by the two groups for the symmetric dissociation energy of diborane. Sinke et al. utilised a Knudsen cell and obtained a value for $D(BH_3-BH_3)$ of $55 \pm 8$ kcal.mole$^{-1}$ from the observed equilibrium between $B_2H_6$ and $BH_3$; Fehlner and Koski, on the other hand, determined the ionisation potential of $BH_3$ and the appearance potential of $BH_3^+$ from $B_2H_6$ using a fast-flow pyrolysis technique, resulting in $D(BH_3-BH_3) = 1.7$ eV or 39 kcal.mole$^{-1}$. The latter workers also observed free $BH_2$ in concentrations of two or three times that of $BH_3$, and postulated the following reactions to account for its formation.

$$B_2H_6 \rightarrow BH_4 + BH_2 \rightarrow 2\,BH_2 + H_2$$

Sinke et al. reported no $BH_2$. Fehlner and Koski continued their investigations with a study of the pyrolysis of $BH_3CO$ in a low-pressure flow system and so obtained[60] $D(BH_3-BH_3) = 37.1 \pm 4$ kcal.mole$^{-1}$, in excellent agreement with their earlier value. Further mass spectrometric studies of the pyrolysis of $B_2H_6$ by Baylis et al.[3] again failed to detect $BH_2$, the only monoborane species observed being $BH_3$. Its mass spectrum was similar to those of the boron trihalides with the most abundant ion being $BH_2^+$. The formation of higher boranes was also followed.

Wilson and McGee[74] also studied the pyrolysis of $B_2H_6$ and observed $BH_3$ but failed to detect $BH_2$. They then completed a thorough study of the energetics of diborane by measuring the appearance potentials of the fragment ions $B_2H_5^+$, $BH_3^+$, $BH_2^+$, $BH^+$, and $B^+$ from $B_2H_6$, and of $BH_3^+$, $BH_2^+$, $BH^+$, and $B^+$ from $BH_3$, and by calculating several bond dissociation energies (Table 6.7). The value obtained for the symmetric dissociation energy of diborane, $D(BH_3-BH_3)$, was 2.56 eV or 59.0 kcal.mole$^{-1}$. Many of the appearance potentials obtained differed considerably from previously reported values (Table 6.5). Confidence was expressed in the new figures, however, because of their self-consistency in that for all fragment ions considered (except $^{10}B^+$) the difference in the appearance potentials of a particular fragment from $BH_3$ and from $B_2H_6$, which corresponds to the

Table 6.7

Appearance Potentials of Fragment Ions from $BH_3$ and $B_2H_6$, and Bond Dissociation Energies Calculated from These Measurements[74]

| Fragment ion | Appearance potential from parent (eV) | |
|---|---|---|
| | $BH_3$ | $B_2H_6$ |
| $^{10}B^+$ | 15.83[a] | 18.39 ± 0.02 |
| $^{10}BH^+$ | 13.66 ± 0.02 | 16.39 ± 0.3 |
| $^{10}BH_2^+$ | 12.95 ± 0.05 | 15.5 ± 0.05 |
| $^{11}BH_3^+$ | 12.32 ± 0.1 | 14.88 ± 0.05 |
| $^{11}B_2H_5^+$ | | 11.84 ± 0.1 |

[a] Calculated as 18.39–2.56 eV.

| Bond dissociation energy | Calculated value (eV) |
|---|---|
| $D(B–H)$ | 3.64 |
| $D(BH–H)$ | 3.90 |
| $D(BH_2–H)$ | 3.58 |
| $D(B^+–H)$ | 2.17 |
| $D(BH^+–H)$ | 4.30 |
| $D(BH_2^+–H)$ | 0.63 |
| $D(BH_3–BH_3)$ | 2.56 |
| $D(BH_3^+–BH_3)$ | < 3.67;  > 3.04 |

symmetric dissociation energy, was essentially constant (2.56 eV). In addition, the value of $D(B^+–H)$ calculated directly from the experimental appearance potentials agreed very well with the value from the spectroscopic results of Bauer et al.[80]. Ganguli and McGee[61] have recently published details of a similar detailed study of the molecular energetics of borane carbonyl which resulted in the confirmation of 59 kcal.mole$^{-1}$ for $D(BH_3–BH_3)$.

Dunbar[81] recently carried out an investigation of the positive and negative ion–molecule chemistry of diborane by ion cyclotron resonance. Since the ion–molecule reaction process may be expected to produce only relatively stable ions, comparisons could be made with the predictions of Lipscomb's topological theory: it appeared from the results that the theory is useful in rationalising the negative ion pattern but is not applicable to the positive ion case. Most of the exothermic reactions observed were of the form

$$(\text{ion})^+ + B_2H_6 \rightarrow (\text{ion} + B_2H_2)^{\ddagger} + 2 H_2$$

and were suggestive of unusual stability for a fragment of structure

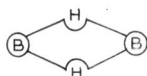

### 4.1.3 Tetraborane-8

This compound has been postulated[82-84] as an intermediate in a number of reactions and interconversions of the boron hydrides. Direct proof of its existence, however, was lacking until it was observed mass spectrometrically in the pyrolysis of $B_2H_6$ (refs. 3, 85), $B_4H_{10}$ (ref. 86), and $B_4H_8CO$ (refs. 62, 62a). The spectrum obtained was similar to that published for $B_4H_8CO$ (ref. 87), prompting the suggestion[86] that the latter compound was pyrolysed in the mass spectrometer ion source with the result that the $B_4H_8$ spectrum was, in fact, observed. The intensity of $B_4H_x^+$ ions was high, and Hollins and Stafford[62a] pointed out that, although one might expect greater fragmentation to the lower mass ions for this transient, "unstable", reactive intermediate, the compound does fit into the "stable" $B_nH_{(n + 4)}$ series and not the "unstable" $B_nH_{(n + 6)}$ series.

### 4.1.4 Tetraborane-10

The mass spectrum of tetraborane was first reported by Shapiro et al.[22] in 1958 but only scant details were presented. Fehlner and Koski[88] published the results of a comprehensive mass spectrometric study of $B_4H_{10}$, $B_4D_{10}$, and the partially deuterated compound $B_4H_8D_2$ in 1963. Considerable attention was devoted to metastable transitions which showed that, initially, the fragment ions of highest mass lose $H_2$ (or $D_2$); this confirmed the postulate of Shapiro et al.[22] which arose from a study of spectra alone. The loss of $2H\cdot$ corresponds to observations made on hydrocarbons[89] but, in addition, strong transitions were observed involving loss of $3H\cdot$ which has no such analogy in the spectra of hydrocarbons.

A comparison of the relative rates of $H\cdot$ and $D\cdot$ loss was made by Fehlner and Koski[88] from the observed metastable transitions; many more were detected in the spectrum of $B_4D_{10}$ than in that of $B_4H_{10}$. In addition, the relative intensities of the monoisotopic spectra of $B_4$-containing ions[4,88] indicated that $B_4H_{10}$ loses H$\cdot$ more readily than $B_4D_{10}$ loses D$\cdot$. This might be expected from a comparison of B–H and B–D bond energies, and led to the conclusion that the rate of loss of D$\cdot$ was approximately $10^{-6}$ sec, the approximate decomposition rate required for observation of metastable species, while the rate of loss of H$\cdot$ was somewhat faster, thus precluding observation of many metastable transitions in $B_4H_{10}$. A review of all reported metastable transitions for tetraborane, however, (Table 6.8) shows little difference in the total numbers involving H$\cdot$ and D$\cdot$, even though both sets of workers observed more involving D$\cdot$. Only one such transition corre-

Table 6.8

Metastable Transitions Observed for Tetraborane[4,69,88]

| Transition | X | Transition | X |
|---|---|---|---|
| $B_4X_9^+ \xrightarrow{*} B_4X_7^+ + X_2$ | H, D | $B_3X_7^+ \xrightarrow{*} B_3X_5^+ + X_2$ | H, D |
| $B_4X_9^+ \xrightarrow{*} B_4X_6^+ + X_2 + X\cdot$ | H | $B_3X_6^+ \xrightarrow{*} B_3X_4^+ + X_2$ | H, D |
| $B_4X_9^+ \xrightarrow{*} B_3X_7^+ + \cdot BX_2$ | H | $B_3X_5^+ \xrightarrow{*} B_3X_3^+ + X_2$ | H, D |
| $B_4X_8^+ \xrightarrow{*} B_4X_6^+ + X_2$ | H, D | $B_3X_4^+ \xrightarrow{*} B_3X_2^+ + X_2$ | H, D |
| $B_4X_8^+ \xrightarrow{*} B_4X_5^+ + X_2 + X\cdot$ | H, D | $B_3X_3^+ \xrightarrow{*} B_3X^+ + X_2$ | D |
| $B_4X_7^+ \xrightarrow{*} B_4X_5^+ + X_2$ | H, D | $B_3X_2^+ \xrightarrow{*} B_3^+ + X_2$ | H, D |
| $B_4X_7^+ \xrightarrow{*} B_4X_4^+ + X_2 + X\cdot$ | H, D | $B_2X_5^+ \xrightarrow{*} B_2X_3^+ + X_2$ | D |
| $B_4X_6^+ \xrightarrow{*} B_4X_4^+ + X_2$ | H, D | $B_2X_4^+ \xrightarrow{*} B_2X_2^+ + X_2$ | D |
| $B_4X_5^+ \xrightarrow{*} B_4X_3^+ + X_2$ | H, D | $BX_2^+ \xrightarrow{*} B^+ + X_2$ | H, D |
| $B_4X_4^+ \xrightarrow{*} B_4X_2^+ + X_2$ | H, D | $BX^+ \xrightarrow{*} B^+ + X\cdot$ | H, D |
| $B_4X_3^+ \xrightarrow{*} B_4X^+ + X_2$ | H, D | $B_4X_6^{2+} \xrightarrow{*} B_3X_5^{2+} + BX$ | H |
| $B_4X_2^+ \xrightarrow{*} B_4^+ + X_2$ | H, D | $B_4X_5^{2+} \xrightarrow{*} B_3X_2^{2+} + BX_3$ | D |
| $B_4X_2^+ \xrightarrow{*} B_3X_2^+ + B\cdot$ | D | $B_4X_4^{2+} \xrightarrow{*} B_3X_3^{2+} + BX$ | D |
| $B_3X_8^+ \xrightarrow{*} B_3X_6^+ + X_2$ | D | | |

References pp. 198–205

sponding to loss of a boron was observed, leading to the conclusion that the loss of B-containing fragments was either very slow or very fast with respect to $10^{-6}$ sec. It was suggested that the latter was unlikely since one would then expect a much higher intensity of the lower mass fragments than is observed. The fact that the intensities of these lower mass fragments were greater in the spectrum of $B_4D_{10}$ than in that of $B_4H_{10}$ (ref. 88) was interpreted as enhancement of the decomposition rate of the B-framework by D. However, this difference in relative intensities of the lower mass fragments is certainly not present in the spectra recorded on the zero source contact mass spectrometer[4].

Table 6.9

Appearance Potentials[88] of Ions Derived from $^{11}B_4H_{10}$ and $^{11}B_4D_{10}$

| Ion | A.P. (eV) from $^{11}B_4H_{10}$ | Calculated $\Delta H°_f$ (eV) (X = H) | A.P. (eV) from $^{11}B_4D_{10}$ |
|---|---|---|---|
| $B_4X_9^+$ | $12.2 \pm 0.2$ | | $11.9 \pm 0.2$ |
| $B_4X_8^+$ | $10.4 \pm 0.1$ | $11.0 \pm 0.1$ | $9.9 \pm 0.1$ |
| $B_4X_7^+$ | $12.5 \pm 0.2$ | | $12.2 \pm 0.1$ |
| $B_4X_6^+$ | $11.2 \pm 0.1$ | $11.8 \pm 0.1$ | $11.1 \pm 0.1$ |
| $B_4X_5^+$ | $12.5 \pm 0.3$ | | $12.5 \pm 0.2$ |
| $B_4X_4^+$ | $12.4 \pm 0.2$ | $13.2 \pm 0.2$ | $12.5 \pm 0.2$ |
| $B_4X_3^+$ | $14.0 \pm 0.4$ | | $14.8 \pm 0.4$ |
| $B_3X_5^+$ | $12.1 \pm 0.2$ | $11.9 \pm 0.2$ | $12.8 \pm 0.3$ |
| $B_3X_3^+$ | $14.2 \pm 0.3$ | $14.0 \pm 0.3$ | $13.8 \pm 0.3$ |
| $B_3X_2^+$ | $17.8 \pm 0.8$ | | |
| $B_3X^+$ | $16.5 \pm 0.8$ | | |

*Metastable peaks*

| Transition | A.P. (eV) from $^{11}B_4H_{10}$ | A.P. (eV) from $^{11}B_4D_{10}$ |
|---|---|---|
| $B_4X_8^+ \xrightarrow{*} B_4X_6^+ + X_2$ | $11.0 \pm 1.0$ | $11.1 \pm 0.5$ |
| $B_4X_7^+ \xrightarrow{*} B_4X_5^+ + X_2$ | $12.3 \pm 1.0$ | |
| $B_4X_8^+ \xrightarrow{*} B_4X_5^+ + X_2 + X\cdot$ | $12.5 \pm 0.2$ | $12.4 \pm 0.5$ |
| $B_4X_7^+ \xrightarrow{*} B_4X_4^+ + X_2 + X\cdot$ | $13.6 \pm 1.0$ | |

On the basis of appearance potential measurements (Table 6.9), observed metastable transitions (Table 6.8), and the spectrum of partially deuterated tetraborane, an energetically favoured series of unimolecular dissociations was proposed[88] to account for $B_4H_x^+$ ions in the spectrum. Structures were also proposed for these ions (and, in addition[69], the ion $B_3H_3D_2^+$) on the basis of the following assumptions: (i) the D atoms in the compound $B_4H_8D_2$ are in the 1,3 positions[90], (ii) $B_4X_8^+$ is $B_4H_6D_2^+$ in the spectrum of this dideuterio compound, (iii) the formation of a "normal" two-centre B–B bond is a part of the process of losing two H atoms, (iv) the $m/e$ 40 peak in the spectrum of $B_4H_8D_2$ is due to $B_3H_3D_2^+$, (v) rearrangement within the partially deuterated compound is not taking place as fast as the primary decompositions, and (vi) the most readily lost H atoms are the closest adjacent ones, and two H atoms do not come from the same B.

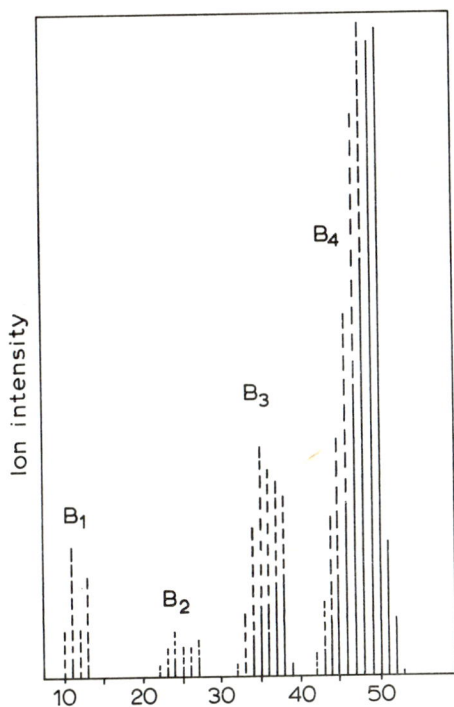

Fig. 6.4. Comparison of the normal isotopic $B_4H_{10}$ mass spectrum recorded under zero ion source contact conditions (solid lines) and conventional conditions where the material may be pyrolysed by contact with the hot ion source (broken lines). (Reproduced with permission from refs. 4, 62.)

Some of these assumptions appear highly speculative. Later, Fehlner and Koski[69] extended their original decomposition scheme slightly on the basis of appearance potential measurements and calculated heat of formation values of ions derived from tetraborane and the higher hydrides $B_5H_9$, $B_5H_{11}$, and $B_6H_{10}$. Norman et al.[4] have since shown by [11]B NMR and gas volumetric techniques that the reaction of $B_2D_6$ with $B_4H_{10}$ at 45 °C did not produce 1,3-dideuteriotetraborane[69,88], but in fact resulted in a mixture of products including $B_5D_xH_{11-x}$ and partially deuterated tetraboranes with the D in both the 1(3) and 2(4) positions.

These workers re-examined the mass spectrum of tetraborane-10 using the isotopically labelled compounds [10]$B_4H_{10}$, [10]$B_4D_{10}$, *B[10]$B_3H_{10}$ (tetraborane labelled specifically with a [11]B isotope in the 2-position), and $\mu$-[10]$B_4H_9D$ containing a single deuterium atom in a bridging position. They realised also that the possibility of isotopic scrambling and/or pyrolysis of the samples in conventional inlet systems and ion sources was high and so employed an inlet system with variable temperature conduits (total contact time of sample ca. 0.5 sec) and a "zero source contact" mass spectrometer. Significant differences existed between the zero source contact and conventional mass spectra of tetraborane (Fig. 6.4). In the latter, peaks of relative intensity 0.5–1.0% (approx.) arising from $B_5H_x^+$ ion fragments were routinely present, and peaks at masses below $m/e$ 46 ([10]$B_4H_{10}$ spectrum) are more intense than in the zero source contact spectrum. A separarate study[86] confirmed that pyrolysis of tetraborane occurs in the conventional mass spectrometer. Another significant feature was the definite observation of a parent molecular ion in the spectrum of [10]$B_4H_{10}$; this ion had not previously been observed[22,88] even at an ionising voltage[91] of 11 eV.

The stepwise dissociation process leading to the formation of $B_4H_x^+$ ions, postulated by Fehlner and Koski[69,88], was substantiated in part by metastable transitions observed[4] in the zero source contact spectra of [10]$B_4H_{10}$ and [10]$B_4D_{10}$ (Table 6.8). The data from the specifically labelled sample $\mu$-[10]$B_4H_9D$ provided no evidence for an energetically favoured process involving loss of H· or D· from specific positions in the molecule, however, but this may have been due to the H and D atoms of the labelled sample becoming isotopically scrambled prior to dissociation. The 2,4-boron atoms were lost in a preferred manner.

The metastable transition involving loss of a B from a $B_4$ fragment ion reported by Fehlner and Koski[88] was not observed by Norman et al.[4], but that corresponding to the decomposition

$$^{10}B_4H_9^+ \xrightarrow[*]{} {}^{10}B_3H_7^+ + {}^{10}BH_2$$

was detected[4] and so indicates a second process which must contribute to dissociation of the boron framework of tetraborane. Further information about this fragmentation mechanism was obtained from the zero source contact spectrum of $*B^{10}B_3H_{10}$. It was concluded[4] that $B_3H_x^+$, $B_2H_x^+$, and $BH_x^+$ ions result, in part, from dissociation and/or ionisation processes in which the ion fragments are lost in a specific fashion and which bear a structural relationship to the original tetraborane molecules. Thus, $B_3H_x^+$ ions are formed by the loss of a monoborane species from the 2 or 4 positions of tetraborane; direct formation of $B_2H_x^+$ ions from tetraborane is a significant process; and preferential loss of $BH_x^+$ ions from 2 or 4 positions of tetraborane must occur.

A combination of all proposed schemes and observed metastable transitions produces the following decomposition paths:

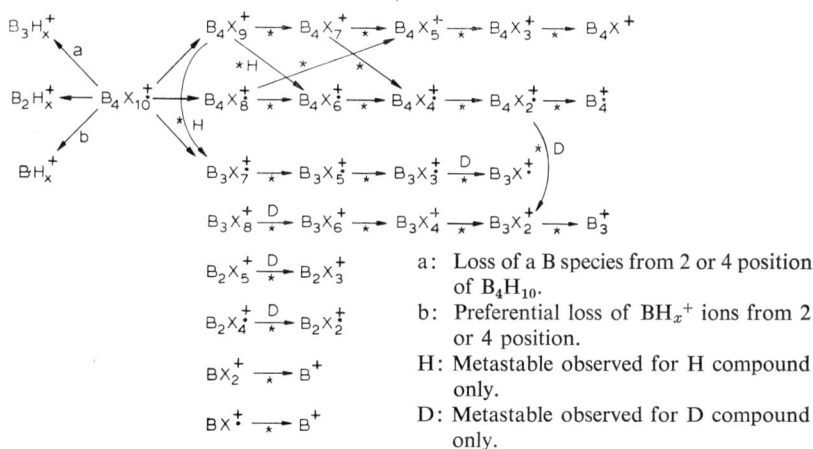

a:  Loss of a B species from 2 or 4 position of $B_4H_{10}$.
b:  Preferential loss of $BH_x^+$ ions from 2 or 4 position.
H:  Metastable observed for H compound only.
D:  Metastable observed for D compound only.

A large number of doubly charged ions was observed[4,69,91] but, because the relative intensities were very low in many cases, some doubt was involved in the assignment of these peaks. Attempts to observe peaks arising from $^{10}B_4D_x^{2+}$ ions were unsuccessful because of their integral $m/e$ values; they could not be sufficiently resolved from the large singly charged ion peaks to be assigned or measured. Metastable transitions corresponding to decomposition of three doubly charged ions were also reported[69].

Pyrolysis of tetraborane-10 was studied[86] in the temperature range 10–

285 °C in a mass spectrometer and the new tetraborane, $B_4H_8$, was identified as an intermediate in the pyrolysis. In addition, $B_2H_6$, $B_5H_9$, $B_5H_{11}$, $B_6H_{12}$, a combination of heptaboranes (most likely $B_7H_{11}$ and $B_7H_{13}$), $B_8H_{12}$, $B_{10}H_{14}$, and possibly $B_9H_{15}$ were also detected. Schaeffer et al.[109] also reported fragments at higher $m/e$ than the parent molecular ion of $B_4H_{10}$ in a spectrometer operating at 185 °C. Monoborane and triborane were not observed, but the possibility of very small amounts being formed and remaining undetected could not be excluded. Other workers[92], however, have suggested the formation of a monoborane in the pyrolysis of tetraborane.

### 4.1.5 Tetraborane carbonyl

It has been shown[62,62a] by the use of a zero source contact mass spectrometer that tetraborane carbonyl is very readily pyrolysed in the source of a conventional mass spectrometer operating at 250 °C; the two spectra so obtained in the two instruments were vastly different (Fig. 6.5). The ion

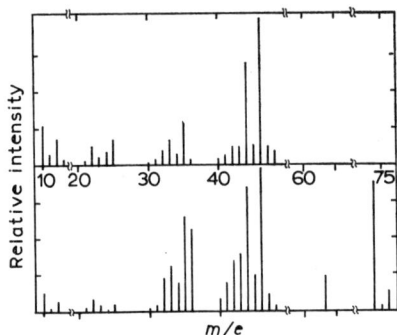

Fig. 6.5. Comparison of the tetraborane carbonyl, $B_4H_8CO$, mass spectrum recorded under zero ion source contact conditions (bottom) and in a conventional mass spectrometer (top). (Reproduced with permission from ref. 62a.)

corresponding to loss of $2H\cdot$ from the molecule was again very pronounced. Appearance potential measurements of a large number of ions were reported[62a].

### 4.1.6 Pentaborane-9

In a similar situation to that of diborane, the mass spectrum of pentaborane-9 was first recorded[23] in an effort to establish the structure of the compound. Two structures had previously been proposed: (*i*) a 5-membered ring compound, $B_5H_5$, with two additional pairs of hydrogen atoms attached to two of the bonds to form protonated double bonds[93] and (*ii*) a methyl-cyclobutane-like structure[94]. The mass spectrum, however, could not be correlated with either. It was noted that the absence of certain ions in the spectrum (Table 6.10) was unlike all $C_5$ hydrocarbons in which loss of

Table 6.10

Monoisotopic Mass Spectra of Pentaborane-9[23] and Pentaborane-11[69] (70 eV)

| Ion | R.I.[a] $B_5H_9$ | R.I.[a] $B_5H_{11}$ | Ion | R.I.[a] $B_5H_9$ | R.I.[a] $B_5H_{11}$ | Ion | R.I.[a] $B_5H_9$ |
|---|---|---|---|---|---|---|---|
| $B_5H_{11}^+$ |      | 0.6  | $B_4H_8^+$ |      | 2.9  | $B_2H_6^+$ | 0.39 |
| $B_5H_{10}^+$ |      | 0.8  | $B_4H_7^+$ |      | 5.4  | $B_2H_5^+$ | 1.28 |
| $B_5H_9^+$ | 78.0 | 35.8 | $B_4H_6^+$ | 4.25 | 14.8 | $B_2H_4^+$ | 0.27 |
| $B_5H_8^+$ | 0.36 | 7.3  | $B_4H_5^+$ | 6.74 | 12.8 | $B_2H_3^+$ | 1.64 |
| $B_5H_7^+$ | 71.7 | 100.0 | $B_4H_4^+$ | 33.0 | 39.8 | $B_2H_2^+$ | 2.54 |
| $B_5H_6^+$ | 8.66 | 6.7  | $B_4H_3^+$ | 14.1 | 21.4 | $B_2H^+$ | 1.54 |
| $B_5H_5^+$ | 100.0 | 59.7 | $B_4H_2^+$ | 12.8 | 15.8 | $B_2^+$ | 0.27 |
| $B_5H_4^+$ | 18.4 | 9.2  | $B_4H^+$ | 7.37 | 9.1  |      |      |
| $B_5H_3^+$ | 18.5 | 8.7  | $B_4^+$ | 4.35 | 4.5  |      |      |
| $B_5H_2^+$ | 20.2 | 12.4 |      |      |      |      |      |
| $B_5H^+$ | 15.6 | 10.4 | $B_3H_6^+$ | 0.81 |      | $BH_2^+$ | 12.2 |
| $B_5^+$ | 20.4 | ~16  | $B_3H_5^+$ | 0.41 |      | $BH^+$ | 3.03 |
|      |      |      | $B_3H_4^+$ | 2.65 |      | $B^+$ | 20.7 |
|      |      |      | $B_3H_3^+$ | 5.04 |      |      |      |
|      |      |      | $B_3H_2^+$ | 6.36 |      |      |      |
|      |      |      | $B_3H^+$ | 2.19 |      |      |      |
|      |      |      | $B_3^+$ | 0.87 |      |      |      |

[a] R.I. = relative intensity.

hydrogen atoms was more of a random process. This phenomenon is now readily understood on consideration of the structure of $B_5H_9$, and boron hydrides in general.

Subsequently published spectra[22,67,95] of $B_5H_9$ agreed very well with that of Dibeler *et al.*[23]. In contrast to diborane and tetraborane, pentaborane-9 gives rise to an intense parent molecular ion. The molecular beam mass spectrum was very similar to those obtained with conventional ion sources[62a]; this might be expected since $B_5H_9$ is the most stable member of the lower boron hydrides. The very low intensity of the ion $B_5H_8^+$ indicates that the preferred mechanism is loss of $2H\cdot$ and not $H\cdot$; the $B_5H_6^+$ ion is likewise of low intensity (Table 6.10). Differences may be noted between the mass spectrum of pentaborane-$d_9$[95,96] and $B_5H_9$, similar to those observed[88] between $B_4H_{10}$ and $B_4D_{10}$ discussed previously. A number of doubly charged ions were again observed.

From the measured appearance potentials of $B_5$ fragment ions derived from $B_5X_9$ (X = H or D), Kaufman et al.[95] calculated ionisation potentials

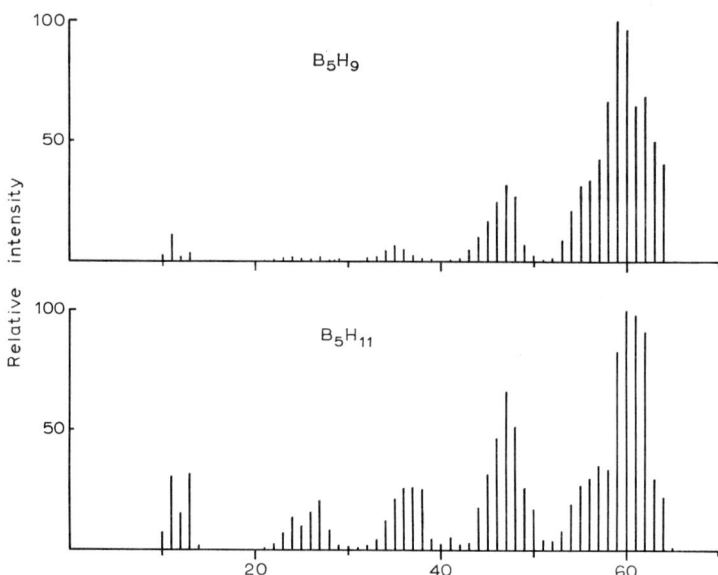

Fig. 6.6. Mass spectra of normal isotopic $B_5H_9$ and $B_5H_{11}$ (refs. 67, 91).

for these fragments employing available bond energy data and processes assumed to account for the formation of the particular ions (Table 6.11). Structures for these ions were also proposed. An almost complete lack of bond energy data, however, made a good deal of speculation and assumption necessary. By considering possible routes by which a fragment ion may be formed, calculating the corresponding heat of formation values for that ion utilising appearance potential measurements (Table 6.11), and then comparing these values with that obtained for the same ion derived from a different boron hydride ($B_4H_{10}$, $B_5H_{11}$, $B_6H_{10}$), Fehlner and Koski[69] were

Table 6.11

Appearance and Ionisation Potentials of Fragment Ions from $B_5X_9$ (X = H, D)

| Ion | Measured A.P. (eV) (ref. 95) | Calculated I.P. (eV) (ref. 95) | Measured A.P. (eV) (ref. 69) | Calculated $\Delta H°_f$ (eV) (ref. 69) |
|---|---|---|---|---|
| $B_5H_9^+$ | $10.3_8 \pm 0.05$ | 10.4 | $10.5 \pm 0.1$ | $11.2 \pm 0.1$ |
|  | $10.8 \pm 0.5^a$ |  |  |  |
| $B_5H_8^+$ | $10.5_0 \pm 0.1$ | 6.5 |  |  |
| $B_5H_7^+$ | $10.4_3 \pm 0.1$ | 10.1 | $11.6 \pm 0.2$ | $12.2 \pm 0.1$ |
| $B_5H_6^+$ | $12.1_8 \pm 0.3$ | 6.8 |  |  |
| $B_5H_5^+$ | $12.6_7 \pm 0.03$ | 10.1 | $12.8 \pm 0.2$ | $13.4 \pm 0.1$ |
| $B_5H_4^+$ | $13.0_1 \pm 0.3$ | 6.4 |  |  |
| $B_4H_6^+$ | $12.2_5 \pm 0.2$ | 10.4 |  | $12.1 \pm 0.2$ |
| $B_4H_5^+$ | $15.0_7 \pm 0.3$ |  |  |  |
| $B_4H_4^+$ | $14.0_6 \pm 0.1$ |  |  | $13.9 \pm 0.1$ |
| $B_4H_3^+$ | $15.9_6 \pm 0.5$ |  |  |  |
| $B_4H_2^+$ | $17.9_9 \pm 0.2$ |  |  |  |
| $B_4H^+$ | $19.9_7 \pm 1.0$ |  |  |  |
| $B_5D_9^+$ | $9.7_7 \pm 0.1$ | 9.8 | $10.0 \pm 0.1$ |  |
| $B_5D_7^+$ | $10.9_3 \pm 0.2$ | 9.6 | $11.2 \pm 0.2$ |  |
| $B_5D_6^+$ |  |  | $12.4 \pm 0.2$ |  |
| $B_5D_5^+$ | $12.9_2 \pm 0.03$ | 10.1 | $12.4 \pm 0.4$ |  |
| $B_4D_6^+$ |  |  | $12.4 \pm 0.2$ |  |
| $B_4D_4^+$ |  |  | $13.7 \pm 0.2$ |  |

[a] Reference 106.

able to suggest, with a fair degree of certainty, processes by which certain ions were produced.

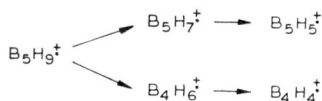

The measured ionisation potential of $B_5D_9$ was 0.6 eV lower than that of $B_5H_9$. It was presumed that an electron was ionised from the boron skeleton rather than from either a terminal or bridge boron–hydrogen bond and it was postulated that the change in boron skeletal energy between regular and deuterated pentaborane, or their fragments, is dependent on whether Hs or Ds are in the bridge positions. Later appearance potential measurements[69] agreed fairly well (Table 6.11) and, in addition, a number of metastable transitions were observed[69], the ones of chief interest corresponding to the decompositions

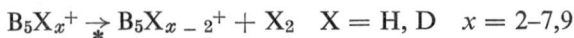

$$B_5X_x{}^+ \underset{*}{\rightarrow} B_5X_{x-2}{}^+ + X_2 \quad X = H, D \quad x = 2\text{--}7,9$$

and

$$B_5H_5{}^{2+} \underset{*}{\rightarrow} B_4H_2{}^{2+} + BH_3$$

In addition, decompositions of fragment ions with lower boron content were observed but not reported in detail.

Fragmentation of $B_5H_9$ has also been accomplished in a mass spectrometer by bombardment with rare gas ions[97] and was found to vary markedly as a function of the recombination energy of the rare gas ion.

The mass spectra of the halogenated pentaboranes $B_5H_8X$ (X = Cl, Br, I)[98,99] indicated that the iodine atom is lost much more readily from the pentaborane nucleus than is either Cl· or Br·, and the general distribution of species in the fragmentation pattern of iodopentaborane is similar to that in $B_5H_9$. Loss of Cl· and Br·, in fact, is accompanied by rupture of the $B_5$ nucleus so that $B_4H_x{}^+$ ions are much more abundant than $B_5H_y{}^+$ ions. From these observations Shapiro and Landesman[99] postulated that the relative ease of breaking B–X bonds in these compounds followed the order BI > BH > BB > BBr ∼ BCl.

Ionisation and appearance potentials of the ions $B_5H_8X^+$ (X = Br, I) and $B_5H_8{}^+$ have been measured[98,100] and are discussed later in connection with decaborane-16.

### 4.1.7 Pentaborane-11

Perhaps because of its instability, very little mass spectrometric work has appeared on $B_5H_{11}$. The polyisotopic spectrum of $B_5H_{11}$ has been reported [22,91] and, in addition, the monoisotopic spectra of $B_5H_{11}$ (refs. 69, 91) and $B_5D_{11}$ (ref. 69) have been compiled (Table 6.10). The intensity of the parent molecular ion was very low. Surprisingly, however, it was not significantly higher at an ionising voltage of 11 eV[91] than it was at 70 eV[69]. Pentaborane-11 is the least stable of the defined lower boron hydrides and rapid decomposition to pentaborane-9, tetraborane-10, and diborane-6 is reported[101,102] to occur at 100 °C. Somewhat surprisingly, there was no discernible parent molecular ion peak in the molecular beam mass spectrum of the compound[62a]. Pyrolysis resulted in dissociation to $B_4H_8$ and $BH_3$ and also $B_2H_6$ and $B_5H_9$.

Table 6.12

Monoisotopic Mass Spectra of Hexaborane-10[69] and Hexaborane-12[107] (70 eV)

| Ion | R.I.[a] $B_6H_{10}$ | R.I.[a] $B_6H_{12}$ | Ion | R.I.[a] $B_6H_{10}$ | R.I.[a] $B_6H_{12}$ | Ion | R.I.[a] $B_6H_{10}$ | R.I.[a] $B_6H_{12}$ |
|---|---|---|---|---|---|---|---|---|
| $B_6H_{12}^{+}$ |  | 1.4 | $B_5H_5^{+}$ | 18.5 | 73.4 | $B_3H_4^{+}$ | 1.4 | 16.7 |
| $B_6H_{11}^{+}$ |  | 7.3 | $B_5H_4^{+}$ | 17.3 | 35.8 | $B_3H_3^{+}$ | 1.3 | 16.8 |
| $B_6H_{10}^{+}$ | 41.0 | 39.8 | $B_5H_3^{+}$ | 11.5 | 21.4 | $B_3H_2^{+}$ | 2.0 | 16.8 |
| $B_6H_9^{+}$ | 0.6 | 2.9 | $B_5H_2^{+}$ | 18.1 | 33.5 | $B_3H^{+}$ | 0.8 | 4.2 |
| $B_6H_8^{+}$ | 47.1 | 56.5 | $B_5H^{+}$ | 15.4 | 14.9 | $B_3^{+}$ | 0.2 | 2.0 |
| $B_6H_7^{+}$ | 2.1 | 5.9 | $B_5^{+}$ | 11.4 | 27.3 |  |  |  |
| $B_6H_6^{+}$ | 100.0 | 100.0 |  |  |  | $B_2H_7^{+}$ |  | 1.6 |
| $B_6H_5^{+}$ | 7.7 | 4.3 | $B_4H_7^{+}$ |  | 3.0 | $B_2H_6^{+}$ |  | 0 |
| $B_6H_4^{+}$ | 34.8 | 23.0 | $B_4H_6^{+}$ | 0.6 | 23.9 | $B_2H_5^{+}$ |  | 21.1 |
| $B_6H_3^{+}$ | 17.7 | 13.9 | $B_4H_5^{+}$ | 2.0 | 16.0 | $B_2H_4^{+}$ | 0.9 | 6.6 |
| $B_6H_2^{+}$ | 14.7 | 13.0 | $B_4H_4^{+}$ | 4.8 | 52.9 | $B_2H_3^{+}$ | 0.6 | 0 |
| $B_6H^{+}$ | 25.6 | 20.2 | $B_4H_3^{+}$ | 5.7 | 30.8 | $B_2H_2^{+}$ | 0.9 | 15.4 |
| $B_6^{+}$ | 4.7 | 4.6 | $B_4H_2^{+}$ | 4.1 | 20.2 | $B_2H^{+}$ | 0.2 | 4.1 |
|  |  |  | $B_4H^{+}$ | 2.8 | 12.2 | $B_2^{+}$ | 0.2 | 2.5 |
| $B_5H_9^{+}$ | 1.8 | 29.6 | $B_4^{+}$ | 3.8 | 6.0 |  |  |  |
| $B_5H_8^{+}$ | 0 | 1.6 |  |  |  | $BH_3^{+}$ | 0.2 |  |
| $B_5H_7^{+}$ | 3.8 | 48.0 | $B_3H_6^{+}$ |  | 17.4 | $BH_2^{+}$ | 9.3 |  |
| $B_5H_6^{+}$ | 1.3 | 7.1 | $B_3H_5^{+}$ | 2.0 | 8.3 | $BH^{+}$ | 1.0 |  |
|  |  |  |  |  |  | $B^{+}$ | 10.4 |  |

[a] R.I. = relative intensity.

As with $B_5H_9$, the hydrogens appeared to be initially lost in pairs. Doubly charged ions[91] and metastable transitions[69] corresponding to the decompositions

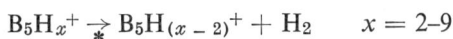

$$B_5H_x^{+} \xrightarrow{*} B_5H_{(x-2)}^{+} + H_2 \qquad x = 2\text{–}9$$

have been reported. From appearance potential measurements and calculated heats of formation of fragment ions (Table 6.12), Fehlner and Koski[69] postulated the following decomposition paths.

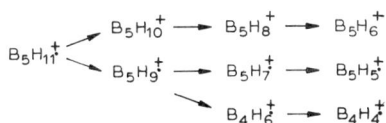

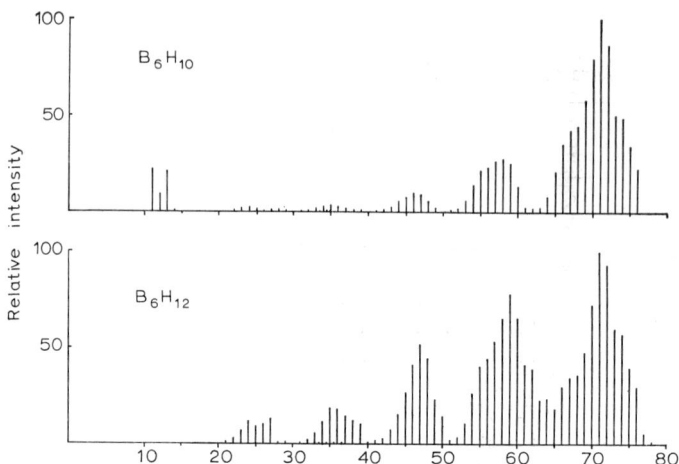

Fig. 6.7. Polyisotopic mass spectra of $B_6H_{10}$ and $B_6H_{12}$ (refs. 103, 107).

### 4.1.8 Hexaborane-10

The mass spectrum of $B_6H_{10}$ has been reported by three groups of workers[22,69,103,104] but agreement between them is not particularly good. The first of these to be published[104] contained a significant peak at $m/e$ 77. This could originate from $B_6H_{10}$ by way of an ion–molecule reaction to produce $B_6H_{11}^+$ but a more likely explanation is the presence of impurity or pyrolysis product which could also account for the observed disagreement (which is particularly noticeable in the $m/e$ 61–64 region) with later spectra.

The hydrogen again appeared to be intitially lost in pairs (Table 6.12). Fehlner and Koski[69] investigated decomposition paths by means of appearance potentials and calculated heats of formation (Table 6.13) for this compound also and postulated

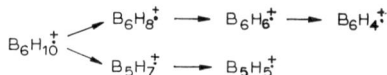

$$B_6H_{10}^+ \begin{cases} \nearrow B_6H_8^{+\cdot} \longrightarrow B_6H_6^{+\cdot} \longrightarrow B_6H_4^{+\cdot} \\ \searrow B_5H_7^+ \longrightarrow B_5H_5^+ \end{cases}$$

Several metastable transitions involving $B_6$ ions were also reported, corresponding to the processes

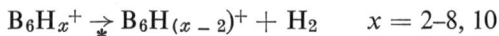

$$B_6H_x^+ \underset{*}{\rightarrow} B_6H_{(x-2)}^+ + H_2 \qquad x = 2\text{–}8, 10$$

A number of doubly charged ion species were recognised[69,103].

Table 6.13

Appearance Potentials[69] of Ions from Pentaborane-11 and Hexaborane-10

| Ion | A.P. (eV) X = H | $\Delta H°_f$ (eV) X = H | A.P. (eV) X = D | $\Delta H°_f$ (eV) X = D |
|---|---|---|---|---|
| *Pentaborane-11* | | | | |
| $B_5X_{10}^+$ | 11.8 ± 0.4 | | 11.3 ± 0.4 | |
| $B_5X_9^+$ | 10.3 ± 0.2 | 11.4 ± 0.2 | 10.4 ± 0.2 | |
| $B_5X_8^+$ | 12.0 ± 0.3 | | 11.4 ± 0.3 | |
| $B_5X_7^+$ | 11.5 ± 0.2 | 12.5 ± 0.2 | 11.1 ± 0.2 | |
| $B_5X_6^+$ | 12.6 ± 0.3 | | 12.2 ± 0.3 | |
| $B_5X_5^+$ | 12.7 ± 0.2 | 13.7 ± 0.2 | 12.3 ± 0.2 | |
| $B_5X_4^+$ | 14.2 ± 0.4 | | | |
| $B_4X_6^+$ | | | 11.4 ± 0.5 | 11.6 ± 0.5 |
| $B_4X_4^+$ | | | 12.4 ± 0.5 | 12.6 ± 0.5 |
| *Hexaborane-10* | | | | |
| $B_6X_{10}^+$ | 9.3 ± 0.1 | 10.2 ± 0.1 | 9.7 ± 0.2 | |
| $B_6X_9^+$ | 11.1 ± 0.4 | | | |
| $B_6X_8^+$ | 11.2 ± 0.1 | 12.0 ± 0.1 | | |
| $B_6X_7^+$ | 11.5 ± 0.3 | | | |
| $B_6X_6^+$ | 11.9 ± 0.1 | 12.8 ± 0.1 | | |
| $B_6X_5^+$ | 12.0 ± 0.3 | | | |
| $B_6X_4^+$ | 13.4 ± 0.3 | | | |
| $B_5X_7^+$ | 12.0 ± 0.2 | 12.1 ± 0.2 | | |
| $B_5X_5^+$ | 13.6 ± 0.2 | 13.7 ± 0.2 | | |

## 4.1.9 Hexaborane-12

Initial identification of trace amounts of this compound was accomplished by mass spectrometry[105]. A preparation resulting in much larger yields was later established by Gaines and Schaeffer[106] who studied many of its properties and reported the polyisotopic mass spectrum of the compound. That obtained in an independent study of the compound by Lutz *et al.*[107] was essentially in agreement and the monoisotopic representation of this spectrum is compared with that of hexaborane-10 in Table 6.12. While there is a marked similarity between the spectra of the two compounds as far as the $B_6$ ions are concerned, the $B_5$ and $B_4$ species are much more abundant in the spectrum of $B_6H_{12}$, suggesting that the boron frameworks in the two compounds are different. The structure of $B_6H_{10}$, I, is known and Lipscomb[82] has postulated that the most likely structures of $B_6H_{12}$ are the two

4212 formulations II and III. The open structure III has been postulated[106] on the basis of NMR evidence. Possible breakdown schemes for $B_4H_{10}$ proposed by Fehlner and Koski[88] have been discussed previously; they assumed that (*i*) closest adjacent hydrogens fragment most readily and two hydrogens are not lost from the one boron and (*ii*) formation of a "normal" two-centre B–B bond is a part of the process. Applying these same assumptions to the decomposition of a $B_6H_{12}$ molecule with structure II, one would expect that the $B_6H_{10}^+$ ion formed would have an identical structure to that of hexaborane-10. Since other unstable boron hydrides such as $B_4H_{10}$ and $B_5H_{11}$ appear to lose hydrogen from the parent molecular ion before breakdown of the boron skeleton, it is expected that the $B_6H_{10}^+$ ion is readily formed and only a small proportion, if any, of the $B_5$ and $B_4$ species would arise from a mechanism such as

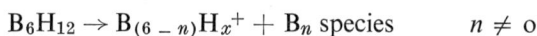

$$B_6H_{12} \rightarrow B_{(6-n)}H_x^+ + B_n \text{ species} \qquad n \neq 0$$

If $B_6H_{12}$ possessed structure II, therefore, one would expect similar cracking patterns for lower fragments in the spectra of the two compounds. Because of the large difference in the intensities of the $B_5$ and $B_4$ ions in the two spectra, it appears that the mass spectral data rules out one of the most likely structures, II, providing at least negative evidence in favour of that postulated on the basis of NMR results, III.

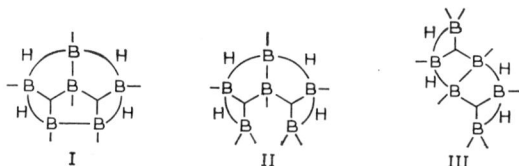

### 4.1.10 Heptaboranes

Various mixtures of boron hydrides have been postulated to contain $B_7$ species on the basis of mass spectral evidence but no heptaborane has been isolated. $B_7H_{11}$ (refs. 67, 108), $B_7H_{13}$ (refs. 105, 107, 108), $B_7H_{15}$ (refs. 105, 110), and $B_7H_{17}$ (refs. 111) have all been reported, but it appears most unlikely that all exist. Ditter et al.[112] reinvestigated one such report and found that the peaks previously thought to arise from a $B_7$ compound [105] were due instead to ethylpentaborane formed from ether used in preparations. However, Baylis et al.[86] have produced convincing evidence for the existence of a

Table 6.14

Partial Monoisotopic Spectra of Octaborane-12 and $n$-Nonaborane-15

| Octaborane-12 | | $n$-Nonaborane-15 | | |
|---|---|---|---|---|
| Ion | Rel. intensity[a] | Ion | Rel. intensity[a] | Rel. intensity[b] |
| $B_8H_{12}^+$ | 37 | $B_9H_{15}^+$ | 0 | 0 |
| $B_8H_{11}^+$ | 0 | $B_9H_{14}^+$ | 0 | 0 |
| $B_8H_{10}^+$ | 38 | $B_9H_{13}^+$ | 22 | 29.5 |
| $B_8H_9^+$ | 8 | $B_9H_{12}^+$ | 0 | 3.5 |
| $B_8H_8^+$ | 58 | $B_9H_{11}^+$ | 67 | 95.4 |
| $B_8H_7^+$ | 0 | $B_9H_{10}^+$ | 4 | 5.1 |
| $B_8H_6^+$ | 100 | $B_9H_9^+$ | 42 | 56.8 |
| $B_8H_5^+$ | 27 | $B_9H_8^+$ | 5 | 2.5 |
| $B_8H_4^+$ | 6 | $B_9H_7^+$ | 100 | 100 |
| $B_8H_3^+$ | 22 | $B_9H_6^+$ | 4 | −0.7 |
| $B_8H_2^+$ | 8 | $B_9H_5^+$ | 18 | 19.8 |
| $B_8H^+$ | 13 | $B_9H_4^+$ | 15 | 1.6 |
| $B_8^+$ | 0 | | | |

[a] Estimated from Fig. 4, ref. 22.
[b] Calculated from the polyisotopic spectrum[112] using a $^{11}B/^{10}B$ ratio of 4.00.

heptaborane, and their spectrum corresponded to that reported[108] for a mixture of $B_7H_{11}$ and $B_7H_{13}$.

### 4.1.11 Octaborane-12

The polyisotopic mass spectrum of this compound has been recorded[22, 113,114] and also calculated[112] from the spectrum of a mixture of $B_8H_{12}$ and $B_9H_{15}$. The monoisotopic spectrum of the $B_8$ ions[22] (Table 6.14) once again indicates that the favoured decomposition path involves loss of $H_2$. Because of the thermal instability of the compound*, considerable decomposition must have occurred in the source of the mass spectrometer with resultant lowering of the intensity of the higher mass ions.

### 4.1.12 Octoborane-18

A crystal structure of this compound has not been published but two

---

* Enrione et al.[115] reported that the compound decomposes "rapidly" near the melting point of approximately −20 °C.

structures are compatible with observed NMR data[116]: two tetraborane-9 units connected by a B–B bond, and a belt-line icosahedral fragment. On the basis of the molecular beam mass spectrum of the compound and the appearance potential of the $B_4H_9^+$ ion, Steck *et al.*[117] have proposed that the former is the correct structure. This mass spectrum, depicted in Fig. 6.8, differs markedly from any other so far observed for the boron hydrides in that the most intense ions are normally those containing the intact boron

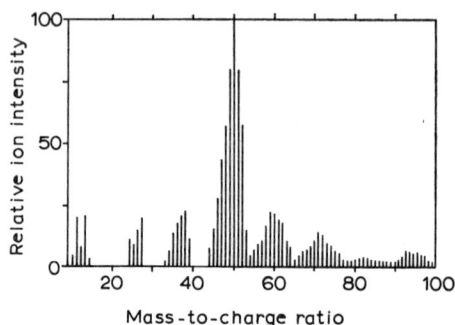

Fig. 6.8. Molecular beam mass spectrum of octaborane-18. (Reproduced with permission from ref. 117.)

skeleton but for this compound the maximum of ion intensity is in the $B_4H_x^+$ region; it was shown that this did not result from thermal decomposition of the compound during inlet into the mass spectrometer.

A separate pyrolysis study of the compound in the mass spectrometer was also undertaken; both tetraborane-8 and tetraborane-10 appeared to be formed in the intitial steps of the decomposition.

### 4.1.13 n-Nonaborane-15

The first reported evidence[118] for the existence of a nonaborane was obtained during a mass spectrometric investigation of decomposition products of $B_5H_{11}$; it was postulated that a compound $B_9H_{13}$ or $B_9H_{15}$ was present with $B_{10}H_{14}$ which had been separated from the reaction products. Three subsequent papers[22,119,120] reported the mass spectrum of the isolated compound followed by a fourth[112] which differed considerably from them. It was realised[112] that this unstable compound was continuously decomposing in the mass spectrometer to $B_8H_{12}$, which in turn itself decomposed, probably to $B_6H_{10}$; small amounts of $B_{10}H_{14}$ also appeared to be formed. Both

$n$-$B_9H_{15}$ and $B_8H_{12}$ were found to decompose by first-order mechanisms: Ditter et al.[112] reported the polyisotopic spectrum of a sample of $n$-$B_9H_{15}$ which, it was believed, contained $<5\%$ $B_8H_{12}$; it was claimed that the previously published spectra were recorded on samples containing up to $60\%$ $B_8H_{12}$. The monoisotopic spectrum of $B_9$ ions has been calculated from their results and is shown in Table 6.14. Neither the parent molecular ion nor the ion $B_9H_{14}^+$ was observed. Once again, ions formed by loss of an uneven number of hydrogens from the parent molecule have a very low relative intensity.

In a study of isotopically substituted $n$-$B_9H_{15}$ and $B_{10}H_{14}$, Maruca et al.[121] recorded and utilised the mass spectra of their various compounds in the determination of isotopic composition. Their spectrum of normal isotopic $n$-$B_9H_{15}$ shows a significant peak at $m/e$ 113 which had not previously been reported. If one assumed that this is due to the ion $^{11}B_9H_{14}^+$ and reduces the figures to a monoisotopic representation (assuming $20\%$ $^{10}B$, $80\%$ $^{11}B$), large negative values are obtained for some fragment ions. The same situation applies if one ignores the $m/e$ 113 peak and assumes that that at 112 is due solely to $^{11}B_9H_{13}^+$, or if one assumes that $B_9H_{12}^+$ is not present and the peak at 111 arises from the ion $^{11}B_8^{10}BH_{13}^+$ only. Apparently some impurity was present in sufficient quantity to affect the spectrum noticeably. It must also be remembered that slight changes in the isotopic abundances of $^{10}B$ and $^{11}B$ would alter the calculated monoisotopic spectrum considerably, as shown by Kaufman et al.[122] for $B_{10}H_{14}$. It is possible, therefore, that these negative values were caused largely by their use of the incorrect $^{10}B$ to $^{11}B$ ratio in this particular case.

### 4.1.14 Isononaborane-15

This unstable compound has been prepared and characterised by Schaeffer and coworkers[123]. Its mass spectrum has not been reported, however, and it appears that, with present day instruments, the spectrum of a relatively pure sample is not possible.

### 4.1.15 Decaborane-14

In 1950, Norton[118] observed that $B_{10}H_{14}$ was formed as a decomposition product of $B_5H_{11}$ and he reported its mass spectrum in the $m/e$ range 85–124. Shapiro et al.[22,124], published spectra of the compound which were in fair agreement. That reported by Quayle[67] was in excellent agreement with Norton's results. The partial mass spectrum of $B_{10}D_{14}$ was also presented[124].

A thorough mass spectral study of decaborane-14 appeared in 1962 and involved[122] both $^{11}B_{10}H_{14}$ (99.68% $^{11}B$) and decaborane containing the normal isotopic abundances of $^{10}B$ and $^{11}B$. A computer programme was used to calculate the complete monoisotopic spectrum. Fragmentation appeared to favour loss of $H_2$ to give $B_{10}H_n^+$ ions of greater abundance for $n$ even than for $n$ odd. In an examination of the spectrum of decaborane containing deuterium in the four bridging position, Shapiro et al.[124] found indications that a high proportion of the hydrogen pairs lost in the initial fragmentation consisted of one terminal hydrogen atom and a deuterium atom from a bridging position. Two other characteristics of the spectrum are of interest: (i) there are appreciable contributions from lower ions (this is contrary to that found for fused-ring hydrocarbons) and (ii) there is a substantial contribution from ions in the $B_9$ range—it was claimed that this was to be expected since a $BH_3$ group is the fragment which is most likely to be withdrawn on rupture of the boron skeleton. Contributions in the lower mass ranges indicated that it was possible to get every conceivable fragment ion from electron-impact induced dissociation of decaborane. Multiply charged ions were also observed.

Table 6.15

Appearance Potentials and Calculated Ionisation Potentials of Decaborane-14 Fragment Ions[122]

| Ion | A.P.[a] (eV) | A.P.[b] (eV) | Calculated A.P.[c] (eV) |
|---|---|---|---|
| $B_{10}H_{14}^+$ | $11.0 \pm 0.5$[d] | $10.26 \pm 0.5$ | 10.26 |
| $B_{10}H_{12}^+$ | $10.90 \pm 0.2$ | $10.87 \pm 0.2$ | 9.61, 9.54 |
| $B_{10}H_{11}^+$ | | $10.81 \pm 0.5$ | 5.51, 5.43 |
| $B_{10}H_{10}^+$ | $11.62 \pm 0.2$ | $11.62 \pm 0.2$ | 9.10, 8.94, 9.02 |
| $B_{10}H_8^+$ | $12.66 \pm 0.3$ | $12.67 \pm 0.3$ | 8.81, 8.73, 8.65 |
| $B_{10}H_7^+$ | | $12.51 \pm 0.5$ | 4.45–4.61 |
| $B_{10}H_6^+$ | $13.56 \pm 0.3$ | $13.14 \pm 0.3$ | 8.36, 8.28 (using A.P.[a]) 7.94, 7.86 (using A.P.[b]) |

[a] Obtained by semilog plot of ion current versus ionising voltage.
[b] Obtained by semilog plot of ion current as a percent of that at 70 eV versus ionising voltage.
[c] More than one value obtained by considering different bonds ruptured and formed during rearrangement.
[d] Value taken from ref. 100.

The appearance potentials of selected ions were measured[122]. Using these appearance potentials together with the limited bond energy data available, a set of apparently self-consistent ionisation potentials for decaborane and various $B_{10}H_n$ fragments were calculated on the basis of the authors' interpretation of the processes taking place on ionisation and fragmentation. These are shown in Table 6.15.

### 4.1.16 Decaborane-16

The structure[125] of this compound is that of two $B_5H_8$ units each derived from $B_5H_9$ by removal of an apical hydrogen atom from the pentagonal prism and joined by a two-centre B–B bond between these apices. The bonding in each $B_5H_8$ unit is assumed to be identical with that in $B_5H_9$. Its mass spectrum was reported[126] in 1962. A large pentaborane ion group is present in the spectrum of $B_{10}H_{16}$ (Fig. 6.9).

Of particular interest is the dissociation energy of the B–B bond which couples the two $B_5H_8$ units. A value of $3.2 \pm 0.2$ eV ($73.8 \pm 4.6$ kcal.mole$^{-1}$) was obtained by first of all determining the ionisation potential of the $B_5H_8$

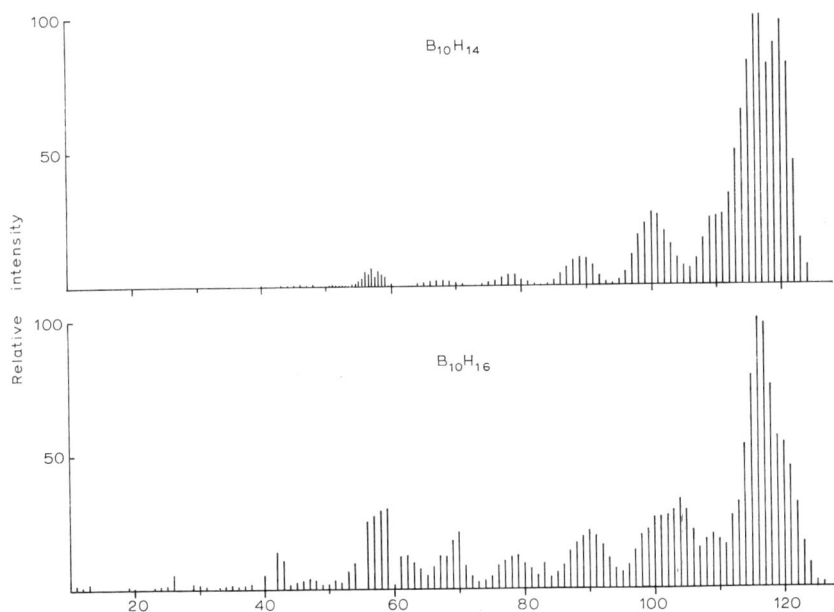

Fig. 6.9. Polyisotopic mass spectra of $B_{10}H_{14}$ and $B_{10}H_{16}$. (Reproduced with permission from ref. 98.)

Table 6.16

Ionisation and Appearance Potentials[98] of Some Ions from $B_{10}H_{16}$, $B_5H_8Br$, and $B_5H_8I$

| Compound | Ion | A.P. (eV) | I.P. (eV) |
|----------|-----|-----------|-----------|
| $B_{10}H_{16}$ | $B_{10}H_{16}^+$ | | $10.1 \pm 0.2$ |
| | $B_5H_8^+$ | $11.6 \pm 0.2$ | |
| $B_5H_8Br$ | $B_5H_8Br^+$ | | $9.5 \pm 0.1$ |
| | $B_5H_8^+$ | $12.0 \pm 0.2$ | $8.4 \pm 0.2$[a] |
| $B_5H_8I$ | $B_5H_8I^+$ | | $9.2 \pm 0.1$[b] |
| | $B_5H_8^+$ | $11.1 \pm 0.1$ | $8.4 \pm 0.2$[a] |

[a] Calculated assuming that $D(B_5H_8–X) \approx D(B–X)$ in $BX_3$ (X = Br, I).
[b] A value of $11.1 \pm 0.5$ eV had previously been reported (ref. 100).

radical from measurements carried out on $B_5H_8Br$ and $B_5H_8I$ and com-
bining this with the appearance potential of the $B_5H_8^+$ ion derived from
$B_{10}H_{16}$. Ionisation and appearance potential values are listed in Table 6.16.
The value thus obtained for the B–B bond dissociation energy in $B_{10}H_{16}$ is
approximately what one would expect for a normal two-centre electron pair
bond. It also has been concluded from an LCAO–MO calculation on the compound that this is not a multiple bond. Other dissociation energies of B–B
bonds which have been reported are: 79.0 and 87.6 kcal.mole$^{-1}$ ($B_2Cl_4$, refs.
127, 128), 72.4 and 103.1 kcal.mole$^{-1}$ ($B_2F_4$, refs. 129, 130), and 79.34
kcal.mole$^{-1}$ ($B_4H_{10}$, ref. 131).

*4.1.17  1,10-$B_{10}H_8(N_2)_2$*
    This diazonium derivative of $B_{10}H_{10}^{-2}$ is the only compound possessing a
closed polyhedral boron skeleton whose mass spectrum has been published.
Middaugh[132] compared its spectrum with that of the open-faced icosahedral
fragment compound $B_{10}H_{14}$ and found that there was less boron fragmentation from the former, as would be expected from their structures. Prominent
ions were $B_{10}H_8N_4^+$ and $B_{10}H_8N_2^+$ with no evidence of similar ions with
fewer than eight hydrogens. Accurate appearance potentials were not
determined but the differences were $0.60 \pm 0.05$ eV between $B_{10}H_8(N_2)_2^+$
and $B_{10}H_8(N_2)^+$ and $1.80 \pm 0.05$ between $B_{10}H_8(N_2)_2^+$ and $B_{10}H_8^+$.
Metastable ions corresponding to the decompositions

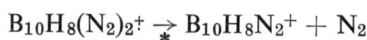

$$B_{10}H_8(N_2)_2^+ \xrightarrow[*]{} B_{10}H_8N_2^+ + N_2$$

$$B_{10}H_8N_2^+ \underset{*}{\rightarrow} B_{10}H_8^+ + N_2$$

were observed, indicating that the stepwise loss of two $N_2$ molecules from the parent molecular ion is an important fragmentation path. Doubly charged ions $B_{10}H_x(N_2)_2^{2+}$, $B_{10}H_8N_2^{2+}$, and $B_{10}H_x^{2+}$ were also detected.

### 4.1.18 Higher boron hydrides

The partial mass spectra of the boranes $B_{16}H_{20}$ (ref. 133), $B_{18}H_{22}$ (ref. 133), and $B_{20}H_{26}$ (ref. 126) have been represented graphically, while that of $B_{20}H_{16}$ has been reported[134] as showing positive ions at every mass up to 236 with those clustered around the base peak at $m/e$ 232 by far the most abundant; no further details were given. Quayle[67] tentatively assigned a formula of $B_{20}H_{24}$ to an impurity present in a sample of decaborane-14 on the basis of its partial mass spectrum. This spectrum, however, is very similar to that published[126] later for $B_{20}H_{26}$ and it therefore appears that this was the compound observed.

### 4.1.19 Summary

The boron hydrides have frequently been classified into a $B_nH_{(n+4)}$ and a $B_nH_{(n+6)}$ series. Another method of classification, however, places those containing a $BH_2$ group into one series and those which do not into another; members of the latter group are more stable chemically and thermodynamically than compounds in the former[83]. Examination of the mass spectra of these compounds confirms that they fall readily into this classification system

Table 6.17

Relative Intensities of the Parent Molecular Ions (p.m.i.) of a Number of the Boron Hydrides

| Containing a BH₂ group | | Containing no BH₂ group | |
|---|---|---|---|
| Compound | Intensity of p.m.i. | Compound | Intensity of p.m.i. |
| $B_2H_6$ | 0–1.5 | $B_5H_9$ | ~80 |
| $B_4H_{10}$ | < 1 | $B_6H_{10}$ | ~40 |
| $B_5H_{11}$ | < 1 | $B_8H_{12}$ | ~40 |
| $B_6H_{12}$ | ~1 | $B_{10}H_{14}$ | ~15 |
| $B_9H_{15}$ | 0 | $B_{10}H_{16}$ | ~15 |

when one considers the relative intensity, and hence the stability, of the parent molecular ions. This is illustrated in Table 6.17.

Three other characteristics stand out. (*i*) The favoured decomposition path of the parent molecular ions in general involves initial loss of pairs of hydrogen atoms; ions formed by loss of an uneven number of hydrogens have a very low relative intensity. (*ii*) The intensity of ions containing the intact boron skeleton is much higher than that of lower fragment ions. (*iii*) The polyisotopic spectra are somewhat complex with peaks appearing at practically every *m/e* value; only the region *m/e* 15–19 is normally devoid of peaks.

Appearance potential studies have been numerous and have contributed greatly to out present knowledge of these compounds, but a full understanding of the bonding involved is still not possible.

Mass spectrometric techniques were also used to follow the oxidation of $B_5H_9$ (refs. 135, 136), $B_4H_{10}$ (ref. 136), and $BH_3CO$ (ref. 136). A kinetic study[137] of the pyrolysis of diborane has been carried out and a reaction scheme proposed for conversion of diborane to higher hydrides.

*4.1.20 Alkyl derivatives of boron hydrides*

The fragmentation patterns of both methyl and ethyl derivatives of diborane have been studied by Shapiro *et al.*[22] and the use of isotopes of boron and hydrogen in determining the particular ionic species present was treated in some detail. For most derivatives, the parent molecular ion was not observed but the compounds $C_2H_5B_2H_5$ and $CH_3{}^{10}B_2H_5$ were exceptions; parent molecular ions of low intensity were shown in the published spectra. This is somewhat puzzling in the latter case since the corresponding ions were apparently not observed in $CH_3B_2H_5$ or $CH_3B_2D_5$, and must have been an instrument characteristic. The mass spectra of 1,1- and 1,2-dimethyldiborane and their ethyl analogs were subsequently investigated in greater detail[138] aided by the use of isotopic labelling. A number of rearrangement ions were observed; of particular interest were the ions $R_2B^+$ ($R = CH_3$, $C_2H_5$) formed from the 1,2-derivatives by rearrangement of the complete alkyl group.

The polyisotopic mass spectra of the somewhat unusual derivatives of diborane, IV and V, which were formed in the vapour phase reaction of 1,3-butadiene and diborane at 100 °C, have been reported[139]. They differ from those of the simple alkyl derivatives and also diborane itself in that the parent molecular ion was present in large quantities.

IV

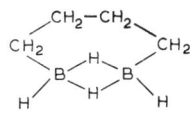

1,2-tetramethylene
diborane

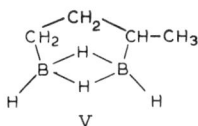

V

1,2-(1'-methyltrimethylene)
diborane

Mass spectrometry has also been used in the characterisation of the compounds $B_4H_9CH_3$, $B_5H_{10}CH_3$, $B_5H_9(CH_3)_2$ (ref. 91), the 1,2, 2,2, and 2,4 isomers of $B_4H_8(CH_3)_2$ (ref. 143), and $B_{10}H_{13}C_2H_5$ (ref. 67), and, in general,

Table 6.18

Comparison of Partial Monoisotopic Spectra of $B_4H_9CH_3$, $B_4H_8(CH_3)_2$, $B_5H_{10}CH_3$ and $B_5H_9(CH_3)_2$ with those of the Parent Hydrides

| $B_4H_9CH_3$[a] | | $B_4H_{10}$[a] | |
|---|---|---|---|
| Ion | R.I.[b] | Ion | R.I.[b] |
| $B_4H_8CH_3^+$ | 0.19 | $B_4H_9^+$ | 0.40 |
| $B_4H_7CH_3^+$ | 7.92 | $B_4H_8^+$ | 13.79 |
| $B_4H_6CH_3^+$ | 4.79 | $B_4H_7^+$ | 5.71 |
| $B_4H_5CH_3^+$ | 100.00 | $B_4H_6^+$ | 100.00 |
| $B_4H_4CH_3^+$ | 3.57 | $B_4H_5^+$ | 2.18 |
| $B_4H_3CH_3^+$ | 27.31 | $B_4H_4^+$ | 15.87 |
| $B_4H_2CH_3^+$ | 1.51 | $B_4H_3^+$ | 0.80 |
| $B_4H_1CH_3^+$ | 2.23 | $B_4H_2^+$ | 0.20 |
| $B_4CH_3^+$ | 0.69 | | |

$B_4H_8(CH_3)_2$ (ref. 143)

| Ion | R.I. (2,4)[d] | R.I. (1,2)[d] | R.I. (2,2)[d] |
|---|---|---|---|
| $B_4H_8(CH_3)_2^+$ | 0.5 | 0.1 | 0.7 |
| $B_4H_7(CH_3)_2^+$ | 0.2 | 0.0 | 0.1 |
| $B_4H_6(CH_3)_2^+$ | 4.6 | 5.6 | 6.9 |
| $B_4H_5(CH_3)_2^+$ | 9.3 | 2.0 | 5.0 |
| $B_4H_4(CH_3)_2^+$ | 100.0 | 100.0 | 100.0 |
| $B_4H_3(CH_3)_2^+$ | 0.0 | 5.9 | 4.6 |
| $B_4H_2(CH_3)_2^+$ | 8.6 | 35.2 | 14.2 |
| $B_4H(CH_3)_2^+$ | 0.2 | 6.2 | 3.3 |
| $B_4(CH_3)_2^+$ | 3.0 | 36.7 | 10.4 |

Table 6.18 (continued)

| $B_5H_{10}CH_3$[a] | | $B_5H_9(CH_3)_2$[a] | | $B_5H_{11}$[a] | |
|---|---|---|---|---|---|
| Ion | R.I.[c] | Ion | R.I.[b] | Ion | R.I.[b] |
| $B_5H_{10}CH_3^+$ | 0.00 | $B_5H_9(CH_3)_2^+$ | 0.00 | $B_5H_{11}^+$ | 0.27 |
| $B_5H_9CH_3^+$ | 0.82 | $B_5H_8(CH_3)_2^+$ | 1.01 | $B_5H_{10}^+$ | 0.47 |
| $B_5H_8CH_3^+$ | 37.68 | $B_5H_7(CH_3)_2^+$ | 46.22 | $B_5H_9^+$ | 51.94 |
| $B_5H_7CH_3^+$ | 4.87 | $B_5H_6(CH_3)_2^+$ | 10.66 | $B_5H_8^+$ | 3.64 |
| $B_5H_6CH_3^+$ | 100.00 | $B_5H_5(CH_3)_2^+$ | 100.00 | $B_5H_7^+$ | 100.00 |
| $B_5H_5CH_3^+$ | 2.59 | $B_5H_4(CH_3)_2^+$ | 5.74 | $B_5H_6^+$ | 1.53 |
| $B_5H_4CH_3^+$ | 32.13 | $B_5H_3(CH_3)_2^+$ | 38.54 | $B_5H_5^+$ | 20.21 |
| $B_5H_3CH_3^+$ | 1.91 | $B_5H_2(CH_3)_2^+$ | 4.42 | $B_5H_4^+$ | 0.83 |
| $B_5H_2CH_3^+$ | 5.91 | $B_5H(CH_3)_2^+$ | 21.58 | $B_5H_3^+$ | 0.87 |
| $B_5H_1CH_3^+$ | 1.63 | $B_5(CH_3)_2^+$ | 4.06 | $B_5H_2^+$ | 0.29 |
| $B_5CH_3^+$ | 0.02 | | | | |

[a] Calculated from reported polyisotopic spectrum[91] using a ratio of 4.00.
[b] Ionising voltage 11 eV.
[c] Ionising voltage 12 eV.
[d] Ionising voltage 24 eV; numbers in parentheses refer to the particular isomer.

the spectrum of each is very similar to that of the parent hydride at least as far as the higher fragments are concerned. This is illustrated in Table 6.18. Calculation of monoisotopic spectra for lower fragments is more difficult. Consider, for example, the compound $B_5H_{10}CH_3$. The $B_4$ ions commence at $m/e$ 66, corresponding to $B_4H_7CH_3^+$, but these ions overlap with the $B_5H_x^+$ ions. Similarly, the subgroups $B_3H_xCH_3^+$ and $B_4H_y^+$ overlap.

The ionisation potential of $B_{10}H_{13}C_2H_5$ was reported[100] to be 9.0 $\pm$ 0.5 e.V.

### 4.1.21 Negative ion mass spectra

In comparison, very little work has been reported on the negative ion mass spectra of the boron hydrides. Enrione and Rosen[140] pointed out that the electron deficiency of these compounds might lead one to expect that negative ions would be readily formed. The first published account[141] of work in this area presented approximate appearance potentials for four ions from pentaborane-9, all of which were formed by resonance capture processes. The compounds diborane, decaborane, triethylborane, and dimethylaminodiborane, $(CH_3)_2NB_2H_5$, were also investigated but no negative ions were observed over an electron energy range of 0–90 V. Later workers were successful in recording the negative ion mass spectra of diborane[140,142], tetraborane[142], and decaborane[140]; the work on pentaborane-9 was repeated[140,142]. Enrione and Rosen demonstrated from clastograms that the spectra were very sensitive to the ionising voltage. While this is undoubtedly partly responsible for differences in the various reported spectra of $B_2H_6$ and $B_5H_9$, it is unlikely that it accounts fully for the poor agreement, shown in Table 6.19.

Table 6.19

Reported Relative Intensities of Ions in the Negative Ion Mass Spectra of Diborane and Pentaborane-9

| Composition of ion | $B_5H_9$ | | | Composition of ion | $B_5H_9$ | | $B_2H_6$ | |
|---|---|---|---|---|---|---|---|---|
| | Ref. 140[a] | Ref. 141[b] | Ref. 142[a] | | Ref. 140[a] | Ref. 142[a] | Ref. 140[a] | Ref. 142[a] |
| $B_5H_9$ | 22.8 | 80 | 2.6 | $B_3H_4$ | 27.9 | 16.7 | | |
| $B_5H_8$ | 100 | 100 | 100 | $B_3H_3$ | 3.19 | 35.2 | | |
| $B_5H_7$ | 8.58 | 57 | 5.3 | $B_3H_2$ | 0.22 | 14.4 | | |
| $B_5H_6$ | 4.83 | 20 | 9.5 | $B_3H$ | 0.43 | 1.3 | | |
| $B_5H_5$ | | | 0.9 | $B_3$ | | 2.7 | | |
| | | | | $B_2H_5$ | 1.93 | 3.1 | | 25.6 |
| | | | | $B_2H_4$ | 0.87 | 1.4 | 3 | 3.1 |
| $B_4H_7$ | | | 2.7 | $B_2H_3$ | 2.78 | 9.1 | 100 | 100 |
| $B_4H_6$ | 16.2 | | 24.6 | $B_2H_2$ | 1.20 | 1.2 | 14 | 16.0 |
| $B_4H_5$ | 7.10 | | 4.4 | $B_2H$ | | 2.2 | | 10.6 |
| $B_4H_4$ | | | 3.3 | $B_2$ | | 2.5 | | 24.5 |
| $B_4H_3$ | 49.2 | | 35.6 | $BH_4$ | | 3.4 | 14 | 96.0 |
| $B_4H_2$ | | | 4.4 | $BH_3$ | | 0.1 | 3 | 15.7 |
| $B_4H$ | | | 1.1 | $BH_2$ | | 0.3 | 8 | 11.9 |
| $B_3H_7$ | | | 13.1 | $BH$ | | 0.7 | | 23.5 |
| | | | | $B$ | | 1.1 | | 35.2 |
| $B_3H_5$ | 6.94 | | 1.8 | | | | | |

[a] Recorded at 70 eV.
[b] Recorded at 90 eV.

It is postulated that the parent molecules undergo a series of resonance captures to form negative ions of different energies, each of which can then decompose by a relatively limited number of pathways such as

$$B_x H_y^- \rightarrow B_x H_{y-2}^- + H_2$$

$$B_x H_y^- \rightarrow B_{x-1} H_{y-3}^- + BH_3$$

Many prominent ions in the spectra, such as $BH_4^-$, could be related to known species. This particular ion is, of course, familiar and is stable in solution and in the solid state. Similarly, one may have expected the well-known $B_3 H_8^-$ ion to be prominent in the spectrum of $B_4 H_{10}$, and perhaps in that of $B_5 H_9$, but it was not observed. However, Hortig et al.[144] bombarded molecules of diborane with positively charged krypton ions, and detected a number of negatively charged boron hydride species including $B_3 H_8^-$; others were $BH_4^-$, $B_2 H_7^-$, $B_2 H_5^-$, $B_3 H_7^-$, $B_4 H_9^-$, $B_4 H_8^-$, and $B_5 H_{10}^-$. The ions $BH_4^-$, $B_2 H_7^-$, $B_3 H_8^-$, $B_4 H_9^-$, $B_5 H_{10}^-$, and the previously unobserved $B_6 H_9^-$ were recently identified by Dunbar[81] during an investigation of the ion–molecule chemistry of diborane by ion cyclotron resonance.

A number of appearance potentials were measured[140] and could be related to the electron affinity of the species involved. The large number of resonance captures mentioned previously complicated the ionisation efficiency curves.

## 4.2 Carboranes

Although the mass spectra of many members of this class of compounds have been recorded, they have principally been used as a means of determining the molecular weight and the composition (usually in conjunction with NMR spectra); further details of the recorded mass spectra in most instances have not been published. Even though rigorous studies similar to those carried out on the boron hydrides have not been attempted, interesting facets and comparisons emerge from complete spectra which have been reported. The parent molecular ion is present in high abundance, and often is responsible for the most intense peak in the spectrum. In addition, the intensity of ions resulting from destruction of the basic skeletal structure is very low. These characteristics are in keeping with the extreme stability of the compounds which appear to be aromatic species[84]. The hydrogens do not appear to be lost in pairs as they are in the boron hydrides, with the possible exception[267] of the three isomeric methyl derivatives of 2-carba-hexaborane-9 (Fig. 6.10, Table 6.20), in each of which the $(P-1)^+$ ion, where P represents the parent molecule, is of low intensity. Ditter et al.[145] have pointed out differences in the mass spectra of closo- and nido-carbo-

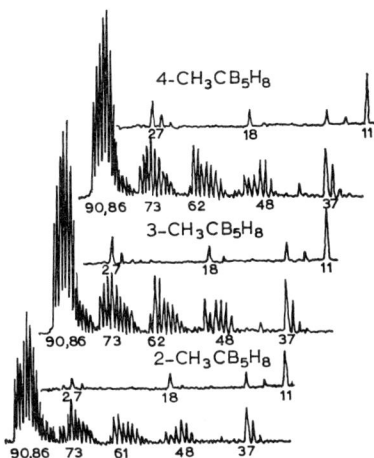

Fig. 6.10. Mass spectra of the methyl derivatives of $CB_5H_9$. The parent molecular ion in each case has an $m/e$ value of 90 while the other $m/e$ values represent the most intense peak in each group. (Reproduced with permission from ref. 267.)

Table 6.20

Partial Monoisotopic Mass Spectra of Three Methyl Derivatives of 2-Carbahexaborane-9[a]

| Ion | Relative intensity | | |
|---|---|---|---|
| | $2\text{-}CH_3CB_5H_8$ | $3\text{-}CH_3CB_5H_8$ | $4\text{-}CH_3CB_5H_8$ |
| $B_5C_2H_{11}^+$ | 100 | 100 | 100 |
| $B_5C_2H_{10}^+$ | 8.3 | 2.3 | 10.5 |
| $B_5C_2H_9^+$ | 52.3 | 88.3 | 86.3 |
| $B_5C_2H_8^+$ | 76.6 | 54.9 | 59.4 |
| $B_5C_2H_7^+$ | 73.1 | 65.6 | 62.5 |
| $B_5C_2H_6^+$ | 38.5 | 14.4 | 17.3 |
| $B_5C_2H_5^+$ | 45.2 | 23.8 | 23.2 |
| $B_5C_2H_4^+$ | 17.9 | 8.4 | 8.3 |
| $B_5C_2H_3^+$ | 0.5 | 1.1 | 3.4 |
| $B_5C_2H_2^+$ | 25.8 | 14.3 | 12.4 |
| $B_5C_2H^+$ | 1.3 | 2.5 | 3.3 |
| $B_5C_2^+$ | 7.0 | 4.6 | 3.6 |

[a] Calculated from published spectra[267].

*References pp. 198–205*

ranes; hydrogen abstraction occurs much more in the latter. The few nido compounds that have been synthesised and analysed mass spectrometrically resemble the boron hydrides more than they resemble the closo-carboranes in their fragmentation patterns. The structures of the closo derivatives are characterised by the absence of hydrogen bridges.

From an investigation of both B-deuterated and C-deuterated sym-carborane-4, Shapiro et al.[146] concluded that fragmentation involving B–X bond rupture occurred more readily than that involving C–X bond rupture (X = H, D). The completely B-methylated compound containing H bonded only to C, showed remarkable resistance to fragmentation[147].

Partial monoisotopic spectra of two carboranes are shown in Table 6.21.

Table 6.21

Partial Monoisotopic Mass Spectra of sym-$B_4C_2H_6$ and $B_3C_2H_5$

$B_4C_2H_6$ (ref. 146)                $B_3C_2H_5$[a]

| Ion | Rel. intensity | Ion | Rel. intensity |
|---|---|---|---|
| $B_4C_2H_6^+$ | 67.2 | $B_3C_2H_5^+$ | 100 |
| $B_4C_2H_5^+$ | 100 | $B_3C_2H_4$ | 46.1 |
| $B_4C_2H_4^+$ | 11.4 | $B_3C_2H_3^+$ | 10.9 |
| $B_4C_2H_3^+$ | 24.1 | $B_3C_2H_2^+$ | 19.6 |
| $B_4C_2H_2^+$ | 7.9 | $B_3C_2H^+$ | 14.5 |
| $B_4C_2H^+$ | 23.9 | $B_3C_2^+$ | 3.4 |
| $B_4C_2^+$ | 4.8 | | |

[a] Calculated from the reported[149] polyisotopic spectrum using a $^{11}B/^{10}B$ ratio of 4.00/1.00 and ignoring contributions from $^{13}C$.

## 4.3 Hydrides of other main Group III elements

The hydride chemistry of aluminium, gallium, and indium is very limited in comparison with that of boron; until recently no uncoordinated mono-meric aluminium trihydride had been observed. In 1964 Breisacher and Siegel[148] reported the identification of both $AlH_3$ and $Al_2H_6$ in a time-of-flight mass spectrometer. Aluminium was slowly evaporated from a hot tungsten filament into hydrogen (0.3 mm pressure) in the spectrometer and peaks at $m/e$ 30 at 1090 °C and 60 at 1170 °C were observed. Confirmation that these were due to the parent molecular ions $AlH_3^+$ and $Al_2H_6^+$ was obtained by the observation of corresponding peaks at $m/e$ 33 and 66 when deuterium was substituted for hydrogen. Besides these two transient compounds, there are no volatile hydrides of aluminium analogous to the boron hydrides.

Further experiments[150] involving the reaction of metal atoms with hydrogen in a fast flow system and the mass spectrometric analysis of the products were reported to show that alane was much more stable than gallane and indane.

## 5. MAIN GROUP III HALIDES

### 5.1 Mass spectra fragmentation

The mass spectra of the boron trihalides have been investigated by a number of workers[151-153]. Because of the simplicity of the molecular structure, the spectra are straightforward; the ions $BX_n^+$ ($n = 0$–3) have all been observed, with the predominant species in each case being $BX_2^+$ (Table 6.22). A number of doubly charged ions were also observed[151,152,154] and evidence was obtained for $BI_3^{n+}$ ions carrying 3+, 4+, and even 5+ charges[151]. The parent molecular ions of all possible mixed halides $BX_2Y$ and $BXYZ$ (X, Y and Z = F, Cl, Br, I) have also been identified in the mass spectrometer when mixtures of the simple trihalides were admitted to the instrument[48,155,156]. There has been considerable study of the question of the existence of the mixed halides of boron and these are only known as mixtures in equilibrium with the simple halides[157]. It is interesting, therefore, that their parent molecular ions are long-lived and are present in the spectra in reasonable abundance. Because of the pronounced affinity of the boron trihalides for water, peaks due to $HX^+$ will normally be present in the spectrum until all traces of moisture have been removed. Both the electron-impact mass spectrum of diboron tetrafluoride[158] and its photoionisation mass spectra[159] and the corresponding tetrachloride[164] have been reported (Table 6.22). The photoionisation and electron impact spectra of $B_2F_4$ differ in that the relative intensities of the $B_2F_3^+$ and $BF_2^+$ ions are reversed. This suggests that some pyrolysis may have been occurring in the electron impact source, causing rupture of the B–B bond. Since no appearance potential measurements were made in the study this cannot be confirmed. Some aspects of the mass spectra of boron sub-chlorides have been discussed briefly by Beynon[160] and by Massey and Urch[161] and the principal ions in

Table 6.22

Mass Spectra of Boron and Aluminium Halide Compounds

References are given in brackets.

| Ion | $BH_3$ | $BF_3$ | $BCl_3$ | $BBr_3$ | $BI_3$ | $AlF_3$[a,b] | $AlCl_3$[c] | $AlBr_3$[a] | $B_2F_4$ | $B_2Cl_4$ | $B_2Cl_4$ |
|---|---|---|---|---|---|---|---|---|---|---|---|
| $M_2X_4^+$ | | | | | | | | | 8.2(158) | 10(159)[d] | 34(164)[d] |
| $M_2X_3^+$ | | | 8.5(159)[d] | 59(164)[d] | | | | | 77.8 | 100 | 57 |
| $M_2X_2^+$ | | | <0.5 | 7.2 | | | | | 0.9 | 14 | |
| $M_2X^+$ | | | | ~0.1 | | | | | 2.2 | | |
| $MX_3^+$ | 31(3) | 3(153) | 36(151) | 37(151) | (151)[f] | 5.7(30) | 43.5(30) | 71(30) | 2.6[e] | 0.8[e] | <0.1[e] |
| $MX_2^+$ | 100 | 100 | 100 | 100 | 100 | 100 | 100 | 100 | 100 | | 100 |
| $MX^+$ | 16 | 1 | 8 | 49 | 22 | 0.7 | 26.1 | 18 | 39.3 | 76 | 30 |
| $M^+$ | 8 | 3 | 3 | 0.5 | 9 | 0.9 | 16.1 | 40 | 3.1 | 30 | 5.4 |

[a] Ions arising from the dimer were also observed.
[b] High-temperature studies.
[c] Ions arising from the dimer and the trimer were also observed.
[d] Photoionisation mass spectrum.
[e] Rearrangement ion.
[f] Reported "out of range".

the spectrum of the compound $C(BCl_2)_4$ have been listed[162]; the most abundant was $BCl_2^+$. The mass spectra of other products of the co-condensation of carbon vapour and various boron halides have recently been reported[163].

The spectrum of the unstable, very reactive compound triboron pentafluoride, $B_3F_5$, has been reported[165] and mass spectrometry was used as an aid in the identification and characterisation of certain volatile boron fluorides and adducts such as $(BF_2)_3BCO$, but few details were published[165]. Ions arising from the compounds $SiCl_3BCl_2$ (refs. 166, 167), $Si_2BF_7$ (ref. 168), $F_2Si(BF_2)_2$ (ref. 169), $FSi(BF_2)_3$ (ref. 169), and MeClB–BClMe (ref. 170) have also been reported.

$AlF_3$ has been previously mentioned in the section on high-temperature studies. Spectra of the corresponding chloride[30,171,172] and bromide[30] have likewise been published. The fragmentation pattern of the monomer (Table 6.22) was, in each case, similar to that of the boron trihalides. In addition, however, ions arising from the dimeric and, in the case of the chloride, from the trimeric species were present. The ion $Al_2Cl_5^+$, in fact, was the most abundant in the spectrum of $AlCl_3$. In addition, Tanaka and Smith[172] observed groups of peaks up to $m/e$ 630 of relative intensities 1–3% of that of the base peak and still others of much weaker intensity above $m/e$ 630. However, no satisfactory assignment could be made although chlorine isotope patterns for six and seven chlorines could be ascertained in some of the groups of peaks. Several weak peaks in the spectrum of $AlCl_3$ were assigned[171] to ions such as $X_2AlCl_4^+$, $XAl_2Cl_7^+$ etc.; they were produced by the reaction of aluminium chloride with compounds (for example $H_2O$) present in the mass spectrometer.

*5.2 Ionisation and appearance potential studies*

A number of ionisation potential studies have been carried out on the simple trihalides but a considerable amount of variation in the values obtained is obvious (Table 6.23). Because of the improvements which have been made in mass spectrometric instrumentation and in the methods of analysing results for calculation of ionisation potentials in recent years, it was believed[156] that confidence could be placed in the most recent electron impact figures, including those on the mixed halides (Table 6.24), especially since the data for the simple boron trihalides (F, Cl, Br) agreed closely with the results obtained very recently from photoelectron spectroscopy[173,174].

Table 6.23

First Ionisation Potentials of the Simple Boron Halides
References are given in brackets.

| Compound | I.P. (eV) | Year | Method[a] |
|---|---|---|---|
| BF$_3$ | 15.5 $\pm$ 0.5 (175) | 1956 | A |
|  | 15.5 $\pm$ 0.3 (176) | 1957 | B |
|  | 15.7 $\pm$ 0.1 (152) | 1957 | B |
|  | 16.4 $\pm$ 0.4 (44) | 1963 | H |
|  | 15.83 (173) | 1967 | C |
|  | 15.95 (174) | 1968 | C |
|  | 15.97 $\pm$ 0.1 (156) | 1968 | D |
|  | 15.89 $\pm$ 0.04 (156) | 1968 | E |
|  | 15.55 $\pm$ 0.04 (159) | 1968 | F |
| BCl$_3$ | 12.0 $\pm$ 0.5 (153) | 1950 | B |
|  | 12.03 $\pm$ 0.02 (152) | 1957 | G |
|  | 10.9 $\pm$ 0.2 (151) | 1959 | H |
|  | 11.69 (173) | 1967 | C |
|  | 11.97 (174) | 1968 | C |
|  | 11.67 $\pm$ 0.1 (156) | 1968 | D |
|  | 11.77 $\pm$ 0.14 (156) | 1968 | E |
|  | 11.60 $\pm$ 0.02 (164) | 1969 | F |
| BBr$_3$ | 9.7 $\pm$ 0.2 (151) | 1959 | H |
|  | 10.64 (173) | 1967 | C |
|  | 10.72 (174) | 1968 | C |
|  | 10.62 $\pm$ 0.1 (156) | 1968 | D |
|  | 10.66 $\pm$ 0.1 (156) | 1968 | E |
| BI$_3$ | 9.0 $\pm$ 0.2 (151) | 1959 | H |
|  | 9.24 $\pm$ 0.06 (156) | 1968 | D |
|  | 9.22 $\pm$ 0.14 (156) | 1968 | E |

[a] Methods used (all except C are mass spectrometric).
    A   Critical slope.
    B   Linear extrapolation.
    C   Photoelectron.
    D   Semi-logarithmic plot.
    E   Energy-distribution difference.
    F   Photoionisation.
    G   Extrapolation voltage difference.
    H   Vanishing current.

Table 6.24

Ionisation Potentials (eV)[156,177,178] of Boron Compounds $BX_3$, $BX_2Y$ and $BXYZ$

| Y | $X_2$ | | | | | | | | | |
|---|---|---|---|---|---|---|---|---|---|---|
| | $F_2$ | $Cl_2$ | $Br_2$ | $(OMe)_2$ | $(OEt)_2$ | $(OPr^n)_2$ | $Et_2$ | $I_2$ | $(SMe)_2$ | $(NMe_2)$ |
| F | 15.97 | 12.18 | 11.11 | 10.92 | | | | 9.69 | 9.79 | |
| Cl | 13.06 | 11.67 | 10.79 | 10.83 | 10.52 | 10.45 | 10.28 | 9.49 | 9.64 | 8.15 |
| Br | 11.95 | 11.13 | 10.62 | 10.62 | | | | 9.40 | 9.55 | 8.05 |
| OMe | 11.97 | 11.55 | 10.68 | 10.62 | | | | 9.21 | | |
| OEt | | 11.27 | | | 10.13 | | | | | |
| $OPr^n$ | | 11.22 | | | | 10.02 | | | | |
| Et | | 10.80 | | | | | 9.66 | | | |
| I | 10.42 | 9.93 | 9.74 | 9.63 | | | | 9.24 | 8.90 | 7.97 |
| SMe | 10.59 | 10.45 | 10.25 | | | | | 9.26 | 9.24 | |
| $NMe_2$ | 9.71 | 9.57 | 9.50 | | | | | 8.99 | | 7.75 |

| Z | XY | | | | |
|---|---|---|---|---|---|
| | FCl | FBr | F(OMe) | ClBr | FI |
| Br | 11.46 | | | | |
| OMe | 11.96 | 11.68 | | 11.07 | |
| I | 10.18 | 10.11 | 9.96 | 9.81 | |
| $NMe_2$ | 9.65 | 9.68 | | | 9.61 |

A contemporary mass spectrometric study[159] employing photoionisation, however, yielded a value for the ionisation potential of $BF_3$ which was about 0.4 eV lower, in agreement with previous electron-impact results. The reason for the discrepancy is not obvious, especially since the corresponding $BCl_3$ values[164] are in fairly close agreement. Only two mass spectral studies[151,156] have provided ionisation potential values for $BBr_3$ and $BI_3$.

Lappert et al.[156] applied various semi-empirical MO theories to calculate the first ionisation potentials for the molecules $BX_3$, $BX_2Y$, and $BXYZ$. Close agreement with the experimental data was found using an extended Hückel approach, modified for the considerable charge-drift occurring in these compounds, which was then employed to calculate the electron distribution. The results indicated that, contrary to the usual assumptions,

$\pi$ back-donation decreases in the series BI $>$ BBr $\approx$ BCl $\gg$ BF; the $\sigma$ charge-drift was the dominant factor in deciding overall bond polarity. It appears appropriate to mention at this stage the extension of this work to other similar species not necessarily containing halogens. A number of parent molecular ions $BX_2Y^+$ and $BXYZ^+$, many of which correspond to molecules which are either unstable or unknown, were obtained by redistribution reactions and characterised mass spectrometrically[155,177,178]. The first ionisation potentials of the parent molecular ions from the simple boron compounds $BX_3$, $BX_2Y$, and $BXYZ$ were measured and are shown in Table 6.24. Comparison of these data were made with a model based on symmetry arguments and on a simple inductive parameter. Interaction between a given $p\pi$ orbital on the boron atom decreases in the order B–N $>$ B–O $\approx$ B–S $>$ B–Hal, and it is also interesting to note that there appeared to be some

Table 6.25

Appearance Potentials of Fragment Ions from Boron Trihalides
References are given in brackets.

| Ion | Appearance potential (eV) | | | |
|---|---|---|---|---|
| | From $BF_3$ | From $BCl_3$ | From $BBr_3$ | From $BI_3$ |
| $BX_2^+$ | 17.0    0.5 (153) | 13.2 $\pm$ 0.5 (153) | 10.8 $\pm$ 0.2 (151) | 9.7 $\pm$ 0.2 (151) |
| | 16.2 $\pm$ 0.2 (175) | 11.8 $\pm$ 0.2 (151) | 11.0 $\pm$ 0.2 (151) | 10.1 $\pm$ 0.2 (151) |
| | 16.17 $\pm$ 0.05 (152) | 12.0 $\pm$ 0.2 (151) | 10.7 $\pm$ 0.2 (151) | |
| | 15.81 $\pm$ 0.04 (159) | 13.01 $\pm$ 0.02 (152) | | |
| | 16.5 $\pm$ 0.8 (35) | 12.3 $\pm$ 0.02 (164) | | |
| | 16.7 $\pm$ 0.4 (44) | | | |
| | 16.25 $\pm$ 0.2 (176) | | | |
| $BX^+$ | 27.2 $\pm$ 1.0 (153) | 18.54 $\pm$ 0.07 (152) | 10.7 $\pm$ 0.2 (151) | 14.4 $\pm$ 0.2 (151) |
| | (1) 11.2 $\pm$ 0.4 (152) | 19.2 $\pm$ 0.5 (153) | 11.8 $\pm$ 0.2 (151) | 14.6 $\pm$ 0.2 (151) |
| | (2) 25.2 $\pm$ 0.2 (152) | 17.2 $\pm$ 0.2 (151) | 15.0 $\pm$ 0.2 (151) | |
| | | 20.0 $\pm$ 0.2 (151) | | |
| | | 18.37 $\pm$ 0.02 (164) | | |
| $B^+$ | 31.5 $\pm$ 1.0 (153) | 19.5 $\pm$ 0.2 (151) | 19.6 $\pm$ 0.2 (151) | 16.6 $\pm$ 0.5 (151) |
| | 30.1 $\pm$ 0.5 (175) | 20.8 $\pm$ 0.5 (151) | | |
| | 31.3 $\pm$ 0.4 (44) | 22.5 $\pm$ 0.5 (153) | | |
| | 30.6 $\pm$ 1 (152) | 19.5 $\pm$ 1.0 (175) | | |
| | | (1) 13.6 $\pm$ 0.2 (152) | | |
| | | (2) 18.4 $\pm$ 0.2 (152) | | |
| | | (3) 22.35 $\pm$ 0.06 (152) | | |

Table 6.26

Threshold Energies of Ions from Photoionisation Study of Diboron Tetrafluoride and Tetrachloride

| Ion process | Threshold energy (eV) | |
|---|---|---|
| | $B_2F_4$ (ref. 159) | $B_2Cl_4$ (ref. 164) |
| $B_2X_4 + h\nu \rightarrow B_2X_4^+ + e$ | $12.07 \pm 0.01$ | $10.32 \pm 0.02$ |
| $\rightarrow B_2X_3^+ + X\cdot + e$ | $15.40 \pm 0.01$ | $11.52 \pm 0.02$ |
| $\rightarrow B_2X_2^+ + 2X\cdot + e$ | | $17.24 \pm 0.03$ |
| $\rightarrow BX_2^+ + BX_2 + e$ | $12.94 \pm 0.01$ | $11.32 \pm 0.02$ |
| $\rightarrow BX_2^+ + BX + X\cdot + e$ | $18.0 \pm 0.05$ | |
| $\rightarrow BX^+ + BX_3 + e$ | $12.75 \pm 0.01$ | $13.71 \pm 0.04$ |

correlation between the ionisation potential and the heats of the redistribution reactions.

A number of appearance potentials of fragment ions from the trihalides have been reported but, in general, agreement is poor here also (Table 6.25). Derived thermodynamic values are summarised in Tables 6.27 and 6.28. Koski et al.[151] sought a two-fold objective: (i) to calculate a value for the B–I bond dissociation energy in $BI_3$ since no thermochemical data were available on the subject and (ii) to attempt to calculate a self-consistent set of ionisation potentials for the boron trihalides, $BCl_3$, $BBr_3$, and $BI_3$ and the fragments formed from these molecules. They found that a plot of bond energy versus bond length in the boron trihalides gave a linear relationship and indications were obtained of a correlation between the calculated ionisation potentials of the $BX_2$ radicals and the nature of the group attached to boron. More than one appearance potential value was obtained for some of the fragment ions and it was postulated that this may be due to thermal dissociation at the source filament, e.g.

$$BX_3 \rightarrow BX + X_2$$
$$BX + e \rightarrow BX^+ + 2e$$

Devyatykh et al.[154] investigated the effect of varying the energy of the ionising electrons on the relative abundances of the various ions produced in the mass spectra of all of the main Group III trichlorides. They observed relationships between the relative abundances of these ions and (i) the dissociation energy

Table 6.27

Calculated Heats of Formation and Ionisation Potentials
References are given in brackets.

| Ion or radical | $\Delta H°_f$ (kcal. mole$^{-1}$) | Radical | Calc. I.P. (eV) |
|---|---|---|---|
| $BF_3^+$ | 87.5 (159) | $\cdot BF_2$ | 8.47 (159) |
| $\cdot BF_2$ | −120.2 (159) | | 9.4 (182) |
| | −130 ± 6 (181) | BF | 10.56 (159) |
| $BF_2^+$ | 75.1 (159) | | 11.5 ± 0.4 (35) |
| BF | − 21.9 (159) | | 11.6 (180)[a] |
| | − 27.4 ± 3 (35) | | 11.0 (183)[a] |
| | − 27.6 ± 4 (35) | $\cdot B_2F_3$ | ∼ 8.1 (159) |
| $BF^+$ | 221.6 (159) | $\cdot BCl_2$ | 7.20 (151) |
| $B_2F_4^+$ | − 65.2 (159) | | 7.52 (164) |
| $B_2F_3^+$ | − 6.8 (159) | BCl | ∼10.44 (151) |
| $BCl_3^+$ | 171.2 (164) | | 10.20 (164) |
| $\cdot BCl_2$ | − 14.7 (164) | $\cdot B_2Cl_3$ | ∼ 6.7 (164) |
| $BCl_2^+$ | 158.6 (164) | $B_2Cl_2$ | ∼ 9.0 (164) |
| $BCl^+$ | 270.0 (164) | $\cdot BBr_2$ | 7.06 (151) |
| $B_2Cl_4^+$ | 120.8 (164) | BBr | 9.25–10.14 (151) |
| $B_2Cl_3^+$ | 119.9 (164) | $\cdot BI_2$ | 7.13 (151) |
| $B_2Cl_2^+$ | ∼223 (164) | BI | 8.96 (151) |

[a] Spectroscopic value.

of the $MCl_x^+$–Cl bonds; (ii) the length of the M–Cl bond; (iii) appearance potentials of the ions; and (iv) the position of the element in the periodic table. Dibeler and associates[159,164] have reported detailed studies of $B_2F_4$ and $BF_3$ and $B_2Cl_4$ and $BCl_3$ using photoionisation mass spectrometry. Ionisation threshold values (Table 6.26) were used to calculate heats of formation of ions and radicals (Table 6.27) and to derive bond dissociation energies (Table 6.28). The values they obtained for the B–B bond dissociation energies, 103.1 and 87.6 kcal.mole$^{-1}$, in $B_2F_4$ and $B_2Cl_4$ were more in keeping with the known relative stabilities and the structural parameters of the two molecules[179] than were previously reported values. On the basis of the relative abundances of the ions $B_2X_3^+$ and $BX_2^+$ in the spectra of these two compounds one might be tempted to postulate that $D(F_2B–BF_2) > D(BF–F)$ and $D(Cl_2B–BCl_2) < D(BCl–Cl)$. However, the reverse was found to be true in each case.

Table 6.28

Bond Dissociation Energies Calculated for Boron–Halogen Compounds
References are given in brackets.

| Molecule | Bond | Dissociation energy (kcal. mole⁻¹) |
|---|---|---|
| $BF_3$ | $BF_2–F$ | 169.3 (159) |
| $B_2F_4$ | $BF–F$ | 117.6 (159) |
| | $B–F$ | 173.6 (159) |
| | $B–B$ | 103.1 (159) |
| | | 72.4 (129)[a] |
| $BCl_3$ | $BCl_2–Cl$ | 110.2 ± 0.5 (164) |
| | $BCl–Cl$ | 78.2 ± 0.5 (164) |
| | $B–Cl$ | 127 ± 1.0 (164) |
| $B_2Cl_4$ | $B–B$ | 87.6 ± 0.5 (164) |
| | | 79.0 (127)[a] |
| $BF$ | $B–F$ | 173.6 ± 1 (159) |
| | | 181 (35) |
| | | 185 (184)[a] |

[a] Calorimetric value.

Hildenbrand and Murad[35] studied the vapour species produced in a Knudsen cell containing a mixture of elemental boron and calcium fluoride. Appearance potentials of the resultant ions $BF^+$ and $BF_2^+$ were measured and it was inferred from these measurements that the neutral precursors were BF and $BF_3$ respectively. The close agreement existing between the spectroscopic[180], photoionisation[159], and electron-impact[35] values of the ionisation potential of boron monofluoride is certainly encouraging.

Porter et al.[48] studied the reactions

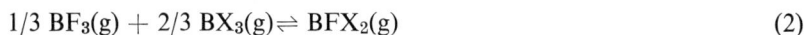

$$2/3 \; BF_3(g) + 1/3 \; BX_3(g) \rightleftharpoons BF_2X(g) \tag{1}$$

$$1/3 \; BF_3(g) + 2/3 \; BX_3(g) \rightleftharpoons BFX_2(g) \tag{2}$$

(where X = Cl, OH) mass spectrometrically at a series of temperatures and calculated equilibrium constants for the reactions and heats of formation of the four mixed species $BF_2X$ and $BFX_2$.

MacNeil and Thynne[185] have studied the formation of negative ions when

$BF_3$ was bombarded with low-energy electrons and observed that the ions $F_2^-$, $BF_2^-$, and $BF_4^-$ were formed. $BF_4^-$ was shown to be a secondary species formed by an ion–molecule reaction

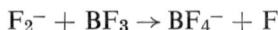

$$F_2^- + BF_3 \rightarrow BF_4^- + F$$

The heat of formation of $BF_4^-$ was calculated.

## 6. MAIN GROUP III–GROUP V COMPOUNDS

### 6.1  Compounds of the types $LBH_3$, $L_2B_2H_4$, $B_2(X_2C_2H_4)_2$, and $(X_2AlNRR')_2$

The fragmentation patterns of exceedingly few simple compounds containing bonds between main Group III and main Group V elements have been reported. A mass spectrometric investigation[186] of the vapour above solid ammonia–borane at room temperature has shown that, in addition to the molecule $BNH_6$, both aminoborane, $BNH_4$, and diborane were present as a result of solid-state decomposition. No evidence was found that either aminoborane or iminoborane, $BNH_2$, was formed during pyrolysis of ammonia–borane. Pyrolysis data for the compound $D_3N.BH_3$ were consistent with an ethylene-like structure, $H_2NBH_2$, for aminoborane rather than the asymmetric structure $H_3NBH$.

In the mass spectrum[187] of the azoalkane adduct of borane,

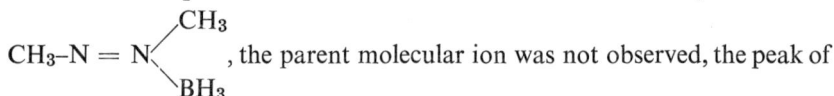

$$CH_3-N = N \underset{BH_3}{\overset{CH_3}{\diagdown}}$$

, the parent molecular ion was not observed, the peak of highest mass corresponding to the ion $(CH_3)_2N_2BH_2^+$. Other borane adducts whose mass spectra have been reported ($L.BH_3$, where $L = NH_3$, $PF_2H$, $P_2F_4$, CO) all exhibit parent molecular ions. In addition, the non-appearance of the ion corresponding to loss of a methyl group from $(CH_3)_2N_2BH_3$ and the occurrence of extensive rearrangement on electron impact, resulting in ions such as $CH_3BH^+$, rendered the mass spectrum of little assistance in structure determination of the compound without additional evidence such as that provided by NMR and IR spectrometry.

The mass spectra of two phosphine-boranes[188,189] have been recorded to assist in the characterisation of the compounds; the monoisotopic spectra have been calculated from the published figures and are shown in Table 6.29.

Table 6.29

Monoisotopic Mass Spectra of Difluorophosphine Borane[188], $HF_2P \cdot BH_3$, and Tetra-fluorodiphosphine Borane[189], $P_2F_4 \cdot BH_3$

| $HF_2P \cdot BH_3$ | | | | $P_2F_4 \cdot BH_3$ | |
|---|---|---|---|---|---|
| Ion[a] | Rel. intensity | Ion[a] | Rel. intensity | Ion[b] | Rel. intensity |
| $BPF_2H_4^+$ | 23.2 | $PF^+$ | 24.5 | $P_2F_4 \cdot BH_3^+$ | 2.1 |
| $BPF_2H_3^+$ | 76.6 | $*BF_2^+$ | 9.2 | $P_2F_4^+$ | 8.15 |
| $BPF_2H_2^+$ | 100 | $BPH_2^+$ | 1.2 | $P_2F_3^+$ | 5.71 |
| $BPF_2H^+$ | 2.3 | $BPH^+$ | 3.4 | $F_2PBH_3^+$ | 10.05 |
| $BPF_2^+$ | 1.5 | $BP^+$ | 1.4 | $F_2PBH_2^+$ | 11.25 |
| $*PF_2H_2^+$ | 3.5 | $*PH_4^+$ | 3.0 | $F_2PBH^+$ | 1.48 |
| $PF_2H^+$ | 13.0 | $*PH_2^+$ | 27.0 | $F_2PB^+$ | 1.55 |
| $PF_2^+$ | 43.6 | $PH^+$ | 16.9 | $F_2P^+$ | 100 |
| $BPFH_2^+$ | 1.9 | $*BFH^+$ | 67.1 | $FP^+$ | 24.5 |
| $BPFH^+$ | 3.1 | $P^+$ | 6.1 | $P^+$ | 25.6 |
| $BPF^+$ | 9.9 | $F^+$ | 0.6 | | |
| $*PH_2F^+$ | 0.9 | $BH_2^+$ | 37.3 | | |
| $PHF^+$ | 34.7 | $BH^+$ | 18.2 | | |

[a] Except for the parent molecular ion, the structures of ions containing H is not known, e.g. $BPF_2H_3^+$ could be either $F_2P \cdot BH_3^+$ or $HF_2P \cdot BH_2^+$. Ions marked with an asterisk were formed by a rearrangement process.
[b] Major ions only were reported[189].

Two points are worth noting. Firstly, the higher stability of difluorophos-phine-borane, compared with that of tetrafluorodiphosphine borane, appears to be reflected in the relative intensities of the ions of highest mass. Fragment ions containing the intact B–P bond account for 40% of the total ion intensity in the former, but only 12%* in the latter. Secondly, a large number of rearrangement ions were detected in the spectrum of the former; the ion $BFH^+$ in particular was present in high abundance.

The mass spectrum of bis(trifluorophosphine)-diborane(4) has recently been reported[190]. The observation of ions formed by loss of one, two, three, and four hydrogens from the parent molecular ion, $B_2H_4P_2F_6^+$, and from $B_2H_4PF_3^+$ is in accord with the proposed structure containing no bridging hydrogens.

---

* This figure is approximate since only the major peaks were listed.

Mass spectra of the diborane(4) derivatives $B_2[N(CH_3)_2]_4$, $B_2(OCH_3)_4$, and $B_2(SCH_3)_4$ were studied and compared with the spectra of heterocyclic systems of composition $B_2(X_2C_2H_4)_2$ in an attempt to differentiate between structures VI and VII for the latter compounds[191]. The technique proved to be more successful than NMR and vibrational spectroscopy and structure VI was assigned. This contradicted the earlier representation of $B_2[N_2(CH_3)_2C_2-H_4]_2$ as a bicyclic compound VII but was in agreement with chemical evidence for the compounds where X = O or S.

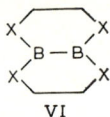

         VI                           VII

No boron–nitrogen cyclic dimers of the type $(R_2B-NR'_2)_2$ appear to have been studied mass spectrometrically, but recently reported spectra of the aluminium compounds $(Et_2AlNHMe)_2$, $(Et_2AlNMe_2)_2$, $(Et_2AlNHBu^t)_2$ (ref. 192), and $(Cl_2AlNMe_2)_2$ (ref. 193) showed that the compounds were dimeric in the vapour state.

*6.2 Aldiminoboranes and related aluminium and gallium compounds*

The aldiminoboranes have attracted some attention because of the possibility of boron–nitrogen $\pi$-bonding accounting for species, represented by VIII, which are isoelectronic with the corresponding allenes IX. In certain cases, however, association to produce the dimeric species with a $B_2N_2$ ring system, X, occurs at least to some extent as shown by their mass spectra [194,195]. The same was found to be true for corresponding aluminium and gallium compounds.

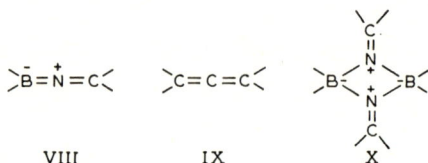

        VIII              IX               X

Groups on the imino carbon atom appear to be intrinsically involved in the electron-impact-induced fragmentation processes of these compounds. Dorokhov and Lappert[194] observed peaks in the mass spectrum of the compound $(Bu^t-HC=N-BCl_2)_2$ which corresponded to loss of $HBCl_2$ fragments from both the monomer and the dimer and the appropriate metastable transitions were detected illustrating that these fragments were lost in a single step. With the corresponding bromo compound $(Bu^t-HC=N-BBr_2)_2$, breakdown occurred by successive loss of Br· and H· from either the monomer or the dimer (presumably due to the weaker N–Br bond) although a peak corresponding to the monomer itself was absent. It appears most likely that the H· originates in both cases from the imino carbon rather than from the butyl groups.

The mass spectra of the compounds $(Ph_2C=N-BX_2)_n$ ($n = 1, 2$; $X = Cl$, Br)[195] were not analogous in that ions corresponding to loss of $PhBX_2$ fragments from the parent molecular ions were not observed. However, the ions $PhBX^+$, $Ph_2B^+$, and $PhB^+$ were present in high abundance so that a similar rearrangement process involving the group on the imino C has therefore occurred. Parent molecular ions were not observed for these compounds, nor for the compound $(Cl_2C=N-BCl_2)_2$[196], but fragment ions containing $B_2N_2$ were present. It was suggested[195] that at the temperature prevailing in the ion source (150–250 °C), it is most likely that considerable dissociation of the neutral dimer molecules occurred before electron impact, especially in the case of the bromo-derivative. The spectrum of the compound $Ph_2C=N-BMe_2$ indicated that no significant amounts of dimer were present in the vapour at 150–250 °C.

The mass spectrum of the fluoride $Ph_2CNBF_2$ differed considerably from those of the corresponding bromide and chloride in that no peaks were detected which could be assigned with certainty to fragments containing more than one boron atom[195]. The structure of the compound is unknown although its properties indicate a polymeric species. Rearrangement ions such as $Ph_2B^+$, $PhBF_2^+$, and $PhBF^+$ were again very prominent.

The mass spectra of similar aluminium[198] and gallium[199] compounds $(Ph_2C=N-MR_2)_2$, where $M = Al$, Ga and $R = Me$, Ph, have also been reported. Weak parent molecular ions were observed in each case together with the moderately abundant $(P-R·)^+$ ions. There is no doubt, therefore, that a considerable proportion of the vapour at 200 °C consists of the dimeric species. Rearrangement ions were very prominent in the spectra of the gallium compounds and, in fact, $PhGaMe^+$ was the most abundant ion in that of $(Ph_2CNGaMe_2)_2$.

## 6.3 s-Diazadiborines

The mass spectra of two compounds postulated to have the s-diazadiborine ring structure XI have been published[200,201]. One of them, however, has apparently since been found to contain a 5-membered ring[202] although the nature of this ring has not been revealed. The mass spectrum of this compound[200] did not reveal a parent molecular ion but the boron isotopic distribution was used to identify fragments containing two B atoms.

```
   -B — N-
    |   |
   -N — B-
      XI
```

A prominent parent molecular ion was evident in the spectrum of the 1,3-di-*tert.*-butyl-2,4-di-*tert.*-butylamino derivative and a favoured fragmentation pathway which was metastable supported involved loss of $CH_3$ from the parent molecular ion followed by successive loss of $(Bu'–H)$ groups with the resultant formation of rearrangement ions[201]. Another feature of the spectrum was the persistence of the $B_2N_2$ nucleus in the fragment ions, but no proof is available that the ring structure was maintained, although this does seem most likely. Simple Hückel MO calculations suggested that this B–N ring system may not be nearly as unstable as the isoelectronic cyclobutadiene ring, although it was stated that the calculations were likely to overestimate stability and did not take ring strain into account[203]. The positive charge on the ions introduces another factor.

## 6.4 Tetrazaborolines

Electron-impact-induced fragmentation of a number of derivatives of tetrazaboroline have been studied by Morris and coworkers and the information which was obtained confirmed the previously proposed structures of the compounds (XII).

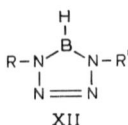

```
           H
           |
           B
         /    \
    R-N        N-R'
       |        |
       N ===== N
          XII
```

A strong parent molecular ion was observed in each case. The N-substituted series first studied[204] underwent an initial fragmentation by loss of $(N_2 + H\cdot)$ from the parent molecular ion and it was suggested on the basis of the subsequent fragmentation pattern that the hydrogen atom originated by cleavage from a sidechain on nitrogen rather than from the boron. This was later confirmed by a study of the B-substituted compounds $R_2N_4BR'$

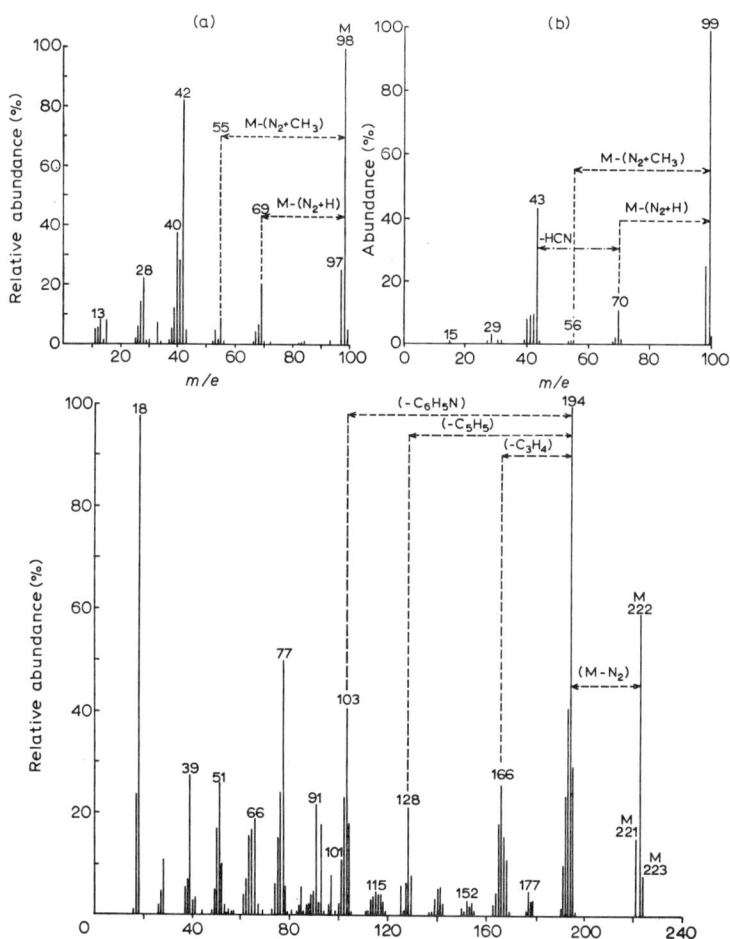

Fig. 6.11. Mass spectra of (a) 2,5-dimethylcyclotetrazenoborane, $Me_2N_4BH$; (b) 1-deutero-2,5-dimethylcyclotetrazenoborane, $Me_2N_4BD$; (c) 2,5-diphenylcyclotetrazenoborane, $Ph_2N_4BH$. (Reproduced with permission from refs. 204, 206.)

(R = Me, R' = D, Me, Et, Vi, Cl, Br; R = Et, R' = Et; R = Ph, R' = Me, Ph)[205,206]. The spectra of $Me_2N_4BH$ and $Me_2N_4BD$ are shown in Fig. 6.11. All major breakdowns in the deuterated compound produce peaks at one mass unit higher than in $Me_2N_4BH$, showing that the D is retained throughout.

Although the initial fragmentation by loss of $(N_2 + H\cdot)$ was virtually complete for 1,4-dimethyl derivatives, for other derivatives loss of $N_2$ was predominant together with a slight contribution from $(N_2 + H\cdot)$. Perhaps the surprising feature was the remarkable stability of the B-substituent bond under electron-impact conditions. The subsequent fragmentation depended heavily on the nature of the 1,4-substituent groups (Fig. 6.11), and the general scheme proposed[207] is shown in Fig. 6.12.

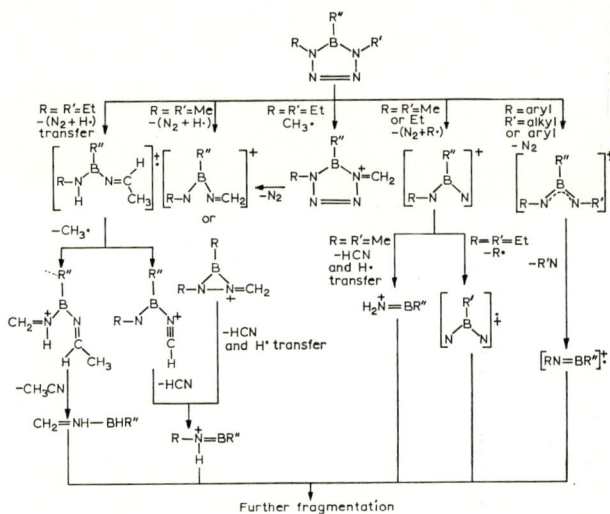

Fig. 6.12. Proposed general fragmentation scheme for tetrazaboroline compounds. (Reproduced with permission from ref. 207.)

## 6.5 Borazines

Many comparisons have been made between borazine and benzene and attention drawn to the similarities existing between the two molecules. It is now generally accepted, however, that the aromaticity of the borazine system has been overemphasized. Niedenzu and Dawson[208] have pointed out that, in general, physical constants of borazine seem to substantiate a resonance structure of the molecule XIII, whereas most chemical evidence tends to support the uncharged formulation of this boron-nitrogen heterocycle XIV.

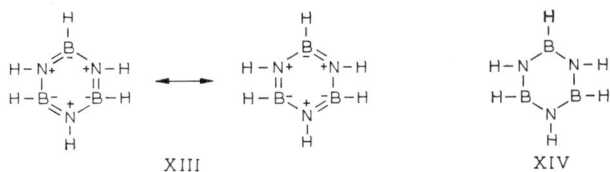

XIII                                              XIV

In many cases where the mass spectra of borazine derivatives have been recorded, full details have unfortunately not been published. Certain aspects, however, are apparent. Immediately obvious is the direct similarity with aromatic compounds in that the most abundant ionic species is usually either the parent molecular ion* or a fragment containing the $B_3N_3$ nucleus. Whether this nucleus maintains the ring structure in these fragment ions is, however, unknown. Evidence has been presented[213] which indicates that electron-impact-induced fragmentation of benzene proceeds through ionised 1,3-hexadiene-5-yne, XV, as an intermediate, so that the structure of the

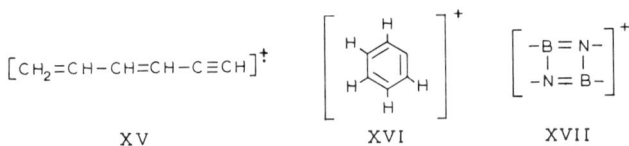

XV                           XVI                  XVII

$C_6H_5^+$ ion should not be the phenyl ion XVI but, instead, an open-chain species. This was substantiated by appearance potential and heat of formation measurements on both benzene and certain derivatives. The possibility of a similar high-energy ring opening reaction rather than a simple ring ionisation occurring in electron-impact-induced fragmentation of borazine

---

* The parent molecular ion of benzene accounts for 43% of the total ion current[209]. In borazine the figure is 15.5% (ref. 210), in $(CH_3)_3B_3N_3H_3$ 12.5% (ref. 211), in $F_3B_3N_3H_3$ 45% considering ions above $m/e$ 44 only (lower $m/e$ values not reported)[212]; in $Cl_3B_3N_3H_3$ 25% (above $m/e$ 30)[209]; in $Cl_3B_3N_3Me_3$ 42% (above $m/e$ 30)[209]; in $Cl_3B_3N_3Ph_3$ 51% (above $m/e$ 90)[209]; in $Ph_3B_3N_3Ph_3$ 27% (above $m/e$ 55)[209]; and for further comparison, in triphenylbenzene 14% (ref. 209). On comparison with fully reported spectra, it is expected that the absence of intensities for the lower fragment ions in some of the above spectra makes little difference to the figures quoted.

compounds has not been investigated. No compounds containing the six-membered chain system containing alternate boron and nitrogen atoms and corresponding to XV have been reported and one might expect that such compounds would be quite unstable compared with borazines. It will be assumed, therefore, that all ions derived from borazine compounds and containing three nitrogens and three borons have the intact borazine ring system. In addition, it seems likely that ions containing two nitrogens and two borons will possess the stable four-membered ring system XVII. It is emphasized, however, that no conclusive proof substantiating these structures has been presented. However, in the case of borazine a difference of

Table 6.30

Partial Monoisotopic Mass Spectra[a] of Various Borazine Derivatives Showing the Relative Intensities of the Parent Molecular Ions and Fragment Ions Formed by Loss of Hydrogen

| Compound | $P^{+}$[b] | $P-1^+$ | $P-2^+$ | $P-3^+$ | $P-4^+$ | $P-5^+$ | $P-6^+$ | $P-Me^+$ | Ref. |
|---|---|---|---|---|---|---|---|---|---|
| $H_3B_3N_3H_3$ | 48.2 | 100 | 17.5 | 11.9 | 8.26 | 5.66 | 0.42 | | 210 |
| $H_2(HO)B_3N_3H_3$ | 80.0 | 100 | 23.0 | 5.1 | 17.2 | 0 | | | 212 |
| $H_2(CH_3O)B_3N_3H_3$ | 100 | 90.1 | 18.9 | 2.8 | 3.8 | 1.0 | | | 215 |
| $H_2(C_2H_5O)B_3N_3H_3$ | 45.9 | 11.2 | 1.1 | 0.8 | | | | | 215 |
| $H_2ClB_3N_3H_3$ | 56.5 | 100 | $\sim$4[c] | | | | | | 216 |
| $H_2ClB_3N_3(CH_3)_3$ | 36.5 | 100 | $\sim$3[c] | | | | | | 216 |
| $HCl_2B_3N_3H_3$ | 77.0 | 100 | 0 | | | | | | 216 |
| $HCl_2B_3N_3(CH_3)_3$ | 25.6 | 100 | $\sim$0[c] | | | | | | 216 |
| $Cl_3B_3N_3H_3$ | 100 | $\sim$6[c] | | | | | | | 209 |
| $F_3B_3N_3H_3$ | 100 | $\sim$0[c] | | | | | | | 217 |
| $(CH_3)_3B_3N_3H_3$ | 29.3 | 0.28 | 0.94 | 0.28 | | | | 100 | 211 |
| $(ClCH_2)_3B_3N_3H_3$ | 30 | $\sim$0[c] | | | | | | 100[d] | 209 |
| $H_2CH_3B_3N_3H_3$ | 47 | 34 | 0 | 3 | 2 | 1 | | 100 | 218 |
| $H(CH_3)_2B_3N_3H_3$ | 36 | 7 | 0 | | | | | 100 | 218 |
| $H_3B_3N_3H_2CH_3$ | 34 | 100 | 3 | 5 | | | | $\sim$1 | 218 |
| $H_3B_3N_3H(CH_3)_2$ | 48 | 100 | 27 | | | | | 5 | 218 |

[a] Calculated using the following isotopic abundances: $^{10}$B 18.8%, $^{11}$B 81.2%, $^{35}$Cl 75.53%, $^{37}$Cl 24.47% except for $H_3B_3N_3H_3$ and $(CH_3)_3B_3N_3H_3$ where the figures $^{10}$B 20% and $^{11}$B 80% were used[210,211].
[b] P = parent molecule.
[c] Some discrepancy existed between the calculated and experimental relative intensities; the $^{10}$B : $^{11}$B ratio quoted in footnote [a] gave the best fit.
[d] Relative intensity of $(P-CH_2Cl)^+$ ion.

only about 1 eV exists between the ionisation potential of the molecule and the appearance potential of the (parent molecule–H·)$^+$ ion, whereas for benzene this difference is about 5 eV[214]. Ring rupture of the type suggested in the formation of the $C_6H_5^+$ ion from benzene is therefore unlikely in the formation of $B_3N_3H_5^+$ from borazine. Partial monoisotopic spectra of borazine and some of its derivatives have been calculated from published data and are shown in Table 6.30. It can be seen that the abundance of fragment ions formed by loss of hydrogen is, in some cases including unsubstituted borazine, much greater than that of corresponding ions resulting from benzene compounds. A closer examination of the various spectra in Table 6.30 reveals that hydrogens attached to boron appear to be lost much more readily than those bonded to nitrogen. In compounds which are tri-boron substituted but with hydrogen bonded to the nitrogens, fragment ions corresponding to loss of hydrogen are of very low intensity if present at all. Such ions are readily formed, however, from compounds containing one or more unsubstituted borons, and in most such cases the base peak corresponds to the (parent molecule-1)$^+$ ion. Methyl groups attached to boron are also lost very readily but not those which are nitrogen substituents.

Another interesting aspect of the mass spectra of benzene derivatives is that only in the case of $\alpha$-branched alkylbenzenes or compounds of the general type $C_6H_5COR$ are the C and H atoms originally present in the phenyl moiety retained in the $C_6H_5^+$ ion. In all other compounds specifically investigated from this point of view, both C and H atoms originating in the substituent groups were found in the $C_6H_5^+$ ions which cannot, therefore, be formed by simple loss of the substituent[213]. No labelling experiments or appearance potential measurements directed towards an understanding of the processes occurring in the electron-impact-induced fragmentation of borazine and its derivatives have been carried out. Obviously a considerable amount of work is required in this field before one can begin to predict these processes and make meaningful comparisons with the fragmentation of aromatic compounds.

Substituents attached to the B–N ring can have a marked effect on fragmentation processes. Snedden[209] investigated two series of compounds, tri-$B$-alkyltri-$N$-phenyl and tri-$B$-alkoxytri-$N$-phenyl derivatives as well as tri-$B$-phenyltri-$N$-phenylborazine, tri-$B$-chloroborazine and tri-$B$-chloromethylborazine (Fig. 6.13). Reduction of the spectra to a monoisotopic representation showed the parent molecular ion to be the most abundant in all except two of the compounds: tri-$B$-chloromethylborazine readily lost one of its chloromethyl groups and the tri-*B-tert.*-butoxide compound exhibited a parent molecular ion with an exceedingly low intensity and a base peak at $m/e$ 41. The lower stability of the former parent molecular ion was attributed by Snedden to lack of substituents such as phenyl groups attached to the N atoms; fewer resonance structures were possible and hence a decrease in stability resulted. In tri-*B*-chloroborazine, he suggested the positive charge on the parent could be shared between the chlorine atoms and the borazine nucleus so that the stability of this ion relative to the fragment ion formed by loss of Cl· was enhanced. As expected, the stability of the parent molecular ion of tri-*B*-chlorotri-*N*-phenylborazine was raised still further.

Rearrangement ions appear to be very common. The *N*-triphenyl derivatives were postulated[209] to undergo reactions of the type (see Fig. 6.13).

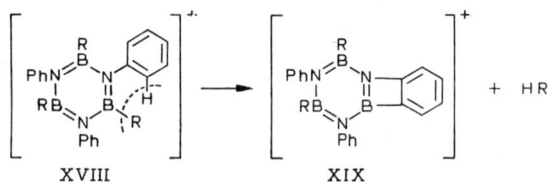

XVIII                    XIX

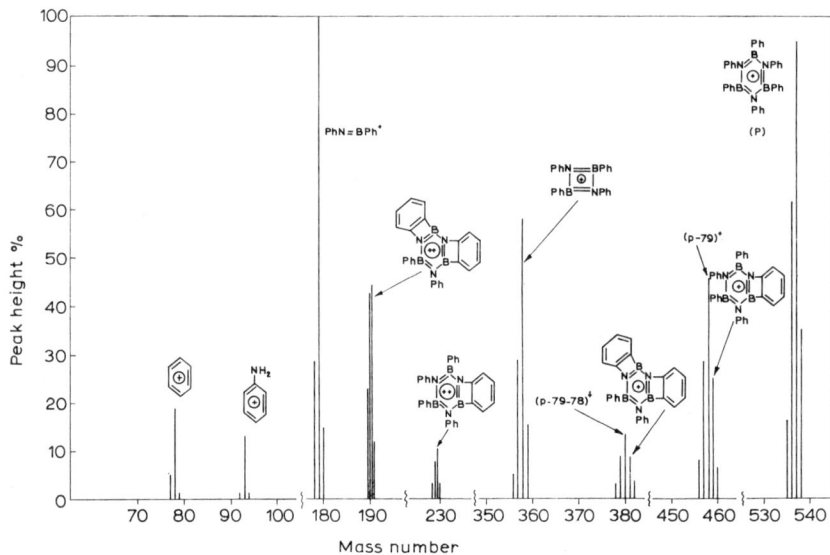

(a)

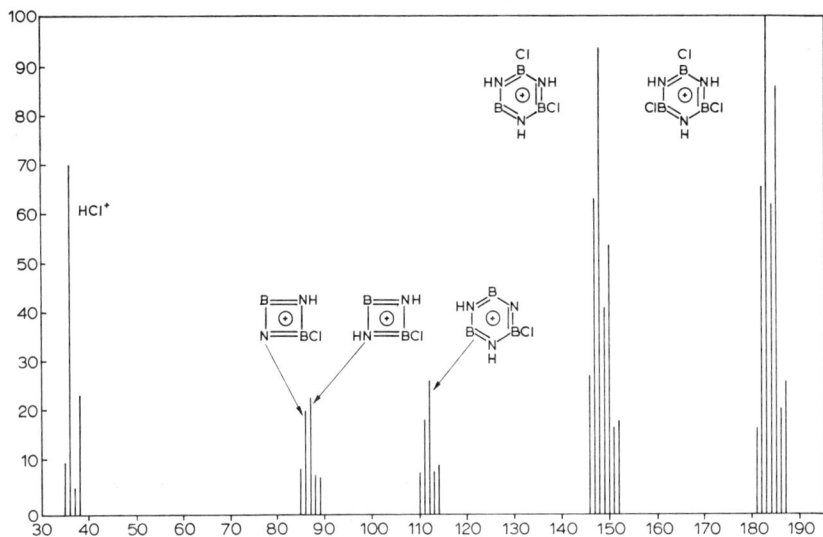

(b)

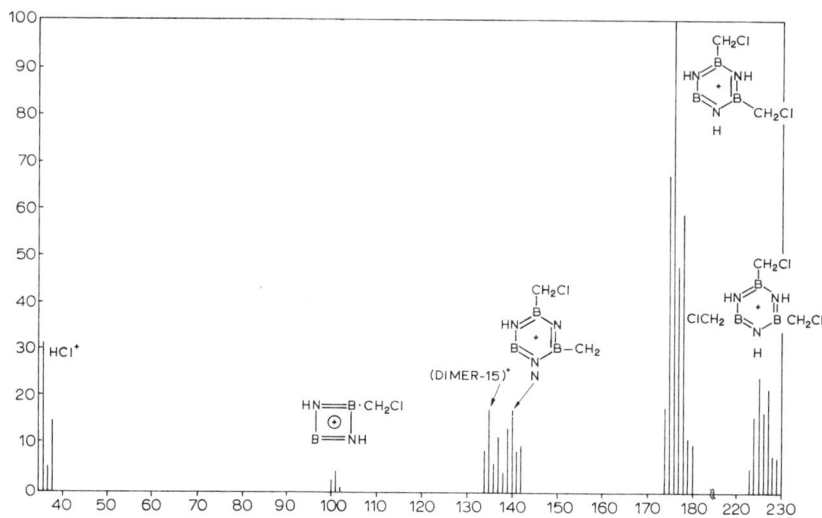

(c)

Fig. 6.13. Mass spectra of (a) tri-*B*-phenyltri-*N*-phenylborazine, (b) tri-*B*-chloroborazine, and (c) tri-*B*-chloromethylborazine. (Reproduced with permission from ref. 209.)

Doubly charged ions with the same composition as XIX possessed a high relative intensity, demonstrating the stability of the structure. In the $B$-alkoxyborazines, fragmentation occurred almost exclusively at the C–O rather than the B–O linkage as expected from a consideration of the thermochemical bond energies of the C–O and B–O single bonds[209]. Laubengayer $et\ al.$[217] found that the major fragment ions for $Cl_3B_3N_3H_3$, $Cl_2FB_3N_3H_3$, and $ClF_2B_3N_3H_3$ were formed by loss of a single Cl· from the parent molecular ion, whereas of $F_3B_3N_3H_3$ (and also $F_3B_3O_3$) the major fragment ion resulted from ring cleavage. These observations were consistent with the high stability of the B–F terminal bond compared with the B–N (and B–O) bond in the ring. In Table 6.31 a comparison is made of the proportion of

Table 6.31

Proportion of Total Ion Current Contributed by Ions Containing the $B_3N_3$ Ring and Those Which Have Lost One or More of the Ring Atoms

| Compound | Ions containing the intact $B_3N_3$ ring (%) | Ions formed by loss of one or more of the ring atoms (%) | Ref. |
|---|---|---|---|
| $F_3B_3N_3H_3$ | 40 | 60 | 217 |
| $Cl_3B_3N_3H_3$ | 92 | 8 | 209 |
| $Cl_3B_3N_3H_3$ | 83 | 17 | 217 |
| $Cl_3B_3N_3Ph_3$ | 69 | 31 | 209 |
| $Cl_3B_3N_3Me_3$ | 80 | 20 | 209 |
| $(MeO)_3B_3N_3Ph_3$ | 69 | 31 | 209 |
| $Me_3B_3N_3H_3$ | 52 | 48 | 211 |
| $H_3B_3N_3H_3$ | 62 | 38 | 210 |

the total ion current contributed by ions containing the intact B–N ring and those formed by ring cleavage and loss of one or more of the ring atoms.

Metastable transitions have been reported for borazine and the $B$-trimethyl derivative only, and correspond to the decompositions[210]

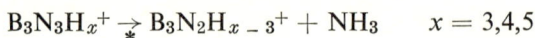

and

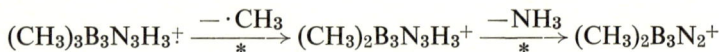

Other borazine compounds whose mass spectra have been recorded are
$N$-tris(diphenylmethyl)borazine[219], $B$-monoaminoborazine[212], $B$-diamino-
borazine[220], diborazinyl ether $(B_3N_3H_5)_2O$ (ref. 212), and the boron–nitrogen
analogs of naphthalene and biphenyl, XX and XXI[220,221].

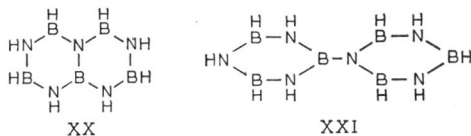

The ionisation potentials of $H_3B_3N_3H_3$, $Me_3B_3N_3H_3$, $H_3B_3N_3Me_3$, and
$Me_3B_3N_3Me_3$ have been measured by an electron-impact technique[222]. It
was observed that methyl substitution lowers the ionisation potential of
benzene more than that of borazine.

*6.6 Cyclotriborazanes and related phosphorus compounds*

No complete mass spectrum of a cyclotriborazane has been published. The
spectra of two isomeric forms of each of five 1,3,5-trialkylcyclotriborazanes
were reported to be very similar[223], with the base peak in all compounds
investigated corresponding to an ion of composition $[P-(\cdot BH_2)-(2H\cdot)]^+$
(P = parent molecule). No further details were given but it appears that, as
expected, the ring is not as stable as that in borazine with its resonance forms.
Two other substituted members of the series were reported to show weak
parent molecular ions[224], while in each case an ion formed by loss of a $\cdot H$
from the parent was eight times more abundant.

Instability of the analogous B–P ring under electron impact was certainly
demonstrated[225] in the case of $(Me_2PBH_2)_3$ which fragmented severely at
70 eV, giving rise to peaks at almost every mass number from 24 to 222. The
ions $Me_6P_3B_2H_3^+$ and $Me_4P_2B_3H_6^+$ were both present in appreciable quantities when 12 eV electrons were employed, indicating that fragmentation of
the ring was a readily occurring process. The most surprising aspect was that
the base peak in the 70 eV spectrum corresponded to the ion $BMe_2^+$; its
formation necessitated transfer of both methyl groups to the boron and its
appearance potential was high ($>20$ eV).

*6.7 Tetrameric hydrazinoboranes*

The mass spectrum of a tetrameric hydrazinoborane derivative (HB–NCH$_3$–NCH$_3$)$_4$ has been reported[226] and the parent molecular ion definitely observed. The postulated structure of the compound is very compact and the absence of ions corresponding to the trimer in the spectrum possibly reflects the difficulty in rearrangement from the tetramer to a plausible structure for such a species.

*6.8 Heteroaromatic boron–nitrogen compounds*

A study of electron-impact fragmentation patterns for derivatives of 10,9-borazarophenanthrene, 2,1-borazaronaphthalene and 4,3-borazaroisoquinoline has been made by Dougherty[227]. The spectra were qualitatively similar to the parent aromatic systems and will not be considered in detail here. The fragmentation patterns of three cyclic amine-boranes

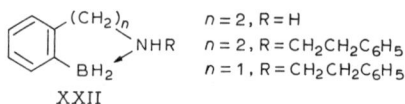

$$n = 2, R = H$$
$$n = 2, R = CH_2CH_2C_6H_5$$
$$n = 1, R = CH_2CH_2C_6H_5$$

XXII

and the catechol derivatives of two of them have recently been discussed in some detail by Catlin and Snyder[228].

7. MAIN GROUP III–GROUP VI COMPOUNDS

*7.1 Alkoxy derivatives*

The mass spectra of the trialkyl borates B(OR)$_3$ (R = Me[229,230], Et[231], $n$-Pr[231], $n$-Bu[231]) have been reported, as has that[229] of the compound HB(OMe)$_2$. Wada and Kiser[230] measured the appearance potentials of the various ions from B(OCH$_3$)$_3$ to assist in determining the processes involved in the fragmentation mechanism. Their proposed mechanism was later confirmed, with the addition of one minor pathway, by the observation of appropriate metastable transitions by Fallon *et al.*[231] (Fig. 6.14). The fragmentation path of the ethyl derivative was also established in detail, and those of the $n$-Pr and $n$-Bu derivatives in part, by observation of the relevant metastable transitions[231].

$$(CH_3O)BO_2^+ \xleftarrow[\ast]{-C_2H_6} (CH_3O)_3B^+ \xrightarrow[\ast]{-\cdot CH_3} (CH_3O)_2BO^+$$

$$-H\cdot \diagdown \quad \diagup -C_2H_5 \qquad \ast \diagdown -CH_3O\cdot$$

$$(CH_3O)BO(OH)^+ \qquad\qquad (CH_3O)_2B^+$$

$$-CH_3OH \diagdown \ast \qquad \ast \diagup -C_2H_6$$

$$BO_2^+$$

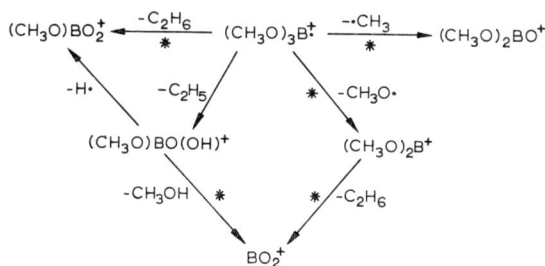

Fig. 6.14. Proposed fragmentation scheme for trimethyl borate, B(OCH₃)₃. (Reproduced with permission from ref. 231.)

A significant feature of the mass spectra of these compounds was the abundance of secondary ions formed by H· transfer; BOH(OR)⁺ was generally the most prominent of these and increased in intensity down the series Me < Et < $n$-Pr ≈ $n$-Bu. Metastable peaks corresponding to the following rearrangements were observed for ethyl, $n$-propyl, and $n$-butyl, but not methyl, borates[231].

$$B(OR)_3^+ \xrightarrow[\ast]{} B(OH)(OR)_2^+ \xrightarrow[\ast]{} BO(OR)_2^+ \xrightarrow[\ast]{} BO(OH)(OR)^+$$

$$B(OR)_2^+ \xrightarrow[\ast]{} B(OH)(OR)^+$$

Although it appeared that no new bond to boron (e.g. B–H, B–R) was formed for the ethyl, propyl, and butyl esters, the ion CH₃BOCH₃⁺ had a relative intensity of 10% in the spectrum of B(OCH₃)₃ (ref. 230).

Equilibrium mixtures of pairs of alkyl borates were investigated by Lockhart and coworkers[232,233] using mass spectrometry and the mixed esters B(OR)(OR')₂ were detected. Approximate spectra of these compounds, which cannot be isolated, were then deduced by subtracting the spectra of the parent tris esters. Metastable transitions corresponding to fragmentation of these mixed esters were also observed.

Wada and Kiser[230] found that their appearance potential values of ions derived from B(OCH₃)₃ differed considerably from those of Law and Margrave[229]. Although the full range of appearance potentials has not been remeasured by other workers, the value for the parent molecular ion obtained by Wada and Kiser has since been confirmed by two other sets of workers (Table 6.32). Heats of formation of the ions from B(OCH₃)₃ were also calculated[230].

Aluminium alkoxides, unlike their boron counterparts, appear to adopt

Table 6.32

Ionisation Potentials (eV) of Trialkyl Borates
References are given in brackets.

| $B(OMe)_3$ | $B(OEt)_3$ | $B(OPr^n)_3$ | $B(OBu^n)_3$ |
|---|---|---|---|
| 8.9 ± 0.2 (229) | 10.47 ± 0.12 (231) | 10.62 ± 0.15 (231) | 10.72 ± 0.74 (231) |
| 10.8 ± 0.3 (230) | 10.13 (178) | 10.02 (178) | |
| 10.62 (178) | | | |

a variety of associated structures. Molecular weight measurements of aluminum isopropoxide, for example, have indicated[234] that the freshly distilled liquid is composed mainly of trimers, while similar measurements[234-237] carried out on solutions of the solid in organic solvents were consistent with a tetrameric structure. Shiner et al.[237] also demonstrated that at higher temperatures, either in the melt or in solution, this tetramer is reconverted to the trimer. However vapour density measurements on the vapour of the freshly distilled compound indicated a dimeric species[238].

Three mass spectral studies of the compound have since been reported [239-241]. In the first two studies the sample was introduced into the instrument as the solid by means of the solid insertion probe and because of the slow attainment of equilibrium in the vapour[242] the resultant spectra should largely reflect the species present in the solid state. The spectra indicated mainly tetramer although no parent molecular ion corresponding to this species was observed. By recording the spectra of both the normal isotopic sample and also one deuterated at the secondary carbon atom of the isopropoxide group, Fieggen et al.[240] were able to propose a fragmentation mechanism based on a number of observed metastable transitions. Low abundance ions which could have originated from either the tetramer, the trimer, or the dimer, were present in the spectrum. No metastable decompositions corresponding to elimination of Al-containing neutral fragments were detected and appearance potentials of the ions were not measured, but it was concluded that the dimer and trimer were possibly present in minor quantities. Daasch and Fales[239] reported that the spectrum depended on the temperature of the probe and only in heated samples were ions corresponding to the dimer observed. It appears, therefore, that at the source temperatures employed by Fieggen et al. (120–150 °C), thermal degradation to the dimer was occurring. Daasch and Fales also reported that samples of the tetramer distilled at atmospheric pressure (and therefore at high temperatures) showed groups of ions at masses above that of the tetramer, indicating the presence of pentameric, hexameric, and heptameric species. Metastable transitions were observed in this study also.

Rearrangement ions involving hydrogen transfer, common in the boron alkoxide spectra, were apparently not formed during electron impact of these aluminium compounds.

## 7.2 Boroxines

The mass spectra of two samples of trifluoroboroxine $(BOF)_3$ formed at high temperatures from (i) $MgF_2 + B_2O_3$ (ref. 243) and (ii) $BF_3 + B_2O_3$ (ref. 244) were in excellent agreement. Under the conditions employed, the former reaction also resulted in the monomeric species BOF so the intensity of the $BOF^+$ ion in this case was quite high, resulting as it did from both simple ionisation of BOF and fragmentation of $(BOF)_3$.

Table 6.33

Comparison of Mass Spectra of $F_3B_3O_3$ and $(CH_3)_3B_3O_3$

| Ion | Rel. intensity[b] | Ion[a] | Rel. intensity[c] |
|---|---|---|---|
| $F_3B_3O_3^+$ | 100 | $(CH_3)_3B_3O_3^+$ | 12.5 |
| $F_2B_3O_3^+$ | 9.6 | $(CH_3)_2B_3O_3^+$ | 100 |
| $F_3B_2O^+$ | 59.5 | | |
| $F_2B_2O_2^+$ | 30.8 | $(CH_3)_2B_2O_2^+$ | 1.6 |
| $FB_2O_2^+$ | 11.3 | $CH_3B_2O_2^+$ | 25 |
| $FBO^+$ | 9.2 | $CH_3BO^+$ | 3 |

[a] Additional low-intensity ions were also reported.
[b] From ref. 244.
[c] From ref. 20.

The corresponding trimethylboroxine exhibits a sharply contrasting fragmentation pattern[20] (Table 6.33). As mentioned previously, B–F terminal bonds appear to be highly stable compared with the B–N and B–O ring bonds in the borazine $F_3B_3N_3H_3$ and in $B_3O_3F_3$ (ref. 217). The B–CH$_3$ bonds in $(CH_3)_3B_3O_3$, on the other hand, were readily ruptured while the highly abundant rearrangement ion $B_2OF_3^+$ present in the spectrum of the fluoro compound has no analogue in that of the methyl derivative. A number of quite low-intensity rearrangement ions were observed for the latter compound, most of which involved H· transfer. Lehmann et al.[20] noted similarities between the spectrum of $(CH_3)_3B_3O_3$ and 1,3,5-trimethylbenzene (mesitylene) on the basis of which they postulated considerable aromatic character for the compound. Appearance potentials of all ions formed by electron impact of $(BOF)_3$ were measured[243].

The mass spectral pattern of boroxine is reported[245] to be analogous to that of borazine rather than to that of trifluoro-boroxine, $B_3O_3F_3$; the most abundant ion in the spectrum of $B_3O_3H_3$ is $B_3O_3H_2^+$ whose appearance potential was only 1 eV greater than the ionisation potential of the molecule. This is reminiscent of the borazine case discussed previously, and quite different from that of benzene.

### 7.3 Other main Group III–oxygen compounds

The mass spectrum of the compound $B_2H_2O_3$, formed as an intermediate in the partial oxidation of pentaborane-9, has been reported[246] as has that of dimethylborinic anhydride[247]. Only a very weak parent molecular ion was observed in this latter compound, but rearrangement ions were again very prominent and, in fact, one such species was the most abundant ion in the spectrum.

The compounds $(Me_2C=NOMMe_2)_n$ (M = Al, Ga, In, Tl) appear to be dimeric in the crystal, solution, and gaseous phases while the boron derivative is monomeric in the gas phase and only partially associated in benzene solution[248]. The mass spectrum of this latter compound is straightforward, showing parent molecular ion and fragment ions formed by bond fission; no rearrangement ions were observed. The spectra of the aluminium and gallium compounds, however, depicted ions formed by loss of one, two, and three methyl groups from the dimer, although the parent molecular ion was not observed in either case. The peaks corresponding to monomer were extremely weak indicating that dissociation of the dimer to the monomer occurred only to a very limited extent if at all under the source conditions. A notable feature was that loss of methyl groups apparently took place much more readily than loss of $Me_2CN$; this was in direct contrast to the situation involving the boron compound and was interpreted[248] as indicating that the N atoms were strongly bound in the molecular skeleton so that the structure

Me₂C=N, O—M, Me₂ / N=CMe₂ M—O Me₂

XXIII

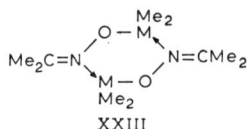

was postulated. In the thallium derivative, however, the group Me₂CNO tended to be lost so that its structure may not be analogous, or else the M–O bond is very weak.

The two aluminium compounds (Me₂Al.NRCOMe)ₙ (R = Ph, Buᵗ) were both dimeric in solution and the eight-membered ring structure

RN—CMe=O
│            ↕
Me₂Al        AlMe₂
↕            │
O=CMe—NR

RN=CMe—O
↕            │
Me₂Al        AlMe₂
│            ↕
O—CMe=NR

was proposed[249]. The mass spectra of these compounds, however, (and also of the compound Me₂Al.NPhCOPh whose low solubility prevented a cryoscopic molecular weight determination) did not reveal any ions which obviously originated from the dimers although that of the latter compound did exhibit peaks of uncertain assignment at $m/e$ values higher than the dimer. It is quite conceivable that at the source temperature of 150–250 °C the Al–O bonds were readily cleaved with the formation of monomer before electron impact occurred.

The mass spectrum of the cage compound $Al_4Cl_4[N(CH_3)_2]_4$ $(NCH_3)_2$ has also been reported[250].

*7.4 Boron–sulphur compounds*

Edwards *et al.*[50,251] identified the ions formed when metathioboric acid, $HBS_2$, was vaporised in the mass spectrometer and also measured their appearance potentials. The results indicated that below 100 °C, $H_2S$ and $(HBS_2)_3$ were the predominant species formed with some $H_3BS_3$ and $H_2B_2S_5$. Metastable transitions were observed to support the following fragmentation reactions.

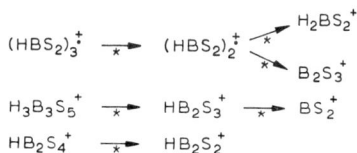

$$(HBS_2)_3^+ \xrightarrow{*} (HBS_2)_2^+ \begin{cases} \nearrow_* H_2BS_2^+ \\ \searrow_* B_2S_3^+ \end{cases}$$

$$H_3B_3S_5^+ \xrightarrow{*} HB_2S_3^+ \xrightarrow{*} BS_2^+$$

$$HB_2S_4^+ \xrightarrow{*} HB_2S_2^+$$

At temperatures above 250 °C, boron sulphide formed a number of polymeric species with molecular weights up to 800[50,51].

The fragmentation pattern of trimeric ethylthioborane, $(H_2BSEt)_3$, has been reported[252] but analogous monoalkoxyboranes have not been isolated. Its behaviour under electron impact differs considerably from that of trimethylboroxine, however, in that there is almost a complete absence of fragments containing the intact B–S six-membered ring. A fragmentation mechanism was proposed and supported, in part, by observed metastable decompositons.

## 8. ORGANO-DERIVATIVES OF MAIN GROUP III ELEMENTS

### 8.1 Trialkyl compounds

The mass spectrum of trimethylborane was first reported by Law and Margrave[253] who also measured the appearance potentials of the principal ions. At about the same time, Tollin et al.[254] investigated a mixture of $B(^{12}CH_3)_3$ and $B(^{13}CH_3)_3$ mass spectrometically and were able to show that methyl migration from one molecule to another did not occur, illustrating that no symmetrical dimer involving methyl bridges was formed. A more detailed mass spectral study of this compound was undertaken by Lehmann et al.[8] who utilised $^{10}B$- and D-labelled samples. These workers were principally concerned with the effects of isotopic labelling on the mass spectra obtained and with the correct value of the $^{10}B/^{11}B$ natural abundance ratio. They recognised that internal rearrangements played an important role in the fragmentation processes of trimethylborane.

A thorough and comprehensive mass spectral study of fourteen trialkyl boron compounds $BR_1R_2R_3$ ($R_1$, $R_2$, $R_3$ = Me, Et, n-Pr, i-Pr) was carried out by Henneberg et al.[255]. By use of high-resolution measurements, metastable transitions, partially deuterated samples such as $B(CH_2CH_2-CH_2D)_3$, molecular models, and recording of spectra at low ionising energies, they were able to deduce detailed fragmentation mechanisms. That for the compound tri-n-propylborane is partially shown in Fig. 6.15. Pertinent information concerning the principal ions observed may be summarised as follows.

(a) The parent molecular ions were all of low abundance and corresponded to about 1 % of the total ionisation.

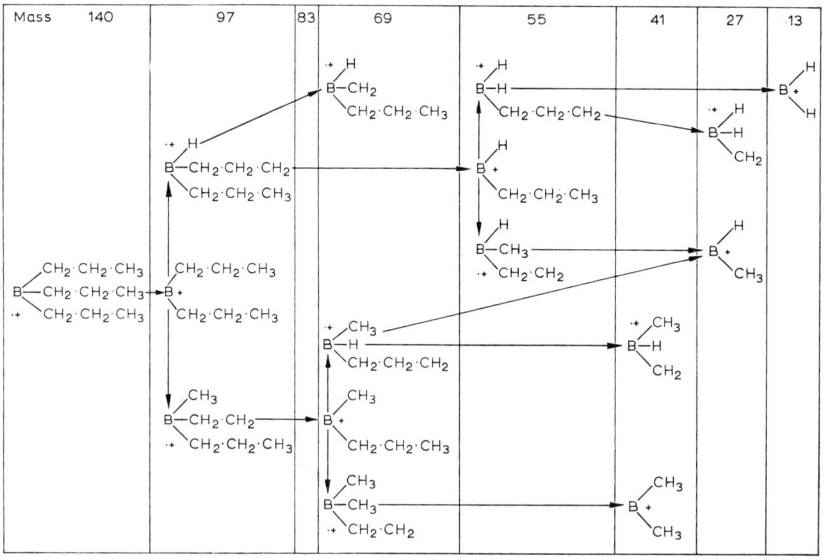

| Mass | 140 | 97 | 83 | 69 | 55 | 41 | 27 | 13 |
|---|---|---|---|---|---|---|---|---|

Fig. 6.15. Proposed fragmentation scheme for tri-*n*-propylborane. (Reproduced with permission from ref. 255.)

(*b*) Ions of the type $BR_2^+$ (10–25% of the total ionisation) were easily recognised and were formed by separation of a complete alkyl group R from $BR_3^+$.

(*c*) Ions of the type $(BC_nH_{(2n + 2)})^+$, not including those of type $BR_2^+$ with this composition, were formed by rearrangement processes and comprised 60–85% of the total ionisation.

(*d*) Ions of the type $(BC_nH_{2n})^+$ and $(BC_nH_{(2n - 2)})^+$ were shown by the observation of metastable transitions and by recording spectra at low ionising energies to have been formed from $(BC_nH_{(2n + 2)})^+$ ions by loss of hydrogen, and so these also originate from rearrangement processes.

(*e*) Hydrocarbon ions contributed 5–10% of the total ionisation.

(*f*) Ions which differed from the types mentioned above, formed by separation of another H atom, constituted a very small fraction of the total ionisation.

Rearrangement processes were again intrinsically involved in the fragmentation of the compounds $(CH_3)_2BCH_2X$ (X = Cl, $N_3$, $NH_2$, OH) in the mass spectrometer[256] as shown by the relative intensities of the predominant ions listed in Table 6.34.

Table 6.34

Relative Intensities of Predominant Ions in the Mass Spectra of Certain Derivatives of Trimethylborane[a]

$(CH_3)_2BCH_2Cl$

| Ion | Rel. intensity |
|---|---|
| $(CH_3)_2BCH_2Cl^+$ | 0.3 |
| $(CH_3)_2BH^{+b}$ | 3.2 |
| $(CH_3)_2B^+$ | 100 |
| $CH_3BH^{+b}$ | 10.3 |

$(CH_3)_2BCH_2N_3$

| Ion | Rel. intensity |
|---|---|
| $(CH_3)_2BCH_2N_3^+$ | 11.5 |
| $CH_3BCH_2N_2^+$ | } 100 |
| $CH_3(H)BCH_2N_3^{+b}$ | |
| $(CH_3)_2BCH_2^+$ | 17.4 |
| $CH_3BH^{+b}$ | 42.0 |
| $CH_3B^+$ | 6.7 |

$(CH_3)_2BCH_2NH_2$

| Ion | Rel. intensity |
|---|---|
| $(CH_3)_2BCH_2NH_2^+$ | 5.5 |
| $CH_3BCH_2NH_2^+$ | 10.8 |
| $(CH_3)_2BH^{+b}$ | |
| $HBCH_2NH_2^{+b}$ | 100 |
| $CH_3BH_2^{+b}$ | 25.2 |
| $CH_3BH^{+b}$ | 4.3 |

$(CH_3)_2BCH_2OH$

| Ion | Rel. intensity |
|---|---|
| $(CH_3)_2BCH_2OH^+$ | 2.1 |
| $CH_3BCH_2OH^+$ | 16.7 |
| $HBCH_2OH^{+b}$ | 100 |

[a] Assignments made from figures in ref. 256.
[b] Rearrangement ion.

In contrast to the boron trialkyls, the trialkyls of aluminium are associated in the vapour phase. The ion $Al_2Me_5^+$ was observed in the mass spectrum of the trimethyl derivative in very low concentrations[172,257,258]. However, a more recent study[259] found the dimeric ions $Al_2Me_3^+$ and $Al_2Me_5^+$ to be quite abundant at source temperatures of about 40 °C but showed that this abundance rapidly diminishes with increasing source temperatures. No dimeric ions have been observed in the spectrum of triethylaluminium even at a source temperature of 45 °C (ref. 259). It had previously been shown[260] that at temperatures of 100–150 °C and at much higher vapour pressures

(90–135 mm) than exist in the mass spectrometer source, the dimers of tri-methyl- and especially triethylaluminium are highly dissociated.

Two additional dimer peaks were reported by Tanaka and Smith[172] who assigned them to the ions $Al_2Me_3H_3^+$ and $Al_2Me_4H^+$. They also reported small peaks corresponding to the ions $Al_3Me_4H_4^+$ and $Al_4Me_6H_5^+$, but their intensities were apparently a function of pressure, indicating that they were formed by ion–molecule reactions. As with many alkyl–metal compounds, hydride ions were formed from the monomer by rearrangement processes, but were present in low abundance. From their constructed clastogram and appearance potential data, Winters and Kiser[257] proposed successive loss of methyl radicals. A number of metastable decompositions have since been reported[172,258,259,261] so that the partial fragmentation shown in Fig. 6.16 may be derived. The heats of formation of the predominant ions formed from trimethylaluminium were calculated from their measured appearance potentials[257].

Dimethylaluminium hydride vapour is believed to consist of a mixture of dimer and trimer[262] and Chambers et al.[259] have observed both trimeric and

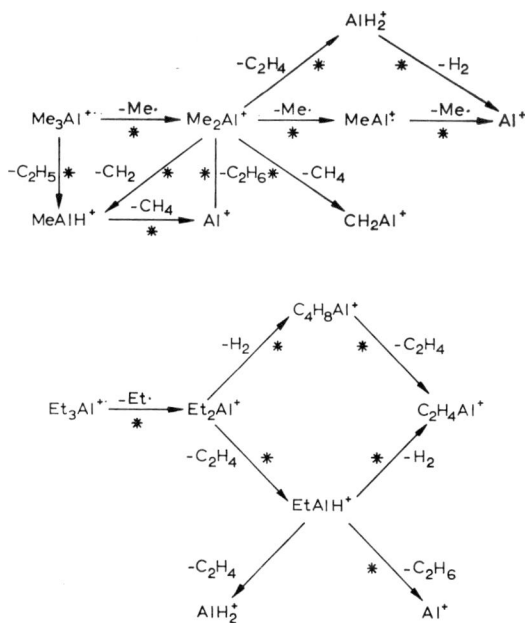

Fig. 6.16. Partial fragmentation schemes for trimethyl- and triethylaluminium.

dimeric ions in the mass spectrum of the compound. Tanaka and Smith[172], however, have also detected polymeric ions corresponding to tetrameric, pentameric, and hexameric species in low abundance; their origin is uncertain. The fragmentation pathways of both this compound and the corresponding ethyl derivative have been established by the observation of metastable transitions[172,259] and are shown in Fig. 6.17. Even-electron ions

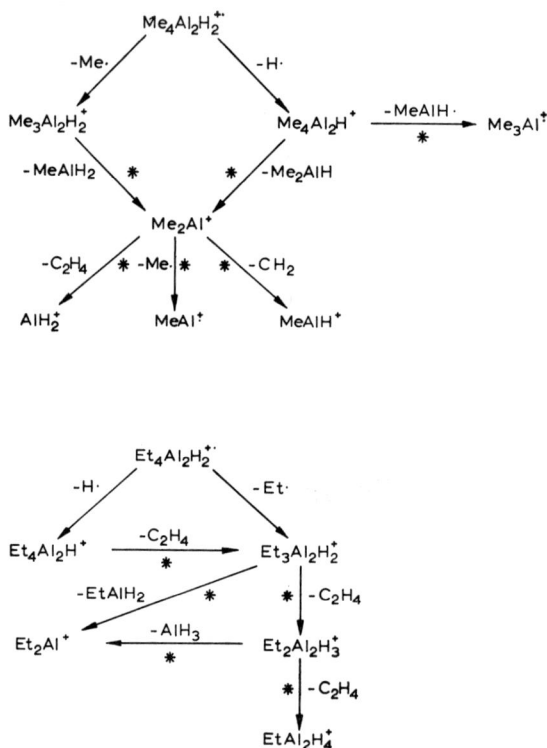

Fig. 6.17. Metastable supported fragmentation paths of dimethyl- and diethylaluminium hydride dimers.

were far more abundant than odd-electron ions in the spectra of all compounds.

Aston employed trimethylindium[263] and ethylthallium[264] in the determination of isotopic abundances but no details of the mass spectra of these compounds have been reported.

Winters and Kiser[257] determined the ionisation potential of $Al(CH_3)_3$ and, in addition, estimated those of the trimethyl derivatives of gallium, indium, and thallium by inteipolation with other metal alkyl data.

### 8.2 $R_nMX_{(3-n)}$ compounds

A comprehensive mass spectral study of methyl-, ethyl-, isopropyl-, and vinylboron difluoride was carried out by Steele et al.[182]. They measured the appearance potentials of the major positive ions formed, enabling them to calculate the B–C bond dissociation energy in each case. The results showed that the B–C bond strength in the vinyl compound was significantly higher than the strengths of the B–C bonds in the alkylboron difluorides (Table 6.35). Such an increase was expected with the change in hybridisation of the

Table 6.35

Bond Dissociation Energies and Heats of Formation of Some Organoboron Difluorides

| Compound | $\Delta H°_f$ (kcal. mole$^{-1}$) | $D(\text{R–BF}_2)$ (kcal. mole$^{-1}$) | $D(\text{H–BF}_2)$ (kcal. mole$^{-1}$) |
|---|---|---|---|
| CH$_3$BF$_2$ | $-199 \pm 3$ | 95–100 | |
| C$_2$H$_5$BF$_2$ | $-209$ | 101 | 106 |
| i-C$_3$H$_7$BF$_2$ | $-212$ | 96 | 99 |
| CH$_2$ : CHBF$_2$ | $-171$ | 111 | 99 |

bonded carbon atom but, in addition, was attributed in part to vinyl–boron $\pi$-interaction. The apparent stability of this molecule was also reflected in the high relative intensity of its parent molecular ion; the spectrum of the compound differed considerably from those of the alkyl derivatives (Table 6.36). Results obtained in the investigation indicated that the process

$$\text{RBF}_2^+ \rightarrow (\text{R–H})^+ + \text{HBF}_2 + 2\,e \quad (\text{R} = \text{Et, i-Pr, Vi})$$

was occurring and an indirect measurement of the B–H bond dissociation energy in the HBF$_2$ molecule was carried out. In addition, the heat of formation of CH$_3$BF$_2$ was calculated and estimates of the other three were made (Table 6.35), and detailed consideration given to the various processes involved in the electron-impact-induced dissociations.

Lockhart and Kelly[265] have recently reported the mass spectrum of phe-

Table 6.36

Relative Intensities of the Principal Boron-Containing Positive Ions Observed in the Mass Spectra of Organoboron Difluorides[182]

| Ion | Relative intensity for R = | | | |
|---|---|---|---|---|
| | $CH_3$ | $C_2H_5$[a] | $i$-$C_3H_7$[a] | $CH_2CH$ |
| $RBF_2^+$ | 13.0 | 17.4 | 9.1 | 100 |
| $(R–CH_3)BF_2^+$ | | | 60.0 | |
| $(R–CH_4)BF_2^+$ | | | 25.6 | |
| $RBF^+$ | 19.7 | 50.5 | | 30.0 |
| $(R–H)BF^+$ | 15.7 | 11.1 | | 27.4 |
| $(R–2H)BF^+$ | | 45.3 | | |
| $(R–3H)BF^+$ | | 25.8 | | |
| $(R–CH_4)BF^+$ | | | 100 | |
| $BF_2^+$ | 100 | 100 | 97.9 | 87.4 |

[a] The most abundant ion in the spectrum was a hydrocarbon ion.

nylboron dichloride. The peak at $m/e$ 121, with an intensity of 84.4% of that of the base peak ($m/e$ 123), was assigned to the ion $C_6H_4{}^{10}B^{35}Cl^+$. If it were due solely to this ion, however, the intensity of $C_6H_4{}^{11}B^{35}Cl^+$ at $m/e$ 122 would be approximately three times that of the base peak. The $m/e$ 121 peak must therefore be largely due to impurity. The appearance potential of the ion at $m/e$ 123, assumed to be $Ph^{11}B^{35} Cl^+$, was measured and combined with the electron affinity of chlorine to give a value for the ionic bond energy $D(PhBCl^+Cl^-)$ of 211 $\pm$ 3 kcal.mole$^{-1}$. Since the spectrum was recorded on a low-resolving-power instrument and no accurate mass measurements were made, the possibility that the previously mentioned impurity present contributes to the peak at $m/e$ 123 cannot be ruled out and the result must therefore be somewhat suspect.

The mass spectra of the compounds dimethylaluminium chloride, methylaluminium chloride, and dimethylaluminium hydride have recently been reported[172]. The existence of a dimer was clearly shown for each of the two chloride compounds while the spectrum of the hydride indicated that, in addition to monomers and dimers, trimers, tetramers, pentamers, and hexamers were present in low concentrations and it was postulated that the associated species were primarily ring or cluster arrangements rather than linear fragments.

A number of metastable transitions were observed[172,258] corresponding to the decompositions

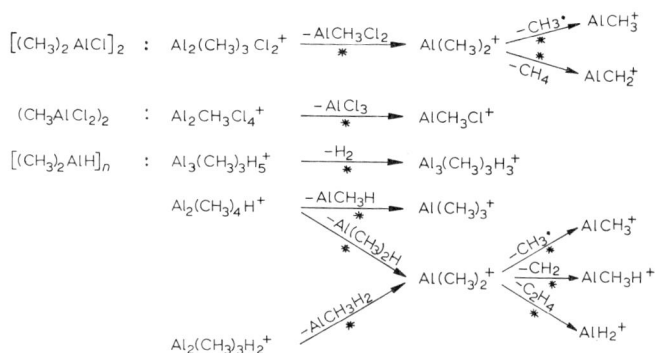

$$[(CH_3)_2AlCl]_2 \; : \; Al_2(CH_3)_3Cl_2^+ \xrightarrow[*]{-AlCH_3Cl_2} Al(CH_3)_2^+ \begin{array}{c} \xrightarrow[*]{-CH_3^{\cdot}} AlCH_3^+ \\ \xrightarrow[*]{-CH_4} AlCH_2^+ \end{array}$$

$$(CH_3AlCl_2)_2 \; : \; Al_2CH_3Cl_4^+ \xrightarrow[*]{-AlCl_3} AlCH_3Cl^+$$

$$[(CH_3)_2AlH]_n \; : \; Al_3(CH_3)_3H_5^+ \xrightarrow[*]{-H_2} Al_3(CH_3)_3H_3^+$$

$$Al_2(CH_3)_4H^+ \xrightleftharpoons[-Al(CH_3)_2H]{-AlCH_3H} Al(CH_3)_3^+$$

$$Al_2(CH_3)_3H_2^+ \xrightarrow{-AlCH_3H_2} Al(CH_3)_2^+ \begin{array}{c} \xrightarrow[*]{-CH_3^{\cdot}} AlCH_3^+ \\ \xrightarrow[*]{-CH_2} AlCH_3H^+ \\ \xrightarrow[*]{-C_2H_4} AlH_2^+ \end{array}$$

The mass spectra of a variety of dimethylthallium derivatives have been reported by Lee[266]: $(CH_3)_2TlX$, where X = Cl, Br, I, CN, $OC_6H_5$, $SCH_3$, $OOCCH_3$, and acac. In contrast to the fragmentation of most organometallic compounds, metal hydride species were not formed except in the case of dimethylthallium phenoxide. Peaks due to dimeric species were observed for the compounds where X = Cl, CN, $OC_6H_5$, and $SCH_3$. A number of metastable transitions were observed where either two methyl groups were lost simultaneously or else a neutral ethane molecule was expelled.

## APPENDIX

*Boron hydride derivatives*

| | |
|---|---|
| $\mu$-HSB$_2$H$_5$ | P. C. Keller, *Inorg. Chem.*, 8 (1969) 2457. |
| $\mu$-H$_3$SiB$_5$H$_8$ | D. F. Gaines and T. V. Iorns, *J. Am. Chem. Soc.*, 90 (1968) 6617. |

*Carboranes*

| | |
|---|---|
| CB$_5$H$_7$, C$_3$B$_5$H$_{13}$ | T. Onak, P. Mattschei and E. Groszek, *J. Chem. Soc. A*, (1969) 1990. |
| CH$_3$GaC$_2$B$_4$H$_6$ | R. N. Grimes and W. J. Rademaker, *J. Am. Chem. Soc.*, 91 (1969) 6498. |

198                                                              THE MAIN GROUP III ELEMENTS

REFERENCES

1  F. J. Norton, *J. Am. Chem. Soc.*, 71 (1949) 3488.
2  J. R. Morrey, A. B. Johnson, Y-C. Fu and G. R. Hill, *Advan. Chem. Ser.*, 32 (1961) 157.
3  A. B. Baylis, G. A. Pressley, Jr. and F. E. Stafford, *J. Am. Chem. Soc.*, 88 (1966) 2428.
4  A. D. Norman, R. Schaeffer, A. B. Baylis, G. A. Pressley, Jr. and F. E. Stafford, *J. Am. Chem. Soc.*, 88 (1966) 2151.
5  R. C. Weast (Ed.), *Handbook of Chemistry and Physics*, The Chemical Rubber Publishing Co., Cleveland, Ohio, U.S.A., 49th edn., 1968–69.
6  J. Drowart and P. Goldfinger, *Angew. Chem. Intern. Ed. Engl.*, 6 (1967) 581.
7  M. G. Inghram, *Phys. Rev.*, 70 (1946) 653.
8  W. J. Lehmann, C. O. Wilson, Jr. and I. Shapiro, *J. Inorg. Nucl. Chem.*, 11 (1959) 91.
9  C. E. Melton, L. O. Gilpatrick, R. Baldock and R. M. Healy, *Anal. Chem.*, 28 (1956) 1049.
10  G. M. Panchenkov and V. D. Moiseev, *Zh. Fiz. Khim.*, 30 (1956) 1118.
11  W. H. McFadden and A. L. Wahrhaftig, *J. Am. Chem. Soc.*, 78 (1956) 1572.
12  V. H. Dibeler, F. H. Mohler and M. de Hemptinne, *J. Res. Natl. Bur. Std.*, 53 (1954) 107.
13  V. H. Dibeler and F. L. Mohler, *J. Res. Natl. Bur. Std.*, 45 (1950) 441.
14  V. H. Dibeler, F. L. Mohler and L. Williamson, *J. Res. Natl. Bur. Std.*, 44 (1950) 489.
15  W. S. Koski, J. J. Kaufman, C. F. Pachucki and F. J. Shipko, *J. Am. Chem. Soc.*, 80 (1958) 3202.
16  D. P. Stevenson, *J. Chem. Phys.*, 19 (1951) 17.
17  O. Beeck, J. W. Otoos, D. P. Stevenson and C. D. Wagner, *J. Chem. Phys.*, 16 (1948) 255.
18  O. A. Schaeffer, *J. Chem. Phys.*, 18 (1950) 1501; 23 (1955) 1305, 1309.
19  J. H. Beynon, *Mass Spectrometry and Its Application to Organic Chemistry*, Elsevier, Amsterdam, 1960, p. 457.
20  W. J. Lehmann, C. O. Wilson, Jr. and I. Shapiro, *J. Inorg. Nucl. Chem.*, 21 (1961) 25.
21  P. J. de Bievre and G. H. Debus, *Intern. J. Mass Spectrom. Ion Phys.*, 2 (1969) 15.
22  I. Shapiro, C. O. Wilson, J. F. Ditter and W. J. Lehmann, *133rd American Chemical Society Meeting, San Francisco, 1958*; *Advan. Chem. Ser.*, 32 (1961) 127.
23  V. H. Dibeler, F. L. Mohler, L. Williamson and R. M. Reese, *J. Res. Natl. Bur. Std.*, 43 (1949) 97.
24  G. Verhaegen and J. Drowart, *J. Chem. Phys.*, 37 (1962) 1367.
25  W. A. Chupka, *Argonne National Lab. Rept. 5667*, 1957, p. 75.
26  R. P. Burns, A. J. Jason and M. G. Inghram, *J. Chem. Phys.*, 46 (1967) 394.
27  G. Verhaegen. F. E. Stafford and J. Drowart, *J. Chem. Phys.*, 40 (1964) 1622.
28  W. A. Chupka, J. Berkowitz, C. F. Giese and M. G. Inghram, *J. Phys. Chem.*, 62 (1958) 611.
29  O. C. Trulson and H. W. Goldstein, *J. Phys. Chem.*, 69 (1965) 2531.
30  R. F. Porter and E. E. Zeller, *J. Chem. Phys.*, 33 (1960) 858.
31  T. C. Ehlert and J. L. Margrave, *J. Am. Chem. Soc.*, 86 (1964) 3901.
32  E. Murad, D. L. Hildenbrand and R. P. Main, *J. Chem. Phys.*, 45 (1966) 263.
33  R. F. Porter, *J. Chem. Phys.*, 33 (1960) 951.

34 F. J. Keneshea and D. Cubicciotti, *J. Phys. Chem.*, 69 (1965) 3910.
35 D. L. Hildenbrand and E. Murad, *J. Chem. Phys.*, 43 (1965) 1400.
36 A. Büchler, E. P. Marram and J. L. Stauffer, *J. Phys. Chem.*, 71 (1967) 4139.
37 J. W. Hastie and J. L. Margrave, *J. Phys. Chem.*, 73 (1969) 1105.
38 A. Büchler and J. B. Berkowitz-Mattuck, *J. Chem. Phys.*, 39 (1963) 286.
39 R. P. Burns, *J. Chem. Phys.*, 44 (1966) 3307.
40 R. P. Burns, G. DeMaria, J. Drowart and M. G. Inghram, *J. Chem. Phys.*, 38 (1963) 1035.
41 M. G. Inghram, R. F. Porter and W. A. Chupka, *J. Chem. Phys.*, 25 (1956) 498.
42 O. T. Nikitin and P. A. Akishin, *Dokl. Akad. Nauk SSSR*, 145 (1962) 1294.
43 P. E. Blackburn, A. Büchler and J. L. Stauffer, *J. Phys. Chem.*, 70 (1966) 2469.
44 D. L. Hildenbrand, L. P. Theard and A. M. Saul, *J. Chem. Phys.*, 39 (1963) 1973.
45 S. K. Gupta and R. F. Porter, *J. Phys. Chem.*, 70 (1966) 871.
46 W. P. Sholette and R. F. Porter, *J. Phys. Chem.*, 67 (1963) 177.
47 D. J. Meschi, W. A. Chupka and J. Berkowitz, *J. Chem. Phys.*, 33 (1960) 530.
48 R. F. Porter, D. R. Bidinosti and K. F. Watterson, *J. Chem. Phys.*, 36 (1962) 2104.
49 K. F. Zmbov and J. L. Margrave, *J. Inorg. Nucl. Chem.*, 29 (1967) 2649.
50 J. G. Edwards and P. W. Gilles, *Advan. Chem. Ser.*, 72 (1968) 211.
51 F. T. Green and P. W. Gilles, *J. Am. Chem. Soc.*, 84 (1962) 3598; 86 (1964) 3964.
52 S. J. Steck, G. A. Pressley, Jr. and F. E. Stafford, *J. Phys. Chem.*, 73 (1969) 1000.
53 J. Berkowitz and W. A. Chupka, *J. Chem. Phys.*, 45 (1966) 1287.
54 J. Drowart, G. DeMaria, R. P. Burns and M. G. Inghram, *J. Chem. Phys.*, 32 (1960) 1366.
55 R. F. Porter, P. O. Schissel and M. G. Inghram, *J. Chem. Phys.*, 23 (1955) 339.
56 C. G. Frosch and C. D. Thurmond, *J. Phys. Chem.*, 66 (1962) 877.
57 R. P. Burns, K. A. Gingerich, L. Malaspina and V. Piacente, *J. Chem. Phys.*, 44 (1966) 2531.
58 E. J. Sinke, G. A. Pressley, Jr., A. B. Baylis and F. E. Stafford, *J. Chem. Phys.*, 41 (1964) 2207.
59 T. P. Fehlner and W. S. Koski, *J. Am. Chem. Soc.*, 86 (1964) 2733.
60 T. P. Fehlner and W. S. Koski, *J. Am. Chem. Soc.*, 87 (1965) 409.
61 P. S. Ganguli and H. A. McGee, Jr., *J. Chem. Phys.*, 50 (1969) 4658.
62 F. E. Stafford, G. A. Pressley, Jr. and A. B. Baylis, *Advan. Chem. Ser.*, 72 (1968) 137.
62a R. E. Hollins and F. E. Stafford, *Inorg. Chem.*, 9 (1970) 877.
63 J. A. Hipple, *Phys. Rev.*, 57 (1940) 350.
64 V. H. Dibeler and F. L. Mohler, *J. Am. Chem. Soc.*, 70 (1948) 987.
65 J. Turkevich, L. Friedman, E. Solomon and F. M. Wrightson, *J. Am. Chem. Soc.*, 70 (1948) 2638.
66 M. W. Evans, N. Bauer and J. Y. Beach, *J. Chem. Phys.*, 14 (1946) 701.
67 A. Quayle, *J. Appl. Chem.*, 9 (1959) 395.
68 J. R. Jennings and K. Wade, *J. Chem. Soc. A*, (1968) 1946.
69 T. P. Fehlner and W. S. Koski, *J. Am. Chem. Soc.*, 86 (1964) 581.
70 Mass spectral data, *American Petroleum Institute Research Project 44, Ser. No. 415*, Natl. Bur. Std., Washington, D.C.
71 J. L. Margrave, *J. Phys. Chem.*, 61 (1957) 38.
72 F. P. Lossing, A. W. Tickner and W. A. Bryce, *J. Chem. Phys.*, 19 (1951) 1254.

73  R. E. Honig, *J. Chem. Phys.*, 16 (1948) 105.

74  J. H. Wilson and H. A. McGee, Jr., *J. Chem. Phys.*, 46 (1967) 1444.

75  W. N. Lipscomb, *Boron Hydrides*, Benjamin, New York, 1963.

76  R. M. Adams, *Boron, Metallo-boron Compounds and Boranes*, Interscience, New York, 1964, Chap. 7.

77  L. Pauling, *The Nature of the Chemical Bond*, Cornell University Press, Ithaca, New York, 1960.

78  M. E. Garabedian and S. W. Benson, *J. Am. Chem. Soc.*, 86 (1964) 176.

79  R. E. McCoy and S. H. Bauer, *J. Am. Chem. Soc.*, 78 (1956) 2061.

80  S. H. Bauer, G. Herzberg and J. W. C. Johns, *J. Mol. Spectry.*, 13 (1964) 256.

81  R. C. Dunbar, *J. Am. Chem. Soc.*, 90 (1968) 5676.

82  W. N. Lipscomb, *Boron Hydrides*, Benjamin, New York, 1963, p. 56.

83  R. M. Adams, *Boron, Metallo-boron Compounds and Boranes*, Interscience, New York, 1964.

84  E. L. Muetterties, *The Chemistry of Boron and Its Compounds*, Wiley, New York, 1967.

85  A. B. Baylis, G. A. Pressley, Jr. and F. E. Stafford, *J. Am. Chem. Soc.*, 86 (1964) 5358.

86  A. B. Baylis, G. A. Pressley, Jr., M. E. Gordon and F. E. Stafford, *J. Am. Chem. Soc.*, 88 (1966) 929.

87  G. L. Brennan, *Dissertation*, Iowa State University, 1960.

88  T. P. Fehlner and W. S. Koski, *J. Am. Chem. Soc.*, 85 (1963) 1905.

89  E. G. Bloom, F. L. Mohler, C. E. Wise and E. J. Wells, *J. Res. Natl. Bur. Std.*, 43 (1949) 65.

90  J. E. Todd and W. S. Koski, *J. Am. Chem. Soc.*, 81 (1959) 2319.

91  C. A. Lutz and D. M. Ritter, *Can. J. Chem.*, 41 (1963) 1344.

92  T. P. Fehlner and W. S. Koski, *J. Am. Chem. Soc.*, 86 (1964) 2733.

93  K. S. Pitzer, *J. Am. Chem. Soc.*, 67 (1945) 1126.

94  S. H. Bauer, *Chem. Rev.*, 31 (1942) 43.

95  J. J. Kaufman, W. S. Koski, L. J. Kuhns and S. S. Wright, *J. Am. Chem. Soc.*, 85 (1963) 1369.

96  I. Shapiro and J. F. Ditter, *J. Chem. Phys.*, 26 (1957) 798.

97  G. R. Hertel and W. S. Koski, *J. Am. Chem. Soc.*, 87 (1965) 404.

98  L. H. Hall, V. V. Subbanna and W. S. Koski, *J. Am. Chem. Soc.*, 86 (1964) 3969.

99  I. Shapiro and H. Landesman, *J. Chem. Phys.*, 33 (1960) 1590.

100  J. L. Margrave, *J. Chem. Phys.*, 32 (1960) 1889.

101  A. B. Burg and H. I. Schlesinger, *J. Am. Chem. Soc.*, 55 (1933) 4009.

102  General Electric Co., *Rept. No. R49A0512*, March 10, 1949.

103  S. G. Gibbins and I. Shapiro, *J. Chem. Phys.*, 30 (1959) 1483.

104  W. V. Kotlensky and R. Schaeffer, *J. Am. Chem. Soc.*, 80 (1958) 4517.

105  S. G. Gibbins and I. Shapiro, *J. Am. Chem. Soc.*, 82 (1960) 2968.

106  D. F. Gaines and R. Schaeffer, *Inorg. Chem.*, 3 (1964) 438.

107  C. A. Lutz, D. A. Phillips and D. M. Ritter, *Inorg. Chem.*, 3 (1964) 1191.

108  T. P. Fehlner and W. S. Koski, *J. Am. Chem. Soc.*, 86 (1964) 1012.

109  R. W. Schaeffer, K. H. Ludlum and S. E. Wiberly, *J. Am. Chem. Soc.*, 81 (1959) 3157.

110  R. Schaeffer, *135th National Meeting American Chemical Society, Boston, Mass., April, 1959*.

111 J. F. Ditter, H. E. Landesman and R. E. Williams, *U.S. Dept. Comm. Office Tech. Serv.*, *AD275,784*, 1961.

112 J. F. Ditter, J. R. Spielman and R. E. Williams, *Inorg. Chem.*, 5 (1966) 118.

113 I. Shapiro and B. Keilin, *J. Am. Chem. Soc.*, 76 (1954) 3864.

114 R. E. Enrione, F. P. Boer and W. N. Lipscomb, *J. Am. Chem. Soc.*, 86 (1964) 1451.

115 R. E. Enrione, F. P. Boer and W. N. Lipscomb, *Inorg. Chem.*, 3 (1964) 1659.

116 J. Dobson, D. Gaines and R. Schaeffer, *J. Am. Chem. Soc.*, 87 (1965) 4072.

117 S. J. Steck, G. A. Pressley, Jr., F. E. Stafford, J. Dobson and R. Schaeffer, *Inorg. Chem.*, 8 (1969) 830.

118 F. J. Norton, *J. Am. Chem. Soc.*, 72 (1950) 1849.

119 R. E. Dickerson, P. J. Wheatley, P. A. Howell and W. N. Lipscomb, *J. Chem. Phys.*, 27 (1957) 200.

120 A. B. Burg and R. Kratzer, *Inorg. Chem.*, 1 (1962) 725.

121 R. Maruca, J. D. Odom and R. Schaeffer, *Inorg. Chem.*, 7 (1968) 412.

122 J. J. Kaufman, W. S. Koski, L. J. Kuhns and R. W. Law, *J. Am. Chem. Soc.*, 84 (1962) 4198.

123 J. Dobson, P. C. Keller and R. Schaeffer, *J. Am. Chem. Soc.*, 87 (1965) 3522; *Inorg. Chem.*, 7 (1968) 399.

124 I. Shapiro, M. Lustig and R. E. Williams, *J. Am. Chem. Soc.*, 81 (1959) 838.

125 R. N. Grimes, F. E. Wang, R. Lewin and W. N. Lipscomb, *Proc. Natl. Acad. Sci. U.S.*, 47 (1961) 996.

126 L. H. Hall and W. S. Koski, *J. Am. Chem. Soc.*, 84 (1962) 4205.

127 S. R. Gunn, L. G. Green and A. I. Von Egidy, *J. Phys. Chem.*, 63 (1959) 1787.

128 V. H. Dibeler and J. A. Walker, *Inorg. Chem.*, 8 (1969) 50.

129 S. R. Gunn and L. G. Green, *J. Phys. Chem.*, 65 (1961) 178.

130 V. H. Dibeler and S. K. Liston, *Inorg. Chem.*, 7 (1968) 1742.

131 S. R. Gunn and L. G. Green, *J. Phys. Chem.*, 65 (1961) 2173.

132 R. L. Middaugh, *Inorg. Chem.*, 7 (1968) 1011.

133 J. Plešek, S. Heřmánek and F. Hanousek, *Collection Czech. Chem. Commun.*, 33 (1968) 699.

134 N. E. Miller and E. L. Muetterties, *J. Am. Chem. Soc.*, 85 (1963) 3506.

135 W. H. Bauer and S. E. Wiberley, *Advan. Chem. Ser.*, 32 (1961) 115.

136 L. Barton, C. Perrin and R. F. Porter, *Inorg. Chem.*, 5 (1966) 1446.

137 J. K. Bragg, L. V. McCarty and F. J. Norton, *J. Am. Chem. Soc.*, 73 (1951) 2134.

138 C. O. Wilson, Jr. and I. Shapiro, *Anal. Chem.*, 32 (1960) 78.

139 H. G. Weiss, W. J. Lehmann and I. Shapiro, *J. Am. Chem. Soc.*, 84 (1962) 3840.

140 R. E. Enrione and R. Rosen, *Inorg. Chim. Acta*, 1 (1967) 169.

141 R. M. Reese, V. H. Dibeler and F. L. Mohler, *J. Res. Natl. Bur. Std.*, 57 (1956) 367.

142 D. F. Munro, J. E. Ahnell and W. S. Koski, *J. Phys. Chem.*, 72 (1968) 2682; *Am. Chem. Soc. Div. Fuel Chem., Prepr.*, 11(2) (1967) 75.

143 W. R. Deever and D. M. Ritter, *Inorg. Chem.*, 8 (1969) 2461.

144 G. Hortig, O. Müller, K. R. Schubert and E. Fluck, *Z. Naturforsch.*, 21b (1966) 609.

145 J. F. Ditter, F. J. Gerhart and R. E. Williams, *Advan. Chem. Ser.*, 72 (1968) 191.

146 I. Shapiro, B. Keilin, R. E. Williams and C. D. Good, *J. Am. Chem. Soc.*, 85 (1963) 3167.

147 H. V. Seklemian and R. E. Williams, *Inorg. Nucl. Chem. Letters*, 3 (1967) 289.

148 P. Breisacher and B. Siegel, *J. Am. Chem. Soc.*, 86 (1964) 5053.

149 I. Shapiro, C. D. Good and R. E. Williams, *J. Am. Chem. Soc.*, 84 (1962) 3837.

150 P. Breisacher and B. Siegel, *J. Am. Chem. Soc.*, 87 (1965) 4255.

151 W. S. Koski, J. J. Kaufman and C. F. Pachucki, *J. Am. Chem. Soc.*, 81 (1959) 1326.

152 J. Marriott and J. D. Craggs, *J. Electron. Control*, 3 (1957) 194.

153 O. Osberghaus, *Z. Physik*, 128 (1950) 366.

154 G. G. Devyatykh, V. G. Rachkov and I. L. Agafonov, *Russ. J. Inorg. Chem.*, 13 (1968) 1497.

155 M. F. Lappert, J. B. Pedley, P. N. K. Riley and A. Tweedale, *Chem. Commun.*, (1966) 788.

156 M. F. Lappert, M. R. Litzow, J. B. Pedley, P. N. K. Riley and A. Tweedale, *J. Chem. Soc., A*, (1968) 3105.

157 J. C. Lockhart, *Chem. Rev.*, 65 (1965) 131.

158 J. N. Gayles and J. Self, *J. Chem. Phys.*, 40 (1964) 3530.

159 V. H. Dibeler and S. K. Liston, *Inorg. Chem.*, 7 (1968) 1742.

160 J. H. Beynon, in *Mass Spectrometry*, R. I. Reed (Ed.), Academic Press, London, 1965, p. 359.

161 A. G. Massey and D. S. Urch, *Chem. Ind. (London)*, (1965) 607.

162 J. E. Dobson, P. M. Tucker, R. Schaeffer and F. G. A. Stone, *Chem. Commun.*, (1968) 452.

163 J. E. Dobson, P. M. Tucker, F. G. A. Stone and R. Schaeffer, *J. Chem. Soc. A*, (1969) 1882.

164 V. H. Dibeler and J. A. Walker, *Inorg. Chem.*, 8 (1969) 50.

165 P. L. Timms, *J. Am. Chem. Soc.*, 89 (1967) 1629.

166 P. L. Timms, *Inorg. Chem.*, 7 (1968) 387.

167 A. G. Massey and D. S. Urch, *Proc. Chem. Soc.*, (1964) 284.

168 J. D. McDonald, C. H. Williams, J. C. Thompson and J. L. Margrave, *Advan. Chem. Ser.*, 72 (1968) 261.

169 R. W. Kirk and P. L. Timms, *J. Am. Chem. Soc.*, 91 (1969) 6315.

170 P. L. Timms, *Chem. Commun.*, (1968) 1525.

171 T. Imanaka, *Bull. Chem. Soc. Japan*, 40 (1967) 2182.

172 J. Tanaka and S. R. Smith, *Inorg. Chem.*, 8 (1969) 265.

173 D. R. Lloyd, *Chemical Society Meeting, Durham, Sept. 1967, Abstr. EG3*.

174 R. J. Boyd and D. C. Frost, *Chem. Phys. Letters*, (1968) 649.

175 R. W. Law and J. L. Margrave, *J. Chem. Phys.*, 25 (1956) 1086.

176 H. Kreuzer, *Z. Naturforsch.*, 12a (1957) 519.

177 J. C. Baldwin, M. F. Lappert, J. B. Pedley, P. N. K. Riley and R. D. Sedgwick, *Inorg. Nucl. Chem. Letters*, 1 (1965) 57.

178 M. F. Lappert, M. R. Litzow, J. B. Pedley, P. N. K. Riley, T. R. Spalding and A. Tweedale, *J. Chem. Soc. A*, (1970) 2320.

179 G. Urry, in *The Chemistry of Boron and Its Compounds*, E. L. Muetterties (Ed.), Wiley, New York, 1967.

180 W. C. Price, T. R. Passmore and D. M. Roessler, *Discussions Faraday Soc.*, 35 (1963) 201.

181 J. L. Margrave, *J. Phys. Chem.*, 66 (1962) 1209.

182 W. C. Steele, L. D. Nichols and F. G. A. Stone, *J. Am. Chem. Soc.*, 84 (1962) 1154.

183 D. W. Robinson, *J. Mol. Spectry.*, 11 (1963) 275.
184 R. F. Barrow, *Trans. Faraday Soc.*, 56 (1960) 952.
185 K. A. G. MacNeil and J. C. J. Thynne, *Inorg. Nucl. Chem. Letters*, 5 (1969) 1009.
186 P. M. Kuznesof, D. F. Shriver and F. E. Stafford, *J. Am. Chem. Soc.*, 90 (1968) 2557.
187 A. Kaldor, I. Pines and R. F. Porter, *Inorg. Chem.*, 8 (1969) 1418.
188 R. W. Rudolph and R. W. Parry, *J. Am. Chem. Soc.*, 89 (1967) 1621.
189 K. W. Morse and R. W. Parry, *J. Am. Chem. Soc.*, 89 (1967) 172.
190 W. R. Deever, E. R. Lory and D. M. Ritter, *Inorg. Chem.*, 8 (1969) 1263.
191 G. L. Brubaker and S. G. Shore, *Inorg. Chem.*, 8 (1969) 2804.
192 K. Gosling, J. D. Smith and D. H. W. Wharmby, *J. Chem. Soc. A*, (1969) 1738.
193 E. Ehrlich, *Inorg. Chem.*, 9 (1970) 146.
194 V. A. Dorokhov and M. F. Lappert, *J. Chem. Soc. A*, (1969) 433.
195 J. R. Jennings, I. Pattison and K. Wade, *J. Chem. Soc. A*, (1969) 565.
196 A. Meller and H. Marecek, *Monatsch. Chem.*, 99 (1968) 1355.
197 I. Pattison and K. Wade, *J. Chem. Soc. A*, (1967) 1098.
198 K. Wade and B. K. Wyatt, *J. Chem. Soc. A*, (1967) 1339.
199 J. R. Jennings, I. Pattison, K. Wade and B. K. Wyatt, *J. Chem. Soc. A*, (1967) 1608.
200 J. Casanova, H. R. Kiefer, D. Kuwada and A. H. Boulton, *Tetrahedron Letters*, (1965) 703.
201 M. F. Lappert and M. K. Majumdar, *Advan. Chem. Ser.*, 42 (1964) 208.
202 J. Casanova, private communication cited in ref. 203.
203 D. S. Matteson, *Organometal. Chem. Rev. B*, 4 (1968) 290.
204 E. F. H. Brittain, J. B. Leach and J. H. Morris, *J. Chem. Soc. A*, (1968) 340.
205 E. F. H. Brittain. J. B. Leach and J. H. Morris, in R. Brymner and J. R. Penney (Eds.), *Mass Spectrometry, Proceedings of the Symposium on Mass Spectrometry, Enfield College of Technology, 6th July, 1967*, Butterworth, London, 1968, p. 195.
206 E. F. H. Brittain, J. B. Leach and J. H. Morris, *Org. Mass Spectrom.*, 1 (1968) 459.
207 A. Finch, J. B. Leach and J. H. Morris, *Organometal. Chem. Rev. A*, 4 (1969) 1.
208 K. Niedenzu and J. W. Dawson, in *The Chemistry of Boron and Its Compounds*, E. L. Muetterties (Ed.), Wiley, New York, 1967.
209 W. Snedden, *Advances in Mass Spectrometry*, Vol. 2, R. M. Elliott (Ed.), Pergamon Press, London, 1963, p. 456.
210 Mass spectral data, *American Petroleum Institute Research Project 44, Ser. No. 1346*, Carnegie Institute of Technology, Pittsburg, Pa.
211 Mass spectral data, *American Petroleum Institute Research Project 44, Ser. No. 1503*, Carnegie Institute of Techqology, Pittsburg, Pa.
212 G. H. Lee, II and R. F. Porter, *Inorg. Chem.*, 6 (1967) 648.
213 H. Budzikiewicz, C. Djerassi and D. H. Williams, *Mass Spectrometry of Organic Compounds*, Holden-Day, San Francisco, 1967.
214 E. D. Loughran, C. L. Mader and W. E. McQuistion, *U.S. At. Energy Comm. Rept. LA-2368*, 1960.
215 M. Nadler and R. F. Porter, *Inorg. Chem.*, 6 (1967) 1739.
216 R. Maruca, O. T. Beachley, Jr. and A. W. Laubengayer, *Inorg. Chem.*, 6 (1967) 575.
217 A. W. Laubengayer, K. Watterson, D. R. Bidinosti and R. F. Porter, *Inorg. Chem.*, 2 (1963) 519.
218 O. T. Beachley, Jr., *Inorg. Chem.*, 8 (1969) 981.

219 I. Pattison and K. Wade, *J. Chem. Soc. A*, (1968) 842.
220 A. W. Laubengayer, P. C. Moews, Jr. and R. F. Porter, *J. Am. Chem. Soc.*, 83 (1961) 1337.
221 G. Mamantov and J. L. Margrave, *J. Inorg. Nucl. Chem.*, 20 (1961) 348.
222 P. M. Kuznesof, F. E. Stafford and D. F. Shriver, *J. Phys. Chem.*, 71 (1967) 1939.
223 M. P. Brown, R. W. Heseltine and L. H. Sutcliffe, *J. Chem. Soc. A*, (1968) 612.
224 M. P. Brown, R. W. Heseltine and D. W. Johnson, *J. Chem. Soc. A*, (1967) 597.
225 R. E. Florin, L. A. Wall, F. L. Mohler and E. Quinn, *J. Am. Chem. Soc.*, 76 (1964) 3344.
226 H. Nöth and W. Regnet, *Chem. Ber.*, 102 (1969) 167.
227 R. C. Dougherty, *Tetrahedron*, 24 (1968) 6755.
228 J. C. Catlin and H. R. Snyder, *J. Org. Chem.*, 34 (1969) 1664.
229 R. W. Law and J. L. Margrave, *J. Chem. Phys.*, 25 (1956) 1086.
230 Y. Wada and R. W. Kiser, *J. Phys. Chem.*, 68 (1964) 1588.
231 P. J. Fallon, P. Kelly and J. C. Lockhart, *Intern. J. Mass Spectrom. Ion Phys.*, 1 (1968) 133.
232 P. J. Fallon and J. C. Lockhart, *Intern. J. Mass Spectrom. Ion Phys.*, 2 (1969) 247.
233 R. Heyes and J. C. Lockhart, *J. Chem. Soc. A*, (1968) 326.
234 R. C. Mehrotra, *J. Indian Chem. Soc.*, 30 (1953) 585.
235 H. Ulich and W. Nespital, *Z. Physik. Chem.*, A165 (1933) 294.
236 R. A. Robinson and D. A. Peak, *J. Phys. Chem.*, 39 (1935) 1125.
237 V. J. Shiner, Jr., D. Whittaker and V. P. Fernandez, *J. Am. Chem. Soc.*, 85 (1963) 2318.
238 R. C. Mehrotra, *J. Indian Chem. Soc.*, 31 (1954) 85.
239 L. W. Daasch and H. M. Fales, paper presented at *A.S.T.M. Committee E-14, 15th Annual Conference on Mass Spectrometry and Allied Topics, Denver, 14th–19th May 1967, No. 153*, p. 498; *Org. Mass Spectrom.*, 2 (1969) 1043.
240 W. Fieggen, H. Gerding and N. M. M. Nibbering, *Rec. Trav. Chim.*, 87 (1968) 377.
241 L. M. Brown and K. S. Mazdiyasni, *Anal. Chem.*, 41 (1969) 1243.
242 R. C. Wilhoit, *J. Phys. Chem.*, 61 (1957) 114.
243 D. L. Hildenbrand, L. P. Theard and A. M. Saul, *J. Chem. Phys.*, 39 (1963) 1973.
244 R. F. Porter, D. R. Bidinosti and K. F. Watterson, *J. Chem. Phys.*, 36 (1962) 2104.
245 W. P. Sholette and R. F. Porter, *J. Phys. Chem.*, 67 (1963) 177.
246 J. F. Ditter and I. Shapiro, *J. Am. Chem. Soc.*, 81 (1959) 1022.
247 G. F. Lanthier and W. A. G. Graham, *Can. J. Chem.*, 47 (1969) 569.
248 J. R. Jennings and K. Wade, *J. Chem. Soc. A*, (1967) 1333.
249 J. R. Jennings, K. Wade and B. K. Wyatt, *J. Chem. Soc. A*, (1968) 2535.
250 H. Nöth and P. Konrad, *Chem. Ber.*, 101 (1968) 3423.
251 J. G. Edwards, H. Wiedmeier and P. W. Gilles, *J. Am. Chem. Soc.*, 88 (1966) 2935.
252 J. W. Gilje, *Intern. J. Mass Spectrom. Ion Phys.*, (1 1968) 500.
253 R. W. Law and J. L. Margrave, *J. Chem. Phys.*, 25 (1956) 1086.
254 B. C. Tollin, R. Schaeffer and H. J. Svec, *J. Inorg. Nucl. Chem.*, 4 (1957) 273.
255 D. Henneberg, H. Damen and R. Köster, *Ann. Chem.*, 640 (1961) 52.
256 R. Schaeffer and L. J. Todd, *J. Am. Chem. Soc.*, 87 (1965) 488.
257 R. E. Winters and R. W. Kiser, *J. Organometal. Chem.*, 10 (1967) 7.
258 D. B. Chambers, F. Glockling and J. R. C. Light, *Quart. Rev.*, 22 (1968) 317.

259 D. B. Chambers, G. E. Coates, F. Glockling and M. Weston, *J. Chem. Soc. A*, (1969) 1712.
260 A. W. Laubengayer and W. F. Gilliam, *J. Am. Chem. Soc.*, 63 (1941) 477.
261 D. B. Chambers, G. E. Coates and F. Glockling, *Discussions Faraday Soc.*, 47 (1969) 157.
262 G. E. Coates, M. L. H. Green and K. Wade, *Organometallic Compounds*, Vol. 1, Methuen, London, 1967.
263 F. W. Aston, *Proc. Roy. Soc. (London), Ser. A*, 149 (1935) 396.
264 F. W. Aston, *Nature*, 128 (1931) 725.
265 J. C. Lockhart and P. Kelly, *Intern. J. Mass Spectrom. Ion Phys.*, 1 (1968) 209.
266 A. G. Lee, *Intern. J. Mass Spectrom. Ion Phys.*, 3 (1969) 239.
267 T. P. Onak, G. B. Dunks, J. R. Spielman, F. J. Gerhart and R. E. Williams, *J. Am. Chem. Soc.*, 88 (1966) 2061.

# Chapter 7

# The Main Group IV Elements

T. R. SPALDING

## 1. INTRODUCTION

A large number of compounds containing main Group IV elements have been studied mass spectrometrically. They have been popular subjects because they are relatively easy to prepare and handle and have convenient volatilities. Although mass spectrometry has been used to characterise entirely inorganic compounds such as main Group IV hydrides and halides, most of the detailed investigations have concerned organometallic derivatives of silicon, germanium, or tin.

The compounds discussed in this chapter are categorised according to the types of bonds between the Group IV element and the surrounding groups. Inorganic compounds are dealt with before organometallic derivatives.

Sufficient work has now been done to be able to make some generalisations about the spectra of main Group IV compounds of the type $ML_4$. Firstly, since the elements of Group IV are generally more electropositive than the ligands attached to them, the ion $(L-ML_n)^+$ will be more likely to dissociate to give $L\cdot + M^+L_n$ than $L^+ + \cdot ML_n$. Secondly, the decomposition of parent molecular (odd-electron) ions will probably occur to give an even-electron (metal-containing) ion by elimination of an odd-electron (radical) fragment. Thirdly, the ternary ion $L_3M^+$ in the mass spectra of $ML_4$ might be expected to be particularly stable because (i) it will have an even number of electrons, (ii) the electron configuration will be that of the corresponding (stable) Group III compound. The radical ions $(L_2M^+)$ might be expected to be unstable for the opposite reasons to (i) and (ii). Fourthly, comparing a series of compounds $L_4M(M = C, Si, Ge, Sn, Pb)$ one might expect that the $M^{II}$ oxidation state would become more important as one descended the series, i.e. $LM^+$ ions would become more important.

Lastly, molecules of the type $L_4M$ exhibit a low (usually 1–10%) parent molecular ion abundance because they are odd-electron ions which can eliminate a radical fairly easily to give the $L_3M^+$ ion.


## 2. THE ELEMENTS

Silicon, germanium, tin, and lead are polyisotopic elements. The naturally occurring isotope abundances[1] and the corresponding isotope patterns are

Table 7.1          .

Natural Abundances of Isotopes of Main Group IV Elements[1]

| Element | Mass no. | Abundance (%) |
|---------|----------|---------------|
| C       | 12       | 98.89         |
|         | 13       | 1.11          |
| Si      | 28       | 92.21         |
|         | 29       | 4.70          |
|         | 30       | 3.09          |
| Ge      | 70       | 20.53         |
|         | 72       | 27.43         |
|         | 73       | 7.76          |
|         | 74       | 36.54         |
|         | 76       | 7.76          |
| Sn      | 112      | 0.96          |
|         | 114      | 0.66          |
|         | 115      | 0.35          |
|         | 116      | 14.30         |
|         | 117      | 7.61          |
|         | 118      | 24.03         |
|         | 119      | 8.58          |
|         | 120      | 32.85         |
|         | 122      | 4.72          |
|         | 124      | 5.94          |
| Pb      | 204      | 1.48          |
|         | 206      | 23.6          |
|         | 207      | 22.6          |
|         | 208      | 52.3          |

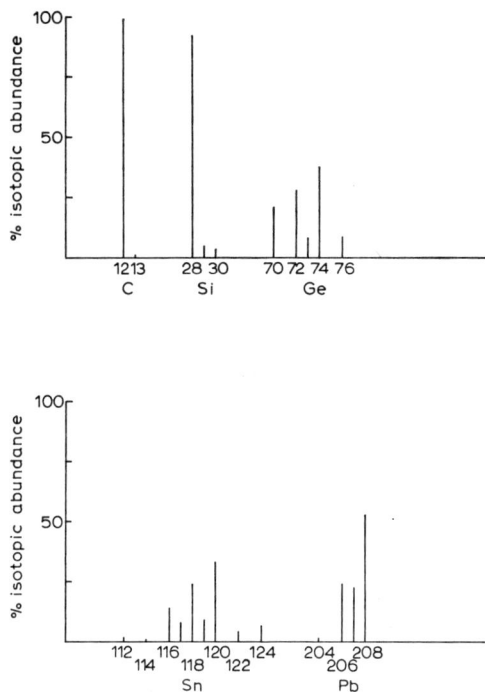

Fig. 7.1. Natural isotopic abundances of main Group IV elements.

given in Table 7.1 and Fig. 7.1. These values apply to all natural (terrestial) sources of silicon, germanium, and tin and their compounds and are generally the ones used in ion abundance calculations. In natural ores of lead, the isotope abundances vary quite considerably with the origin of the sample, consequently isotope abundances in any lead-containing compound have to be calculated from the mass spectrum of a suitable compound[2]. Both lead alkyls[3–5] and lead halides[5,6] have been used to this end. The values quoted above should be regarded as typical for lead from a commercial source *i.e.* a mixture of many naturally occurring ores.

Lead isotope analysis has been used in dating geological specimens[2,3,7] and numerous publications on the abundances of silicon, germanium, tin, and lead in terrestial and meteoric samples have appeared, (see, for example, ref. 8). Both of these aspects are outside the scope of this book and therefore will not be discussed further.

## 3. HIGH-TEMPERATURE STUDIES

Mass spectrometric studies at high temperatures have shown that the main Group IV elements vaporise to produce monatomic and polyatomic species. The degree of polymerisation, $n$, depends on the element. Parent ions containing up to seven atoms of silicon[9] or germanium[10], five atoms of tin[10], and two of lead[9] have been observed.

Ionisation potential studies[9,10,15,17] of the polyatomic species $M_n$ have shown that the first ionisation potential is the same within experimental error irrespective of the value of $n$. That is to say, parent molecular ions, $M_n^+$, are being observed, not ions resulting from fragmentation reactions. Using data from a combination of mass spectrometry and effusion techniques, Kant and Strauss[17] have calculated the heats of atomisation of the germanium species $M_n$ and the free energies of the gas phase reactions.

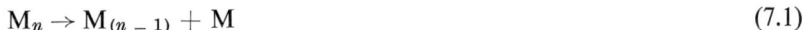

$$M_n \rightarrow M_{(n-1)} + M \tag{7.1}$$

They have tabulated bond dissociation energies for the $M-M_{(n-1)}$ bonds for up to $n = 4$ for silicon and tin, $n = 7$ for germanium, and $n = 6$ for carbon (Table 7.2). In all cases except carbon, the bond dissociation energy when $n$ equals two is smaller than when $n$ is less than two; when $n$ is greater than two the values are fairly constant. The trends for Si, Ge, and Sn are in sharp contrast to those in the analogous carbon systems where the carbon–carbon bond dissociation energies alternate with $n$, being greatest for $n = 3$ or 5 and least when $n = 4$. The structure of these molecules is still in doubt

Table 7.2

Bond Dissociation Energies $D(M_{(n-1)}-M)$ in Group IV Polyatomics[17]

| Molecule $M_n$ | $D(M_{(n-1)}-M)$ (kcal. mole$^{-1}$) for M = | | | |
|---|---|---|---|---|
| | C | Si | Ge | Sn |
| $M_2$ | 143 | 75 | 65 | 46 |
| $M_3$ | 179 | 101 | 84 | 74 |
| $M_4$ | 118 | 98 | 89 | 72 |
| $M_5$ | 170 | | 83 | |
| $M_6$ | <132 | | 86 | |
| $M_7$ | | | 75 | |

but calculations of the heats of atomisation of the polymeric germanium species based on the second and third laws of thermodynamics have indicated that the structures of the polymers were more likely to be linear than cyclic. However, this could not be proved unambiguously because of the size of the possible errors involved in the calculations. The molecule $C_3$ is known to be linear from spectroscopic studies[18]. Some molecular orbital calculations on $C_n$ molecules imply that linear structures are the important structures and under the conditions used[20] for a mass spectrometric study at 1900–2200 °C, $C_3$ and $C_5$ would be particularly stable. This was in agreement with the ion abundance measurements. However, to make assumptions on these grounds about the structure of the other Group IV polyatomics is dangerous, especially in view of the observed differences in bond dissociation energy trends.

In a study of the evaporation of silicon and germanium, Honig[9] has calculated values for the heat of sublimation of silicon, $105 \pm 12$ kcal.mole$^{-1}$ at 1300 °C, and germanium, $89 \pm 2$ kcal.mole$^{-1}$ at 900 °C, and reported the heat of evaporation of germanium as $79 \pm 2$ kcal.mole$^{-1}$ at 1150 °C. He was also able to measure the heat of sublimation of $Si_2$, $Si_3$, and $Si_4$ species and the heat of evaporation of $Ge_2$, $Ge_3$, and $Ge_4$. His result for the difference in the heats of sublimation and evaporation of germanium ($10 \pm 4$ kcal.mole$^{-1}$)[9] agreed well with 8.1 kcal.mole$^{-1}$ from a direct measurement[11] and satisfactorily with results from effusion techniques[12]. The result for silicon agreed within experimental error with previously published values (112 (ref. 13) and 95 (ref. 14)) from vapour pressure experiments and has been confirmed by Drowart et al.[15,16] who obtained[16] a value of $\Delta H°(sub) = 108.4 \pm 3$ kcal.mole$^{-1}$. Drowart et al.[21] studied the species in equilibrium with the hexagonal form of SiC and calculated the following heats of formation for the gaseous species, $\Delta H°_f Si = 125 \pm 3$; $\Delta H°_f Si_2 = 172 \pm 5$; $\Delta H°_f Si_2C = 168 \pm 5$ and $\Delta H°_f SiC_2 = 162 \pm 5$ kcal.mole$^{-1}$. They also observed ions corresponding to gaseous SiC, $Si_2C_2$, $Si_2C_3$, $Si_3C$, and $Si_3$ molecules, and calculated the heats of reactions involved in their formation. A similar study of Ge–C, Ge–Si, and Ge–Si–C systems was reported by Drowart et al.[15]. They observed a number of polyatomic molecules at temperatures between 1495–1650 °C, including the mixed species GeC, $Ge_2C$, $GeC_2$, $Ge_3C$, GeSi, $Ge_2Si$, $Ge_3Si$, GeSiC, and $Ge_2SiC$. Dissociation energies of the inter-Group IV molecules were calculated.

The oxides of main Group IV elements have been studied by several workers[22]. Porter et al.[23] studied the Si–SiO$_2$ system and observed that the major ions were due to $SiO^+$, $SiO_2^+$, and $Si_2O_2^+$ under nearly neutral oxidising conditions. They calculated the heat of sublimation of $SiO_2$, $\Delta H°$(sub) $= 136 \pm 8$ kcal.mole$^{-1}$. Hildenbrand and Murad[24] have studied the SiO molecule and reported its heat of formation $\Delta H°_f = -18.3 \pm 3$ kcal.mole$^{-1}$. A study of the oxides of germanium $GeO_2$ (hexagonal) and GeO (amorphous) was carried out by Drowart et al.[25]. From the amorphous oxide they observed polymeric species $(GeO)_n$ ($n = 2$ or 3) in the gas phase. The heat of sublimation of GeO (amorphous) was calculated to be $53.1 \pm 1$ kcal. mole$^{-1}$, and its heat of formation $\Delta H°_f(GeO) = -60.8 \pm 1.4$ kcal. mole$^{-1}$. Tin oxide and mixtures of tin and its oxide were studied by Colin et al.[26]. They observed the ion due to the SnO molecule and polymeric species $(SnO)_n$ ($n = 2$–4). A value of the heat of sublimation of SnO of $\Delta H°$(sub) $= 71.9$ kcal.mole$^{-1}$ was reported. Drowart et al.[27] reported a study of lead oxide (yellow, $\beta$-rhombic form). They observed ions due to species PbO, $(PbO)_n$ ($n = 2$–6) and calculated the heat of sublimation of PbO, $\Delta H°$(sub) $= 69.2 \pm 1.3$ kcal.mole$^{-1}$.

Properties such as metal–oxygen bond dissociation energies[28] and appearance and ionisation potentials[29] of oxide species have been listed for all the Group IV oxides (Table 7.3).

In contrast to the oxides, the main Group IV sulphides, selenides, and tellurides vaporise to give monomeric MX molecules as the predominant species. Several studies of such systems have provided quite precise values

Table 7.3

Bond Dissociation Energies and Ionisation Potentials of Group IV Oxides

| Element (M) | $D$(M–O) (kcal.mole$^{-1}$) (Refs. 28, 29) | Ionisation or appearance potential (eV) | | | | |
|---|---|---|---|---|---|---|
| | | MO | $MO_2$ | $M_2O_2$ | $M_3O_3$ | $M_4O_4$ |
| C | $256.2 \pm 0.15$ | $14.01 \pm 0.01$ | $13.78 \pm 0.01$ | | | |
| Si | $190.4 \pm 1.5$ | $10.8 \pm 0.5^a$ | $11.7 \pm 0.5^a$ | $10.1 \pm 1.0^a$ | | |
| Ge | $156.4 \pm 1.5^b$ | $10.1 \pm 0.8$ | | $8.7 \pm 1.0$ | $8.6 \pm 1.0$ | |
| Sn | $126.5 \pm 2^c$ | $10.5 \pm 0.5$ | | $9.8 \pm 1.0$ | $9.8 \pm 1.0$ | $9.2 \pm 1.0$ |
| Pb | $88.4 \pm 1.4^d$ | $9.6 \pm 0.5$ | | $9.8 \pm 0.5$ | | $8.5 \pm 0.7$ |

[a] Reference 23.
[b] Reference 25.
[c] Reference 26.
[d] Reference 27.

of heats of vaporisation, since the complication of the presence of polymeric species is not nearly as important as in the oxides.

Germanium monosulphide has been studied by Coppens *et al.*[28]. They calculated the bond dissociation energy $D(Ge-S) = 130.8 \pm 1.3$ kcal.mole$^{-1}$, comparable to $132.2 \pm 1.6$ kcal.mole$^{-1}$ derived from thermochemical data. Karbanov *et al.*[30] have reported the heats of formation of both the mono- and disulphides of germanium, $\Delta H°_f GeS = 27 \pm 10$ and $\Delta H°_f GeS_2 = 71 \pm 16$ kcal.mole$^{-1}$. The tin and lead monosulphides were studied by Colin and Drowart[31]. They calculated the bond dissociation energies $D(Sn-S) = 110.1 \pm 3.0$ and $D$ (Pb-S) $= 79.1 \pm 2.8$ kcal.mole$^{-1}$, in good agreement with previously published work. The mass spectrometric obser- vation of ions due to the species $SiS_2$ and $SiSe_2$ has been reported[39].

The heats of formation of germanium selenides have been reported[30] as $\Delta H°_f GeSe = 29 \pm 10$ and $\Delta H°_f GeSe_2 = 38 \pm 10$ kcal.mole$^{-1}$. Colin and Drowart[32] studied the vapour in equilibrium with solid SnSe and were able to calculate the heat of formation of SnSe as $\Delta H°_f = -21.5 \pm 1.7$ and the heat of dimerisation of SnSe going to $Sn_2Se_2$ as $\Delta H°(dim) = 46.5 \pm 5.0$ kcal.mole$^{-1}$. They reported a value for the bond dissociation energy $D(Sn-Se)$[32]. A study of PbSe by Porter[33] provided a value for the bond dissociation energy $D(Pb-Se)$ of $61.5 \pm 2.5$ kcal.mole$^{-1}$.

The tellurides of silicon, germanium, tin, and lead have all been studied mass spectrometrically .Exsteen *et al.*[34] observed that the vapour over $Si_2Te_3$ consisted mainly of $Te_2$, SiTe, and a small amount of $SiTe_2$. They were able to calculate the heat of formation of $Si_2Te_3$, $\Delta H°_f Si_2Te_3 = -18.3 \pm 4$, and the bond dissociation energy $D(Si-Te) = 103 \pm 2$ kcal.mole$^{-1}$. The germanium monotelluride has been studied by Colin and Drowart[35] and by Ban and Knox[36]. The former workers observed that the major species in the gas phase were $Te_2$, GeTe, and $GeTe_2$, with some evidence for the polymeric species $(GeTe)_n$ being formed in extremely small amounts. Ban and Knox, using a laser beam to vaporise or vaporise and ionise GeTe observed similar species, $GeTe_2$ being the most abundant, and polymeric species $(GeTe)_n$ with ($n = 2-4$). Values of the heat of formation of solid GeTe, $\Delta H°_f GeTe = -6.0 \pm 2.3$ and the bond dissociation energy $D(Ge-Te) = 92.5 \pm 2$ kcal.mole$^{-1}$ were obtained by Colin and Drowart[35]. The studies of solid SnTe (ref. 32) and PbTe (ref. 33) have provided values for the heat of forma- tion of SnTe, $\Delta H°_f SnTe = -14.6 \pm 1.3$ and the bond dissociation ener- gies $D(Sn-Te) = 79.9 \pm 1.5$ and $D(Pb-Te) = 51.4 \pm 2.0$ kcal.mole$^{-1}$. Sokolov *et al.*[37] observed ions due to $SnPbTe_2$ from mixtures of SnTe and

PbTe, and $PbTeSe_2$ and $PbTe_2Se$ from mixtures of PbTe and PbSe. Luybinov *et al.*[38] reported ions due to PbTe in the spectra of both *p*- and *n*-doped semiconductor-grade lead monotelluride.

Several systems with main Group IV–Group V elements have been studied. Zmbov and Margrave have reported the observation of the $Si_2N$ species in a silicon–nitrogen mixture[40]. They calculated the heat of formation $\Delta H^\circ_f Si_2N = 93 \pm 5$ kcal.mole$^{-1}$ and measured the ionisation potential ($9.4 \pm 0.3$ eV). Muenow and Margrave[41] have reported a value for the heat of formation of SiCN gas, $\Delta H^\circ_f SiCN = 90.3 \pm 4$ kcal.mole$^{-1}$ and ionisation potentials of $8.7 \pm 0.5$ eV and $7.4 \pm 0.5$ eV respectively for the species SiCN and OSiCN. The bond dissociation energy $D(Pb-Bi) = 33.0 \pm 3.5$ kcal.mole$^{-1}$ has been calculated from a study by Rovner *et al.*[42].

Apart from the thermochemical studies already described, mass spectrometers have been employed in the study of gas evolution from silica surfaces[43,44] and in an apparatus[45] designed to monitor the products of thermolytic reactions such as the evolution of nitrogen from heated $PbN_6$.

## 4. MAIN GROUP IV MONO- AND DIHALIDES

The characterisation of several halo compounds of Group IV with the metals in oxidation states I or II has been achieved mass spectrometrically. Margrave and his co-workers have been particularly active in this field. Ehlert and Margrave[46] identified $SiF_2$ and $\cdot SiF$ as products of the fluorination of silicon with calcium fluoride. The ionisation potentials of these species were $11.0 \pm 0.5$ eV and $7.5 \pm 0.4$ eV respectively. Later, a value of I.P. $SiF_2 = 11.29 \pm 0.1$ eV and the heats of formation of $SiF_2$, $\Delta H^\circ_f$-$SiF_2(g) = -136 \pm 5$ and of $\cdot SiF$, $\Delta H^\circ_f \cdot SiF(g) = -2 \pm 3$ kcal.mole$^{-1}$ and bond dissociation energies $D(SiF-F) = 153 \pm 5$ and $D(Si-F) = 130 \pm 3$ kcal.mole$^{-1}$ have been reported[47]. Margrave and coworkers also concluded that stable polymeric species of $SiF_2$ existed in the gas phase and these caused the observed discrepancy between their value for $\Delta H^\circ_f SiF_2$ and one obtained from high-pressure studies. This conclusion was verified by Kana'an and Margrave from transpiration studies[48]. Work on the germanium–calcium fluoride system provided values of ionisation potentials of $GeF_2$ and $\cdot GeF$ of $11.6 \pm 0.3$ eV and $7.8 \pm 0.4$ eV respectively[46]. Ehlert and Margrave were able to calculate the heats of formation, $\Delta H^\circ_f GeF_2(g) = -113 \pm 17$ and $\Delta H^\circ_f \cdot GeF(g) = -7.4 \pm 5$ kcal.mole$^{-1}$ and bond dissociation energies $D(\cdot\text{GeF–F}) = 115.0 \pm 18.4$ and $D(\text{Ge–F}) = 116.2 \pm 4$ kcal.mole$^{-1}$. The above values may be compared with $11.8 \pm 1$ eV for the ionisation potential of GeF$_2$ (ref. 49) and $\varDelta H^{\circ}{}_{\text{f}}\text{GeF}_2(\text{g}) = -99.5 \pm 1.50$ kcal.mole$^{-1}$ reported in later work by Margrave et al. From the reaction of germanium with germanium tetrafluoride at 360–400 °C, evidence for polymeric species (GeF$_2$)$_n$ ($n = 2$–4) was found[49] as well as the formation of GeF$_2$ and $\cdot$GeF. The ionisation potential of Ge$_2$F$_4$ was $10.6 \pm 0.3$ eV[49]. Both germanium dichloride and germanium dibromide have been observed[51] as products of the reaction between germanium and the corresponding tetrahalide at temperatures in the region of 400 °C, viz.

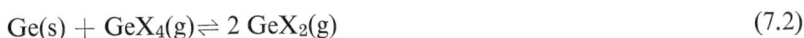

$$\text{Ge(s)} + \text{GeX}_4(\text{g}) \rightleftharpoons 2\ \text{GeX}_2(\text{g}) \tag{7.2}$$

The ionisation potentials of GeCl$_2$ and GeBr$_2$ were $10.4 \pm 0.3$ eV and $9.5 \pm 0.3$ eV respectively. Heats of formation $\varDelta H^{\circ}{}_{\text{f}}\text{GeCl}_2(\text{g}) = -42 \pm 1$ and $\varDelta H^{\circ}{}_{\text{f}}\text{GeBr}_2(\text{g}) = -13 \pm 1$ kcal.mole$^{-1}$ and bond dissociation energies $D(\text{Ge–Cl}) = 90$ and $D(\text{Ge–Br}) = 80$ kcal.mole$^{-1}$ were calculated. Corresponding values of $\varDelta H^{\circ}{}_{\text{f}}\text{GeI}_2 = 13 \pm 2$ (ref. 52a) and $D(\text{Ge–I}) = 70$ kcal.mole$^{-1}$ (ref. 52b) have been reported from studies using methods other than mass spectrometry.

The thermodynamics of the vaporisation of SnF$_2$ have been studied by Zmbov et al.[53]. They found evidence for polymeric species (SnF$_2$)$_n$ ($n = 2,3$) in the vapour phase and measured the ionisation potentials of these species. They also used the reaction between tin and tin difluoride at 300–360 °C to produce $\cdot$SnF and measured its ionisation potential. Their results were as follows: Sn$_3$F$_6 = 10.0 \pm 0.5$ eV, Sn$_2$F$_4 = 10.5 \pm 0.5$ eV, SnF$_2 = 11.5 \pm 0.2$ eV, and $\cdot$SnF $= 8.5 \pm 0.3$ eV. The bond dissociation energies $D(\text{Sn–F}) = 111.5 \pm 3$, and $D(\text{SnF–F}) = 80.5 \pm 7$ kcal.mole$^{-1}$ were reported and the average bond energy between –SnF$_2$– units was about $20 \pm 2$ kcal.mole$^{-1}$. From a study of tin dichloride–tin tetrachloride mixtures, Buchanan et al.[54] found the ionisation potential of SnCl$_2$ to be $10.1 \pm 0.4$ eV and calculated that of $\cdot$SnCl to be about 6.8 eV.

The trend in the importance of polymerisation of the Group IV difluorides which increases in going from silicon to tin is not continued at lead. Instead of polymerising, PbF$_2$ disproportionates[53] to give Pb and PbF$_4$. The thermodynamics of this system was studied by Zmbov et al. They reported the ionisation potentials of PbF$_4 = 10.4 \pm 0.3$ eV, PbF$_2 = 11.6 \pm 0.3$ eV, and $\cdot$PbF $= 7.5 \pm 0.3$ eV and calculated the heat of atomisation of PbF$_4$,

$\Delta H°$(atom) $PbF_4 = 382 \pm 2$ kcal.mole$^{-1}$, and the bond dissociation energy $D$(Pb–F) $= 85.0 \pm 2$ kcal.mole$^{-1}$.

An electron-impact study of $PbCl_2$, $PbBr_2$, and $PbBrCl$ has been reported by Hastie *et al.*[55]. A quadrupole mass spectrometer with which the investigators could observe both positive and negative ions was used. From their appearance potential data they concluded that they were observing both simple ion formation processes of type (7.3), and processes involving ion pair production, type (7.4).

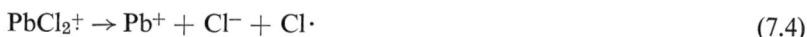

$$PbCl_2^{\dagger} \rightarrow PbCl^+ + Cl \cdot \tag{7.3}$$

$$PbCl_2^{\dagger} \rightarrow Pb^+ + Cl^- + Cl \cdot \tag{7.4}$$

They calculated the following ionisation potentials; $PbCl_2 = 10.3 \pm 0.1$ eV, $\cdot PbCl = 7.5 \pm 0.2$ eV, $\cdot PbBr_2 = 10.2 \pm 0.2$ eV, $\cdot PbBr = 7.8 \pm 0.2$ eV and $\cdot PbClBr = 10.4 \pm 0.2$ eV and electron affinities; $\cdot PbCl = 1.0 \pm 0.2$ eV, $\cdot PbBr = 0.9 \pm 0.2$ eV. The mixed halide $PbClBr$ was produced[56] in the vapour of a mixture of $PbCl_2$ and $PbBr_2$ at 420–495 °C. This reaction was shown to be thermoneutral with $\Delta H_{(react)} = 1 \pm 3$ kcal.mole$^{-1}$. The heat of vaporisation $\Delta H_{(vap)}PbClBr(g) = 31 \pm 2$ kcal.mole$^{-1}$ was calculated[56]. A study of the thermal and photolytic decomposition of $PbCl_2$ has been published[57]. At or above 225 °C, the compound decomposes thermally to give $Cl_2(g)$, Cl radicals and Pb(s). The same products are observed from the photolytic decomposition carried out between 25–225 °C.

The lead halides $PbBr_2$, $PbI_2$ have been studied at high temperature in a mass spectrometer in an investigation of the recombination rates of gaseous ions produced by photolysis[58]. The reactions studied were

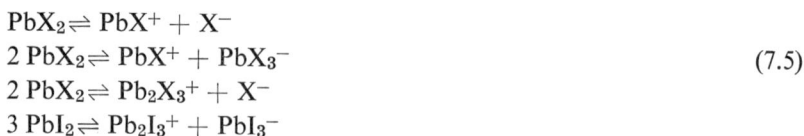

$$
\begin{aligned}
PbX_2 &\rightleftharpoons PbX^+ + X^- \\
2\,PbX_2 &\rightleftharpoons PbX^+ + PbX_3^- \\
2\,PbX_2 &\rightleftharpoons Pb_2X_3^+ + X^- \\
3\,PbI_2 &\rightleftharpoons Pb_2I_3^+ + PbI_3^-
\end{aligned}
\tag{7.5}
$$

for X = Br or I, and were carried out in xenon and argon.

Vapour-phase equilibria (7.6) have been studied and some enthalpies and entropies of reaction calculated.

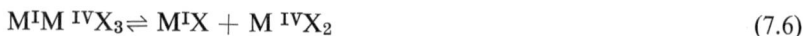

$$M^IM^{IV}X_3 \rightleftharpoons M^IX + M^{IV}X_2 \tag{7.6}$$

where

$M^I = $ Na, K; $M^{IV} = $ Sn; X $= $ F (ref. 60)
$M^I = $ Cs, Rb; $M^{IV} = $ Pb; X $= $ Cl (ref. 61)
$M^I = $ K; $M^{IV} = $ Pb; X $= $ Br (ref. 59)

In the study[60] of $M^ISnF_3$, ions derived from polymeric species were observed. This was in contrast to the behaviour of $M^IPbCl_3$ (ref. 61) and $KPbBr_3$ (ref. 59) which do not polymerise in the gas phase but are in simple equilibrium with the corresponding Group I and IV halides.

## 5. MAIN GROUP IV HYDRIDES

### 5.1 Hydrides of the type $MH_4$

The mass spectra of the Group IV hydrides are well documented. Silane and germane were first reported[62] in 1954, but it appears that incorrect mass assignments were made and the reported ion abundances are wrong[63]. This was pointed out by Saalfeld and Svec when they published their work on the $MH_4$ molecules M $=$ Si, Ge, Pb (ref. 64) and Sn (refs. 63, 64) (Table 7.4). Saalfeld and Svec's results for the ion abundances of singly charged ions have been satisfactorily confirmed by other workers for silane[65,66], ger-

Table 7.4

Ion Abundances from $MH_4$ Compounds

| Composition of ion | Singly charged | | | | Doubly charged Sn[b] |
|---|---|---|---|---|---|
| | M = Si[a] | Ge[a] | Sn[b] | Pb[a] | |
| $MH_4$ | 0.45 | 0.2 | 0.04 | Trace | |
| $MH_3$ | 35.6 | 34.7 | 37.6 | 31.8 | 1.6 |
| $MH_2$ | 45.4 | 37.0 | 20.2 | 52.5 | 2.1 |
| MH | 9.2 | 10.5 | 8.1 | 2.6 | 3.8 |
| M | 9.4 | 17.6 | 26.6 | 13.1 | Trace |

[a] Reference 64.
[b] Reference 63.

mane[67-70], and stannane[71]. The results for the doubly charged ions in the mass spectrum of stannane have been questioned on the grounds of wrong mass assignment[71]. The reported ion abundances from plumbane[64] must be viewed with reservations because of the difficulty in handling the compound, and comparison with the results from the other compounds is difficult because the measurements were made under different experimental conditions.

There is considerable disagreement in the reported values of appearance potentials of ions from $MH_4$ molecules. Whilst the differences are in some part due to the different machines and methods used, the disagreements

Table 7.5

Appearance Potentials for Ions from $MH_4$ Compounds and Ionic Bond Dissociation Energies $D(H_nM-H)^+$

| Composition of ion | Appearance potential (eV) for M = | | | | |
|---|---|---|---|---|---|
| | Si[a] | Si | Ge | Sn | Pb |
| $MH_4$ | | 11.4[b] | 10.5[b] | 9.2[b] | 9.1[b] |
| $MH_3$ | 12.3±0.03 | 11.8±0.2 | 10.8±0.3 | 9.4±0.3 | 9.6±? |
| $MH_2$ | 11.9±0.02 | 12.1±0.2 | 11.8±0.2 | 9.5±0.3 | |

[a] Ref. 66.
[b] Estimated values of Saalfeld and Svec[64].
Processes forming ions from silane, germane, and stannane were postulated to be

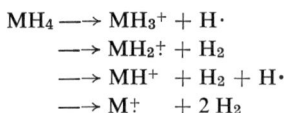

$MH_4 \longrightarrow MH_3^+ + H\cdot$
$\longrightarrow MH_2^{+} + H_2$
$\longrightarrow MH^+ + H_2 + H\cdot$
$\longrightarrow M^{+} + 2 H_2$

Other processes were suggested for $PbH_4$. The appearance potential of $Pb^+$ (11.2 eV) was reported and this ion was postulated to be formed *via*
$PbH_4 \longrightarrow Pb^+ + H_2 + 2 H\cdot$

| Bond | $D(H_nM-H)^+$ (kcal.mole$^{-1}$) for M = | | | |
|---|---|---|---|---|
| | Si | Ge | Sn | Pb |
| $(H_3M-H)^{+}$ | 9.2 | 6.9 | 4.6 | 11.5 |
| $(H_2M-H)^+$ | 108.2 | 105.9 | 103.6 | 11.5 |
| $(HM-H)^{+}$ | 89.8 | 32.2 | 20.7 | 23.1 |
| $(M-H)^+$ | 9.2 | 34.5 | 50.6 | |

cannot be explained entirely by experimental conditions. We shall consider here the results of Saalfeld and Svec for silane, germane, and stannane[64] since they were obtained on one machine by the same method, and the results of Potzinger and Lampe[66] for silane (Table 7.5). The former workers modified the ion source of their machine and were able to show that the ions from silane, germane, and stannane contained negligible excess kinetic energy. They claimed to identify the following processes

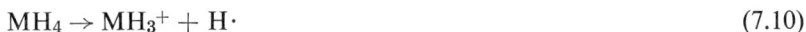

$$MH_4 \rightarrow M^+ + 2\,H_2 \qquad\qquad\qquad (7.7)$$

$$MH_4 \rightarrow MH^+ + H_2 + H\cdot \qquad\qquad\qquad (7.8)$$

$$MH_4 \rightarrow MH_2^+ + H_2 \qquad\qquad\qquad (7.9)$$

$$MH_4 \rightarrow MH_3^+ + H\cdot \qquad\qquad\qquad (7.10)$$

Using published values for the heats of formation of hydrogen and $M^+(g)$, Saalfeld and Svec calculated the heats of formation of the $MH_4$ compounds (7.7) from the equation

$$\text{A.P.}_{M^+ \text{ from } MH_4} = \Delta H°_f M^+(g) + 2\Delta H°_f H_2(g) - \Delta H°_f MH_4(g) \quad (7.11)$$

Their values were $\Delta H°_f(SiH_4)g = 7.8 \pm 4$, $\Delta H°_f(GeH_4)(g) = 20.8 \pm 4$, $\Delta H°_f(SnH_4)g = 35.0 \pm 4$ and $\Delta H°_f(PbH_4)(g) = 59.7 \pm 4$ kcal.mole⁻¹. The results for silane, germane, and stannane are in good agreement with the thermochemically determined values of $\Delta H°_f SiH_4 = 7.3$, $\Delta H°_f GeH_4 = 21.6$ and $\Delta H°_f SnH_4 = 38.9$ kcal.mole⁻¹, respectively of Gunn and Green[72]. The work of Potzinger and Lampe reported the same reactions ((7.7)–(7.10)) for silane but their value for the appearance potential of $SiH_2^+$ was the only one in reasonable agreement with Saalfeld and Svec (Table 7.5).

Successive ionic bond dissociation energies $D(H_nM–H)^+$, $(n = 0–3)$ were reported by Saalfeld and Svec (Table 7.5) and they may be interpreted as reflecting the general trends in ion abundances for each compound.

Pullen et al.[73] have reported the vertical and adiabatic ionisation potentials of $SiH_4$ and $GeH_4$ determined with a photoelectron spectrometer. The values were, for $SiH_4$, 11.66 eV (adiabatic), 12.36 and 12.85 eV (vertical) and for $GeH_4$ 11.31 eV (adiabatic), 11.98 and 12.46 eV (vertical). The study of the pyrolysis of $SiH_4$ and mixtures of $SiH_4/SiD_4$, $SiD_4/H_2$ by Ring et al.[74] lead to a bond dissociation energy $D(H_3Si–H)$ of about 94 kcal.mole⁻¹.

Table 7.6

Ion Abundance from the Negative Ion Spectrum of $SiH_4$ and Appearance Potentials

| Composition of ion | % Abundance at | | | Appearance potential (eV) |
|---|---|---|---|---|
| | 20 eV | 50 eV | 70 eV | |
| $SiH_3$ | 56.5 | 37.5 | 39.8 | 6.7 |
| $SiH_2$ | 18.7 | 14.3 | 13.2 | 7.7 |
| $SiH$ | 22.1 | 28.5 | 25.5 | 7.7 |
| $Si$ | 2.7 | 19.6 | 21.5 | 2.5 |

Potzinger and Lampe[66] have reported the negative ion spectrum of $SiH_4$ at 20, 50, and 70 eV and suggested reaction schemes. Their results are given in Table 7.6 together with the appearance potentials of the fragment ions. At all electron voltages the most abundant ion was $SiH_3^-$ and $SiH^-$ was the second most abundant.

### 5.2 Hydrides of the type $H_{(2n + 2)}M_n$ (n = 2 or more)

The mass spectra of the hydrides $Si_2H_6$, $Ge_2H_6$, $Ge_3H_8$, $Sn_2H_6$ (ref. 75), and $H_3SiGeH_3$ (ref. 76) have been reported by Saalfeld and Svec. Their ion abundances for $Si_2$-containing ions from disilane agree well with other published work but those for Si-containing ions do not[66,77]. The ion abundances from digermane are in agreement with those reported in a previous study[68]. The ion abundances from $M_2H_6$ compounds reported by Saalfeld and Svec and from $Si_2H_6$ reported by Potzinger and Lampe are shown in Table 7.7 together with the appearance potentials of the ions from $Si_2H_6$.

As in the work on $MH_4$ compounds, there is a great deal of disagreement in the reported values of appearance potentials. It is extremely difficult to draw satisfactory conclusions from these values. For instance, Saalfeld and Svec have calculated the heats of formation $\Delta H°_fSi_2H_6 = 15.1 \pm 4$, $\Delta H°_f$-$Ge_2H_6 = 39.1 \pm 4$, and $\Delta H°_fSn_2H_6 = 65.6 \pm 4$ kcal.mole$^{-1}$. The values for disilane and digermane are in excellent agreement with the thermochemically determined values of 17.1 and 38.7 kcal.mole$^{-1}$ respectively[76]. However, later work on $H_3Si–GeH_3$ gave results in disagreement by $\sim$20 kcal.mole$^{-1}$, i.e. $\sim$1 eV. There is no obvious reason why there should be such disagreement. Saalfeld and Svec[76] and Gunn and Kindsvater[78] used the same

Table 7.7

Ion Abundances from $M_2H_6$ Compounds (M = Si, Ge, Sn) and Appearance Potentials for Ions from $Si_2H_6$

| Composition of ion | % Abundances (appearance potential eV) for M = | | | |
|---|---|---|---|---|
| | $Si^{75}$ | $Si^{66}$ | $Ge^{75}$ | $Sn^{75}$ |
| $M_2H_6$ | 14.5 (10.6) | 13.2 (10.15±0.1) | 8.8 | 0.8 |
| $M_2H_5$ | 5.7 | 7.8 (11.4 ±0.1) | 5.6 | 1.6 |
| $M_2H_4$ | 31.7 | 25.7 (10.85±0.1) | 14.3 | 2.0 |
| $M_2H_3$ | 5.8 | 5.3 (12.5 ±0.1) | 6.0 | 10.5 |
| $M_2H_4$ | 21.8 | 16.5 (11.8 ±0.1) | 19.1 | 52.2 |
| $M_2H$ | 11.3 | 10.6 (12.9 ±0.2) | 11.8 | 15.6 |
| $M_2$ | 7.7 (12.2) | 7.6 (13.0 ±?) | 12.6 | 14.9 |
| $MH_3$ | 0.2 | 4.8 (11.95±0.15) | 5.3 | 0.1 |
| $MH_2$ | 0.3 | 2.5 (11.95±0.1) | 5.7 | 0.8 |
| $MH$ | 0.2 | 6.0 | 3.2 | 1.0 |
| $M$ | 0.6 (15.2) | | 7.0 | 0.5 |
| $MH_4^*$ | 0.1 | | 0.8 | Trace |

* The production of $MH_4$ was observed as a secondary process[64]. The intensity of the corresponding $MH_4^+$ ion was proportional to the square of the pressure in the source.

techniques in each of their studies for disilane and digermane. The results of Gunn and Kindsvater are compatible with the physical properties of $H_3Si-GeH_3$, being intermediate between $Si_2H_6$ and $Ge_2H_6$, but Saalfeld and Svec's results are not. We are therefore inclined to take the former's value of heat of formation namely $\Delta H°_f(H_3Si-GeH_3) = 27.8$ kcal.mole$^{-1}$.

From their study of silane and disilane, Potzinger and Lampe[66] calculated the heat of formation of $\cdot SiH_3$ as 51 kcal.mole$^{-1}$ and the ionisation potential of this radical as 8.16 eV. These results are in excellent agreement with the work of Steele et al.[79] who obtained values of $\Delta H°_f SiH_3 \cdot (g) = 49.5 \pm 2$ kcal.mole$^{-1}$, the ionisation potential of the $SiH_3$ radical (= 8.32 ± 0.15 eV) and $D(H_3Si-SiH_3) = 81.3 \pm 4$ kcal.mole$^{-1}$. The agreement between the values of $D(H_3Si-SiH_3)$ reported by Steele et al. and Saalfeld and Svec[75,80] (83.7) is apparently fortuitous since they used values of the ionisation potential of $\cdot SiH_3$ and the appearance potential of the $H_3Si^+$ ion from $Si_2H_6$ differing by 0.6–0.7 eV. The pyrolysis studies reported by Ring et al.[74] support a bond dissociation energy $D(H_3Si-SiH_3)$ of about 81 kcal.mole$^{-1}$.

Similar work to that described above has been carried out by Saalfeld

Table 7.8

Other Hydride Compounds of Group IV Elements Studied Mass Spectrometrically[a]

| Compound | Comments | Ref. |
|---|---|---|
| $D_6Si_2$ | Mass spectrometric characterisation of | 82 |
| $D_5Si_2H$ | these deutero hydrides produced by | |
| $D_2HSiSiHD_2$ | electrical discharge in a mixture of | |
| $D_2HSiSiH_2D$ | $SiH_4/SiD_4$. | |
| $D_2HSiSiH_3$ | | |
| $DSi_2H_5$ and $Si_2H_6$ | | |
| $D_6Si_2$ | P.m.i. and expected ions | 83 |
| $H_8Si_3$ | P.m.i., base peak $Si^+$ | 84, 77 |
| $H_{10}Si_4$ | P.m.i. | 312 |
| $H_3SiSiH_2GeH_3$ | P.m.i., $H_4SiGe^+$ base peak | 85 |
| $H_3SiGeH_2GeH_3$ | P.m.i., $H_2Ge_2^+$ base peak | 85 |
| $H_3SiGeH_2SiH_3$ | P.m.i. | 85 |
| $H_8Ge_3$ | P.m.i. | 312 |
| $H_8Si_2Ge$ | | 312 |
| $H_8SiGe_2$ | | 312 |
| $H_3GeNH_2$ | | 86 |
| $H_3GePH_2$ | | 86 |
| $H_3GeAsH_2$ | | 86 |
| $H_2Ge(NH_2)_2$ | | 86 |
| $(H_3Si)_2NH$ | P.m.i., $SiH_2^+$ base peak | 87 |
| $H_3SiNR_2$ | R = Me | 88 |
| $H_3SiNHPh$ | P.m.i. | 89 |
| $H_2Si(PH_2)_2$ | P.m.i.s; base peaks due to | 99 |
| $HSi(PH_2)_3$ | (p.m.i. − ·$PH_2$)$^+$ ion | 99 |
| $(H_3Ge)_3P$ | P.m.i. | 91 |
| $H_3SiOPh$ | P.m.i. | 92 |
| $H_3GeOPh$ | P.m.i. | 92 |
| $H_3GeOCH_3$ | P.m.i. | 93 |
| $H_3GeSCH_3$ | P.m.i. | 94 |
| $(H_3Ge)_2O$ | | 95 |
| $(H_3Ge)_2CN_2$ | | 95 |
| $H_3MSH$ | M = Si or Ge | 312 |
| $H_3MSMe$ | | 312 |
| $(H_3M)_2S$ | | 312 |

[a] P.m.i. = parent molecular ion.

and Svec[80] on $H_3Si-PH_2$, and by Saalfeld and McDowell[81] on $H_3Si-AsH_2$. These compounds showed molecular ions and all the expected ions due to H· losses. The base peak in the spectrum of $H_5SiP$ was due to the $HSiP^+$ ion and from $H_5SiAs$, the $H_3SiAs^+$ ion was the most abundant. Saalfeld and Svec reported the ionisation potential of $H_3SiPH_2$ as 10.0 eV and calculated $\Delta H°_f(H_5SiP)(g) = 1.9$, $\Delta H°_f(H_3Si·)g = 50.4$ and $D(H_3Si-PH_2) = 88.3$ kcal.mole$^{-1}$. The ionisation potential of $H_3SiAsH_2$ was 10.1 ± 0.1 eV and its heat of formation, $\Delta H°_f(H_5SiAs)(g) = 37$ kcal.mole$^{-1}$. The bond dissociation energy $D(H_3Si-AsH_2) = 73$ kcal.mole$^{-1}$ was calculated.

Other hydride compounds in this category for which mass spectral data has been reported are given in Table 7.8.

*5.3 Kinetic studies involving silane*

For a long time ion–molecule reactions of the types (7.12) and (7.13) have been of interest to kineticists and theoretical chemists.

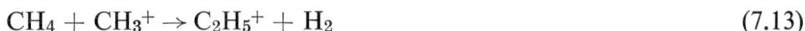

$$CH_4 + CH_4^+ \rightarrow CH_5^+ + ·CH_3 \qquad (7.12)$$

$$CH_4 + CH_3^+ \rightarrow C_2H_5^+ + H_2 \qquad (7.13)$$

Several workers have now studied ion–molecule reactions involving silane. Hess and Lampe[96] showed that the important ionic reactant in ionised silane is $SiH_2^+$. They studied the reaction of this ion (and its deuterated analogue) with silane (or deuterosilane) at pressures of between $4 \times 10^{-3}$ and $12 \times 10^{-3}$ torr. They demonstrated that the major reactions were (7.14)–(7.19) and calculated rate constants for these reactions.

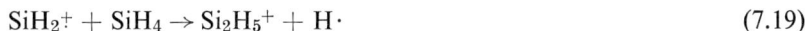

$$SiH_2^+ + SiH_4 \rightarrow SiH_3^+ + ·SiH_3 \qquad (7.14)$$

$$SiH_2^+ + SiH_4 \rightarrow Si_2H^+ + H· + 2 H_2 \qquad (7.15)$$

$$SiH_2^+ + SiH_4 \rightarrow Si_2H_2^+ + 2 H_2 \qquad (7.16)$$

$$SiH_2^+ + SiH_4 \rightarrow Si_2H_3^+ + H· + H_2 \qquad (7.17)$$

$$SiH_2^+ + SiH_4 \rightarrow Si_2H_4^+ + H_2 \qquad (7.18)$$

$$SiH_2^+ + SiH_4 \rightarrow Si_2H_5^+ + H· \qquad (7.19)$$

Interestingly, the $SiH_5^+$ ion was not observed as the product of primary ion reaction, but in later work Beggs and Lampe[97] observed $SiH_5^+$ produced in a secondary ion reaction (7.20) in mixtures of $CH_4$ and $SiH_4$.

$$CH_4^{\dot+} + SiH_4 + M \rightarrow SiH_5^+ + \cdot CH_3 + M \qquad (7.20)$$

or

$$CH_4^{\dot+} + SiD_4 + M \rightarrow SiD_4H^+ + \cdot CH_3 + M$$

where $M = CH_4$ or $SiH_4$. Beggs and Lampe[98] also studied the ion–molecule reactions in mixtures of $SiH_4$–$CH_4$ and $SiH_4$–$C_6H_6$. The major reacting ions in the methane–silane system were $CH_4^{\dot+}$, $CH_3^+$, $SiH_2^{\dot+}$, and $SiH_3^+$ and the major reactions were (7.21) giving the ionic products shown.

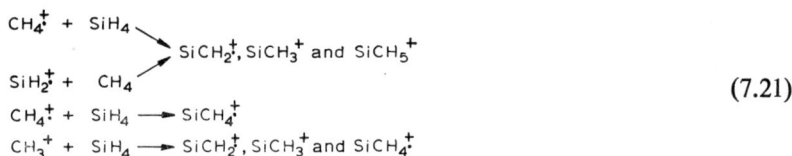

$$
\begin{aligned}
&\left.\begin{array}{l} CH_4^{\dot+} + SiH_4 \\ \\ SiH_2^{\dot+} + CH_4 \end{array}\right\} \rightarrow SiCH_2^{\dot+}, SiCH_3^+ \text{ and } SiCH_5^+ \\[6pt]
&CH_4^{\dot+} + SiH_4 \longrightarrow SiCH_4^{\dot+} \\
&CH_3^+ + SiH_4 \longrightarrow SiCH_2^{\dot+}, SiCH_3^+ \text{ and } SiCH_4^{\dot+}
\end{aligned}
\qquad (7.21)
$$

The ionic products of a reaction between $SiH_3^+$ and $CH_4$ could not be ascertained but were energetically limited to $SiCH_3^+$ and $SiCH_5^+$. The reaction between $CH_4^{\dot+}$ and $SiH_4$ was by far the fastest and in addition to the products above, an observable amount of $SiH_5^+$ ions was produced. In the benzene–silane mixtures, the $C_6H_6^{\dot+}$ ion was the only reactant ion of importance. The ionic products were $SiC_6H_5^+$, $SiC_6H_6^{\dot+}$, $SiC_6H_7^+$, $SiC_6H_8^{\dot+}$, and $SiC_6H_9^+$. Beggs and Lampe calculated rate constants for the various reactions observed. These same workers have also studied the ion–molecule reaction in mixtures of acetylene–silane[99,100] and ethylene–silane[100]. The principal reactions observed were (7.22) and (7.23).

$$
\begin{aligned}
&\left.\begin{array}{l} C_2H_2^{\dot+} + SiH_4 \\ \\ SiH_2^{\dot+} + C_2H_2 \end{array}\right\} \rightarrow SiCH_3^+ \\[6pt]
&SiH_2^{\dot+} + C_2H_2 \longrightarrow SiC_2H_3^+ + H\cdot \\
&C_2H_2^{\dot+} + SiH_4 \longrightarrow SiH_2^{\dot+} + C_2H_4
\end{aligned}
\qquad (7.22)
$$

$$
\begin{aligned}
&C_2H_4^{\dot+} + SiH_4 \longrightarrow SiH_2^{\dot+} + C_2H_6 \\
&SiH_2^{\dot+} + C_2H_4 \longrightarrow SiCH_3^+ + \cdot CH_3
\end{aligned}
\qquad (7.23)
$$

## 5.4 Analytical studies involving hydrides

Schmidt[101] studied the radiolysis reactions of silane and mixtures of silane and ethylene using mass spectrometry qualitatively to analyse the products which included disilane from both reactions. The products of the reaction between methyl radicals and silane, namely methane and the silyl radical, have been reported[102]. Mass spectrometry has also been used in the study of the $^3P_1$ Hg photosensitised decomposition of silane[103] and germane[104] and the radiation-induced addition of digermane to ethylene[105]. However, these applications will not be discussed here.

Meinrenken has studied the natural abundances of silicon isotopes from samples of silane[106]. The analysis of mixtures of silane and germane, arsine, phosphine, diethyl ether, ethylene[107], or silane and methane[108] by mass spectrometry has been reported. Similarly, mixtures of germane and phosphine, diethyl ether, ethane, ethylene[107], or germane and diethyl ether[108] have been quantitatively analysed.

### 6. MAIN GROUP IV HALIDES

## 6.1 Halides of the type $MX_4$ and $M_nX_{(2n + 2)}$

Halides of Group IV elements have received considerable attention, particularly those of silicon. The positive ion spectra of $SiF_4$, $SiCl_4$, $SiBr_4$, $SiI_4$ (ref. 109), $GeCl_4$, and $SnCl_4$ (ref. 110) have been reported in detail (Table 7.9). Previously reported spectra for $SiF_4$ (refs. 111–113) were in good agreement with Svec and Sparrow's results[109]. However, the agreement between published spectra of $SiCl_4$ was not so good. Svec and Sparrow suggested that, in some cases, reaction between $SiCl_4$ and water in the mass spectrometer had occurred and the resulting spectrum contained peaks not directly derived from $SiCl_4$. Their spectrum was in quite good agreement with that obtained by Agafonov et al.[116] but not such good agreement with the work of Devyatykh et al.[110] and Osberghaus[117]. The two published monoiso-topic spectra of $GeCl_4$ are in reasonable agreement, and agree with the polyisotopic spectrum reported by Larin et al.[118].

In all cases the parent molecular ions are observed and $MX_3^+$ ions produce the base peaks. The expected $MX_n^+$ ions ($n = 0$–3) are observed in all spectra; so too are the $X_2^+$ and $X^+$ ions (excepting $GeCl_4$ which only exhibits the $Cl^+$ ion). The % ion abundances in the series $MCl_4$ (M = Si, $Ge_2$ or Sn) reflect to some extent the decrease in M–Cl bond strengths as the atomic number of M increases. The interpretation of the series of silicon compounds in terms of decreasing Si–X bond strengths is less obvious. It does appear, however, that in the spectrum of $SiF_4$, the ion $SiF_3^+$ is particularly stable and the increased abundances of the parent molecular ions for X = Cl, Br, or

Table 7.9

Ion Abundances from Group IV Halides, $MX_4$

| Composition of ion | % Abundance | | | | | |
|---|---|---|---|---|---|---|
| | $SiF_4$ | $SiCl_4$ | SiBr | $SiI_4$ | $GeCl_4$ | $SnCl_4$ |
| *Singly charged* | | | | | | |
| $MX_4$ | 1.9 | 26.3 | 29.9 | 28.7 | 9.7 | 6.0 |
| $MX_3$ | 88.3 | 43.4 | 42.7 | 46.8 | 56.1 | 51.8 |
| $MX_2$ | 0.4 | 3.4 | 1.9 | 2.5 | 2.7 | 12.5 |
| MX | 2.1 | 13.5 | 8.1 | 12.0 | 14.5 | 17.2 |
| M | 1.1 | 6.0 | 5.7 | 2.1 | 10.0 | 12.5 |
| $X_2$ | 0.3 | 0.4 | 0.3 | 0.7 | | 1.0 |
| X | 0.6 | 5.3 | 10.0 | 6.9 | 5.5 | 6.6 |
| *Doubly charged* | | | | | | |
| $(MX_4)$ | Trace | 0.1 | 0.3 | 0.1 | | |
| $(MX_3)$ | 1.1 | 1.0 | 0.4 | 0.1 | 0.4 | Trace |
| $(MX_2)$ | 3.4 | 0.1 | 0.1 | Trace | | 0.1 |
| (MX) | 0.6 | 0.3 | Trace | Trace | 1.1 | 1.9 |
| (M) | 0.2 | 0.1 | Trace | Trace | | 0.4 |
| (X) | Trace | 0.1 | 0.6 | 0.1 | | |

I may be partly due to the increased ease with which these halogen atoms can lose an electron compared with fluorine. Doubly charged ions were observed for all the singly charged species found (except $X_2^{2+}$) in most cases (Table 7.9).

Svec and Sparrow detected many metastable supported reactions which are summarised in reaction scheme (7.24).

$SiX_4^{+} \rightarrow SiX_3^{+} + X\cdot$
$\rightarrow SiX_2^{+} + X_2$
$\rightarrow SiX^{+} + X_2 + X\cdot$
$\rightarrow Si^{+} + 2 X_2$
$\rightarrow X_2^{+} + SiX_2$
$\rightarrow X^{+} + SiX_2 + X\cdot$
$SiX_3^{+} \rightarrow SiX_2^{+} + X\cdot$
$\rightarrow SiX^{+} + X_2$
$\rightarrow Si^{+} + X_2 + X\cdot$              (7.24)
$\rightarrow X_2^{+} + \cdot SiX$
$\rightarrow X^{+} + SiX_2$
$SiX_2^{+} \rightarrow SiX^{+} + X\cdot$
$\rightarrow Si^{+} + X_2$
$\rightarrow X_2^{+} + Si$
$\rightarrow X^{+} + \cdot SiX$
$SiX^{+} \rightarrow Si^{+} + X\cdot$
$\rightarrow X^{+} + Si$

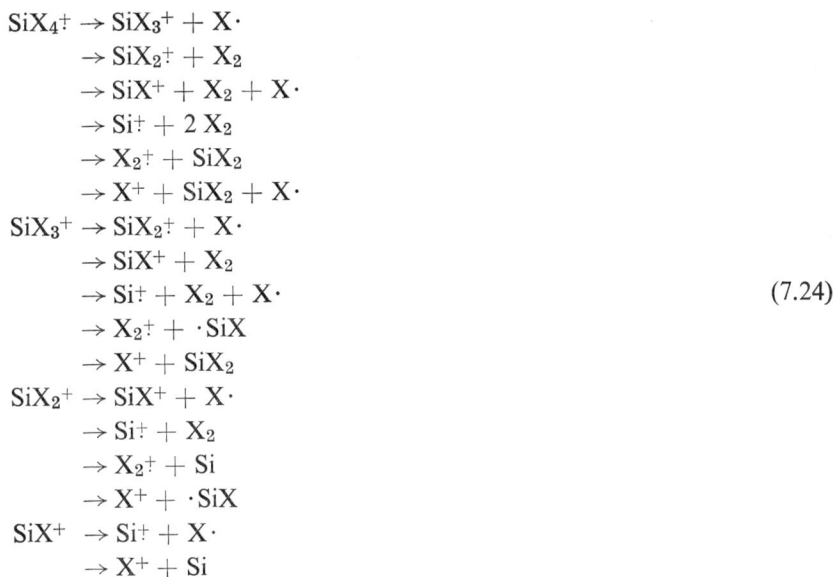

They concluded that silicon tetrahalides dissociate primarily by consecutive loss of $X\cdot$ atoms. However, dissociation to halogen ions was also observed. The ionisation potentials of $SiF_4$ ($15.7 \pm 0.1$ (ref. 47), $15.6 \pm 0.1$ (ref. 120), and 15.4 eV[121])), $SiCl_4 = 11.7 \pm 0.1$ eV, $GeCl_4 = 11.6 \pm 0.3$ eV, $GeBr_4 = 10.8 \pm 0.3$ eV[51], $SnCl_4 = 11.5 \pm 0.4$ eV[54] have been measured mass spectrometrically and the ionisation potentials of $\cdot SiF$ ($7.3 \pm 0.2$ eV), $\cdot SiF_3$ ($8.5 \pm 1$ eV)[47] and $\cdot SnCl_3$ (about 9.5 eV)[54] have been calculated. Appearance potentials of $GeX_n^{+}$ ions ($n = 0\text{–}3$) from $GeCl_4$ and $GeBr_4$ have been reported[51]. Bull et al.[122] reported the adiabatic and vertical ionisation potentials of $SiF_4$ determined by photoelectron spectroscopy as 15.19 and 16.46 eV respectively. Values of bond dissociation energies $D(F_2Si\text{–}F) = 118 \pm 10$ and $D(F_3Si\text{–}F) = 170 \pm 10$ kcal.mole$^{-1}$ have been calculated[47].

Margrave et al.[119] have reported characterising the mixed halides $ISiF_3$ and $I_2SiF_2$. Both showed parent molecular ions, $I^{+}$ being the base peak in the former, and the parent molecular ion that in the latter spectrum. All the expected ions $SiF_nI^{+}$ ($n = 1\text{–}3$) from $ISiF_3$ and $SiF_nI_2^{+}$ ($n = 1,2$), $SiF_nI^{+}$ ($n = 1,2$) from $I_2SiF_2$ were observed.

Several negative ion studies of $SiF_4$ and $GeF_4$ have been reported. Macneil and Thynne studied negative ions formed in $SiF_4$ as a function of electron energy[123]. At 70 eV the three major ions observed were $SiF_3^{-}$ (3%), $F_2^{-}$

(97%), and $F^-$ (trace) but at 20 eV these ions were found in the following abundance $SiF_3^-$ (68%), $F_2^-$ (30.5%), and $F^-$ (1.5%). Macneil and Thynne also observed $SiF_5^-$, presumably formed in a secondary ion process

$$SiF_3^- + SiF_4 \to SiF_5^- + SiF_2 \tag{7.25}$$

Using a value for the heat of formation of $SiF_3^-$, $\Delta H^\circ{}_f(SiF_3^-) = -328$ kcal.mole$^{-1}$, Thynne and Macneil[124] have calculated $\Delta H^\circ{}_f(SiF_5^-) = \leqslant -583$ kcal.mole$^{-1}$. The negative ion spectrum of $GeF_4$ showed[125] as the major germanium-containing ions $GeF_2^-$ and $GeF_3^-$. The ion $GeF_5^-$ was observed as the product of an ion–molecule reaction

$$GeF_3^- + GeF_4 \to GeF_5^- + GeF_2 \tag{7.26}$$

The polynuclear species $Ge_2F_4^-$ and $Ge_2F_8^-$ were observed in low abundance and it was suggested that they too were produced by the ion–molecule reactions

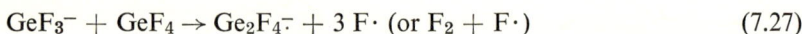

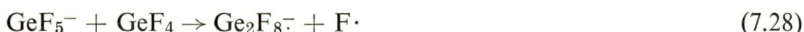

$$GeF_3^- + GeF_4 \to Ge_2F_4^- + 3\ F\cdot\ (\text{or } F_2 + F\cdot) \tag{7.27}$$

$$GeF_5^- + GeF_4 \to Ge_2F_8^- + F\cdot \tag{7.28}$$

McDonald et al.[47] used a time-of-flight mass spectrometer to measure the ionisation potential of $Si_3F_8$ (10.84 $\pm$ 0.1 eV) and appearance potentials of various ions from $Si_2F_6$, $Si_3F_8$, and $Si_2BF_7$. From their results they calculated $\Delta H^\circ{}_f Si_2F_6 = -565 \pm 5$, $\Delta H^\circ{}_f Si_3F_8 = -754 \pm 10$, and $\Delta H^\circ{}_f Si_2BF_7 = -637 \pm 25$ kcal.mole$^{-1}$, respectively, and the following values of mean bond dissociation energies ($\bar{D}$), $\bar{D}(Si-F) = 139 \pm 3$, $\bar{D}(Si-Si) = 55 \pm 10$, and $\bar{D}(Si-B) = 55 \pm 15$ kcal.mole$^{-1}$. They were able to estimate the ionisation potentials of $Si_2F_6$ (10.6 $\pm$ 1.0 eV) and $Si_2BF_7$ (10.6 $\pm$ 1.0 eV) although the parent molecular ions of these compounds were not observed.

*6.2 Other studies involving halides*

Meinrenken[106] has used $SiF_4$ and $SiCl_4$ to study the natural abundances of silicon isotopes. Arshakuni[127,128] has discussed the use of $SiF_4$ and $GeF_4$ in the mass spectral analysis of silicon and germanium compounds and

Table 7.10

Ion Abundances from $Me_4M$ Compounds

| Composition of ion | M | | | | |
|---|---|---|---|---|---|
| | C | Si | Ge | Sn | Pb |
| $(CH_3)_4M$ | Trace | 0.4 | 0.3 | 0.3 | 0.1 |
| $(CH_3)_3MH$ | | | Trace | 0.2 | 0.3 |
| $(CH_3)_3M$ | 40.2 | 55.0 | 60.5 | 50.1 | 31.5 |
| $(CH_3)_2MCH_2$ | 1.5 | 0.5 | 0.4 | 0.5 | 3.1 |
| $(CH_3)_2MCH$ | 1.3 | 0.2 | | | |
| $(CH_3)_2MC$ | Trace | 0.1 | | | |
| $MC_3H_5$ | 0.6 | Trace | | | |
| $MC_3H_4$ | 0.1 | Trace | | | |
| $MC_3H_3$ | 0.6 | 0.1 | | | |
| $MC_3H_2$ | 0.5 | 0.1 | | | |
| $MC_3H$ | 0.1 | 0.1 | | | |
| $MC_3$ | Trace | Trace | | | |
| $(CH_3)_2MH$ | 0.6 | 1.2 | 0.9 | 0.7 | 0.1 |
| $(CH_3)_2M$ | 0.7 | 1.0 | 3.8 | 11.3 | 8.0 |
| $CH_3MCH_2$ | 17.6 | 0.9 | 0.6 | 0.1 | 0.2 |
| $CH_3MCH$ | 0.5 | 0.1 | Trace | | |
| $CH_3MC$ | 7.0 | 1.4 | 0.6 | | Trace |
| $MC_2H_2$ | 0.8 | 0.2 | Trace | | |
| $MC_2H$ | 0.3 | 1.4 | 0.7 | | Trace |
| $MC_2$ | Trace | 0.9 | Trace | | Trace |
| $CH_3MH_2$ | 14.4 | 6.5 | 3.2 | | |
| $CH_3MH$ | 1.0 | 10.2 | 0.3 | 0.1 | 0.3 |
| $CH_3M$ | 6.9 | 8.7 | 13.7 | 18.1 | 28.5 |
| $MCH_2$ | 0.8 | 2.3 | 2.8 | 3.7 | 2.1 |
| $MCH$ | 0.1 | 0.5 | 0.3 | 0.8 | 1.1 |
| $MC$ | | Trace | 0.3 | 0.2 | 0.3 |
| $MH_3$ | 3.7 | 1.9 | 0.3 | | |
| $MH_2$ | 0.5 | 0.1 | Trace | | |
| $MH$ | 0.1 | 4.2 | 6.7 | 10.1 | 2.0 |
| $M$ | Trace | 2.0 | 4.6 | 3.8 | 22.4 |

Devyatykh et al.[129] have similarly discussed the use of the chlorides of silicon and germanium.

Mass spectrometry has been used to characterise compounds such as $H_3GeSiF_2H$, $H_3GeSi_2F_4H$ and $H_3GeSi_3F_6H$. The ion appearing at highest

$m/e$ was always due to (parent molecule $-1$)$^+$. A parent molecular ion was observed from $(F_2ClSi)_2O$ but $(HF_2Si)_2O$ showed the ion due to (parent molecule $-HF_2$)$^+$ at highest $m/e$ (ref. 131). Timms has reported the parent molecular ion from $Cl_3SiPCl_2$ and the base peak was due to the $SiCl_3^+$ ion[132]. Likewise, $Cl_3SiBCl_2$ produced a parent molecular ion and the base peak was due to the $BCl_2^+$ ion.

## 7. MAIN GROUP IV ORGANOMETALLIC COMPOUNDS

### 7.1 Symmetrical tetraalkyl compounds $MR_4$

#### 7.1.1 $R = Me$

Although Aston et al.[133] used tetramethylsilane in a study of the isotope abundances of silicon in 1941, it was not until 1952 that Dibeler[134] first reported the abundances of ions from the tetramethyl compounds of silicon and the other main Group IV elements. Dibeler calculated the monoisotopic spectra on the basis of the isotope abundances he observed in his spectra[134]. The major ions ($\geqslant 0.1 \%$ total ion current) reported are shown in Table 7.10. Other workers who have reported the mass spectra of $Me_4M$ compounds are in reasonably good agreement with Dibeler's results[135-141]. The spectra of the organometallic compounds were qualitatively similar. In each case the (parent molecule $- Me \cdot$)$^+$ ion was the base peak and parent molecular ions were in low abundance. As we have seen previously in $MX_4$ compounds (X = H or halogen), these observations are general for Group IV molecules with this kind of configuration. The abundances of some of the major fragment ions such as $C_2H_6M^+$, $CH_3M^+$, and $M^+$ vary fairly regularly across the series from carbon to lead, increasing in abundance as the M–Me bond becomes weaker. An interesting feature in these spectra was the formation of hydride ions. These are formed by H rearrangements involving the elimination of $:CH_2$ from one methyl group or $C_2H_4$ from two methyl groups, viz.

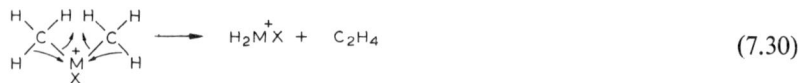

$$H-\underset{\underset{H}{|}}{\overset{\overset{H}{|}}{C}}\!\!-\!\!\overset{+}{M}X \longrightarrow H\overset{+}{M}X + :CH_2 \qquad (7.29)$$

$$\longrightarrow H_2\overset{+}{M}X + C_2H_4 \qquad (7.30)$$

These reactions are metastable supported in some cases[135,141]. Monohydride ions ($C_2H_6MH^+$, $CH_3MH^+$, and $MH^+$) were observed for all tetramethyl compounds. Trimethyl hydride ions ($C_3H_9MH^+$) were observed for $M = Ge$, Sn, Pb and other hydride ions ($MH_2^+$, $MH_3^+$, and $CH_3MH_2^+$) were found in the spectra of the silicon and germanium compounds.

De Ridder and Dijkstra[135] have published a general fragmentation scheme of metastable supported decomposition pathways (7.31).

$$(CH_3)_4M^{+\cdot} \xrightarrow[\ast]{-Me\cdot} (CH_3)_3M^+ \begin{array}{c} \xrightarrow{-C_2H_4} H_2M^+CH_3 \xrightarrow[\ast]{-H_2} CH_3M^+ \\[4pt] \xrightarrow[\ast]{-:CH_2} HM^+C_2H_6 \xrightarrow[\ast]{-C_2H_4} H_3M^+ \\[4pt] \xrightarrow[\ast]{-Me\cdot} \end{array} \left. \rule{0pt}{40pt}\right\} \quad (7.31a)$$

$$C_2H_6M^{+\cdot} \xrightarrow[\ast]{-Me\cdot} CH_3M^+ \xrightarrow[\ast]{-Me\cdot} M^{+\cdot} \qquad (7.31b)$$

These workers interpreted ion abundances and observed metastable peaks as showing that reactions (7.31a) were more favourable for $M = C$, Si, and Ge and reactions (7.31b) were more favourable for $M = Sn$, Pb. They suggested that the differences in the observed reactions could be explained in terms of the different $(XM-C)^+$ and $(XM-H)^+$ bond dissociation energies relative to each other for each element and the other Group IV elements. Although this simple qualitative picture is useful, one must bear in mind that these bond dissociation energies are, for the most part, unknown. Tamas and Ujszaszy[141] have reported a similar scheme to (7.31) for $Me_4Si$. They also observed the following metastable supported reactions.

$$\begin{array}{l} (CH_3)_4Si^{+\cdot} \xrightarrow[\ast]{-CH_4} (C_2H_6)Si^+CH_2 \\[4pt] (CH_3)_3Si^+ \xrightarrow[\ast]{-CH_4} C_2H_5Si^+ \\[4pt] \phantom{(CH_3)_3Si^+} \xrightarrow[\ast]{-C_2H_6} CH_3Si^+ \\[4pt] CH_4Si^{+\cdot} \xrightarrow[\ast]{-CH_4} Si^{+\cdot} \\[4pt] C_2H_6SiCH_2^+ \xrightarrow[\ast]{-H_2} C_3H_6Si^{+\cdot} \\[4pt] CSiCH_3^+ \xrightarrow[\ast]{-H_2} C_2HSi^+ \end{array} \qquad (7.32)$$

The major ions in the negative ion mass spectra of $Me_4Si$ and $Me_4Sn$ have been reported[143,147]. Several ionising energies were used[142]. At 4, 7.8, and 30 eV, the base peak in the $Me_4Si$ spectrum was $(CH_3)_3SiCH_2^-$. At 8 eV, the base peak in the $Me_4Sn$ spectrum was $(CH_3)_3SnCH_2^-$ but at 30 eV this is no longer the case and the base peak was $(CH_3)_3Sn^-$ (or $(CH_3)_2HSnCH_2^-$). The 30 eV spectrum of $Me_4Si$ showed the base peak in 68% abundance and the major ions $C_3H_7Si^-$ (13.5%), $C_2H_5Si^-$ (7.5%), $CHSi^-$ (6%), $CH_3Si^-$ (2%), $C_2H_4Si^-$ (2%), and $CH_2Si^-$ (1%). At the same electron energy, $Me_4Sn$ showed the base peak in 45% abundance and $C_2H_5Sn^-$ (23%) $(CH_3)_3SnCH_2^-$ (12%), $CH_3Sn^-$ (9%), and $C_2H_4Sn^-$ (11%)[143].

Several workers have measured ionisation potentials and appearance potentials of ions from $Me_4M$ compounds. The results of their work are discussed in Section 8 of this chapter on energetics of Group IV organometallic compounds.

### 7.1.2 $R = Et$

The tetraethyl compounds of Group IV have been reported by de Ridder and Dijkstra[135]. The ion abundances found by these workers are in general agreement with other published values for $SiEt_4$ (ref. 144), $GeEt_4$ (refs. 68, 145), $SnEt_4$ (refs. 137, 138, 145, 147), and $PbEt_4$ (refs. 147, 148). As with the $Me_4M$ compounds, the parent molecular ions are in low abundance and $(C_2H_5)_3M^+$ is a prominent ion being the base peak from $Et_4Sn$ and $Et_4Pb$. The base peak in the spectra of $Et_4Si$ and $Et_4Ge$ is due to the $(C_2H_5)_2MH^+$ ion. In the carbon compounds' spectrum the base peak is due to the $C_4H_9^+$ ion, formed from $(C_2H_5)_3C^+$ by a metastable supported decomposition reaction. The ion $C_4H_9^+$ probably has the structure $(CH_3)_3C^+$ of the tertiary butyl ion which is known to be particularly stable[149]. In the spectra of the other elements, the $(CH_3)_3M^+$ ions are not particularly abundant, the elimination of ethylene from $(C_2H_5)_3M^+$ being more favourable (7.33). This reaction was metastable supported[135] and probably involves a $\beta$-H atom, since no corresponding hydride ions were observed in the spectra of (I), $M = Sn$ (ref. 147) or Pb (ref. 150).

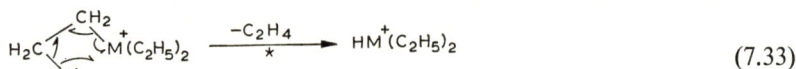

$$(7.33)$$

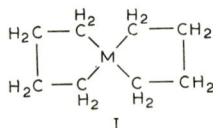

I

de Ridder and Dijkstra[135] have reported a generalised, often metastable supported, fragmentation scheme for $Et_4M$ compounds, viz.

$$(C_2H_5)_4M^{+\cdot} \xrightarrow{-C_2H_5\cdot} (C_2H_5)_3M^+ \xrightarrow{-C_2H_4} HM^+(C_2H_5)_2 \xrightarrow{-C_2H_4} H_2\overset{+}{M}C_2H_5 \quad \begin{matrix} \nearrow^{-C_2H_4} \searrow^{H_3M^+} \\ \searrow_{-H_2} \end{matrix} \qquad \Bigg\} \quad (7.34\text{a})$$

$$\Big\downarrow_{-C_2H_5\cdot} \qquad\qquad\qquad HM^+ \xleftarrow{-C_2H_4} C_2H_5M^+ \xrightarrow{-C_2H_5\cdot} M^{+\cdot}$$

$$(C_2H_5)_2M^{+\cdot} \xrightarrow{-C_2H_5\cdot} C_2H_5M^+ \xrightarrow{-C_2H_5\cdot} M^{+\cdot} \qquad \Bigg\} \quad (7.34\text{b})$$

$$\searrow_{-C_2H_4} \quad M\overset{\cdot}{H}{}^+$$

They suggested that reactions (7.34a) were more favourable for $M = Si$ or Ge and reactions (7.34b) were favoured for $M = Sn$ or Pb for similar reasons to those given for $Me_4M$ compounds above[135]. de Ridder and Dijkstra also reported ionisation potentials and appearance potentials for the tetraethyl compounds. These values are probably high because they were calculated using the linear extrapolation method, and will not be discussed here. Glockling and coworkers[147] in their study of $Et_4Sn$ and $Et_4Pb$ at high resolution, confirm the formation of $(C_2H_5)MH^+$ ions and were able to identify the elimination of methylene from $(C_2H_5)_3M^+$ ions,

$$(C_2H_5)_3M^+ \xrightarrow[*]{} (C_2H_3)_2MCH_3^+ + :CH_2 \qquad (7.35)$$

This process was supported by a weak metastable peak.

### 7.1.3 Other symmetrical tetraalkyl compounds

The positive and negative ion mass spectrum of $(F_3C \cdot C \equiv C-)_4Si$ has been reported[151]. No parent molecular ion was observed in the positive ion spectrum, the ion at highest $m/e$ was due to (parent molecule $- \cdot F$)$^+$. Other fragment ions were produced by loss of $\cdot CF_3$, $:CF_2$, and $C_3F_3$ groups. The formation of the fluorocarbon ion $C_{12}F_9^+$ by loss of $\cdot SiF_3$ from the parent molecular ion was observed. The negative ion spectrum exhibited a parent molecular ion which was the base peak. Fragmentation by loss of $\cdot CF_3$ was reported.

In the cases of germanium and tin, several studies of the higher tetraalkyls have been reported.

Ion abundances and metastable peaks in the spectra of $GePr^n_4$, $GeBu^n_4$, $Ge(n\text{-pentyl})_4$ and $Ge(n\text{-hexyl})_4$ have been reported by de Ridder and Dijkstra[152]. Glockling $et\ al.$ have studied $GePr^i_4$ (ref. 153), $GeBu^i_4$, and Ge (benzyl)$_4$ (ref. 70). The main characteristics of the spectra observed by de Ridder and Dijkstra were that as the alkyl chain length increases ($a$) the

Table 7.11

Variation of Ion Abundances with R group in $R_4Ge$ Compounds

| Composition of ion | % Abundance (of total ion current) from $R_4Ge$ for R = | | | | | |
|---|---|---|---|---|---|---|
| | Me | Et | $Pr^n$ | $Bu^n$ | $n$-$C_5H_{11}$ | $n$-$C_6H_{17}$ |
| $R_3Ge$ | 71.1 | 26.6 | 20.4 | 14.7 | 9.1 | 11.6 |
| $R_2GeH$ | 0.9 | 35.9 | 39.3 | 31.1 | 26.9 | 24.2 |
| $RGeH_2$ | 2.9 | 15.0 | 13.3 | 19.1 | 20.8 | 19.2 |
| $RGeH$ | 2.8 | 2.9 | 1.2 | 0.9 | 0.5 | 0.3 |
| $RGe$ | 1.6 | 1.5 | 1.4 | 1.1 | 0.6 | 0.3 |
| $\dfrac{\Sigma \text{ C–H ions}^a}{\Sigma \text{ Ge–C–H ions}}$ | 3.5 | 3.0 | 7.4 | 18.0 | 28.3 | 30.5 |

a Expressed as a % of total ion current, *i.e.* of hydrocarbon and organometallic ions.

abundance of $R_3Ge^+$ decreases, (b) the $HGeR_2^+$ ion formed by elimination of alkene increases till R = $Pr^n$ and decreases to R = $n$-hexyl, (c) overall, the abundance of $H_2GeR^+$ ions increases, (d) the total of germanium-containing ions which have been formed with carbon-carbon bond fission increases, (e) the abundance of $Ge^+$ and $GeH^+$ ions decrease, and finally, (f) the % abundance of hydrocarbon fragments increases as shown in Table 7.11. They also found that the ion $H_2Ge^+R$ (R = $Bu^n$, $n$-pentyl, $n$-hexyl) decomposed by two paths, either by the elimination of $H_2$ or the alkene[154], *viz.*

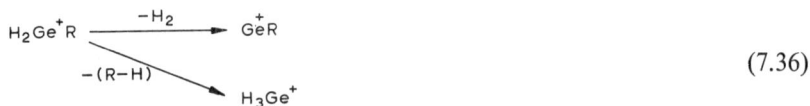

$$H_2Ge^+R \xrightarrow[\phantom{xx}-(R-H)\phantom{xx}]{-H_2} \begin{array}{c} Ge^+R \\ H_3Ge^+ \end{array} \tag{7.36}$$

The work by Glockling *et al.* showed similar results. From tetraisopropyl-germane loss of $\cdot C_3H_7$, $C_3H_6$, $C_2H_4$, and $H_2$ was observed. The base peak was due to the $C_6H_{15}Ge^+$ ion. The isobutyl derivative showed similar fragmentation paths and $C_8H_{19}Ge^+$ was the base peak. Tetrabenzylgermane lost a benzyl radical from the parent molecular ion to give the base peak ion $(C_6H_5CH_2)_3Ge^+$. Metastable supported decompositions of this ion by the loss of $H_2$, $C_6H_5CH_3$, and $(C_6H_5CH)_2$ were reported, *viz.*

$$(C_6H_5CH_2)_3Ge^+ \begin{array}{l} \xrightarrow[*]{-H_2} C_{21}H_{19}Ge^+ \\ \xrightarrow[*]{-C_6H_5CH_3} \\ \xrightarrow[*]{} C_6H_5CH_2Ge^+C_7H_6 \\ -(C_6H_5CH)_2* \\ \searrow \\ C_6H_5CH_2GeH_2^+ \end{array} \qquad (7.37)$$

Another interesting decomposition path was the loss of Ge from $C_6H_5CH_2$-$Ge^+$ to give the very stable $C_7H_7^+$ ion.

$$C_6H_5CH_2Ge^+ \xrightarrow[*]{-Ge} C_7H_7^+ \qquad (7.38)$$

The spectra of the symmetrical tetraalkyltin compounds $R_4Sn$ ($R = Pr^n$ (refs. 146, 154), $Pr^i$ (ref. 146), $Bu^n$ (refs. 137, 154, 155), $Bu^i$ (ref. 146), vinyl (ref. 137), and cyclohexyl[154]) have been reported.

The spectra of the compounds $Pr^n{}_4Pb$ and $Bu^i{}_4Pb$ have been reported[332].

### 7.2 Unsymmetrical tetraalkyl compounds

Generally, the fragmentation of compounds containing different alkyl groups attached to the Group IV element occur by the same pathways as for the symmetrical $MR_4$ compounds. We will discuss them according to the central metal atom.

### 7.2.1 Silanes

A large number of $R_3Si$– derivatives have been studied mass spectrometrically mostly in order to obtain the molecular weight of the compound. However, in a few cases, the fragmentation behaviour has been discussed. One such case is the compound $Me_3Si$–$CH_2$–$SiMe_3$. The spectrum has been reported in detail by Tamas et al.[156,157] who found a parent molecular ion in very low abundance (0.07%) and the base peak ion (parent molecule $- \cdot CH_3)^+$ in 51.1% abundance. A large number of metastable supported decompositions were observed corresponding to fragmentation by loss of $\cdot CH_3$, $CH_4$ and $C_2H_4$ groups, e.g.

$$(CH_3)_5Si_2CH_2^+ \xrightarrow[*]{-CH_3\cdot} (CH_3)_4Si_2CH_2^+ \begin{array}{l} \swarrow -CH_4 \\ * \end{array} \begin{array}{l} \searrow -C_2H_4 \\ * \end{array} \qquad (7.39)$$
$$(CH_3)_2Si_2(CH_2)_2^+ \qquad (CH_3)_3Si_2H^+$$

The loss of the $SiCH_2$ group was observed from $(CH_3)_3Si_2CH_2^+$ and $(CH_3)_3Si_2^+$ ions. The decompositions shown in (7.39) were also found with the species shown doubly charged. Chernyak et al.[158] studied a series of compounds, $Me_3Si-(CH_2)_n-SiMe_3$ ($n = 1-6$) and $Me_3SiCH_2CH(Me)-SiMe_3$. Parent molecular ions were reported for all the series and when there was more than one methylene group, the base peak was due to the $(CH_3)_3Si^+$ ion. The loss of larger neutral fragments including $SiC_3H_{10}$, $SiC_4H_{10}$, and $(CH_3)_4Si$ was a feature of the fragmentation of the higher members of the series, e.g.

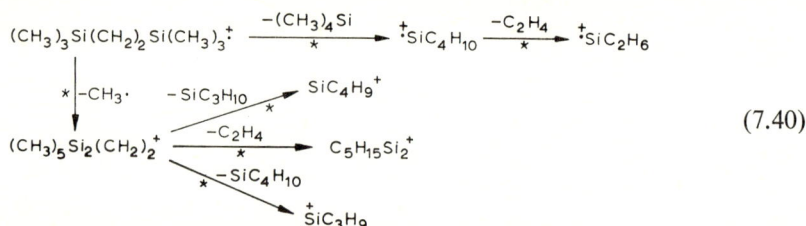

$$(CH_3)_3Si(CH_2)_2Si(CH_3)_3^+ \xrightarrow[\ *\ ]{-(CH_3)_4Si} \ ^+_.SiC_4H_{10} \xrightarrow[\ *\ ]{-C_2H_4} \ ^+_.SiC_2H_6$$

(7.40)

$$*\Big|-CH_3 \cdot \quad -SiC_3H_{10} \quad \nearrow SiC_4H_9^+$$

$$(CH_3)_5Si_2(CH_2)_2^+ \xrightarrow[\ *\ ]{-C_2H_4} \quad C_5H_{15}Si_2^+$$

$$\searrow -SiC_4H_{10}$$

$$^+_.SiC_3H_9$$

The tendency to lose large groups from the parent molecular ion became less as the number of methylene groups increased, until at $n = 5$ or 6 ions due to the loss of $SiC_3H_{10}$ and $(CH_3)_4Si$ were no longer observed. Instead, loss of these fragments occurred from the (parent molecule $- \cdot CH_3)^+$ ion. The branched chain derivative showed a parent molecular ion, $(CH_3)_3Si^+$, as base peak and fewer ions containing both silicon atoms.

Weber et al.[159] reported that the spectrum of II was quite simple. Four major ions were observed, namely, in order of their abundance, $(CH_3)_3Si^+$, $C_6H_5(CH_2)_2Si(CH_3)_2^+$, $C_6H_5Si(CH_3)_2^+$, and the parent molecular ion. The formation of the $C_6H_5Si(CH_3)_2^+$ ion from the (parent molecule $- \cdot CH_3)^+$ ion was shown by studying 1-phenyl-1, 1-dideuterio-2-trimethylsilane to go according to reaction (7.41).

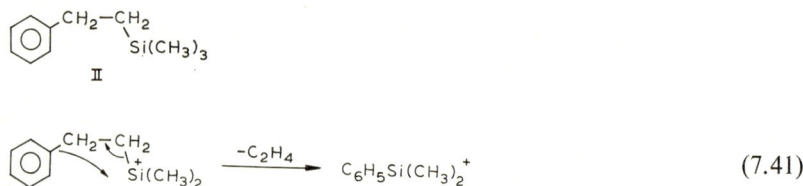

$$\begin{array}{c} \text{CH}_2\text{---CH}_2 \\ \diagdown \\ \text{Si(CH}_3)_3 \end{array}$$

II

$$\begin{array}{c} \text{CH}_2\text{---CH}_2 \\ \diagdown + \\ \text{Si(CH}_3)_2 \end{array} \xrightarrow{-C_2H_4} C_6H_5Si(CH_3)_2^+$$

(7.41)

No analogous rearrangement was observed in the spectrum of the corresponding carbon compound. These workers reported a somewhat similar rearrangement from methyl-3-trimethylsilyl propionate, the acid, 3-trimethyl-silylpropionic acid, and trimethylsilylpropionyl chloride[160], *viz.*

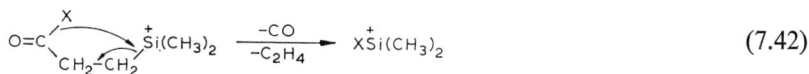

$$O=C\overset{X}{\underset{CH_2-CH_2}{\diagdown}}\overset{+}{S}i(CH_3)_2 \quad \xrightarrow[-C_2H_4]{-CO} \quad X\overset{+}{S}i(CH_3)_2 \qquad (7.42)$$

where X = OCH₃, OH, or Cl.
The formation of the ion $C_2H_6SiO_2CH_3^+$ was observed from the ester.

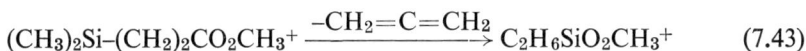

$$(CH_3)_2Si-(CH_2)_2CO_2CH_3^+ \xrightarrow{-CH_2=C=CH_2} C_2H_6SiO_2CH_3^+ \qquad (7.43)$$

With three methylene groups, a rearrangement still occurred with the loss of $C_2H_4$.

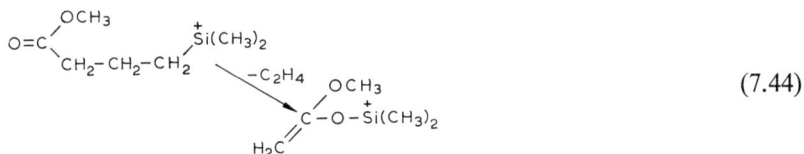

$$O=C\overset{OCH_3}{\underset{CH_2-CH_2-CH_2}{\diagdown}}\overset{+}{S}i(CH_3)_2 \quad \searrow^{-C_2H_4} \quad \overset{OCH_3}{\underset{H_2C}{\diagdown}}C-O-\overset{+}{S}i(CH_3)_2 \qquad (7.44)$$

Some unsymmetrical tetraalkylsilanes containing methyl, ethyl, or propyl groups have been studied by Chernyak *et al.*[144]. They reported similar fragmentations to those observed for the symmetrical compounds, *i.e.* the major ions were due to loss of an alkyl radical or an alkene elimination. These workers pointed out that the unique ion abundances and decomposition pathways of the silanes in a mass spectrometer could be used to identify them from mixtures.

Workers in the USSR have published brief reports[155,161,162] of the mass spectra of some silicon-containing enynic hydrocarbons, $R_3SiC \equiv C-CH = CHR'$ (R or R′ = Me or Et), and some corresponding germanium compounds[163]. The major ions are formed by the loss of R· and then if R or R′ = Et by elimination of alkene from R· or $H_2$ from $(H_2MC \equiv C-CH = CHR')^+$. Ions due to loss of the conjugated grouping are in low abundance.

Other compounds containing alkyl groups or substituted alkyl groups attached to silicon are given in Table 7.12.

### 7.2.2 Germanes

Glockling *et al.*[70] have published detailed mass spectra of $(C_6H_5CH_2)_3GeEt$

Table 7.12

Other Compounds Containing $R_3M-R'$ or $R_2M{\Large\langle}\genfrac{}{}{0pt}{}{R'}{R''}$ Groups

| Compound | Comments | Ref. |
|---|---|---|
| $Me_3SiCH_2SiMe_2Et$ | P.m.i. | 164 |
| $Me_3SiCH_2SiMe_2(CH=CH_2)$ | | 164 |
| $Me_3Si(CH_2)_2SiMe_3$ | | 164 |
| $Me_3Si-C{\equiv}C-SiMe_3$ | | 164 |
| $Me_3Si-CH=CH=CH_2SiMe_3$ | | 164 |
| $Me_3MCH_2C_6H_5$ | | 165 |
| (M = Si or Sn) | | |
| $Me_3M(C_9H_7)$ | P.m.i.s, $(CH_3)_3M^+$ | 166 |
| (M = Si, Ge or Sn) | base peaks | |
| $Me_3SiCH_2SCH_3$ | P.m.i.; $(CH_3)_3Si^+$ base peak | 167 |
| $Me_3Si(CH_2)_2X(CH_3)_2$ | P.m.i.s | 168 |
| $Me_2Si[(CH_2)_2X(CH_3)_2]_2$ | | 168 |
| (X = N or P) | P.m.i.s | |
| $Me_3SiCH_2SiMe_2CH_2C-(SiMe_3)_3$ | No p.m.i., but (parent molecule $-\cdot Me)^+$ | 169 |
| $(Me_3Si)_2CHSiMe_2CH_2C-(SiMe_3)_3$ | at highest $m/e$ | 169 |
| $Me_3SiCH_2F$ | No p.m.i., but $(CH_3)_2Si^+$ and $(CH_3)_2SiF^+$ | 170 |
| | observed | |
| $Me_3Si(CH_2)_2N(CF_3)_2$ | No p.m.i., but $(CH_3)_3Si^+$ | 170 |
| | $(CH_3)_2SiF^+$ and $(CH_2)_2N(CF_3)_2{}^+$ observed | |
| $Et_3SiCCl_2D$ | P.m.i., $(C_2H_5)_3Si^+$ base peak | 171 |

| | | |
|---|---|---|
| $Me_2Si{\Large\langle}\genfrac{}{}{0pt}{}{CH_2}{CH_2}{\Large\rangle}GeMe_2$ | P.m.i. | 172 |

| | | |
|---|---|---|
| structure (Me2Si, SiHMe, C=C, H, R ring) | P.m.i.<br>R = H or Me | 164 |

| | | |
|---|---|---|
| structure (Me2Si, SiMe2, C=C, H, Me ring) | P.m.i. | 164 |

| | | |
|---|---|---|
| structure (Me2Si, CH2(=C), CH2, C-SiMe2, H2 ring) | P.m.i. | 164 |

Table 7.12 (continued)

| Compound | Comments | Ref. |
|---|---|---|
| | P.m.i. | 164 |
| | P.m.i. | 164 |
| | P.m.i. | 164 |
| | P.m.i. | 164 |
| | P.m.i. | 174 |
| | P.m.i. | 185 |
| $Pr^i_3GePr^n$ | P.m.i. | 297 |
| $Me_3MCMe_3$ | P.m.i., $Me_3M^+$ base peaks for M = Ge, Sn, Pb | 139 |
| $Me_2Pr^i$ (cyclohexyl)Sn | P.m.i., $C_2H_7Sn^+$ base peak | 179 |

Table 7.12 (continued)

| Compound | Comments | Ref. |
|---|---|---|
| MeEtPr$^i$(cyclohexyl)Sn | P.m.i., $C_3H_9Sn^+$ base peak | 179 |
| MeEtPr$^n$Pr$^i$Sn | P.m.i., $C_3H_9Sn^+$ base peak | 179 |
| MePr$^n$Pr$^i$Bu$^s$Sn | P.m.i., $C_4H_{11}Sn^+$ base peak | 179 |
| MePr$^n$Pr$^i$Bu$^t$Sn | P.m.i., $C_4H_{11}Sn^+$ base peak | 179 |
| ($\sigma C_5H_5$)MePr$^i$RSn | P.m.i. | 180 |
| $\quad$ R = –CH=C=CH$_2$ or | | |
| $\quad$ –CH$_2$–C≡CH | | |
| Me$_3$PbR | P.m.i. | 332 |
| $\quad$ R = Et, Bu$^n$, Bu$^s$, Bu$^t$ | | |
| Me$_2$PbEt$_2$ | P.m.i. | 332 |
| MePbEt$_3$ | P.m.i. | 332 |
| Pr$^n_4$Pb | P.m.i. | 332 |
| Bu$^i_4$Pb | P.m.i. | 332 |

and $(C_6H_5CH_2)_2GeMe_2$ and described the major features in the spectra of $(C_6H_5CH_2)_3GeMe$ (ref. 70) and $Et_3GeBu^n$ (ref. 173). The parent molecular ions were observed in low abundance and both types of ion $R_2Ge^+R'$ and $R_3Ge^+$ (or $RGe^+R_2$) were found although their abundance varied greatly depending on R and R'. For instance, from $(C_6H_5CH_2)_2GeMe_2$ the $(C_6H_5-CH_2)_2GeMe^+$ ion was in 2.3% abundance but $(C_6H_5CH_2)GeMe_2^+$ was in 63.8% abundance. The most important fragmentation paths were loss of radicals R or R' and elimination of alkenes. However, several interesting rearrangements were reported, some of which are given in the following scheme.

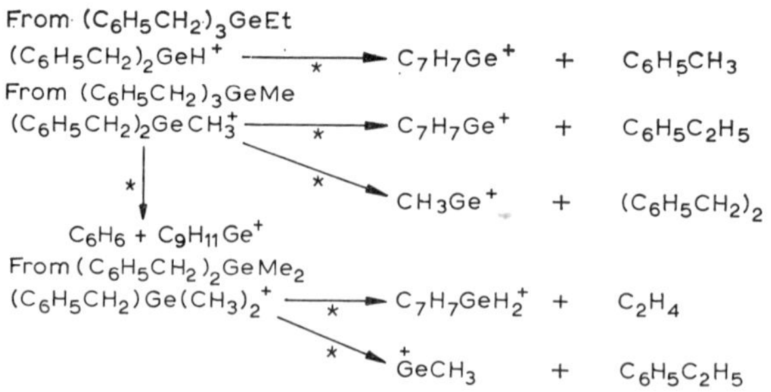

$$\text{(7.45)}$$

Duffield et al.[175] reported some studies of substituted germacyclopentanes, III.

$$Me \quad R''$$

$$\overset{Ge}{\underset{R \quad R'}{\bigvee}}$$

R = Et, Me, Et
R' = Et, Me, Et
R'' = H, Me, Me

III

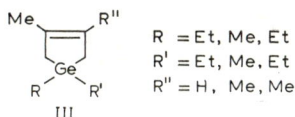

All the compounds showed parent molecular ions. The Ge–Et compounds lost an ethyl radical, then the elimination of ethylene occurred. The base peak was due to the $C_2H_5Ge^+$ ion. The entirely methylated compound showed $^+GeCH_3$ as base peak ion, produced from the parent molecular ion *via* the reaction.

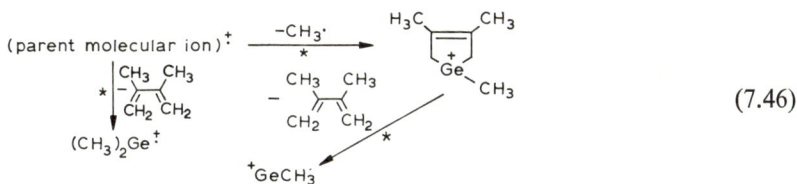

$$(7.46)$$

Other compounds in this category are given in Table 7.12.

### 7.2.3 Stannanes

Heldt *et al.*[138] have studied the mass spectra of the compounds $Me_{4-n}SnEt_n$ ($n = 0$–4). As one would expect, the fragmentation behaviour of these mixed alkyl compounds is markedly similar to that of the symmetrical compounds, *i.e.* both Me· and Et· are lost and hydride ions are more common in the spectra of compounds containing a Sn–Et bond because of the facile elimination of $C_2H_4$ (Table 7.13). Metastable supported eliminations of methylene or ethylene were common features. In a later paper, Heldt *et al.*[176] reported using mass spectrometry to identify the products of the radiolysis of $Et_2SnMe_2$ which were predominantly $Me_4Sn$, $Me_3SnEt$, and $Et_3SnMe$. Boué *et al.*[154] have reported studies on a series of $Me_3SnR$ (R = Et, $Pr^n$, $Pr^i$, $Bu^n$, $Bu^t$, cyclohexyl, *p*-methylbenzyl, *o*-methylbenzyl, and *m*-methylbenzyl) and $Et_3SnR$ (R = Me, $Pr^n$, $Pr^i$, $Bu^n$). Their spectra of $Me_3SnEt$ and $MeSnEt_3$ agree reasonably well with Heldt *et al.*[138]. Boué *et al.* attempted to discuss their ion abundance results in terms of "the ease of cleavage of a carbon–tin bond in the molecular ion", which they suggested parallelled "the possible stabilisation of the leaving radical by hyperconjugation". Such discussions are usually not very satisfactory because there is not enough data about stabilisation effects in ions and the discussions ignore the dynamic nature of mass spectra and ion abundances.

Table 7.13

Ion Abundances for Hydride Ions from $Me_nSnEt_{(4-n)}$

| Composition of ion | % Abundance from | | | | |
|---|---|---|---|---|---|
| | $Me_4Sn$ | $Me_3SnEt$ | $Me_2SnEt_2$ | $MeSnEt_3$ | $Et_4Sn$ |
| $(CH_3)_3SnH$ | | 0.5 | | | |
| $(CH_3)_2SnH$ or $C_2H_5SnH_2$ | | 17.1 | 26.4 | 1.0 | 14.6 |
| $CH_3SnH_2$ | | | 3.1 | 9.1 | |
| $CH_3SnH$ | | | 0.7 | 0.5 | |
| $SnH_2$ | | | | 1.1 | 1.8 |
| $SnH$ | 3.2 | 4.8 | 3.1 | 5.4 | 16.7 |

The mass spectrum of $Bu_2{}^nSn(vinyl)_2$ has been briefly reported[137]. As expected, the molecule fragmented by loss of butyl, and to a lesser extent vinyl, radicals and by the elimination of butene. The following processes were identified by metastable peaks.

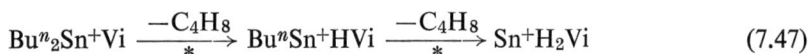

$$Bu^n{}_2Sn^+Vi \xrightarrow[*]{-C_4H_8} Bu^nSn^+HVi \xrightarrow[*]{-C_4H_8} Sn^+H_2Vi \qquad (7.47)$$

$$Bu^nSn^+(Vi)_2 \xrightarrow[*]{-C_4H_8} SnH(Vi)_2$$

Kuivila et al.[177] have reported some features in the spectra of the compounds $Me_3Sn(CH_2)_nCOR$ ($n = 2$ or 3; R = Me or Ph). Weak parent molecular ions and base peaks due to the $(CH_3)_3Sn^+$ ion were found from all compounds. Major fragmentation paths involved the loss of ·H, ·CH₃, $C_2H_6$, $C_3H_9$, and $C_2H_4$ groups, e.g.

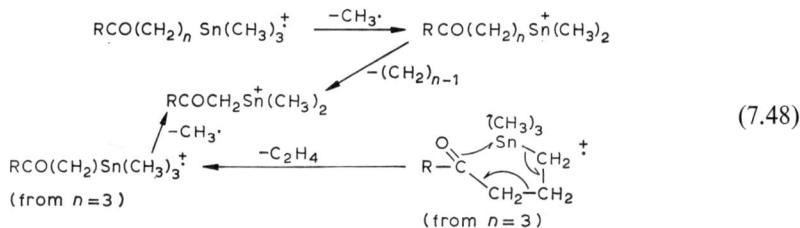

$$(7.48)$$

The rearrangement ion $C_6H_5Sn(CH_3)_2{}^+$ was observed from the phenyl compounds, viz.

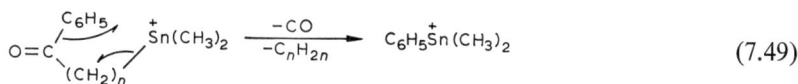

$$(7.49)$$

Other trialkyltin compounds studied[178] include $R_3SnCH_2CO_2Me$, $R_3SnCH_2$-COMe ($R$ = Me, Et), and $Et_3Sn\ CH_2CH_2COMe$. Although fragmentation proceeded predominantly by cleavage of Sn–C bonds, some fragments were formed by cleavages at $\alpha$-carbon (7.50) or $\beta$-carbon (7.51) atoms.

$$(7.50)$$

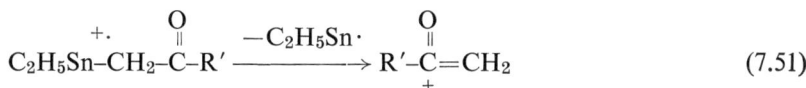

$$(7.51)$$

The rearrangement ion $(C_2H_5)_2SnOCH_3{}^+$ was produced by the metastable supported elimination of $CH_2{=}C{=}O$ from $(C_2H_5)_2SnCH_2CO_2CH_3{}^+$.
Other compounds in this category are given in Table 7.12.

## 7.3 Symmetrical tetraaryl compounds $(Ar)_4M$

### 7.3.1 Ar = Ph

The mass spectra of the tetraphenyl compounds of carbon[147,181,182], silicon[120,181,182], germanium[70,181,182], tin[147,188], and lead[147], have been reported. The ion abundances of the major ions ($\geqslant 0.1\%$ of the total ion current) are shown in Table 7.14. There appears to be reasonable agreement between the spectra observed by different workers in the references given above, but another spectrum of $Ph_4Sn$ reported by Occolowitz[137] shows slight disagreements with that of Glockling et al.[70] in the abundances of $(C_6H_5)_2Sn^+$, $C_6H_5Sn^+$, $C_2HSn^+$ and $C_6H_5{}^+$ ions. However, the work of Glockling et al. is more detailed and their spectrum is given in Table 7.14. The spectra of the tetraphenyl compounds of silicon, germanium, tin, and lead were, in general, much simpler than the spectra of tetraalkyl compounds of these elements. Ions of the type $R_2M^+$ are more abundant and hydride ions are less abundant in the spectra of $Ph_4M$ compounds compared to those of tetraalkyl compounds. The major fragmentation modes involve the loss of phenyl radicals or elimination of $C_6H_6$, $C_{12}H_{10}$, and $C_2H_2$ groups. Parent molecular ions are observed in low abundance and the base peaks

Table 7.14

Ion Abundances from $Ph_4M$ Compounds

| Composition of ion | M | | | | |
|---|---|---|---|---|---|
| | C (ref. 147) | Si (ref. 120) | Ge (ref. 70) | Sn (ref. 143) | Pb (ref. 147) |
| $M(C_6H_5)_4$ | 14.4 | 17.8 | 2.0 | (0.1) | (0.05) |
| $M(C_6H_5)_3C_6H_4$ | 2.9 | 0.1 | | | |
| $M(C_6H_5)_3$ | 37.6 | 29.9 | 43.0 | 42.7 | 34.1 |
| $M\ C_{18}H_{14}$ | 1.0 | 2.1 | | | |
| $M\ C_{18}H_{13}$ | 2.9 | 4.4 | 0.9 | | |
| $M\ C_{18}H_{11}$ | 2.7 | 0.5 | | | |
| $M\ C_{17}H_{12}$ | 2.0 | | | | |
| $M\ C_{17}H_{10}$ | 1.0 | | | | |
| $M\ C_{16}H_{11}$ | 1.6 | | | | |
| $M\ C_{15}H_{10}$ | 1.0 | | | | |
| $M\ C_{13}H_{10}$ | 1.6 | | | | |
| $M\ C_{12}H_{11}$ | 1.6 | | | | |
| $M\ (C_6H_5)_2$ | 2.7 | 19.4 | 27.7 | 16.9 | |
| $M\ C_{12}H_9$ | 15.5 | 8.9 | 4.8 | 1.5 | 0.6 |
| $M\ C_{12}H_8$ | 0.8 | 3.7 | | | |
| $M\ C_{12}H_7$ | 0.7 | 0.7 | 0.6 | | |
| $M\ C_{11}H_{10}$ | 0.7 | | | | |
| $M\ C_{11}H_8$ | 0.6 | 0.1 | | | |
| $M\ C_{10}H_7$ | 0.7 | 1.9 | 1.4 | | |
| $M\ C_8H_7$ | 1.4 | 0.1 | | | |
| $M\ C_8H_5$ | | 2.3 | 1.3 | | |
| $M\ C_6H_5$ | | 4.7 | 12.4 | 17.8 | 33.4 |
| $M\ C_6H_4$ | | 0.1 | | 0.9 | |
| $M\ C_6H_3$ | | 0.7 | 0.5 | | |
| $M\ C_5H_6$ | 3.8 | | | | |
| $M\ C_5H_3$ | 1.9 | | | | |
| $M\ C_4H_3$ | | 1.3 | 2.2 | | |
| $M\ C_3H_4$ | 0.7 | | | | |
| $M\ C_3H_3$ | 1.0 | | | | |
| $M\ C_3H_2$ | 0.6 | | | | |
| $M\ C_2H_3$ | 1.0 | | | | |
| $M\ C_2H$ | | 1.2 | 2.4 | 1.1 | |
| M | | 0.1 | 1.2 | 18.9 | 31.6 |

are due to $(C_6H_5)_3M^+$ ions. Typical metastable supported decomposition reactions are shown in scheme (7.52). These include some interesting eliminations from odd-electron ions.

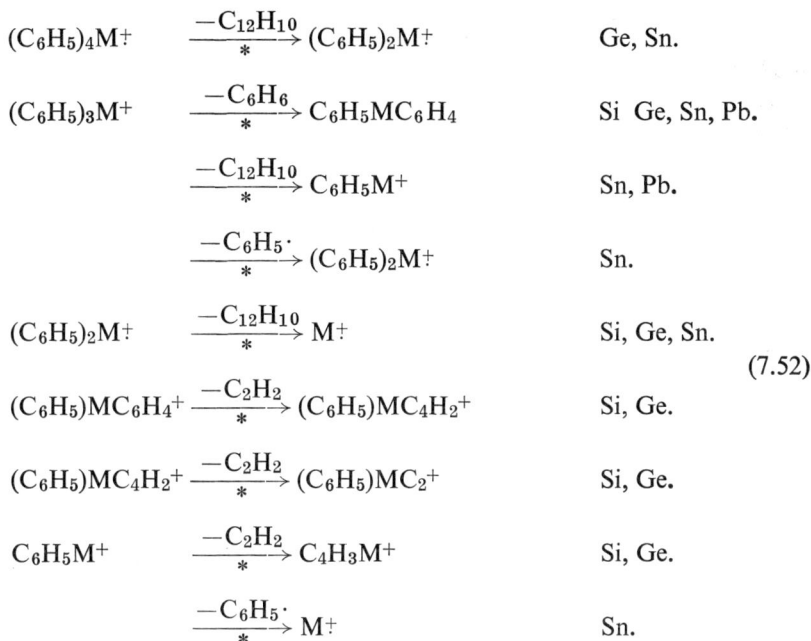

$$(C_6H_5)_4M^{\dagger} \xrightarrow[*]{-C_{12}H_{10}} (C_6H_5)_2M^{\dagger} \qquad \text{Ge, Sn.}$$

$$(C_6H_5)_3M^+ \xrightarrow[*]{-C_6H_6} C_6H_5MC_6H_4 \qquad \text{Si Ge, Sn, Pb.}$$

$$\xrightarrow[*]{-C_{12}H_{10}} C_6H_5M^+ \qquad \text{Sn, Pb.}$$

$$\xrightarrow[*]{-C_6H_5\cdot} (C_6H_5)_2M^{\dagger} \qquad \text{Sn.}$$

$$(C_6H_5)_2M^{\dagger} \xrightarrow[*]{-C_{12}H_{10}} M^{\dagger} \qquad \text{Si, Ge, Sn.}$$

(7.52)

$$(C_6H_5)MC_6H_4^+ \xrightarrow[*]{-C_2H_2} (C_6H_5)MC_4H_2^+ \qquad \text{Si, Ge.}$$

$$(C_6H_5)MC_4H_2^+ \xrightarrow[*]{-C_2H_2} (C_6H_5)MC_2^+ \qquad \text{Si, Ge.}$$

$$C_6H_5M^+ \xrightarrow[*]{-C_2H_2} C_4H_3M^+ \qquad \text{Si, Ge.}$$

$$\xrightarrow[*]{-C_6H_5\cdot} M^{\dagger} \qquad \text{Sn.}$$

Glockling et al.[147] have pointed out that the processes involving fragmentation of the phenyl groups attached to a metal require considerable energy. In the study of SnPh$_4$, they found that ions from processes involving phenyl ring fragmentation disappeared at electron voltages below 20 eV. The ion $(C_6H_5)M^+C_6H_4$ observed in all Group IV spectra, was postulated to have a structure IV in which the metal remains tri-coordinate e.g. from SnPh$_4$,

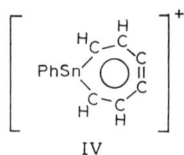

IV

It is not difficult to imagine that ions of type IV would lose acetylene easily to give PhM$^+$C$_4$H$_2$ and PhM$^+$C$_2$ ions.

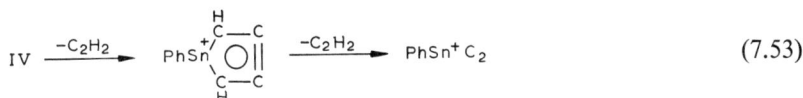

(7.53)

Moreover, the tri-coordinate structure of IV would explain its relatively high abundance because of the possibility of extensive delocalisation of the postive charge. It is perhaps significant that the fragmentation of phenyl groups on $Ph_3M^+$ by loss of acetylene was not observed. This may be due to the (possibly) large amount of charge delocalisation in $Ph_3M^+$. Another contrasting feature of the spectra of tetraphenyl and tetraalkyl compounds (other than the tetrabenzylgermanium)[70] is the loss of H radical from the parent molecular ion from $Ph_4Ge$, *viz.*

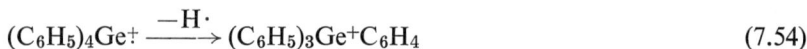

$$(C_6H_5)_4Ge^+ \xrightarrow{\ -H\cdot\ } (C_6H_5)_3Ge^+C_6H_4 \qquad (7.54)$$

Corresponding reactions have also been observed[70] with *m*- or *p*-(tolyl)$_4$Ge (see below). In their work on $Ph_4M$ compounds, Bowie and Nussey also reported the metastable supported loss of $C_{12}H_{10}$ from $(C_6H_5)_3M^+$ ions for $M = Ge$, Sn, or Pb but not for $Si^{182}$. However, the silicon compound lost $C_{12}H_{10}$ from the parent molecular ion and by studying the deuterated compounds, V, and $(C_6H_5)_3Si(C_6D_5)$, they established that H–D scrambling

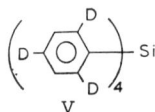

prior to $C_{12}H_{10}$ loss did not occur. Bowie and Nussey reported some features in the negative ion spectra of the organometallic tetraphenyls. The base peaks were due to $(C_6H_5)_2Si^-$ and $(C_6H_5)_3M^-$ for $M = Ge$, Sn, Pb. The spectra of the deuterated compounds showed that loss of $C_6H_5\cdot$ or $C_{12}H_{10}$ from $(C_6H_5)_4Si^-$ was not accompanied by H–D randomisation but complete randomisation did occur in the elimination of $H_2$ from the $(C_6H_5)_2Si^-$ ion.

$$(C_6H_5)_2Si^- \xrightarrow{\ -H_2\ } \qquad\qquad (7.55)$$

Similar conclusions concerning the spectra of $Ph_3M$ ($M = P$, As, Sb) were found (see Chapter 8, Section 7.4).

### 7.3.2 Other aryl groups
Tetraperfluorophenyl derivatives of Group IV elements have been studied mass spectrometrically by Miller[183,184]. Glockling *et al.*[70] questioned the

Table 7.15

Ion Abundances from $(C_6F_5)_4M$ Compounds

| Composition of ion | % Abundance for M = | | | |
|---|---|---|---|---|
| | Si | Ge | Sn | Pb |
| $(C_6F_5)_4M$ | 56.6 | 32.3 | 9.1 | 1.7 |
| $(C_6F_5)_3MC_6F_4$ | 0.5 | 0.2 | | 0.1 |
| $(C_6F_5)_3M$ | 2.1 | 40.6 | 25.1 | 24.9 |
| $(C_6F_5)_2MC_5F_3$ | 7.6 | 1.6 | | |
| $(C_6F_5)_2MF$ | | 5.4 | 13.2 | |
| $(C_6F_5)_2M$ | 5.6 | | | |
| $C_{11}F_9M$ | 10.2 | | | |
| $C_6F_5MF_2$ | | 5.8 | 3.5 | |
| $C_6F_5MF$ | 6.6 | | 0.3 | 0.1 |
| $C_6F_5M$ | | | 1.6 | 21.3 |
| $C_5F_5M$ | 1.7 | | | |
| $MF_3$ | 6.6 | | | |
| MF | 1.5 | 12.2 | 43.9 | 31.5 |
| M | | | 2.9 | 20.4 |

$(C_6F_5)_4M^{2+}$ was also observed in the abundances Si 1.0, Ge 1.9, Sn 0.3, and Pb 0.1%.

ion abundances from $Ge(C_6F_5)_4$ reported first by Miller[184]. The results of the later work[183] are given in Table 7.15 with the spectra of the other Group IV compounds. Parent molecular ions were observed, that of the silicon compound being the base peak of the spectrum. The other compounds showed base peaks due to $(C_6F_5)_3Ge^+$ and $MF^+$ (M = Sn or Pb). Fragmentation mainly occurred by loss of $F\cdot$, $C_6H_5\cdot$, $C_{12}F_{10}$, $C_6H_4$, and $MF_2$ groups. Rearrangement ions with M–F bonds were common. Some metastable supported decompositions are shown in schemes (7.56)–(7.58).

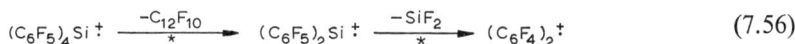

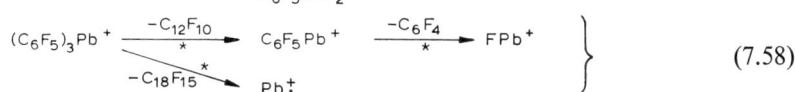

$$(C_6F_5)_4Si^{\ddagger} \xrightarrow[\;\;*\;\;]{-C_{12}F_{10}} (C_6F_5)_2Si^{\ddagger} \xrightarrow[\;\;*\;\;]{-SiF_2} (C_6F_4)_2^{\ddagger} \quad (7.56)$$

$$(C_6F_5)_4Sn^{\ddagger} \xrightarrow[\;\;*\;\;]{-C_6F_5\cdot} (C_6F_5)_3Sn^+$$

$$(C_6F_5)_2SnF^+ \begin{array}{c} \xrightarrow[\;\;*\;\;]{-SnF_2} C_{12}F_9{}^+ \\ \xrightarrow[\;\;*\;\;]{-C_6F_4} C_6F_5SnF_2{}^+ \end{array} \qquad \left.\right\} \quad (7.57)$$

$$(C_6F_5)_3Pb^+ \begin{array}{c} \xrightarrow[\;\;*\;\;]{-C_{12}F_{10}} C_6F_5Pb^+ \xrightarrow[\;\;*\;\;]{-C_6F_4} FPb^+ \\ \xrightarrow[\;\;*\;\;]{-C_{18}F_{15}} Pb^{\ddagger} \end{array} \qquad \left.\right\} \quad (7.58)$$

The spectra of the *ortho*-, *meta*-, and *para*-tetratolylgermanes have been discussed by Glockling *et al.*[70]. These spectra were in many respects similar to that of tetraphenylgermane. All the tolyl compounds exhibited loss of $H\cdot$, $(C_7H_7)_2$, $C_7H_8$, and $C_2H_2$ groups.

$$(C_7H_7)_4Ge^{\ddagger} \xrightarrow{\phantom{xx}-H\cdot\phantom{xx}} (C_7H_7)_3GeC_7H_6^+ \tag{7.59}$$

$$\xrightarrow{-C_{14}H_{14}} (C_7H_7)_2Ge^{\ddagger} \tag{7.60}$$

$$(C_7H_7)_3Ge^+ \xrightarrow[*]{-C_7H_8} (C_7H_7)GeC_7H_6^{\ddagger} \tag{7.61}$$

$$(C_7H_7)Ge^+ \xrightarrow[*]{-C_2H_2} C_5H_5Ge^+ \underset{\text{(for } m\text{- and } p\text{-only)}}{\searrow} \xrightarrow[*]{-C_2H_2} C_3H_3Ge^+ \tag{7.62}$$

The spectra of the tolyl compounds also show some differences. For example, both the *m*- and *p*-compounds lose methyl radicals, (7.63), and the *o*- and *m*-compounds show decomposition pathways involving elimination of germanium or $GeH_2$, (7.64) and (7.65).

$$(C_7H_7)_4Ge^{\ddagger} \xrightarrow{-CH_3\cdot} (C_7H_7)_3GeC_6H_4^+ \tag{7.63}$$

$$(C_7H_7)Ge^+ \xrightarrow[*]{-Ge} C_7H_7^+ \tag{7.64}$$

$$(C_7H_7)GeC_7H_6^+ \xrightarrow[*]{-GeH_2} C_{14}H_{11}^+ \tag{7.65}$$

Generally, the spectra of *m*- and *p*-(tolyl)$_4$Ge are very similar and differ somewhat from *o*-(tolyl)$_4$Ge. This may be seen easily in the table of abun-

Table 7.16

Ion Abundances of Some Ions from *o*, *m*- and *p*-(tolyl)$_4$Ge

| Composition of ion | % Abundance | | |
|---|---|---|---|
| | *o* | *m* | *p* |
| $(C_7H_7)_3Ge$ | 65.0 | 46.0 | 44.0 |
| $(C_7H_7)_2Ge$ | 2.5 | 28.0 | 29.0 |
| $C_7H_7Ge$ | 9.0 | 6.0 | 8.0 |
| $C_{14}H_{11}Ge$ | 3.5 | 1.0 | 1.0 |

dances of the major ions (Table 7.16). The relatively small abundance of $(C_7H_7)_2Ge^+$ from the o-compound is probably due to the o-methyl groups inhibiting the formation of the C–C bond in $C_{14}H_{14}$ from the parent molecular ion. Another difference is the formation of $Ge^+$ and $C_{14}H_{14}$ from $(C_7H_7)_2Ge^+$ only in the cases of m- and p-tolyls. It is interesting to note that $C_6H_5Ge^+$ was formed in the spectra of both o- and p-(tolyl)$_4$Ge, but its formation followed different routes in each case, i.e. reaction (7.66) for the o-compound and reaction (7.67) for the p-compound.

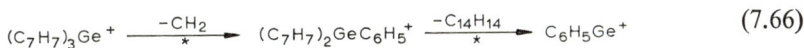

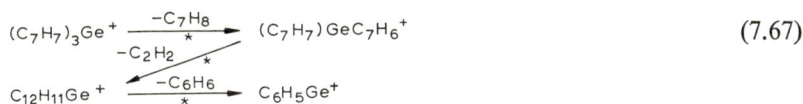

$$(C_7H_7)_3Ge^+ \xrightarrow[\;*\;]{-CH_2} (C_7H_7)_2GeC_6H_5^+ \xrightarrow[\;*\;]{-C_{14}H_{14}} C_6H_5Ge^+ \qquad (7.66)$$

$$(C_7H_7)_3Ge^+ \xrightarrow[\;*\;]{-C_7H_8} (C_7H_7)GeC_7H_6^+ \qquad (7.67)$$

$$C_{12}H_{11}Ge^+ \xleftarrow[\;*\;]{-C_2H_2} \;\; \xrightarrow[\;*\;]{-C_6H_6} C_6H_5Ge^+$$

### 7.4 Mixed alkyl–aryl compounds

The spectra of these compounds contain ions with both aryl and alkyl groups attached to the metal. Generally, ions containing most aryl groups are most abundant and those which contain most alkyl groups are least abundant. This is probably because of the stronger (Ar–M) bonds compared with the (R–M) bonds, the chance of extensive delocalisation of charge in $(Ar–M)^+$ ions, the lower ionisation potentials of (Ar–M)-containing species, and the fact that the fragmentation of an aryl group attached to the metal requires a large amount of energy[147]. The fragmentation of the groups attached to M occurs by the same routes as in the tetraalkyl and tetraaryl compounds.

### 7.4.1 Silanes

Relatively few studies of mixed aryl–alkyl silanes have been reported. The spectra of some trimethylsilyl derivatives $Me_3SiAr$ (Ar = 1-naphthyl, 2-naphthyl, p-methoxyphenyl) have been compared with the corresponding tin compounds[186]. In all cases the base peaks were due to $(CH_3)_2MAr^+$ ions, the parent molecular ions were in greater abundance for the silicon compounds and a larger abundance of $M^+$ ions was observed from the tin compounds. Miller[184] has reported some decomposition paths observed from $Me_3SiC_6X_5$ (X = F or Cl). The most interesting rearrangement reaction reported was (7.68), involving the transfer of three X· groups on to silicon. The reaction was metastable supported.

$$C_6X_5Si(CH_3)_2^+ \xrightarrow[*]{-X_3SiCH_3} C_6X_2CH_3^+ \qquad (7.68)$$

Kinstle et al.[187] have also studied rearrangements in $ArSiR'R^+$ ions generated from $ArRR'Si-SiRR'Ar$ compounds. They studied the $\gamma$-H transfer reaction

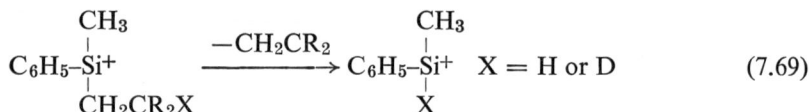

$$\underset{\underset{CH_2CR_2X}{|}}{\overset{\overset{CH_3}{|}}{C_6H_5-Si^+}} \xrightarrow{-CH_2CR_2} \underset{\underset{X}{|}}{\overset{\overset{CH_3}{|}}{C_6H_5-Si^+}} \quad X = H \text{ or } D \qquad (7.69)$$

The rearranged ion was the base peak in the spectra of the compounds studied. From deuterium labelling experiments, they showed that reaction (7.70) went with greater than 97% specifity. Confirmation that it was a $\gamma$-H atom involved was obtained from reaction (7.71) which went with greater than 95% specifity.

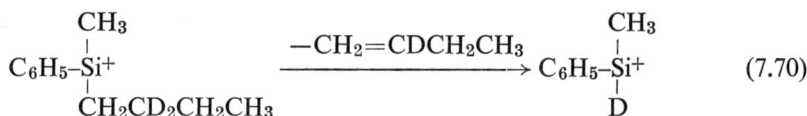

$$\underset{\underset{CH_2CD_2CH_2CH_3}{|}}{\overset{\overset{CH_3}{|}}{C_6H_5-Si^+}} \xrightarrow{-CH_2=CDCH_2CH_3} \underset{\underset{D}{|}}{\overset{\overset{CH_3}{|}}{C_6H_5-Si^+}} \qquad (7.70)$$

$$\underset{\underset{D}{|}}{\overset{\overset{C_2H_5}{|}}{C_6H_5-Si^+}} \xrightarrow{-C_2H_4} \underset{\underset{H}{|}}{\overset{\overset{D}{|}}{C_6H_5-Si^+}} \qquad (7.71)$$

However, the fragmentation of ions with a phenyl group on the $\gamma$-carbon atom was more complicated. Reaction (7.72) went cleanly with R = H, R' = D but scrambling of H–D occurred with R = D, R' = H and the elimination of $C_2H_4$ was observed.

$$\underset{\underset{C_6H_5}{|}}{\overset{\overset{H}{|}}{C_6H_5CR_2CR'_2-Si^+}} \xrightarrow{-CR_2CR'_2} \underset{\underset{C_6H_5}{|}}{\overset{\overset{H}{|}}{C_6H_5-Si^+}} \qquad (7.72)$$

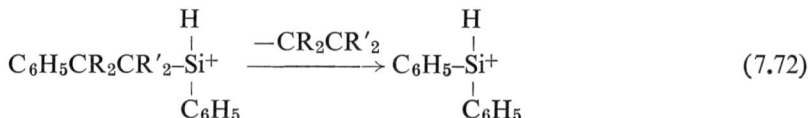

Preston et al.[188] have reported the spectra of the compounds $Ph_3M-CH=CH_2$ (M = C, Si, Ge, Sn). Parent molecular ions were observed in all the spectra; those of the carbon and silicon compounds were in higher abundance than from the germanium or tin compounds. As well as some of the metastable supported decompositions of the $(C_6H_5)_3Si^+$ ion previously discussed in the Section 7.3.1. above, Preston *et al.* reported the loss of $H_2$, *viz.*

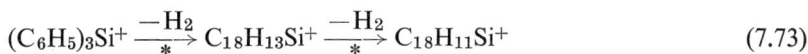

$$(C_6H_5)_3Si^+ \xrightarrow[*]{-H_2} C_{18}H_{13}Si^+ \xrightarrow[*]{-H_2} C_{18}H_{11}Si^+ \qquad (7.73)$$

Other metastable supported decompositions were the loss of benzene from the parent molecular ion, reaction (7.74), and acetylene from the (parent molecule $-C_6H_5 \cdot)^+$ ion, reaction (7.75).

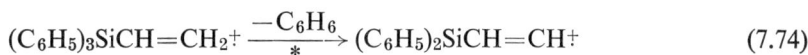

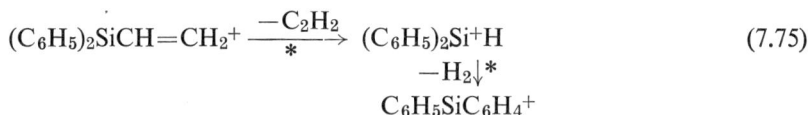

$$(C_6H_5)_3SiCH=CH_2^+ \xrightarrow[*]{-C_6H_6} (C_6H_5)_2SiCH=CH^+ \qquad (7.74)$$

$$(C_6H_5)_2SiCH=CH_2^+ \xrightarrow[*]{-C_2H_2} (C_6H_5)_2Si^+H \qquad (7.75)$$
$$-H_2\downarrow*$$
$$C_6H_5SiC_6H_4^+$$

The carbon compound behaved differently, losing $\cdot CH_3$ and $C_6H_5CH=CH_2$ as well as $C_6H_6$ from the parent molecular ion. The germanium and tin compounds behaved differently again in that loss of $\cdot CH=CH_2$, reaction (7.76), from $(C_6H_5)_2M^+CH=CH_2$ was preferred to loss of $C_2H_2$ and then $H_2$ as found for $M = Si$, reaction (7.75).

$$(C_6H_5)_2M^+CH=CH_2 \xrightarrow[*]{-C_2H_3 \cdot} (C_6H_5)_2M^+ \quad M = Ge \text{ or } Sn \qquad (7.76)$$

The germanium compound showed the loss of $C_{12}H_{10}$, reaction (7.77), from the parent molecular ion in contrast to the loss of benzene observed from the silicon compound, reaction (7.74).

$$(C_6H_5)_3GeCH=CH_2^+ \xrightarrow[*]{-C_{12}H_{10}} (C_6H_5)GeCH=CH_2^+ \qquad (7.77)$$

### 7.4.2 Germanes

Glockling *et al.*[70] have published detailed mass spectra of $Ph_3GeEt$ obtained at high resolution and $Ph_3GeMe$, $Ph_3Ge(CH_2Ph)$, $Ph_2GeEt_2$ and $PhGeEt_3$ at low resolution. Parent molecular ions were observed in each case and ions containing Ar–Ge bonds were favoured over ions containing R–Ge bonds. Some interesting elimination reactions involving the loss of $H_2$ or $C_6H_6$ observed in the spectra of these compounds are given in scheme (7.78).

From $Ph_3GeEt$

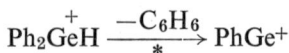

$$\overset{+}{Ph_2GeH} \xrightarrow[*]{-C_6H_6} PhGe^+$$

From $PhGeEt_3$

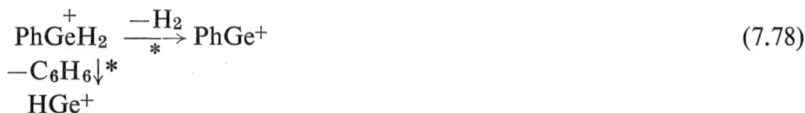

$$\overset{+}{PhGeH_2} \xrightarrow[*]{-H_2} PhGe^+ \qquad\qquad (7.78)$$
$$-C_6H_6\downarrow*$$
$$HGe^+$$

From $Ph_3GeMe$

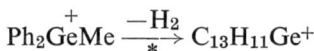

$$\overset{+}{Ph_2GeMe} \xrightarrow[*]{-H_2} C_{13}H_{11}Ge^+$$

The ion abundances of ions from the compounds in the series $Ph_{(3-n)}GeEt_n$ ($n = 0$–3) parallelled those from the corresponding tin compounds which were studied in more detail[147] and are discussed below. Exceptions to this generalisation were the abundances of $Ge^+$ (0.6 %) compared to $Sn^+$ (13.4 %) from $Ph_3MEt$ compounds and the greater abundance of $(C_6H_5)_2Ge^+$ ions compared with $(C_6H_5)_2Sn^+$. Also the ions $(C_6H_5)_3Ge^{2+}$ (from $Ph_3GeEt$), $C_6H_5Ge(C_2H_5)^{2+}$, and $C_6H_5GeC_2H_6^{2+}$ (from $PhGeEt_3$) and $(C_6H_6)_2GeH^{2+}$ from $Ph_2GeEt$ had no counterparts in the spectra of the corresponding tin compounds.

Comparison of the spectra of $Ph_3GeR$ (R = Me, Et, Bu$^n$)[70], shows that as the chain length increases, the % abundance of ions containing the R group decreases. This is probably because of the decreasing Ge–R bond dissociation energies and more favourable reaction paths for parent ion decomposition by elimination of an alkene. In the spectrum of $Ph_3GeBu^n$, the only ions not from $(C_6H_5)_3Ge^+$ were $(C_6H_5)_2Ge(C_4H_9)_2^+$, $(C_6H_5)_2Ge^+$ $C_4H_9$, and $(C_6H_5)_2Ge^+H$ in very low abundance.

Glockling et al.[70] also mention some observations from the spectrum of p-(tolyl)$_3GeCO_2H$. They report ions formed by the elimination of H$\cdot$, a rearrangement ion $(C_7H_7)_2Ge^+OH$, and the elimination of $H_2O$ from this ion, viz.

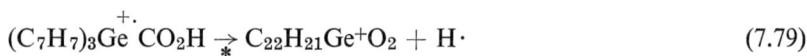

$$(C_7H_7)_3\overset{+\cdot}{Ge}\,CO_2H \xrightarrow[*]{} C_{22}H_{21}Ge^+O_2 + H\cdot \qquad\qquad (7.79)$$

$$(C_7H_7)_2Ge^+OH \xrightarrow[*]{} (C_7H_7)Ge^+C_7H_6 + H_2O$$

### 7.4.3 Stannanes

Gielen and Nasielski[186] reported the mass spectra of a series of $Me_3SnAr$ compounds (Ar = 9-phenanthryl, $p$-bromophenyl, $p$-trimethylsilylphenyl, $p$-neopentylphenyl, 1-naphthyl, 2-naphthyl, $p$-methoxyphenyl, $m$-methoxyphenyl, 2,6-dimethylphenyl, $p$-tolyl, $m$-tolyl, $o$-tolyl, and phenyl groups). Parent molecular ions were observed for all compounds and the base peaks were due to $ArSn(CH_3)_2^+$ ions. The spectra were essentially very simple, the major ions being due to loss of methyl groups. Some doubly charged ions of the type $ArSn(CH_3)_2^{2+}$ were observed. These workers tried to deduce a series of "the ease of cleavage of the aryl–Sn bond" from their ion abundances, (presumably in the parent ions) and then attempted to explain their results "in terms of bond energies" and "electron delocalisation". They tried unsuccessfully to correlate their reactivity sequence with other data such as Taft $\sigma^+$ functions. This was not surprising since $\sigma^+$ functions are ground-state functions and their observed sequence was deduced for ions, some in excited states. The spectra of the compounds $PhSnR_3$ (R = Me, Et, $Pr^n$, $Pr^i$ and $Bu^n$) were reported[186]. Weak parent molecular ions were observed and rearrangement ions involving the loss of alkenes were common, particularly for ·R other than methyl.

Glockling et al.[145,147] have reported the spectra of $Ph_nSnEt_{(4-n)}$ ($n$ = 0–4). The abundance of ions containing ethyl groups was less than for ions containing phenyl groups, for example from $Ph_2SnEt_2$ the $(C_6H_5)_2Sn^+C_2H_5$ ion was 50.8% abundant and the $C_6H_5Sn^+(C_2H_5)_2$ ion 1.1%. This may be expected on simple bond strength arguments. Also, the Sn–Et fragments have a more favourable decomposition path by loss of ethylene to the hydride ions than do Ph–Sn fragments. The formation of hydride ions increases with increasing number of ethyl groups.

Lengyel and Aaronson[189,190] have studied a series of 10,10-dialkyl-9-oxo 10-stanna-9,10-dihydroanthracenes, VI.

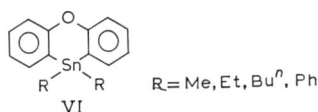

R = Me, Et, $Bu^n$, Ph

VI

The spectra of the methyl and phenyl compounds showed some interesting differences in the modes of fragmentation compared to the ethyl and butyl compounds. However, all compounds exhibited parent molecular ions and (parent molecule $-R\cdot)^+$ ions. The ethyl and butyl compounds then lost alkene to give a hydride ion, *e.g.* the butyl compound showed the reaction sequence[190].

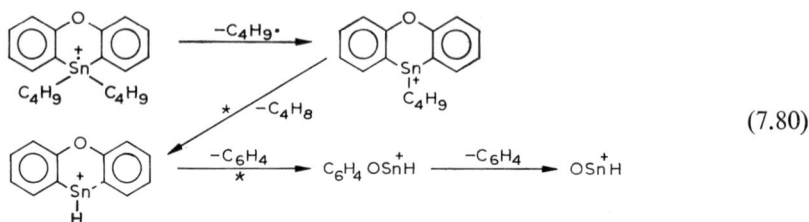

$$(7.80)$$

The methyl and phenyl compounds, after losing an R· group, preferred to fragment further by elimination of $C_6H_4O$ and $C_{12}H_8O$, or $C_6H_4$ from the remaining phenyl group.

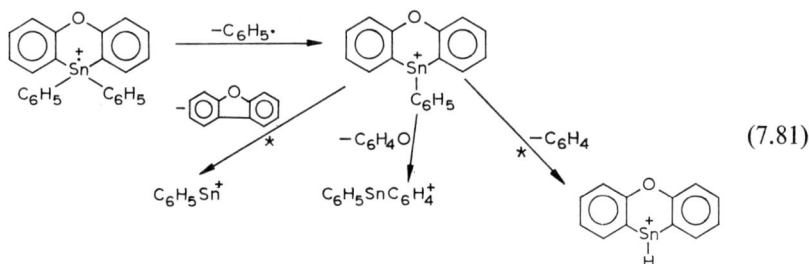

$$(7.81)$$

Table 7.17

Other Mixed Aryl–Alkyl Compounds of Group IV Elements

| Compound | Comments | Ref. |
|---|---|---|
| PhSnMe₃ | | 165 |
| (Me₃SiPh)₂Cr | P.m.i., base peak Cr⁺ | 191 |
| (structure: Ge ring with D₂, D₂, R, R') | R = R′ = Ph; p.m.i., (C₆H₅)₂Ge⁺ and C₆H₅Ge⁺ most abundant ions | 175 |
| | R = Buⁿ, R′ = Ph, p.m.i., (C₆H₅)Ge⁺ base peak. | |
| | Both compounds lose C₆H₅ radicals and eliminate CH₂CD₂ | |
| Ph₃SnCOPh | P.m.i., (C₆H₅)₃Sn⁺ base peak | 192 |
| MePrⁱ(cyclohexyl)Sn–C₆H₄–OMe | P.m.i., CH₃OC₆H₄SnCH₄⁺ base peak. | 179 |

Other compounds containing aryl and alkyl groups bonded to Group IV atoms which have been studied are shown in Table 7.17.

## 7.5 Compounds containing alkyl and or aryl groups and hydrogen bonded to main Group IV elements

### 7.5.1 Silanes

Van der Kelen et al.[136] have reported the spectra of methylsilanes $Me_nSiH_{(4-n)}$ ($n = 0–4$). Their ion abundance values for $SiH_4$ and $SiMe_4$ are in good agreement with previously published work. Parent molecular ions and some metastable supported elimination reactions of $\cdot H, H_2, \cdot CH_3$, $CH_4$ and $C_2H_4$ were observed. Some examples from the spectra of $MeSiH_3$ and $Me_3SiH$ are given below.

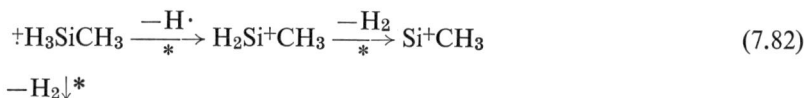

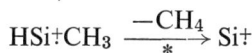

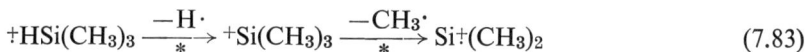

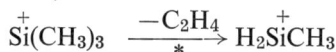

$$\dagger H_3SiCH_3 \xrightarrow[*]{-H\cdot} H_2Si^+CH_3 \xrightarrow[*]{-H_2} Si^+CH_3 \qquad (7.82)$$

$$-H_2\!\downarrow *$$

$$HSi\dagger CH_3 \xrightarrow[*]{-CH_4} Si\dagger$$

$$\dagger HSi(CH_3)_3 \xrightarrow[*]{-H\cdot} {}^+Si(CH_3)_3 \xrightarrow[*]{-CH_3\cdot} Si\dagger(CH_3)_2 \qquad (7.83)$$

$$-H\cdot\!\downarrow *$$

$$\overset{+}{Si}(CH_3)_3 \xrightarrow[*]{-C_2H_4} \overset{+}{H_2SiCH_3}$$

An interesting point in the spectrum of $H_3SiCH_3$ was the formation of $Si\dagger CH_4$, the most abundant ion (32%). In the spectrum of $C_2H_6$, the corresponding ion $C_2H_4{}^+$ probably has the structure VII.

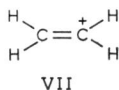

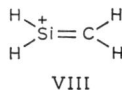

VII                          VIII

It is tempting to write the silicon-containing species as VIII and explain its apparent stability in terms of possible resonance stabilisation.

The spectra of $(C_6X_5)_3SiH$ ($X = H$ or $F$) have been studied by Schrieke and West[193]. The phenyl compound exhibited a parent molecular ion which was the base peak ion. Several interesting metastable supported decompositions were observed including the successive loss of two phenyl groups from the parent molecular ion.

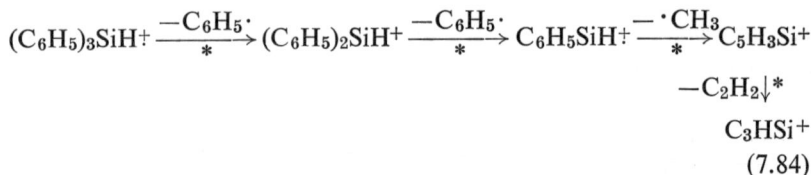

$$(C_6H_5)_3SiH^{\ddagger} \xrightarrow[*]{-C_6H_5^{\cdot}} (C_6H_5)_2SiH^+ \xrightarrow[*]{-C_6H_5^{\cdot}} C_6H_5SiH^{\ddagger} \xrightarrow[*]{-\cdot CH_3} C_5H_3Si^+$$

$$-C_2H_2\!\downarrow^*$$

$$C_3HSi^+$$

$$(7.84)$$

The fluorocarbon derivative, on the other hand, showed rather different decomposition pathways (7.85) losing $\cdot C_6F_5$ and $H\cdot$ rather than two $\cdot C_6F_5$ groups from the parent molecular ion and fragmenting further by elimination of $SiF_4$, $:SiF_2$, and fluorocarbon species.

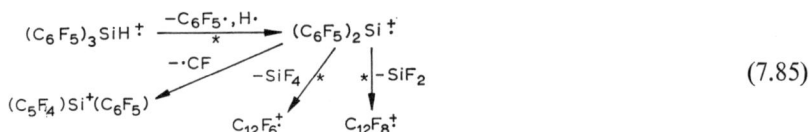

$$(7.85)$$

The base peak in the spectrum of $(C_6F_5)_3SiH$ was due to the parent molecular ion. Rearrangement ions which contained Si–F bonds, such as $F_3Si^+$ and $F_2Si^{\ddagger}$, ions due to $F\cdot$ loss, and ions such as $F_3CSiH^{\ddagger}$ and $F_2CSiH^+$ which retained the Si–H bond even after cleavage of the fluorophenyl ring, were quite abundant.

The spectra of silacyclobutane[194], IX, and silacyclopentane[195], X, have

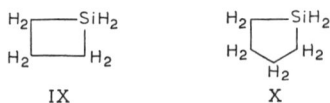

IX

X

been discussed. Both showed strong parent molecular ions, base peaks due to (parent molecule $-28$)$^+$ ions, and ions due to the loss of $\cdot H$ and $H_2$ were abundant. The spectrum of deuterium-labelled silacyclobutane $(CH_2)_3SiD_2$ confirmed that (parent molecule $-1$)$^+$ and (parent molecule $-2$)$^+$ ions were produced by cleavage of Si–H bonds[194]. Differences in the spectra of IX and X were the type of ions $SiH_n{}^+$ formed ($n = 2,3$ for silacylobutane but $n = 0$ or 1 for silacylopentane) and the rearrangement reaction (7.86) observed only from the silacylobutane.

$$\begin{array}{c} \mathrm{H_2} \boxed{\begin{array}{c} \!\!\!\!\!\!\!\!\!\!\!\!\!\!\!\!\! \mathrm{SiH_2} \\[6pt] \end{array}} \!\!\mathrm{H_2} \\ \mathrm{H_2} \qquad\quad \mathrm{H_2} \end{array} \xrightarrow{-\,:\mathrm{CH_2}} \mathrm{H_2C{=}CH{-}SiH_3^+} \qquad\qquad (7.86)$$

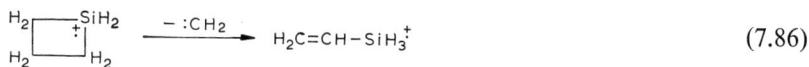

Several studies involving methylsilane in connection with the kinetics of ion–molecule reactions or as a precursor for $CH_3SiH_2$ radicals have been reported. Potzinger and Lampe[196] studied the formation of both positive and negative ions and using a value for $\varDelta H°_f SiH_2^+$, they calculated $\varDelta H°_f Me\text{-}SiH_3 = 0 \pm 2$ kcal.mole$^{-1}$. The ion–molecule reactions between $CH_3SiH^+$ and $CH_3SiH_3$ which produced a large number of Si- and Si$_2$-containing species were also studied[196]. The generation of $CH_3SiH_2$ radicals from $CH_3SiH_3$ by irradiation from a $^3P_1$, Hg lamp and their reactions with nitric oxide have been studied[197]. The major products are siloxanes and $N_2O$. The reactions of methyl radicals with $CH_3SiH_3$ have been studied by Morris and Thynne[102]. Methane and $C_3H_9Si$ radicals were observed as products. Meinrenken[106] has used methylsilane to study the natural abundances of silicon.

Other compounds in this category are listed in Table 7.18.

### 7.5.2 Germanes

The spectra of $Me_3GeH$, $Et_3GeH$, $(PhCH_2)_3GeH$, and $Ph_3GeH$ have been reported in detail by Glockling et al.[70]. All compounds showed parent molecular ions and the base peaks were due to the ions $(CH_3)_2GeH^+$, $C_2H_5GeH_2^+$, $C_7H_7Ge^+$, and $(C_6H_5)_2Ge^+$ respectively. Some metastable supported decompositions which were reported are given in reaction schemes (7.87)–(7.90)

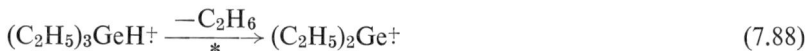

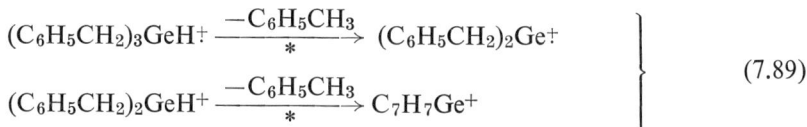

$$(CH_3)_2GeH^+ \xrightarrow[*]{-C_2H_4} H_3Ge^+ \qquad\qquad (7.87)$$
$$\begin{array}{c} -CH_4\!\downarrow\!* \\ CH_3Ge^+ \end{array}$$

$$(C_2H_5)_3GeH^{\underline{+}} \xrightarrow[*]{-C_2H_6} (C_2H_5)_2Ge^{\underline{+}} \qquad\qquad (7.88)$$

$$(C_6H_5CH_2)_3GeH^{\underline{+}} \xrightarrow[*]{-C_6H_5CH_3} (C_6H_5CH_2)_2Ge^{\underline{+}}$$

$$(C_6H_5CH_2)_2GeH^+ \xrightarrow[*]{-C_6H_5CH_3} C_7H_7Ge^+ \qquad\qquad \left.\right\} \quad (7.89)$$

$$(C_6H_5)_2GeH^+ \xrightarrow[*]{-C_6H_6} C_6H_5Ge^+ \qquad\qquad (7.90)$$

Table 7.18

Other Alkyl or Aryl Group IV Hydrides

| Compound | Comments | Ref. |
|---|---|---|
| $H_3SiCH_2OCH_3$ | Base peak $H_3SiCH_2O^+$ | 198 |
| $H_3SiCH_2X$ | P.m.i., base peaks were $SiCl^+$, and | |
| X = Cl,Br,I | $H_3SiCl^+$ respectively | 199 |
| $R_3SiH$ | R = Et or $Pr^i$, p.m.is., $(C_2H_5)_2SiH^+$ and | |
| | $(CH_3)_2SiH^+$ respectively, base peaks | |
| $H_3SiCH_2GeH_3$ | | 200 |
| $H_3GeSiH_2CH_3$ | | 200 |
| $(H_3SiCH_2)_2GeH_2$ | | 200 |
| $CH_3Si_2H_5$ | | 83 |
| $CH_3SiH_2SiD_2H$ | | 83 |
| $C_2H_5Ge_2H_5$ | } P.m.i. | 201 |
| $CH_3Ge_2H_5$ | | 201 |
| $(CH_3)_2GeHGeH_3$ | | 202 |
| $CH_3GeH_2GeH_2CH_3$ | } P.m.i. and expected fragment ions | 202 |
| $(CH_3)_2GeHGeH_2CH_3$ | | 202 |
| $(CH_3)_2GeHGeH_2CH_3$ | | 202 |
| $C_6H_5H_2SiSiH_2C_6H_5$ | } P.m.i. | 203 |
| $C_6H_5Si_2H_5$ | | 203, 204 |
| $C_6H_5SiH(SiH_3)_2$ | | 204 |
| $(C_6H_5)_2SiHSiH_3$ | | 204 |
| $(C_6H_5)GeH_2SiH_3$ | } P.m.i. | 204 |
| $(C_6H_5)_2GeHSiH_3$ | | 204 |
| $(C_6H_5)Ge(SiH_3)_3$ | | 204 |
| $(CH_3)_3SiSiH_2Si(CH_3)_3$ | P.m.i. | 205 |
| $(CH_3)_3SiGeH_3$ | | 206 |
| $(CH_3)_3SiGeD_3$ | } P.m.i. and expected fragment ions | 206 |
| $(CH_3)_3GeGeD_3$ | | 206 |

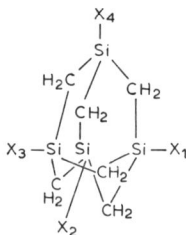

$X_1 = X_2 = X_3 = X_4 = H$ or D, 207
p.m.is., metastable supported loss of
$\cdot CH_3$ or $\cdot CH_2D$ and $\cdot CH_3$ from p.m.i.

$X_1 = X_2 = X_3 = X_4 = H$, p.m.i.  174

$X_1 = X_2 = X_3 = H, X_4 = Me$, p.m.i.  174

| $Me_3SnH$ | No p.m.i. observed, $(CH_3)_2SnH^+$ base peak |

In the spectrum of triphenylgermane, the only ions with the Ge–H bond intact were $(C_6H_5)_2GeH^+$ (4%) and $C_6H_5GeH^+$ (0.4%). These abundances may be compared to $(C_6H_5)_3Ge^+$ (6.1%) and $(C_6H_5)_2Ge^+$ (31.5%) from $Ph_3GeH$.

Duffield et al.[175] reported the spectra of some substituted germacyclopentanes, XI. They all exhibited parent molecular ions and (parent molecule $-C_2H_4)^+$ ions.

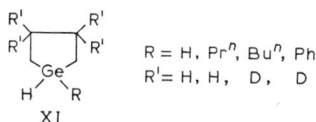

R = H, Pr$^n$, Bu$^n$, Ph
R' = H, H, D, D

XI

By labelling the R' positions with deuterium, Duffield et al. showed that $C_2$ and $C_3$ only were involved in the lost $C_2H_4$ group, viz.

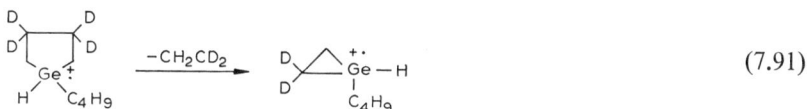

$$-CH_2CD_2 \qquad (7.91)$$

The base peaks in these spectra were due to the $Ge^+$ ion for all compounds except the phenyl derivative which showed $C_6H_5Ge^+$ ion as base peak.

Other compounds which are in this category are listed in Table 7.18.

### 7.6 Compounds containing aryl and or alkyl groups, and halogens bonded to a main Group IV element

#### 7.6.1 Silicon compounds

Accurate spectra of organosilicon halides are often difficult to obtain because the silicon–X bonds hydrolyse readily even under the conditions found in mass spectrometer sources. Trimethylsilyl chloride, for instance, will hydrolyse immediately to hexamethyldisiloxane and only after all the water present in the source has reacted can an accurate spectrum of the $Me_3SiCl$ be obtained. It is therefore advisable to dry out the mass spectrometer with a water scavenging substance such as $BCl_3$ before attempting to record the spectra of organosilicon halides. Furthermore, as Foltz et al.[208] have pointed out, there is a possibility that trimethylsilyl compounds of esters, amines, and amides will react with fluorocarbon substances commonly used in gas chromatographic apparatus linked to mass spectrometers and produce spectra due to $Me_3SiF$ and $Me_6Si_2O$.

Jutzi[209] has reported the elimination of HCl from the parent molecular ion of XII. He also observed ions due to (parent molecule $-CH_3\cdot$)$^+$ and the doubly charged species (parent molecule $-HCl$)$^{2+}$.

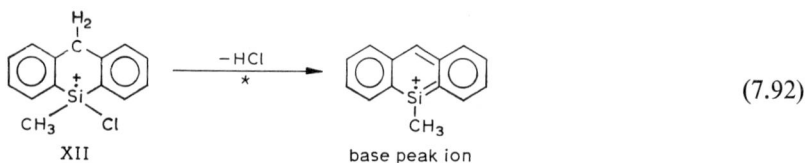

(7.92)

XII                                                    base peak ion

Van Mourik and Bickelhaupt[210] reported some interesting features in the spectra of XIII a and b.

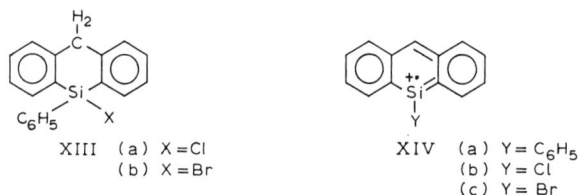

XIII  (a) X=Cl                      XIV  (a) Y=C$_6$H$_5$
      (b) X=Br                           (b) Y=Cl
                                          (c) Y=Br

They found parent molecular ions of XIIIa and b and metastable supported eliminations of HX or C$_6$H$_6$ to produce silaanthracene ions a, b or c.

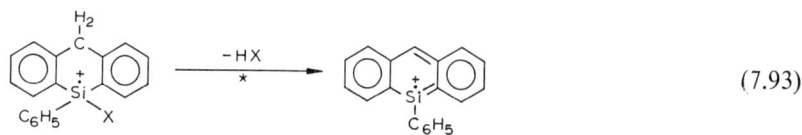

(7.93)

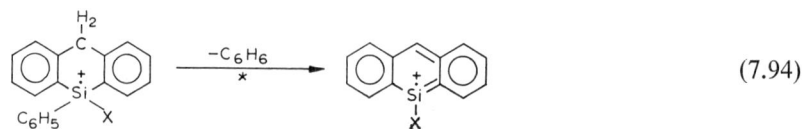

(7.94)

The silaanthracene ion (parent molecule $-HCl$)$^+$ was the base peak ion in the spectrum of XIIIa and the corresponding ion from XIIIb was very abundant. Although the formation of silaanthracene ions was apparently facile, attempts to prepare compounds of the parent silaanthracenes were not conclusive.

Other compounds in this category which have been reported are listed in Table 7.19.

---

Table 7.19

Other Alkyl or Aryl Group IV Halides

| Compound | Comments | Ref. |
|---|---|---|
| $(CH_3)_3SiF$ | From the reaction of $SiF_2$ with $(CH_3)_3SiCl$ | 211 |
| $(CH_3)_3SiF$, $(CH_3)_3Si_2F_3$, $(CH_3)_3SiSi(CH_3)_2F$, $F_3SiOCH_3$ and $(CH_3)_3SiSiF_2OCH_3$ | From the reaction of $SiF_2$ with $(CH_3)_3SiOCH_3$ | 211 |
| $(C_2H_5)_3SiX$   X = F or Cl | P.m.i., $(C_2H_5)_2SiX^+$ base peaks | 212 |
| $(C_2H_5)(CH_3)_2SiF$ | P.m.i., $(CH_3)_2SiF^+$ base peak | 170 |
| $(CF_3)_2N(CH_2)_2SiCl_2(CH_3)$ | No p.m.i., but $(CF_3)_2N(CH_2)_2^+$, $CH_3SiCl_2^+$ and $(CH_2)_2SiFCl_2^+$ | 170 |
| $(CF_3)_2N(CH_2)_2SiX_3$   X = F or Cl | No p.m.i., but $(CF_3)_2N(CH_2)_2^+$, and $(CH_2)_2SiX_3^+$ | 170 |
| $(CH_3)_3SiSiCl_3$ | P.m.i., $(CH_3)_3Si^+$ base peak | 213 |
| $(CH_3)_2ClSiSiCl(CH_3)_2$ | P.m.i. | 218 |
| $(CH_3)Cl_2SiSiCl_2(CH_3)$ | P.m.i. | 213 |
| $(CH_3)Cl_2Si \cdot Si(CH_3)Cl$ $-SiCl_2(CH_3)$ | P.m.i. | 213 |
| $ClC{\equiv}CSiCl_3$ | P.m.i., $C_2SiCl_3^+$ base peak | 214 |
| 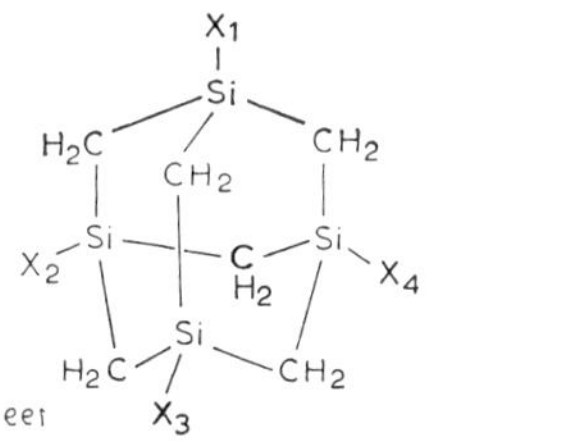 | $X_1 = X_2 = X_3 = X_4 = Cl$, p.m.i., metastable supported loss of $\cdot CH_3$ observed | 207 |
| | $X_1 = X_2 = X_3 = X_4 = Cl$, p.m.i. | 174 |
| | $X_1 = X_2 = X_3 = CH_3, X_4 = Cl$, p.m.i. | 174 |
| | $X_1 = X_2 = CH_3, X_3 = X_4 = Cl$, p.m.i. | 174 |
| | $X_1 = CH_3, X_2 = X_3 = X_4 = Cl$, p.m.i. | 174 |
| $(C_6H_5)_3SiCl$ | P.m.i., loss of $C_6H_5SiCl$ to give $C_{12}H_{10}^+$ metastable supported. $C_6H_5SiCl^+$ also observed | 182 |
| $CH_3GeX_3$   X = Cl, Br | P.m.i. | 215 |
| 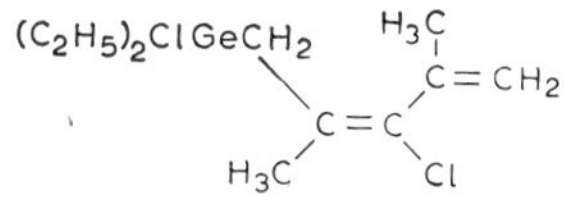 | P.m.i. | 185 |
| 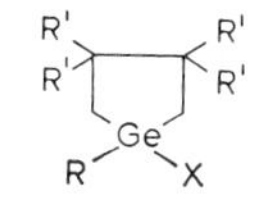 | $R = R' = H, X = Cl$, p.m.i., $GeCl^+$ base peak, loss of $C_2H_4$ from p.m.i. reported | 175 |
| | $R = C_4H_9, R' = D, X = Br$, p.m.i., $GeBr^+$ base peak | 175 |

*References pp. 310–319*

(Table 7.19 continued)

| Compound | Comments | Ref. |
|---|---|---|
| $Cl_3CGeCl_3$ | P.m.i., $CGeCl^+$ base peak | 216 |
| $Cl_2C(GeCl_3)_2$ | P.m.i., $GeCl^+$ base peak | 216 |
| $Cl_2C=CClGeCl_3$ | P.m.i., $GeCl^+$ base peak | 216 |
| $(CH_3)_2(C_3H_7{}^i)SnX$ | P.m.i., $(CH_3)_2SnX^+$ base peak | 179 |
| $\quad$ X = Br, I | | |
| $(CH_3)(C_3H_7{}^n)(C_3H_7{}^i)SnBr$ | P.m.i., $(CH_3)(C_3H_7)SnBr^+$ base peak | 179 |
| $(CH_3)(cyclo\text{-}C_6H_{11})(C_3H_7{}^i)SnBr$ | P.m.i., $(CH_3)(C_6H_{11})SnBr^+$ base peak | 179 |
| $(CH_3)_2(cyclo\text{-}C_6H_{11})SnBr$ | P.m.i., $(CH_3)_2SnBr^+$ base peak | 179 |
|  | P.m.i., loss of $SnCl_2$ from p.m.i. | 189 |

### 7.6.2 Germanium compounds

Detailed mass spectrometric studies of $Me_3GeCl$ and $Ph_3GeX$ (X = Cl, Br, I) have been reported by Glockling et al.[70]. The methyl compound showed a parent molecular ion and the base peak was due to $(CH_3)_2{}^+GeCl$. Fragmentation occurred by loss of $\cdot CH_3$, $C_2H_4$, $C_2H_6$, and groups containing Cl such as HCl and $CH_3Cl$. Some metastable supported decompositions of the base peak ion are shown in scheme (7.95).

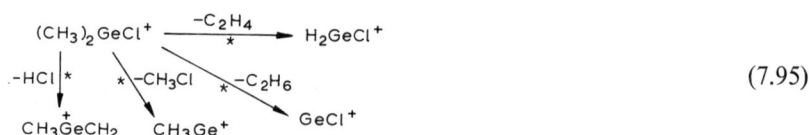

(7.95)

The spectra of the phenyl compounds show ion abundances of Ge–X containing ions and $(C_6H_5)_3Ge^+$ which may be interpreted as reflecting the effect of the decreasing Ge–X bond strength across the series from GeCl to GeI, Ge–Cl > Ge–Br ≫ Ge–I, (Table 7.20). In contrast to the chloride and bromide, the parent molecular ion from $(C_6H_5)_3GeI$ was not observed and only 4 % of the ions of the total ion abundance contained the Ge–I bond. Several interesting metastable supported decompositions were observed; some are shown in scheme (7.96). The decomposition corresponding to the first reaction has been reported for $Ph_3SiCl$.

Table 7.20

Ion Abundances from Triphenylgermanium and Triphenyltin Halides

| Compound | % Abundance of ions | | | |
|---|---|---|---|---|
| | $(C_6H_5)_3MX^+$ | $(C_6H_5)_2MX^+$ | $(C_6H_5)_3M^+$ | $MX^+$ |
| $Ph_3GeCl$ | 10.2 | 53.7 | 1.9 | 11.2 |
| $Ph_3GeBr$ | 7.5 | 12.1 | 52.8 | |
| $Ph_3GeI$ | | 2.5 | 51.9 | 1.4 |
| $Ph_3SnF$ | 0.8 | 44.4 | 3.6 | 25.0 |
| $Ph_3SnCl$ | 1.2 | 47.0 | 1.5 | 27.6 |
| $Ph_3SnBr$ | 1.4 | 46.3 | 5.2 | 26.2 |
| $Ph_3SnI$ | 0.05 | 3.7 | 52.6 | 5.6 |

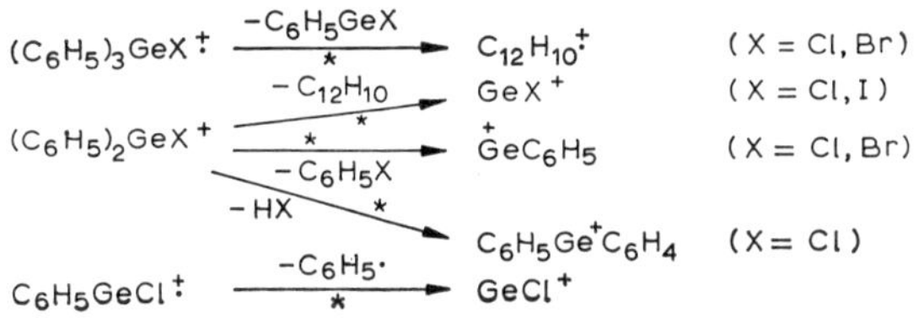

(7.96)

Other compounds in this category are listed in Table 7.19.

### 7.6.3 Tin compounds

Carrick and Glockling[217] have produced a computer programme to calculate combinations of isotope ratios for a compound $a_{0-10}$ $b_{0-10}$ $c_{0-10}$ $d_{0-10}$ containing four polyisotopic elements with up to ten isotopes each. They used the programme to calculate the isotopic pattern for $Et_2SnBr_2$.

Gielen and Mayence[146] have reported spectra for all the compounds $R_3SnX$ (R = Et, $Pr^n$, $Bu^n$, and $Bu^i$; X = F, Cl, Br, I). Parent molecular ions were found for all compounds except $Pr_3SnI$ and $Bu^i_3SnF$. Base peaks were due to ions of the type $R_2SnX^+$ in all cases except $Pr_3SnI$ which gave $(C_3H_7)_2$-$Sn^+H$ and $Bu^n_3SnF$ which produced $(C_2H_6)SnF^+$ as base peak ions. Fragmentation by the loss of R· groups was preferred to the loss of X· except where a facile elimination of an olefin was possible. Some rearrangement ions which were commonly observed were of the types $CH_3Sn^+$, $R_2SnH^+$, $RSnH_2^+$, $RSnXH^+$ and $SnH_m^+$ ($m = 3$ or 1). Several metastable supported decompositions were reported, viz.

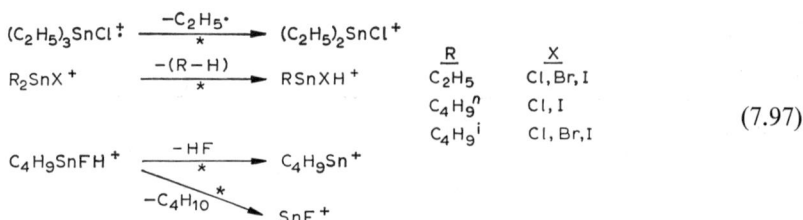

$$(C_2H_5)_3SnCl \cdot \xrightarrow[\quad *\quad]{-C_2H_5 \cdot} (C_2H_5)_2SnCl^+$$

$$R_2SnX^+ \xrightarrow[\quad *\quad]{-(R-H)} RSnXH^+$$

| R | X |
|---|---|
| $C_2H_5$ | Cl, Br, I |
| $C_4H_9^n$ | Cl, I |
| $C_4H_9^i$ | Cl, Br, I |

$$C_4H_9SnFH^+ \xrightarrow[\quad *\quad]{-HF} C_4H_9Sn^+$$

$$\xrightarrow[\quad\quad]{-C_4H_{10}}{}^* SnF^+$$

(7.97)

The mass spectra of the triphenyltin halides (X = F, Cl, Br, I) have been reported by Glockling et al.[147] The abundances of the parent molecular ions, $(C_6H_5)_3Sn^+$, $(C_6H_5)_2SnX^+$, and $SnX^+$ are shown in Table 7.20.

From these figures, one may predict a Sn–X bond strength series of Sn–F ∼ Sn–Cl ∼ Sn–Br ≫ Sn–I using the same reasoning as in the case of the germanium compounds[70]. This perhaps demonstrates the danger of using ion abundances as the criteria of bond strengths, which for these compounds almost certainly fall into the series Sn–F > Sn–Cl > Sn–Br ≫ Sn–I. The conclusions based on ion abundances ignored the fact that the compounds fragmented by different routes.

Several metastable supported fragmentation reactions of the parent molecular ions were reported, viz.

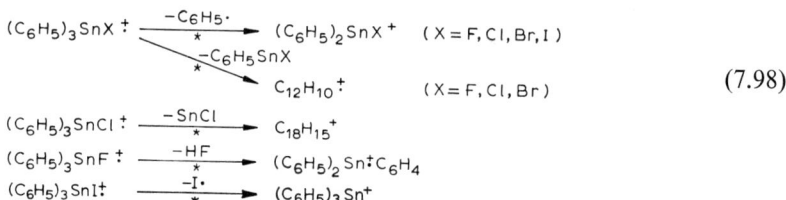

$$(C_6H_5)_3SnX \cdot \xrightarrow[\quad *\quad]{-C_6H_5 \cdot} (C_6H_5)_2SnX^+ \quad (X=F,Cl,Br,I)$$

$$\xrightarrow[\quad *\quad]{-C_6H_5SnX} C_{12}H_{10} \cdot{}^+ \quad (X=F,Cl,Br)$$

$$(C_6H_5)_3SnCl \cdot \xrightarrow[\quad *\quad]{-SnCl} C_{18}H_{15}^+$$

$$(C_6H_5)_3SnF \cdot \xrightarrow[\quad *\quad]{-HF} (C_6H_5)_2Sn \cdot C_6H_4$$

$$(C_6H_5)_3SnI \cdot \xrightarrow[\quad *\quad]{-I \cdot} (C_6H_5)_3Sn^+$$

(7.98)

The ion $C_{12}H_{10} \cdot^+$ accounted for about 10% of the total (metal- and non-metal-containing ions) ion current in the spectra of the fluoride, chloride, and bromide, a much bigger % than in the corresponding germanium chloride and bromide compounds. Other metastable supported decompositions observed are given in the scheme

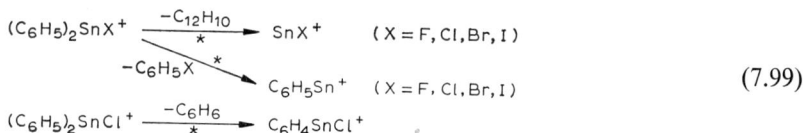

$$(C_6H_5)_2SnX^+ \xrightarrow[\quad *\quad]{-C_{12}H_{10}} SnX^+ \quad (X=F,Cl,Br,I)$$

$$\xrightarrow[\quad\quad]{-C_6H_5X}{}^* C_6H_5Sn^+ \quad (X=F,Cl,Br,I)$$

$$(C_6H_5)_2SnCl^+ \xrightarrow[\quad *\quad]{-C_6H_6} C_6H_4SnCl^+$$

(7.99)

Other compounds in this category are listed in Table 7.19.

## 7.7 Compounds with aryl or alkyl groups and bonds between main Group IV and Group V elements

Although a number of detailed reports of spectra of silicon–nitrogen-containing systems have been published, very little data are available for other Group IV–nitrogen organometallics, (Table 7.21). Perhaps the simplest compound to have been studied is $Me_6Si_2NH$. Features in the spectrum

Table 7.21

Other Alkyl or Aryl Group IV Compounds Containing Bonds to Group V Atoms

| Compound | Comments | Ref. |
|---|---|---|
| $Me_3SiN(CF_3)_2$ | P.m.i. | 224 |
| $Me_3Si-N=N-SiMe_3$ | P.m.i. base peak | 225 |
| $(Me_3Si)_2N-N(SiMe_3)_2$ | P.m.i., $(CH_3)_3Si^+$ base peak | 226 |
| $(Ph_3Ge)_2NH$ | P.m.i., $(C_6H_5)_3Ge^+$ base peak | 227 |
| $(Pr_3^nSn)_2NH$ | P.m.i., $(C_2H_5)_3Sn^+$ base peak | 227 |
| $(Pr_3^nSn)_3N$ | P.m.i., $(C_3H_9)Sn^+$ base peak | 227 |
| $[(PhNH)RSi-NPh]_2$ | $R = Et,Pr^n$ and $Bu^n$; p.m.i. | 228 |
| $[(Pr^nNH)Pr^nSi-NPr^n]_2$ | P.m.i. | 228 |
| $[(R'NH)RSi-NR]_5$ | $R = Et,Pr^n,Bu^n$, $R' = Ph$, p.m.i. | 229 |
| $(Me_3Si-N)_3S$ | P.m.i., $(CH_3)_3Si^+$ base peak | 230 |
| $(Me_3Si-N)_2SF_2$ | P.m.i., $(CH_3)_3Si^+$ base peak | 230 |
| $Me_3SiNHPSF_2$ | P.m.i., $C_3H_6SiNPS^+$ base peak | 231 |
| $Me_3SiNHPSFCl$ | P.m.i., $(CH_3)_3Si\dot{+}Cl$ and $NPS^+$ base peaks | 231 |
| $(Me_3Si)_2N-AuPMe_3$ | P.m.i. | 232 |
| $(Me_2Ge)_6P_4$ | P.m.i., a "cage" molecule | 233 |
| $Me_2Si(PH_2)_2$ | P.m.i., $(CH_3)_2SiPH_2^+$ base peak | 234 |
| $Me_3GePH_2$ | P.m.i., $(CH_3)_3Ge^+$ base peak | 234 |
| $Me_2Ge(PH_2)_2$ | P.m.i., $(CH_3)_2GePH_2^+$ base peak | 234 |

were first reported by Starkey et al.[218] and later work by Tamas et al.[156,157] elucidated some decomposition pathways. The parent molecular ion $(1.1\%)$ and a base peak due to (parent molecule $-\cdot CH_3)^+$ $(46\%)$ were observed. Fragmentation occurred by loss of $\cdot CH_3$, $CH_4$, $C_2H_4$, and also nitrogen-containing species $NH_3$, $HCN$, and $SiNH$. Several interesting metastable supported decompositions were reported[156,157] and they are shown in scheme (7.100).

Tamas et al.[156,157] reported the spectrum and metastable supported decompositions from $Me_6Si_2NMe$. The parent molecular ion $(4.1\%)$ and base

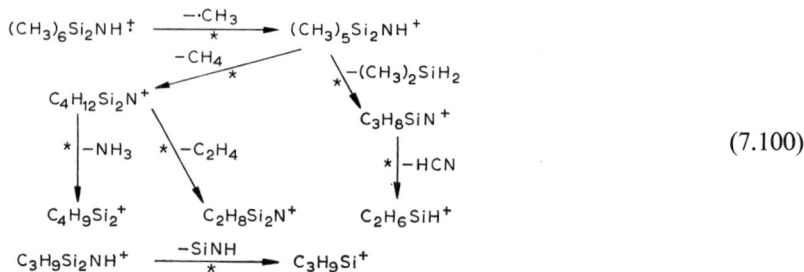

$$
\begin{array}{c}
(CH_3)_6Si_2NH^{+\cdot} \xrightarrow[\;*\;]{-\cdot CH_3} (CH_3)_5Si_2NH^{+} \\
\Big\downarrow{\scriptstyle -CH_4\;*} \qquad \Big\downarrow{\scriptstyle *\;-(CH_3)_2SiH_2} \\
C_4H_{12}Si_2N^{+} \qquad\qquad C_3H_8SiN^{+} \\
{\scriptstyle *\,\big|\,-NH_3} \quad {\scriptstyle *\,\big\backslash\,-C_2H_4} \qquad {\scriptstyle *\,\big|\,-HCN} \\
C_4H_9Si_2^{+} \qquad C_2H_8Si_2N^{+} \qquad C_2H_6SiH^{+} \\
C_3H_9Si_2NH^{+} \xrightarrow[\;*\;]{-SiNH} C_3H_9Si^{+}
\end{array}
\qquad (7.100)
$$

peak ion ($C_6H_{18}Si_2N^+$, 29.7%) were observed and fragmentation proceeded similarly to that found in $Me_6Si_2NH$. Among the metastable supported decompositions observed were those given in scheme (7.101).

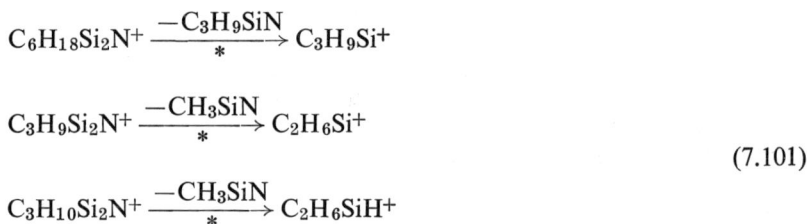

$$C_6H_{18}Si_2N^+ \xrightarrow[\;*\;]{-C_3H_9SiN} C_3H_9Si^+$$

$$C_3H_9Si_2N^+ \xrightarrow[\;*\;]{-CH_3SiN} C_2H_6Si^+$$

$$C_3H_{10}Si_2N^+ \xrightarrow[\;*\;]{-CH_3SiN} C_2H_6SiH^+$$

$$(7.101)$$

Both compounds showed doubly charged ions due to (parent molecule $- \cdot CH_3)^{2+}$ and (parent molecule $-2\cdot CH_3)^{2+}$ and some metastable supported decompositions from these ions, $viz.$

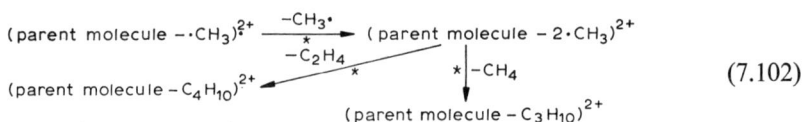

$$
\begin{array}{c}
(parent\ molecule - \cdot CH_3)^{2+} \xrightarrow[\;*\;]{-CH_3\cdot} (parent\ molecule - 2\cdot CH_3)^{2+} \\
{\scriptstyle -C_2H_4}\;\nwarrow{\scriptstyle *} \qquad\qquad {\scriptstyle *\,\big|\,-CH_4} \\
(parent\ molecule - C_4H_{10})^{2+} \qquad (parent\ molecule - C_3H_{10})^{2+}
\end{array}
\qquad (7.102)
$$

Draffen $et\ al.$[219] reported that the spectrum of $Me_3SiN(H)(CH_2)_{10}(H)$-$NSiMe_3$ showed a weak parent molecular ion and the only ions in significant abundance were $H_2C{=}\overset{+}{N}HSi(CH_3)_3$ (base peak), $(CH_3)_3Si^+$, and the (parent molecule $- \cdot CH_3)^+$ ion.

Diekman $et\ al.$[220] have studied compounds containing the $Me_3Si$–$N$ group (XV–XVIII).

Ph(CH$_2$)$_n$NHSiMe$_3$               PhCHNHSiMe$_3$
          XV                                      |
      ($n$ = 1 or 2)                             Me
                                                 XVI


PhCH$_2$CHNHSiMe$_3$            PhCH$_2$N(R)SiMe$_3$
         |                             XVIII
        Me                          (R = Me or Et)
       XVII

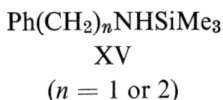

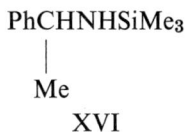

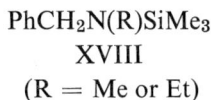


Compound XV with $n$ = 1 showed the parent molecular ion, (parent molecule $-CH_3 \cdot$)$^+$, $CH_2 \overset{+}{=} NHSi(CH_3)_3$ and $(CH_3)_3Si^+$ as major ions and a base peak due to the rearrangment ion $C_6H_5Si(CH_3)_2{}^+$. The formation of $(CH_3)_3$-$Si^+$ from $CH_2 \overset{+}{=} NHSi(CH_3)_3$ was metastable supported.

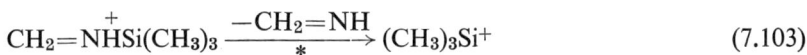

$$CH_2 \overset{+}{=} NHSi(CH_3)_3 \xrightarrow[*]{-CH_2 = NH} (CH_3)_3Si^+ \qquad (7.103)$$

With $n$ = 2 the rearrangement ion $C_6H_5Si(CH_3)_2{}^+$ was very much less abundant and did not exist in the spectra of the branched chain derivatives XVI and XVII. The latter compound gave a very weak parent molecular ion and a weak (parent molecule $-CH_3 \cdot$)$^+$ ion and the formation of $(CH_3)_3Si^+$ from the base peak ion $CH_3CH \overset{+}{=} NHSi(CH_3)_3{}^+$ was metastable supported, a reaction comparable with (7.101). The $N$-methyl and $N$-ethyl compounds were similar, showing the rearrangement ion $C_6H_5Si(CH_3)_2{}^+$ in low abundance. Diekman et al.[221] have also reported features in the spectra of XIX and XX.

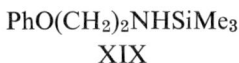

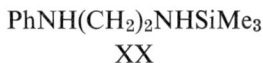

Pho(CH$_2$)$_2$NHSiMe$_3$          PhNH(CH$_2$)$_2$NHSiMe$_3$
         XIX                              XX

Compound XIX showed a parent molecular ion, (parent molecule $- \cdot CH_3$)$^+$, $(CH_3)_3Si^+$, $CH_2 = NHSi(CH_3)_3{}^+$ (base peak), and the rearrangement ion $C_6H_5OSi(CH_3)_2{}^+$, formed via the decomposition in scheme (7.104).

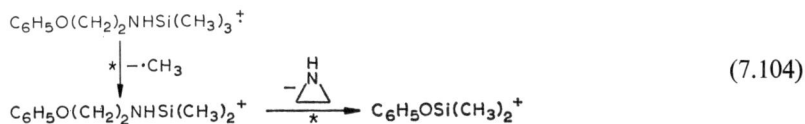

$$C_6H_5O(CH_2)_2NHSi(CH_3)_3 \!{}^{\ddagger}$$
$$\Big\downarrow {}^{*}|{-\cdot CH_3} \qquad \overset{H}{\underset{-\,\triangle}{N}} \qquad (7.104)$$
$$C_6H_5O(CH_2)_2NHSi(CH_3)_2{}^+ \xrightarrow{*} C_6H_5OSi(CH_3)_2{}^+$$

The compound XX showed no rearrangement comparable with (7.104), instead $\alpha$-cleavage was preferred, to give $C_6H_5\overset{+}{N}H=CH_2$ and $(CH_3)_3\overset{+}{S}i=N-H=CH_2$ ions. The ion $(CH_3)_3Si^+$ was also abundant.

A series of compounds of type XXI has been studied by Das et al.[222].

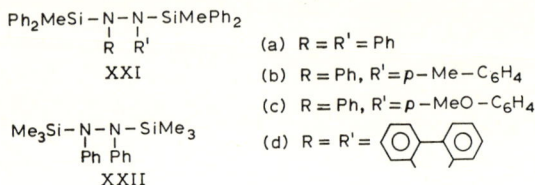

$$Ph_2MeSi-\underset{\underset{R}{|}}{N}-\underset{\underset{R'}{|}}{N}-SiMePh_2$$
XXI

(a) $R = R' = Ph$

(b) $R = Ph,\ R' = p\text{-}Me-C_6H_4$

(c) $R = Ph,\ R' = p\text{-}MeO-C_6H_4$

$$Me_3Si-\underset{\underset{Ph}{|}}{N}-\underset{\underset{Ph}{|}}{N}-SiMe_3$$
XXII

(d) $R = R' = $ 

The silylhydrazines XXI a,b,c, all have the $(C_6H_5)_2SiCH_3^+$ ion as base peak and show metastable peaks for its formation from the parent molecular ion, reaction (7.105). This base peak ion further fragments by loss of $H_2, CH_4$. and $C_6H_6$ species, all metastable supported decompositions.

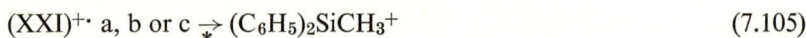

$$(XXI)^{+\cdot}\ a,\ b\ or\ c\ \underset{*}{\rightarrow}\ (C_6H_5)_2SiCH_3^+ \tag{7.105}$$

Decompositions of the parent molecular ions by loss of $\cdot C_6H_5$ or $\cdot CH_3$ are apparently not very important. Cleavage of the N–N bond takes place with H transfer as in scheme (7.106).

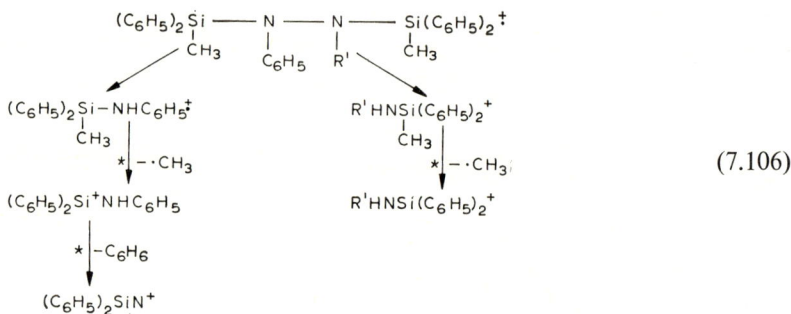

(7.106)

The spectrum of XXId showed ions due to the loss of two $(C_6H_5)_2SiCH_3$ radicals and no peak corresponding to the $(C_6H_5)_2Si(CH_3)NHC_6H_5^+$ ion, presumably because the H-transfer reaction was unfavourable.

Beck et al.[311] have discussed some features in the spectra of the bis(trimethylsilyl) derivatives, XXIII and XXIV.

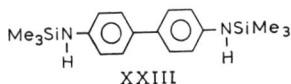

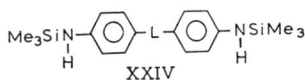

Me$_3$SiN—(O)—(O)—NSiMe$_3$
  |                    |
  H                    H
              XXIII

(a) L = O

Me$_3$SiN—(O)—L—(O)—NSiMe$_3$     (b) L = CO
  |                        |
  H         XXIV           H        (c) L = CH$_2$

(d) L = (CH$_2$)$_2$

All these compounds showed strong parent molecular ions, the base peaks in every spectrum except XXIVd. However, perhaps the most interesting features in the spectra of these compounds were the fairly abundant doubly charged ions due to (parent molecule $-2\times \cdot CH_3)^{2+}$ found in all the spectra. These ions were ten to thirty times as abundant as the corresponding singly charged species (parent molecule $-2\times \cdot CH_3)^+$.

Cyclic silicon–nitrogen compounds of the silazane and tetraza-3,6-disilacyclohexane types have been studied in some detail. Silbiger et al.[223] reported spectra for compounds XXV and XXVI.

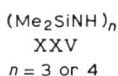

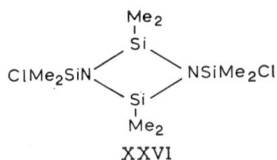

(Me$_2$SiNH)$_n$
     XXV
   n = 3 or 4

                    Me$_2$
                     |
                     Si
                   /    \
    ClMe$_2$SiN            NSiMe$_2$Cl
                   \    /
                     Si
                     |
                    Me$_2$
                   XXVI

Both XXV, $n = 3$ or 4, showed weak parent molecular ions and (parent molecule $- \cdot CH_3)^+$ ions as base peaks. Metastable supported elimination of NH$_3$ groups were a feature of these spectra, viz.

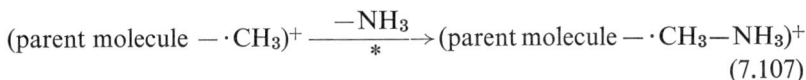

$$\text{(parent molecule } - \cdot CH_3)^+ \xrightarrow[\quad * \quad]{-NH_3} \text{(parent molecule } - \cdot CH_3 - NH_3)^+ \tag{7.107}$$

The spectrum of XXVI is similar to XXV; however, the doubly charged ion (parent molecule $-2\times \cdot CH_3)^{2+}$ was the second most abundant ion in the spectrum. This is possibly because of stabilisation in the Si–N ring system produced by donation of lone pairs of electrons from the N atoms on to the Si ring atoms, XXVII, and the absence of a hydrogen atom on N which precludes $\beta$-elimination reactions of CH$_4$ or HCl species.

$$\text{(CH}_3)_2\text{ClSi}-\overset{+}{\underset{\overset{|}{\text{N}}}{\text{N}}}\overset{\overset{\text{CH}_3}{|}\overset{+}{\underset{\text{Si}}{\diagup}}}{\diagdown}\overset{+}{\underset{\overset{|}{\text{Si}}{\underset{|}{\text{CH}_3}}}{}}\overset{}{\text{N}}-\text{SiCl(CH}_3)_2$$

XXVII

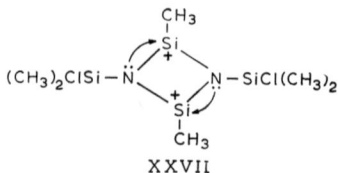

The spectra of some tetraza-3,6-disilacyclohexanes, XXVIII, have been studied[222].

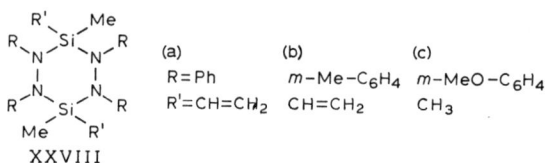

R, R', Me structure (XXVIII)

| | (a) | (b) | (c) |
|---|---|---|---|
| | R=Ph | m-Me-C$_6$H$_4$ | m-MeO-C$_6$H$_4$ |
| | R'=CH=CH$_2$ | CH=CH$_2$ | CH$_3$ |

XXVIII

The parent molecular ions were the most abundant ions in the spectra. Fragmentation by the loss or successive loss of $\cdot$CH$_3$, $\cdot$CH=CH$_2$ or $\cdot$C$_6$H$_5$ groups was unfavourable, but elimination of groups such as arylnitrene (RN:)$_2$ from the parent molecular ions was very important.

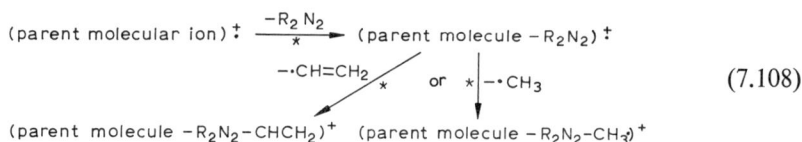

$$\text{(parent molecular ion)}^{\ddagger} \xrightarrow[*]{-R_2N_2} \text{(parent molecule}-R_2N_2)^{\ddagger}$$

$$\begin{array}{cc} \swarrow\;{-\cdot\text{CH=CH}_2} & \text{or} \quad *\Big|{-\cdot\text{CH}_3} \end{array} \qquad (7.108)$$

$$\text{(parent molecule}-R_2N_2-\text{CHCH}_2)^+ \quad \text{(parent molecule}-R_2N_2-\text{CH}_3)^+$$

An interesting metastable supported decomposition of the doubly charged parent molecular ion of XXVIIIa was observed to produce two singly charged species.

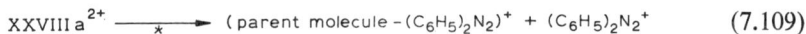

$$\text{XXVIIIa}^{2+} \xrightarrow[*]{} \text{(parent molecule}-(C_6H_5)_2N_2)^+ + (C_6H_5)_2N_2^+ \qquad (7.109)$$

Other organometallic compounds containing Group IV–Group V bonds are given in Table 7.21.

## 7.8 Compounds with aryl and or alkyl groups and a main Group IV element bonded to Group VI elements

The preparation of trimethylsilyl derivatives of many organic compounds has facilitated their characterisation by GLC, mass spectrometry, and GLC–mass spectrometry. The reasons for preparing these derivatives, namely their increased chemical stability and volatility, are well known. Most studies have been concerned with the identification of the compound and the determination of the positions of the trimethylsilyl substituted functional groups. Substances which have been reported include trimethylsilyl derivatives of carbohydrates[235-240], lipids[241-246], steroids[247-250], vitamins[251,252], amino compounds[253-256], alcohols[257,258], nucleosides and nucleotides[259], and aminoacids[260]. The spectra of these compounds will not, for the most part, concern us here. The reader who is interested in them is referred to the above references and the book by Budzikiewicz et al.[261] We will, however, discuss some simpler compounds whose trimethylsilyl derivatives have been studied.

### 7.8.1 Silanes

The spectrum of $Me_3SiOH$ has been reported by Orlov[262] who observed a weak parent molecular ion and a base peak due to (parent molecule $-CH_3\cdot$)$^+$. He reported a number of metastable supported decompositions including those shown in scheme (7.110).

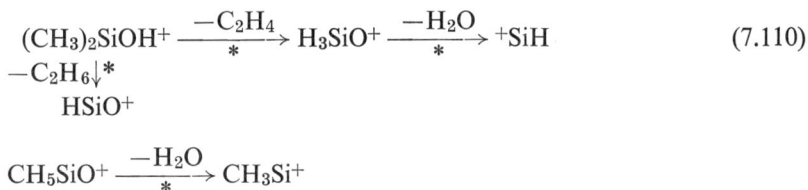

$$(CH_3)_2SiOH^+ \xrightarrow[*]{-C_2H_4} H_3SiO^+ \xrightarrow[*]{-H_2O} {}^+SiH \qquad (7.110)$$
$$-C_2H_6\!\downarrow\!* $$
$$HSiO^+$$

$$CH_5SiO^+ \xrightarrow[*]{-H_2O} CH_3Si^+$$

Starkey et al.[218] reported the spectra of the trimethylsilyl ethers of twenty six aliphatic alcohols, $Me_3Si-OR$. Primary, secondary and tertiary alkyl groups were studied. The primary straight chain ethers (R $=$ Me, Et, Pr$^n$, Bu$^n$, n-pentyl, n-hexyl, n-heptyl, n-octyl, n-nonyl, and n-decyl) all showed parent molecular ions in low abundance. Base peaks in the spectra of the compounds with $C_1$, $C_2$ and $C_6$ to $C_{10}$ alkyl groups were due to the $(CH_3)_3SiO^+$ ion; for the $C_3$ to $C_5$ groups, the base peaks were due to the rearrangement ion $(CH_3)_2Si(H)O^+$. Intense peaks corresponding to (parent molecule $-\cdot CH_3$)$^+$, $(CH_3)_3Si-OCH_2^+$, $(CH_3)_3Si^+$, $CH_3Si(H)OH^+$, $(CH_3)$-$SiH_2^+$ or $SiOH^+$, and $CH_3Si^+$ ions were observed in the spectra of most of these compounds. Some branched-chain primary alcohol derivatives (R $=$ 2-methylpropyl, 2-methylbutyl, 3-methylbutyl, 2,2-dimethylpropyl, 2-methylpentyl, 3-methylpentyl, 2-ethylbutyl, and 2-ethylhexyl) were studied. Parent molecular ions were weak for these compounds. The rearrangement ion $(CH_3)_2Si(H)O^+$ was the base peak in the spectra of all these compounds except the 3-methylbutyl, 2,2-dimethylpropyl and 3-methylpentyl derivatives which showed (parent molecule $-CH_3\cdot$)$^+$, $(CH_3)_3Si^+$, and $(CH_3)_2Si(H)O^+$ base peak ions, respectively. Other intense ions were due to $(CH_3)_3SiOCH_2^+$, $(CH_3)Si(H)OH^+$, and $CH_3Si^+$. The secondary alcohol derivatives studied contained R = 2-propyl, 2-butyl, 2-pentyl, 3-methyl-2-butyl, 2-hexyl, and 3-heptyl groups. Parent molecular ions were weak. The ion $(CH_3)_3Si^+$ was observed in all the spectra but $(CH_3)_2Si(H)O^+$ was less important in these compounds. The five ethers derived from straight-chain R groups all showed intense peaks due to ions formed by cleavage of the larger fragment of the R group ($R_2 > R_1$) as shown in (7.111).

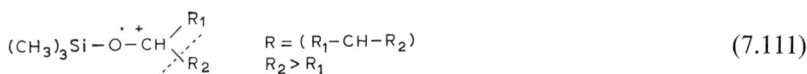

$$(CH_3)_3Si-O-CH\overset{\overset{R_1}{\diagup}}{\underset{\diagdown R_2}{\cdot}} \qquad R = (R_1-CH-R_2) \qquad R_2 > R_1 \tag{7.111}$$

The ions corresponding to $(CH_3)_3SiO–CHR_1^+$ ($R_1 = CH_3$) were the base peaks in the spectra of the 2-pentyl and 2-hexyl compounds (and also the branched chain 3-methyl-2-butyl derivatives). For the 2-butyl and 3-heptyl compound, $(CH_3)_3Si^+$ was the base peak ion and for the 2-propyl compound, $(CH_3)_2Si(H)O^+$ was the base peak ion. Compared with the primary R group derivatives, the secondary R group compounds showed less intense (parent molecule $-\cdot CH_3$)$^+$ ions. Derivatives of tertiary R groups (R = 2-methyl-2-propyl and 2-methyl-2-butyl) showed base peak ions due to $(CH_3)_2-Si(H)O^+$ and $(CH_3)_3Si^+$ respectively. No parent molecular ion was observed from the 2-methyl-2-propyl compound but a very weak parent molecular ion was found in the spectrum of the 2-methyl-2-butyl compound. Other intense ions occurred at $m/e$ values corresponding to ions with the formulae $(CH_3)_3SiO\overset{+}{C}(CH_3)_2$, $(CH_3)SiH_2^+$ or $SiOH^+$, $CH_3Si^+$, and (parent molecule $-\cdot CH_3$)$^+$. Starkey et al. also reported a few features in the spectra of $Me_3SiSBu^n$, $Me_3SiOPh$, and the derivatives of o-, m-, and p-cresol, and propylene glycol. The compounds all showed strong parent molecular ions and (parent molecule $-\cdot CH_3$)$^+$ ions except the glycol derivative which showed a very weak parent molecular ion and intense $(CH_3)_5Si_2O^+$ and $(CH_3)_3Si_2O^+$ ions. Base peak ions for the derivatives containing aryl groups were due to (parent molecule $-\cdot CH_3$)$^+$ ions but the propylene glycol compound showed $(CH_3)_3Si^+$ as the base peak ion.

Dube and Kriegsmann[263] studied the spectra of a number of trimethylsilyl ether derivatives of alcohols XXIX and XXX, some of which had been previously studied by Starkey et al.

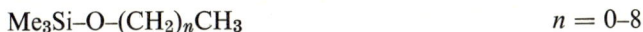

$$\text{Me}_3\text{Si–O–(CH}_2)_n\text{CH}_3 \qquad\qquad n = 0\text{–}8$$
$$\text{XXIX}$$

$$\text{Me}_3\text{Si–O–(CH}_2)_n\text{CH(CH}_3)_2 \qquad\qquad n = 0\text{–}3$$
$$\text{XXX}$$

The major fragment ions observed were the same as those found in the earlier report. Several metastable supported decompositions were reported; some of these have been studied in more detail by Diekmann et al.[220] The compounds studied by Diekmann et al. were trimethylsilyl ethers of penta-nols and deuterated derivatives and a triethylsilyl ether XXXI.

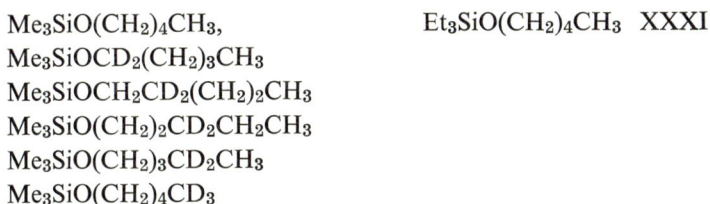

$$\text{Me}_3\text{SiO(CH}_2)_4\text{CH}_3, \qquad\qquad \text{Et}_3\text{SiO(CH}_2)_4\text{CH}_3 \quad \text{XXXI}$$
$$\text{Me}_3\text{SiOCD}_2(\text{CH}_2)_3\text{CH}_3$$
$$\text{Me}_3\text{SiOCH}_2\text{CD}_2(\text{CH}_2)_2\text{CH}_3$$
$$\text{Me}_3\text{SiO(CH}_2)_2\text{CD}_2\text{CH}_2\text{CH}_3$$
$$\text{Me}_3\text{SiO(CH}_2)_3\text{CD}_2\text{CH}_3$$
$$\text{Me}_3\text{SiO(CH}_2)_4\text{CD}_3$$

and

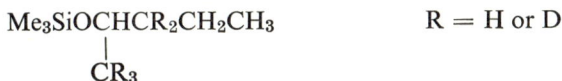

$$\text{Me}_3\text{SiOCHCR}_2\text{CH}_2\text{CH}_3 \qquad\qquad R = H \text{ or } D$$
$$\underset{\displaystyle \text{CR}_3}{|}$$

The spectra of these compounds were in broad agreement with those found by Starkey et al. Parent molecular ions were weak and base peaks were due to $(\text{CH}_3)_2\overset{+}{\text{Si}}\text{OH}$ from the straight-chain compounds and $(\text{CH}_3)_3\text{Si}^+$ from the branched-chain derivative. The spectra of the deuterated compounds confirmed that the methyl group lost from the parent molecular ion of $\text{Me}_3\text{SiO(CH}_2)_4\text{CH}_3$ came from the $\text{Me}_3\text{Si}$ group and the ion formed rear-ranged on elimination of butene, viz.

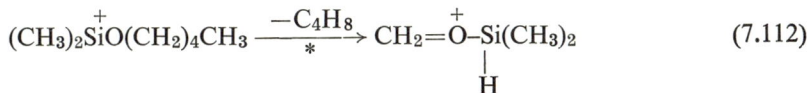

$$(\text{CH}_3)_2\overset{+}{\text{Si}}\text{O(CH}_2)_4\text{CH}_3 \xrightarrow[*]{-\text{C}_4\text{H}_8} \text{CH}_2{=}\overset{+}{\text{O}}\text{-Si(CH}_3)_2 \qquad (7.112)$$
$$\underset{\displaystyle \text{H}}{|}$$

The triethylsilyl compound showed a strong peak due to $(\text{C}_2\text{H}_5)_2\overset{+}{\text{Si}}\text{OH}$ and metastable supported elimination of three molecules of ethylene from the (parent molecule $- \cdot\text{C}_5\text{H}_{11})^+$ ion.

House et al.[266] have characterised a number of trimethylsilyl enol ethers. They observed similar fragmentation reactions to some reported by Diekmann et al.[221], for example from $Bu^nCH=C(Me)OSiMe_3$ and $Bu^nCH_2-C(:CH_2)OSiMe_3$, schemes (7.113) and (7.114).

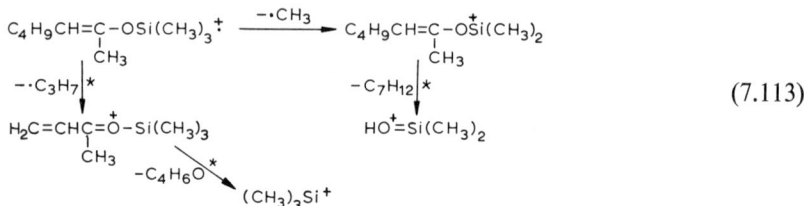

$$(7.113)$$

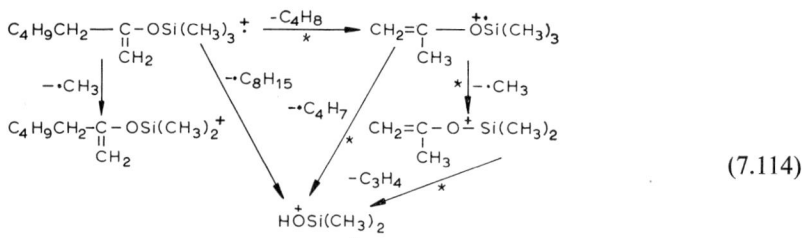

$$(7.114)$$

Compounds containing more than one alkoxy group have not been studied much. Tamas et al.[264] have reported that the spectrum of $(MeO)_2SiMe_2$ showed a weak parent molecular ion. The base peak was due to the $(CH_3)_2Si\overset{+}{O}CH_3$ ion and fragmentation occurred by the loss of $\cdot CH_3$, $\cdot OCH_3$, $CH_2O$, $CH_4$, $C_2H_4$, $H_2$, and $SiO$ groups. The positive and negative ions from organotriptych-siloxazolidines, XXXII, have been interpreted as supporting the assumption that a coordinate Si–N bond exists in these compounds and Si is therefore pentavalent[265]. Whilst the interpretation of this evidence must be viewed with a great amount of reservation, it was true that a large number

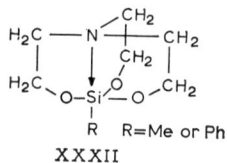

XXXII

of Si–N-containing ions were observed in the positive ion spectra, including species such as XXXIII.

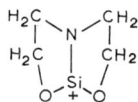

$$H_2C \diagdown_{\underset{|}{N}} \diagup CH_2$$
$$H_2C \quad | \quad CH_2$$
$$\diagdown \underset{O \diagup^+\diagdown O}{Si} \diagup$$

XXXIII

A large number of trimethylsilyl ether derivatives of benzylic and similar compounds have been studied by Diekmann et al.[220] (Table 7.22). The spectrum of the compound $Me_3SiOCH_2Ph$ was simple, with only five major silicon-containing peaks. They were the parent molecular ion, (parent molecule $- \cdot CH_3)^+$, (parent molecule $- \cdot CH_3 - CH_2O)^+$, $(CH_3)_2SiOH^+$, and $(CH_3)_3Si^+$ ions. The base peak was due to $(CH_3)_2SiOCH_2C_6H_5^+$ and the elimination of $CH_2O$ from this was shown to be a clean process by deuterium-labelling experiments. All the phenyl-substituted ethers showed (parent molecule $- \cdot CH_3 - CH_2O)^+$ ions but there was no apparent correlation between the % abundance of this ion and the substituent X. A correlation had previously been suggested by Teeter for spectra of trimethylsilyl benzoates in which a similar ion (parent molecule $- \cdot CH_3 - CO_2)^+$ had appeared[267]. The compound with $X = NO_2$ showed a strong parent mole-

Table 7.22

Trimethylsilyl Derivatives of Benzylic and Similar Compounds

*A. Benzylic compounds*

$Me_3SiOCR_2 -\hspace{-2pt}\langle\bigcirc\rangle\hspace{-2pt}- X$

R = H, H,　H,　　H, D, H, H
X = H, $NO_2$, $NMe_2$, OMe, H, F, Cl
$Me_3SiOCHPh$
　　|　　　　　　R = Me, Ph
　　R
$Me_3SiOCMe_2Ph$　　R = Me

*B. Other compounds which are similar*
　　$Me_3SiO(CH_2)_nPh$　　$n = 2$ or 3
　　$Me_3SiOCD_2(CH_2)_2Ph$
　　$Me_3SiOCH_2CD_2CH_2Ph$
　　$Me_3SiO(CH_2)_2CD_2Ph$
　　$Me_3SiS(CH_2)_nPh$　　$n = 1$ or 2

cular ion in contrast to all the other compounds. Fragmentation by loss of
$\cdot CH_3$ was observed in all spectra. The halo compounds showed (parent
molecule $-\cdot X)^+$ ions and the dimethylamino derivative showed a (parent
molecule $-H\cdot)^+$ ion. Although both the dimethylamino and methoxy groups
are *para*-directing, this did not appear to influence their spectra much and a
number of striking differences were noted. The methoxy derivative showed
no (parent molecule $-\cdot H)^+$ ion but did show (parent molecule $-\cdot CH_3)^+$ in
high abundance and the rearrangement ion $(CH_3)_2SiC_6H_5^+$ in low abundance. The dimethylamino compound did not exhibit the rearranged ion. The
spectra of compounds with more than one methylene group (Table 7.22B)
showed the $(CH_3)_2SiC_6H_5^+$ ion in very low abundance and base peaks due
to $(CH_3)_3Si^+$ and $(CH_3)_2Si(H)\overset{+}{O}CH_2$ for $n = 2$ or 3, respectively. Replacement of oxygen by sulphur in the benzyl compounds produced some signif-
icant changes in the spectra. The compound $Me_3SiSCH_2Ph$ showed a weak
parent molecular ion, (parent molecule $-\cdot CH_3)^+$ ion, and a very weak
$(CH_3)_2SiC_6H_5^+$ ion but the $(CH_3)_3Si^+$ ion remained very strong. With two
methylene groups, the rearrangement ion was no longer observed and the
parent molecular ion and (parent molecule $-\cdot CH_3)^+$ ion were again weak.
The branched-chain derivatives also showed weak or no $(CH_3)_2SiC_6H_5^+$
ions.

VandenHeuvel *et al.*[268] have observed doubly charged species corresponding to the (parent molecule $-2\times\ \cdot CH_3)^{2+}$ ions in quite high abundance
in the spectra of a number of compounds XXXIV–XXXVI.

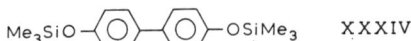

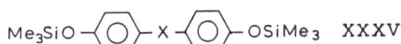

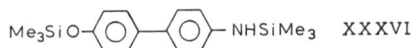

From compound XXXIV the $[(CH_3)_2SiO-C_6H_4-C_6H_4-OSi(CH_3)_2]^{2+}$ ion
was the fourth most abundant species after the parent molecular ion (base
peak), $(CH_3)_3Si^+$, and (parent molecule $-\cdot CH_3)^+$, and was some seventeen
times as abundant as the singly charged counterpart. Similar results were
observed from compounds of type XXXV and XXXVI. Other doubly
charged ions were found with either very low or zero abundance singly
charged counterparts.

Table 7.23

Trimethylsilyl Derivatives of Polymethylene Glycols and related Compounds

---

*Polymethylene glycols*

$Me_3SiO(CH_2)_nOSiMe_3$, $n = 2$–$8$ (ref. 221); $n = 10$ or $22$ (ref. 219)

*Related compounds*

$MeO(CH_2)_nOSiMe_3$, $n = 2, 4, 5$ (ref. 221)
$EtO(CH_2)_2OSiMe_3$ (ref. 221)
$PhO(CH_2)_nOSiMe_3$, $n = 2$–$7$ (ref. 221)
  and $Ph^{18}O(CH_2)_2OSiMe_3$ (ref. 221)
$PhOCH_2CMe_2OSiMe_3$ (ref. 221)
$PhO(CH_2)_2OSiEt_3$ (ref. 221)
$Me_3SiO(CH_2)_2O(CH_2)_2OR$, $R = Me_3Si$ or $Et$ (ref. 221)
$Me_3SiS(CH_2)_nSSiMe_3$, $n = 2$ (ref. 221); $n = 10$ (ref. 219)
$PhS(CH_2)_2SSiMe_3$ (ref. 221)
$Me_2N(CH_2)_2OSiMe_3$ (ref. 221)

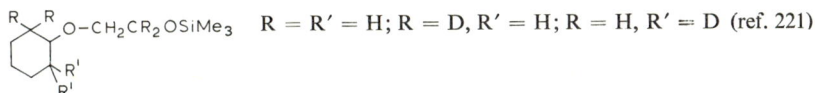 $R = R' = H$; $R = D, R' = H$; $R = H, R' = D$ (ref. 221)

 (ref. 221)

 (ref. 221)

---

Diekmann *et al.*[221] and Draffen *et al.*[219] have studied the spectra of some trimethylsilyl ethers of polymethylene glycols and related compounds (Table 7.23). From the polymethylene glycol derivatives, parent molecular ions were weak but a characteristic ion $(CH_3)_5Si_2O^+$ due to the expulsion of the $(CH_2)_nO$ part of the molecule from the (parent molecule $- \cdot CH_3)^+$ ion was observed in all cases[219,221]. Furthermore, the corresponding $ROSi(CH_3)_2{}^+$ ions from the monotrimethylsilyl ethers ($R = Me$, Et, or Ph) were observed in reasonably high abundance. The decomposition modes for the formation of the $(CH_3)_5Si_2\overset{+}{O}$ ion were as shown in (7.115) for $n = 2$ and (7.116) for $n = 3$–$8$.

$$(CH_3)_3Si-O \overset{\displaystyle H_2C \diagdown CH_2}{\underset{\displaystyle {}^+Si(CH_3)_2}{\diagup\diagdown O\diagdown}} \quad \xrightarrow[\ast]{-\triangle O} \quad (CH_3)_3Si-O-Si(CH_3)_2^+ \qquad (7.115)$$

$$(CH_3)_3Si-O \overset{\displaystyle (CH_2)_n}{\underset{\displaystyle CH_3 \diagup Si \diagdown CH_3}{\diagup\diagdown O}} \quad \xrightarrow[\ast]{-C_{n-1}H_{2(n-1)}} \quad \begin{array}{l} (CH_3)_3SiOSi(CH_3)_2-\overset{+}{O}=CH_2 \\[4pt] \quad\quad -CH_2O \Big| \\[4pt] (CH_3)_3SiOSi(CH_3)_2^+ \end{array} \qquad (7.116)$$

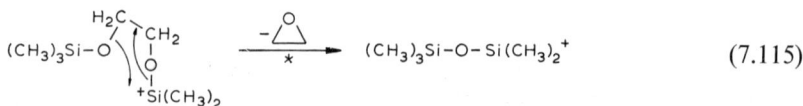

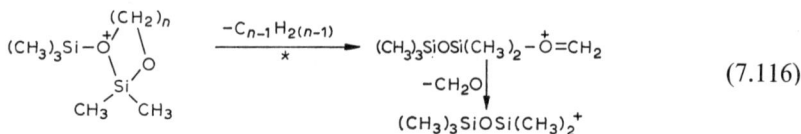

The ethyl ether $EtO(CH_2)_2OSiMe_3$ showed the ion of formula $C_4H_{11}SiO^+$ to be formed from both an $\alpha$-cleavage reaction giving $(CH_3)_3Si\overset{+}{O}CH_2$ and a rearrangement ($cf.$ (7.115)) to give $(CH_3)_2Si\overset{+}{O}CH_2CH_3$. The phenyl derivatives behaved analogously to (7.115) and (7.116) giving $C_6H_5OSi(CH_3)_2^+$. The labelling of the phenoxy oxygen by $^{18}O$ enabled Diekmann $et~al.$ to show that it was this oxygen atom incorporated in the rearranged ion. The formation of the $(CH_3)_3Si^+$ ion from the parent molecular ion in the spectra of the phenoxy compounds was a metastable supported process, $viz.$

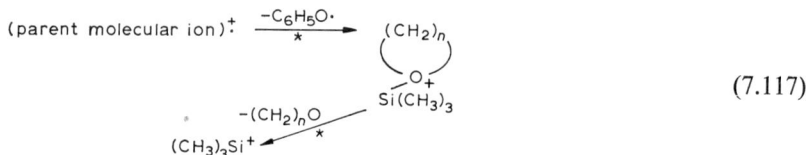

$$(\text{parent molecular ion})^+ \xrightarrow[\ast]{-C_6H_5O\cdot} \overset{\displaystyle (CH_2)_n}{\underset{\displaystyle Si(CH_3)_3}{\diagdown O^+ \diagup}}$$
$$\xleftarrow[\ast]{-(CH_2)_nO} (CH_3)_3Si^+ \qquad\qquad (7.117)$$

The branched-chain compound $PhOCH_2C(Me_2)OSiMe_3$ showed a metastable supported $\alpha$-cleavage decompostion pathway (7.118) to give the base peak on $(CH_3)_2C=\overset{+}{O}Si(CH_3)_3$.

$$C_6H_5OCH_2\overset{\displaystyle CH_3}{\underset{\displaystyle CH_3}{\overset{|}{\underset{|}{C}}}}-OSi(CH_3)_3^+ \xrightarrow[\ast]{-C_6H_5OCH_2\cdot} (CH_3)_2C=\overset{+}{O}Si(CH_3)_3 \,(7.118)$$

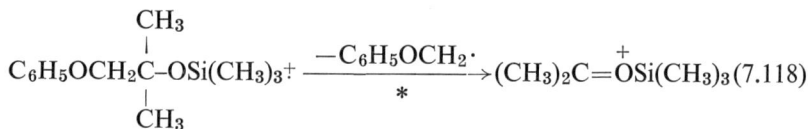

The triethylsilyl ether $PhO(CH_2)_2OSiEt_3$ showed the rearranged ion $(C_2H_5)_2SiO^+C_6H_5$ as the base peak. Another prominent ion was $H_2SiO^+C_6H_5$ formed by the elimination of two ethylene molecules from the base peak ion. The spectrum of the cyclohexyl derivative showed the rearrangement ion formed by loss of ethylene oxide in only low abundance but other rearrangement ions such as $C_6H_{11}OCH_2Si(CH_3)_2^+$ and $(CH_3)_2Si^+OH$ were observed in reasonable abundance. The formation of the latter ion was a metastable supported process from the (parent molecule $-\cdot CH_3)^+$ ion.

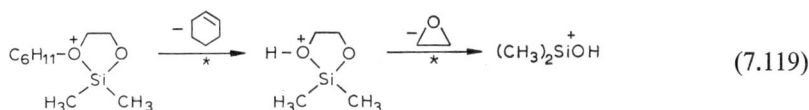

$$(7.119)$$

From the spectra of the deuterium-labelled compounds it was found that the H transferred from the cyclohexyl group came approximately 50% of the time from either the 2 or 6 positions and 50% from other positions in the ring. The dimethylamino derivative $Me_2N(CH_2)_2OSiMe_3$ showed the rearranged ion $(CH_3)_2Si=N(CH_3)_2{}^+$ and ion formed by an $\alpha$-cleavage reaction, $(CH_3)_2N^+=CH_2$, similar to the ethoxy compound.

The replacement of O by S produced a strong peak from $Me_3SiS(CH_2)_2S$-$SiMe_3$ corresponding to the rearrangement ion $(CH_3)_5Si_2S^+$. The base peak ion was $(CH_3)_3Si^+$ produced in a metastable supported decomposition of $C_2H_4SSi(CH_3)_3{}^+$, viz.

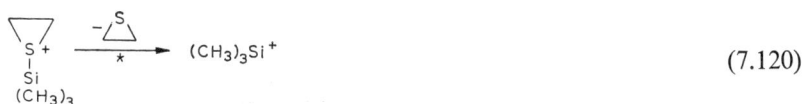

$$(7.120)$$

With the longer-chain compound $Me_3SiS(CH_2)_{10}SSiMe_3$, fragment ions (parent molecule $-\cdot SSi(CH_3)_3)^+$, $(CH_3)_5Si_2S^+$, $(CH_3)_3Si^+$, and $(CH_3)_2Si=SH^+$ were prominent. A metastable supported reaction (7.121) involving the loss of the SH radical from the parent molecular ion was reported.

$$(CH_3)_3SiS(CH_2)_{10}SSi(CH_3)_3{}^+ \xrightarrow[*]{-\cdot SH} (CH_3)_3SiS(CH_2)_{11}{}^+Si(CH_3)_2 \quad (7.121)$$

The phenoxy derivative $PhO(CH_2)_2SSiMe_3$ showed a parent molecular ion and (parent molecule $-\cdot CH_3)^+$, $C_2H_4SSi(CH_3)_3{}^+$, and $(CH_3)_3Si^+$ ions. The compound $PhS(CH_2)_2OSiMe_3$ also showed a parent molecular ion and (parent molecule $-\cdot CH_3)^+$ ion, and the $CH_2=\overset{+}{O}Si(CH_3)_3$ ion formed by $\alpha$-cleavage. The base peak ion was $(CH_3)_3Si^+$.

Timethylsilyl esters of phenoxyalkanoic acids have been studied by Diekmann et al.[269] and derivatives of some dicarboxylic acids, hydroxy acids, and cyano acids have been studied by Draffan et al.[219] (Table 7.24). In general, the phenoxyalkanoic acid derivatives $PhO(CH_2)_nCO_2SiMe_3$ showed intense parent molecular ions (the base peak from the $n = 10$ compound), and

Table 7.24

Phenoxyalkanoic Acid Derivatives and Related Compounds[296]

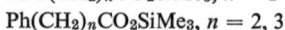

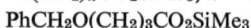

---

$PhO(CH_2)_nCO_2SiMe_3$, $n = 1-6$ and 10
$Ph(CH_2)_nCO_2SiMe_3$, $n = 2, 3$
$PhCH_2O(CH_2)_3CO_2SiMe_3$
$RO(CH_2)_nCO_2SiMe_3$, $R = Me$, $n = 3, 4$; $R = Et$, $n = 3$

Dicarboxylic Acid Derivatives[219]
$RO_2C(CH_2)_nCO_2R'$, $R = R' = Me_3Si$; $n = 8, 10, 16, 22$
$R = Me_3Si$, $R' = CH_3$;
$n = 8, 10, 22$

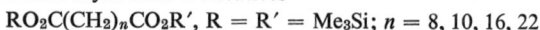

Hydroxy Acid Derivative[219]
$Me_3SiO_2C(CH_2)_{10}CH(CH_2)_5CH_3$
$\quad\quad\quad\quad\quad |$
$\quad\quad\quad\quad OSiMe_3$

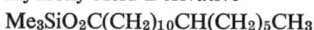

Cyano Acid Derivatives[219]
$Me_3SiO_2C(CH_2)_nCN$, $n = 8, 10$

---

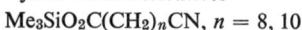

weak (parent molecule $- \cdot CH_3)^+$ ions. This was unusual; most $Me_3SiO$ compounds show the reverse trend. Characteristic ions at $m/e = 75$ $(CH_3)_2$-$SiOH^+$ and 73 $(CH_3)_3Si^+$ (base peak from the $n = 1$ and 6 compounds) were observed and all compounds except $n = 1$ showed the ions (parent molecule $-93)^+$ and (parent molecule $-109)^+$. The formation of these last two ions were metastable supported processes.

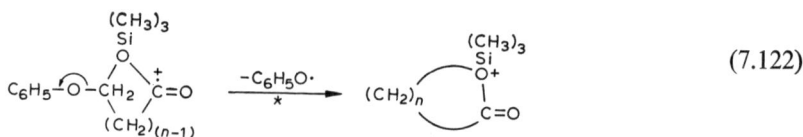

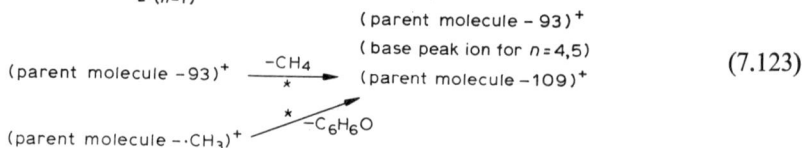

$$(7.122)$$

$$(7.123)$$

Two rearrangement ions occurred in all the spectra at $m/e = 166$ and 151. Their formation was metastable supported in some cases.

$$\text{(parent molecular ion)}^{\ddagger} \xrightarrow[n=2,3]{-(CH_2)_nCO_2} C_6H_5OSi(CH_3)_3^{\ddagger}$$

$$m/e = 166$$

$$\xrightarrow{-\cdot CH_3}$$

$$C_6H_5OSi(CH_3)_2^+ \quad m/e = 151 \quad \text{(base peak for } n=2\text{)}$$

$$-(CH_2)_nCO_2$$

$$C_6H_5O(CH_2)_nCO_2Si(CH_3)_2^+$$

(7.124)

The spectra of the compounds $Ph(CH_2)_nCO_2SiMe_3$ showed significant differences from those of the phenoxyalkanoic acid derivatives in that there were no ions produced by interaction of the terminal functional groups. For instance, no rearrangements from the parent molecular ion or the (parent molecule $-\cdot CH_3)^+$ ion involving the rearrangement of the $-SiMe_3$ moiety and expulsion of the $-(CH_2)_n-$ portion of the molecule were observed. The most important ions were formed by the independent decomposition of the two terminal functional groups e.g. $(CH_3)_2SiOH^+$, $(CH_3)_3Si^+$, and $C_nH_n^+$ (n = 7 or 8). The compound $PhCH_2O(CH_2)_3OSiMe_3$ showed several similarities to the phenoxyalkanoic acid derivatives including the loss of the benzoyl radical from the parent molecular ion, comparable with the loss of the phenoxy group from the phenoxyalkanoic compounds and ions produced by interaction between the benzoyl ether and trimethylsilyl ester groups. Compounds of the type $RO(CH_2)_nCO_2SiMe_3$, $(R = Me$ or $Et)$ showed weak parent molecular ions but intense peaks corresponding to the ions $(CH_3)_2$-$SiOC(CH_2)OH^+$, $(CH_3)_2SiOH^+$, and $(CH_3)_3Si^+$ (base peak for $R = Me$, n = 4 and $R = Et$, n = 3). The base peak ion in the spectrum of the $R = Me$, $n = 3$ compound was $CH_3OSi(CH_3)_2^+$ formed via metastable supported decompositions from the parent molecular ion, viz.

$$RO(CH_2)_3CO_2Si(CH_3)_3^+ \xrightarrow{-\cdot CH_3} RO\begin{matrix}(CH_2)_3\\ \backslash C=O \\ Si-O \\ (CH_3)_2\end{matrix}$$

$$\xrightarrow{-(CH_2)_3CO_2}$$

$$R\overset{+}{O}Si(CH_3)_2$$

$(R = CH_3$ or $C_2H_5)$

(7.125)

The compound $MeO(CH_2)_4CO_2SiMe_3$ showed a strong peak due to an ion at $m/e = 157$ whose formation from the (parent molecule $-\cdot CH_3)^+$ ion by loss of $CH_4O$ was a metastable supported process. The derivatives of dicarboxylic acids showed many ions in common with the phenoxyalkanoic compounds discussed previously, particularly those derived from $\alpha$-cleavage reactions adjacent to the $Me_3Si-$ group, $(CH_3)_2SiOH^+$, and $(CH_3)_3Si^+$. The hydroxy acid derivative showed a parent molecular ion, (parent molecule $- \cdot CH_3)^+$, and (parent molecule $- \cdot CH_3-CH_4)^+$ ions as well as the rearrangement ion $(CH_3)_3SiOC(CH_2)_{10}OSi(CH_3)_3^+$ and some ions produced by cleavage of the groups $\alpha$ to the $Me_3Si$ groups. The cyano acid derivatives showed similar fragmentation behaviour and also a decomposition which appeared to involve the transfer of an $-SiMe_3$ group from O to N, viz.

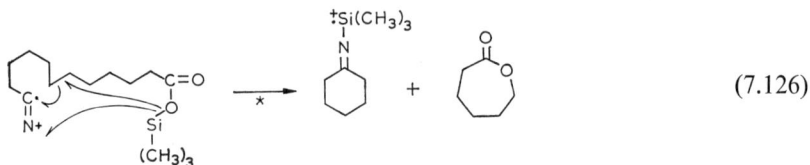

$$(7.126)$$

Mass spectral rearrangements in some acycloxytriphenylmethanes, -silanes, and -germanes have been reported[270]. An important fragmentation path in the decomposition of ions from $Ph_3MOCOR$ (M = Si, Ge, R = alkyl or aryl) involves the metastable supported rearrangement of R onto M.

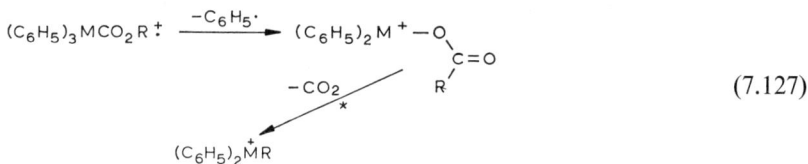

$$(7.127)$$

This is in contrast to the behaviour of the carbon analogues which did not show the loss of $\cdot Ph$ from the parent molecular ion or ions of the type $(C_6H_5)_2C^+R$ (Table 7.25).

The bis-trimethylsilyl derivative, XXXVII, showed three major peaks due to the parent molecular ion, (parent molecule $- \cdot CH_3)^+$, and $(CH_3)_3Si^+$ (base peak) ions[268]. Other ions were produced by the loss of $\cdot CH_3$, $CO_2$, CO, $\cdot CO_2SiMe_3$, and $\cdot OSiMe_3$ groups.

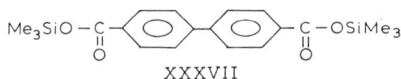

XXXVII

A number of fairly intense doubly charged ions was observed including (parent molecule $-2 \times \cdot CH_3)^{2+}$, (parent molecule $-2 \times \cdot CH_3-CO)^{2+}$ and (parent molecule $-2 \times \cdot CH_3-CO_2)^{2+}$.

Table 7.25

Ion Abundances of Some Ions from Ph₃MOCOR Compounds

| Compound Ph₃MOCOR | Relative abundance of ions (base peaks = 100) | | | |
|---|---|---|---|---|
| | $\left[\begin{array}{c}\text{(Parent molecule)} \\ - \cdot C_6H_5\end{array}\right]^+$ | $(C_6H_5)_2MR^+$ | $(C_6H_5)_3M^+$ | $(C_6H_5)_3CO^+$ |
| Ph₃COCOMe | | | 75.5 | 100 |
| Ph₃SiOCOMe | 100 | 8.4 | 2.6 | |
| Ph₃COCOPh | | | 61.3 | 100 |
| Ph₃SiOCOPh | 100 | 10 | <5 | |
| Ph₃GeOCOPh | 100 | 43 | 43 | |

*7.8.1.1 Siloxanes* Hexamethyldisiloxane has been the subject of a number of studies[156,157,218,262,264,291]. This compound exhibits a weak parent molecular ion $(0.01\%)^{[156]}$ and the (parent molecule $- \cdot CH_3$)$^+$ ion is the base peak ion $(65.4\%)^{[156]}$. A detailed examination of the fragmentation behaviour of Me₆Si₂O was published by Tamas *et al.*[157].

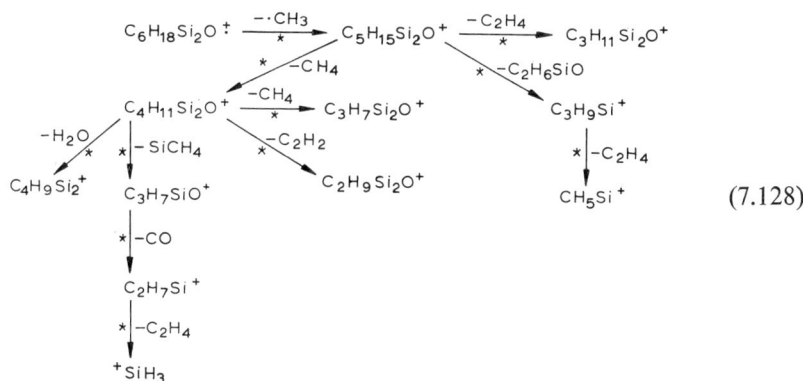

$$(7.128)$$

Decompositions by loss of $\cdot CH_3$, $CH_4$, $C_2H_4$, $C_2H_2$, CO, SiCH₄, Si(CH₃)₂O, and SiO groups were observed, many of them metastable supported reactions. The decompositions of doubly charged species were also reported. It is particularly interesting to note the formation of $(CH_3)_3Si^+$ ion from $(CH_3)_3Si_2O^+$ by elimination of SiO. Comparable processes were observed in the spectra of Me₆Si₂CH₂ (losing SiCH₂), Me₆Si₂NH (losing SiNH) and

*References pp. 310–319*

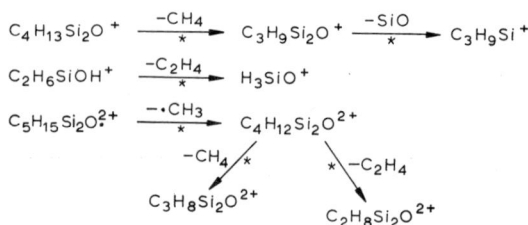

$$C_4H_{13}Si_2O^+ \xrightarrow[*]{-CH_4} C_3H_9Si_2O^+ \xrightarrow[*]{-SiO} C_3H_9Si^+$$

$$C_2H_6SiOH^+ \xrightarrow[*]{-C_2H_4} H_3SiO^+$$

$$C_5H_{15}Si_2O^{2+} \xrightarrow[*]{-\cdot CH_3} C_4H_{12}Si_2O^{2+}$$

$$C_3H_8Si_2O^{2+} \xleftarrow[*]{-CH_4} \quad \xrightarrow[*]{-C_2H_4} C_2H_8Si_2O^{2+}$$

$Me_6Si_2NMe$ (losing $SiNCH_3$). Orlov[262] reported some metastable supported decompositions also found by Tamas $et\ al.$[157] and some others, $viz.$

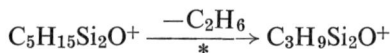

$$C_2H_7Si_2O^+ \xrightarrow[*]{-SiO} C_2H_7Si^+$$
(7.129)
$$C_5H_{15}Si_2O^+ \xrightarrow[*]{-C_2H_6} C_3H_9Si_2O^+$$

Features in the spectra of $(H_2C{=}CH)Me_2SiOSiMe_2(CH{=}CH_2)$ were reported by Tamas $et\ al.$[264] The parent molecular ion was in low abundance and the base peak was due to the (parent molecule $-\cdot CH_3)^+$ ion. The loss of $C_4H_6$ (or two vinyl groups) from the base peak ion was a metastable supported process.

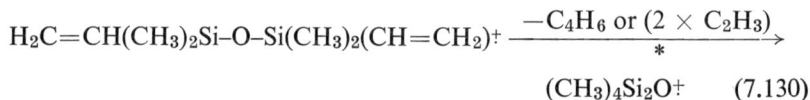

$$H_2C{=}CH(CH_3)_2Si{-}O{-}Si(CH_3)_2(CH{=}CH_2)^+ \xrightarrow[*]{-C_4H_6\ or\ (2\times C_2H_3)} (CH_3)_4Si_2O^+ \quad (7.130)$$

He studied other siloxanes of the general formulae $Me_3Si(OSiMe_2)_n\,OSiMe_3$ ($n = 1$–$4$)[262,271]. These compounds showed weak parent molecular ions and increasingly complicated fragmentation behaviour as the series was ascended. The first member, $Me_3SiOSiMe_2OSiMe_3$, showed the (parent molecule $-\cdot CH_3)^+$ ion as the base peak and $(CH_3)_3Si^+$ in high abundance. The next member, $Me_{10}Si_4O_3$, had the (parent molecule $-\cdot CH_3 - Si(CH_3)_4)^+$ ion as the base peak and $Me_{12}Si_5O_4$ and $Me_{14}Si_6O_5$ had $C_5H_{15}Si_2O^+$ and $(CH_3)_3Si^+$ ions, respectively, as their base peak ions. The loss of silicon-containing species becomes more important in decomposition pathways as the parent molecule becomes more complex. Some metastable supported decompositions from the (parent molecule $-\cdot CH_3)^+$ ion of $Me_{14}Si_6O_5$ are shown in scheme (7.131).

$$C_{13}H_{39}Si_6O_5^+ \xrightarrow[\ast]{-(CH_3)_4Si} C_9H_{27}Si_5O_5^+$$

$$\xrightarrow[\ast]{-C_6H_{18}Si_2O} C_7H_{21}Si_4O_4^+$$

$$\xrightarrow[\ast]{-C_8H_{24}Si_3O_2} C_5H_{15}Si_3O_3^+$$

$$\xrightarrow[\ast]{-C_6H_{18}Si_3O_3} C_7H_{21}Si_3O_2^+ \qquad (7.131$$

$$\xrightarrow[\ast]{-C_8H_{24}Si_4O_4} C_5H_{15}Si_2O^+$$

$$\xrightarrow[\ast]{-C_{10}H_{30}Si_5O_5} C_3H_9Si^+$$

$$\xrightarrow[\ast]{-C_2H_6SiO} C_{11}H_{33}Si_5O_4^+$$

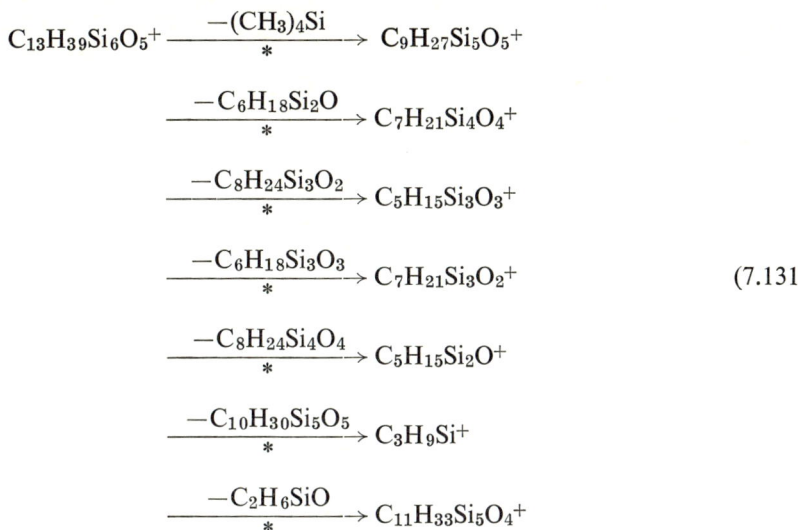

Some interesting multiply charged ions from $Me_8Si_3O_2$ and $Me_{10}Si_4O_3$, (parent molecule $-2 \times \cdot CH_3)^{2+}$, and (parent molecule $-3 \times \cdot CH_3)^{3+}$ were reported[262,271]. Metastable supported fragmentations of the doubly charged species were detected which showed both decomposition to other doubly charged ions and decomposition to two singly charged species.

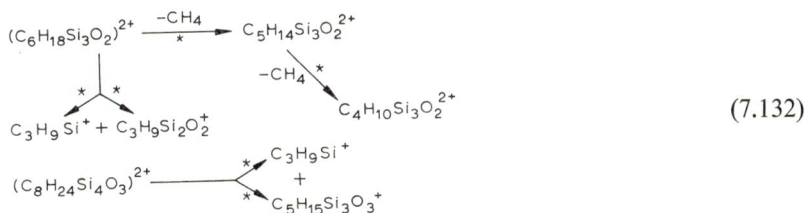

$$(7.132)$$

7.8.1.2 Cyclic siloxanes. Orlov[262] has studied several cyclic siloxanes $(Me_2SiO)_n$ ($n = 3$–5). Parent molecular ions were weak but the (parent molecule $- \cdot CH_3)^+$ ion was intense, being the base peak ion in the spectra of the $n = 3$ and 4 compounds; $Me_{10}Si_5O_5$ showed the $(CH_3)_3Si^+$ ion as base peak. Fragmentations by loss of $:CH_2$, $\cdot CH_3$, $CH_4$, $C_2H_6$, $C_2H_6SiO$, and SiO groups were observed. The metastable supported decompositions in the spectra of $Me_6Si_3O_3$ and $Me_{10}Si_5O_5$ are shown in schemes (7.133) and (7.134).

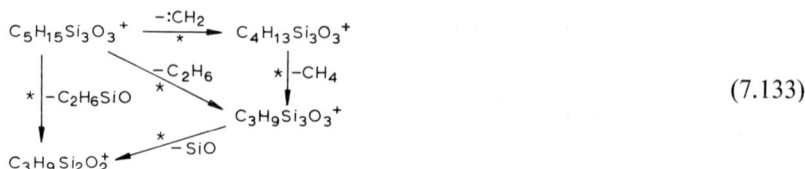

$$(7.133)$$

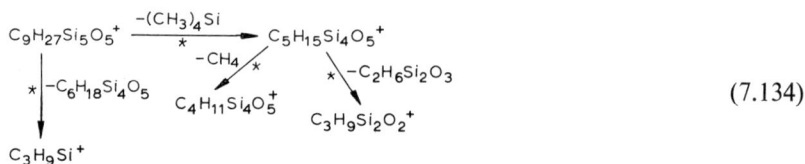

$$(7.134)$$

The decomposition of a doubly charged ion $(C_6H_{18}Si_4O_4)^{2+}$ from $Me_8Si_4O_4$ to two singly charged species was reported[271], *viz.*

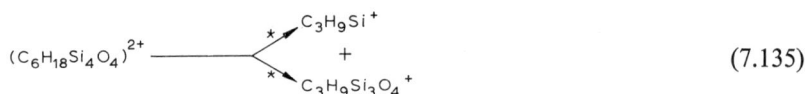

$$(7.135)$$

Orlov pointed out that many of the ions observed from linear siloxanes containing four or more silicon atoms and the corresponding cyclic siloxanes appeared at the same $m/e$ values and probably had the same structure. He suggested that cyclic structures would be more stable than linear ones and therefore these common ions were probably cyclic.

Orlov *et al.*[272] studied XXVIII and a series of cyclocarbosiloxanes, XXIX–XLI, and the corresponding eight-membered ring systems.

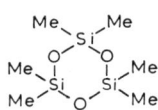

XXXVIII

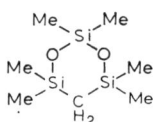

XXXIX

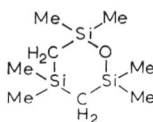

XL

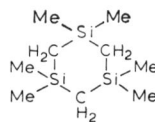

XLI

The base peaks were always due to the (parent molecule $- \cdot CH_3$)$^+$ ions. Initially, fragmentation by loss of hydrocarbon groups such as $CH_4$ and $C_2H_4$ occurred leaving the cyclic framework intact. There was an increase in the number of ions produced by cleavage of the cyclic framework as the number of $-CH_2-$ groups was increased from zero to 3 or 4. Other fragmentation routes involved the loss of silicon-containing groups such as SiO, $(CH_3)_2SiO$, $(CH_3)_2SiCH_2$, and $(CH_3)_4Si$, for instance from $Me_8Si_4O_2(CH_2)_2$.

$$C_9H_{25}Si_4O_2{}^+ \xrightarrow[\quad*\quad]{-(CH_3)_2SiCH_2} C_6H_{17}Si_3O_2{}^+$$

$$*\big|-CH_4 \qquad *\diagdown -(CH_3)_4Si \qquad *\big|-CH_4 \qquad\qquad (7.136)$$

$$C_8H_{21}Si_4O_2{}^+ \xrightarrow[\quad*\quad]{-(CH_3)_2SiCH_2} C_5H_{13}Si_3O_2{}^+$$

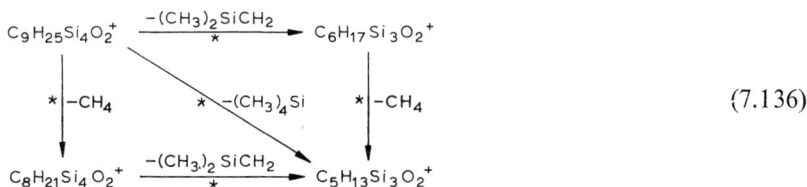

A number of doubly and triply charged ions were found of the types (parent molecule $-2 \times \cdot CH_3)_2{}^+$ and (parent molecule $-3 \times \cdot CH_3)^{3+}$. Other compounds which have been studied are shown in Table 7.26.

### 7.8.2 Germanes

Dube[273] has studied the series of methoxymethylgermanes $Me_{(4-n)}Ge(OMe)_n$ ($n = 1$–4). These compounds show weak parent molecular ions and fragment predominantly by the loss of $\cdot CH_3$ and $CH_2O$ groups. Several metastable supported decompositions were reported and these are shown in scheme (7.137).

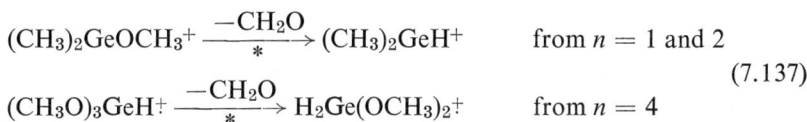

$$(CH_3)_2GeOCH_3{}^+ \xrightarrow[\quad*\quad]{-CH_2O} (CH_3)_2GeH^+ \qquad \text{from } n = 1 \text{ and } 2$$

$$\qquad\qquad\qquad\qquad\qquad\qquad\qquad\qquad (7.137)$$

$$(CH_3O)_3GeH^+ \xrightarrow[\quad*\quad]{-CH_2O} H_2Ge(OCH_3)_2{}^+ \qquad \text{from } n = 4$$

Base peak ions were $(CH_3)_2GeH^+$, $CH_3Ge(OCH_3)_2{}^+$, $(CH_3O)_2GeH^+$, and $CH_3OGe^+$ from $n = 1, 2, 3$, and 4, respectively.

#### 7.8.2.1 Germoxanes. Glockling et al.[70,153] have reported several interesting features in the spectra of some germoxanes of the formulae $R_3GeOGeR_3$ ($R = Me$, Et (ref. 70), $Pr^i$ (ref. 153) and Ph (ref. 70)). The methyl compound showed two metastable supported decompositions from the (parent molecule $-\cdot CH_3)^+$ ion which involved the loss of $OGeMe_2$ and $H_2O$ groups (reaction (7.138)). The corresponding ion from the ethyl compound preferred to decompose by the successive loss of $C_2H_4$ molecules (reaction (7.139)) and the isopropyl compound showed similar fragmentation behaviour.

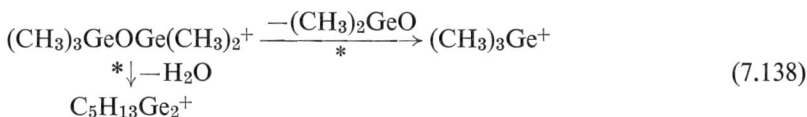

$$(CH_3)_3GeOGe(CH_3)_2{}^+ \xrightarrow[\quad*\quad]{-(CH_3)_2GeO} (CH_3)_3Ge^+$$

$$*\big\downarrow -H_2O \qquad\qquad\qquad\qquad\qquad (7.138)$$

$$C_5H_{13}Ge_2{}^+$$

Table 7.26

Other Organometallic Compounds of Group IV Elements Containing Bonds to Elements of Group VI

| Compound | Comments | Ref. |
|---|---|---|
| Me$_3$SiOMe | P.m.i., base peak (CH$_3$)$_2$SiOH$^+$ | 274 |
| Me$_3$SiCO$_2$R | R = CF$_3$, C$_3$F$_7$; p.m.i. | 275 |
| Me$_3$SiCHOPh(C$_6$X$_5$) | X = F, Cl; p.m.i. | 276 |
| | P.m.i. | 277 |
| | R = Me$_3$Si | 278 |
| (CH$_3$)$_3$SiOCH$_2$CH$_2$OCH$_3$ | Base peak (CH$_3$)$_3$SiO(CH$_2$)$_2$OCH$_2^+$ (may be (CH$_3$)$_5$Si$_2$O$^+$?) | 279 |
| | P.m.i. | 280 |
| Me$_3$SiO–AuPMe$_3$ | P.m.i. | 232 |
| (Me$_3$SiO)$_3$P=O | P.m.i. | 281 |
| | P.m.i. for X = Y = OSiMe$_3$, OSi(OEt)$_3$, OSi(OBu$^t$)$_3$ and X = OH, Y = OSi(OEt)$_3$ | 282 282 282 |
| | P.m.i. base peak, fragments by loss of :CH$_2$ and C$_6$H$_5$· groups | 283 |
| [–O–SiMe–CH$_2$–SiMe–O–]$_4$ | P.m.i. | 283 |

Table 7.26 (continued)

---

| | | |
|---|---|---|
| $[-Me_2SiOC(CF_3)_2-SC(CF_3)_2-O-]_3$ | P.m.i. | 284 |
| $[Me_2SiOC(CF_3)(CF_2Cl)SC(CF_3)CF_2Cl-O]$ | | 284 |
| $[Me_2SiOC(CF_2Cl)_2SC(CF_2Cl)_2-O]$ | | 284 |
| $PhO(CH_2)_2OGeMe_3$ | P.m.i., $(CH_3)_3Ge^+$ base peak rearrangement ion $C_6H_5OGe(CH_3)_2^+$ from (cf. 7.113) $C_6H_5O(CH_2)_2OGe^+(CH_3)_2 \overset{\triangle}{\rightarrow} C_6H_5OGe(CH_3)_2^+$ | 221 |
| $Me_3GeO-\underset{\overset{\|}{O}}{C}-GeMe_2Ph$ | P.m.i. | 285 |

| | | |
|---|---|---|
| 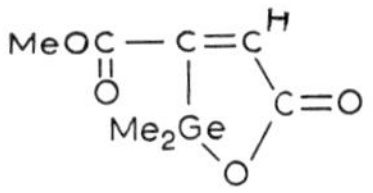 | P.m.i. | 285 |

| | | |
|---|---|---|
| 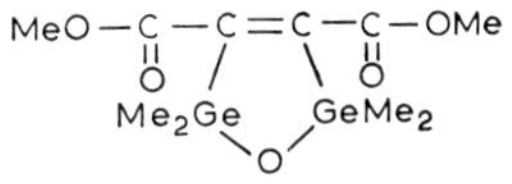 | P.m.i. | 285 |
| $Me_4Ge_4S_6$ | P.m.i., suggested "adamantane" type structure on NMR evidence | 286 |

| | | |
|---|---|---|
| 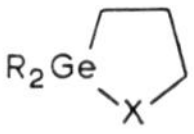 | $R = Bu^n, X = S$; p.m.i. $R = Bu^n, X = Se$; p.m.i. Base peaks due to (parent molecule $-\cdot Bu^n)^+$ | 287 |
| $Et_3SnOMe$ | P.m.i. | 77 |
| $Me_3SnO-\underset{\overset{\|}{O}}{S}-Me$ | P.m.i. | 288–290 |
| $Me_3SnO-\underset{\overset{\|}{O}}{C}-Me$ | No p.m.i. | 289 |
| $Me_3SnO-\underset{\overset{\|}{O}}{C}-CD_3$ | No p.m.i. | 289 |
| $Ph_3SnO-\underset{\overset{\|}{O}}{S}-Ph$ | P.m.i., $(C_6H_5)_3Sn^+$ base peak | 291 |
| $Me_3SnON=C_6H_{10}$ | P.m.i., $(CH_3)_3Sn^+$ base peak | 292 |

---

Table 7.26 (contuined)

| | | |
|---|---|---|
| $Me_3SnON{=}C\langle^{Me}_{Me}$ | P.m.i., $(CH_3)_3Sn^+$ base peak | 292 |
| $Me_3PbO{-}\underset{\underset{\displaystyle O}{\Vert}}{S}{-}Me$ | P.m.i., $(CH_3)_3Pb^+$ base peak, ions at higher *m/e* due to polymeric species | 288 |
| $Ph_3PbO{-}\underset{\underset{\displaystyle O}{\Vert}}{S}{-}CH_2CH{=}CH_2$ | P.m.i. | 288 |

$$
\begin{array}{l}
(C_2H_5)_5Ge_2O^+ \xrightarrow[*]{-C_2H_4} (C_2H_5)_4Ge_2(H)O^+ \\
\qquad\qquad\qquad\qquad\qquad -C_2H_4 \downarrow * \\
\qquad\qquad\qquad\qquad\quad (C_2H_5)_3Ge_2(H)_2O^+
\end{array}
\qquad (7.139)
$$

$$
(C_2H_5)_3Ge_2O^+ \xrightarrow[*]{-C_2H_4} (C_2H_5)_2Ge_2(H)O^+
$$

$$
\begin{array}{l}
(C_2H_5)_2Ge(H)O^+ \xrightarrow[*]{-C_2H_4} (C_2H_5)Ge(H)_2O^+ \\
\qquad\qquad\qquad\qquad\qquad\qquad * \downarrow -C_2H_4 \\
\qquad\qquad\qquad\qquad\qquad\quad H_2Ge^+OH
\end{array}
$$

The loss of ethylene from doubly charged ions from the ethyl compound was also observed.

$$
[(C_2H_5)_4Ge_2O]^{2+} \xrightarrow[*]{-C_2H_4} [(C_2H_5)_3Ge_2(H)O]^{2+} \qquad (7.140)
$$

The base peak ion in the spectrum of $Ph_6Ge_2O$ was $(C_6H_5)_5Ge_2O^+$ and apparently all other ions except $(C_6H_5)_2GeOH^+$, $GeOH^+$, and $(C_6H_5)_2GeH^+$ were derived from the base peak ion and contained the Ge–O–Ge linkage. Several metastable supported decompositions were reported including those shown in scheme (7.141).

$$
\begin{array}{l}
(C_6H_5)_6Ge_2O^{+\cdot} \xrightarrow[*]{-H_2O} C_{36}H_{28}Ge_2^{+\cdot} \\
\\
(C_6H_5)_5Ge_2O^+ \xrightarrow[*]{-C_6H_6} (C_6H_5)_3Ge_2(C_6H_4)O^+ \\
\qquad * \downarrow -(C_6H_5)_2GeO \\
\quad (C_6H_5)_3Ge^+
\end{array}
\qquad (7.141)
$$

The $(C_6H_5)_3Ge^+$ ion was also produced from the rearrangement ion $(C_6H_5)_5Ge_2^+$.

Other compounds in the above category are shown in Table 7.26.

## 7.9 Organometallic compounds containing main Group IV metal–metal bonds

In this section we are concerned with the mass spectra of metal–metal bonded organometallic compounds which give rise to ions with the metal–metal bond intact, and ions which result from the cleavage of the metal–metal bond. The fragmentation of ions containing one metal atom has been discussed previously and will not be dealt with in detail here.

Table 7.27

Ion Abundances for Ions from the Hexamethyl Compounds $Me_6M_2$

| Composition of ion | % Abundance for M–M = | | | | |
|---|---|---|---|---|---|
| | C–C | Si–Si | Ge–Ge | Sn–Sn | Pb–Pb |
| $C_6H_{18}MM$ | | 12 | 7 | 5 | 2 |
| $C_5H_{15}MM$ | 3 | 20 | 18 | 16 | 12 |
| $C_4H_{12}MM$ | | 1 | 1 | 2 | 2 |
| $C_3H_9MM$ | 1 | | 2 | 1 | 2 |
| $C_2H_6MM$ | | | 1 | 1 | 1 |
| $C_2H_5MM$ | | | 1 | 1 | 1 |
| $CH_3MM$ | 2 | | | 1 | 3 |
| $CH_2MM$ | | | | 1 | 1 |
| MM | | | | 2 | 6 |
| $C_3H_{10}M$ | 2 | | | | |
| $C_3H_9M$ | 44 | 42 | 57 | 46 | 38 |
| $C_3H_8M$ | 12 | | | | |
| $C_3H_7M$ | 2 | | | | |
| $C_2H_7M$ | 6 | 3 | 2 | 1 | |
| $C_2H_6M$ | 1 | 5 | 2 | 4 | 3 |
| $C_2H_5M$ | 13 | | | | |
| $CH_5M$ | 7 | 9 | 1 | 1 | |
| $CH_4M$ | 2 | 2 | | | |
| $CH_3M$ | 3 | 6 | 7 | 15 | 23 |
| $H_3M$ | 2 | | | | |
| HM | | | 1 | 1 | |
| M | | | | 2 | 6 |

*References pp. 310–319*

### 7.9.1 Non-cyclic compounds

A number of reports of the spectra of methyl compounds $Me_6M_2$ and ethyl compounds $Et_6M_2$ have appeared. All the methyl compounds of Group IV elements have been studied by Lappert et al.[139]. Vandendunghen[293] has reported spectra for the silicon, germanium, and tin compounds and several other workers have discussed $Me_6Si_2$ (refs. 156, 157, 294) and $Me_6Ge_2$ (refs. 295, 163). The abundance of the major ions ($\geqslant 1\%$ abundance) reported by Lappert et al. are shown in Table 7.27. All the compounds showed parent molecular ions and had base peaks due to $(CH_3)_3M^+$ ions. Important decomposition pathways were by the successive loss of $\cdot CH_3$ groups and these increased in importance with increasing atomic number of M. In each case, about 30% of the total ion current is carried by ions with $M_2$ species intact. The (parent molecule $- \cdot CH_3)^+$ ion was the most abundant $M_2$-containing ion.

The spectrum of $Me_6Si_2$ reported by Tamas et al.[157] is more detailed than those reported by Vandendunghen[293] and Lappert et al.[139] There is reasonable agreement between these spectra. A number of metastable supported decompositions were observed by Tamas et al.[137] and these are shown in scheme (7.142).

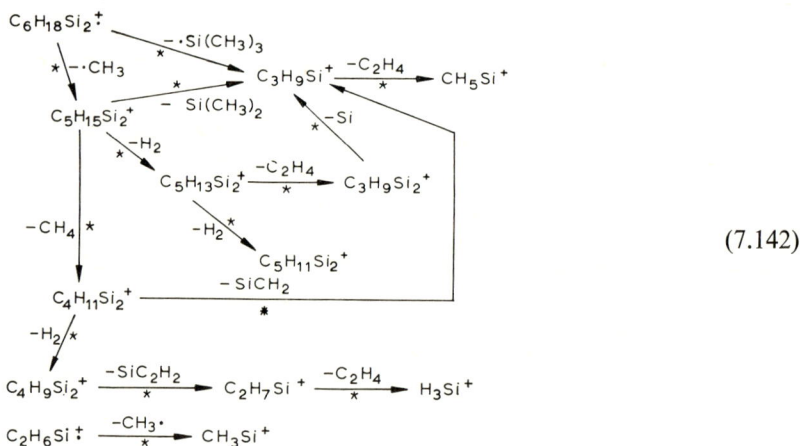

$$(7.142)$$

Fragmentation of the ions containing both silicon atoms was by loss of $H_2$, $CH_4$, $C_2H_4$, and the silicon-containing species $\cdot SiC_3H_9$, $SiC_2H_6$, $SiC_2H_2$, and $SiCH_2$. Gohlke[294] has observed that the negative ion spectrum of $Me_6Si_2$ shows a parent molecular ion but $Me_6Si_2O$ does not.

The ion abundance data reported by Glockling et al.,[173] and de Ridder and Dijkstra[295] for $Me_6Ge_2$ was in fair agreement with that of Lappert et al.[139] (Table 7.27). According to de Ridder and Dijkstra, decomposition of some of the $Ge_2$-containing ions was by loss of $H_2$, $CH_4$, $^-CH_3$, $:CH_2$, $C_2H_6$, and germanium-containing species. Their reported metastable supported decompositions of $Ge_2$-containing ions are given in scheme (7.143).

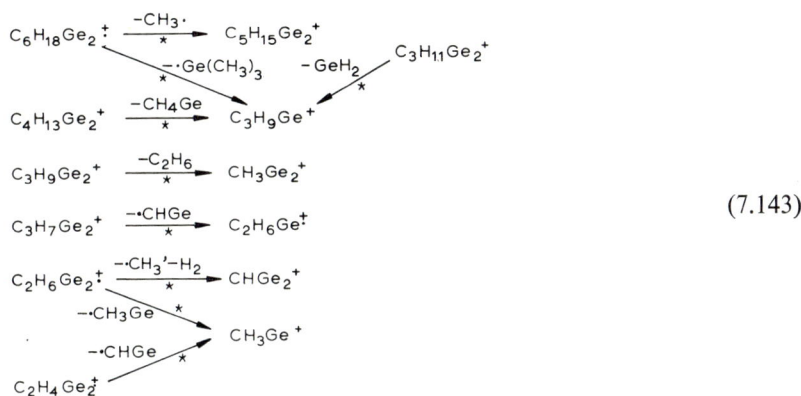

$$(7.143)$$

The spectra of $Me_6Sn_2$ reported by Lappert et al. and Vandendunghen were only in fair agreement. No metastable supported decompositions were reported.

Hexamethyl compounds with two different Group IV elements bonded together, $Me_6MM'$, have been studied by Lappert et al. (M = Si, M' = Ge, Sn; M = Ge, M' = Sn) and Vandendunghen ($Me_3SiSnMe_3$). The same trends in ion abundances are found as for the $M_2$ compounds. Parent molecular ions were observed and the base peaks were $(CH_3)_3M^+$ ions (M = Si for the silicon compounds and M = Ge for $Me_3GeSnMe_3$). The number of ions containing the $MM'$ species was intermediate between those from the corresponding $M_2$ compounds; $(CH_3)_5MM'^+$ ions were the most abundant $MM'$-containing ions.

The hexaethyl compounds $Et_6M_2$ have been studied by Vandendunghen (M = Si, Ge or Sn)[293]. Glockling and Light[70] and de Ridder and Dijkstra[295] have studied $Et_6Ge_2$. Parent molecular ions were observed, that of $Et_6Sn_2$ being much weaker than those of the silicon and germanium compounds. The base peak ions were $(C_2H_5)_3M^+$ for M = Si (23 %) or Sn (27 %) but $(C_2H_5)_3Ge_2H_2^+$ (21 %) for $Et_6Ge_2$. The spectra of these compounds showed quite substantial differences in the type and abundance of the ions found[293]. The silicon compound exhibited the ions $(C_2H_5)_nSi_2H_{(5-n)}^+$ ($n = 1–5$) and $Si_2H_3^+$ in moderate abundance (2–9 %). The ions $(C_2H_5)_n$-$M_2H_{(5-n)}^+$ were also found in the spectrum of $Et_6Ge_2$ but in quite high abundance (3–21 %) and in $Et_6Sn_2$ in low abundance (0–4 %). The ion $(C_2H_5)_2MH^+$ was found in high abundance for $M = Si$ (22 %) and $M = Sn$ (26 %) and only (8.5 %) for $M = Ge$. The ions $CH_3M_2H_4^+$, $M_2^+$, and $MH_3^+$ were reported only in the spectrum of $Et_6Ge_2$ whilst $(C_2H_5)_4M^+$, $(C_2H_5)_2$-$MCH_3^+$, $(CH_3)_3M^+$, and $CH_3MH_2^+$ were reported only for $M = Sn$. The proportion of the ion current from $Et_6M_2$ compounds carried by ions containing the $M_2$ species was much larger for $M = Si$ (47 %) and Ge (75 %) compared with the corresponding hexamethyl compounds but for $M = Sn$ it was very much lower (4 %). Vandendunghen's spectrum of $Et_6Ge_2$ is in overall good agreement with those of Glockling $et\ al.$ and de Ridder and Dijkstra. De Ridder and Dijkstra reported metastable supported decompositions from $Et_6Ge_2$ (7.144) some of which were also observed by Glockling $et\ al.$

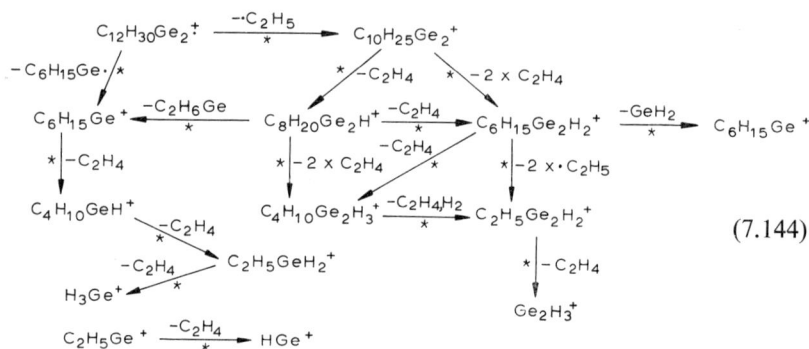

(7.144)

Other compounds which have been studied include $Pr^i_6Si_2$ (ref. 293), $Pr^i_6Ge_2$ (ref. 97), $Bu^i_6Ge_2$, and $(C_6H_5CH_2)_6Ge_2$ (ref. 70). The silicon compound showed a parent molecular ion and $(C_3H_7)_3Si^+$ as the base peak ion. Numerous moderately abundant hydride ions were found including the series $(C_3H_7)_nSi_2H_{(5-n)}^+$ ($n = 1–5$). The silicon compound fragmented mainly by the loss of $\cdot C_3H_7$, and $C_2H_4$ or $C_3H_6$ olefins. The germanium compounds with $R = Pr^i$ or $Bu^i$ behaved similarly, but hexabenzyldigermane showed several differences.

No parent molecular ion was observed from $(PhCH_2)_6Ge_2$ but the ion $(PhCH_2)_5Ge_2^+$ was abundant (26.8 %) and all ions of the type $(PhCH_2)_nGe_2^+$ ($n = 1$–4) were observed. The base peak was $C_7H_7Ge^+$ (35.6 %). Elimination of $PhCH_3$ and $(PhCH_2)_2Ge$: from $(PhCH_2)_5Ge_2^+$ were metastable supported processes.

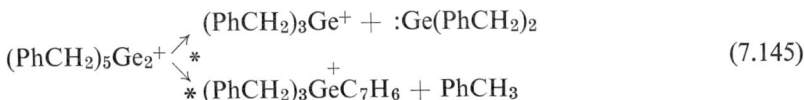

$$(PhCH_2)_5Ge_2^+ \Big\langle \begin{matrix} {}^{\nearrow *} (PhCH_2)_3Ge^+ + :Ge(PhCH_2)_2 \\ {}_{\searrow *} \ \ \overset{+}{(PhCH_2)_3Ge}C_7H_6 + PhCH_3 \end{matrix} \tag{7.145}$$

The spectrum of $Et_3GeGeMe_3$ has been described by de Ridder and Dijkstra and by Vandendunghen who also studied the tin compound $Et_3SnSn$-$Me_3$. Parent molecular ions were observed from both the compounds. The reported spectra of the germanium compound were not in very good agreement, in particular the base peak was reported as $C_3H_{11}Ge_2^+$ (13.9 %) by de Riddel and Dijkstra but as $C_5H_{15}Ge_2^+$ (16.7 %) by Vandendunghen. However, there was agreement that the three most abundant ions were $C_7H_{19}$-$Ge_2^+$, $C_5H_{15}Ge_2^+$, and $C_3H_{11}Ge_2^+$, all in moderately high (11–17 %) a-bundance. The tin compound showed the $C_7H_{19}Sn_2^+$ ion as the base peak. The correct constitution of the ions in these spectra could not always be assigned since for example, an ion containing $-MC_2H_6$ species may be in the form $(CH_3)_2M-$ or $C_2H_5M(H)-$ and therefore the ion $C_2H_7M_2^+$ may have been $C_2H_5HM-MH^+$ or $C_2H_5MMH_2^+$ or $(CH_3)_2MMH^+$ etc. The identification of a number of metastable supported decompositions helped to reveal the fragmentation behaviour of the germanium compound, including those shown in scheme (7.146), but in interpreting these the possibility of rearrangement of $\cdot H$, $\cdot CH_3$ and $\cdot C_2H_5$ groups should not be ignored.

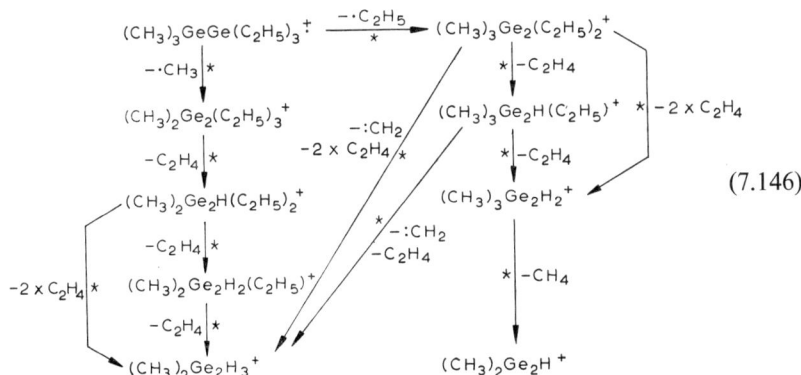

$$(7.146)$$

Such rearrangements have been shown[296] to occur in the spectrum of $(C_6H_5-CH_2)_3GeSiMe_3$. This compound showed a parent molecular ion and a base peak due to the ion $(C_6H_5CH_2)_2GeSi(CH_3)_3^+$ (28%). The corresponding ion due to the loss of $\cdot CH_3$, $(C_6H_5CH_2)_3GeSi(CH_3)_2^+$ was also observed but in only 0.7% abundance. Several interesting metastable supported decompositions were reported (scheme 7.147), including some which involved the formation of rearranged ions.

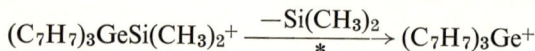

$$(C_7H_7)_3GeSi(CH_3)_2^+ \xrightarrow[*]{-Si(CH_3)_2} (C_7H_7)_3Ge^+$$

(comparable to the elimination of $(C_7H_7)_2Ge$ from $(C_6H_5CH_2)_5Ge_2^+$ in (7.145)).

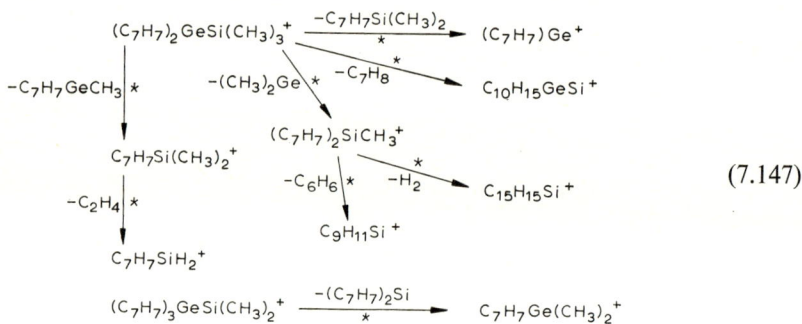

(7.147)

The abundances of the rearrangement ions are shown in Table 7.28A.

Electron-impact studies of the aryl compounds $Ph_6Si_2$ (refs. 296, 197), $Ph_6Ge_2$ (refs. 70, 297), $Ph_6Sn_2$ (refs. 293, 297), and $m$ (tolyl)$_6Ge_2$ (ref. 70) have shown that the spectra of these compounds are relatively simple. Parent molecular ions were observed and the base peaks were due to $(Ar)_3M^+$ ions in every case, being (70%), (57%), (51%), and (68%) abundant for $Ph_6Si_2$, $Ph_6Ge_2$, $Ph_6Sn_2$, and $m$(tolyl)$_6Ge_2$ respectively. Ions containing the $M_2$ species were fewer and less abundant than in the spectra of the alkyl compounds and carried about 10–12% of the total ion current. In particular, the $Ar_5M_2^+$ ions were less abundant than the parent molecular ions, in sharp contrast to the $Me_6M_2$ and $Et_6M_2$ compounds. Fragmentation by cleavage of the M–M bond was preferred to the cleavage of C–M or C–C bonds. The formation of the base peak ions from parent molecular ions were metastable supported processes.

Table 7.28

Ion Abundances of Some Rearranged Ions from Compounds with Metal–Metal Bonds

(A) From $(C_6H_5CH_2)_3GeSi(CH_3)_3$

| Composition of ion | % Abundance |
|---|---|
| $(CH_3)_2SiC_7H_7$ | 12.2 |
| $(CH_3)Si(C_7H_7)_2$ | 3.9 |
| $(C_7H_7)_2GeCH_3$ | 0.2 |
| $(C_7H_7)Ge(CH_3)_2$ | 0.04 |
| $Ge(CH_3)_3$ | 0.9 |

(B) From $(C_6H_5)_3MM'(CH_3)_3$ $(M,M' = Sn \ and \ Ge)$

| Composition of ion | % Abundance from | |
|---|---|---|
| | M–M' $Ph_3SnGeMe_3$ | M–M' $Ph_3GeSnMe_3$ |
| $M(CH_3)_3$ | 0.39 | 0.90 |
| $C_6H_5M(CH_3)_2$ | 0.65 | 1.47 |
| $(C_6H_5)_2M(CH_3)$ | 0.65 | 13.00 |
| $(C_6H_5)_3M'$ | 0.97 | 0.12 |
| $(C_6H_5)_2M'CH_3$ | 5.50 | 0.48 |
| $C_6H_5M'(CH_3)_2$ | 9.02 | 1.67 |

The mass spectra of several mixed aryl–alkyl compounds have been reported by Chambers and Glockling[296] including $Ph_3MMMe_3$ (M = C, Si, Ge, Sn), $Et_3Si–SiPh_3$, $Et_3Si–GePh_3$, $Me_3Ge–SnPh_3$, and $Me_3Sn–GePh_3$. $Ph_3MMMe_3$ (M = Si, Ge), $Ph_3Ge–SiMe_3$ (ref. 297) and $Ph_3SnSnMe_3$ (ref. 293) have also been observed by other workers. Parent molecular ions were observed in all cases and the base peaks from the organometallic compounds studied by Chambers and Glockling[296] were respectively $(C_6H_5)_3M^+$ (M = Si 40.6%; Ge 40.5%), $C_6H_5Sn^+$ (27.8%), $(C_6H_5)_3Si^+$ (34.8%), $(C_6H_5)_2Ge^+$ (28.9%), $Sn^+$ (19.0%) and $(C_6H_5)_3Ge^+$ (26.8%).

In all cases, the ion due to the loss of an alkyl group from the parent molecular ion was in greater abundance than that due to phenyl loss. Always the $(C_6H_5)_3M^+$ ion was more abundant than $R_3M^+$, presumably because there could be more delocalisation of charge in $(C_6H_5)_3M^+$, the decomposition routes of $(C_6H_5)_3M^+$ require more energy than those of $R_3M^+$, and the ionisation potentials of the $Ph_3M \cdot$ radicals are lower than those of $R_3M \cdot$. Loss of radicals $R \cdot$ or $Ph \cdot$ and $\cdot MR_3$ (but not $Ph_3M \cdot$) was a common reaction, $e.g.$ the metastable supported decompositions from $Ph_3Ge$-$SnMe_3$.

$$(C_6H_5)_3GeSn(CH_3)_3^{\ddagger} \begin{array}{c} \xrightarrow[\quad *\quad]{-\cdot CH_3} (C_6H_5)_3GeSn(CH_3)_2^+ \\ \\ \xrightarrow[\quad *\quad]{-\cdot C_6H_5} (C_6H_5)_2GeSn(CH_3)_3^+ \\ \\ \searrow^{-\cdot Sn(CH_3)_3} {}_* \\ \\ (C_6H_5)_3Ge^+ \end{array}$$  (7.148)

In each of the compounds $Ph_3M\text{-}M'R_3$, all six possible ions due to rearrangement processes were observed, (7.147), although the relative abundances of these ions varied considerably, for example from $Ph_3MM'Me_3$ (M,M' = Sn and Ge), Table 7.28B.

$$(C_6H_5)_3MM'R_3^{\ddagger} \to \begin{array}{c} (C_6H_5)_2M^+R + (C_6H_5)M^+R_2 + {}^+MR_3 \\ \\ R_2^+M'(C_6H_5) + R^+M'(C_6H_5)_2 + {}^+M'(C_6H_5)_3 \end{array}$$  (7.149)

Glockling $et\ al.$ suggested that these rearrangements occurred either just before the decomposition of parent molecular ion or as a synchronous process. They considered that the reactions probably proceeded $via$ a five-coordinate intermediate involving the participation of $d$ orbitals. This could then explain the absence of rearrangement ions from $Ph_3CCMe_3$ (ref. 291). Some of the interesting metastable supported elimination reactions including those involving rearrangement ions are listed in scheme (7.150).

$$(C_6H_5)_3SiSi(C_2H_5)_3^{\ddagger} \xrightarrow[\quad *\quad]{-\cdot C_6H_5Si(C_2H_5)_2} (C_6H_5)_2SiC_2H_5^+$$

$$(C_6H_5)_3SiSi(CH_3)_2^+ \xrightarrow[\quad *\quad]{-C_6H_5SiCH_3} (C_6H_5)_2SiCH_3^+$$

$$\begin{array}{c} (C_6H_5)_3GeSn(CH_3)_2^+ \xrightarrow[\quad *\quad]{-C_6H_5SnCH_3} (C_6H_5)_2GeCH_3^+ \\ -(C_6H_5)_2Ge(CH_3)_2\!\downarrow\!* \\ C_6H_5Sn^+ \end{array}$$  (7.150)

$$(C_6H_5)_3GeSi(C_2H_5)_3^{\ddagger} \xrightarrow[\quad *\quad]{-C_6H_5Si(C_2H_5)_3} (C_6H_5)_2Ge^{\ddagger}$$

$$(C_6H_5)_3SnGe(CH_3)_3^{\ddagger} \xrightarrow[\quad *\quad]{-C_6H_5Ge(CH_3)_3} (C_6H_5)_2Sn^{\ddagger}$$

Other compounds in this category are listed in Table 7.29.

Table 7.29

Other Metal–Metal Bonded Organometallic Compounds of Group IV Elements

| Compound | Comments | Ref. |
|---|---|---|
| $Me_5Ge_2Et$ | P.m.i., fragmentation by loss of $\cdot R$ | 163 |
| $Me_5Ge_2Pr$ | groups and elimination of olefins | 163 |
| $Et_5Ge_2C_6H_{13}$ | | 163 |
| $Bu_5Ge_2I$ | | 163 |
| $Me_8Ge_3$ | | 163 |
| $Et_8Ge_3$ | | 163 |
| $Me_5Ge_2CH_2GeMe_3$ | | 163 |
| $(Me_3Ge)_3GeMe$ | | 163 |
| $(Me_3Ge)_2GeMeGeMe_2Et$ | | 163 |
| $Me_{10}Ge_4CH_2$ | | 163 |
| $Me_9Ge_4Pr$ | | 163 |
| $Me_8Ge_4Et_2$ | | 163 |
| $(Me_3Ge)_4Ge$ | | 163 |
| $(Me_3Ge)_2Ge(Me)Ge_2Me_5$ | | 163 |
| $(Me_3Ge)_2Ge(Et)Ge_2Me_5$ | | 163 |
| $Me_{14}Ge_6$ | | 163 |
| $Me_{16}Ge_7$ | | 163 |
| $Me_{18}Ge_8$ | | 163 |
| $Me_{20}Ge_9$ | | 163 |
| $Me_{22}Ge_{10}$ | | 163 |
| $Ph_6Pb_2$ | No p.m.i., highest $m/e$ $(C_6H_5)_3Pb_2^+$, $(C_6H_5)Pb^+$ base peak | 298 |
| $Ph_3Pb_2(O_2CCH_3)_3$ | No p.m.i., highest $m/e$, $(C_6H_5)_3Pb_2$ $(O_2CCH_3)_2^+$; $Pb(O_2CCH_3)^+$ base peak | 298 |
| $\begin{array}{ccc} & CH_2 & \\ & \diagup \quad \diagdown & \\ Me_2Si & & SiMe_2 \\ | & & | \\ Me_2Si & & SiMe_2 \\ & \diagdown \quad \diagup & \\ & CH_2 & \end{array}$ | P.m.i., $(CH_3)_3Si^+$ base peak | 303 |
| $[p(tolyl)_2Si]_n$ | P.m.i., $n = 4–6$ | 304 |
| $[p(tolyl)_2Ge]_n$ | P.m.i., $n = 4–6$ | 304 |

## 7.9.2 Cyclic compounds

The methylcyclopolysilanes $(Me_2Si)_n$ ($n = 5–7$) have been studied by Kinstle et al.[299] and $(Me_2Si)_5$ was reported by Carberry and West[300]. Fairly intense parent molecular ions were observed and the base peak ions were due to $(CH_3)_3Si^+$ in all cases. The fragmentation of these compounds was mainly by the loss of $\cdot CH_3$, $H_2$, $\cdot Si(CH_3)_3$, $Si(CH_3)_4$, $(CH_3)_2SiH_2$, and $(CH_3)_3SiH$ groups. Some metastable supported decompositions from $(Me_2Si)_6$ are shown in scheme (7.151).

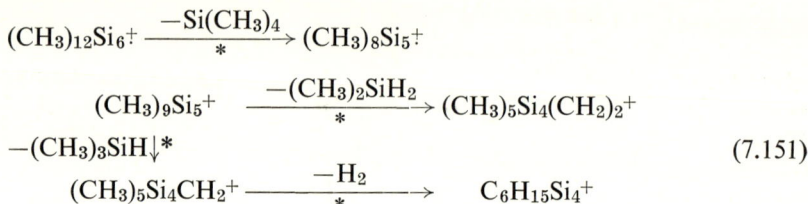

$$(CH_3)_{12}Si_6^+ \xrightarrow[*]{-Si(CH_3)_4} (CH_3)_8Si_5^+$$

$$(CH_3)_9Si_5^+ \xrightarrow[*]{-(CH_3)_2SiH_2} (CH_3)_5Si_4(CH_2)_2^+$$

$$-(CH_3)_3SiH\downarrow * \qquad\qquad (7.151)$$

$$(CH_3)_5Si_4CH_2^+ \xrightarrow[*]{-H_2} \qquad C_6H_{15}Si_4^+$$

The compounds $(Me_2Si)_5$ and $(Me_2Si)_7$ behaved similarly but the five-membered silicon compound preferred to lose $(CH_3)SiH_3$ rather than $(CH_3)_3$-SiH and the seven-membered system also lost $\cdot Si_2(CH_3)_5$ from the parent molecular ion. Gohlke[294] observed that $(Me_2Si)_6$ showed a parent molecular ion in its negative ion spectrum.

Kinstle et al.[299] reported some features in the spectra of the phenyl compounds $(Ph_2Si)_n$ ($n = 4$–$6$). Parent molecular ions were moderately abundant and the rearrangement ion $(C_6H_5)_3Si^+$ was the base peak from all the compounds. Fragmentation occurred mainly by the loss of $\cdot C_6H_5$, $Si(C_6H_5)_2$, and $\cdot Si(C_6H_5)_3$ groups. These findings supported the earlier observations of Kuhlein and Neumann[301] who reported some metastable supported decompositions.

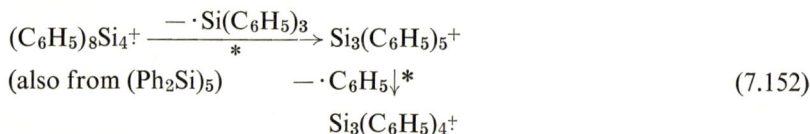

$$(C_6H_5)_8Si_4^+ \xrightarrow[*]{-\cdot Si(C_6H_5)_3} Si_3(C_6H_5)_5^+$$

(also from $(Ph_2Si)_5$)  $\qquad -\cdot C_6H_5\downarrow *$  $\qquad\qquad (7.152)$

$$Si_3(C_6H_5)_4^+$$

The germanium compounds $(Ph_2Ge)_n$ ($n = 4$–$6$) behaved analogously[301] but the tin derivative $Ph_{12}Sn_6$ decomposed[301,302] under the conditions used for $Ph_6Sn_2$ and $Ph_4Sn$.

Other compounds which have been studied are listed in Table 7.29.

*7.10 Organometallic compounds with more than one type of main Group IV–other element bond*

Compounds in this category which have not been discussed previously in connection with other sections above are listed in Table 7.30.

Table 7.30

Other Organometallic Compounds of Group IV Elements

| Compound | Comments | Ref. |
|---|---|---|
| $(C_5H_5)_2SnBF_3$ | No p.m.i., highest $m/e$, (parent molecule $- 2 \cdot H)^+$; $C_5H_5Sn^+$ base peak | 305 |
| $MeGeHX_2$ | P.m.i. for X = Cl, Br, I | 215 |
| $MeGeH_2X$ | | 215 |
| $H_3GeCH_2SiH_2Cl$ | | 200 |
| $H_3GeCH_2SiHCl_2$ | | 200 |
| $MeOSiH_2Me$ | P.m.i. | 306 |
| $MeOSiHMe_2$ | P.m.i. | 274 |
| $MeOSiF_3$ | No p.m.i., hydrolysis products observed in mass spectrometer $e.g.$ $(CH_3O)Si_2OF_4^+$ | 307 |
| (OC(CF₃)₂OEt)<br>\|<br>EtO–Si–H<br>\|<br>(OC(CF₃)₂OEt) | No p.m.i., highest $m/e$, (parent molecule $- \cdot CF_3)^+$ | 308 |
| Me<br>\|<br>Me₃SiO–Si–OSiMe₃<br>\|<br>X | P.m.i. for X = H; no p.m.i. for X = F but (parent molecule $- \cdot F)^+$ and (parent molecule $- \cdot CH_3)^+$ ions | 309 |
| $(Ph(H)BrSi-)_2$ | P.m.i. and rearranged ions $(C_6H_5)_2SiBr^+$, $(C_6H_5)_2SiH^+$ | 202 |
| $Me_2Si(PH_2)H$ | P.m.i., base peaks were $(CH_3)_2MH^+$ | 234 |
| $Me_2Ge(PH_2)H$ | ions | 234 |

## 8. IONISATION AND APPEARANCE POTENTIAL MEASUREMENTS ON MAIN GROUP IV ORGANOMETALLIC COMPOUNDS

Several research schools have been interested in the energetics of dissociation of organometallic molecules. Usually, the intention is to obtain data on bond dissociation energies in molecules and ions and heats of formation of molecules and ions. Sometimes enough information is available for bond energies to be calculated. The measurement of ionisation and appearance potentials has been discussed in Part I of this book (Chapter 3). In this section we will discuss the results of the work on Group IV organometallic compounds. Most of the studied reported have concerned compounds of the

*References pp. 310–319*

types $R_3MX$ (R = alkyl or aryl group; M = Si, Ge, or Sn). These will be discussed first before the reported work on alkylsilanes and trichlorosilanes.

## 8.1 Trimethyl-M derivatives (Me₃MX)

Most investigations have tried to measure or estimate the ionisation potential of the $Me_3M$ radical and then from the appearance potentials of the $(CH_3)_3M^+$ ion from $Me_3MX$, to calculate bond dissociation energies and heats of formation.

$$A.P._{Me_3M^+} = I.P._{Me_3M} \cdot + D(Me_3M-X) \qquad (7.153)$$

$$= \Delta H°_f Me_3M^+ + \Delta H°_f X \cdot - \Delta H°_f Me_3MX \qquad (7.154)$$

The dissociation process is always assumed to be straightforward, (7.155). Unfortunately very few attempts have been made to check this assumption. In some cases the production of an ion pair may conceivably be important, (7.156), and therefore some care is needed in the interpretation of appearance potentials.

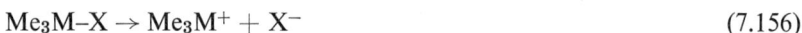

$$Me_3M-X \rightarrow Me_3M^+ + X \cdot \qquad (7.155)$$

$$Me_3M-X \rightarrow Me_3M^+ + X^- \qquad (7.156)$$

Values of the ionisation potentials of $Me_3M \cdot$ (M = C, Si, Ge, Sn, Pb) which have been reported are shown in Table 7.31. The remarkable spread of the results, (1.3 eV for $Me_3Si \cdot$) is testimony to the difficulty in obtaining reliable thermochemical data by mass spectrometry. However, all the results should not be dismissed completely, at least until a careful examination of the methods used has been made. The most common method used was the combination of appearance potential data with thermochemically determined (or estimated) heat of formation data. As far as the silicon compounds are concerned, nearly all the thermochemically determined heat of formation data for silicon alkyls are notoriously inaccurate (see footnote j; Table 7.32, p. 307) and results relying solely upon such data may be treated with reservation. However, there is reliable heat of formation data for a number of $Me_3C-X$, $Me_3Sn-X$, and $Me_3Pb-X$ compounds. The method adopted by Lappert et al.[139] to determine the ionisation potentials of all the $Me_3M$

Table 7.31

Ionisation Potentials of the Me₃M Radicals

| Radical | I.P. (eV) | Method[a] | Ref. |
|---------|-----------|-----------|------|
| Me₃C· | 7.42 ± 0.07 | A | 328 |
| | 7.45 ± 0.1 | B | 325 |
| | 6.9 ± 0.1 | C | 329 |
| | 7.45 ± 0.1 | C | 139 |
| Me₃Si· | 7.10 ± 0.15 | C | 315,324 |
| | 6.94 ± 0.43 | C | 312,313 |
| | 7.8 | D | 330 |
| | 7.26 ± 0.1 | C | 139 |
| | 7.31 ± 0.18 | C | 331 |
| | 7.54 ± 0.1 or | C | 318 |
| | 7.95 ± 0.1 | | |
| | 6.67 ± 0.05 | C | 140 |
| Me₃Ge· | 8.0 | D | 330 |
| | 7.08 ± 0.1 | C | 139 |
| | 7.11 ± 0.18 | C | 331 |
| Me₃Sn· | 6.54 ± 0.15 | C | 322 |
| | 6.80 ± 0.3 | C | 321 |
| | 7.10 ± 0.05 | B | 325 |
| | 7.6 | D | 330 |
| | 6.83 ± 0.1 | C | 139 |
| | 6.93 ± 0.26 | C | 331 |
| Me₃Pb· | 7.6 | D | 330 |
| | 6.66 ± 0.1 | C | 139 |

[a] A, pyrolysis to produce Me₃C radicals in the mass spectrometer; B, Me₃M radicals produced and ionised in a double-beam mass spectrometer; C, combination of thermochemical and mass spectrometric data; D, not stated (these values were quoted without references or probable errors).

radicals was to use all the reliable data for $Me_4M$ and $Me_6M_2$ compounds and measure the appearance potentials of the $(CH_3)_3M^+$ ions from these compounds and some mixed derivatives $Me_6MM'$. Their results were self-consistent to ± 2 kcal.mole⁻¹ and agree quite well with the work of Yergey and Lampe[321] who used only Me₃Sn– compounds. This appears to be entirely fortuitous since the values of the appearance potentials of the $(CH_3)_3Sn^+$ ions from $Me_4Sn$, $Me_3SnBu'$, and $Me_6Sn_2$ were all higher by 0.2–0.3 eV than the values obtained by Lappert et al. and the differences

Table 7.32

Ionisation Potentials of $R_3MX$, Appearance Potentials of $R_3M^+$ ions, Heat of Formation and Bond Dissociation Energies $D(R_3M-X)$

| M | X | I.P.(eV) | A.P.(eV) | Method* | $\Delta H^\circ_f$(g) (kcal.mole$^{-1}$) | $D(R_3M-X)$ (kcal.mole$^{-1}$) | Ref. |
|---|---|---|---|---|---|---|---|
| *A* | *R = Me* | | | | | | |
| Si | Me | 9.98 ± 0.13 | 10.63 ± 0.13 | RPD | (−69)[a] | 85 or 79 | 312, 313 |
| | | 9.8 ± 0.15 | 11.3 ± 0.15 | EVD | (−63)[b] | | 314 |
| | | 9.9 ± 0.1 | 10.4 ± 0.1 | MIB | −68 ± 2 | 76 ± 2 | 315, 316 |
| | | 9.85 ± 0.16 | 10.53 ± 0.2 | SLP | −48.3 ± 2 | 74.4 ± 2 | 139 |
| | | 11.2 ± 0.2 | 12.0 ± 0.2 | IB | | | 295 |
| | | 9.81 ± 0.1 | | PI | | | 317 |
| Si | H | | 10.53 ± 0.1 | PI | | 78.9 | 318, 319 |
| | | 9.86 ± 0.02 | 10.09 ± 0.02 | RPD | (−60)[a] | | 140 |
| | | 9.8 ± 0.3 | 10.78 ± 0.07 | EVD | (−54)[b] | 88 or 83 | 312, 313 |
| | | (9.6?) | 10.9 ± 0.2 | MIB | −55 | | 320 |
| | | | 10.6 ± 0.1 | | | 81 | 315, 316 |
| | | | 10.72 ± 0.1 | | | | 318 |
| Si | $F^h$ | 10.55 ± 0.06 | 11.7 ± 0.5 | RPD | | 193 | 312 |
| Si | Cl | 10.58 ± 0.04 | 12.4 ± 0.06 | RPD | −117 (or −118) | 126 (or 120) | 312, (313) |
| | | 9.9 ± 0.1 | 10.9 ± 0.1 | MIB | (−84.7)[c] | 88 ± 2 | 315, 316 |
| | | | 11.5 ± 0.2 | | | | 318 |
| Si | Br | 10.24 ± 0.02 | 10.69 ± 0.06 | RPD | −80 | 86 | 312 |
| | | 9.8 ± 0.1 | 10.5 ± 0.1 | MIB | −77 ± 2 | 78.5 ± 2 | 315, 316 |
| Si | I | 8.9 ± 0.1 | 10.1 ± 0.1 | MIB | −69 ± 2 | 69 ± 2 | 315, 316 |
| Si | Et | 9.70 ± 0.01 | 10.53 ± 0.09 | RPD | −77 | 83 (or 77) | 312 or (313) |
| | | | 10.34 ± 0.1 | | | | 318 |
| Si | $Pr^i$ | 9.50 ± 0.03 | 10.56 ± 0.16 | RPD | −91 | 84 (or 77) | 312, (313) |
| Si | $Bu^t$ | 9.34 ± 0.06 | 10.53 ± 0.09 | RPD | −98 | 83 (or 77) | 312, (313) |

| Element | Substituent | | | Method | | | Refs. |
|---|---|---|---|---|---|---|---|
| Si | CH$_2$Ph | 9.79 ± 0.04 | 10.05 ± 0.1 | RPD | | | 318 |
| Si | OMe | 9.59 ± 0.04 | 12.43 ± 0.18 | RPD | −143 | 127 | 312 |
| Si | OSiMe$_3$[h] | 8.06 ± 0.02 | 15.36 ± 0.13 | RPD | −186[d] | 194 | 312 |
| Si | NEt$_2$ | 8.79 ± 0.08 | 12.61 ± 0.03 | RPD | −113 | 131 | 312 |
| Si | SiMe$_3$ | | 10.69 ± 0.04 | MIB | −126 (or −129)[d] | 86 (or 81) | 312 (313) |
| Si | | | 10.0 ± 0.1 | | −118 | (67 ± 2)[g] | 315, 316 |
| Si | | | 10.03 ± 0.1 | | | (49 ± 6)[g] | 319 |
| Si | GeMe$_3$ | 8.35 ± 0.12 | 10.22 ± 0.18 | SLP | −86.4 ± 2 | 68.0 ± 2 | 139 |
| Si | SnMe$_3$ | 8.31 ± 0.10 | 10.19 ± 0.12 | SLP | −74.1 ± 2 | 67.7 ± 2 | 139 |
| Ge | Me | 8.18 ± 0.14 | 10.18 ± 0.26 | SLP | −49.7 ± 2 | 68.3 ± 2 | 139 |
| | | 9.2 ± 0.2 | 10.2 ± 0.1 | EVD | −35[d] | | 320 |
| | | 11.2 ± 0.2 | 11.4 ± 0.2 | IB | | | 295 |
| Ge | Bu$^t$ | 9.29 ± 0.14 | 10.05 ± 0.14 | SLP | −32 ± 2 | 69.0 ± 2 | 139 |
| Ge | SiMe$_3$ | 8.98 ± 0.12 | 9.91 ± 0.22 | SLP | −55.7 ± 2 | 64.7 ± 2 | 139 |
| Ge | GeMe$_3$ | 8.31 ± 0.10 | 9.99 ± 0.14 | SLP | −74.1 ± 2 | 67.7 ± 2 | 139 |
| | | 8.5 ± 0.1 | 11.3 ± 0.1 | IB | | | 295 |
| Ge | GeEt$_3$ | 8.18 ± 0.11 | 9.96 ± 0.16 | SLP | −62.5 ± 2 | 67.0 ± 2 | 139 |
| | | 7.6 ± 0.1 | 13.2 ± 0.1 | IB | | | 295 |
| Ge | SnMe$_3$ | 8.20 ± 0.10 | 10.01 ± 0.18 | SLP | −39.7 ± 2 | 69.1 ± 2 | 139 |
| Sn | Me | 8.25 ± 0.15 | 9.9 ± 0.15 | EVD | | | 314 |
| | | 9.1 ± 0.2 | 10.8 ± 0.2 | IB | | | 295 |
| | | 8.76 ± 0.02 | 9.72 ± 0.03 | RPD | −4.6 ± 0.6[e] | 69 ± 6 | 321, 322 |
| | | | 9.7 ± 0.2 | | | 55 | 137 |
| | | | 9.58 ± 0.19 | SLP | −3.4 ± 2 | 65.4 ± 2 | 139 |
| Sn | Br | | | | | 76 | 326 |
| Sn | I | | | | | 62 | 326 |
| | | | | | | | 326 |
| Sn | Et | 8.76 ± 0.12 | 9.49 ± 0.07 | RPD | −7.1 ± 0.7[e] | 64 ± 6 | 321, 322 |
| | | | | | | 50 | 326 |

Table 7.32 (continued)

| M | X | I.P.(eV) | A.P.(eV) | Method* | $\Delta H^\circ_f(g)$ (kcal.mole⁻¹) | $D(R_3M-X)$ (kcal.mole⁻¹) | Ref. |
|---|---|---|---|---|---|---|---|
| Sn | Pr$^n$ | 8.54 ± 0.01 | 9.50 ± 0.06 | RPD | −11.6 | 66 ± 7 | 321 |
| Sn |  |  |  |  |  | 50 | 326 |
| Sn | Pr$^i$ | 8.28 ± 0.01 | 9.17 ± 0.14 | RPD | −11.2 | 60 ± 6 | 321 |
| Sn | Bu$^n$ |  | 9.80 ± 0.04 | RPD | −16.5 | 65 ± 6 | 321 |
| Sn | Bu$^s$ | 8.27 ± 0.01 | 9.20 ± 0.05 | RPD | −15.9 | 60 ± 6 | 321 |
| Sn | Bu$^i$ | 8.34 ± 0.02 | 9.79 ± 0.12 | RPD | −18.5 | 65 ± 6 | 321 |
| Sn | Bu$^t$ |  | 9.50 ± 0.10 | RPD | −16.0 | 55 ± 6 | 321 |
| Sn | Bu$^t$ | 8.34 ± 0.11 | 9.32 ± 0.16 | SLP | −24.9 ± 2 | 58.9 ± 2 | 139 |
| Sn | $-(CH=CH_2)$ |  | 10.44 ± 0.11 | RPD | 22.1 | 75 ± 7 | 321 |
| Sn |  |  |  |  |  | 60 | 326 |
| Sn | $C_3H_5$ |  | 8.68 ± 0.02 | RPD |  |  | 321 |
| Sn | $CH_2Ph$ |  |  |  |  | 39 | 326 |
| Sn | $C_6H_5$ |  |  |  |  | 63 | 326 |
| Sn | $SnMe_3$ | 8.08 ± 0.02 | 9.85 ± 0.16 | RPD | −5.4 | 69 ± 8 | 321 |
| Sn |  |  |  |  |  | 57 | 326 |
| Sn | $GeMe_3$ | 8.02 ± 0.15 | 9.51 ± 0.22 | SLP | −7.1 ± 2 | 61.6 ± 2 | 139 |
| Sn | $SiMe_3$ | 8.20 ± 0.10 | 9.85 ± 0.22 | SLP | −39.7 ± 2 | 69.1 ± 2 | 139 |
| Sn |  | 8.18 ± 0.14 | 9.80 ± 0.24 | SLP | −49.7 ± 2 | 68.3 ± 2 | 139 |
| Pb | Me | 8.0 ± 0.4 | 8.9 ± 0.1 | EVD | 22.6$^t$ |  | 314 |
| Pb |  | 9.3 ± 0.2 | 10.1 ± 0.2 | IB |  |  | 295 |
| Pb | Bu$^t$ | 8.26 ± 0.17 | 8.77 ± 0.16 | SLP | 32.6 ± 2 | 48.8 ± 2 | 139 |
| Pb |  | 7.99 ± 0.13 | 8.67 ± 0.21 | SLP | 6.9 ± 2 | 46.5 ± 2 | 139 |
| Pb | $PbMe_3$ | 7.41 ± 0.10 | 9.02 ± 0.14 | SLP | 38.7 ± 2 | 54.6 ± 2 | 139 |

| $B$ | $R = Ph^k$ | | |
|---|---|---|---|
| Sn | Ph | 10.1 ± 0.2 | 137 |
| | | 9.5 ± 0.2 | 147 |
| Sn | Et | 8.6 ± 0.2 | 147 |

Thermochemical data ($\Delta H°f$ values) used in the above work:

a S. Tannenbaum, *J. Am. Chem. Soc.*, 76 (1954) 1027.

b S. Tannenbaum, S. Kaye and G. F. Lewenz, *J. Am. Chem. Soc.*, 75 (1953) 3753.

c A. E. Beezer and C. T. Mortimer, *J. Chem. Soc. A*, (1966)( 514.

d Estimated by the investigators who obtained the appearance potential results.

e H. A. Skinner, *Advan. Organometal. Chem.*, 2 (1964) 49.

f W. D. Good, W. D. Scott, J. L. Lacina and J. P. McGullough, *J. Phys. Chem.*, 63 (1959) 1139.

g Determined kinetically.

h Authors state that reproducible values could not be obtained. In their opinions, their results for bond dissociation energies are "probably too high".

i Some workers (notably in refs. 312 and 321) report the appearance potentials of the $Me_2SnX^+$ ions.

j It is generally recognised that thermochemically determined $\Delta H°f$ values of organosilicon compounds which were published up to 1968 are probably unreliable—see footnote e and J. D. Cox and G. Pilcher, *Thermochemistry of Organic and Organometallic Compounds*, Academic Press, New York, 1970.

k Some other appearance potential data on Ph–Sn compounds are:
(i) from $SnPh_4$, A.P. $^{\ddagger}SnPh_2$ = 9.1 ± 0.2; A.P. $^{+}SnPh$ = 16.1 ± 0.5; A.P. $^{\ddagger}Sn$ = 9.4 ± 0.2 (ref. 137),
(ii) from $Ph_3SnEt$, A.P. $EtSn^{+}Ph_2$ = 9.0 ± 0.2 (ref. 147),
(iii) from $PhSnEt_3$, A.P. $Et_3Sn^{+}$ = 8.5 ± 0.2 (ref. 147).

* Methods: RPD, retarded potential difference; EVD, extrapolated voltage difference; MIB, modified initial break; SLP, semi-logarithmic plot; IB, initial break; PI, photoionisation.

were therefore cancelled out. However, the higher appearance potentials obtained by Yergey and Lampe did, of course, give rise to higher values of bond dissociation energies $D(Me_3Sn-X)$.

Davidson et al.[324] have used a slightly different approach in their work. They have determined the bond dissociation energy $D(Me_3Si-SiMe_3)$ from a kinetic study of the gas phase dissociation of hexamethyldisilane. From an electron-impact study of this compound, they calculated[315] the ionisation potential of $Me_3Si\cdot$ from eqn. (7.153). Their value of $7.1 \pm 0.1$ eV is in agreement with that obtained by Lappert et al.[139] within experimental error.

A third method has been the direct ionisation of $Me_3M$ radicals produced either by pyrolysing a suitable compound or by dissociation of a compound by electron impact. Lampe and Niehaus[325] have used a double-beam mass spectrometer to measure the ionisation potentials of $(CH_3)_3C\cdot$, $(CH_3)_3Sn\cdot$, and $(CH_3)_2Sn$:. Their value for the tert.-butyl radical agreed well with a previous value obtained using a pyrolysis–mass spectrometric technique and that for the trimethylstannyl radical was in satisfactory agreement with other results obtained by Yergey and Lampe and Lappert et al.

In connection with the determination of the appearance potentials of $Me_3M^+$ ions, it should be mentioned that several different methods of obtaining the appearance potentials have been used. A number of workers have used the retarding potential difference (RPD) method[312,313]. Their results are generally higher than those obtained with the semi-logarithmic plot method[139] or modified initial break method[315,316]. The reason for this may be because the relative insensitivity of the RPD-type source obscures the tails of the ionisation efficiency curves which are usually quite long for $(CH_3)_3M^+$ ions. Distefano[140] using a photoionisation source obtained a value of the appearance potential of $(CH_3)_3Si^+$ from $Me_4Si$ which was much lower (by 0.3–0.4 eV) than any reported value using a conventional source. His value for the ionisation potential of $Me_4Si$ was, however, in good agreement with that of Lappert et al. who used a conventional source and the semi-logarithmic plot method of calculation.

The ionisation and appearance potentials of $Me_3MX$ compounds and thermochemical data derived therefrom are given in Table 7.32A.

Bock and Seidl have reported the ionisation potentials of a number of $Me_3Si$-substituted ethylenes in a study of the bonding of the $Me_3Si$ and $Me_3SiCH_2$ groups associated with an olefinic bond[327]. The compounds studied and ionisation potentials are listed below.

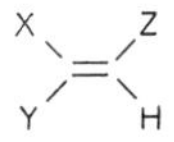

| X | Y | Z | Ionisation potential (eV $\pm$ 0.1) |
|---|---|---|---|
| SiMe$_3$ | H | H | 9.82 |
| CH$_2$SiMe$_3$ | H | H | 8.85 |
| SiMe$_3$ | H | SiMe$_3$ | 9.32 |
| H | CMe$_3$ | SiMe$_3$ | 9.08 |
| H | CH$_2$SiMe$_3$ | CH$_2$SiMe$_3$ | 7.95 |
| SiMe$_3$ | SiMe$_3$ | H | 9.25 |
| SiMe$_3$ | SiMe$_3$ | SiMe$_3$ | 8.85 |
| SiMe$_3$ | SiMe$_3$ | CMe$_3$ | 8.72 |

*8.2 Triphenyl-M derivatives*

Glocking *et al.*[147] have reported the appearance potentials of (C$_6$H$_5$)$_3$Sn$^+$ ions from some Ph$_3$SnX compounds (X = Ph, Et, Table 7.32B). Very little other work on phenyl compounds has been reported but Occolowitz[137] obtained appearance potentials for (C$_6$H$_5$)$_n$M$^+$ ions ($n$ = 0-3) from Ph$_4$Sn. His result for (C$_6$H$_5$)$_3$Sn$^+$ was 0.6 eV higher than that obtained by Glockling *et al.*, Table 7.32B).

Table 7.33

Ionisation Potentials, Appearance Potentials of X$_3$Si$^+$ Ions, Heats of Formation and Bond Dissociation Energies from X$_3$Si–R Compounds[79,323]

*A Derivatives RSiH$_3$*

| R | I.P. (eV) | A.P. H$_3$Si$^+$ (eV) | $\Delta H^\circ_f$ RSiH$_3$[a] (kcal.mole$^{-1}$) | $D$(R–SiH$_3$) (kcal.mole$^{-1}$) |
|---|---|---|---|---|
| H | | 12.4 $\pm$ 0.02 | (7.5) | 94 $\pm$ 3 |
| Me | | 12.8 $\pm$ 0.1 | $-$ 4 $\pm$ 4 | 86 $\pm$ 4 |
| Et | 10.18 $\pm$ 0.05 | 12.8 $\pm$ 0.2 | $-15 \pm 4$ | 89 $\pm$ 2 |
| Pr$^i$ | 9.85 $\pm$ 0.1 | 13.1 $\pm$ 0.2 | $-14 \pm 4$ | 79 $\pm$ 3 |
| Bu$^t$ | 9.5 $\pm$ 0.2 | 13.7 $\pm$ 0.2 | $-11 \pm 4$ | 65 $\pm$ 3 |
| SiH$_3$[b] | | 11.85 $\pm$ 0.05 | | 81.3 $\pm$ 4 |

*B   Derivatives RSiCl₃*

| R | I.P. (eV) | A.P. $H_3Si^+$ (eV) | $\Delta H°_f RSiH_3$ [a] (kcal.mole)$^{-1}$ | $D(R–SiH_3)$ (kcal.mole$^{-1}$) |
|---|---|---|---|---|
| Cl | | $12.48 \pm 0.02$ | $(-145.7)$ | $106 \pm 4$ |
| H | | $11.91 \pm 0.03$ | $-110 \pm 5$ | $93 \pm 4$ |
| Me | $11.36 \pm 0.03$ | $11.90 \pm 0.08$ | $-129 \pm 5$ | $93 \pm 4$ |
| Et | $10.74 \pm 0.04$ | $12.10 \pm 0.03$ | $-140 \pm 5$ | $95 \pm 4$ |
| Pr$^i$ | $10.28 \pm 0.1$ | $13.1 \pm 0.2$ | $-133 \pm 5$ | $80 \pm 4$ |
| Bu$^t$ | | $13.0 \pm 0.1$ | $-140 \pm 5$ | $76 \pm 4$ |
| SiCl₃ [b] | | $11.55 \pm 0.1$ | | $\leqslant 85 \pm 6$ |

[a] Heat of formation data in brackets are taken from previously published thermochemical work.
[b] Reference 323.

## 8.3 Other compounds studied

Steele et al.[79,323] studied a series of alkylsilanes (RSiH₃) and trichlorosilanes (RSiCl₃) (R = Me, Et, Pr$^i$, Bu$^t$) and SiH₄, HSiCl₃, SiCl₄, Si₂H₆, and Si₂Cl₆. The appearance potentials of the $H_3Si^+$ and $Cl_3Si^+$ ions were measured (Table 7.33) and heats of formation and bond dissociation energies were calculated.

REFERENCES

1 R. C. Weast (Ed.), *Handbook of Chemistry and Physics*, The Chemical Rubber Publishing Co., Cleveland, Ohio, U.S.A., 49th edn., 1968–69.
2 A. O. Nier, R. W. Thompson and B. F. Murphy, *Phys. Rev.*, 60 (1941) 112.
3 C. B. Collins, J. R. Freeman and J. T. Wilson, *Phys. Rev.*, 82 (1951) 966.
4 V. H. Dibeler and F. L. Mohler, *J. Res. Natl. Bur. Stand.*, 47 (1951) 337.
5 R. M. Farquar, G. H. Palmer and K. L. Aitken, *Nature*, 172 (1953) 860.
6 J. R. White and A. E. Cameron, *Phys. Rev.*, 74 (1948) 991.
7 E. Welin and G. Blomqvist, *Geol. Foren. Stolkholm Forh.*, 88 (1966) 3; *Chem. Abstr.*, 65 (1966) 8599e.
8 M. Shima, *J. Geophys. Res.*, 68 (1963) 4289.
9 R. E. Honig, *J. Chem. Phys.*, 22 (1954) 1610.
10 R. E. Honig, *J. Chem. Phys.*, 21 (1953) 573.
11 E. S. Greiner, *J. Metals*, 4 (1952) A1044.
12 A. W. Searcy, *J. Am. Chem. Soc.*, 74 (1952) 4789.
13 O. Ruff and M. Konshak, *Z. Electrochem.*, 34 (1926) 68.
14 E. Baur and R. Brunner, *Helv. Chim. Acta*, 17 (1934) 958.
15 J. Drowart, G. de Maria, A. J. H. Boerboom and M. G. Inghram, *J. Chem. Phys.*, 30 (1959) 308.
16 G. Verhaegen, F. E. Stafford and J. Drowart, *J. Chem. Phys.*, 40 (1964) 1622.

17  A. Kant and B. H. Strauss, *J. Chem. Phys.*, 45 (1966) 822.
18  G. Herzberg, *Molecular Spectra and Molecular Structure*, Vol. III, Van Norstrand, Princeton, N.J., 1965, p. 498.
19  K. Pitzer and E. Clementi, *J. Am. Chem. Soc.*, 81 (1959) 4477.
20  W. A. Chupka and M. G. Inghram, *J. Phys. Chem.*, 59 (1955) 100.
21  J. Drowart, G. de Maria and M. G. Inghram, *J. Chem. Phys.*, 29 (1958) 1015.
22  J. Drowart and P. Goldfinger, *Quart. Rev.*, (1966) 545.
23  R. F. Porter, W. A. Chupka and M. G. Inghram, *J. Chem. Phys.*, 23 (1955) 216.
24  D. L. Hildenbrand and E. Murad, *J. Chem. Phys.*, 51 (1969) 807.
25  J. Drowart, F. Degrève, G. Verhaegen and R. Colin, *Trans. Faraday Soc.*, 61 (1965) 1072.
26  R. Colin, J. Drowart and G. Verhaegen, *Trans. Faraday Soc.*, 61 (1965) 1364.
27  J. Drowart, R. Colin and G. Exsteen, *Trans. Faraday Soc.*, 61 (1965) 1376.
28  P. Coppens, S. Smoes and J. Drowart, *Trans. Faraday Soc.*, 63 (1967) 2140.
29  J. Drowart, *Bull. Soc. Chim. Belges*, 73 (1964) 451.
30  S. G. Karbanov, V. I. Belusov, V. P. Zlomenov and A. V. Novoziolova, *Vestn. Mosk. Univ.*, *Ser. II*, 5 (1968) 93.
31  R. Colin and J. Drowart, *J. Chem. Phys.*, 37 (1962) 1120.
32  R. Colin and J. Drowart, *Trans. Faraday Soc.*, 60 (1964) 673.
33  R. F. Porter, *J. Chem. Phys.*, 34 (1961) 583.
34  G. Exsteen, J. Drowart, A. Vander Auwera-Mahieu and R. Callaerts, *J. Phys. Chem.*, 71 (1967) 4130.
35  R. Colin and J. Drowart, *J. Phys. Chem.*, 68 (1964) 428.
36  V. S. Ban and B. E. Knox, *Intern. J. Mass Spectrom. Ion Phys.*, 3 (1969) 131.
37  V. V. Sokolov, V. I. Belousov, B. V. Shol'ts and L. N. Sidovov, *Russ. J. Phys. Chem.*, 40 (1966) 885.
38  A. P. Luybinov, I. I. Bespal'tseva and Ya. G. Politanskii, *Russ. J. Phys. Chem.*, 42 (1968) 1458.
39  J. Drowart and R. Colin, *Techn. Note No. 14*, A.F. 61 (052)-225, 15 July, 1963.
40  K. F. Zmbov and J. L. Margrave, *J. Am. Chem. Soc.*, 89 (1967) 2492.
41  D. W. Muenow and J. L. Margrave, *J. Phys. Chem.*, 74 (1970) 2577.
42  L. Rovner, A. Drowart and J. Drowart, *Trans. Faraday Soc.*, 63 (1967) 2906.
43  Yu. P. Sitonite, F. S. Zimyin and I. V. Krglova, *Russ. J. Phys. Chem.*, 44 (1970) 1023.
44  L. T. Zhuravlev, A. V. Kiselev and V. P. Naidinia, *Russ. J. Phys. Chem.*, 42 (1968) 1200.
45  G. T. Shechkov, V. A. Kaplii, Yu. A. Kakharov and E. N. Svobodin, *Russ. J. Phys. Chem.*, 44 (1970) 296.
46  T. C. Ehlert and J. L. Margrave, *J. Chem. Phys.*, 41 (1964) 1066.
47  J. D. McDonald, C. H. Williams, J. C. Thompson and J. L. Margrave, *Advan. Chem. Ser.*, 72 (1968) 261.
48  A. S. Kana'an and J. L. Margrave, *Inorg. Chem.*, 3 (1964) 1037.
49  K. F. Zmbov, J. W. Hastie, R. Hauze and J. L. Margrave, *Inorg. Chem.*, 7 (1968) 608.
50  T. V. Charlu, G. A. P. Adams and J. L. Margrave, unpublished results given as ref. a, Table 6 in ref. 51.
51  O. M. Uy, D. W. Muenow and J. L. Margrave, *Trans. Faraday Soc.*, 65 (1969) 1296.
52  (a) F. Jona, R. F. Lever and H. R. Wendt, *J. Electrochem. Soc.*, 111 (1964) 413; D. F.

Evans and R. E. Richards, *J. Chem. Soc.*, (1952) 1292; (b) R. C. Feber, *USAEC Rept. LA 3164*, 1965.

53 K. Zmbov, J. W. Hastie and J. L. Margrave, *Trans. Faraday Soc.*, 64 (1968) 861.

54 A. S. Buchanan, D. J. Knowles and D. L. Swingler, *J. Phys. Chem.*, 73 (1969) 4394.

55 J. W. Hastie, H. Bloom and J. D. Morrison, *J. Chem. Phys.*, 47 (1967) 1580.

56 H. Bloom and J. W. Hastie, *J. Phys. Chem.*, 71 (1967) 2360.

57 A. Kaldor and G. A. Somorjai, *J. Phys. Chem.*, 70 (1967) 3538.

58 G. A. Fisk, B. H. Mahan and E. K. Parks, *J. Chem. Phys.*, 46 (1967) 2649.

59 H. Bloom and J. W. Hastie, *Australian J. Chem.*, 21 (1968) 583.

60 K. F. Zmbov, J. W. Hastie and J. L. Margrave, *J. Inorg. Nucl. Chem.*, 30 (1968) 729.

61 H. Bloom and J. W. Hastie, *Australian J. Chem.*, 19 (1966) 1003.

62 H. Neuert and H. Clasen, *Z. Naturforsch.*, 7a (1954) 410.

63 F. E. Saalfeld and H. J. Svec, *J. Inorg. Nucl. Chem.*, 8 (1961) 98.

64 F. E. Saalfeld and H. J. Svec, *Inorg. Chem.*, 2 (1963) 46.

65 D. C. Frost, University of British Columbia, Vancouver, B.C., quoted in ref. 64.

66 P. Potzinger and F. W. Lampe, *J. Phys. Chem.*, 73 (1969) 3912.

67 M. N. de Mévergnies, *Ann. Soc. Sci. Bruxelles, Ser. I.*, 64 (1950) 188.

68 G. P. Van der Kelen and D. F. van de Vondel, *Bull. Soc. Chim. Belges*, 69 (1960) 504.

69 N. V. Larin and I. L. Agafonov, *Tr. Khim. i Khim. Technol.*, 1 (1967) 90.

70 F. Glockling and J. R. C. Light, *J. Chem. Soc. A*, (1968) 717.

71 N. V. Lavin, I. L. Agafonov and S. M. Vlsov, *Russ. J. Inorg. Chem.*, 13 (1968) 1.

72 S. R. Gunn and L. G. Green, *J. Phys. Chem.*, 65 (1961) 779.

73 B. P. Pullen, T. A. Carbon, W. E. Moddeman, G. K. Schweitzer, W. E. Bull and F. A. Grimm, *J. Chem. Phys.*, 53 (1970) 768.

74 M. A. Ring, M. J. Puentes and H. E. O'Neal, *J. Am. Chem. Soc.*, 92 (1970) 4845.

75 F. E. Saalfeld and H. J. Svec, *Inorg. Chem.*, 2 (1963) 50.

76 F. E. Saalfeld and H. J. Svec, *J. Phys. Chem.*, 70 (1966) 1753.

77 J. D. Puperzin and K. F. Zmbov, *Bull. Inst. Nucl. Sci. Boris Kidrich (Belgrade)*, 8 (1959) 89.

78 S. R. Gunn and J. H. Kindsvater, *J. Phys. Chem.*, 70 (1966) 1750.

79 W. C. Steele, L. D. Nichols and F. G. A. Stone, *J. Am. Chem. Soc.*, 84 (1962) 4441.

80 F. E. Saalfeld and H. J. Svec, *Inorg. Chem.*, 3 (1964) 1442.

81 F. E. Saalfeld and M. V. McDowell, *Inorg. Chem.*, 6 (1967) 96.

82 M. A. Ring, G. D. Beverly, F. H. Koester and R. P. Hollandsworth, *Inorg. Chem.*, 8 (1969) 2033.

83 P. Estacio, M. D. Sefvik, E. K. Chan and M. A. Ring, *Inorg. Chem.*, 9 (1970) 1068.

84 P. P. Gasper, C. A. Levy and G. M. Adair, *Inorg. Chem.*, 9 (1970) 1272.

85 K. M. Mackay, S. T. Horsfield and S. R. Stobart, *J. Chem. Soc. A*, (1969) 2937.

86 P. Royen and C. Rocktaschel, *Z. Anorg. Allgem. Chem.*, 346 (1966) 290.

87 B. J. Aylett and M. J. Hakim, *J. Chem. Soc. A*, (1969) 639.

88 B. J. Aylett and J. Emsley, *J. Chem. Soc. A*, (1967) 652.

89 B. J. Aylett and M. J. Hakim, *J. Chem. Soc. A*, (1969) 800.

90 A. D. Norman and D. C. Wingeleth, *Inorg. Chem.*, 9 (1970) 98.

91 S. Craddock, E. A. V. Ebsworth,,G. Davidson and L. A. Woodward, *J. Chem. Soc. A*, (1967) 1229.

92 C. Glidewell and D. W. H. Rankin, *J. Chem. Soc. A*, (1969) 753.

93 C. H. Van Dyke, G. A. Gibbon and J. T. Wang, *Inorg. Chem.*, 6 (1967) 1989.
94 J. T. Wang and C. H. Van Dyke, *Inorg. Chem.*, 7 (1968) 1319.
95 S. Craddock and E. A. V. Ebsworth, *J. Chem. Soc. A*, (1968) 1423.
96 G. G. Hess and F. W. Lampe, *J. Chem. Phys*, 44 (1966) 2257.
97 D. P. Beggs and F. W. Lampe, *J. Chem. Phys.*, 49 (1968) 4230.
98 D. P. Beggs and F. W. Lampe, *J. Phys. Chem.*, 73 (1969) 4194.
99 D. P. Beggs and F. W. Lampe, *J. Phys. Chem.*, 73 (1969) 3307.
100 D. P. Beggs and F. W. Lampe, *J. Phys. Chem.*, 73 (1969) 3315.
101 J. F. Schmidt, *Ph.D. Thesis*, Pennsylvania State University; *Dissertation Abstr.*, 30B (1970) 3105-B.
102 E. R. Morris and J. C. J. Thynne, *J. Phys. Chem.*, 73 (1969) 3294.
103 H. Niki and G. J. Mains, *J. Phys. Chem.*, 68 (1964) 304.
104 Y. Rousseau and G. J. Mains, *J. Phys. Chem.*, 70 (1966) 3158.
105 K. M. Mackay and R. Watt, *J. Organometal. Chem.*, 14 (1968) 123.
106 J. Meinrenken, *Z. Physik. Chem. (Frankfurt)*, 63 (1969) 205.
107 N. V. Larin, G. G. Devyatykh and I. L. Agafonov, *J. Anal. Chem.*, USSR 22 (1967) 245.
108 I. L. Agafonov, N. V. Larin and V. K. Afonskii, *Tr. Khim. I. Khim. Technol.*, 1 (1966) 112.
109 H. J. Svec and G. R. Sparrow, *J. Chem. Soc. A*, (1970) 1163.
110 G. G. Devyatykh, V. G. Rachkov and I. L. Agafonov, *Russ. J. Inorg. Chem.*, 13 (1968) 1497.
111 P. L. Timms, R. A. Kent, T. C. Ehlert and J. L. Margrave, *J. Am. Chem. Soc.*, 87 (1965) 2824.
112 V. H. Dibeler and F. L. Mohler, *J. Res. Natl. Bur. Std.*, 40 (1948) 24.
113 R. E. Rummel, J. R. Sites and R. Baldock *ORNL Rpt. 1406*, Oak Ridge National Laboratory, Oak Ridge, Tenn., 1952.
114 N. N. Sokolov, *Zh. Obshch. Khim.*, 25 (1955) 675.
115 R. H. Vought, *Phys. Rev.*, 71 (1947) 93.
116 I. L. Agafonov, G. G. Devyatykh and N. V. Larin, *Russ. J. Inorg. Chem.*, 8 (1963) 810.
117 O. Osberghaus, *Z. Physik*, 128 (1950) 366.
118 N. V. Larin, I. L. Agafonov and G. G. Devyatykh, *Russ. J. Inorg. Chem.*, 12 (1967) 1194.
119 J. L. Margrave, K. G. Sharpe and P. W. Wilson, *J. Inorg. Nucl. Chem.*, 32 (1970) 1813.
120 M. F. Lappert, M. R. Litzow and T. R. Spalding, unpublished results, 1969.
121 J. T. Herron and V. H. Dibeler, *J. Chem. Phys.*, 33 (1960) 1595.
122 W. E. Bull, B. P. Pullen, F. A. Grimm, W. E. Moddeman, G. K. Schweitzer and T. A. Carlson, *Inorg. Chem.*, 9 (1970) 2474.
123 K. A. G. Macneil and J. C. J. Thynne, *Intern. J. Mass Spectrom. Ion Phys.*, 3 (1970) 455.
124 J. C. J. Thynne and K. A. G. Macneil, *Inorg. Chem.*, 9 (1970) 1946.
125 S. Cradock, P. W. Harland and J. C. J. Thynne, *Inorg. Nucl. Chem. Letters*, 6 (1970) 425.
126 J. L. Margrave, D. L. Williams and P. W. Wilson, *Inorg. Nucl. Chem. Letters*, 7 (1971) 103.
127 R. G. Arshakuni, *Geokimiya*, 1 (1967) 121.

128 R. G. Arshakuni, *J. Anal. Chem. USSR*, 23 (1968) 251.
129 G. G. Devyatykh, V. G. Rachov and I. L. Agafonov, *J. Anal. Chem. USSR*, 240 (1969) 320.
130 D. Sloan and P. L. Timms, *Inorg. Chem.*, 7 (1968) 2157.
131 J. L. Margrave, K. G. Sharpe and P. W. Wilson, *J. Am. Chem. Soc.*, 92 (1970) 1530.
132 P. L. Timms, *Inorg. Chem.*, 7 (1968) 387.
133 J. G. Aston, R. M. Kennedy and G. H. Messerly, *J. Am. Chem. Soc.*, 63 (1941) 2343.
134 V. H. Dibeler, *J. Res. Natl. Bur. Std.*, 49 (1952) 235.
135 J. J. de Ridder and J. Dijkstra, *Rec. Trav. Chim.*, 86 (1967) 737.
136 G. P. Van der Kelen, O. Volders, H. van Onckelen and Z. Eecklaut, *Z. Anorg. Allgem. Chem.*, 338 (1965) 106.
137 J. L. Occolowitz, *Tetrahedron Letters*, 43 (1966) 5291.
138 V. E. Heldt, K. Hoppner and K. H. Krebs, *Z. Anorg. Allgem. Chem.*, 347 (1966) 95.
139 M. F. Lappert, J. B. Pedley, J. Simpson and T. R. Spalding, *J. Organometal. Chem.*, 29 (1971) 195.
140 G. Distefano, *Inorg. Chem.*, 9 (1970) 1919.
141 J. Tamas and K. Ujszaszy, *Acta Chim. Acad. Sci. Hung.*, 56 (1968) 125.
142 R. G. Kostyanovsky, *Tetrahedron Letters*, 22 (1968) 2721.
143 R. B. Kostyanovskii, *Bull. Acad. Sci. USSR*, 12 (1967) 2659.
144 N. Ya Chernyak, R. A. Khmel'nitskii, E. V. D'yakova and V. M. Volovin, *J. Gen. Chem. USSR*, 36 (1966) 93.
145 D. B. Chambers, F. Glockling, J. R. C. Light and M. Weston, *Chem. Commun.*, (1966) 281.
146 M. Gielen and G. Mayence, *J. Organometal. Chem.*, 12 (1968) 363.
147 D. B. Chambers, F. Glockling and M. Weston, *J. Chem. Soc. A*, (1967) 1759.
148 M. R. Ghate and K. N. Bhide, *Indian J. Chem.*, 2 (1964) 243.
149 F. W. McLafferty (Ed.), *Mass Spectrometry of Organic Ions*, Academic Press, New York, 1963.
150 K. C. Williams, *J. Organometal. Chem.*, 19 (1969) 210.
151 B. C. Pant and R. E. Sacher, *Inorg. Nucl. Chem. Letters*, 5 (1969) 549.
152 J. J. de Ridder and G. Dijkstra, *Rec. Trav. Chim.*, 86 (1967) 1325.
153 A. Carrick and F. Glockling, *J. Chem. Soc. A*, (1966), 623.
154 S. Boué, M. Gielen and J. Nasielski, *Bull. Soc. Chim. Belges*, 77 (1968) 43.
155 I. L. Agafonov, B. I. Faerman, Yu. N. Tsinovoi, N. V. Larin and V. A. Umutin, *Izvt. Akad. Nauk SSSR, Ser. Khim.*, 6 (1968) 1289.
156 J. Tamas, K. Ujszazy, T. Szekely and G. Bujtas, *Acta Chim. Acad. Sci. Hung.*, 62 (1969) 335.
157 J. Tamas, K. Ujszazy, T. Szekely and G. Bujtas, *Magy. Khem. Folyoirat*, 75 (1969) 148
158 N. Ya Chernyak, R. A. Khmel'nitskii, T. V. D'yakova, K. S. Pushchevaya and V. M. Vdovin, *J. Gen. Chem. USSR*, 37 (1967) 867.
159 W. P. Weber, R. A. Felix and A. K. Willard, *Tetrahedron Letters*, 12 (1970) 907.
160 W. P. Weber, R. A. Felix and A. K. Willard, *J. Am. Chem. Soc.*, 91 (1969) 6544.
161 A. A. Polykova, K. I. Zimina, A. A. Petrov and R. A. Khmel'nitskii, *Dokl. Akad. Nauk. SSSR*, 134 (1960) 833.
162 A. A. Khmel'nitskii, A. A. Polyakova and A. A. Petrov, *Tr. Komis. po Analit. Khim. Akad. Nauk SSSR*, 13 (1963) 482.

163 R. A. Khmel'nitskii, A. A. Polyakova, A. A. Petrov, F. A. Medvedev and M. D. Stadnichuck, *Zh. Obshch. Khim.*, 35 (1965) 773.
164 G. Tritz and N. Gotz, *Z. Anorg. Allgem. Chem.*, 375 (1970) 171.
165 R. Alexander, C. Eaborn and T. G. Traylor, *J. Organometal. Chem.*, 21 (1969) 65.
166 P. E. Rakita and A. Davison, *Inorg. Chem.*, 8 (1969) 1164.
167 D. J. Peterson, *J. Org. Chem.*, 32 (1967) 1717.
168 J. Grobe and U. Möller, *J. Organometal. Chem.*, 17 (1969) 263.
169 L. C. Quass, R. West and G. R. Husk, *J. Organometal. Chem.*, 21 (1970) 65.
170 E. S. Alexander, R. N. Hazeldine, M. J. Newlands and A. E. Tipping, *J. Chem. Soc. A*, (1970) 2285.
171 D. Seyferth, R. Damraeuer, J. Yick-Pui Mui and T. F. Jula, *J. Am. Chem. Soc.*, 90 (1968) 2944.
172 D. Seyferth and C. J. Attridge, *J. Organometal. Chem.*, 21 (1970) 103.
173 F. Glockling, J. R. C. Light and R. G. Stafford, *J. Chem. Soc. A*, (1970) 426.
174 G. Fritz, W. Konig and H. Scheer, *Z. Anorg. Allgem. Chem.*, 377 (1970) 241.
175 A. M. Duffield, C. Djerassi, P. Mazerolles, J. Dubac and G. Manuel, *J. Organometal. Chem.*, 12 (1968) 123.
176 E. Heldt, K. Hoppner and K. H. Krebs, *Z. Anorg. Allgem. Chem.*, 348 (1966) 113.
177 H. G. Kuivila, K.-H. Tsai and D. G. I. Kington, *J. Organometal. Chem.*, 23 (1970) 129.
178 V. V. Khrapov, V. I. Goldanskii, A. K. Prokof'ev, V. Ya. Rochev and R. G. Kostyanovskii, *Bull. Acad. Sci. USSR*, 6 (1968) 1192.
179 S. Boué, M. Gielen, J. Nasielski, J.-P. Lieutenant and R. Spielman, *Bull. Soc. Chim. Belges*, 78 (1969) 135.
180 C. W. Fong and W. Kitching, *J. Organometal. Chem.*, 22 (1970) 107.
181 P. N. Preston and N. A. Weir, *J. Inorg. Nucl. Chem. Letters*, 4 (1968) 279.
182 J. H. Bowie and B. Nussey, *Org. Mass Spectrom.*, 3 (1970) 933.
183 J. M. Miller, *Can. J. Chem.*, 47 (1969) 1613.
184 J. M. Miller, *J. Chem. Soc. A*, (1967) 828.
185 D. Seyferth, T. F. Jula, D. C. Mueller, P. Mazerolles, G. Manuel and F. Thoumas, *J. Am. Chem. Soc.*, 92 (1970) 657.
186 M. Gielen and J. Nasielski, *Bull. Soc. Chim. Belges*, 77 (1968) 5.
187 T. H. Kinstle, P. J. Ihrig and E. J. Goettert, *J. Am. Chem. Soc.*, 92 (1970) 1780.
188 P. N. Preston, P. J. Rice and N. A. Weir, *Intern. J. Mass Spectrom. Ion Phys.*, 1 (1968) 303.
189 I. Lengyel and M. J. Aaronson, *Angew. Chem. Intern. Ed. Engl.*, 9 (1970) 161.
190 I. Lengyel and M. J. Aaronson, *J. Chem. Soc. D*, (1970) 129.
191 Ch. Elschenbroich, *J. Organometal. Chem.*, 22 (1970) 677.
192 G. J. D. Peddle, *J. Organometal. Chem.*, 14 (1968) 139.
193 R. R. Schrieke and B. O. West, *Australian J. Chem.*, 22 (1969) 49.
194 J. Laane, *J. Am. Chem. Soc.*, 89 (1967) 1144.
195 A. M. Duffield, H. Budzikiewicz and C. Djerassi, *J. Am. Chem. Soc.*, 87 (1965) 2920.
196 P. Potzinger and F. W. Lampe, *J. Phys. Chem.*, 74 (1970) 587.
197 E. Kamarato and F. W. Lampe, *J. Phys. Chem.*, 74 (1970) 2267.
198 R. Varma, *Inorg. Nucl. Chem. Letters*, 6 (1970) 9.
199 J. M. Bellama and A. G. MacDairmid, *J. Organometal. Chem.*, 18 (1969) 275.

200 G. A. Gibbon, E. W. Kifer and C. H. Van Dyke, *Inorg. Nucl. Chem. Letters*, 6 (1970) 617.
201 K. M. Mackay, R. D. George, P. Robinson and R. Watt, *J. Chem. Soc. A*, (1968) 1920.
202 R. D. George and K. M. Mackay, *J. Chem. Soc. A*, (1969) 2122.
203 F. Feher, P. Plinchta and R. Guillery, *Tetrahedron Letters*, 33 (1970) 2889.
204 F. Feher, P. Plichta and R. Guillery, *Tetrahedron Letters*, 51 (1970) 4443.
205 P. S. Skell and P. W. Owen, *J. Am. Chem. Soc.*, 89 (1967) 3933.
206 R. D. George, K. M. Mackay and S. R. Stobart, *J. Chem. Soc. A*, (1970) 3250.
207 R. S. Gohlke and P. J. Robinson, *Org. Mass Spectrom.*, 3 (1970) 967.
208 R. L. Foltz, M. B. Neher and E. R. Hinnenkamp, *Anal. Chem.*, 39 (1967) 1338.
209 P. Jutzi, *J. Organometal. Chem.*, 16 (1969) p. 71.
210 G. L. Van Mourik and F. Bickelhaupt, *Rec. Trav. Chim.*, 88 (1969) 868.
211 J. L. Margrave, D. L. Williams and P. W. Wilson, *Inorg. Nucl. Chem. Letters*, 7 (1971) 103.
212 H. Bock and H.-J. Semmler, *Z. Anal. Chem.*, 230 (1967) 161.
213 C. R. Bettler, J. C. Sendra and G. Urry, *Inorg. Chem.*, 9 (1970) 1060.
214 J. Binenboym and R. Schaeffer, *Inorg. Chem.*, 9 (1970) 1578.
215 J. E. Dale, R. T. Hemmings and C. Riddle, *J. Chem. Soc. A*, (1970) 3359.
216 M. J. McGlinchey, J. D. Odom, T. Reynoldson and F. G. A. Stone, *J. Chem. Soc. A*, (1970) 31.
217 A. Carrick and F. Glockling, *J. Chem. Soc. A*, (1967) 40.
218 A. G. Starkey, R. A. Friedel and S. H. Langer, *Anal. Chem.*, 29 (1957) 770.
219 G. H. Draffen, R. N. Stillwell and J. A. McCloskey, *Org. Mass Spectrom.*, 1 (1968) 669.
220 J. Diekman, J. B. Thomson and C. Djerassi, *J. Org. Chem.*, 32 (1967) 3904.
221 J. Diekman, J. B. Thomson and C. Djerassi, *J. Org. Chem.*, 33 (1968) 2271.
222 K. G. Das, P. S. Kulkami, V. Kalyanaraman and M. V. George, *J. Org. Chem.*, 35 (1970) 2140.
223 J. Silbiger, C. Lifshitz, J. Fuchs and A. Mandelbaum, *J. Am. Chem. Soc.*, 89 (1967) 4306.
224 H. G. Ang, *J. Chem. Soc. A*, (1970) 2734.
225 N. Wiberg, W-Ch. Joo and W. Uhlenbrock, *Angew. Chem. Intern. Ed. Engl.*, 7 (1968) 640.
226 K. Seppelt and W. Sandermeyer, *Chem. Ber.*, 102 (1969) 1247.
227 R. E. Highersmith and H. H. Sisler, *Inorg. Chem.*, 8 (1969) 996.
228 G. Kannengiesser, F. Damm, A. Deluzarche and A. Maillard, *Bull. Soc. Chim. France*, 3 (1969) 894.
229 G. Kannengiesser and F. Damm, *Bull. Soc. Chim. France*, 3 (1969) 891.
230 O. Glemser and J. Wegener, *Angew. Chem. Intern. Ed. Engl.*, 9 (1970) 309.
231 O. Glemser, U. Biermann and S. P. v. Halasz, *Inorg. Nucl. Chem. Letters*, 5 (1969) 643.
232 A. Shiotani and H. Schmidbaur, *J. Am. Chem. Soc.*, 92 (1970) 7003.
233 A. R. Dahl and A. D. Norman, *J. Am. Chem. Soc.*, 92 (1970) 5525.
234 A. D. Norman, *Inorg. Chem.*, 9 (1970) 870.
235 D. C. de Jongh, T. Radford, J. D. Ilabir, S. Hanessian, M. Bieber, G. Dawson and C. C. Sweeley, *J. Am. Chem. Soc.*, 91 (1969) 1728.

236 B. Arreguin and J. Taboada, *J. Chromatog. Sci.*, 8 (1970) 187.
237 G. Peterson, *Tetrahedron*, 25 (1969) 4437.
238 G. Peterson. O. Samuelson, K. Anjou and E. v. Sydow, *Acta Chem. Scand.*, 23 (1969) 3597.
239 N. N. Kochetkov, O. S. Chizhov and N. V. Molodtsov, *Tetrahedron*, 24 (1968) 5587.
240 S. A. Barker, M. J. How, P. V. Peplow and P. J. Somers, *Anal. Biochem.*, 26 (1968) 219.
241 F. A. J. M. Leemans and J. A. McCloskey, *J. Am. Oil Chemists' Soc.*, 44 (1967) 11.
242 P. Capella, C. Galli and R. F. Fumagalli, *Lipids*, 3 (1968) 341.
243 C. J. Argoudelis and E. G. Perkins, *Lipids*, 3 (1968) 379.
244 C. J. Argoudelis and E. G. Perkins, *Lipids*, 4 (1969) 619.
245 G. Eglinton and D. H. Hunneman, *Phytochemistry*, 7 (1968) 313.
246 J. Esselmann and C. O. Clagett, *J. Lipid Res.*, 10 (1969) 234.
247 J. A. Gustafsson, R. Ryhage, J. Sjövall and R. M. Moriarty, *J. Am. Chem. Soc.*, 91 (1969) 1234.
248 W. Vetter, W. Walker, M. Vecchi and M. Cereghetti, *Helv. Chim. Acta*, 52 (1969) 1.
249 B. E. Gustafsson, J. A. Gustafsson and J. Sjövall, *Acta Chem. Scand.*, 20 (1966) 1827.
250 T. Laatikainen and R. Vihko, *European J. Biochem.*, 10 (1969) 165.
251 M. Vecchi, W. Vetter, W. Walther, S. F. Jermstad and G. W. Schutt, *Helv. Chim. Acta*, 50 (1967) 1243.
252 W. Vetter, M. Vecchi, H. Gutmann, R. Ruegg, W. Walther and P. Meyer, *Helv. Chim. Acta*, 50 (1967) 1866.
253 A. J. Polito, J. Naworal and C. C. Sweeley, *Biochemistry*, 8 (1969) 1811.
254 K. A. Karlsson, *Acta Chem. Scand.*, 19 (1965) 2425.
255 B. Samuelsson and K. Samuelsson, *J. Lipid Res.*, 10 (1969) 41.
256 M. Hamburg, W. G. Niehaus and B. Samuelsson, *Anal. Biochem.*, 22 (1968) 145.
257 M. Capella and C. M. Zorzut, *Anal. Chem.*, 40 (1968) 1459.
258 R. T. Gray, J. Diekmann, G. L. Larson, W. K. Musker and C. Djerassi, *Org. Mass Spectrom.*, 3 (1970) 973.
259 J. A. McClosky, A. M. Lawson, K. Tsuboyama, P. M. Kruger and R. N. Stillwell, *J. Am. Chem. Soc.*, 90 (1968) 4182.
260 K. Bergström, J. Gurtler and R. Blomstrand, *Anal. Biochem.*, 34 (1970) 74.
261 H. Budzikiewicz, C. Djerassi and D. H. Williams, *Mass Spectrometry of Organic Compounds*, Holden-Day Inc., New York, 1967.
262 V. Yu. Orlov, *J. Gen. Chem. USSR*, 37 (1967) 2188.
263 G. Dube and H. Kriegsmann, *Org. Mass Spectrom.*, 1 (1968) 891.
264 J. Tamas, K. Ujszazy and Gy. Bujtas, *Magy. Khem. Folyoirat*, 76 (1970) 131.
265 R. Muller and H. J. Frey, *Z. Anorg. Allgem. Chem.*, 368 (1969) 113.
266 H. O. House, L. J. Czuba, M. Gall and H. D. Olmstead, *J. Org. Chem.*, 34 (1969) 2324.
267 R. M. Teeter, *10th Conf. on Mass Spectrom.*, *N. Orleans, 1962*, Am. Chem. Soc., p. 51.
268 W. J. A. VandenHeuvel, J. L. Smith and J. L. Beck, *Org. Mass Spectrom.*, 4 (1970) 563.
269 J. Diekmann, J. B. Thompson and C. Djerassi, *J. Org. Chem.*, 34 (9169) 3147.
270 A. G. Brook, A. G. Harrison and P. F. Jones, *Can. J. Chem.*, 46 (1968) 2862.
271 V. Yu Orlov, *Russ. J. Phys. Chem.*, 43 (1969) 13.
272 V. Yu Orlov, N. S. Nametkin, L. E. Gusel'nikov and T. H. Isalamov, *Org. Mass Spectrom.*, 4 (1970) 195.
273 G. Dube, *Z. Chem.*, 9 (1969) 316.

274 N. Viswanathan and C. H. Van Dyke, *J. Chem. Soc. A*, (1968) 487.
275 C. S. Wang, K. E. Pullen and J. M. Shreeve, *Inorg. Chem.*, 9 (1970) 90.
276 A. F. Webb, D. S. Sethi and H. Gilman, *J. Organometal. Chem.*, 21 (1970) 61.
277 G. Stork and P. F. Hudrlik, *J, Am. Chem. Soc.*, 90 (1968) 4462.
278 H. Paulen and G. Steinert, *Chem. Ber.*, 103 (1970) 475.
279 W. E. Newton and E. G. Rochow, *Inorg. Chem.*, 9 (1970) 1076.
280 S. P. Narula, *Indian J. Chem.*, 5 (1967) 346.
281 M. Zinbo and W. R. Sherman, *Tetrahedron Letters*, 33 (1969) 2811.
282 D. B. Boylan and M. Calvin, *J. Am. Chem. Soc.*, 89 (1967) 5472.
283 D. J. Cook, N. C. Lloyd and W. J. Owen, *J. Organometal. Chem.*, 22 (1970) 55.
284 E. W. Abel, D. J. Walker and J. N. Wingfield, *J. Chem. Soc. A*, (1968) 2642.
285 J. G. Zavistoski and J. J. Zuckermann, *J. Am. Chem. Soc.*, 90 (1968) 6612.
286 K. Moedritzer, *Inorg. Chem.*, 6 (1967) 1248.
287 P. Mazerolles, J. Dubac and M. Lesbre, *J. Organometal. Chem.*, 12 (1968) 143.
288 C. W. Fong and W. Kitching, *J. Organometal. Chem.*, 21 (1970) 365.
289 C. W. Fong and W. Kitching, *J. Organometal. Chem.*, 22 (1970) 95.
290 E. Lindner, U. Kunze, G. Ritter and A. Haag, *J. Organometal. Chem.*, 24 (1970) 119.
291 E. Lindner, U. Kunze, G. Ritter, A. Haag and G. Vitzthum, *J. Organometal. Chem.*,
    24 (1970) 131.
292 P. G. Harrison and J. J. Zuckerman, *Inorg. Chem.*, 9 (1970) 175.
293 G. Vandendunghen, *Ph.D. Thesis*, Université Libre, Brussells, 1969.
294 R. S. Gohlke, *J. Am. Chem. Soc.*, 90 (1968) 2713.
295 J. D. de Ridder and G. Dijkstra, *Org. Mass Spectrom.*, 1 (1968) 647.
296 D. B. Chambers and F. Glockling, *J. Chem. Soc. A*, (1968) 735.
297 M. F. Lappert, J. Simpson and T. R. Spalding, unpublished results, 1969.
298 V. G. Kumar Das and P. R. Wells, *J. Organometal. Chem.*, 23 (1970) 143.
299 T. H. Kinstle, I. Haiduc and H. Gilman, *Inorg. Chim. Acta*, 3 (1969) 373.
300 E. Carberry and R. West, *J. Organometal. Chem.*, 6 (1966) 582.
301 K. Kuhlein and W. P. Neumann, *J. Organometal. Chem.*, 14 (1968) 317.
302 J. J. de Ridder and J. G. Noltes, *J. Organometal. Chem.*, 20 (1969) 287.
303 C. R. Bettler and G. Urry, *Inorg. Chem.*, 9 (1970) 2372.
304 M. Richter and W. P. Neumann, *J. Organometal. Chem.*, 20 (1969) 81.
305 P. G. Harrison and J. J. Zuckerman, *J. Am. Chem. Soc.*, 92 (1970) 2577.
306 J. T. Wang and C. H. Van Dyke, *Inorg. Chem.*, 6 (1967) 1741.
307 W. Airey and G. M. Sheldrick, *J. Inorg. Nucl. Chem.*, 32 (1970) 1827.
308 R. A. Braun, *Inorg. Chem.*, 5 (1966) 1831.
309 A. D. Britt and W. B. Moriz, *J. Am. Chem. Soc.*, 91 (1969) 6204.
310 J. E. Drake and C. Riddle, *J. Chem. Soc. A*, (1970) 3134.
311 J. L. Beck, W. J. A. VandenHeuvel and I. L. Smith, *Org. Mass Spectrom.*, 4 (Supple-
    mentary) (1970) 237.
312 G. G. Hess, F. W. Lampe and L. H. Sommer, *J. Am. Chem. Soc.*, 87 (1965) 5327.
313 G. G. Hess, F. W. Lampe and L. H. Sommer, *J. Am. Chem. Soc.*, 86 (1964) 3174.
314 B. G. Hobrock and R. W. Kiser, *J. Phys. Chem.*, 65 (1961) 2186.
315 S. J. Band, I. M. T. Davidson and C. A. Lambert, *J. Chem. Soc. A*, (1968) 2068.
316 S. J. Band, I. M. T. Davidson, C. A. Lambert and T. L. Stephenson, *Chem. Commun.*,
    (1967) 723.

317 T. N. Radwan, *Ph.D. Thesis*, Imperial College, 1966, London.
318 J. A. Connor, G. Finney, G. H. Leigh, R. N. Hazeldine, P. J. Robinson, R. D. Sedgwick and R. F. Simmons, *Chem. Commun.*, (1966) 178.
319 J. A. Connor, R. N. Hazeldine, G. J. Leigh and R. D. Sedgwick, *J. Chem. Soc. A*, (1968) 768.
320 B. G. Hobrock and R. W. Kiser, *J. Phys. Chem.*, 66 (1962) 155.
321 A. L. Yergey and F. W. Lampe, *J. Organometal. Chem.*, 15 (1968), 339.
322 A. L. Yergey and F. W. Lampe, *J. Am. Chem. Soc.*, 87 (1965) 4204.
323 W. C. Steele and F. G. A. Stone, *J. Am. Chem. Soc.*, 84 (1962) 3599.
324 I. M. T. Davidson and I. L. Stephenson, *J. Chem. Soc. A*, (1968) 282.
325 F. W. Lampe and A. Niehaus, *J. Chem. Phys.*, 49 (1968) 2949.
326 W. P. Neumann, *The Organic Chemistry of Tin*, Wiley, London, 1970.
327 H. Bock and H. Seidl, *J. Organometal. Chem.*, 13 (1968) 87.
328 F. P. Lossing and J. B. De Sousa, *J. Am. Chem. Soc.*, 81 (1959) 281.
329 D. P. Stevenson, *Discussions Faraday Soc.*, 10 (1951) 35.
330 R. W. Kiser, *Introduction to Mass Spectrometry and Its Applications*, Prentice-Hall, New York, 1965.
331 M. F. Lappert, J. Simpson and T. R. Spalding, *J. Organometal. Chem.*, 17 (1969) 1.
332 E. I. Quinn, V. H. Dibeler and F. L. Mohler, *J. Res. Natl. Bur. Std.*, 57 (1956) 41.

# Chapter 8

# The Main Group V Elements

T. R. SPALDING

## 1. INTRODUCTION

Most of the discussion in this chapter will concern the elements phosphorus, arsenic, antimony, and bismuth and their compounds and compounds of nitrogen with elements other than carbon. The organic compounds of nitrogen will not be discussed in any detail since there is ample literature on them[1-3].

An extremely large number of compounds has been studied. In the case of phosphorus they extend from phosphine and the phosphorus trihalides through to complicated organophosphorus pesticides. The discussion will be mainly concerned with studies which have reported spectra or fragmentation behaviour. Compounds which have been studied merely to determine the molecular weight or to ascertain if a certain group was present are generally tabulated without further discussion in the appropriate section.

Compounds are categorised according to the types of main Group V to other element bonds involved. For example, $POF_2H$ will be found after sections on Group V hydrides, Group V halides and Group V–main Group VI compounds. This categorisation is somewhat flexible in that a compound such as $F_3PNSO_2F$ will be found under compounds of Group V elements bonded to halogens and Group VI elements even though neither Group V element is bonded to *both* a halogen and a Group VI element as in for example the compound $F_3PNPF_2(S)$.

In some sections there are sufficient data on a series of compounds to furnish a complete subsection. For example, in the section on compounds with only main Group V–carbon bonds, there is a subsection on ylides and their derivatives.

*References pp. 439–447*

Table 8.1

Natural Abundances of Isotopes of Main Group V Elements[4]

| Element | Mass no. | % Abundance |
|---------|----------|-------------|
| N       | 14       | 99.63       |
|         | 15       | 0.37        |
| P       | 31       | 100         |
| As      | 75       | 100         |
| Sb      | 121      | 57.25       |
|         | 123      | 42.75       |
| Bi      | 209      | 100         |

2. THE ELEMENTS

Nitrogen and arsenic each have two naturally occurring isotopes but phosphorus, antimony, and bismuth are monoisotopic (Table 8.1 and Fig. 8.1)[4].

A number of publications concerning molecular nitrogen or ions derived therefrom has appeared. While these are not a primary concern of this book, it is still worth pointing out the types of studies which have been undertaken. Several workers have reported studies of the energetic states in nitrogen ions produced by electron-impact[5,6] or by photoionisation[7] or photoelectron techniques[8]. The energy distributions for energetic ions produced by electron impact on $N_2$ have been reported[9] and Vance[10] has studied the relative populations of $N^+$ and $N_2^{2+}$ in a $m/e = 14$ ion beam.

Reactions involving $N_2$ or $N_2^+$ ions have been extensively studied. Some

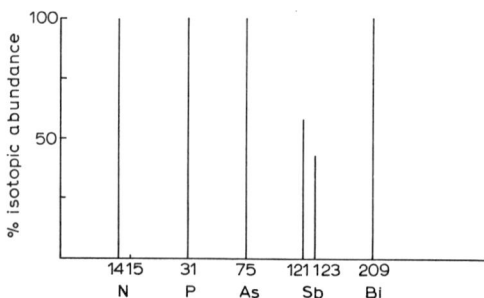

Fig. 8.1. Natural isotopic abundances of main Group V elements.

controvery arose as to whether the reaction between $N_2$ and $N_2^+$ to produce $N^+$ ions proceeded according to equation (8.1) (ref. 11) or (8.2) (ref. 12).

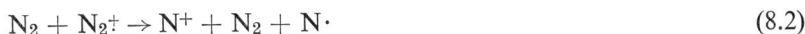

$$N_2 + N_2^+ \rightarrow N^+ + N_3 \cdot \tag{8.1}$$

$$N_2 + N_2^+ \rightarrow N^+ + N_2 + N \cdot \tag{8.2}$$

The formation of $N_3^+$ and $N_4^+$ species[13] has been shown to occur by equations (8.3) and (8.4).

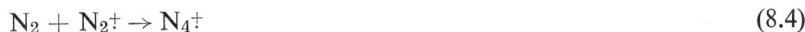

$$N_2 + N_2^+ \rightarrow N_3^+ + N \cdot \tag{8.3}$$

$$N_2 + N_2^+ \rightarrow N_4^+ \tag{8.4}$$

Although $N_3^+$ ions are formed at pressures of $10^{-4}$ torr, $N_4^+$ ions are only found at substantially higher pressures[13,14]. Other reactions involving nitrogen which have been studied mass spectrometrically include[15]

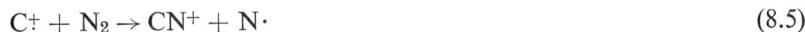

$$C^+ + N_2 \rightarrow CN^+ + N \cdot \tag{8.5}$$

The composition and reactions of nitrogen-containing species in the upper and lower atmosphere have been the subjects of several studies, for instance Johnson et al.[16] have observed some ion–molecule reactions involving $N_2^+$ and $N^+$ ions and Donahue[17] studied nitrogen-containing species and their reactions in the ionosphere.


3. HIGH-TEMPERATURE STUDIES

The Group V elements from phosphorus to bismuth vaporise to give molecules which are predominantly diatomic or tetratomic[18]. Higher polyatomic species have been reported in some studies, for instance ions derived from $M_8$ molecules have been observed for phosphorus[19,20] and arsenic[21,22] and $Sb_5^+$ ions were observed in laser-produced antimony vapour[22]. Usually, all $M_n^+$ ions ($n = 1–4$ or 8) are observed from $M_n$ molecules. With conventional techniques and temperatures in the region of $1000°$ C, the most abundant ionic species were $M_4^+$ ions in all cases except bismuth which had $Bi^+$ in highest abundance[21,23,24]. The most abundant ions in the laser-produced vapour of arsenic, antimony, and bismuth were[22] the monometallic ions $As^+$, $Sb^+$, and $Bi^+$.

From mass spectrometric studies using Knudsen cell techniques, the following reactions have been postulated for phosphorus[19,21,24,25], arsenic[21,24], antimony[21,26,27], and bismuth[28,29].

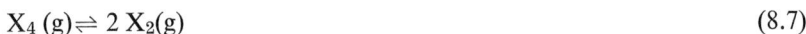

$$4 X \cdot (s) \rightleftharpoons X_4(g) \tag{8.6}$$

$$X_4 (g) \rightleftharpoons 2 X_2(g) \tag{8.7}$$

Workers who have studied these systems have sometimes reported relevant thermodynamic data such as heats of atomisation, heats of sublimation or bond dissociation energies. Kane and Reynolds[21] report the heat of sublimation by reaction (8.6) to be 56 kcal.mole$^{-1}$ for phosphorus and 41 kcal.-mole$^{-1}$ for arsenic. Another value for arsenic of $\Delta H^\circ$ (sub) $= 43.0 \pm 0.4$ kcal.mole$^{-1}$ was found by Westmore et al.[36,37] in good agreement with Kane and Reynolds, but these values disagree with that of 34.5 kcal.mole$^{-1}$ obtained by using equilibrium methods[35]. The heat of evaporation of $Sb_4$ was found to be $\Delta H^\circ$(vap) $= 49.78 \pm 0.3$ kcal.mole$^{-1}$ by Boerboom et al.[26]. For bismuth, the following values of heats of sublimation have been calculated[32] $\Delta H^\circ$(sub)(Bi) $= 50.0 \pm 1.5$, $(Bi_2) = 53.0 \pm 1.8$, $(Bi_3) = 64.1 \pm 4.3$, and $(Bi_4)$ 59.0 $\pm$ 3.4 kcal.mole$^{-1}$. Kohl et al.[28] reported the heat of atomisation of $Bi_4$ $\Delta H^\circ$(atom) $= 139.7 \pm 1.9$ kcal.mole$^{-1}$.

Some dissociation energies of diatomic species have been calculated. Gingerich[30] obtained a value of $D$(P-P) of $117.0 \pm 2.5$ kcal.mole$^{-1}$ and values of $D$(Sb-Sb) $= 70.6 \pm 1.5$ (ref. 27) and $D$(Bi-Bi) $= 47.0 \pm 1.8$ kcal.mole$^{-1}$ (ref. 32) have been reported. Dissociation energies corresponding to equation (8.7) have been obtained for $D(P_2-P_2) = 55.7 \pm 0.5$, $D(As_2-As_2) = 71.8 \pm 0.5$, and $D(Sb_2-Sb_2) = 60.7 \pm 0.5$ by Drowart and Goldfinger[24]. These can be compared with the corresponding values of 54.43, $>61.5$, and 63.8 reported by Stull and Sinke[35]. Other values which have been published are $D(As_2-As_2) = 69.6 \pm 2.5$, $D(Sb_2-Sb_2) = 63.4 \pm 2.5$ from the work of Goldfinger and Jeunehomme[40], $D(Sb_2-Sb_2) = 67.5 \pm 2$ found by Boerboom et al.[26], $D(Bi_2-Bi_2) = 47.0$ (ref. 32), and $D(Bi_2-Bi_2) = 68$ kcal.mole$^{-1}$ (ref. 29). Overall, the agreement in these results is reasonable except for that between the last two mentioned.

Several systems of two main Group V elements have been studied[18,28,30]. One such[28] was the evaporation of $P_3N_5$ at 850° C. Careful examination of the appearance potentials of ions showed that the neutral molecules in equilibrium with $P_3N_5$ were $P_4$, $P_2$, PN, and $N_2$. The reactions were formulated as

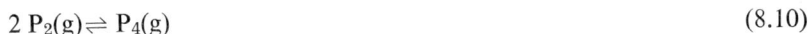

$$P_3N_5(s) \rightleftharpoons 3\ PN(g) + N_2(g) \tag{8.8}$$

$$2\ PN(g) \rightleftharpoons P_2(g) + N_2(g) \tag{8.9}$$

$$2\ P_2(g) \rightleftharpoons P_4(g) \tag{8.10}$$

Reaction (8.9) was postulated as the most important between nitrogen and phosphorus vapour to produce PN, but another reaction was also thought to be involved[30], *viz.*

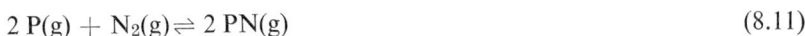

$$2\ P(g) + N_2(g) \rightleftharpoons 2\ PN(g) \tag{8.11}$$

The heat of formation of PN was found[30] to be $\Delta H^\circ_f(PN) = 45.3 \pm 4$ kcal.mole$^{-1}$ and $D(P-N) = 147.5 \pm 5$ kcal.mole$^{-1}$ according to equation (8.11).

A study of phosphorus–arsenic-containing species produced from either a mixture of the elements or from the alloy $In(P_{0.56}As_{0.44})$ has been reported[28]. The same ions were observed in each system. All the tetratomic ions $P_4^+$, $P_3As^+$, $P_2As_2^+$, $PAs_3^+$, and $As_4^+$ were found together with the expected triatomic, diatomic, and monatomic species. The following reactions were suggested as the most important in producing the mixed species.

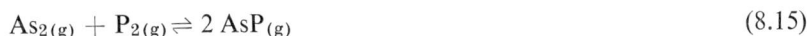

$$As_{2(g)} + P_{2(g)} \rightleftharpoons As_2P_{2(g)} \tag{8.12}$$

$$3\ As_{2(g)} + P_{2(g)} \rightleftharpoons 2\ As_3P_{(g)} \tag{8.13}$$

$$As_{2(g)} + 3\ P_{2(g)} \rightleftharpoons 2\ AsP_{3(g)} \tag{8.14}$$

$$As_{2(g)} + P_{2(g)} \rightleftharpoons 2\ AsP_{(g)} \tag{8.15}$$

A similar study[31] of the antimony–bismuth system and the arsenic–bismuth system showed the presence of the expected fourteen ionic species. The monoatomic, diatomic, and tetratomic ions had ionisation potentials which were the same as those of the corresponding element of lowest ionisation potential (*i.e.* Sb in $Sb_4^+$, but Bi in $BiSb_3^+$). In contrast, the triatomic species always had appearance potentials several volts higher and were therefore considered to be fragment ions from tetratomic species. There has been a brief report[31] of the study of a system of three Group V elements, arsenic-antimony–bismuth, from which thirty four ionic species may be observed.

A large number of systems containing main Group V–Group III or main Group V–Group VI elements have been studied; they are listed in Table 8.2.

Table 8.2

Systems of Group V–Group III and Group V–Group VI Elements
References are given in brackets.

*Group V–Group III*

| P–Al | (33) | As–Al | (34) | | |
|------|------|-------|------|-------|------|
| P–Ga | (18) | As–Ga | (28) | Sb–Ga | (36,22 ‡) |
| P–In | (28) | As–In | (22 ‡) | Sb–In | (27,24)[b] |

*Group V–Group VI*[a]

| As–O | (22 ‡,39 ‡) | Sb–O | (22 ‡,39 ‡) | Bi–O | (22 ‡,43,45)[c] |
|------|------|------|------|------|------|
| As–S | (22 ‡,39 ‡) | Sb–S | (22 ‡,39 ‡,42) | Bi–S | (22 ‡,43,44,45)[c] |
| As–Se | (22 ‡,39 ‡) | Sb–Se | (22 ‡,39 ‡,41,42) | Bi–Se | (22 ‡,43,41,45)[c] |
| As–Te | (22 ‡,39 ‡) | Sb–Te | (22 ‡,39 ‡,41) | Bi–Te | (22 ‡,43,41,45)[c] |

and

| $BiSbTe_3$ | (43,22 ‡) |
|------|------|
| $BiSe_{1.5}Te_{1.5}$ | (43,22 ‡) |
| $BiSbSeTe_2$ | (43,22 ‡) |
| $As_2S_{2.5}Se_{2.5}$ | (22 ‡) |

‡ References marked by ‡ indicate that laser-induced vaporisation techniques were used.
[a] A study of phosphorus oxides has been reported[38].
[b] A value of $\Delta H°_f(In–Sb) = 35.4 \pm 2.5$ kcal.mole$^{-1}$ was reported[27].
[c] Values of $D(Bi–O) = 81.0 \pm 1.4$, $D(Bi–S) = 74.5 \pm 1.1$, $D(Bi–Se) = 66.0 \pm 1.4$, and $D(Bi–Te) = 54.6 \pm 2.7$ kcal.mole$^{-1}$ were reported[45].

Most of the studies have involved the use of mass spectrometric–Knudsen cell techniques but recently compounds containing Group V elements have been vaporised or simultaneously vaporised and ionised by using laser beams. The neutral species and the ions which are also produced by the laser can be distinguished by mass spectrometric techniques. Originally it was suggested[39,43] that laser-induced vaporisation produced species which were more complex than those found in the more conventional mass spectrometric studies. However, by using molecular beam techniques and higher temperatures it has been shown that the same species are produced, at least in the cases of antimony sulphides and antimony selenides[42].

The work with the laser beam clearly demonstrated that the number and types of ions produced was, to some extent, dependent on the structure of the (solid) material which was being examined. For instance, two forms of $As_2O_3$ were examined[39], one with a structure in the solid based on a tetrahedron of arsenic atoms (arsenolite) and the other with a layer structure (claudite). The spectra of these compounds differed markedly. Arsenolite produced neutral fragments and primary ions of higher complexity than claudite but claudite produced more types of ions and neutral fragments. Further, the base peak of both the neutral and primary ion spectra of arsenolite was due to the $As_8O_8$ fragment but from claudite the base peaks were due to the $As_4O_4$ fragment. Results were obtained from two forms of $Sb_2O_3$, senamontite and valententite, but the differences were much less marked, both compounds showing the same fragment at highest $m/e$ and the same base peaks due to $^+Sb_3O_4$. The reason for this was rather obscure but may possibly have been related to the fact that the two forms of $Sb_2O_3$ interconvert thermally much more readily than the forms of $As_2O_3$.

4. MAIN GROUP V HYDRIDES

The mass spectra of all the hydrides $MH_3$ of Group V elements have been reported by workers in two laboratories, one in America[46] and the other in USSR[47]. Their results are shown in Table 8.3, those from ref. 46 being given first. The spectra are very simple as expected and the agreement is excellent. The similarities in the spectra of the hydrides of phosphorus, arsenic, antimony, and bismuth are very striking and these spectra differ quite considerably from $NH_3$. Particularly noticeable is that the base peak for ammonia is the parent molecular ion (53 %) but for the other compounds $MH^+$ ions (36–45 %) are base peaks. The overall trends in decreasing abundance of $MH_3^+$ ions and increasing abundance of $M^+$ ions from N to Bi may be explained in terms of decreasing M–H bond strengths. The bond dissociation energies in ions $D(H_nM-H)^+$, were measured by Saalfeld and Svec[46] for the compounds of phosphorus to bismuth (Table 8.4). These values seem to concur with the bond strength arguments for the similarities in the spectra of the phosphorus to bismuth compounds.

Table 8.3

Ion Abundances from $MH_3$ Compounds

| Ion | % Abundance | | | | | | | | | |
|---|---|---|---|---|---|---|---|---|---|---|
| | $NH_3$ | | $PH_3$ | | $AsH_3$ | | $SbH_3$ | | $BiH_3$ | |
| | Ref.46 | Ref.47 | Ref.46 | Ref.47[a] | Ref.46 | Ref.47[a] | Ref.46 | Ref.47[b] | Ref.46 | Ref.47 |
| $MH_3^+$ | 53 | 53 | 31 | 34 | 29 | 32 | 27 | 28 | 25 | 25 |
| $MH_2^+$ | 42 | 41 | 11 | 10 | 10 | 9 | 12 | 14 | 18 | 19 |
| $MH^+$ | 4 | 4 | 45 | 42 | 44 | 44 | 37 | 36 | 38 | 37 |
| $M^+$ | 1 | 2 | 13 | 14 | 17 | 15 | 24 | 22 | 19 | 19 |

[a] These results also appear in ref. 48.
[b] These results also appear in ref. 49.

Several other workers[50-53] have reported the mass spectrum of $PH_3$. There is reasonably good agreement with the results in Table 8.3 except, perhaps, where the parent molecular ion (27%) appears rather low and the $P^+$ ion (16%) is rather high[51] and where the opposite pertains with (38%) and (8%) respectively[53].

The spectrum of $AsH_3$ has also been reported by Saalfeld and McDowell[54]. The agreement with the results in Table 8.3 is excellent. The formation of doubly charged ions in low abundance $(AsH_3)^{2+}$ (trace), $(AsH_2)^{2+}$ (0.1%), $(AsH)^{2+}$ (1%), and $As^{2+}$ (0.7%) was observed in two studies[46,54].

The appearance potentials of the ions in the spectrum of phosphine have been reported by various workers (Table 8.5).

The disagreement in the appearance potential measurements is at once

Table 8.4

Bond Dissociation Energies[46] (kcal.mole$^{-1}$) in Ions from $MH_3$

| Bond | M | | | |
|---|---|---|---|---|
| | P | As | Sb | Bi |
| $(H_2M-H)^+$ | 67 | 55 | 44 | 53 |
| $(HM-H)^+$ | 46 | 53 | 55 | |
| $(M-H)^+$ | 101 | 60 | 58 | 28 |

Table 8.5

Appearance Potentials (eV) of Ions from $PH_3$

| Ion | Appearance potential | | | | |
|---|---|---|---|---|---|
| | Ref. 48 | Ref. 50 | Ref. 51 | Ref. 52 | Ref. 55 |
| $PH_3^+$ | 11.5 ± 0.3 | 10.0 ± 0.3 | 10.05 ± 0.05 | 10.2 ± 0.2 | 10.4 ± 0.3 |
| $PH_2^+$ | 14.4 ± 0.2 | 13.9 ± 0.3 | 13.2 ± 0.2 | 13.2 ± 0.2 | 14.0 ± 0.2 |
| $PH^+$ | 12.2 ± 0.2 | 12.0 ± 1 | 12.6 ± 0.2 | 13.3 ± 0.2 | 13.1 ± 0.2 |
| $P^+$ | 16.5 ± 0.2 | 16.7 ± 1 | 15.9 ± ? | 17.2 ± ? | 16.0 ± 1 |

Note: (a) Various methods were used including linear extrapolation, vanishing current and the semi-log plot method. (b) Other values for the ionisation potential of $PH_3$ are
  (i) 9.98 eV (adiabatic I.P.) using a photon-impact source[56] and
  (ii) 10.13 eV (?) (adiabatic (?)) and 10.49 eV (vertical I.P.) using a photoelectron spectrometer[55].

obvious. It is difficult to assess which are correct. To illustrate this point let us take the results from ref. 46, 51, 52, 56, and 57. There is reasonable agreement for the values for $PH_3^+$ and $PH_2^+$ from Fehlner and Callen[51] and Wada and Kiser[52]. The ionisation potentials are further substantiated by the results of Price and Passmore[56] and Branton et al.[57]. However, the results of Saalfeld and Svec[46] would then appear to be too high. But Saalfeld and Svec calculated a value for the heat of formation of $PH_3$ from their appearance potential of $P^+$ assuming the reaction (8.16)

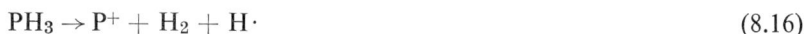

$$PH_3 \rightarrow P^+ + H_2 + H \cdot \qquad (8.16)$$

of $-2.3$ kcal.mole$^{-1}$, in good agreement with the thermochemically determined value of $-1.4$ kcal.mole$^{-1}$ by Gunn and Green[58]. Saalfeld and Svec have calculated heats of formation of the other Group V hydrides except $NH_3$, assuming similar reactions to (8.16) in each case[46]. Their results are shown in Table 8.8 (p. 331) together with the values obtained by Gunn and Green. Excellent agreement for $AsH_3$ and $SbH_3$ was found.

The ionisation potential of $AsH_3$ has been reported as 12.1 ± 0.2 eV (ref. 46), 10.03 eV (ref. 56), 10.5 ± 0.1 eV (ref. 54), 10.06 ± 0.03 eV (adiabatic), 10.58 ± 0.02 eV (vertical) (ref. 57), and 10.6 eV (ref. 59). One value of 9.58 eV has been reported for the ionisation potential of $SbH_3$ (ref. 56) and another

Table 8.6

Negative Ion Spectra and Appearance Potentials from $PH_3$ and $AsH_3$

| Ion | $PH_3$ | | $AsH_3$ | |
|---|---|---|---|---|
| | % Abundance | A.P. (eV) | % Abundance | A.P. (eV) |
| $MH_2^-$ | 63 | 2.3 and 5.2 | 52 | 2.0 and 5.2 |
| $MH^-$ | 32 | (2.1)?, 6.3 | 34 | (2.0)?, 6.2 |
| | | and 8.1 | | |
| $M^-$ | 5 | 5.7 and 8.3 | 14 | 5.3 and 7.8 |

of $9.9 \pm 0.3$ eV (ref. 46). The ionisation potential of $BiH_3$ was reported[46] as 10.1 eV.

The negative ion spectra of $PH_3$ and $AsH_3$ had $MH_2^-$ as the most abundant ion[60] (Table 8.6).

Other hydrides of Group V elements which have been studied include $M_2H_4$ (M = N, P, As, Sb), $P_2H_2$, and $P_3H_5$. The spectra of $M_2H_4$ compounds were first reported by Saalfeld and Svec[61]. Later, other workers reported

Table 8.7

Ion Abundances from $M_2H_4$ Compounds

| Ion | M | | | | | |
|---|---|---|---|---|---|---|
| | N | P[a] | P[b] | P[c] | As | Sb |
| $M_2H_4^+$ | 32.5 | 6.0 | 16.0 | 35.0 | 2.0 | 1.0 |
| $M_2H_3^+$ | 15.0 | 1.0 | 1.0 | 3.0 | 7.0 | 1.0 |
| $M_2H_2^+$ | 9.5 | 5.0 | 11.0 | 18.0 | 20.0 | 4.0 |
| $M_2H^+$ | 13.0 | 5.0 | 9.0 | 16.0 | 6.0 | 15.0 |
| $M_2^+$ | 7.0 | 78.0 | 12.0 | 21.0 | 58.0 | 69.0 |
| $MH_3^+$ | 8.5 | | 11.0 | | | 1.0 |
| $MH_2^+$ | 9.5 | 4.0 | 4.0 | 1.0 | 6.0 | 4.0 |
| $MH^+$ | 2.5 | | 21.0 | 3.0 | | 4.0 |
| $M^+$ | 2.0 | 1.0 | 14.0 | 3.0 | 1.0 | 1.0 |

[a] From ref. 61.
[b] From ref. 52.
[c] From ref. 51.

$P_2H_4$ (refs. 51, 52) and $P_2D_2H_2$ (ref. 52). The results of these studies are shown in Table 8.7. There is very great disagreement between comparable results (*i.e.* for $P_2H_4$). Fehlner and Callen[51] seriously questioned the earlier work of Saalfeld and Svec on the grounds that their spectra of $P_2H_4$ and $Sb_2H_4$ were actually those of decomposition products. Fehlner and Callen used a zero-contact (*i.e.* collision free) source to obtain their spectrum. This disagreement must cast doubt on the spectra of $As_2H_4$ and $Sb_2H_4$.

Saalfeld and Svec[61] calculated heats of formation of $M_2H_4$ compounds, assuming the equation

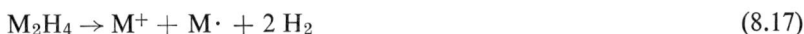

$$M_2H_4 \rightarrow M^+ + M \cdot + 2 H_2 \tag{8.17}$$

Table 8.8

Heat of Formation and Bond Dissociation Energy Data for Group V Hydrides

| Compound | Heat of formation (kcal.mole$^{-1}$) | Ref. |
|---|---|---|
| $PH_3$ | $-$ 2.3 | 46 |
|  | $-$ 1.4 | 58 |
| $AsH_3$ | 15.2 | 46 |
|  | 15.9 | 58 |
| $SbH_3$ | 34.6 | 46 |
|  | 34.7 | 58 |
| $BiH_3$ | 66.4 | 46 |
| $P_2H_4$ | 9.9 | 61 |
|  | 5.0 | 62, 58 |
| $As_2H_4$ | 35.2 | 61 |
| $Sb_2H_4$ | 57.2 | 61 |

| Bond | Bond dissociation energy (kcal.mole$^{-1}$) for M = | | |
|---|---|---|---|
|  | P | As | Sb |
| $(H_2M-MH_2)^+$ | 69 | 30 | 7 |
| $(H_2M-MH_2)$ | 43.7[a] or 46.8[b] | 44.7 | 30.7 |

[a] Reference 61.
[b] Reference 62.

and the bond dissociation energy for the $(H_2M-MH_2)^+$ bond. Their results are given in Table 8.8 together with values of heats of formation determined thermochemically by Gunn and Green[58] and by Fehlner[62]. A value of $D(H_2P-PH_2)$ of 54.5 kcal.mole$^{-1}$ has been reported by Grishin et al.[63].

Mass spectrometry has been used to identify hydrides of Group V elements produced by hydrolysing mixtures of nitrides, phosphides, and arsenides of calcium and magnesium[64]. As well as the simple hydrides, the following species were found (in less than 1 % of the total yield); $PH_2NH_2$, $PH(NH_2)_2$, $P_2H_3NH_2$, $P_3H_5$, $PH_2AsH_2$, $AsH_2NH_2$, $AsH(NH_2)_2$, $As(NH_2)_3$, and $As_3H_5$.

Fehlner[65] has studied the pyrolysis of $P_2H_4$ at temperatures between 570° and 650°C. He showed that zero-order kinetics were obeyed and the products from 1 mole of $P_2H_4$ were initially $PH_3$ and $P_2H_2$. A second reaction of $P_2H_2$ gave tetratomic phosphorus and more $PH_3$.

$$P_2H_4 \rightarrow PH_3 + {}^1/_2\,P_2H_2 \qquad\qquad\qquad\qquad (8.18)$$
$${}^1/_2\,P_2H_2 \rightarrow {}^1/_6\,P_4 + {}^2/_6\,PH_3$$

The intensities of $P_2$-containing ions from $P_2H_2$ have been reported by Fehlner and Callen[51]. They found $P_2H_2^+$ (30%), $P_2H^+$ (26%), and $P_2^+$ (40%). These authors also reported the appearance potentials of $P^+$ and/or $P_2^+$ from $PH_3$, $P_2H_4$, and $P_2H_2$ and postulated the reactions

|  | A.P. (eV) |  |
|---|---|---|
| $PH_3 \rightarrow P^+ + H\cdot + H_2$ | $15.9 \pm ?$ | (8.19) |
| $P_2H_4 \rightarrow P_2^+ + 2\,H_2$ | $13.2 \pm 0.2$ | (8.20) |
| $\rightarrow P^+ + \cdot P + 2\,H_2$ | $19.4 \pm 0.5$ | (8.21) |
| $P_2H_2 \rightarrow P_2^+ + H_2$ | $11.9 \pm 0.4$ | (8.22) |

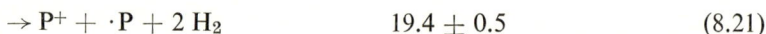

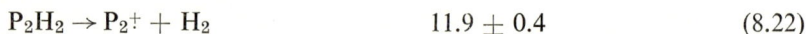

Then, using Gunn and Green's values[58] for the heats of formation of $PH_3$ and $P_2H_4$, Fehlner and Callen calculated the heat of formation of $P_2H_2$ as $26 \pm 8$ kcal.mole$^{-1}$.

Some studies of the mass spectra of $P_3H_5$ have been reported[62]. The compound is very unstable, decomposing at room temperature. A molecular beam sampling device was used to obtain the spectrum. Although % ion abundances are not quoted, there were large abundances of $P^+$, $P_2H^+$, and

$P_2H_3^+$ observed and the formation of $PH_4^+$ was reported. By comparing the appearance potentials of the $P_2H_3^+$ ion from $P_2H_4$ and $P_3H_5$ and using previously published values of the bond energies $E$(P–H) and $E$(P–P), the following bond dissociation energies were estimated,

$D(H_2P-H) = 77$ kcal.mole$^{-1}$

$D(H_2P-PH_2) = 47$ kcal.mole$^{-1}$

$D(PH=PH) = 80$ kcal.mole$^{-1}$

The ionisation potential of $P_3H_5$ was reported[62] as $8.7 \pm 0.1$ eV.

From the appearance potential of $PH_4^+$, the proton affinity of $PH_3$ was calculated[62] to be 172 kcal.mole$^{-1}$. Previously, figures of 207 kcal.mole$^{-1}$ (calculated) and 153 kcal.mole$^{-1}$ (indirect measurement) had been reported. These figures may be compared with a value of $186 \pm 3$ obtained by Eyler from studies of ion–molecule reactions of $PH_3$ in an ion cyclotron resonance mass spectrometer[66]. Eyler showed that at pressures of about $2 \times 10^{-5}$ torr and using 23 eV ionising electrons the ion $PH_4^+$ was produced by the reaction

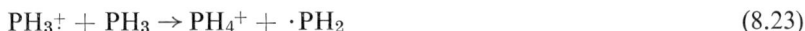

$$PH_3^+ + PH_3 \rightarrow PH_4^+ + \cdot PH_2 \qquad (8.23)$$

Other ion–molecule reactions occurred to produce the ions $P_2H_n^+$ ($n = 0$–5), $P_3H_n^+$ ($n = 0$–2), and $P_4^+$. The ion–molecule reaction between $PH_3$ and $PH_3^+$ was also studied by Halmann and Platzner and shown to be second-order[67]. The rate constant was reported to be $9.8 \times 10^{-10}$ cm.molecules$^{-1}$. sec.$^{-1}$.

Platzner has studied the reaction of $PH_3$ with $H_2O$ in the gas phase by mass spectrometry[68]. He found $PO^+$, $PO_2^+$, $PO_3^+$, and possibly $PO_4^+$ ions as well as hydride species such as $HPO_3^+$, $H_2PO_3^+$, and $H_3PO_3^+$.

### 5. MAIN GROUP V HALIDES

*5.1 Trihalides, $MX_3$*

The spectra of $NF_3$ (ref. 69) and $ClNF_2$ (ref. 70) have been reported and appearance potentials of the five most important positive ions from $NF_3$ have been measured. In their study of $NF_3$, Reese and Dibeler[69] also observed two negative ions, $F_2^-$ and $F^-$. The spectra of these halides were very simple. Parent molecular ions were observed $NF_3^+$ (28.5%) $NF_2Cl^+$ (18.5 %)and all the expected ions including those from the elements. The compound $NF_2Cl$ also showed an ion due to $Cl_2^+$ (3.0%). The base peak in both spectra was due to the $NF_2^+$ ion (48%) and (31%) from $NF_3$ and $NF_2Cl$ respectively. The ionisation potential of $NF_3$ was reported to be 13.2 $\pm$ 0.2 eV and the appearance potential of the $NF_2^+$ ion was 14.2 $\pm$ 0.3 eV. Reese and Dibeler used Rossini's value [71] for the heat of formation of $NF_3$ ($-27.2$ kcal mole$^{-1}$) and assumed values for bond dissociation energies of $D(F_2N-F) = 73$ kcal.-mole$^{-1}$, $D(FN-F) = 62$ kcal.mole$^{-1}$ and $D(N-F) = 62$ kcal.mole$^{-1}$ to calculate a value for the ionisation potential of the $NF_2$ radical (11.0 eV)[69]. This value would seem to be low (by 0.7–1.0 eV) in the light of later work on the dissociation of $N_2F_4$.

The ionisation potential of $PF_3$ has been reported as 9.71 eV by Price and Passmore[56] and as 12.31 eV by Green et al.[72] using a photoelectron spectrometer. The latter value would seem more likely in view of the value for the ionisation potential of $NF_3$ of 13.2 $\pm$ 0.2 eV obtained by Reese and Dibeler[69] and that of $PCl_3$ of 9.91 eV obtained by Price and Passmore themselves[56].

Several studies of phosphorus trichloride have been published. The ion

Table 8.9

Ion Abundances from $PX_3$

| Ion | X = Cl | | | | X = Br |
|---|---|---|---|---|---|
| | Ref. 73 | Ref. 74 | Ref. 75 | Ref. 76 | Ref. 77 |
| $PX_3^+$ | 18 | 19.3 | 20 | 27.9 | 28.5 |
| $PX_2^+$ | 51 | 51.7 | 55 | 50.7 | 48.0 |
| $PX^+$ | 10 | 11.2 | 10 | 10.6 | 14.5 |
| $X^+$ | 16 | 10.8 | 9 | | Trace |
| $P^+$ | 5 | 5.8 | 6 | 10.0 | 9 |
| $PX_3^{2+}$ | | Trace | | | |
| $PX_2^{2+}$ | | 0.9 | | 0.7 | |
| $PX^{2+}$ | | Trace | | Trace | |
| $PX_2^-$ | | | 96 | | |
| $PX^-$ | | | 4 | | |
| $X^-$ | | | | | |

abundances of the singly and doubly charged positive ions and the negative ions are given in Table 8.9.

Apart from the fact that Tate and coworkers[76] did not report the % abundance of $Cl^+$, the overall results are in good agreement. The parent molecular ion was observed in quite high abundance and the $PCl_2^+$ ion was the base peak (50% of the total ion abundance in the spectra of the positively charged ions). The species $PCl_2^{2+}$ was the most abundant doubly charged ion in about 1% abundance[74,78]. Only Halmann and Klein reported the negative ion spectrum[75]; $PCl_2^-$ was by far the most abundant negative ion. This is not surprising since the $PCl_2^-$ ion will have an electron configuration equivalent to a neutral compound $SCl_2$ of Group VI. Appearance potentials for the major singly charged positive ions have been measured by some workers[73,75,76]. Their results are given in Table 8.10.

Table 8.10

Appearance Potentials for Ions from $PX_3$ (eV)

| Composition of ion | X = Cl | | | X = Br |
|---|---|---|---|---|
| | Ref. 73 | Ref. 75 | Ref. 76 | Ref. 77 |
| $PX_3^+$ | $10.75 \pm 0.2$ | $10.6 \pm 0.2$ | $12.2 \pm ?$ | $10.0 \pm 0.2$ |
| $PX_2^+$ | $12.3 \pm 0.2$ | $11.8 \pm 0.5$ | $12.5 \pm ?$ | $11.4 \pm 0.2$ |
| $PX^+$ | $16.8 \pm 0.3$ | $16.5 \pm 0.5$ | $17.5 \pm ?$ | $15.6 \pm 0.3$ |
| $X^+$ | $20.2 \pm 0.5$ | $19.8 \pm 0.4$ | | $13.1 \pm 0.5$ |
| $P^+$ | $21.2 \pm 0.5$ | $21.0 \pm 0.5$ | $22.2 \pm ?$ | $20.1 \pm 0.5$ |

The work of Kiser and coworkers[73] and Halmann and Klein[75] agrees very well but the results of Tate and coworkers[76] appear to be too high. The values for the ionisation potential of $PCl_3$ may be compared with 9.91 eV reported by Price and Passmore[56] using a photon impact method. Kiser and coworkers[73] attempted to study the reactions producing some of the ions observed. For example, from evidence from clastograms and the lack of corresponding metastable peaks, they concluded[73] that the $PCl_2^+$ ion was produced mainly by the reaction

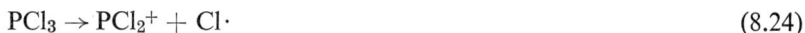

$$PCl_3 \rightarrow PCl_2^+ + Cl \cdot \qquad (8.24)$$

However, many of their conclusions were questioned by Halmann and Klein[75] in the light of their own appearance potential data and data on the negative ions $PCl_2^-$ and $PCl^-$. The latter workers concluded, for instance, that the $PCl_2^+$ ion was produced mainly by reaction (8.25) involving the formation of a pair of ions.

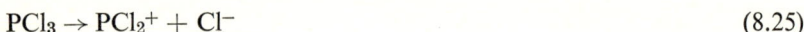

$$PCl_3 \rightarrow PCl_2^+ + Cl^- \qquad\qquad (8.25)$$

Obviously further work is necessary to clarify this situation.

The mass spectrum and appearance potentials of the singly charged positive ions of phosphorus tribromide have been reported by Kiser et al.[77]. Their results are given in Tables 8.9 and 8.10. The spectrum has strong similarities with that of phosphorus trichloride in that the parent molecular ion is quite abundant and the base peak is the $PBr_2^+$ ion in about 50% abundance. However, the abundance of the halide ion $Br^+$ is much less than for the corresponding $Cl^+$ ion. The ionisation potential of $PBr_3$ of $10.0 \pm 0.2$ eV appears to be reasonable in view of the results for $PCl_3$ (Table 8.10).

Although no ion abundance data were reported for $PI_3$ by Finch et al.[78], they measured the appearance potentials for the ions $PI_2^+$ and $I^+$ and postulated their formation by the reactions

$$PI_3 \rightarrow PI_2^+ + I\cdot \qquad A.P. = 11.9 \pm 0.15 \text{ eV} \qquad (8.26)$$

$$PI_3 \rightarrow I^+ + \cdot PI_2 \qquad A.P. = 12.7 \pm 0.1 \text{ eV} \qquad (8.27)$$

From reaction (8.27), Finch et al. calculated the bond dissociation energy $D(I_2P-I) = 52$ kcal.mole$^{-1}$. This result may be compared to a previously published value of 44 kcal.mole$^{-1}$ and suggests that the measured appearance potential for reaction (8.27) may be too high. Perhaps, on the basis of "Stevenson's rule", the appearance potential of the $PI_2^+$ ion should be used for the calculation of $D(I_2P-I)$ for preference. Taking the values of 52 and 44 kcal.mole$^{-1}$ for $D(I_2P-I)$, one may calculate the ionisation potential of the $PI_2$ radical as 9.65 and 10.0 eV, respectively.

Of the arsenic trihalides, arsenic trichloride has received most attention. Both Agafonov and coworkers[74] and Tate and coworkers[76] have published the mass spectrum of $AsCl_3$. Their results are given in Table 8.11 together with the appearance potentials of the major ions found by Tate and coworkers. Apart from the discrepancy for the % of parent molecular ion and the

Table 8.11

Ion Abundances and Appearance Potentials from $MCl_3$

| Ion | As | | | Sb | | | Bi |
|---|---|---|---|---|---|---|---|
| | % Abundance | | A.P.(eV) | % Abundance | | A.P.(eV) | % Abundance |
| | Ref. 74 | Ref. 76 | Ref. 76 | Ref. 74 | Ref. 76 | Ref. 76 | Ref. 74 |
| $MCl_3^+$ | 21.9 | 31.9 | 12.3 | 14.5 | 27.2 | 11.4 | 7.7 |
| $MCl_2^+$ | 52.3 | 52.3 | 13.0 | 44.7 | 46.5 | 12.3 | 25.0 |
| $MCl^+$ | 8.8 | 8.7 | 17.0 | 9.4 | 8.2 | 16.4 | 17.0 |
| $M^+$ | 6.8 | 5.8 | 21.6 | 15.0 | 10.9 | 17.0 | 32.6 |
| $Cl^+$ | 8.1 | | | 12.3 | 4.5 | 20.7 | 15.0 |
| $MCl_3^{2+}$ | | | | Trace | | | |
| $MCl_2^{2+}$ | 1.7 | 1.2 | 33.1 | 1.6 | 1.5 | 32.1 | Trace |
| $MCl^{2+}$ | 0.4 | 0.2 | | 2.2 | 1.5 | 34.8 | 1.0 |
| $M^{2+}$ | Trace | | | 0.3 | Trace | 38.0 | 1.7 |

fact that Tate and coworkers did not report the % of $Cl^+$ ion found, the spectra agree quite well. The base peak is due to the $AsCl_2^+$ ion (52%) and doubly charged ions make up between 1 and 2% of the total ion abundance. The ionisation potential of $AsCl_3$ has been reported by Cullen and Frost[59] as 11.7 eV, 0.6 eV lower than the value obtained by Tate and coworkers. It is likely that the other appearance potentials reported by Tate and coworkers are also rather high.

There has been a brief reference[79] to the study of the gas phase equilibrium between $AsCl_3$ and $AsF_3$, viz.

$$AsF_3 + AsCl_3 \rightleftharpoons AsFCl_2 + AsF_2Cl \qquad (8.28)$$

The mass spectrum of $AsF_3$ was not reported in any detail.

The spectrum of antimony trichloride has been obtained by Agafonov and coworkers[74] and Tate and coworkers[76] (Table 8.11) and appearance potentials of the major ions were measured by the latter. There is not such good agreement between these spectra as there was for the phosphorus and arsenic compounds. However, both sets of workers agree that the parent molecular ion appears in quite good abundance and the $AsCl_2^+$ ion is the base peak. Previous comments about the appearance potential data being rather high apply here also.

Agafonov and coworkers[74] have published a mass spectrum of bismuth trichloride (Table 8.11). The parent molecular ion is observed but in lower % abundance than for the corresponding phosphorus, arsenic, and antimony compounds; also, the $BiCl_2^+$ ion is not the base peak, the $Bi^+$ ion being more abundant. Overall, the spectra of the trichlorides of phosphorus, arsenic, antimony, and bismuth (Tables 8.9 and 8.11) show a decrease in abundance of the ions $MCl_3^+$ and $MCl_2^+$ but an increase in the abundance of the $M^+$ ion as the atomic number of M increases. This is consistent with the reduction of M–Cl bond strengths in going from phosphorus to bismuth.

*5.2 Pentahalides, $MX_5$*

The spectrum of phosphorus pentafluoride has been published by Schmutzler and coworkers[80]. No parent molecular ion was detected but all the expected $PF_n^+$ fragment ions were reported. They were observed in the following % abundance, $PF_4^+$ (87.3), $PF_3^+$ (3.8), $PF_2^+$ (3.3), $PF^+$ (4.3), $P^+$ (1.3). One might expect a Group V element containing ions of the formulae $ML_4^+$ to be particularly stable since this ion would have the electron configuration of the corresponding main Group IV element compound. Other pentavalent phosphorus compounds which have been studied[81] are "$PCl_5$" and "$P_2Cl_9F$". Unfortunately, no firm conclusions as to the form of $PX_5$ species in the gas phase could be drawn for these compounds. Ion abundances were not reported, but peaks were observed due to the ions $P_2Cl_n^+$ ($n = 0$–9) and all $PCl_n^+$ ($n = 0$–5) ions from "$PCl_5$" and $P_2FCl_n^+$ ($n = 0$–9) and all $PFCl_n^+$ ($n = 0$–4) from "$P_2Cl_9F$". Rogowski and Cohn[82] reported that parent molecular ions were not observed from $PF_4Cl$ or $PF_4Br$.

*5.3 $M_2X_4$ compounds*

Several groups have studied $N_2F_4$ mass spectrometrically. The spectrum and appearance potentials originally reported by Loughran and Mader[83] have been criticised by Colburn and Johnson[84] who were able to show that the earlier spectrum was largely due to the $NF_2$ radical and not $N_2F_4$. Colburn and Johnson also showed that the spectrum of $N_2F_4$ has a marked dependence on temperature (Table 8.12) and this was one reason for Loughran and Mader not observing the correct spectrum of $N_2F_4$. A reduction in temperature from 170° to ~20° C increased the % of $N_2F_n$-containing

Table 8.12

Spectrum of $N_2F_4$ and Effect of Temperature

| Ion | Temperature of source chamber (°C) | |
|---|---|---|
| | Room temp. ($\sim$20) | 170 |
| $N_2F_4^+$ | 1 | Trace |
| $N_2F_3^+$ | 3.5 | 0.5 |
| $N_2F_2^+$ | 4 | Trace |
| $NF_2^+$ | 56 | 39 |
| $N_2F^+$ | 4.5 | Trace |
| $NF^+$ | 22.5 | 52 |
| $N_2^+$ | 3 | 1 |
| $F^+$ | 2 | 2.5 |
| $N^+$ | 3.5 | 5 |

ions from about 0.5% to about 13% of the total ion abundance. The base peak also changed from the $NF^+$ ion at 170° to the $NF_2^+$ ion at 20 °C. Kennedy and Colburn[85] used heat of formation data[86] for $N_2F_4$ (2.0 $\pm$ 2.5 kcal.mole$^{-1}$) and the heat of dissociation[87] of $N_2F_4$ into two $NF_2$ radicals (19.3 $\pm$ 1.0 kcal.mole$^{-1}$) to calculate the heat of formation of the $NF_2$ radical (8.9 $\pm$ 2.5 kcal.mole$^{-1}$). Kennedy and Colburn[85] also calculated the bond dissociation energy $D(F_2N-NF_2)$ as 20.7 kcal.mole$^{-1}$ and the bond dissociation energies $D(F_2N-F) = 57.1 \pm 2.5$ kcal.mole$^{-1}$, $D(FN-F) = 71$ kcal.mole$^{-1}$, and $D(N-F) = 71$ kcal.mole$^{-1}$ from $NF_3$. These values for $NF_3$ are not in good agreement with those assumed by Reese and Dibeler[69] in their study of $NF_3$. However, later work by Herron and Dibeler[88] in which they measured the appearance potentials of a series of ions from $N_2F_4$, $N_2F_2$, and $NF_3$, and the $NF_2$ radical, is in much better agreement with the results of Kennedy and Colburn. Herron and Dibeler found the ionisation potential of the $NF_2$ radical to be 12.0 $\pm$ 0.1 eV, which was 1 eV higher than that reported in an earlier paper[69], and calculated $D(F_2N-F) = 52$ kcal.mole$^{-1}$ and $D(F_2N-NF_2) = 21.5$ kcal.mole$^{-1}$. The value for the bond dissociation energy $D(F_2N-NF_2)$ is in good agreement with that obtained by Kennedy and Colburn[85]. Johnson[89] has reported using mass spectrometry to characterise $N_2F_2$.

Evidence for the formation of the $PF_2$ radical from $P_2F_4$ has been sought using a mass spectrometer[90]. The ten-fold increase in the ratios of the ions $PF_2^+/P_2F_3^+$ and $PF_2^+/P_2F_4^+$ over the temperature range 350–700° C was suggested to be due to the formation of the $PF_2$ radicals. Further evidence was obtained from the pyrolysis of $P_2F_4$ at 900° and 3 torr. The product was $P_4F_6$ which was characterised by its mass spectrum ($P_4F_6^+$ and $PF_2^+$ being the most abundant ions) and which possibly had the structure $P(PF_2)_3$.

The mass spectrum of $P_2Cl_4$ has been reported in detail by Kiser and co-workers[73]. These workers measured appearance potentials and calculated the heats of formation of the ions and postulated possible reaction processes. Apart from the ions $PCl^+$ (22 %) and $PCl_2^+$ (61 %) (base peak) all the other ions contained two phosphorus atoms. A parent molecular ion (I.P. = 9.36 ± 0.2 eV) was observed and $P_2Cl_n^+$ ion species accounted for 13 % of the total ion abundance. By assuming the bond energy $E(P–Cl) = 78$ kcal.mole$^{-1}$, which was equal to the bond dissociation energy $D(Cl_2P–Cl)$, Kiser and coworkers estimated the bond dissociation energy $D(Cl_2P–PCl_2) = 58$ kcal.mole$^{-1}$. A value of $D(Cl_2P–PCl_2) = 58$ kcal.mole$^{-1}$ has also been reported by Grishin et al.[63].

Cowley and Cohen[91] have studied the reaction of $P_2I_4$ and $Br_2$ mass spectrometrically and observed the ionic species $PI_3^+$ $PBrI_2^{·+}$, and $PBr_2I^+$, and their expected fragment ions, but not $PBr_3^+$ or $P_2I_4^+$. They also reported that they obtained "similar results" in mass spectrometric and NMR studies of the $PI_3$–$PBr_3$ system where they observed all $PI_nBr_{(3-n)}$ ($n = 0$–3) species in the NMR spectrum.

Finch et al.[78] report the appearance potential of $PI_2^+$ from $P_2I_4$ and assume the reaction.

$$P_2I_4^+ \rightarrow PI_2^+ + \cdot PI_2 \qquad \text{A.P.} = 12.8 \pm 0.15 \text{ eV} \qquad (8.29)$$

They calculate $D(I_2P–PI_2)$ as 64.8 kcal.mole$^{-1}$. From the appearance potential of $I^+$ from $P_2I_4$ they calculate $D(I_2P–PI_2)$ as 73 kcal.mole$^{-1}$. Probably the first value is more correct but even this seems high in the light of the work on $P_2Cl_4$.

## 5.4 Phosphonitriles $(PNX_2)_n$

Several detailed studies of phosphonitriles have been published with $X = F$, Cl, Br or $X = Cl$ and $X' = Br$. The chlorophosphonitrile compounds will be discussed first.

Table 8.13

Ion Abundances from $P_4N_4Cl_8$ and $P_3N_3Cl_6$

| Ion | % Abundance | | Ion | % Abundance | |
|---|---|---|---|---|---|
| | $P_4N_4Cl_8$ | $P_3N_3Cl_6$ | | $P_4N_4Cl_8$ | $P_3N_3Cl_6$ |
| $(P_4N_4Cl_8)^+$ | 4.3 | | $(P_3N_2Cl_7)^{2+}$ | 0.7 | |
| $(P_4N_4Cl_8)^{2+}$ | 0.3 | | $P_3N_2Cl_6^+$ | 0.3 | |
| $(P_4N_4Cl_7)^+$ | 19.9 | | $P_3N_2Cl_5^+$ | 0.1 | |
| $(P_4N_4Cl_7)^{2+}$ | 0.1 | | $(P_3N_2Cl_5)^{2+}$ | 1.1 | 1.3 |
| $(P_4N_4Cl_6)^+$ | 0.1 | | $P_3N_2Cl_4^+$ | 1.1 | 0.4 |
| $(P_4N_4Cl_6)^{2+}$ | 5.0 | | $(P_3N_2Cl_3)^{2+}$ | 1.3 | 1.4 |
| $(P_4N_4Cl_5)^+$ | 0.3 | | $(P_3N_2Cl_2)^+$ | 0.1 | 0.1 |
| $(P_4N_4Cl_5)^{2+}$ | | | $(P_3N_2Cl)^{2+}$ | 1.4 | 1.3 |
| $(P_4N_4Cl_4)^+$ | | | $P_3N_2^{2+}$ | 1.2 | 1.1 |
| $(P_4N_4Cl_4)^{2+}$ | 1.4 | | $P_2N_2Cl_3^+$ | | 0.7 |
| $(P_3N_3Cl_6)^+$ | | 5.5 | $P_2N_2Cl_2^+$ | 0.1 | 0.4 |
| $(P_3N_3Cl_6)^{2+}$ | | 0.1 | $P_2N_2Cl^+$ | | 0.3 |
| $P_3N_3Cl_5^+$ | 4.5 | 27.5 | $P_2N_2^+$ | 0.5 | |
| $(P_3N_3Cl_5)^{2+}$ | | 0.3 | $PN_2Cl^+$ | 0.1 | 0.3 |
| $P_3N_3Cl_4^+$ | 0.3 | 0.5 | $PN_2^+$ | 0.8 | 1.1 |
| $(P_3N_3Cl_4)^{2+}$ | 1.7 | 4.7 | $PNCl_2^+$ | 0.3 | 0.7 |
| $P_3N_3Cl_3^+$ | 2.0 | 2.6 | $PNCl^+$ | 2.3 | 3.0 |
| $(P_3N_3Cl_3)^{2+}$ | | 0.1 | $(PNCl)^{2+}$ | | 0.1 |
| $P_3N_3Cl_2^+$ | | 0.1 | $PN^+$ | 8.2 | 9.0 |
| $(P_3N_3Cl_2)^{2+}$ | 0.1 | 0.3 | $PCl_4^+$ | 0.3 | 0.3 |
| $P_3N_3Cl^+$ | 0.3 | 0.7 | $PCl_3^+$ | 0.4 | 0.4 |
| $(P_3N_3Cl)^{2+}$ | | | $PCl_2^+$ | 9.3 | 11.3 |
| $(P_4N_3Cl_7)^{2+}$ | 0.3 | | $PCl^+$ | 9.1 | 10.8 |
| $(P_4N_3Cl_5)^{2+}$ | 0.1 | | $P^+$ | 14.1 | 11.2 |
| $(P_4N_3Cl_3)^{2+}$ | 0.1 | | $NCl^+$ | 0.3 | 0.1 |
| $(P_4N_3Cl)^{2+}$ | 0.1 | | $P_2Cl_3^+$ | 0.1 | |
| | | | $P_2Cl_2^+$ | 0.1 | |
| | | | $P_2Cl^+$ | 0.8 | 1.2 |
| | | | $P_2^+$ | 2.2 | 1.1 |

Traces of $HCl^+$, $NHCl^+$, and $Cl_2^+$, $Cl^+$ were also found, the first two ions being due presumably to hydrolysis of the phosphonitriles.

The mass spectra of hexachlorophosphonitrile and octachlorophosphonitrile were reported in full by Schmulbach et al.[92]. Their results are given in Table 8.13. A very large number of ions was observed ranging from the parent molecular ions down to ions containing only one or two atoms of P, N, or Cl. In both cases the base peak was due to the parent molecule less one chlorine atom. Apart from the $[(PNCl_2)_n - \cdot Cl]^+$ ions, the only other ions found in greater than $10\%$ abundance were $P^+$ from both $P_4N_4Cl_8$ and $P_3N_3Cl_6$ and $PCl_2^+$, $PCl^+$ from $P_3N_3Cl_6$. The unusually high abundance of doubly charged ions ($14.9\%$) from $P_4N_4Cl_8$ and ($10.7\%$) from $P_3N_3Cl_6$ is noteworthy. Of the doubly charged species, even electron ions are much more abundant than odd-electron ions; the $(P_4N_4Cl_6)^{2+}$ ion is the sixth most abundant ion in the spectrum of $P_4N_4Cl_8$ and $(P_3N_3Cl_4)^{2+}$ is the seventh most abundant in the spectrum of $P_3N_3Cl_6$.

An attempt was made to explain the unusal stability of some of the even-electron ions, for example $P_4N_4Cl_7^+$, $P_4N_4Cl_6^{2+}$, $P_4N_4Cl_4^{2+}$, and $P_4N_4Cl_2^{2+}$ from $P_4N_4Cl_8$ and $P_3N_3Cl_5^+$ and $P_3N_3Cl_4^{2+}$ from $P_3N_3Cl_6$, in terms of increased $\pi$ bonding between phosphorus and nitrogen in the $(PN)_x$ rings[92]. Let us take as an example the formation of the $P_3N_3Cl_5^+$ and $P_3N_3Cl_4^{2+}$ ions from $P_3N_3Cl_6$,

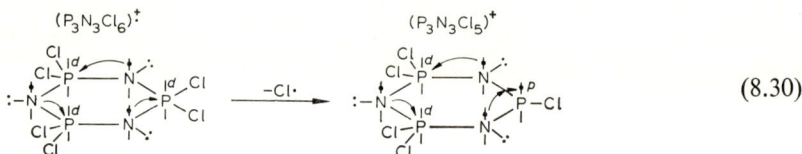

$$(8.30)$$

where arrows indicate some $p\pi$–$d\pi$ $(N \rightarrow P)$ bonding and double-headed arrows indicate formation of a $p\pi$–$p\pi$ $(N \rightarrow P)$ bond. With the loss of the Cl radical from $P_3N_3Cl_6^+$ the $p\pi$ orbital on a phosphorus atom becomes "free" and is able to participate in $N \rightarrow P$ $p\pi$–$p\pi$ bonding. Further loss of a $Cl^-$ or $Cl \cdot$ and an electron gives $(P_3N_3Cl_4)^{2+}$ in which two $p\pi$–$p\pi$ $(N \rightarrow P)$ bonds are possible, viz.

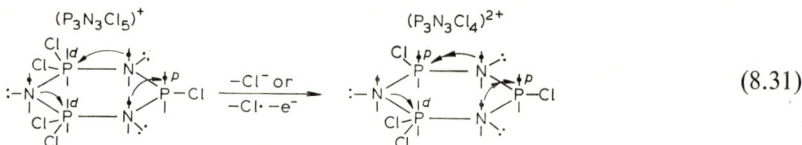

$$(8.31)$$

Schmulbach et al. put forward tentative arguments from Molecular Orbital theoretical calculations of orbital overlap populations in $d\pi$–$p\pi$ and $p\pi$–$p\pi$ $P \leftarrow N$ bonds to support these suggestions. However, some negative ion data are probably more convincing evidence in that $Cl^-$ ions are formed in reactions of the types[92]

$$P_nN_nCl_{2n} \rightarrow P_nN_nCl_{(2n-1)}^+ + Cl^- \qquad (8.32)$$

and

$$P_4N_4Cl_7^+ \rightarrow P_4N_4Cl_6^{2+} + Cl^- \qquad (8.33)$$

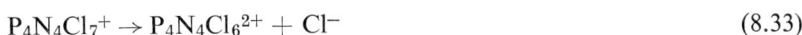

In their study, Schmulbach *et al.* noticed several rearrangement ions *e.g.* $PCl_3^+$, $PCl_4^+$, $P_2Cl_2^+$, $P_3N_2Cl_7^{2+}$, $P_2NCl_5^+$, $P_2NCl_5^{2+}$, $P_4^+$, and $P_3N^+$. Metastable peaks were observed for halogen elimination reactions and elimination of PN.

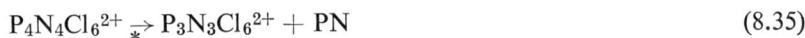

$$P_4N_4Cl_5^+ \underset{*}{\rightarrow} P_3N_3Cl_5^+ + PN \qquad (8.34)$$

$$P_4N_4Cl_6^{2+} \underset{*}{\rightarrow} P_3N_3Cl_6^{2+} + PN \qquad (8.35)$$

In another study of chlorophosphonitriles, Brion and Paddock[93] reported the spectra of compounds $(PNCl_2)_n$ for $n = 3$–$8$. Overall, the spectra showed many similarities with those of Schmulbach *et al.* Two types of ions were predominant, namely $(P_nN_nCl_y)^{a+}$ where $y = 2n$–$0$ and $(P_nN_{(n-1)}Cl_z)^{b+}$ where $z = (2n+1)$–$0$. The authors suggested that since ions of the general type $(P_xN_{(x-1)}Cl_{(2x+2)})^+$ are known from preparative phosphonitrile chemistry to have linear P–N structures, *e.g.* where $x = 2$ or $3$, $(P_2NCl_6)^+$, and $(P_3N_2Cl_8)^+$; so too $(P_nN_{(n-1)}Cl_y)^{n+}$ ions are linear but $(P_nN_nCl_y)^{n+}$ ions are cyclic.

A comparison of the ion abundances found by Brion and Paddock and Schmulbach *et al.* is difficult because the former workers do not give the complete spectra of $P_3N_3Cl_6$ or $P_4N_4Cl_8$. However, if one compares the ratios of abundances of prominent ions there appears to be quite large disagreements. For example, in the spectrum of $P_3N_3Cl_6$, Schmulbach *et al.* find $P_3N_3Cl_6^+/P_3N_3Cl_5^+ = 1/5.03$ but Brion and Paddock find $P_3N_3Cl_6^+/P_3N_3Cl_5^+ = 1/3.03$. These differences may be due, at least in some part, to different machines and conditions being used.

Several other interesting points emerge from the study of Brion and Paddock. First, with regard to the stabilities of linear and cyclic ions. Some results for $P_3N_3Cl_6$ and $P_6N_6Cl_{12}$ are shown in Table 8.14. A greater % of the ion current is carried by cyclic type ions than by linear ones. Generally,

Table 8.14

Ion Abundances of Linear and Cyclic Ions from $P_3N_3Cl_6$ and $P_6N_6Cl_{12}$

| Fragment type | $P_3N_3Cl_6$ | | $P_6N_6Cl_{12}$ | |
|---|---|---|---|---|
| | Cyclic | Linear | Cyclic | Linear |
| $P^a$ | | 5.6 | | 6.2 |
| $PN^a$ | 1.7 | | 0.4 | |
| $P_2N$ | | 12.0 | | 9.0 |
| $P_2N_2$ | 0.9 | | 0.0 | |
| $P_3N_2$ | | 1.4 | | 3.8 |
| $P_3N_3$ | 78.4 | | 40.6 | |
| $P_4N_3$ | | | | 2.3 |
| $P_4N_4$ | | | 10.3 | |
| $P_5N_4$ | | | | 0.1 |
| $P_5N_5$ | | | 15.1 | |
| $P_6N_5$ | | | | 0.0 |
| $P_6N_6$ | | | 12.0 | |

[a] For convenience, P-containing ions are included in the "linear" series and PN ions in the "cyclic" series.

$P_3N_3$, $P_4N_4$, and $P_5N_5$ rings are quite stable but after $n = 5$ fragmentation occurs more easily to produce smaller rings; for example, it was found that $P_6N_6Cl_x$ gives $P_3N_3Cl_y$ rings, $P_7N_7Cl_x$ gives about equal amounts of

Table 8.15

Abundances of doubly and triply charged ions from $(PNCl_2)_n$ ($n = 3-8$)

| Fragment type | $(P_nN_nCl_y)^{2+}$ (%) | $(P_nN_nCl_y)^{3+}$ (%) |
|---|---|---|
| $P_3N_3^a$ | 3.8 | |
| $P_4N_4^b$ | 8.2 | |
| $P_5N_5$ | 9.8 | |
| $P_6N_6$ | 17.7 | 0.14 |
| $P_7N_7$ | 15.2 | 0.28 |
| $P_8N_8$ | 8.5 | 1.02 |

[a] Schmulbach et al. found values of 5.5%.
[b] Schmulbach et al. found values of 6.8%.

$P_3N_3Cl_y$ and $P_4N_4Cl_y$ rings, and $P_8N_8Cl_x$ gives about equal amounts of $P_4N_4Cl_y$ and $P_5N_5Cl_y$ rings. The parent molecular ions from the $(PNCl_2)_n$ ($n = 6$–8) compounds are also remarkably abundant. The second point was that ions containing $P_nN_{(n+1)}$ nuclei were found, e.g. $P_7N_8$ from $P_8N_8$, $P_6N_7$ from $P_8N_8$ and $P_7N_7$, and $P_5N_6$ from $P_8N_8$, $P_7N_7$ and $P_6N_6$. The reaction producing these ions was suggested to be

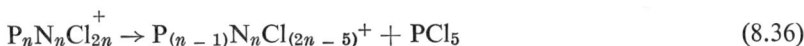

$$P_nN_nCl_{2n}^+ \rightarrow P_{(n-1)}N_nCl_{(2n-5)}^+ + PCl_5 \qquad (8.36)$$

with the product possibly having a condensed ring structure, i.e. $P_6N_7Cl_9^+$ may be

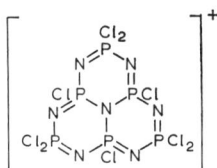

The third point was the high abundance of doubly and triply charged ions (Table 8.15). From $P_8N_8Cl_{16}$ the following triply charged ions were observed.

| $P_nN_nCl_y^{3+}$ | % | $P_nN_nCl_y^{3+}$ | % |
|---|---|---|---|
| $P_8N_8Cl_{15}$ | 0.03 | $P_6N_6Cl_9$ * | 0.15 |
| $P_8N_8Cl_{13}$ * | 0.25 | $P_6N_6Cl_7$ | 0.04 |
| $P_7N_7Cl_{11}$ * | 0.38 | $P_6N_5Cl_{12}$ | 0.08 |
| $P_7N_7Cl_9$ | 0.09 | | |

Lastly, Brion and Paddock reported metastable peaks corresponding to loss of Cl radicals for the reactions

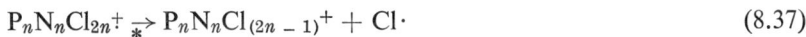

$$P_nN_nCl_{2n}^+ \underset{*}{\rightarrow} P_nN_nCl_{(2n-1)}^+ + Cl\cdot \qquad (8.37)$$

for $n = 3$–8,

---

* According to Schmulbach et al.[92] these ions should contain 3 $p\pi$–$p\pi$ P–N bonds.

$$P_5N_5Cl_{10}{}^+ \underset{*}{\rightleftharpoons} P_5N_5Cl_9{}^+ + Cl\cdot \tag{8.38}$$

for $n = 6, 7$

$$P_3N_2Cl_5{}^+ \underset{*}{\rightleftharpoons} P_3N_2Cl_4{}^{\ddagger} + Cl\cdot \tag{8.39}$$

for $n = 6$.

Brion and Paddock[94] have reported the mass spectra of several fluoro-phosphonitriles, $(PNF_2)_n$ where $n = 3$–16. Generally, the fragmentation schemes are similar to those of the corresponding chlorides. Two major series of ions $P_nN_nF_y{}^+$ ($y = 0$–$2n$) and $P_nN_{(n-1)}F_x{}^+$ ($x = 0$–$(2n + 1)$) were observed and ions of the type $P_nN_{(n+2)}F_{(2n+6)}{}^+$ were prominent for $n = 4$–12. The fluoro compounds had a smaller % of the ion current carried by doubly and triply charged ions compared to the chloro compounds and, as with the chloro compounds, the cyclic $P_nN_nF_x{}^+$ ions of $n = 3, 4, 5$ were important features in the spectra. Another interesting feature was that once $n$ became greater than 5, the % of the parent ion steadily increased e.g. from 23.5% for $n = 6$ to 67.7% for $n = 9$. This was explained in terms of greater electron delocalisation in the larger rings thus making the removal of a single electron more facile.

Although the cyclic and linear species accounted for between 98 and 99% of the ions in these spectra, other fragments which were identified were $N_2F_4{}^{\ddagger}$, $N_2F_3{}^+$, $N_2F^+$, and $NF_2{}^+$ where $n = 6$ and also $PN_2F_2{}^+$, $PN_2F^{\ddagger}$, and $PN_2{}^+$ where $n \leqslant 12$. Table 8.16 contains reactions for which metastable peaks were observed. Brion and Paddock suggest that the cyclisation of the smaller cyclic ions is by a transannular attack e.g. of $P_3N_3F_6$ from $P_6N_6F_{12}$ via

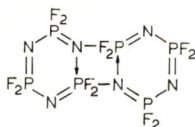

Bromophosphonitriles have been studied by Sowerby and coworkers[95]. The mass spectra of $(PNBr_2)_n$ where $n = 3$–6 were reported. The fragmentation patterns were similar to the results from the fluoro and chloro compounds but there was a lower % abundance of parent ions and a greater % abundance of doubly charged ions.

Table 8.16

Metastable Supported Decompositions from Fluoro- and Bromophosphonitriles

(a) *Halogen atom elimination*

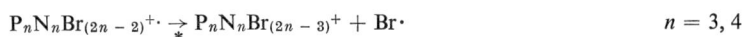

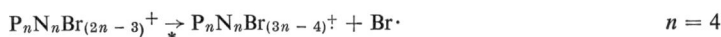

$P_nN_nF_{2n}^+ \underset{*}{\rightarrow} P_nN_nF_{(2-n)}^+ + F\cdot$     $n = 3\text{--}7$

$P_nN_nBr_{2n}^+ \underset{*}{\rightarrow} P_nN_nBr_{(2n-1)}^+ + Br\cdot$     $n = 3\text{--}6$

$P_nN_nBr_{(2n-1)}^+ \underset{*}{\rightarrow} P_nN_nBr_{(2n-2)}^{+\cdot} + Br\cdot$     $n = 3, 4$

$P_nN_nBr_{(2n-2)}^{+\cdot} \underset{*}{\rightarrow} P_nN_nBr_{(2n-3)}^+ + Br\cdot$     $n = 3, 4$

$P_nN_nBr_{(2n-3)}^+ \underset{*}{\rightarrow} P_nN_nBr_{(3n-4)}^+ + Br\cdot$     $n = 4$

(b) *PN elimination*

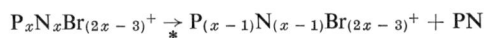

$P_xN_xBr_{(2x-3)}^+ \underset{*}{\rightarrow} P_{(x-1)}N_{(x-1)}Br_{(2x-3)}^+ + PN$

for $x = 4$ or $5$ from $n = 5$, and $x = 4$ from $n = 4$.

(c) *Other decomposition paths*

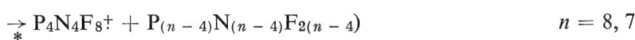

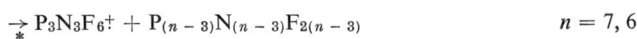

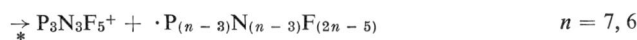

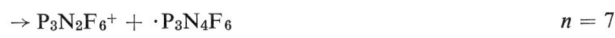

$P_nN_nF_{2n}^+ \underset{*}{\rightarrow} P_nN_{(n-1)}F_{2n}^+ + N\cdot$     $n = 3\text{--}7$

$P_nN_nF_{2n}^+ \underset{*}{\rightarrow} P_{(n-1)}N_{(n-2)}F_{(2n-2)}^+ + \cdot PN_2F_2$     $n = 3\text{--}7$

$P_nN_nF_{2n}^+ \underset{*}{\rightarrow} P_{(n-1)}N_{(n-1)}F_{(2n-3)}^+ + \cdot PNF_3$     $n = 4\text{--}7$

$P_nN_nF_{2n}^+ \underset{*}{\rightarrow} P_{(n-1)}N_{(n-1)}F_{(2n-1)}^+ + \cdot PNF$     $n = 6, 7$

$P_4N_3F_7^+ \underset{*}{\rightarrow} P_3N_3F_5^+ + \cdot PF_2$     $n = 4, 5$

$P_nN_nF_{2n}^+ \underset{*}{\rightarrow} P_6N_6F_{12}^+ + P_3N_3F_6$     $n = 9$

$\underset{*}{\rightarrow} P_5N_5F_9^+ + \cdot P_4N_4F_9$     $n = 9$

$\underset{*}{\rightarrow} P_4N_4F_8^+ + P_{(n-4)}N_{(n-4)}F_{2(n-4)}$     $n = 8, 7$

$\underset{*}{\rightarrow} P_4N_4F_7^+ + \cdot P_3N_3F_7$     $n = 7$

$\underset{*}{\rightarrow} P_4N_3F_8^+ + P_3N_4F_6$     $n = 7$

$\underset{*}{\rightarrow} P_3N_3F_6^+ + P_{(n-3)}N_{(n-3)}F_{2(n-3)}$     $n = 7, 6$

$\underset{*}{\rightarrow} P_3N_3F_5^+ + \cdot P_{(n-3)}N_{(n-3)}F_{(2n-5)}$     $n = 7, 6$

$\rightarrow P_3N_2F_6^+ + \cdot P_3N_4F_6$     $n = 7$

*References pp. 439–447*

| *Ions from* | $P_3N_3Br_6$ | $P_4N_4Br_8$ | $P_5N_5Br_{10}$ | $P_6N_6Br_{12}{}^*$ |
|---|---|---|---|---|
| *% of doubly* | | | | |
| *charged ions* | 4.2 | 16.2 | 32.5 | (32.5) |

Several triply charged ions and $P_nN_{(n-1)}$ $Br_x{}^+$ ions were observed. A number of reactions were metastable peak supported, (Table 8.16).

The mass spectra of mixed bromochlorophosphonitriles, $P_3N_3Br_xCl_{(6-x)}$ ($x = 0$–6) have been reported[96]. The fragmentation patterns were essentially similar to $P_3N_3Cl_6$ and $P_3N_3Br_6$ and it was suggested that differences were due to the bond strengths P–Cl $>$ P–Br. For example, base peaks $P_nN_n$-$(XX')_{(2n-1)}{}^+$ were due to ions with most Cl attached, *i.e.* $P_3N_3Cl_5{}^+$ from $P_3N_3Cl_6$, $P_3N_3Cl_5{}^+$ from $P_3N_3BrCl_5$, $P_3N_3BrCl_4{}^+$ from $P_3N_3Br_2Cl_4$ and so on; $P_3N_3Br_4Cl^+$ from $P_3N_3Br_5Cl$ and $P_3N_3Br_5{}^+$ from $P_3N_3Br_6$. Doubly charged species accounted for between 9.5 and 12.7% of the ion abundance except for $P_3N_3Br_4Cl_2$ which had 5.3% of the doubly charged ions and an anomolously high abundance of $P_2NX^+$ and $P_2N^+$ ions. All the metastable peaks observed related to the loss of X· radicals, *viz.*

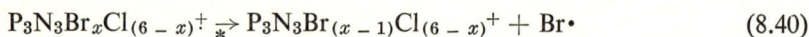

$$P_3N_3Br_xCl_{(6-x)}{}^+ \xrightarrow[*]{} P_3N_3Br_{(x-1)}Cl_{(6-x)}{}^+ + Br\cdot \qquad (8.40)$$

$x = 1$ or 5

$$P_3N_3Br_yCl_{(5-y)}{}^+ \xrightarrow[*]{} P_3N_3Br_{(y-1)}Cl_{(5-y)}{}^+ + Br\cdot \qquad (8.41)$$

$y = 1$ or 4

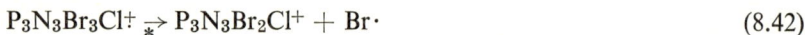

$$P_3N_3Br_3Cl^+ \xrightarrow[*]{} P_3N_3Br_2Cl^+ + Br\cdot \qquad (8.42)$$

Mass spectrometry has been used to characterise several derivatives of phosphonitriles. The results are summarised in Table 8.17.

Several workers have reported ionisation potential measurements on phosphonitriles. The results are given in Table 8.18.

In all cases the ionisation potential of the compound $P_3N_3X_6$ is larger than for the $P_4N_4N_8$ compound by about 0.4 eV, except for X = F or OPh where the difference is about 0.75 eV and 0.1 eV respectively. Where a large series of compounds has been studied, *i.e.* for X = F or Cl, the ionisation poten-

---

* The $P_6N_6$ compound was run at a different temperature (100 °C) from the other com - pounds (250 °C).

Table 8.17

Derivatives of Phosphonitriles Characterised by Mass Spectrometry

| Compound | Comments | Ref. |
|---|---|---|
| $P_3N_3F_5(NPF_3)$ | Parent molecular ion was base peak; fragment ions were those expected and the rearrangement ion $PF_4^+$ was observed | 97 |
| $P_3N_3F_5(NSF_2)$ | Parent molecular ion: $P_3N_3F_5NSF^+$ base peak | 98 |
| $P_3N_3F_5(NSO)$ | Parent molecular ion: $P_3N_3F_5^+$ base peak | 98 |
| $P_3N_3F_5(NSCl_2)$ | Parent molecular ion: $P_3N_3F_5^+$ base peak | 98 |
| $(P_3N_3F_5N)_2S$ | Parent molecular ion: no base peak given | 98 |
| $P_3N_3Cl_4(N=PPh_3)_2$ | Parent molecular ion and expected fragment ions | 99 |
| $P_3N_3Cl_4(Ph)(N=PPh_3)$ | No parent molecular ion reported but expected fragment ions found | 100 |
| $P_4N_4Cl_4(Ph)_4$ | No parent molecular ion reported but expected fragment ions found | 100 |

tials appear to alternate with $n$ even or odd. Branton et al.[102] have interpreted their results with the aid of Molecular Orbital theory as showing that there is a dual $\pi$ system which involves $3d$ orbitals on phosphorus and valence shell orbitals on nitrogen, and which supplements the existing $\sigma$ framework (Fig. 8.2(a) and (b)). A comparison of the experimental results with simple Molecular Orbital calculations suggests that the highest $\pi$ system is of the homomorphic type (Fig. 8.2(a)) involving mainly the $3d_{x^2-y^2}$ orbital on phosphorus and that it is accompanied by a lower and partly overlapping heteromorphic system to which the $3d_{xz}$ orbital makes the major contribution (Fig. 8.2(b)). Sowerby and coworkers[96] suggest that their results are consistent with the views of Brion et al.[101] (and presumably Branton et al.[102])

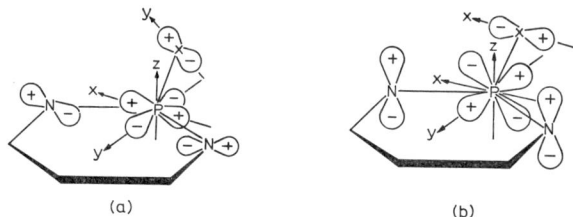

(a)                          (b)

Fig. 8.2. Overlap schemes of the atomic orbitals in the (a) homomorphic $\pi$ system and (b) heteromorphic $\pi$ system. (Reproduced with permission from ref. 102.)

Table 8.18

Ionisation Potentials of Phosphonitriles

| X in $(PNX_2)_n$ | Ionisation potential (eV) for $n =$ | | | | | | Ref. |
|---|---|---|---|---|---|---|---|
| | 3 | 4 | 5 | 6 | 7 | 8 | |
| $F^a$ | 11.64 | 10.86 | ~11.1 | ~11.1 | | | 101 |
| $F^b$ | 11.4 | 10.7 | 11.4 | 10.9 | 11.3 | 10.9 | 102 |
| $Cl^a$ | 10.26 | 9.8 | 9.83 | 9.81 | 9.80 | | 101, 102 |
| $OCH_2CF_3{}^a$ | 10.43 | 10.01 | | | | | 102 |
| $OMe^a$ | 9.29 | 8.83 | | | | | 102 |
| $OPh^a$ | 8.83 | 8.70 | | | | | 102 |
| $NMe_2{}^a$ | 7.85 | 7.45 | | | | | 102 |
| $Me^a$ | 8.35 | 7.99 | | | | | 102 |

| Compound | Ionisation potential[b,c] (eV) | Ion | Appearance potential[b,c] (eV) |
|---|---|---|---|
| $P_3N_3Cl_6$ | 10.27 | $(P_3N_3Cl_5)^+$ | 11.06 |
| $P_3N_3BrCl_5$ | 9.83 | $(P_3N_3Cl_5)^+$ | 10.49 |
| $P_3N_3Br_2Cl_4$ | 9.80 | $(P_3N_3BrCl_4)^+$ | 10.54 |
| $P_3N_3Br_3Cl_3$ | 9.72 | $(P_3N_3Br_2Cl_3)^+$ | 10.32 |
| $P_3N_3Br_4Cl_2$ | 9.60 | $(P_3N_3Br_3Cl_2)^+$ | 10.22 |
| $P_3N_3Br_5Cl$ | 9.47 | $(P_3N_3Br_4Cl)^+$ | 10.01 |
| $P_3N_3Br_6$ | 9.56 | $(P_3N_3Br_5)^+$ | 10.29 |
| $P_4N_4Br_8$ | 9.21 | | |

[a] $\pm 0.05$ eV.
[b] $\pm 0.1$ eV.
[c] Reference 96.

that the ionisation corresponds to the removal of a $\pi$-type electron. However, Sowerby and coworkers could not explain the apparently anomolous position of $P_3N_3Br_6$ in their series; in the case of the bromo derivatives the interaction of a bromine lone pair may have a significant effect on the ionisation potential. Using the thermochemical value for the heat of formation of $P_3N_3Cl_6$ and the appearance potential of $P_3N_3Cl_5{}^+$ from $P_3N_3Cl_6$ and $P_3N_3BrCl_5$, Sowerby and coworkers calculated the ionisation potential of

$P_3N_3Cl_5$ as 8.1 eV. Assuming the $D(P_3N_3Cl_5-Cl)$ as 80 kcal.mole$^{-1}$, a value for $D(P_3N_3Cl_5-Br)$ of 67 kcal.mole$^{-1}$ was obtained.

### 5.5 Compounds containing $[MF_6]^-$ anions

Some compounds containing $MF_6^-$ anions have been studied mass spectrometrically. These include $[NF_4]^+$ $[AsF_6]^-$ (ref. 103), $[Br_3]^+$ $[AsF_6]^-$ (ref. 104), and $[O_2]^+$ $[MF_6]^-$ M = As or Sb (ref. 105). The ion with the Group V element at highest $m/e$ was always $MF_4^+$; parent molecular ions were not observed. This behaviour is also reported to be observed[105] for $AsF_5$.

## 6. OTHER COMPOUNDS OF MAIN GROUP V ELEMENTS NOT CONTAINING A M–CARBON BOND

### 6.1 Compounds containing a Group V element bonded to elements of main Group VI

The study of nitrogen oxides has been undertaken by several workers, usually in connection with ion–molecule reactions or in the study of the excited states of the ions. The collision-induced dissociation of molecular ions of NO and $N_2O$ have been studied by Cheng et al.[106] and negative ion–neutral molecule reactions in $N_2O$ were observed by Paulson[107]. Isotopic exchange reactions in the nitrogen oxides were investigated by Sharma et al.[108] and a similar study of the exchange between $N_2O$ and $N_2$ over tungsten was reported by Gasser et al.[109]. The reactions of NO and $N_2O$ in the ionosphere were studied by Bates[110]. A possible lower limit for the electron affinity of NO obtained from a study of the collision of monoenergetic electrons with $NO_2$ was reported by Stockdale et al.[111]. Mathix et al.[112] have studied the abundance of excited ions in an $NO^+$ ion beam. A detailed study of the mass spectrum of $NO_2$ has been reported by Lindholm and Sahlstrom[113]. Four ions were found; $NO_2^+$ (22%), $NO^+$ (59%), $O^+$ (13%), and $N^+$ (6%). The spectrum was interpreted with the aid of Molecular Orbital theory and information on $NO_2$ from photoelectron spectra.

Several workers have reported ionisation potentials for the nitrogen oxides. The published values (in eV) include 12.89 (adiabatic)[114] for $N_2O$ and 9.26 (adiabatic)[114] for NO. The value for $NO_2$ is still open to discussion but values between 9.78 and 13.92 have been reported[114].

Compounds with alkoxy groups or thioalkoxy groups should, strictly, be dealt with here. However, it is more convenient to discuss them after later sections dealing with compounds containing Group V–carbon bonds. Other than the simple compounds of nitrogen and oxygen, the more complicated compounds with Group V elements bonded directly to Group VI elements are summarised in Table 8.19.

Table 8.19

Other Group V–Group VI Compounds Reported

| Compound | Comments[a] | Ref. |
|---|---|---|
| $F_2S(=NSF_5)_2$ | P.m.i. and expected fragment ions, $SF_5^+$ ion base peak | 195 |
| $F_2S=NSO_2F$ | P.m.i., $SOF^+$ ion base peak and expected fragment ions | 196 |
| $F_2S=NSO_2Cl$ | P.m.i., $SO_2^+$ ion base peak and expected fragment ions | 116 |
| $F_2S=NSO_2CF_3$ | P.m.i., $CF_3^+$ ion base peak and expected fragment ions | 117 |
| $F_2S=NSCF_3$ | P.m.i. and expected fragment ions | 118 |
| $(CF_3S-N=)_2S$ | P.m.i. and expected fragment ions | 114 |
| $(CF_3S-N=S=O)$ | P.m.i. and expected fragment ions | 114 |
| $FSO_2N=S(O)F_2$ | P.m.i. and expected fragment ions | 119 |
| $ClSO_2N_3$ | P.m.i., $SO_2N_3^+$ ion base peak | 120 |

[a] P.m.i. = parent molecular ion.

### 6.2 Compunds containing a Group V element bonded to hydrogen and halogens

Relatively few compounds of this type have been studied. Rudolph and Parry [121] identified $PHF_2$, and Treichel et al.[122] characterised $H_2PF_3$ and $HPF_4$ by their mass spectra. Lustig and Roesky[123] reported that the spectrum of $F_3P(NH_2)_2$ showed no parent molecular ion but other expected fragment ions were present; the ion $F_3PNH_2^+$ was the base peak.

### 6.3 Compounds containing a Group V element bonded to hydrogen and a main Group VI element

Very few compounds of this type have been reported. Schenk and Leutner[124] identified $H_2PSH$, $D_2PSH$, and $H_2PSCH_3$ by their mass spectra. Shozda and Vernon[120] reported that $H_2NSO_2N_3$ showed a parent molecular ion and the ion $H_2NSO_2^+$ was the base peak.

### 6.4 Compounds containing a Group V element bonded to a halogen and a main Group VI element

Although detailed spectra of the compounds $POCl_3$, $PSF_3$, and $PSF_2I$ have been published, the majority of compounds in this category have been studied merely to obtain information about the molecular weight and the constitution of the molecule. The spectrum of $POCl_3$ and the appearance potentials of most of the positive ions have been reported by Kiser et al.[77] and Halmann and Klein[75] (Table 8.20).

Table 8.20

Ion Abundances and Appearance Potentials of Ions from $POCl_3$

| Ion | Ref. 77 | | Ref. 75 | |
|---|---|---|---|---|
| | % Abundance | A.P. (eV) | % Abundance | A.P. (eV) |
| $POCl_3^+$ | 19.4 | $11.4 \pm 0.3$ | 19 | $13.1 \pm 0.2$ |
| $POCl_2^+$ | 43.8 | $12.8 \pm 0.3$ | 46 | $13.3 \pm 0.2$ |
| $POCl^+$ | 1.8 | $15.6 \pm 0.3$ | | |
| $PCl_3^+$ | | | 1 | $12.3 \pm 0.5$ |
| $PCl_2^+$ | | | 4 | $13.3 \pm 0.5$ |
| $PCl^+$ | 3.0 | $20.2 \pm 0.4$ | 5 | $17.0 \pm ?$ |
| $PO^+$ | 16.3 | $16.6 \pm 0.4$ | 13 | $14.5 \pm 0.5$ |
| $P^+$ | 5.5 | $28.1 \pm 0.5$ | 5 | |
| $Cl^+$ | 10.2 | | 7 | |

Several negative ions, $POCl_2^-$, $POCl^-$, $PO^-$, $PCl\cdot^-$, $Cl^-$, and $PCl_2^-$ (ref. 77) and $O^-$ (ref. 75) were reported. The later work of Kiser et al. questions the validity of the spectrum obtained by Halmann and Klein. Kiser et al. suggested that some impurity ($PCl_3$) was present in the earlier study. They also question the value for the ionisation potential of $POCl_3$ found by Halmann and Klein and quote values for $POF_3$ (13.4 eV) and $POBr_3$ (10.46 eV) which are consistent with the value of 11.4 eV (ref. 74) but not 13.1 eV (ref. 75). There is considerable disagreement between these sets of workers about proposed decomposition paths. For instance, for the production of the $POCl_2^+$ ion from the parent molecular ion, Kiser *et al.* favour the reaction

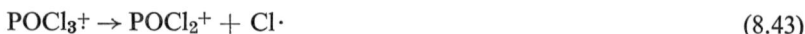

$$POCl_3^+ \rightarrow POCl_2^+ + Cl \cdot \qquad (8.43)$$

whereas, Halmann and Klein suggest

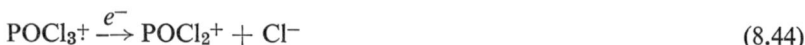

$$POCl_3^+ \xrightarrow{e^-} POCl_2^+ + Cl^- \qquad (8.44)$$

Both sets of workers do agree, though, that the ion $POCl_2^+$ is the base peak and the parent molecular ion is found in quite high abundance.

Kiser *et al.*[77] reported the spectrum of $PSF_3$ and the appearance potentials of the positive ions $PSF_3^+$, $PSF_2^+$, $PS^+$, and $PF_n^+$ ($n = 1$–3). The parent molecular ion was the base peak (41.1 %). A greater % abundance of $PX_n^+$ ions in the case of $PSF_3$ compared to $POCl_3$ was explained as being due largely to the stronger P–F compared to P–Cl bonds. A relatively low ionisation potential $11.1 \pm 0.3$ eV for $PSF_3$ was suggested to indicate the loss of an electron located mainly about the P=S system.

The spectrum of $PSF_2I$ has been reported[125]. The parent molecular ion was characterised by accurate mass measurement (the $^{32}S$ isotope was taken; all the other elements are monoisotopic). As with $PSF_3$, the parent molecular ion was the base peak (55.6 %) and this was the only P–I-containing ion. However, the reaction to produce the ion $PSF_2^+$ from the parent molecular ion was metastable supported.

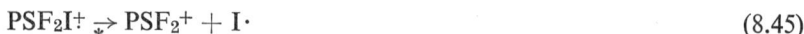

$$PSF_2I^+ \xrightarrow[*]{} PSF_2^+ + I \cdot \qquad (8.45)$$

Accurate mass measurement of the parent molecular ions from the compounds $(SPF_2)O$, $(SPF_2)S$, and $(SPF_2)O(OPF_2)$ has been reported[126]. The base peak for the first compound was the parent molecular ion, but that for the second compound was the ion $PF_2^+$. Reactions (8.46) and (8.47) were observed to be metastable supported.

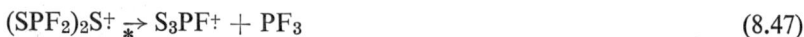

$$(SPF_2)_2O^+ \xrightarrow[*]{} S_2^+ + (PF_2)_2O \qquad (8.46)$$

$$(SPF_2)_2S^+ \xrightarrow[*]{} S_3PF^+ + PF_3 \qquad (8.47)$$

The formation of the rearrangement ion $S_2^+$ (8.46) is particularly interesting.

This ion was also formed[127] from $F_2P(S)SH$ by a metastable supported reaction

$$F_2P(S)SH^{\ddagger} \underset{*}{\rightarrow} S_2^+ + PF_2H \qquad (8.48)$$

The base peak in the spectrum of the thiol was due to the $PF_2S^+$ ion and its production from the parent molecular ion was metastable supported.

$$F_2P(S)SH^{\ddagger} \underset{*}{\rightarrow} PF_2S^+ + \cdot SH \qquad (8.49)$$

Other compounds related to those discussed above which have been studied[126] are $(OPF_2)_2NMe$ and $(SPF_2)_2NMe$. Parent molecular ions and the following metastable supported reactions were observed.

$$(OPF_2)_2NMe^{\ddagger} \begin{array}{c} \xrightarrow[*]{-\cdot NCH_2} O_2P_2F_4H^+ \\ \xrightarrow[*]{-CH_2O} PF_2NP(O)F_2^{\ddagger} \end{array} \qquad (8.50)$$

$$(SPF_2)_2NCH_3^{\ddagger} \xrightarrow[*]{-\cdot SPF_2} SPF_2NCH_3^+ \qquad (8.51)$$

The base peaks were due to the ions $OPF_2NCH_3^+$ and $(SPF_2)_2NCH_3^{\ddagger}$.

Compounds in the present category which have been studied but less extensively are given in Table 8.21.

*6.5 Compounds containing a Group V element bonded to hydrogen, a halogen and a main Group VI element*

The mass spectra of $P(O)F_2H$, $P(S)F_2H$ and $P(Se)F_2H$ have been published in detail. Two sets of workers published spectra for $P(O)F_2H$ which are not in good agreement (Table 8.22) and Treichel *et al.*[122] have observed some of the ions reported by the other workers. Both agree that the parent molecular ion is the base peak and the $PF_2^+$ ion is abundant but Centofanti and Parry[148] report the $PO^+$ ion in high abundance whereas Charlton and Cavell[147] do not report it at all. The latter workers reported a metastable reaction involving the elimination of HF from the parent molecular ion, *viz.*

$$POF_2H^{\ddagger} \underset{*}{\rightarrow} POF^{\ddagger} + HF \qquad (8.52)$$

Many features in the spectrum of $P(S)F_2H$ resemble that of $P(O)F_2H$. The

Table 8.21

Other Compounds Which Contain Group V Elements Bonded to Group VI Elements
and Halogen

| Compound | Comments | Ref. |
|---|---|---|
| Br–N=S=N–Br | P.m.i., $SN^+$ base peak | 128 |
| I–N=S=N–I | P.m.i., base peak | 128 |
| $ClNSF_2$ | P.m.i., $SF_2^+$ base peak | 129 |
| $FNSF_2O$ | P.m.i., $SFO^+$ base peak | 129 |
| $ClNSF_2O$ | P.m.i., $NCl^+$ base peak | 129 |
| $BrNSF_2$ | P.m.i., $SF_2^+$ base peak | 130 |
| $(CN)NSF_2$ | P.m.i. | 131 |
| $F_2NOSO_2F$ | P.m.i. | 132 |
| $(F_5S)FNSO_2F$ | $FSONFSF_5^+$ fragment at highest $m/e$; $SO_2NSF_3^+$ base peak | 133 |
| $S_3N_3Cl_3$ | No p.m.i. but expected fragment ions e.g. $S_3N_3^+$ and $SNCl_2^+$, also some ions due to pyrolysis effects ? e.g. $S_4N_4^+$ | 134 |
| $S_3N_3O_3Cl_3$ | P.m.i., $N_3S_3O_3Cl_2^+$ base peak | 136 |
| $S_3N_3O_3Ph_2X$ | X = F p.m.i. base peak; X = Cl, no p.m.i. | 136 |
| $F_3PNSO_2F$ | P.m.i. base peak | 97 |
| $F_2P(S)(CN)$ | P.m.i., $PF_2^+$ base peak | 145 |
| $F_2P(S)(NC)$ | P.m.i., base peak for X = O or S | 145 |
| $(Me_2N)F_2PNSO_2F$ | P.m.i. | 135 |
| $(Me_2N)_2FPNSO_2F$ | P.m.i. | 135 |
| $F_3PNSO_2Cl$ | P.m.i. | 137 |
| $F_3PNPF_2(S)^a$ | P.m.i., base peak | 138 |
| $F_3PNPF(Cl)S^a$ | P.m.i., $SPFNPF_3^+$ base peak | 138 |
| $F_3PNPCl_2(S)^a$ | P.m.i., $SPCl_2^+$ base peak and ions due to $F_3PNPFCl(S)$ compound | 138 |
| $F_2(Cl)PNPF_2(S)$ | P.m.i., $SPCl_2^+$ base peak | 138 |
| $F_2(Cl)PNPF(Cl)S$ | P.m.i. and ions due to rearrangement or impurities e.g. $SPCl_3^+$, $SPF_3^+$, $PF_3^+$ | 138 |
| $F_2(Cl)PNPCl_2(S)$ | P.m.i., $SPClNPF_2Cl^+$ base peak, some rearrangement ions e.g. $PF_2Cl_2^+$ and $SPF_2N^+$ | 138 |
| $F_2(Br)PNPF_2(S)$ | P.m.i., $PF_2^+$ base peak and PO-containing ions observed, presumably due to impurity | |
| $F_2(Br)PNPF(Cl)S$ | P.m.i., $POF_2^+$ base peak due to a PO-containing impurity | 139 |
| $Cl_3PNPF_2(S)$ | P.m.i., $PF_2^+$ base peak | 140 |
| $Cl_3PNPF(Cl)S$ | P.m.i. | 140 |
| $Cl_3PNPF_2NPF_2(S)$ | P.m.i. and expected fragment ions | 141 |
| $Cl_3PNPF_2NPF(Cl)S$ | P.m.i., $SP(Cl)FNPF_2NPCl_2^+$ base peak | 141 |

Table 8.21 (continued)

| Compound | Comments | Ref. |
|---|---|---|
| $F_2P(O)NCS$ | P.m.i., base peak | 142 |
| $F_2PNP(O)F_2$ | P.m.i. | 143 |
| $F_2PNP(S)F_2$ | P.m.i. | 143 |
| $(Me_2N)S(O)ClNP(O)Cl_2$ | P.m.i. | 144 |
| $(Et_2N)S(O)ClNP(O)Cl_2$ | P.m.i. | 144 |
| $F_2SNP(O)F_2$ | P.m.i. | 119 |
| $F_2P(S)N_3$ | P.m.i., $PF_2^+$ base peak and impurity? giving | |
| | $S_2^+$ ions | 145 |
| $(F_2PS)_2O$ | P.m.i., $S_2^+$ base peak, a rearrangement ion | 145 |
| $F_2P(S)OSO_2F$ | P.m.i., $PF_2^+$ base peak | 146 |
| $(SCN)PF_2NPF_2(S)$ | P.m.i., base peak | 139 |
| $(Me_2N)PF_2NPF_2(S)$ | P.m.i., base peak | 139 |
| $(Me_2N)PF_2NPF(Cl)(S)$ | P.m.i., base peak | 139 |
| $(Me_2N)_2PFNPF_2(S)$ | P.m.i., base peak | 139 |

[a] Note that these compounds all had the rearrangement ion $PF_4^+$ in their spectra.

Table 8.22

Ion Abundances from $PXF_2H$ Compounds

| Ion | X = O | | X = S | X = Se |
|---|---|---|---|---|
| | Ref. 147 | Ref. 148 | Ref. 147 | Ref. 149 |
| $PXF_2H^+$ | 44.0 | 26.6 | 52.9 | 16.3 |
| $PXF_2^+$ | 11.4 | 7.8 | 6.9 | 42.0 |
| $PF_2^+$ | 21.1 | 19.0 | 28.0 | 7.6 |
| $PXFH^+$ | 9.3 | 11.5 | 1.6 | 0.4 |
| $PXF^+$ | 11.0 | 12.5 | 3.2 | 7.1 |
| $PX^+$ | | 18.0 | 4.2 | 6.3 |
| $PF^+$ | 1.7 | 3.7 | 3.2 | 3.4 |
| $P^+$ | 1.5 | | | 1.9 |
| $PF_2H^+$ | | | | 1.0 |
| $PFH^+$ | | 0.4 | | 0.7 |
| $PH^+$ | Trace | 0.5 | | 0.4 |
| $XH^+$ | | | | 0.5 |
| $X^+$ | | | | 12.2 |

base peak is due to the parent molecular ion which has been characterised by accurate mass measurement and $PF_2^+$ is very abundant. However, for $P(Se)F_2H$, the parent molecular ion is observed but the base peak is due to the $PSeF_2^+$ ion. A number of hydrogen-containing ions were reported in this case which apparently did not have counterparts from the corresponding sulphur compound and, similarly, the ion $Se^+$ was the third most abundant in the spectrum of $P(Se)F_2H$ but the $S^+$ ion was not reported from $P(S)F_2H$. Compounds which may be conveniently discussed here also include $(OPF_2)_2NH$, $(SPF_2)_2NH$, and their trimethylamine adducts[126]. Several metastable supported reactions were observed in the spectra of the adducts and some of these are shown in reaction schemes (8.53) and (8.54).

$$(OPF_2)_2N\overset{H}{\overset{}{}}{}^{\ddagger} \xrightarrow[\underset{*}{-HF}]{\overset{-OH\cdot}{\underset{*}{}}} \begin{array}{l} PF_2NP(O)F_2^+ \\ \\ OPFNP(\dot{O})F_2^{\ddagger} \end{array} \qquad (8.53)$$

$$(SPF_2)_2NH^{\ddagger} \xrightarrow[\underset{*}{-HF}]{\overset{-(PF_2)_2NH}{\underset{*}{}}} \begin{array}{l} S_2^{\ddagger} \\ \\ SPFNP(S)F_2^{+\cdot} \end{array} \qquad (8.54)$$

Parent molecular ions were observed from the adducts and the base peaks were due to $(OPF_2)_2NHN(CH_3)_3^+$ and $S_2^+$ and $PFN^+$ from the sulphur-containing compound.

Other compounds in this category are given in Table 8.23.

## 7. ORGANO-DERIVATIVES OF MAIN GROUP V ELEMENTS

### 7.1 Trialkyl compounds

Several studies of the trimethyl compounds of nitrogen, phosphorus, arsenic, antimony, and bismuth have been reported. Where the same compound has been studied by different workers the ion abundance figures are not usually in good agreement. The reported spectra of $Me_3P$ are given in Table 8.24.

Overall, the results of Wada and Kiser[154] are in much better agreement with those of Halmann[53] than with those of Kostyanovsky and Yakshin[153] and fewer types of ions were reported by the Russian workers. However, all three groups do agree that the base peak is due to the $(CH_3)_2P^+$ ion. The fragmentation patterns of the other trimethyl–Group V compounds are not

Table 8.23

Other Compounds Reported Containing Group V Element–Hydrogen–Halogen and Group VI Element Bonds

| Compound | Comments | Ref. |
|---|---|---|
| $H_2NPF_2S$ | P.m.i., base peak | 140 |
| $H_2NPF(Cl)S$ | P.m.i. but $SPF_2NH^+$ was the most abundant ion observed | 140 |
| $H_2NPCl_2S$ | P.m.i. but $SPCl_3^+$ ion also observed | 140 |
| $H_2NPF(Br)S$ | P.m.i. but $SPF_2NH_2^+$ was the most abundant ion; the ion $SPFBr_2^+$ was also observed | 138 |
| $(H_2N)_2P(O)F$ | P.m.i., base peak | 151 |
| $(P(O)F_2)NH(P(S)F_2)$ | P.m.i. and $PF_2^+$ base peaks; rearrangement ion $PF_4^+$ observed | 152 |
| $(P(O)F_2)ND(P(S)ClF)$ | P.m.i., base peak; $PF_4^+$ observed | 152 |
| $(P(O)F_2)NH(P(S)Cl_2)$ | P.m.i., (parent molecule − Cl·)$^+$ ion base peak | 152 |
| $(P(O)F_2)ND(P(S)Cl_2)$ | P.m.i., (parent molecule − Cl·)$^+$ ion base peak | 152 |
| $HN(F)SO_2F$ | P.m.i., $FSNHF^+$ base peak | 133 |

Table 8.24

Ion Abundances from $(CH_3)_3P$

| Composition of ion | % Abundance | | |
|---|---|---|---|
| | Ref. 153 | Ref. 154 | Ref. 53 |
| $(CH_3)_3P$ | 8.5 | 19 | 20 |
| $(CH_3)_2PCH_2$ | | 5 | 6 |
| $(CH_3)_2PC$ | | | 1.5 |
| $(CH_3)_2PH$ | | 1 | 1 |
| $(CH_3)_2P$ | 43 | 24 | 30 |
| $C_2H_5P$ | | 2.5 | 2 |
| $C_2H_4P$ | 27 | 16 | 20 |
| $C_2H_3P$ | | 3 | 2 |
| $C_2H_2P$ | 10 | 8 | 1.5 |
| $HPCH_3$ | | 3 | 5 |
| $PCH_3$ | 4 | 2.5 | 2 |
| $PCH_2$ | 7.5 | 13.5 | 9 |
| $PCH$ | | 2.5 | 2 |
| $(CH_3)_3PH$ | | 1 | |

given here in full because of the discrepancies in the reported spectra. Instead, the composition of the four highest abundance ions for these compounds from the work of Kostyanovsky and Yakshin are listed in Table 8.25.

Winters and Kiser[155] have reported a more detailed spectrum of $Me_3Sb$ and found the four most abundant ions were $(CH_3)_2Sb^+$ (30%), $(CH_3)_3Sb^+$ (23.5%), $(CH_3)Sb^+$ (10.5%), and $Sb^+$ (10.5%). This spectrum is probably more reliable than the one given in ref. 153.

Table 8.25

The Four Most Abundant Ions from $Me_3M$ Compounds
The % abundance of each ion is given in brackets.

| Compound | Composition of the most abundant ions | | | |
|---|---|---|---|---|
| $Me_3N$ | $Me_3N$ (55.5) | $(CH_2)_2N$ (37) | $(CH_2)N$ (4) | $Me_2N$ (2) |
| $Me_3P$ | $Me_2P$ (43) | $(CH_2)_2P$ (27) | $(CH)_2P$ (10) | $Me_3P$ (8.5) |
| $Me_3As$ | $(CH_2)As$ (47.5) | $Me_2As$ (37.5) | $(CH_2)As$ (10) | $MeAs$ (2.5) |
| $Me_3Sb$ | $Me_2Sb$ (44) | $MeSb$ (21) | $(CH_2)_2Sb$ (13.5) | $Sb$ (13) |
| $Me_3Bi$ | $Bi$ (42) | $Me_2Bi$ (32) | $MeBi$ (26) | |

Gillis and Long[156] have published a study of metastable supported fragmentations from $Me_3N$ and $Me_3P$. They were able to draw up reaction scheme (8.55). Where both compounds showed the same fragmentation, the pathway is marked ** where $Me_3P$ alone was observed to follow a particular reaction, the pathway is marked as usual and the product ion is written as the phosphorus-containing entity.

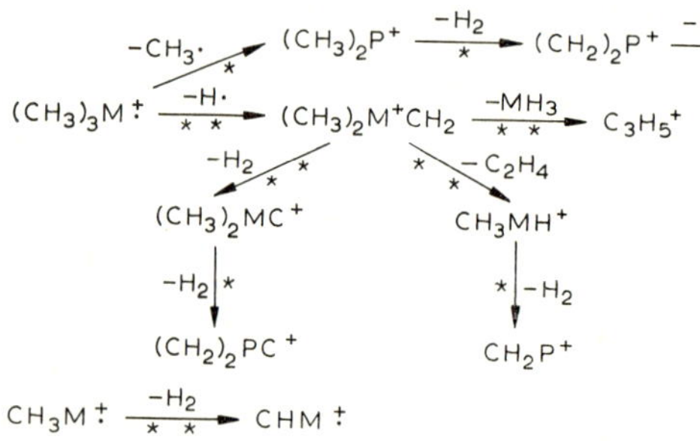

(8.55)

The spectra of triethylphosphine and triethylantimony have been published and are given in Table 8.26. The replacement of methyl groups by ethyl groups gives rise to a large number of hydride-containing ions which include

Table 8.26

Ion Abundances from $(C_2H_5)_3M$ (M = P,Sb) and Appearance Potentials of Ions from $(C_2H_5)_3Sb$

| Composition of ion | % Abundance for M = | | Appearance potential[a] for M = Sb |
|---|---|---|---|
| | P (ref. 154) | Sb (ref. 157) | |
| $(C_2H_5)_3M$ | 9.5 | 18.5 | 9.2 |
| $(C_2H_5)_2MCH_2$ | 4.0 | | |
| $(C_2H_5)_2MH$ | 14.0 | 1.6 | 10.0 |
| $(C_2H_5)_2M$ | 2.5 | 6.5 | 10.7 |
| $m/e = 76\ (?)^b$ | 1.5 | | |
| $(C_2H_5)MCH_3$ | 3.5 | | |
| $m/e = 63\ (?)^b$ | 1.0 | | |
| $(C_2H_5)MH_2$ | 18.0 | 0.6 | 12.6 |
| $(C_2H_5)MH$ | 13.0 | 29.7 | 12.6 |
| $(C_2H_5)M$ | 2.5 | 9.1 | 12.4 |
| $C_2H_4M$ | 8.5 | 14.5 | 14.8 |
| $C_2H_3M$ | 4.0 | Trace | |
| $C_2H_2M$ | 8.5 | 0.6 | 13.0 |
| $CH_3MH$ | 2.0 | Trace | 17.4 |
| $CH_3M$ | 1.0 | 0.8 | 16.5 |
| $CM$ | 5.5 | | |
| $MH_3$ | 1.0 | | |
| $MH_2$ | | 3.9 | 17.2 |
| $MH$ | | 8.4 | 18.3 |
| $M$ | | 5.8 | 19.2 |

[a] $\pm$ 0.3 eV.
[b] Two ions in the spectrum of Et$_3$P, at $m/e = 76$ and $m/e = 63$ were not assigned formulae by Wada and Kiser[154].

the base peaks of both compounds $(C_2H_5)PH_2^+$ and $(C_2H_5)SbH^+$. Other abundant hydride ions include $(C_2H_5)_2MH^+$ and $SbH^+$ which does not appear to have an equivalent in the spectrum of Et$_3$P.

Table 8.27

Other Reported Aliphatic Derivatives of Phosphorus

| Compound | Comments | Ref. |
|---|---|---|
| $(CH_3)_2PC_3H_7{}^n$ | P.m.i. | 158 |
| $(CH_3)_2PCH_2CH_2F$ | P.m.i. | 158 |
| $(CH_3)_2PCH_2CHF_2$ | P.m.i. | 158 |
| $(CH_3)_2P-C=C\overset{CF_3}{\underset{H}{\diagdown}}$ with F on C | P.m.i. | 158 |
| and *trans*-isomer | | 158 |
| $CH_3-C\overset{O}{\underset{PMe_2}{\diagdown}}$ | P.m.i. | 159 |
| $CH_3P\overset{CH_2}{\underset{CH_2}{\diagup\diagdown}}$ | Rearrangement ion $C_3H_6{}^+$ observed | 160 |
| $H_3CP\overset{C=CH}{\underset{C=CH}{\diagup\diagdown}}$ with H | P.m.i. base peak and ion due to $(p.m. + 1)^+$ peak 14% of base peak. Other abundant ions were due to $C_5H_6P^+$, $C_4H_4P^+$ and $C_2H_2P^+$ | 161 |

## 7.2 Other aliphatic derivatives containing trivalent phosphorus

Compounds which can be listed under this heading are shown in Table 8.27 with relevant features of their spectra.

## 7.3 Alkyl compounds containing bonds between Group V elements

### 7.3.1 Compounds of the type $R_4M_2$

The spectrum of $Me_4P_2$ has been reported by Seel and Rudolph[162]. The base peak was due to the $CH_3P^+$ ion (19.8%) and several ions formed by cleavage of the P–P bond and or rearrangement processes were observed including $CH_3PCH_2{}^+$, $CH_2PCH^+$, and $CH_3PH^+$. Ions containing two phosphorus atoms accounted for about 38% of the total ion abundance of which (17.6%) was due to the $P_2{}^+$ ion. The other ions containing the $P_2$ entity were the parent molecular ion and the $(CH_3)_3P_2{}^+$, $CH_2P_2{}^+$, and

$CHP_2^+$ ions. The corresponding arsenic compound $Me_4As_2$ is reported to show the parent molecular ion and the ion $(CH_3)_3As_2^+$ as the major ions[163].

Although very little ion abundance data were reported for $Et_4P_2$ other than that the base peak was due to the ion $(C_2H_5)_2P_2H^+$, Grishin et al.[63] have measured the appearance potential of the $(C_2H_5)_2P^+$ fragment ion and, using a value of the ionisation potential of the $(C_2H_5)_2P$ radical, calculated the bond dissociation energy $D(Et_2P-PEt_2) = 86$ kcal.mole$^{-1}$. The ionisation potential of the $(C_2H_5)_2P$ radical was found to be 9.56 eV by measuring the appearance potential of the ion from $(C_2H_5)_3P$ and using a value of $\bar{D}(P-C) = 65.5$ kcal.mole$^{-1}$. These workers also reported the bond dissociation energies $D(H_2P-PH_2) = 54.5$ kcal.mole$^{-1}$ and $D(Cl_2P-PCl_2) = 58$ kcal. mole$^{-1}$. They state that the fact that the P–P bond in $Et_4P_2$ is stronger than the P–C bond "is in full accord with photolytic and thermal stability of diphosphines". The apparently high value of the P–P bond dissociation energy in $Et_4P_2$ still seems very surprising compared with that in $P_2H_4$ and $P_2Cl_4$.

The mass spectrum of $Et_4Sb_2$ and appearance potentials of some of the ions observed have been reported by Boglyubov et al.[157]. The base peak was due to the parent molecular ion and a large percentage (81 %) of the ion abundance was due to ions containing two antimony atoms. As with the spectrum of triethylantimony, hydride-containing ions were a prominent feature in the spectrum of $Et_4Sb_2$. Such ions included $(Sb_2H_n)^+$ ($n = 0$–3) and some which were also found from $Et_3Sb$, e.g. $C_2H_5SbH^+$ and $(SbH_n)^+$ ($n = 0$ or 2). The ion $(C_2H_5)_2MH^+$ which was the base peak in the spectrum of $Et_4P_2$ was also observed in $Et_4Sb_2$, being the second most abundant ion.

*7.3.2 Compounds of the type $[RM]_x$*

Cowley and Pinnell have characterised some cyclopolyphosphines $(RP)_x$ ($R = Me$, Et, $Pr^n$, and $Bu^n$; $x = 5$) by observing their parent molecular ions[164]. They also reported the five most abundant peaks for the methyl and ethyl compounds. These were, for the methyl compound, the parent molecular ion, $(CH_3)_4P_5^+$, $(CH_3P)^+_4$, $(CH_3)P_2^+$ base peak, and $(CH_3)_2P^+$ and, for the ethyl compound, the parent molecular ion, $(C_2H_5)_4P_5^+$ base peak $(C_2H_5P)^+_4$, $C_2H_5PH^+$, and $(C_2H_5P)_5P_3^+$. The appearance of ions at higher $m/e$ values than for the parent molecules which occurred for $R = Et$, $Pr^n$ was explained as due to some pyrolytic decomposition reaction in the spectrometer. Evidence for the conversion of $(EtP)_4$ into $(EtP)_5$ in the sample chamber on heating was reported by Ang and West[165]. However, the redistribution reaction (8.56) which is observed[156] for $R = Me$, $R' = Et$ at 130° C under normal pressure conditions did not apparently proceed inside a mass spectrometer source at 150 °C.

$$2\,(RP)_5 + 2\,(R'P)_5 \rightarrow R_4R'P_5 + R_3R'_2P_5 + R_2R'_3P_5 + RR'_4P_5 \quad (8.56)$$

Wells et al.[167] were the first to report a parent molecular ion from the compound $(AsMe)_5$. Other ions observed included $(AsCH_3)_3^+$, $(CH_3)_2As_3^+$, and some $As_2$- and As-containing species. Subsequently this compound has been studied in more detail together with the ethyl and n-propyl analogues by Elmes et al.[168] Although no ion abundances were given, it was stated that all compounds showed parent molecular ions and the base peaks for the methyl and ethyl compounds were due to $As(CH_3)_2^+$ and $As_3^+$, repectively. Other features in the spectra were the production of ions by loss of methyl radicals, $AsR_n$ and $As_2$ species, the ions $As_x(CH_3)_y^+$ ($x = 4$ or $5$, $y = 0-x$) from the methyl compound, and the formation of hydride-containing ions $As_5H_2^+$, $As_3H_n^+$ ($n = 1-3$) from the ethyl compound. Metastable supported fragmentations are shown in reaction schemes (8.57)–(8.62).

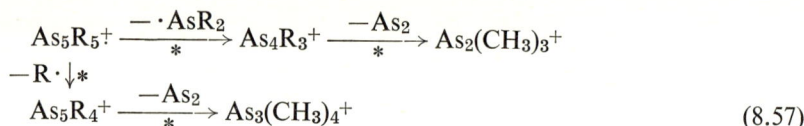

$$\text{(8.57)}$$

where R = Me, Et, Pr''

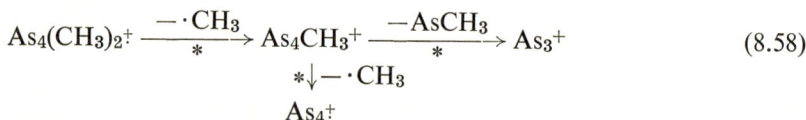

$$\text{(8.58)}$$

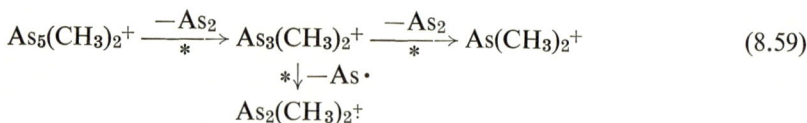

$$\text{(8.59)}$$

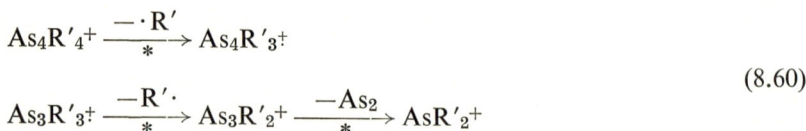

$$\text{(8.60)}$$

where R' = Et or Pr''

and olefin elimination reactions

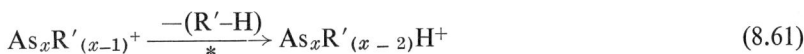

$$As_xR'_{(x-1)}{}^+ \xrightarrow[*]{-(R'-H)} As_xR'_{(x-2)}H^+ \tag{8.61}$$

where $x = 5, 4$; $R' = Pr^n$, $x = 3$; $R' = Et$

$$As_5HC_2H_5{}^+ \xrightarrow[*]{-C_2H_4} As_5H_2{}^+$$

$$As_4C_2H_5{}^+ \xrightarrow[*]{-C_2H_4} As_4H^+$$

$$\text{(8.62}$$

$$As_2(C_2H_5)_3{}^+ \xrightarrow[*]{-C_2H_4} As_2H(C_2H_5)_2{}^+$$

$$As(C_2H_5)_2{}^+ \xrightarrow[*]{-C_2H_4} AsHC_2H_5{}^+$$

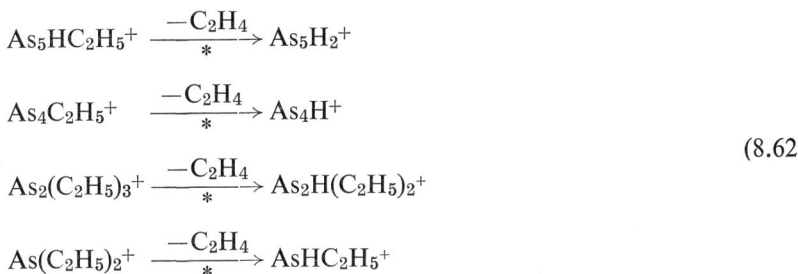

### 7.3.3 Compounds with two different Group V elements

Several workers have studied tris(dimethylamino)phosphine, $(Me_2N)_3P$. The spectrum has been published by Borer and Cohn[169], (Table 8.28). Metastable supported fragmentation reactions (8.63) have been reported by these workers and by Braterman[170] and Kostyanovskii et al.[171]. The parent molecular ion was observed in reasonable abundance; the base peak was due to the $(C_2H_6N)_2P^+$ ion but the ion $C_2H_6NPH^+$ was almost as abundant.

$$(C_2H_6N)_3P^+ \xrightarrow[*]{-C_2H_6N\cdot} (C_2H_6N)_2P^+ \xrightarrow[*]{-CH_2NCH_3} C_2H_6N\overset{+}{P}H \tag{8.63}$$

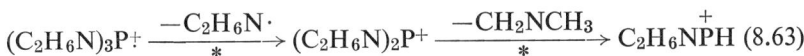

The Russian workers reported metastable supported fragmentations from

Table 8.28

Ion Abundances from $(Me_2N)_3P$

| Composition of ion | % Abundance |
| --- | --- |
| $[(CH_3)_2N]_3P$ | 7.1 |
| $[(CH_3)_2N]_2P$ | 25.9 |
| $(CH_3)_2NP$ | 25.2 |
| $CH_3NP$ | 10.8 |
| $(CH_3)_2N$ | 22.1 |
| $(CH_2)_2N$ | 8.9 |

compounds I, reaction (8.64), II, reaction (8.65), and $EtP(NMe_2)_2$, reaction (8.66).

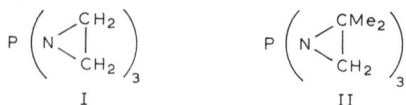

These three compounds all showed loss of a complete group from the parent moleculair ion $\left( \cdot N{<}^{CH_2}_{CH_2} , \cdot\cdot N{<}^{CMe_2}_{CH_2} \text{ or, } \cdot Et \text{ respectively} \right)$ to give the ions $^+P(NR_2)_2$. The base peaks were due to $\overset{+}{P}\left( N{<}^{CH_2}_{CH_2} \right)_2$, $H\overset{+}{P}\left( N{<}^{C(CH_3)_2}_{CH_2} \right)$ and $HP^+NC_2H_6$ respectively.

$$(8.64)$$

$$(8.65)$$

$$EtP(NC_2H_6)_2^{\overset{+}{\cdot}} \xrightarrow[*]{-C_2H_5\cdot} {}^+P(NC_2H_6)_2 \xrightarrow[*]{-CH_2=NCH_3} HP^+NC_2H_6$$

$$EtPNC_2H_6^+ \xrightarrow[*]{-C_2H_4} HP^+NC_2H_6 \qquad\qquad (8.66)$$

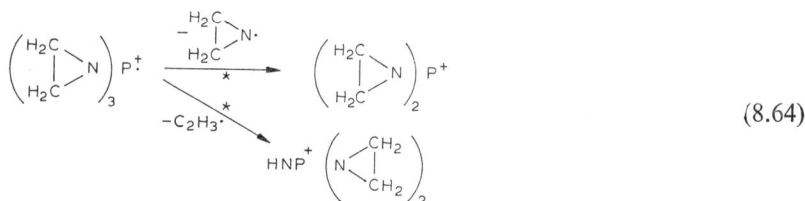

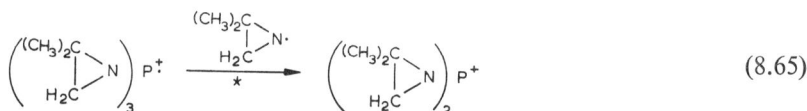

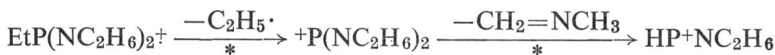

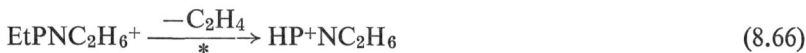

Whigan *et al.*[172] report that compound III, X = N–NMe_2, shows a parent molecular ion at 20 eV; the base peak in the spectrum is due to the $P_3N_4$-$C_2H_6^+$ ion. Other derivatives containing the $>$N–NMe_2 group, [ClPNNMe_2]_3 and $Cl_2PNNMe_2P(Cl)(NNMe_2)PCl_2$ also showed parent molecular ions and base peaks due to the ions $PN_4C_3H_{10}^+$ and $ClPN_2C_2H_6^+$, respectively.

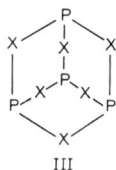

III

## 7.4 Triaryl compounds

The simple compounds $Ph_3M$ (M = N, P, As, Sb, and Bi) have all been reported and studied and those of phosphorus, arsenic, and antimony have been reported in detail. The mass spectra are relatively simple. Parent molecular ions were observed for all the compounds. For those of N (refs. 173, 180) and P (refs. 173–177, 180) the base peak is due to the parent molecular ion. The As (refs. 173, 175, 177–180) and Sb (refs. 173, 177–180) compounds have the $C_6H_5M^+$ ion as base peak, but triphenylbismuth produces the $Bi^+$ ion in largest abundance according to Bublitz and Baker[173] and Zeeh and Thomson[178] but $C_6H_5Bi^+$ according to Bowie and Nussey[180]. The trends in the % abundance of these ions can be explained in terms of the decreasing M–phenyl bond strength with increasing atomic number of M. Table 8.29 shows the ion abundances for the phosphorus, arsenic, and antimony triphenyls found by Rake and Miller[177,179]. Previously reported figures by Miller[174] and by Bublitz and Baker[173] for triphenylphosphine are only in

Table 8.29

Ion Abundances from $(C_6H_5)_3M$ Compounds

| Composition of ion | % Abundance for M = | | |
|---|---|---|---|
| | P | As | Sb |
| $(C_6H_5)_3M$ | 45.6 | 9.7 | 9.0 |
| $(C_6H_5)_2MC_6H_4$ | 4.5 | 0.1 | |
| $C_6H_5M(C_6H_4)_2$ | 0.6 | | |
| $(C_6H_5)_2M$ | 4.2 | 3.9 | 7.0 |
| $C_{12}H_9M$ | 4.9 | | |
| $(C_6H_4)_2M$ | 22.9 | 17.5 | 4.4 |
| $C_6H_5M$ | 13.0 | 61.2 | 72.1 |
| $C_6H_4M$ | 4.3 | 7.6 | 5.8 |
| M | | | 1.7 |

fair agreement with those in Table 8.29 and some ions reported by Rake and Miller were not found in significant abundance in the earlier studies.

General fragmentation schemes (8.67) and (8.68) were suggested by Rake and Miller based on the metastable peaks they observed[177,179].

$$(C_6H_5)_3M^{\ddagger} \xrightarrow[\ast]{-C_{12}H_{10}} C_6H_5M^{\ddagger} \xrightarrow[\ast]{-C_6H_5 \cdot} M$$

$$\downarrow -H \cdot \bigg|^{\ast} \qquad \searrow^{-C_6H_5M}_{\ast} \qquad \text{(Sb only)}$$

$$(C_6H_5)_2MC_6H_4^+ \qquad \searrow (C_6H_5)_2^{\ddagger} \qquad \qquad (8.67)$$

$$(C_6H_5)_2M^+ \xrightarrow[\ast]{-H_2} (C_6H_4)_2M^+ \xrightarrow[\ast]{-M\cdot} \text{[structure]} \quad m/e = 152 \qquad (8.68)$$

The earlier work of Miller and Bublitz and Baker also reported some of the metastable supported reactions in the above scheme for M = P. However, the most detailed report of the fragmentation of Ph₃P appeared in work by Williams et al.[176] who studied a number of deuterated triphenylphosphines, $(C_6D_5)_2P(C_6H_5)$, IV, $(C_6D_5)_3P$, V,

$$\left( \text{[structure]} \right)_3 P \quad \text{and} \quad \left( \text{[structure]} \right)_2 P(C_6H_5)$$
$$\qquad \text{VI} \qquad \qquad \qquad \text{VII}$$

All the fragmentation paths corresponding to the reaction schemes (8.67) and (8.68), except that giving the biphenylene ion $m/e = 152$, $(C_6H_5)_2^{\ddagger}$, were observed as metastable supported processes and four other reactions were reported, viz.

$$(C_6H_5)_3P^{\ddagger} \xrightarrow[\ast]{-C_6H_6} \text{[structure]} \xrightarrow[\ast]{-H\cdot} \text{[structure]} \qquad (8.69)$$
$$\qquad \qquad \qquad \qquad \qquad \qquad m/e = 183$$

$$\text{[structure]} \xrightarrow[\ast]{-C_6H_6} m/e = 183 \qquad (8.70)$$

$$C_6H_5P^{\ddagger} \xrightarrow[\ast]{-H\cdot} C_6H_4P^+ \qquad (8.71)$$

Several interesting features were noted from the spectra of the deuterated compounds. Peaks in the spectrum of IV at $m/e = 113$, $(C_6D_5)P^{\ddagger}$, and $m/e =$

108, $(C_6H_5)P^+$, in the ratio of 2:1 abundance showed that randomisation of hydrogen and deuterium did not occur. These ions decomposed solely by the loss of a D or H atom. Strong evidence for ions containing bridged aromatic groups (e.g. of the types at $m/e = 183$ and $m/e = 152$ for $Ph_3P$) was found in the spectrum of IV where the following reactions were possible for the decomposition of $C_6D_5P^+C_6H_5$.

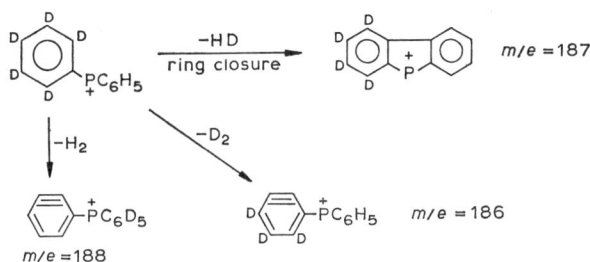

The abundance of the ions at $m/e = 186:187:188$ was $1.0:3.8:1.6$ and therefore showed that ring closure was the preferred fragmentation path. (The difference in the abundances at 188 and 186 was regarded as reflecting the smaller energy required to break a C–D compared with a C–H bond)[176]. Similar work on compounds VIII and $PhAs(C_6D_5)_2$ was reported by Bowie and Nussey[180] for reaction (8.73).

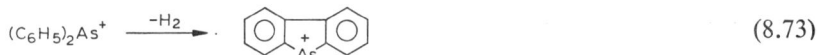

$$(C_6H_5)_2As^+ \xrightarrow{-H_2} \qquad\qquad (8.73)$$

However, these workers interpreted their results for the elimination of $H_2$, HD, and $D_2$ from VIII and $PhAs(C_6D_5)_2$ as showing that randomisation of

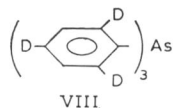

VIII

H and D did take place in reaction (8.73). The agreement between calculated (expected) ratios of abundances for the relevant ions and those found was

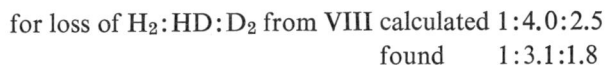

for loss of $H_2:HD:D_2$ from VIII calculated $1:4.0:2.5$
found $\qquad 1:3.1:1.8$

from PhAs$(C_6D_5)_2$              calculated $1:2.05:1$
                                   found      $1:2.05:0.7$

Bowie and Nussey[181] drew similar conclusions regarding reaction of type (8.74) from a study of the negative ions in the spectrum of VIII.

$$\text{(structure)} \xrightarrow{-H_2} \text{(structure)} \tag{8.74}$$

Loss of $H_2$, HD, and $D_2$ from the $(C_6H_2D_3)_2As^-$ ion was found in the ratios $1:3.0:1.9$ comparable with the calculated ratios $1:4:2.5$, indicating randomisation of the hydrogen atoms prior to reaction (8.74).

The negative ion spectra of $Ph_3M$ (M = P, As, Sb, and Bi) showed[180,181] parent molecular ions only for M = Sb and Bi. The base peaks in all cases were due to the $(C_6H_5)_2M^-$-type ion. The reported spectra are given in Table 8.30. Very few triaryl compounds with substituted phenyl groups have

Table 8.30

Negative Ion Mass Spectrum of $(C_6H_5)_3M$ Compounds

| Composition of ion | % Abundance for M = | | | |
| --- | --- | --- | --- | --- |
|  | P | As | Sb | Bi |
| $(C_6H_5)_3M^-$ |  |  | 7 | 1 |
| $(C_6H_5)_2M^-$ | 60 | 80 | 86 | 76 |
| $(C_6H_4)_2M^-$ | 40 | 20 | 7 | 1 |
| $(C_6H_5)M^-$ |  |  |  | 19 |
| $C_6H_3M^-$ |  |  |  | 3 |

been studied. The fluoro compound $(4\text{-}FC_6H_4)_3P$ has been reported by Hawthorne et al.[182] and by De Ketelaere et al.[183]. These reports appear to conflict substantially in that the base peak is $(FC_6H_4)P^+$ according to Hawthorne et al. but $(FC_6H_4)_2P^+$ according to De Ketelaere et al. The latter workers also studied $(3\text{-}FC_6H_4)_3P$ and reported the same ion as base peak as in the 4-fluoro derivative. A metastable supported decomposition (8.75) was observed by Hawthorne et al.

$$(FC_6H_4)_3P^{\ddagger} \xrightarrow[\ast]{-C_6H_4F\cdot} (FC_6H_4)_2P^+ \tag{8.75}$$

The compound IX is reported to show a parent molecular ion[185].

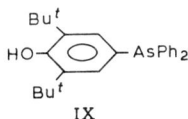

IX

Rather unusual arsenic and antimony compounds containing the metals bonded to diphenyl groups (*e.g.* X) have been reported to show parent molecular ions. From the arsenic compound there is a base peak at $m/e = 531$ and the metastable supported fragmentations are shown in scheme (8.76)[184]. The antimony compound shows the same types of ions except the ion corresponding to the base peak in the arsenic compound's spectrum[186]. The base peak is now due to $(C_6H_4)_4Sb^+$ and the other fragment ions are apparently derived from it.

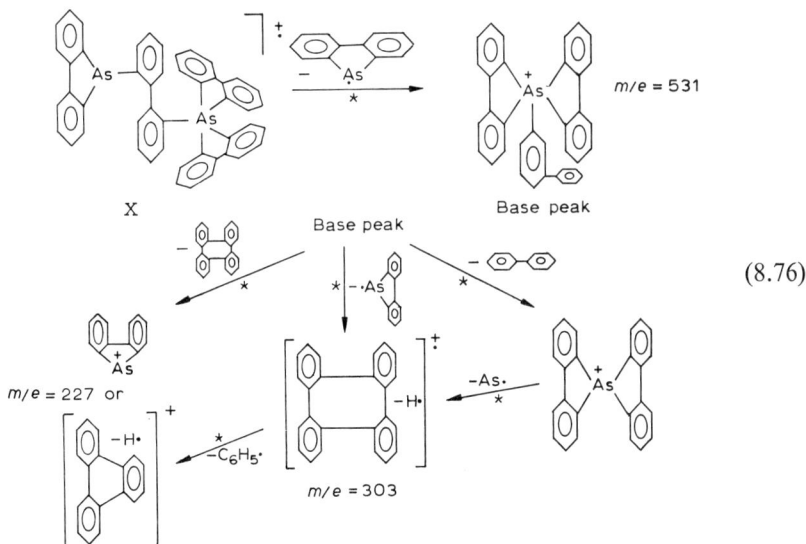

$$(8.76)$$

Features in the mass spectra of tri-cyclopentadienyl derivatives of As and Sb have been discussed by Müller[187]. The abundances of parent molecular ions were very low (1 % or less for M = Sb) and the base peak ions $(C_5H_5)_2M^+$

(47.1% for M = As and 45.5% for M = Sb) did not appear to lose $H_2$ as readily as the equivalent $(C_6H_5)_2M^+$ ions from the triphenyl derivatives. Presumably fragmentation of $(C_5H_5)_2M^+$ ions by a ring closure mechanism is unfavourable and the further loss of a $C_5H_5$ group is preferred. This decomposition was metastable supported in the case of the antimony compound (8.81). Ions formed by the disruption of the $C_5H_5$ ring were quite numerous and included types such as $C_nH_nM^+$ ($n$ = 4, 3, 2 for M = As; $n$ = 3 for M = Sb) and $C_nH_{(n-1)}M$ ($n$ = 5, 3, 2 for M = As; $n$ = 3, 2 for M = Sb). Müller pointed out that the ionisation potentials of As, Sb, and several hydrocarbon species such as the cyclopentadienyl radical were quite similar 9.81 eV, 8.64 eV, and 8.69 eV respectively. Thus in the arsenic compounds particularly, there was a reasonable probability that some fragmentations would proceed with the formation of a positively charged hydrocarbon ion and the elimination of a metal-containing species. He observed the reactions

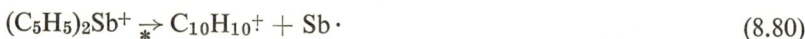

$$(C_5H_5)_2As^+ \xrightarrow{*} C_{10}H_9^+ + AsH \tag{8.77}$$

$$C_9H_7As^+ \xrightarrow{*} C_9H_7^+ + As\cdot \tag{8.78}$$

$$C_5H_5As^+ \xrightarrow{*} C_5H_5^+ + As\cdot \tag{8.79}$$

$$(C_5H_5)_2Sb^+ \xrightarrow{*} C_{10}H_{10}^+ + Sb\cdot \tag{8.80}$$

The following metastable supported decompositions were observed(* — M = As; ‡ — M = Sb).

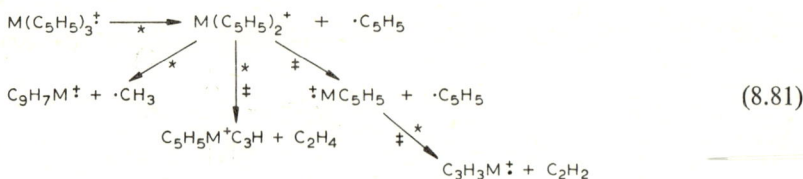

(8.81)

## 7.5 Aryl compounds containing bonds between Group V elements

Mass spectrometry has been used to characterise the cyclopolyphosphines $(PPh)_4$ (ref. 188) and $(PPh)_5$ (ref. 166). The tetramer showed a parent molecular ion and the ion $(C_6H_5)_3P^+$ was the base peak. Line diagrams of the spectrum of the pentamer indicated that major peaks were due to $(C_6H_5)_nP_5^+$

($n = 5$ or 4), $(C_6H_5)_nP_4^+$ ($n = 4$ or 3), $(C_6H_5)_nP_3^+$ ($n = 3$–1), $(C_6H_5)_nP_2^+$ ($n = 4$–1), $(C_6H_5)_nP^+$ ($n = 3$–1), and the cyclised ion $(C_6H_4)_2P^+$. The base peak was the $(C_6H_5)_3P^{+}$ ion as in the case of the tetramer.

The arsenic compound $(PhAs)_6$ has been studied by Elmes et al.[168]. The parent molecular ion and the ions $(C_6H_5)_nAs_5^+$ ($n = 5$ or 4), $(C_6H_5)_nAs_4^+$ ($n = 4$ or 3) were prominent species. Ions formed by the fragmentation of the phenyl ring were much less abundant than ions formed by fragmentation of alkyl groups in the spectra of $(AsMe)_5$, $(AsEt)_5$, and $(AsPr^n)_5$.

*7.6 Group V compounds containing alkyl and aryl groups*

No simple mixed alkyl–aryl compounds appear to have been studied. However, the compounds $Ph_2M$–$CH_2$–$MPh_2$ (M $=$ N, P, As) and $Ph_2M$–$(CH_2)_2$–$MPh_2$ (M $=$ P, As) have been discussed[175]. Major differences in the behaviour of the N compound and the P and As analogues were found to be due to the phosphorus and arsenic compounds undergoing rearrangements which the nitrogen compound did not. The spectrum of $(Ph_2N)_2CH_2$ was simple and had only 6 major ions, $(C_6H_5)_2NH^{+}$, $C_4H_3^+$, $C_6H_5^+$ and those shown in the scheme (8.82)

$$(C_6H_5)_2N \\ \quad\quad CH_2^{+} \xrightarrow[*]{-(C_6H_5)_2N\cdot} (C_6H_5)_2N^+=CH_2 \xrightarrow[*]{-C_6H_6} C_6H_5\overset{+}{N}\equiv CH \\ (C_6H_5)_2N$$

$$(8.82)$$

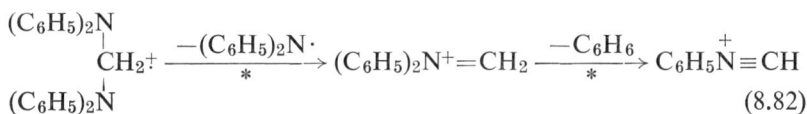

The spectrum of the corresponding phosphorus compound showed much more extensive fragmentation to produce 16 major ions some of which were analogous to those from the N compound (e.g. $(C_6H_5)_2P^+=CH_2$ and $C_6H_5$-$P^+\equiv CH$). However, the base peak was $(C_6H_5)_3P^{+}$ formed by a 1,3 phenyl migration, viz.

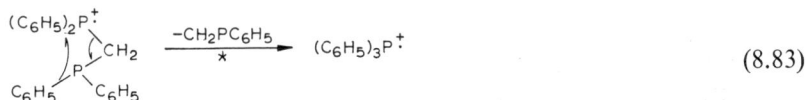

$$(C_6H_5)_2P^{+} \\ \quad\quad CH_2 \xrightarrow[*]{-CH_2PC_6H_5} (C_6H_5)_3P^{+} \\ C_6H_5\quad C_6H_5$$

$$(8.83)$$

The $(C_6H_5)_3P^{+}$ ion produced fragmented as described above in the section on triarylphosphines. The As compound reproduced nearly all the features found for the P compound, but the relative abundances of ions containing As to C multiple bonds were less.

$(C_6H_5)_3M^{\dot{+}}{=}CH_2$          As(1:9)P

$C_6H_5M^+{\equiv}CH$             As(1:4)P

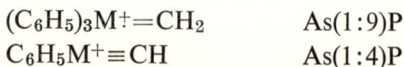

The following metastable supported decompositions were observed for the P compound (no metastable transitions were observed in the spectrum of the As compound).

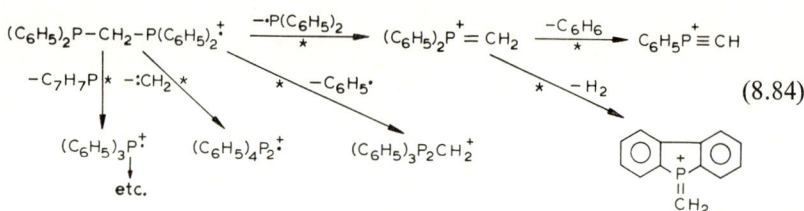

(8.84)

The mass spectrum of $Ph_2P(CH_2)_2PPh_2$ showed a large number of rearrangement ions. The As compound in general resembled it. The base peak of the P compound was $m/e = 183$, probably $(C_6H_4)_2P^+$, and ions due to the production of $(C_6H_5)_3P^{\dot{+}}$ by a 1:4 phenyl shift were prominent. Some metastable supported decompositions were

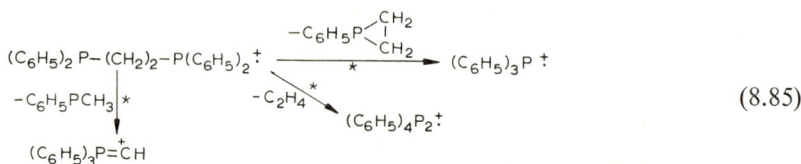

(8.85)

Compounds of the type XI and XII are reported to show parent molecular ions[189].

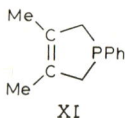

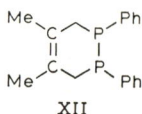

XI                    XII

## 7.7 Ylides and their derivatives

The spectra of the compound $Ph_3P{=}CH_2$ and some deuterated species have been reported[176]. The spectra were essentially similar to those of

$Ph_3P$ (and $Ph_3P{=}O$), with ions which contained bridged structures being important. From the study of a deuterated compound it was shown that elimination of $H\cdot$ (or $D\cdot$) was specifically from the $o$-position to P, the initial ion formed then rearranging to a $CH_2D$-containing species.

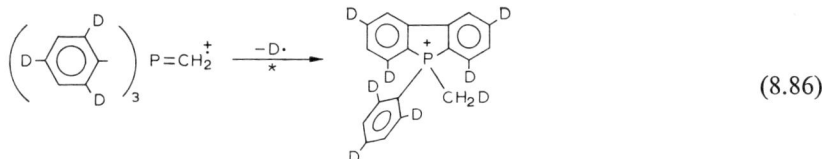

$$\left( D{-}\!\!\bigcirc\!\!{-}{\overset{D}{\underset{D}{\bigcirc}}} \right)_3 \overset{+}{P}{=}CH_2^{\cdot} \quad \xrightarrow[\ast]{-D\cdot} \quad \tag{8.86}$$

A number of alkyl- and arylidenetriphenylphosphoranes have been studied by several workers. The compounds for which fairly detailed data has been reported are listed in Table 8.31.

Ion abundances were not always given but in cases where they were, parent molecular ions were usually quite abundant, being the base peak in some cases. Peaks due to (parent molecule $-1)^+$ were also prominent even in the spectrum of $Ph_3P{=}C(CN)_2$. This led Cooks et al.[193] to conclude that the hydrogen atoms in the aromatic groups were involved, a conclusion which was supported by evidence from the spectra of the deuterated compounds.

Alpin et al.[195] and Cooks et al.[193] found similar behaviour for the $\beta$-ketoalkylidenephosphoranes. Alpin et al. suggested reaction scheme (8.87) for these compounds based on observed metastable peaks.

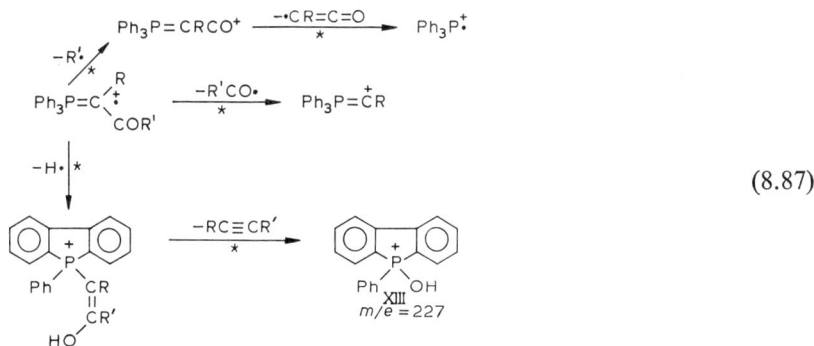

$$\tag{8.87}$$

Gara et al.[194] also reported the metastable supported reactions producing XIII, $m/e = 227$ but wrote a different structure for the ion formed by loss of a H atom, viz.

Table 8.31

Ylide Derivatives Which Have Been Reported in Some Detail

| R | R' | Ref. |
|---|---|---|

(A) $(C_6H_5)_3P=C\diagup^R_{R'}$

| R | R' | Ref. |
|---|---|---|
| H | COMe | 193–195 |
| H | COEt | 193–195 |
| H | $COBu^i$ | 193–195 |
| H | $COBu^t$ | 193–195 |
| H | cyclo-$C_3H_5CO$ | 193–195 |
| H | $COCH_2Cl$ | 193–195 |
| H | COPh | 193–195 |
| D | COPh | 193–195 |
| H | $CO_2Me$ | 193, 195 |
| Me | $CO_2Et$ | 193, 195 |
| $CO_2Et$ | COMe | 193, 195 |
| $CO_2Et$ | $CO_2Et$ | 193, 195 |
| $CO_2Me$ | $CO_2Me$ | 193, 195 |
| $CO_2Me$ | Ph | 193, 195 |
| $CO_2Me$ | $CH_2CO_2Me$ | 195 |
| $CO_2Me$ | $COCH_2COMe$ | 195 |
| CN | CN | 193 |
| CN | $CO_2Me$ | 193 |
| CN | $CO_2Et$ | 193 |

(B) $(C_6D_5)_3P=C\diagup^R_{R'}$

| R | R' | Ref. |
|---|---|---|
| H | $CO_2Me$ | 193 |
| H | Me | 194 |
| H | Ph | 194 |

(C) $\left(D-\bigcirc\diagdown^D_D\right)_3 P=C\diagup^R_{R'}$

| R | R' | Ref. |
|---|---|---|
| H | $CO_2Me$ | 193 |
| $CO_2Et$ | $CO_2Et$ | 193 |

$$Ph_3P=CH \cdot \overset{+}{\underset{\underset{O}{\overset{|}{C}}\diagdown R}{|}} \xrightarrow[\;\;\;\;\ast\;\;\;\;]{-H\cdot} \overset{Ph\diagdown \underset{P}{\diagup} Ph}{\boxed{\phantom{xx}}} \xrightarrow[\;\;\;\;\ast\;\;\;\;]{-RC\equiv CH} m/e = 227 \qquad (8.88)$$

These workers observed another ion containing a P–O bond, $(C_6H_5)_2P$–OH$^+$, in the spectra of the $\beta$-ketoalkylidenephosphoranes were R = H and R' = COMe, COCH$_2$Cl, cyclo-C$_3$H$_5$CO and COPh. They were also able to differentiate between the Bu$^i$ and Bu$^t$ derivatives by observing that reaction (8.89) occurred only with the Bu$^i$ compound.

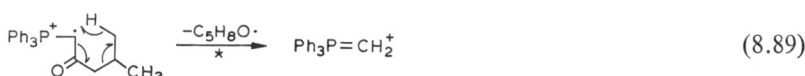

$$Ph_3\overset{+}{P} \xrightarrow[\;\;\;\;\ast\;\;\;\;]{-C_5H_8O\cdot} Ph_3P=CH_2^+ \qquad (8.89)$$

The ester derivatives exhibited similar fragmentation reactions to the keto compounds in that loss of the R radical was very common and often supported by a metastable peak. This was followed by elimination of the R'C=C=O entity to give the $(C_6H_5)_3PO^+$ ion. This second reaction was often metastable supported. Alpin et al.[195] reported the loss of the CH$_3$O radical from the parent molecular ions of the compounds with R = CO$_2$Me, R' = COCH$_2$COCH$_3$ and CH$_2$CO$_2$Me, and the loss of CH$_3$CO$_2$ and CH$_3$COCH$_2$ radicals and CH$_3$CO$_2$H from the latter compound as metastable supported reactions. A number of other interesting decomposition routes were observed by Cooks et al. including those shown in reaction schemes (8.90)–(8.92).

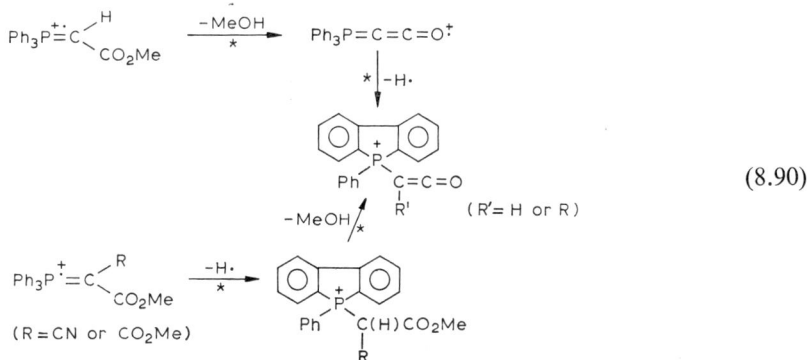

$$Ph_3\overset{+\cdot}{P}=C\diagup{\overset{H}{\diagdown CO_2Me}} \xrightarrow[\;\;\;\;\ast\;\;\;\;]{-MeOH} Ph_3P=C=C=O^+$$

$$\ast\Big|-H\cdot$$

(8.90)

$$Ph_3\overset{+}{P}=C\diagup{\overset{R}{\diagdown CO_2Me}} \xrightarrow[\;\;\;\;\ast\;\;\;\;]{-H\cdot}$$

(R = CN or CO$_2$Me)

$$\underset{\substack{(\text{R = alkyl or alkoxy})}}{\overset{\substack{H_2C-CH_2 \\ H \\ Ph_3\overset{+}{P}-\overset{\cdot\cdot}{C}-\overset{O}{C} \\ | \quad \Vert \\ COR \quad O}}{}} \xrightarrow[-CO_2]{-C_2H_4} \quad \underset{\substack{| \\ COR}}{Ph_3\overset{+\cdot}{P}=CH} \xrightarrow[\ast]{-R\cdot} \underset{\substack{| \\ C\equiv O^+}}{Ph_3P=C(H)} \tag{8.91}$$

$$\underset{\substack{CO_2Et}}{Ph_3\overset{+\cdot}{P}=C{\Big\langle}_{CO_2Et}} \xrightarrow[\substack{-CO_2 \\ \ast}]{-C_2H_4} \xrightarrow[\substack{-CO_2 \\ \ast}]{-C_2H_4} Ph_3\overset{+\cdot}{P}=CH_2$$

$$\ast \Big\vert -H\cdot$$

$$\tag{8.92}$$

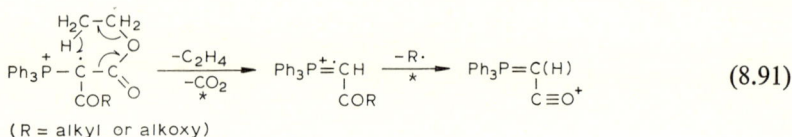

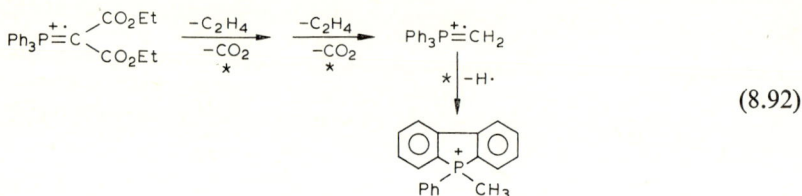

Birum and Matthews[196] have reported some features in the spectrum of $Ph_3P{=}C{=}C{=}O$ including the appearance of a parent molecular ion and ions formed by rearrangements, such as $(C_6H_5)_3PO^{+}_{\cdot}$. Other derivatives of phosphorus ylides which have been studied are given in Table (8.32).

### 7.8 Compounds containing the $(Ar_3P{=}N)$ entity

The spectra of several $N$-phenyliminotriphenylphosphoranes[203] have been reported for

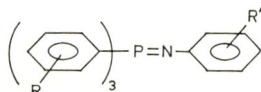

$$\left({\Big\langle}\!\!\!\bigcirc\!\!\!{\Big\rangle}\right)_3\!\!\!\underset{R}{}\!\!\!P{=}N{-}{\Big\langle}\!\!\!\bigcirc\!\!\!{\Big\rangle}\!\!\!\overset{R'}{}$$

R = H,   H,      H,      H,      H,      H,      H,    4-Cl, 3-Me
R'= H, 4-OMe, 4-Me, 4-Br, 4-NO$_2$, 3-NO$_2$, 3-Cl,  H,    H

All compounds showed parent molecular ions, ions due to (parent molecule $-1)^+$, $(RC_6H_4)_3P^{+}_{\cdot}$, and ions of the $(RC_6H_3)_2P^+$ type. The structure of the (parent molecule $-1)^+$ ion was suggested to be XIV.

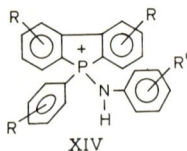

XIV

The only other P–N-containing ion in reasonable abundance was due to (parent molecule $-\cdot RC_6H_4)^+$.

Table 8.32

Other Ylide Derivatives Reported

| Compounds | Comments | Ref. |
|---|---|---|
| 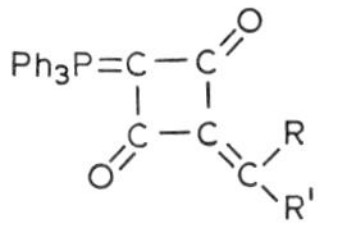 $Ph_3P=C$ (cyclic structure with $C=O$ groups and substituents R, R′) | P.m.i. and expected fragment ions for R = R′ = H; R = R′ = Ph; R = CN, R′ = Ph; R = H, R′ = 4-CNC$_6$H$_4$, 4-NO$_2$C$_6$H$_4$, 4-CHOC$_6$H$_4$; R = R′ = CF$_3$; R = CF$_3$, R′ = 4-ClC$_6$H$_4$ | 196 |
| 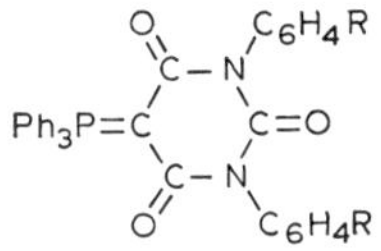 $Ph_3P=C$ (ring structure with C$_6$H$_4$R substituents and C=O) | P.m.i. and expected fragment ions for R = H, Me, F, CN, NO$_2$ | 196 |
| 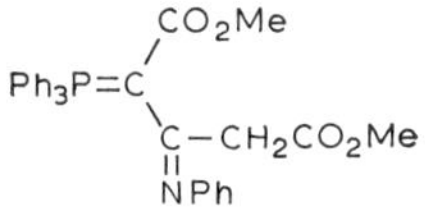 $Ph_3P=C$ with $CO_2Me$ and $C-CH_2CO_2Me$, $NPh$ | P.m.i. | 197 |
| 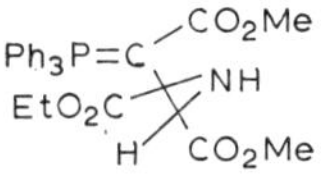 $Ph_3P=C$ with $CO_2Me$, $EtO_2C$, $NH$, $H$, $CO_2Me$ | P.m.i. | 197 |
| 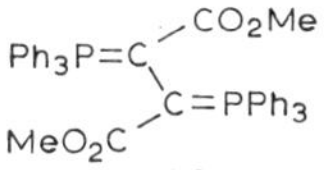 $Ph_3P=C$ with $CO_2Me$, $MeO_2C$, $C=PPh_3$ | P.m.i., $(C_6H_5)_3P^{\underline{+}}$ base peak | 198 |
| 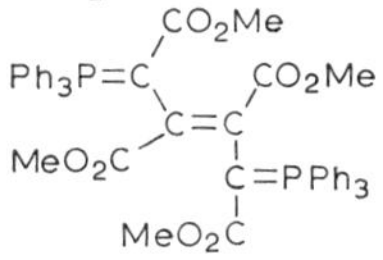 $Ph_3P=C$ with $CO_2Me$, $MeO_2C$, $C=C$, $CO_2Me$, $C=PPh_3$, $MeO_2C$ | P.m.i.; 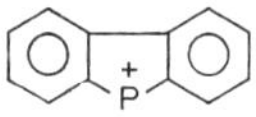 base peak | 199 |
| 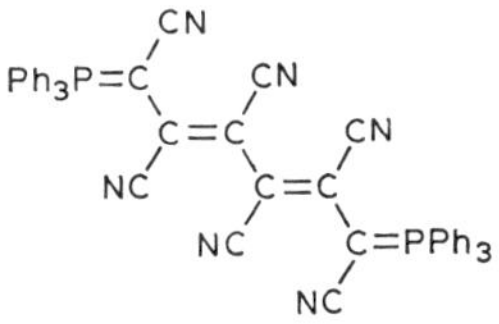 $Ph_3P=C$ with $CN$, $NC$, $C=C$, $CN$, $NC$, $C=C$, $CN$, $C=PPh_3$, $NC$ | P.m.i., high resolution | 200 |
| 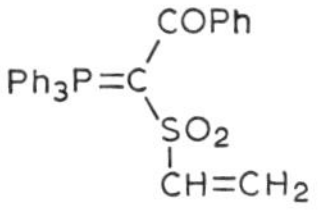 $Ph_3P=C$ with $COPh$, $SO_2$, $CH=CH_2$ | P.m.i. | 201 |
| 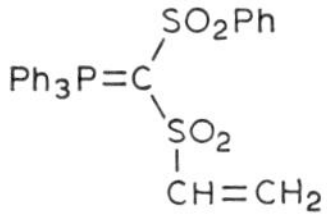 $Ph_3P=C$ with $SO_2Ph$, $SO_2$, $CH=CH_2$ | P.m.i. | 201 |

*References pp. 439–447*

Table 8.32 (continued)

| Compounds | Comments | Ref. |
|---|---|---|
| $Ph_3P=C$ with $SO_2Ph$ and $SO_2Ph$ | P.m.i. | 201 |
|  | X = H, p.m.i. base peak fragmentation of C==PPh₃ bond to give Ph₃P⁺ and reactions analogous to those observed by Cooks *et al.* X = Me, or CN, p.m.i. base peaks | 196 |
|  | P.m.i. in low abundance Ph₃P⁺ base peaks | 202 |
|  | P.m.i. in low abundance Ph₃P⁺ base peaks | 202 |

A study of several phosphazines has been reported by Zeeh and Beutler[204].

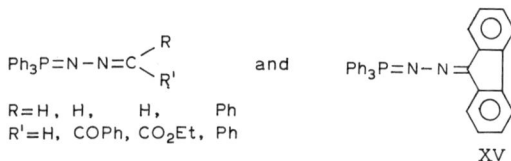

Ph₃P≡N–N≡C with R and R'     and     Ph₃P≡N–N≡

R=H, H, H, Ph
R'=H, COPh, CO₂Et, Ph

XV

Parent molecular ions were observed for the first, fourth and fifth compounds
mentioned above. The base peak from the compound R = R' = H was due

to the $Ph_3P=NH^{\ddagger}$ ion. The ion derived from triphenylphosphine was the base peak for compounds where $R = H$, $R' = COPh$; $R = R' = Ph$; and XV. Where $R = H$, $R' = CO_2Et$ the base peak was reported at $m/e = 78$ ($C_6H_6$?). The elimination of a nitrogen molecule from compounds $R = R' = Ph$ and XV was a metastable supported process giving rise to the corresponding ylide-type ions $Ph_3P=CPh_2^{\ddagger}$ and $Ph_3P=C_{13}H_8^{\ddagger}$.

Baldwin et al.[205] report the parent molecular ion from $Ph_3P=N-P(O)$-MePh.

### 7.9 Heterocyclic compounds containing phosphorus or arsenic

#### 7.9.1 Heterocyclic compounds containing phosphorus
Cyclic phosphinic acids or esters are not included here.

Several interesting cyclic phosphorus compounds have been studied including azaphosphatriptycene[206], XVI, dibenzo (b,d) phosphorin[207], XVII, and the P-chloro derivative, and compound XVIII[208]. All the above produced parent molecular ions.

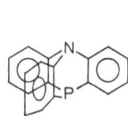

XVI

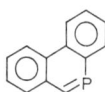

XVII

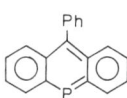

XVIII

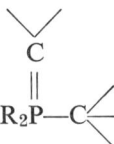

Some data have been published on compounds containing the $R_2P$—C group. Parent molecular ions were observed[209-211] for compounds XIX–XXII).

XIX

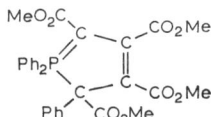

XX

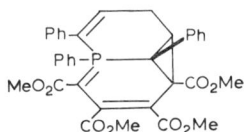

XXI

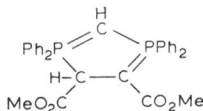

XXII

References pp. 439–447

Compound XIX showed the (parent molecule $-\cdot CO_2Me)^+$ ion and $(C_6H_4)_2P^+$ both as base peaks[209]. Several rearrangement ions were reported including $(C_6H_5)_4P_2^+$.

Granoth et al.[212] have observed parent molecular ions for compounds of type XXIII.

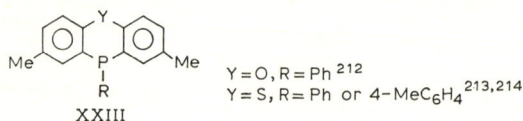

$Y = O, R = Ph$ [212]
$Y = S, R = Ph$ or $4-MeC_6H_4$ [213,214]

XXIII

The metastable supported elimination of an ArPH species from the parent molecular ion (Y = S) of XXIII was reported, viz.

$XXIII^{\dot{+}} \xrightarrow[*]{-Ar\dot{P}H}$

XXIV                                                                    (8.93)

Oxide derivatives of XXV[213] namely the compounds

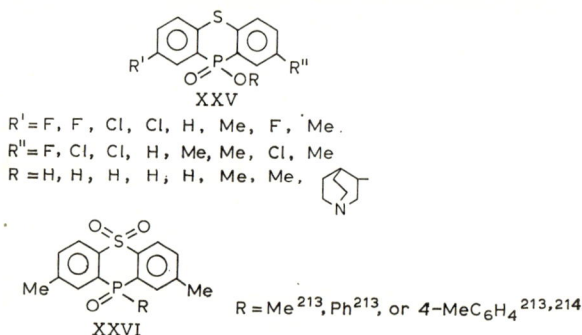

XXV

R' = F, F, Cl, Cl, H, Me, F, Me.
R" = F, Cl, Cl, H, Me, Me, Cl, Me
R = H, H, H, H, H, Me, Me,

XXVI                    $R = Me$ [213], $Ph$ [213], or $4-MeC_6H_4$ [213,214]

are reported to give very abundant parent molecular ions which are the base peaks in some cases. Compounds of the type (R = H) lost an SH radical readily and the elimination of $PO_2H$ was a metastable supported reaction, viz.

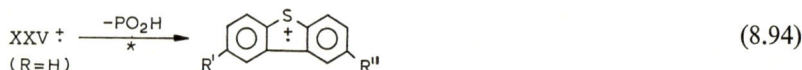

$XXV^{\dot{+}} \xrightarrow[*]{-PO_2H}$
(R=H)

(8.94)

For the corresponding esters (R ≠ H), the loss of $\cdot SH$ was **not** so important;

loss of an alkoxy (OR) radical followed by loss of an S atom or PO entity was now favoured. In the case of **XXVI** (R = Me), the (parent molecule — · Me)⁺ ion was almost as abundant as the parent molecular ion base peak but ions of the type (parent molecule — · OR)⁺ were prominent only for (R = Ph or 4-MeC₆H₄). Ions of the type **XXIV** were found to be present in all spectra. Compounds which can be better classified as cyclic phosphinic acids or esters will be dealt with in Section 8.

A few compounds containing the grouping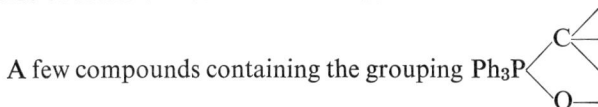

have been reported. A parent molecular ion was found only for **XXVII**[197]. For **XXVIII** the peak at highest *m/e* was due to (parent molecule — · Ph)[215] and for compounds of type **XXIX** the highest *m/e* ion was that due to O=P[(CH₂)₃CH₃]₃⁺ (ref. 216).

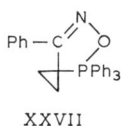

XXVII

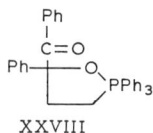

XXVIII

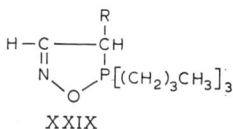

R = Ph or 4-BrC₆H₄.

XXIX

Parent molecular ions have been reported for a few compounds containing an R—P=O linkage:

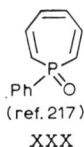

(ref. 217)

XXX

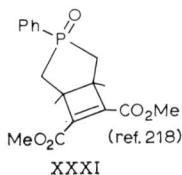

MeO₂C    (ref. 218)

XXXI

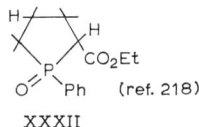

(ref. 218)

XXXII

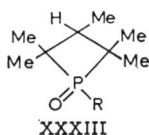

R=Me,Ph or N(H) CH₂Ph    (ref. 218)

XXXIII

XXXIV      XXXV

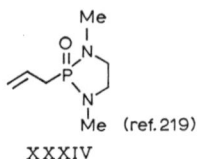

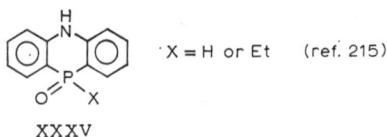

Finally, the compound $(PhPS)_n$ ($n = 3$) was reported to show a parent molecular ion and expected fragment ions[220].

### 7.9.2 Heterocyclic compounds containing arsenic

Jutzi and Deuchert[221] have used mass spectrometry to characterise XXXVI. They observed a parent molecular ion and a metastable peak corresponding to the reaction

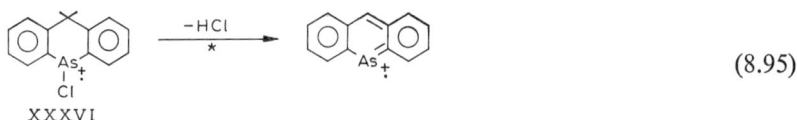

$$(8.95)$$

XXXVI

A compound which was originally assigned the structure XXXVII was shown to be a dimeric species ($n = 2$) in the gas phase by Vermeer and Bickelhaupt[222]. The parent molecular ion at $m/e = 604$ was formulated as XXXVIII. The base peak was $m/e = 227$, an ion which probably has the structure XXXIX.

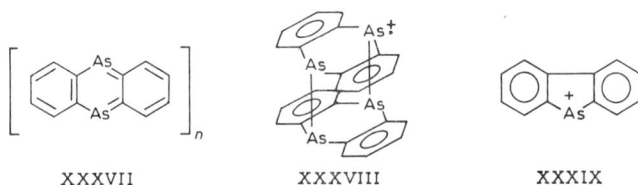

XXXVII      XXXVIII      XXXIX

Apart from the two compounds XXXVII and XXXVIII mentioned above, other heterocyclic arsenic containing compounds which have been studied may be classified as of the following types: derivatives of XL and XLI, derivatives of phenarsazines, derivatives of phenoxarsines, and derivatives of the general type $(RCO)_2AsR$.

Allen et al.[223] have reported identifying compounds XL–XLI mass spectrometrically.

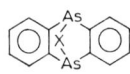 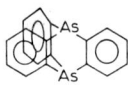 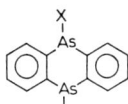

X=O,S,Se,Te          X    X = (a)Cl or (b)Me
XL              XL a            XLI

All the compounds produced parent molecular ions. The ion XXXIX was the base peak from XL and XLIa but XLa and XLIb had ions corresponding to $C_{12}H_8^+$ and the parent molecule less one methyl group respectively as base peaks. Metastable peaks were observed for the formation of the base peak from XLI b and the reactions

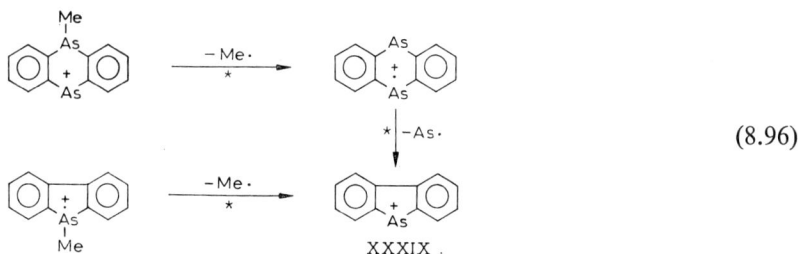

(8.96)

*7.9.2.1 Derivatives of phenarsazines.* The mass spectrum of 5,10-dihydro-5,10-σ-benz-exophenarsazine XLII and the oxide derivative were studied by Earley and Gallagher[224,225]. Parent molecular ions were observed in both spectra; in the case of the oxide it was the base peak. The base peak of XLII was due to the parent molecule less one hydrogen atom and in this respect the spectrum resembled that of the entirely hydrocarbon compound triptycene. Elimination of the arsenic atom from the base peak ion was a metastable supported process, *viz.*

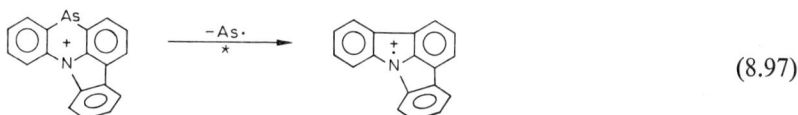

(8.97)

The oxide derivative was observed to lose a CO molecule from the parent molecular ion to give an ion of suggested structure XLIII.

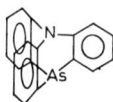         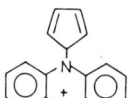

XLII          XLIII

Other compounds studied by Earley and Gallagher were the derivatives XLIV

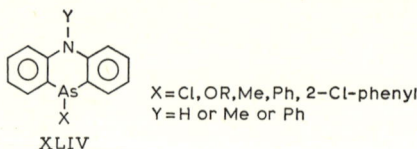

X=Cl,OR,Me,Ph, 2–Cl-phenyl
Y=H or Me or Ph

XLIV

These compounds were reported to all give parent molecular ions and have the metastable supported reactions

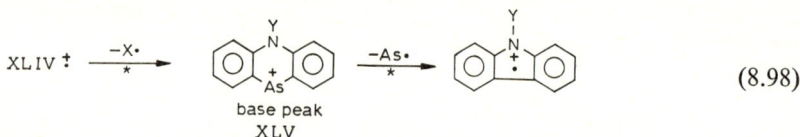

base peak
XLV

(8.98)

However, although they obtained similar results for X = Me, Y = H, Buu-Hoi et al.[226] reported, in the case of X = Cl, Y = H, different % ion abundances from Earley and Gallagher and a base peak corresponding to the ion from the parent molecule less an HCl molecule. Another study reported by Tou and Wang[227] confirmed that the base peak from X = Cl, Y = H was in fact XLV as found by Earley and Gallagher. Tou and Wang also observed the elimination of AsCl from XLIV (X = Cl, Y = H) as a metastable peak supported process. It is interesting to note that Tou and Wang studied XLIV where X = SCN, Y = H and observed metastable peaks for the loss of HSCN from the parent molecular ion as well as for the loss of the SCN radical. Indeed, in this case the base peak was the ion XLVI which decomposed by the loss of ·As, viz.

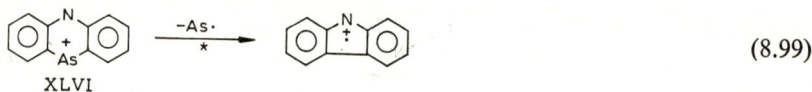

XLVI

(8.99)

Buu-Hoi et al.[226] reported features in the mass spectra of XLVII a, b, and c, and XLVIII.

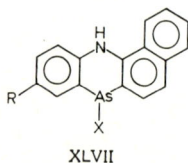

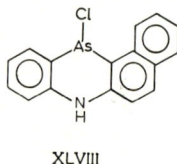

XLVII

XLVIII

Parent molecular ions were observed in all cases and the base peaks corresponded to the ion produced by loss of HX.

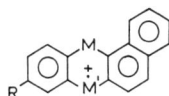

base peak ions
for XLVII a,b and c
$M = N, M' = As$,
for XLVIII, $M = As$,
$M' = N$

The mass spectra of compounds of type XLIX have been studied by Earley and Gallagher (for X = Me, Ph, and 2Cl-phenyl) and by Tou and Wang[227] (for X = OH). Parent molecular ions were observed from all compounds. The base peak was due to the ion formed by loss of the X radical in the cases of X being methyl, phenyl, or *o*-chlorophenyl, but corresponded to XLVI in the case where X = OH.

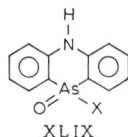

XLIX

Other phenarsazine derivatives to have been studied were XLIV where X = L and LI, Y = H.

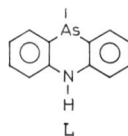

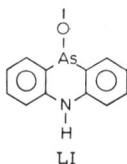

L                  LI

Both compounds showed parent molecular ions. In the case of XLIV (X = L), a metastable peak corresponded to the formation of the base peak ion XLV (Y = H) from the parent molecular ion and other metastable peaks corresponding to loss of As· and loss of As· and H· from the base peak were observed. For XLIV (X = LI), the base peak was XLVI and reaction (8.99) was a metastable supported process.

Earley and Gallagher[225] have briefly reported a study of LII and LIII. Both compounds showed parent molecular ions; in the case of LIII, the most abundant ions were the parent molecular ion and an ion corresponding to the parent molecule less a phenyl group, in about equal intensity.

LII                    LIII

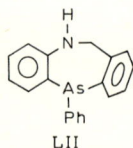

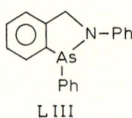

*7.9.2.2 Derivatives of phenoxarsines.* Tou and Wang[228] have studied a number of phenoxarsine derivatives, LIVa–i.

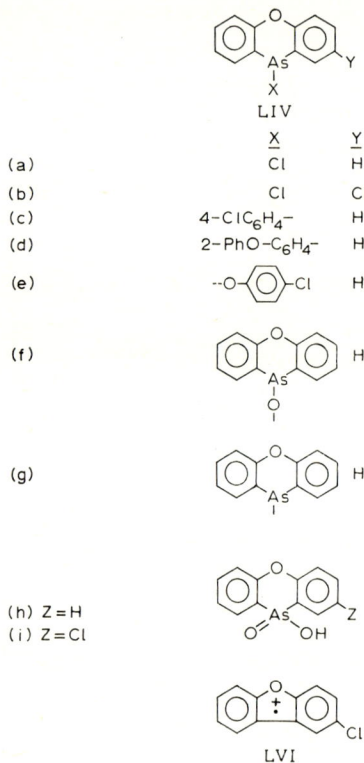

LIV

| | X | Y |
|---|---|---|
| (a) | Cl | H |
| (b) | Cl | Cl |
| (c) | 4-ClC$_6$H$_4$– | H |
| (d) | 2-PhO-C$_6$H$_4$– | H |
| (e) | --O-⟨O⟩-Cl | H |

(f)

  H

(g)

  H

(h) Z = H
(i) Z = Cl

LVI

All compounds showed parent molecular ions. Compound LIVa lost AsCl from the parent molecular ion to give an ion $m/e = 168$ as the base peak ion.

$m/e = 168$

(8.100)

The same reactions were observed in the pyrolysis at 600 °C of the unionised molecules. The ion $m/e = 168$ was also formed by elimination of As from the ion of $m/e = 243$, viz.

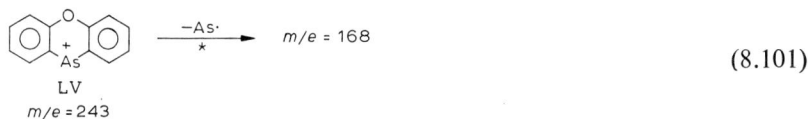

$$\text{(8.101)}$$

Reactions similar to (8.100) and (8.101) were observed for LIVb. Reaction (8.101) was a metastable supported process for ions from LIVf, g, and h also. The ion $m/e = 243$ was the most abundant for compounds LIV c, d, e, f, and h. The production of this ion by the loss of the X radical from the parent molecular ion was a metastable supported process for LIV d, e, and g. Compound LIVc showed the intitial loss of a Cl radical from the p-chlorophenyl group as a metastable supported reaction but not a metastable peak for the loss of the complete X radical. The pentavalent arsenic compounds LIVh and i showed some interesting temperature effects. At a source temperature of 350 °C, both compounds showed behaviour which paralleled that of the other compounds discussed, i.e. parent molecular ions and base peaks at $m/e = 243$ and $m/e = 202$ corresponding to LV and LVI, respectively. However, at 390 °C two molecules of compound LIVh condensed with the elimination of water to give a new compound LVII with a parent molecular ion at $m/e = 534$, reaction (8.102). Metastable peaks were observed corresponding to the loss of $C_6H_4O_2$ and $AsO_2H$ from the parent molecular ion to give two ions at $m/e = 426$. The formulae to these species were established by high-resolution mass measurement and found to be $C_{18}H_{12}As_2O_3^+$ and $C_{24}H_{15}AsO_3^+$, reaction (8.103). Further fragmentation of both ions to produce LV at $m/e = 243$, reaction (8.104), and of $C_{24}H_{15}AsO_3^+$ to give LX, reaction (8.105) were observed.

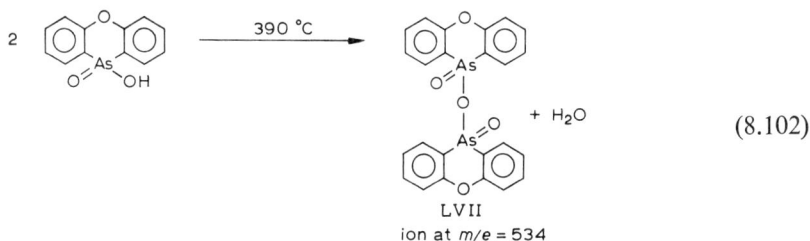

$$\text{(8.102)}$$

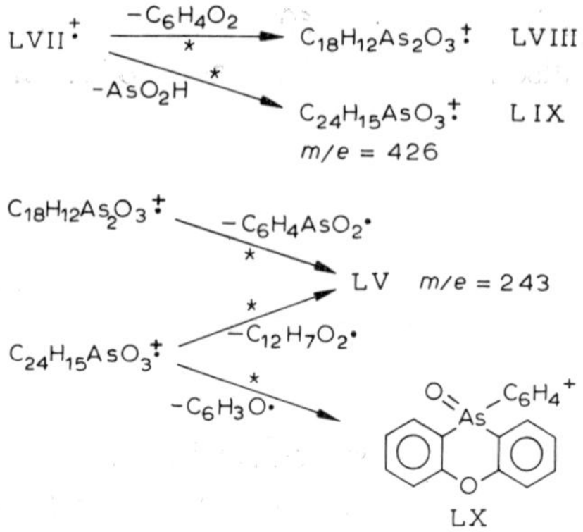

$$\text{LVII}^{\stackrel{+}{\bullet}} \xrightarrow[\substack{-\text{AsO}_2\text{H} \\ *}]{\substack{-\text{C}_6\text{H}_4\text{O}_2 \\ *}} \begin{array}{l} \text{C}_{18}\text{H}_{12}\text{As}_2\text{O}_3{}^{\stackrel{+}{\bullet}} \quad \text{LVIII} \\[8pt] \text{C}_{24}\text{H}_{15}\text{AsO}_3{}^{\stackrel{+}{\bullet}} \quad \text{LIX} \\ m/e = 426 \end{array}$$

(8.103)

$$\text{C}_{18}\text{H}_{12}\text{As}_2\text{O}_3{}^{\stackrel{+}{\bullet}} \xrightarrow[*]{-\text{C}_6\text{H}_4\text{AsO}_2{}^{\bullet}} \text{LV} \quad m/e = 243$$

(8.104)

LX

(8.105)

Compound LVIi behaved similarly; at 450 °C, two molecules of LVIi condensed with the elimination of water to give the corresponding chloro derivative of LVII which showed a parent molecular ion at $m/e = 602$. Metastable peaks supported the formation of chloro derivatives of LVIII and LIX, $\text{C}_{18}\text{H}_{10}\text{As}_2\text{O}_3\text{Cl}_2{}^+$ and $\text{C}_{24}\text{H}_{15}\text{AsO}_3\text{Cl}^+$, respectively, and the chloro derivatives of LV and LVI. The latter ion was the base peak in the spectrum of LVIi at 450 °C.

Mass spectrometry has been used[230] to characterise compounds of type LXI. Parent molecular ions were observed and the base peaks were ions corresponding to LXII.

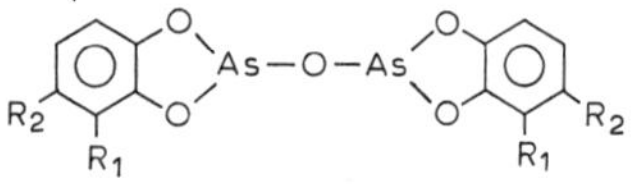

LXI

LXII

R$_1$ = H, Me, H, OH
R$_2$ = H, H, Me, H.

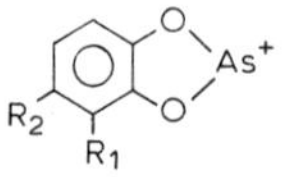

Parent molecular ions have been observed[231] for compounds of the type LXIII

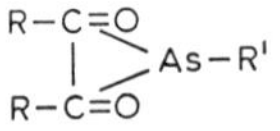

LXIII

R = Me, H, Me, Me, H
R' = OH, OH, Ph, OMe, OMe

*7.10  Compounds of Group V elements bonded to alkyl and/or aryl groups and hydrogen*

The mass spectrum of MePH$_2$ has been reported by Wada and Kiser[154] and by Halmann[53] who also reported the spectrum of Me$_2$PH. Their results for MePH$_2$ are not consistent. Wada and Kiser find a parent ion in 22% abundance and other major ions, CH$_3$P$\overset{+}{\cdot}$ (35%) base peak, CH$_2$P$^+$ (20%), CHP$\overset{+}{\cdot}$ (15%), and CH$_3$PH$^+$ (5.5%). Halmann found only a trace of the parent ion (0.1%) and CH$_3$P$\overset{+}{\cdot}$ (3.1%), CH$_2$P$^+$ (63%) base peak, CHP$\overset{+}{\cdot}$ (1%) ,and CH$_3$PH$^+$ (1%). Because there was reasonable agreement in the spectra of PMe$_3$ reported in the same work, it is difficult to determine which spectrum of MePH$_2$ is the more accurate. Moreover, the divergence in the results for MePH$_2$ calls into question the spectra of EtPH$_2$ and HPMe$_2$ (ref. 51) also reported in this work.

An interesting comparison of the fragmentation behaviour of alkylamines and alkylphosphines (with R group not Me) has been pointed out by Colton and Porter[175]. For example, the ethyl compounds fragment according to

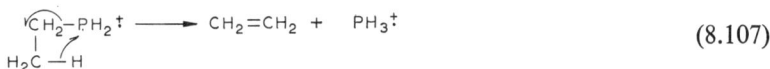

$$CH_3 \overset{\frown}{-} CH_2 - NH_2 \overset{+}{\cdot} \longrightarrow CH_2 = \overset{+}{N}H_2 + \cdot CH_3 \qquad (8.106)$$

$$\overset{\frown}{C}H_2 - PH_2 \overset{+}{\cdot} \longrightarrow CH_2 = CH_2 + PH_3 \overset{+}{\cdot} \qquad (8.107)$$
$$\underset{H_2C \overset{|}{-} H}{}$$

The preferred process with amines is the formation of a C=N double bond and elimination of an (R–CH$_2$) radical. Phosphines prefer to eliminate olefins (this route is very common with other organometallics containing an alkyl group other than Me) and the ion PH$_3$$\overset{+}{\cdot}$ is produced. This may reflect the better ability of nitrogen to double bond to carbon compared with phosphorus and the lower ionisation potential of phosphine compared with ammonia.

Ionisation potentials either calculated by a unified atom method or measured experimentally have been reported for methyl- and ethylphosphines[154] and ethylstibenes[157], and some related radicals. The values obtained (in eV) were for Me$_2$PH 9.7 (found) and 9.02 (calc.); MePH$_2$ 9.72 (found) and 9.53 (calc.); EtPH$_2$ 9.61 (found) and 9.40 (calc.); and from calculations only, Et$_2$PH (8.87), Et$_2$SbH (9.4), EtSbH$_2$ (10.1), Et(H)Sb· (8.5). A value of 8.2 eV has been calculated from appearance potential and thermochemical data for the Et$_2$Sb radical.

The mass spectrum of the phosphiran

$$\underset{CH_2-CH_2}{\overset{PH}{\diagup\!\!\diagdown}}$$

has been reported[190]. The parent molecular ion has often been postulated to have a structure

$$\underset{H_2C}{\overset{H_2C}{\diagdown}}\!\!\Big|\!\!\underset{}{\diagup}PH\!\!\overset{+}{\cdot}$$

and is common in the spectra of alkylphosphines. It was the base peak in the spectrum of $C_2H_4PH$ (Table 8.33). Unfortunately, a second report of the

Table 8.33

Ion Abundances from Phosphiran

| Composition of ion | % Abundance |
|---|---|
| $C_2H_4PH$ | 35.0 |
| $C_2H_3P$ | 21.5 |
| $C_2H_2P$ | 19.5 |
| $C_2HP$ | 5.5 |
| $CH_2P$ | 10.5 |
| $CHP$ | 3.5 |
| $PH_3$ | 4.5 |

spectrum of phosphiran, although listing the same four most abundant peaks as in Table 8.33, differs from the first in that ions $PH_3^+$ and $CHP^+$ are not included and the $CH_3P^+$ ion is[160]. The later work briefly discusses features in the spectra of other phosphirans, LXIX.

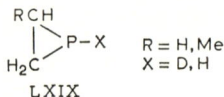

$$\underset{H_2C}{\overset{RCH}{\diagdown}}\!\!\Big|\!\!\overset{}{\diagup}P-X \qquad \begin{array}{l} R = H, Me \\ X = D, H \end{array}$$

LXIX

From the spectra of these compounds it was deduced that the P–H bond remains intact in $C_2PH_n^+$-type ions. Peaks corresponding to the ions $C_2H_5PH_2^+$, $C_2H_5PH_3^+$ from 2-methylphosphiran were suggestive of extensive rearrangements. Tavs[191] has characterised the cyclic compound LXX

LXX

by its mass spectrum.
Hays and Logan[192] studied compounds of the type LXXI by mass spectrometry.

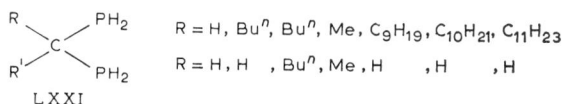

$R = H$, $Bu^n$, $Bu^n$, $Me$, $C_9H_{19}$, $C_{10}H_{21}$, $C_{11}H_{23}$

$R = H,H$ , $Bu^n$, $Me$, $H$ , $H$ , $H$

LXXI

The spectra of $PhPH_2$, $Ph_2PH$, and $Ph_2PD$ exhibit base peaks due to the $(C_6H_5)P^+$ ion[178]. The fragmentation of the parent molecular ions of $Ph_2PH$ or $Ph_2PD$ by loss of $m/e = 3$ ($H_2 + H\cdot$) and $m/e = 4$ ($H_2 + D\cdot$) respectively was reported as giving the ion $(C_6H_4)_2P^+$.

### 7.11 Alkyl and aryl halides

#### 7.11.1 Compounds of trivalent Group V elements, $R_2MX$ and $RMX_2$

Comparatively few compounds of these types have been studied in detail and data for aryl compounds are particularly lacking. The spectra of $Me_2PX$ ($X = F$ or $Cl$) have been reported by Seel and Rudolph[162,232]. Parent molecular ions were very abundant, being $33.0\%$ for the fluoride and $36\%$ (the base peak) for the chloride. The $CH_3PF^+$ ion was the base peak ($46\%$) from the fluoride. Ions containing the P–X bond were much more abundant for $X = F$ than for $X = Cl$ and ions of the types $^+PHX$ and $PX^+$ were reported only for $X = F$.

The spectra of several dimethylaminophosphorus halides have been published by Borer and Cohn[169] including those given in Table 8.34. Parent molecular ions were observed decreasing in abundance going from the fluoride to the bromide. The base peak for bis(dimethylamino)phosphorus chloride was due to the $[(CH_3)_2N]PCl^+$ ion; for the bromide it was $[(CH_3)_2N]_2P^+$ and for the fluoride $(CH_2)_2N^+$ derived from the dimethylamino

Table 8.34

Ion Abundances from Some $(Me_2N)_nPX_{(3-n)}$ Compounds

| Composition of ion | % Abundance | | |
|---|---|---|---|
| | $(Me_2N)_2PCl$ | $(Me_2N)_2PBr$ | $(Me_2N)PF_2$ |
| $[(CH_3)_2N]_2PX$ | 10.9 | 1.8 | |
| $[(CH_3)_2N]PX_2$ | | | 26.0 |
| $(CH_3)_2NPX$ | 29.5 | 7.9 | 6.0 |
| $[(CH_3)_2N]_2P$ | 10.1 | 29.5 | |
| $(CH_3)_2NPH$ | 13.1 | 24.7 | |
| $(CH_3)_2NP$ | 10.5 | 13.0 | |
| $(CH_3)_2N$ | 11.7 | 6.2 | 2.0 |
| $(CH_2)_2N$ | 13.4 | 12.3 | 48.0 |
| $PX_2$ | | | 14.0 |
| $PX$ | | 2.9 | 3.1 |
| $P$ | 0.8 | 1.7 | 0.9 |

entity. The spectra of the compounds $(Me_2N)PX_2$ ($X = Cl$ or $Br$) were also reported by Borer and Cohn. Because ions were found which were not due to the parent molecules but apparently were due to a disproportion or pyrolysis reaction in the spectrometer these spectra will not be discussed here. Some metastable supported reactions were reported including (8.108). Other compounds for which data have been reported are listed in Table (8.35).

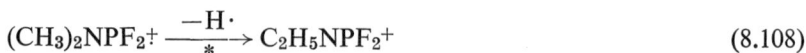

$$(CH_3)_2NPF_2^+ \xrightarrow[\quad *\quad]{-H\cdot} C_2H_5NPF_2^+ \qquad (8.108)$$

As far as aryl derivatives are concerned, relatively few data have been published. The fluoro derivatives $(4\text{-}FC_6H_4)_2PCl$ and $(3\text{-}FC_6H_4)_2PCl$ show parent molecular ions and $(FC_6H_4)PCl^+$ as base peak[183]. Loss of PCl from the parent molecular ions was a metastable supported process in both spectra but only the $(3\text{-}FC_6H_4)_2PCl$ compound exhibited a (parent molecule $-F_2)^+$ ion. The dichlorides were also studied; they gave parent molecular ions and $(FC_6H_4)PCl^+$ ions as base peaks.

*7.11.2  Compounds of pentavalent Group V elements $R_nMX_{(5-n)}$ (n = 1-4)*

Group V element compounds of the type $RMX_4$ which have been studied

Table 8.35

Other Compounds of the Type $R_nMX_{(3-n)}$ Reported

| Compound | Comments | Ref. |
|---|---|---|
| $(CH_3)_2NF$ | P.m.i., $(CH_3)_2N^+$ base peak | 233 |
| $(Bu^tNH)_2PF$ | P.m.i. | 234 |
| $(CH_3)_2NPFBr$ | P.m.i., $(CH_3)_2NPF^+$ base peak | 235 |
| $(C_2H_5)_2NPF_2$ | P.m.i. | 241 |
| $(Bu^nNH)PF_2$ | P.m.i. | 234 |
| $MeCO_2PF_2$ | P.m.i. | 236 |
| $NC(O)PF_2$ | P.m.i. | 236 |
| $CH_3PBr_2$ | P.m.i., base peak | 237 |
| $CH_3P(CN)_2$ | P.m.i., $HCN^+$ base peak | 237 |
| $CH_3P(Br)CN$ | P.m.i., observed in a mixture of $MePBr_2$ and $MeP(CN)_2$ | 237 |
| $(C_6H_{11})_2PCl$ | P.m.i. | 238 |
| $C_6H_{10}\big\langle{\overset{PCl_2}{PCl_2}}$ | P.m.i. | 238 |
| benzene ring with $PCl_2$ Cl | P.m.i. | 238 |
| $Cl_3CNF_2$ | No p.m.i. but $F_2NCl_2^+$ and $FNCCl_3^+$; $CCl_3^+$ base peak | 239 |
| $ClSCCl_2NF_2$ | P.m.i., $CS^+$ base peak | 239 |
| $Cl_3CPCl_2$ | P.m.i., $CCl_3^+$ base peak and $PCl_3^{\dagger}$ rearrangement ions | 240 |
| $Cl_2C=C(Cl)-PCl_2$ | P.m.i., $CCl_3^+$ base peak and other rearrangement ion $PCl_3^{\dagger}$ | 240 |
| $Cl_2C(PCl_2)_2$ | P.m.i., $PCl_2^+$ base peak and $PCl_3^{\dagger}$ and $P_2^{\dagger}$ rearrangement ions | 240 |
| $Ph_2PCl$ | $PhP^{\dagger}$ observed | 179 |
| $PhPCl_2$ | $PhP^{\dagger}$ observed | 179 |

are nearly all derivatives of $XPF_4$. The spectra of $RPF_4$ (R = Me, Et, Ph) and $(R_2N)PF_4$ derivatives have been reported in detail by Schmutzler and coworkers[80]. Parent molecular ions were not observed for $MePF_4$ or $EtPF_4$ but were found in reasonable abundance in the other cases. The presence of an amino group apparently stabilised the parent molecular ion considerably. The base peak for all the compounds studied was due to the $PF_4^+$ ion. Fragmentation by loss of a fluorine atom from the parent molecule appeared to be favourable. The loss of ethylene from compounds containing ethyl groups was unimportant; the only ion observed which possibly contained a P–H bond was the $(C_2H_6)NPF_3^+$ ion from $(Et_2N)PF_4$.

Other compounds of the $RMX_4$ type reported are $Me_2NP(CN)F_3$ (ref. 235) and LXXII (ref. 241). Both are reported to show parent molecular ions.

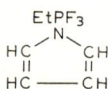

LXXII

Several compounds of the type $R_2PF_3$ have been studied. Schmutzler and coworkers[80] report spectra for $R_2PF_3$ (R = Me, Ph), $(R_2N)_2PF_3$ (R = Me, Et), $EtPF_3(NR_2)$ (R = Me, Et), $PhPF_3(NEt_2)$, and compounds LXXIII and LXXIV.

EtPF₃                                          PhPF₃

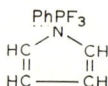

LXXIII                                         LXXIV

With the exception of $Me_2PF_3$, all the compounds examined gave parent molecular ions in moderate abundance. The base peaks were always due to an ion of the type $RPF_3^+$. In compounds with two different R groups, the base peak was always the ion with R = alkyl or aryl and not a P–N-containing ion. The spectrum of the $Me_2PF_3$ has also been reported by Seel and Rudolph[162]. It is not in very good agreement with that of Schmutzler and coworkers. The % abundances of $MePF_3^+$, $Me_2PF_2^+$, and $PF_2^+$ are 51%, 16%, and 15% respectively according to Seel and Rudolph, but 57%, 24% and 11% respectively according to Schmutzler and coworkers. Other $R_2PF_3$ compounds to have been characterised by their mass spectra are $(CH_3NPF_3)_2$ (ref. 242) and the cyclic compounds LXXV and LXXVI.

⬡PF₃        ⬠PF₃

LXXV         LXXVI

Some interesting compounds which may be classed as derivatives of $R_2PX_3$ are LXXVII

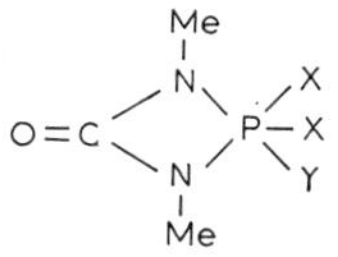

$$X = F, Cl, \quad \left(\text{and also } F, \ F, \ F, \ Me\atop Me, \ Et, \ Ph, \ F\right)^{244}$$
$$Y = F, Cl,$$

LXXVII

Two metastable supported reactions from the parent molecular ion producing rearrangement ions were observed, viz.

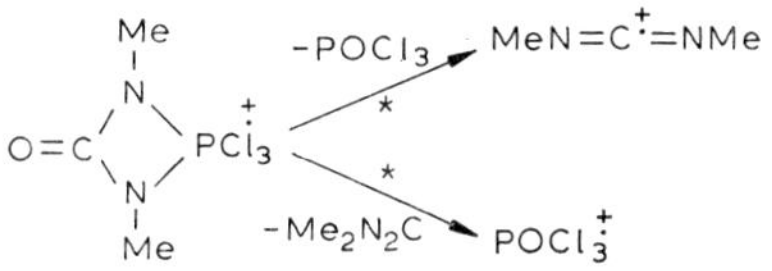

$$MeN=C^{+\cdot}=NMe \qquad (8.109)$$

$$POCl_3^+ \qquad (8.110)$$

The corresponding reactions were observed for X = F, Y = Ph and X = Me, Y = F. When X = F and Y = Me or Et, only reaction (8.110) was observed.

Difluorophosphoranes have been studied by Schmutzler and coworkers[80]. They included $R_3PF_2$, (R = Me, $Bu^n$, Ph); $R_2R'PF_2$ (R = Me, R = Ph; R' = Me), $R_2PF_2(NR_2)$, (R = Me or Ph, R' = Me) and $Ph_2PF_2$–$CH_2$–$F_2PPh_2$. The parent molecular ions were either in very low abundance or not detected. Base peaks corresponding to ions of the type $R_2PF_2^+$ were observed, namely from $Me_3PF_2$ and $Me_2PF_2(NMe_2)$, $(CH_3)_2PF_2^+$ and from $Bu^n_3$-$PF_2$, $(C_4H_9)_2PF_2^+$. Schmutzler and coworkers could not identify the base peak ions in the spectra of $Me_2(Ph)PF_2$, $Ph_2P(F_2)$–$CH_2$–$(F_2)PPh_2$, and $Ph_3PF_2$, but from $Ph_2(Me)PF_2$ and $Ph_2PF_2(NMe_2)$ they found $Ph_2PFCH_2^+$ and $(CH_3)_2N^+$.

A few compounds of the formulae $[R_4P]^+X^-$ have been reported. Parent molecular ions were observed[245] for $[(PhNH)(R_2N)(H_2N)_2P]^+Cl^-$ for R = Et, $Bu^n$ but not for compounds of the type $[R–C–CH_2–P(Bu^n)_3]^+X^-$

$$\overset{\|}{\underset{NOH}{}}$$

(R = Ph, 4-$BrC_6H_4$; X = Cl, Br)[216]

or

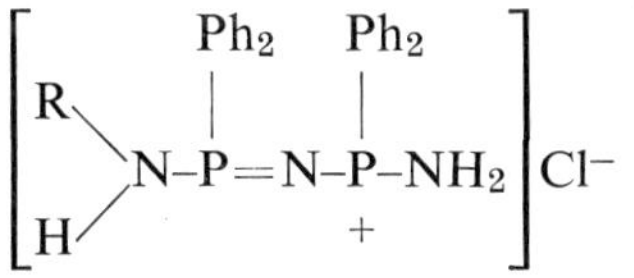

$Cl^-$ (R = Me or $C_6H_5CH_2$[217].)

The highest $m/e$ ions were the (parent molecule $-HX)^{\ddagger}$ and (parent molecule $-\cdot Cl)^{+}$ respectively.

### 7.12 Compounds containing both M–H and M–X bonds

Very few compounds of this type have been reported. Seel and Rudolph[162] published a spectrum of $Me_2PHF_2$. The parent molecular ion was observed in very low abundance but the (parent molecule $-\cdot H)^{+}$ ion was in reasonable abundance (10%) and the $(CH_3)_2PFH^{+}$ ion was found in about the same abundance. The base peak was due to $CH_3(H)PF_2^{+}$ (41.7%). Parent molecular ions have been reported[234] from $(MeNH)_2PF_2H$ and $(EtNH)_2PF_2H$.

### 7.13 Compounds containing P=O or P=S, or P–OR or P–SR groups

A very large number of studies of compounds falling into this category have been reported. It is intended to classify each compound in terms of one of a few generalised formulae. For instance, compounds of the general formulae $RR'P(L)L'R''$ (where $L, L' = O$ or $S$) will include $Ph(Me)P(O)OMe$, $Ph_2P(S)OH$, $Ph(H)P(O)OMe$, $Me(F)P(O)SH$, etc., i.e. R or R' may be organic entities, hydrogen, or halogen, and $L'R''$ may be an alkoxy or thioalkoxy, or an OH or SH group. The discussion of compounds of a particular formula will be in the order organic derivatives before hydrides, before halides. Compounds containing two or more different types of groups will be discussed after the groups themselves have been discussed; for example, $Ph(H)P(O)OMe$ will be discussed after the sections on organic and hydride derivatives of $RR'P(O)(OR'')$.

#### 7.13.1 Compounds of the formulae $R_3M=L$ ( R = organic derivative, H or halide; L = O or S)

Williams et al.[176] have shown that the mass spectra of $Ph_3P=O$ and $Ph_3P=S$ are very similar. Strong parent ions and (parent molecule $-1)^{+}$ ions were detected. They predicted that the (parent molecule $-1)^{+}$ ion would be important for all $R_3P=X$ compounds and this was borne out in the spectra of such compounds as $Ph_3P=CH_2$ and $Ph_3P=C=O$. From their studies on the triphenyl compounds and the deuterated compounds

they were able to suggest fragmentation pathways some of which were metastable supported, *e.g.* reaction scheme (8.111) from $Ph_3P=O$.

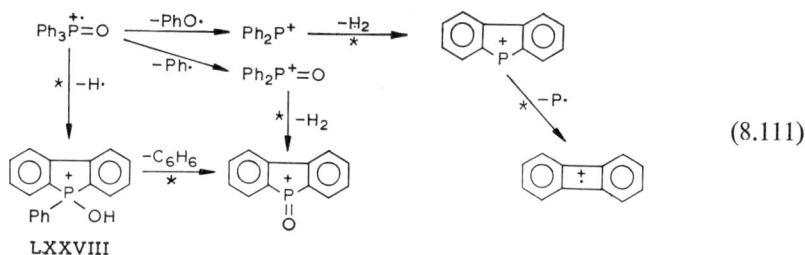

(8.111)

LXXVIII

There was strong evidence that the base peak **LXXVIII** contained a closed ring in that,

(*i*) the ion was extremely stable, hence its abundance and

(*ii*) in the spectrum of  the corresponding peak was

formed overwhelmingly (93%) by loss of a D atom to give **LXXIX** which then lost $C_6D_4H_2$, *viz.*

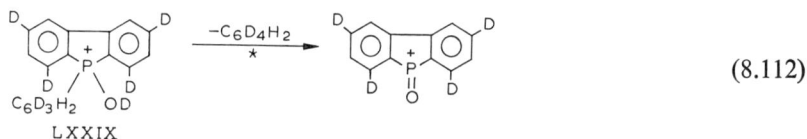

(8.112)

LXXIX

Evidence of scrambling of ·H and ·D in the reaction

(8.113)

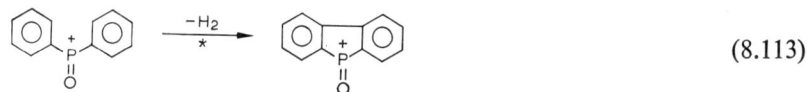

was found in the spectrum of the compound

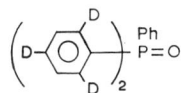

which lost $H_2$, HD, and $D_2$ apparently randomly.

Other compounds in this category are included in Table 8.36.

Table 8.36

Other Compounds of the Type $R(R')(R'')P=L$ Reported

| Compound | Comments | Ref. |
|---|---|---|
| $Ph_2(C_6H_5CH_2)P=O$ | P.m.i. | 247 |
| $Ph_2(4-MeC_6H_4)P=O$ | P.m.i. | 247 |
| $R-C{\overset{P(O)Ph_2}{\underset{P(O)Ph_2}{-OH}}}$ | P.m.i. for $R = H$ or $PhCH_2CH_2$ | 248 |
| pyridine ring–$CH_2P(O)Ph_2$ | P.m.i. | 248 |
| cyclopentene ring with Ph, Ph, Ph, Ph, H, C=O, $P(O)Ph_2$ | P.m.i. | 249 |
| cyclopentene ring with Ph, R, R', Ph, C=O, $P(O)Ph_2$, R'' | P.m.i. for<br>$R = H, Ph, H$<br>$R' = Ph, H, Ph$<br>$R'' = Ph, Ph, Me$ | 249 |
| $H_2N-\underset{S}{\overset{\parallel}{C}}-P(L)Ph_2$ | P.m.i. and expected fragment ions for $L = O$ or $S$ | 250 |
| $\left(Ph_2P(L)-\underset{NH}{\overset{\parallel}{C}}\right)_2 S$ | P.m.i. and expected fragment ions for $L = O$ or $S$ | 250 |
| $R_2P(O)-\underset{NH}{\overset{\parallel}{C}}-S-\underset{NH}{\overset{\parallel}{C}}-(S)PR_2$ | P.m.i. and expected fragment ions for $R = Ph$ or $4-MeC_6H_4$ | 250 |
| $HO-\text{(}Bu^t, Ph, Ph\text{ benzene ring)}-As=L$ | P.m.i. and expected fragment ions, $L = O$ or $S$ | 185 |
| $C\left[CH_2-\underset{S^-}{\overset{Ph}{\underset{\mid}{P}}}=C=S,Na^+\right]_4 2\,Na$ | No p.m.i. | 251 |

## 7.13.2  Compounds of the formulae $R_2XP=L$ and $RX_2P=L$

Compounds of the general formulae $R_2XP=L$ and $RX_2P=L$ for which parent molecular ions have been reported in their spectra are given in Table 8.37. Nearly all reports discuss neither % abundances nor fragmentation reactions.

## 7.13.3  Compounds of the formula $(RL)_3P$

The mass spectra of $(RO)_3P$ compounds ($R = Me$, $Et$, $Pr^n$, $Pr^i$, $Bu^n$, and

Table 8.37

Other Compounds of the Types $R_2XP=L$ and $RX_2P=L$ Reported

| Compound | Comments | Ref. |
|---|---|---|
| $R_2XP=L$ | | |
| $Bu^t_2P(O)X$ | P.m.i. for $X$ = Cl or H | 252, 253 |
| $Ph_2P(S)X$ | P.m.i. for $X$ = Cl, base peak; p.m.i. for $X$ = NCS, $Ph_2P=S^+$ base peak | 254 |
| $Ph_2P(O)NHR$ | P.m.i. for $R$ = Ph, (parent molecule $-1)^+$ base peak; p.m.i. for $R$ = $(CH_2)_3CH_3$, $Ph_2P=O^+$ base peak | 254 |
| $Ph_2P(S)NHR$ | P.m.i. for $R$ = H, $PhPNH_2^+$ base peak; p.m.i. for $R$ = $(CH_2)_3CH_3$, $C_3H_7CHNH_2^+$ base peak; p.m.i. for $R$ = Ph, $Ph_2P=S^+$ base peak | 254 |
| $Ph_2P(S)NH$ $\mid$ $C=S$ $\mid$ $NHR$ | P.m.i. for $R$ = H, $Ph_2P=S^+$ base peak; p.m.i. for $R$ = $(CH_2)_2CH_3$, $Ph_2P=S^+$ base peak; p.m.i. for $R$ = Ph, $PhNH_2\dagger$ base peak | 254 |
| $(Me_2N)_2P(O)F$ | P.m.i. | 255 |
| $Ph(Me_2N)P(O)F$ | P.m.i. | 241 |
| $Ph(Et_2N)P(S)F$ | P.m.i. | 241 |
| | P.m.i. | 241 |
| $RX_2P=L$ | | |
| | P.m.i. | 241 |
| $R(H_2N)P(O)F$ | P.m.i. for $R$ = Me, $NCHMe^+$ base peak; $R$ = Et, $H_3CNP(O)FNH_2^+$ base peak; $R$ = Ph and $C_6H_{11}$ | 151 |
| $(Me_2N)P(O)F_2$ | P.m.i. | 255 |
| $NC(O)P(O)F_2$ | | 257 |
| $(RNH)(H_2N)P(O)F$ | P.m.i. for $R$ = Me, Et; $(H_2N)P(O)F^+$ base peaks | 151 |
| $(EtNH)(H_2N)P(S)F$ | P.m.i. and expected fragment ions | 256 |

Ph) have briefly been discussed[258]. Numerous ions were observed formed by simple P–OR bond cleavages, loss of R or Me radicals, elimination of olefins and hydrogen rearrangement reactions. Fewer rearrangement ions were found in the spectra of $(MeO)_3P$ and $(PhO)_3P$ than in the others. Parent molecular ions were observed in all cases and base peaks from the alkyl compounds were as follows, given in the same order as above, $(CH_3O)_2P^+$, $(HO)_2P^+$, $(HO)_3PH^+$, $(HO)_3P^+$, and $(HO)_3PH^+$. Triphenyl phosphite had the ion $(C_6H_5O)_2P^+$ as the base peak for P-containing fragments. Braterman[170] has suggested some decomposition paths in the fragmentation of $(MeO)_3P$, viz.

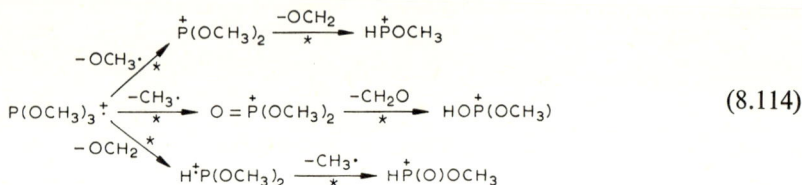

$$
\begin{array}{c}
\overset{+}{P}(OCH_3)_2 \xrightarrow[\ *\ ]{-OCH_2} H\overset{+}{P}OCH_3 \\[4pt]
-OCH_3\cdot\Big/_{*} \\
P(OCH_3)_3^{+} \xrightarrow[\ *\ ]{-CH_3\cdot} O = \overset{+}{P}(OCH_3)_2 \xrightarrow[\ *\ ]{-CH_2O} HO\overset{+}{P}(OCH_3) \\
-OCH_2\Big\backslash^{*} \\
H\overset{+}{P}(OCH_3)_2 \xrightarrow[\ *\ ]{-CH_3\cdot} H\overset{+}{P}(O)OCH_3
\end{array}
\tag{8.114}
$$

Several interesting derivatives of trialkyl phosphites have been studied. Hendricker[259] reported that parent molecular ions were observed for compounds of the type LXXX

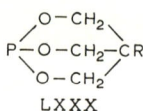

$$
P\overset{O-CH_2}{\underset{O-CH_2}{\overbrace{\phantom{-}}\!\!-O-CH_2-CR}}
$$

LXXX

for R = Me, Et, Pr and $NO_2$.
He studied the fragmentation of these compounds and suggested the generalised reaction scheme 8.115. For some reactions he observed metastable peaks.

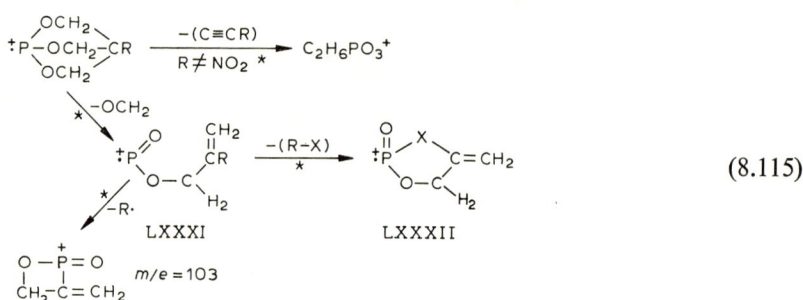

$$
\begin{array}{c}
\overset{+}{P}\overset{OCH_2}{\underset{OCH_2}{\overbrace{\phantom{-}}\!\!-OCH_2-CR}} \xrightarrow[R \ne NO_2\ *]{-(C\equiv CR)} C_2H_6PO_3^{+} \\[6pt]
\Big\backslash_{*}\,-OCH_2 \\
\overset{+}{P}\overset{O}{\underset{O-C}{\diagup}}\overset{CH_2}{\underset{H_2}{\diagdown\!CR}} \xrightarrow[\ *\ ]{-(R-X)} \overset{+}{P}\overset{O}{\underset{O-C}{\diagup}}\overset{X}{\underset{H_2}{\diagdown\!C=CH_2}} \\
\Big\backslash_{*}\,-R\cdot \\
\quad\quad LXXXI \quad\quad\quad\quad\quad LXXXII
\end{array}
\tag{8.115}
$$

$$
\underset{CH_2-C=CH_2}{\overset{\overset{+}{O}-P=O}{|\quad\quad|}}\quad m/e = 103
$$

where, in the formation of LXXXII, R = $C_2H_5$, X = $C_2H_4$ and R = $C_3H_7$, X = $CH_2$ or $C_2H_4$. Other metastable supported reactions observed were

$$\overset{+}{P}\begin{matrix} \diagup O \\ \diagdown O-CH_2 \end{matrix}\overset{CH_2}{\underset{}{\overset{\parallel}{CR}}} \xrightarrow[*]{-HPO_2} C_3H_3R^{+} \quad (R=Me, Pr) \tag{8.116}$$

$$m/e = 103 \xrightarrow[*]{-HPO} C_3H_3O^{+} \quad (R=NO_2, Et) \tag{8.117}$$

$$\overset{+}{P}\begin{matrix} \diagup O \\ \diagdown O-C-CCH_3 \\ \quad H_2 \end{matrix}\overset{CH_2}{\underset{}{\overset{\parallel}{}}} \begin{matrix} \xrightarrow[*]{-PO\cdot} C_4H_7O^{+} \\ \xrightarrow[]{-H_2O} PC_4H_5O^{+} \end{matrix} \quad (R=Me) \tag{8.118}$$

$$\overset{+}{P}\begin{matrix} \diagup OCH_2 \\ \diagdown OCH_2 \\ \diagdown OCH_2 \end{matrix}CNO_2 \xrightarrow[*]{-NO\cdot} PC_4H_6O_4^{+} \tag{8.119}$$

$$\overset{+}{P}\begin{matrix} \diagup O \\ \diagdown O-C \end{matrix}\overset{O}{\underset{H_2}{\overset{}{C=CH_2}}} \xrightarrow[*]{-CO} {}^{+}PC_2H_4O_2 \xrightarrow[*]{-C_2H_2} {}^{+}P(OH)_2 \quad (R=NO_2) \tag{8.120}$$

$$P\begin{matrix} \diagup OCH_2 \\ \diagdown OCH_2 \\ \diagdown OCH_2 \end{matrix}C^{+} \begin{matrix} \xrightarrow[*]{-CO} PC_3H_6O_2^{+} \\ \xrightarrow[]{-CH_2O} PC_3H_4O_2^{+} \end{matrix} \quad (R=NO_2) \tag{8.121}$$

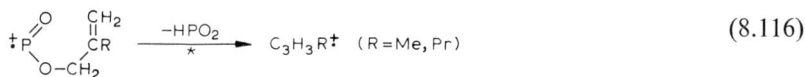

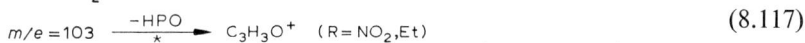

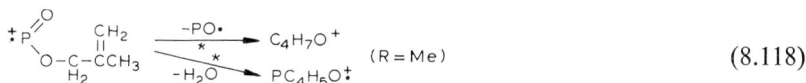

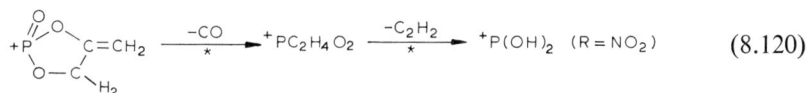

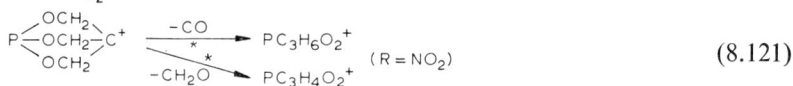

Reaction (8.119) is a particularly interesting fragmentation of the parent molecular ion; loss of $\cdot NO$ is typical of nitro compounds and in this case competes with the loss of $CH_2O$, reaction (8.115). Base peaks were due to the ions LXXXI for R = Me and Et and LXXXII for R = $NO_2$ (X = O) and Pr(X = $C_2H_4$).

Mass spectrometry has been used to identify, by the appearance of the parent molecular ions, other cyclic phosphites LXXXIII[260] and 1,6 diphosphahexaosadiamantane, LXXXIV[261].

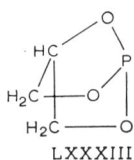

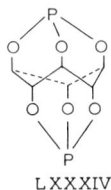

LXXXIII          LXXXIV

### 7.13.4 Compounds of the formulae (RO)₂MX and (RO)MX₂

Compounds of the types $(RO)_2MX$ and $(RO)MX_2$ have scarcely been studied. Parent molecular ions were reported for

$$(RO)_2PF \; (R=Me \text{ or } Et)^{241} \; , \; (MeO)(Et_2N)PF^{241} \; , \; (RO)P\left(N\overset{\diagup CH_2}{\underset{\diagdown CH_2}{|}}\right)_2$$

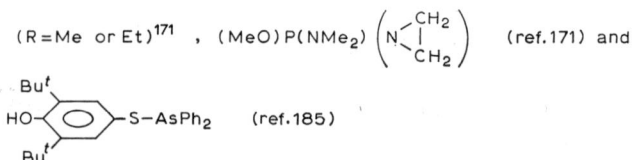

$(R=Me \text{ or } Et)^{171}$ , $(MeO)P(NMe_2)\left(N\begin{smallmatrix}CH_2\\|\\CH_2\end{smallmatrix}\right)$  (ref.171) and

HO–(Bu$^t$)(Bu$^t$)C$_6$H$_2$–S–AsPh$_2$   (ref.185)

## 7.13.5 Compounds of the formulae (PO)₃P=L and (RO)₂(R'L')P=L

*7.13.5 Compounds of the formulae (PO)$_3$P=L and (RO)$_2$(R'L')P=L*

McLafferty[262] was the first to discuss some points relating to the spectra of $(RO)_3P=O$ compounds. He observed that $(EtO)_3PO$ gave a parent molecular ion and fragmentation occurred by simple P–OR bond cleavage and elimination of ethylene and/or $:CH_2$ to give the ions $(HO)_2P(OEt)_2^+$ (base peak), $(HO)_3POEt^+$, $(HO)_3P^+$, $(HO)_2PO^+$, and $(MeO)_2PO^+$. He was able to show[263] that the decomposition of the parent molecular ion from trimethyl phosphate to give the $HP(OH)_3^+$ ion was by the reaction

$$(C_2H_5O)_3P=O^+ \rightarrow HP(OH)_3^+ + 2\,C_2H_4 + \cdot C_2H_3 \qquad (8.122)$$

Other alkoxy compounds ($R = Pr^n$ or $Bu^n$) were reported to fragment similarly to $(EtO)_3PO$ and to fragment by alkyl chain fission to give ions of the $CH_2OP^+(O)OH$ type.

Cooks and Gerrard have studied the compounds $(PhO)_2(RO)PS$, ($R = Me$ or Et), and $(PhO)_2(MeS)PO$ in some detail[264]. Parent molecular ions were observed as base peaks in the spectra of these compounds. Fragmentation by the loss of $\cdot R$ and $PhO\cdot$ groups was important and loss of MeOH from $R = Me$, and $C_2H_4$ from $R = Et$ was observed. Several interesting decomposition reactions were suggested and in some cases deuterated species were used to confirm them. The loss of an SH radical from the parent molecular ion was found for $R = Me$ or Et compounds but not $(PhO)_2(MeS)PO$. The H atom involved came from the methyl group when $R = Me$; this was confirmed in the spectrum of the $R = CD_3$ compound. In the case of $R = Et$, however, loss of $\cdot SH$ competed with decomposition *via* a McLafferty-type rearrangement, *viz.*

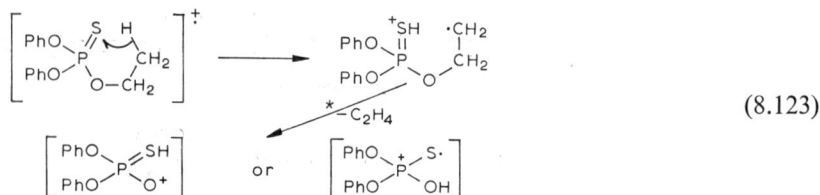

$$\left[\begin{smallmatrix}PhO\\PhO\end{smallmatrix}P\begin{smallmatrix}S & H\\ \\ O-CH_2\end{smallmatrix}CH_2\right]^{\ddagger} \longrightarrow \begin{smallmatrix}PhO\\PhO\end{smallmatrix}P\begin{smallmatrix}^+SH & \cdot CH_2\\ \\ O\end{smallmatrix}CH_2$$

$$\swarrow ^*{-C_2H_4}$$

$$\left[\begin{smallmatrix}PhO\\PhO\end{smallmatrix}P\begin{smallmatrix}SH\\ \\ O^+\end{smallmatrix}\right] \quad \text{or} \quad \left[\begin{smallmatrix}PhO\\PhO\end{smallmatrix}P\begin{smallmatrix}S\cdot\\ \\ OH\end{smallmatrix}\right]^+ \qquad (8.123)$$

and the latter was apparently more favourable in this case. Rearrangement of organic groups from OR to SR apparently occurred when R = Me or Et, (8.124).

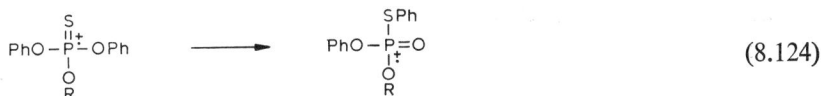

$$\begin{array}{c} \text{S} \\ \| + \\ \text{PhO} - \text{P} - \text{OPh} \\ | \\ \text{O} \\ | \\ \text{R} \end{array} \quad \longrightarrow \quad \begin{array}{c} \text{SPh} \\ | \\ \text{PhO} - \text{P} = \text{O} \\ | + \\ \text{O} \\ | \\ \text{R} \end{array} \qquad (8.124)$$

Ions due to both (parent molecule — ·PhS) and (parent molecule — ·MeS) were observed from R = Me.

This type of behaviour, (8.124), was also observed by Jörg et al.[265] in a study of some organo-phosphorus pesticides and plant growth regulators. These workers reported relatively few spectral features in detail but were more interested in identifying the compounds from their fragmentation patterns. Similar reports on similar compounds have been published by

Table 8.38

Other Compounds of the Type (RO)(R′O)(R″L′)P=L Reported

| Compound | | | Comments | Ref. |
|---|---|---|---|---|
| (RO)(R′O)(R″O)P=O | | | | |
| R = Me | R′ = Me | R″ = CH=CCl$_2$ | P.m.i. | 265 |
| Me | Me | CH(OH)CCl$_3$ | P.m.i. | 265 |
| Me | Me | CH$_2$CO$_2$C=CH$_2$ <br> $\quad\quad$ \| <br> $\quad\quad$ C$_3$H$_7$ | P.m.i. | 265 |
| Me | Me | $\begin{array}{c}\text{N} \quad \text{Pr}^i \\ \text{N}\end{array}$ | P.m.i. | 265 |
| (Et) | (Et) | | P.m.i. | (266) |
| Me | Me | $\begin{array}{c}\quad\quad\quad \text{O} \\ \quad\quad\quad \| \\ -\text{C}=\text{C(Cl)}-\text{C}-\text{NEt}_2 \\ | \\ \text{Me}\end{array}$ | P.m.i. | 266 |
| Me | Me | $\begin{array}{c}\quad\quad\quad \text{O} \\ \quad\quad\quad \| \\ -\text{C}=\text{C(H)}-\text{C}-\text{OMe} \\ | \\ \text{Me}\end{array}$ | P.m.i. | 266 |

Table 8.38 (continued)

| Compounds | Comments | Ref. |
|---|---|---|

Et          Et          $C_6H_4NO_2$                                                    226, 267

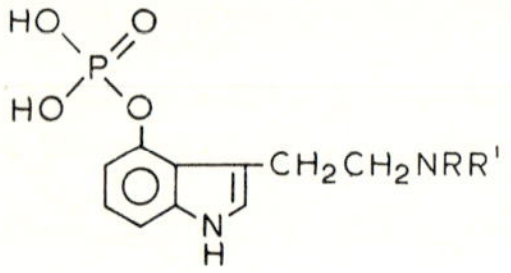

[structure: $(HO)_2P(=O)O$– attached to indole ring with $CH_2CH_2NRR'$ side chain, NH]

$R = R' = Me$ (psilycybin) p.m.i.?        268, 269

$R = R' = H$ (norbaecystin) p.m.i.        268

$R = Me$, $R' = H$ (baecystin); p.m.i.        268

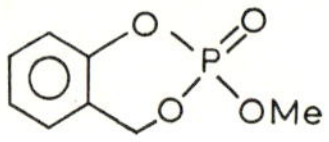

[structure: benzo-fused cyclic phosphate $O$–$P(=O)$–$OMe$ with $O$–$CH_2$]

P.m.i.        270

$(RO)(R'O)(R''O)P{=}S$

| R = Me | R' = Me | R'' = 3-Me, 4-$NO_2C_6H_3$ | P.m.i. | 265–267 |
| Me | Me | 3-Cl, 4-$NO_2C_6H_3$ | P.m.i. | 267 |
| Me | Me | 4-$NO_2C_6H_4$ | P.m.i. | 265–267 |
| Me | Me | 3-Me, 4-MeS$C_6H_3$ | P.m.i. | 265 |
| Me | Me | 2,4,5,$Cl_3C_6H_2$ | P.m.i. | 266 |
| Et | Et | 4-$NO_2C_6H_4$ | P.m.i. | 266, 267 |

Et          Et          [pyrimidine ring with $Pr^i$, Me substituents]        P.m.i.        266, 267, 272

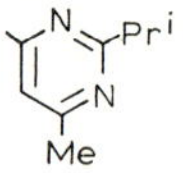

Et          Et          [chromen-2-one ring with Cl, Me]        P.m.i.        266

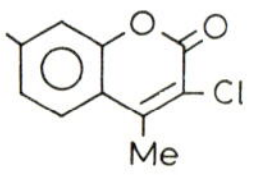

Et          Et          $(S)P(OEt)_2$        P.m.i.        266

Et          Et          [pyrimidine ring with $C(Me)_2$, OH, Me]        P.m.i.        272

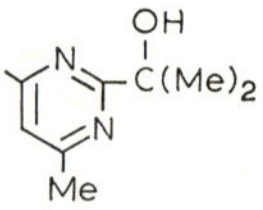

[structure: benzo-fused cyclic thiophosphate $S{=}P$–$OMe$ with $O$–$CH_2$]        P.m.i.        270

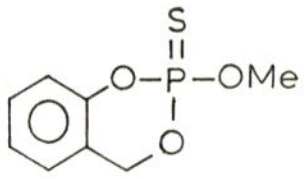

$(RO)(R'O)(R''S)P{=}O$

| R = Me | R' = Me | R'' = $(CH_2)_2SEt$ | P.m.i. | 265 |
| Me | Me | $CH_2$–(O)C–N(H)(Me) | P.m.i. | 266, 267 |

Table 8.38 (continued)

| Compounds | | | Comments | Ref. |
|---|---|---|---|---|
| Me | Me | –CH–(O)C–OEt<br>\|<br>CH₂–(O)C–OEt | P.m.i. | |
| Me | Me | | P.m.i. | 266 |
| | | | P.m.i. | 270 |
| (RO)(R′O)(R″S)P=S | | | | |
| R = Me  R′ = Me | | R″ = –CH –C–OEt<br>(O)<br>\|<br>CH₂–C–OEt<br>(O) | P.m.i. | 265–267 |
| Me | Me | | P.m.i. | 265–267 |
| Me | Me | (CH₂)₂SEt | P.m.i. | 265 |
| (Et) | (Et) | | P.m.i. | (265–267) |
| Et | Et | CH₂SEt | P.m.i. | 265 |
| Et | Et | –CH₂–C–N⟨H/Me (O) | P.m.i. | 265 |
| | | | P.m.i.?<br>p.m.i. by field<br>ionisation | 265, 266 |
| Me | Me | –CH₂S(4-ClC₆H₄) | P.m.i. | 266 |
| (Et) | (Et) | | P.m.i. | 266 |
| Me | Me | | P.m.i. | 266 |
| Et | Et | –CH₂–S–P(S)(OEt)₂ | P.m.i. | 266, 267 |

Damico *et al.*[266,267] who used both electron-impact and field ionisation sources. The compounds studied are given in Table 8.38, together with other compounds of the general formulae which have been characterised by mass spectrometry.

The mass spectrum of tetraethyl pyrophosphate has been reported to show the parent molecular ion[271]. Several metastable supported fragmentation reactions of this ion were observed.

$$\begin{bmatrix} (EtO)_2P \begin{smallmatrix} \nearrow O \\ \searrow O \end{smallmatrix} \\ (EtO)_2P \begin{smallmatrix} \nearrow \\ \searrow O \end{smallmatrix} \end{bmatrix}^{\dot{+}} \xrightarrow[\;*\;]{-C_2H_3\cdot} (EtO)_3(OH)_2P_2\overset{+}{O}_2 \tag{8.125}$$

$$(EtO)_3(HO)_2P_2O_2{}^+ \xrightarrow[\;*\;]{-3 \times C_2H_4} (HO)_3\overset{+}{P}-O-\overset{\overset{O}{\|}}{P}-(OH)_2 \tag{8.126}$$

$$\xrightarrow[\;*\;]{-H_2O} \begin{smallmatrix} O \\ \diagdown \\ HO \end{smallmatrix}\overset{+}{P}-O-\underset{\underset{O}{\|}}{P}(OH)_2$$

There have been two reports of the mass spectrum of psilocybin, an halucinogenic naturally occurring compound. Leung and Paul[268] observed a parent molecular ion, but Picker and Richards[269] reported the absence of this ion in the spectrum. Of course, the difference in these reports may be due to some instrumental feature or possibly the nature of the compound being studied. Two other similar compounds were reported by Leung and Paul; they are given in Table 8.38 with psilocybin.

### 7.13.6 Compounds of the formula RP(L)(OR')(OR'')

Compounds having the general formula RP(O)(OR')(OR'') include those with R, R', and R'' all organic entities or with one or two either hydrogen or a halogen. We will discuss the ester derivatives of phosphinic acids first *i.e.* with R, R', and R'' organic entities.

A large number of esters of phosphinic acid has been studied (Table 8.39). Generally, all compounds showed parent molecular ions and in some cases (parent molecule $+1)^+$ ions were observed. Although ion abundances were not often quoted, fragmentation behaviour was usually discussed. Fragmentation by simple bond cleavages (P–OR, PO–R, or PO(CH$_2$)$_n$–R), eliminations of olefins or water, were common decomposition paths. Major peaks were due to species such as RP(O)(OR')(OX)$^+$ where X = (R''–Me),

Table 8.39

Other Compounds of the Type $R(OR')_2P=O$ Reported

| R | R' | R'' | Base peak (parent molecular ion observed unless stated otherwise) | Ref. |
|---|----|-----|------------------------------------------------------------------|------|
| Me | $Pr^i$ | $Pr^i$ | $MeP(OH)_3^+$ | 258 |
| Et | Et | Et | $EtP(OH)_3^+$ | 258 |
| Me | Et | Et | $MeP(OH)_3^+$ | 273 |
| $n\text{-}C_5H_{11}$ | Et | Et | $C_5H_{11}P(O)(OH)_2^+$ | 273 |
| $n\text{-}C_8H_{17}$ | Et | Et | | 273 |
| $CH_2Cl$ | Me | Me | $P(O)(OR')_2^+$ | 273 |
| $CH_2Cl$ | Et | Et | $P(O)(OR')OH^+$ | 273 |
| $CH_2Cl$ | $CH_2(Cl)CH_3$ | $CH_2(Cl)CH_3$ | $R'^+$ | 273 |
| $CH_2I$ | Et | Et | $RP(O)(OR')(OH)^+$ | 273 |
| $CH(Cl)CH_3$ | Et | Et | $P(O)(OR')OH^+$ | 273 |
| $CH_2CH_2Cl$ | Et | Et | $HP(O)(OR')_2^+$ | 273 |
| $CH_2(CH_2)_2Br$ | Et | Et | | 273 |
| $CH_2CH(Cl)C_8H_{17}{}^n$ | Et | Et | | 273 |
| $CH_2CH=CH_2$ | Et | Et | $P(O)(OR')(OH)^+$ | 273 |
| $CH_2CH=CH_2$ | $Pr^i$ | $Pr^i$ | $R'^+$ | 273 |
| $CH_2OEt$ | Et | Et | | 273 |
| $CH_2C(Cl)=CHC_5H_{11}{}^n$ | Et | Et | $HP(OR)(OR')_2^+$ | 273 |
| $CH=CH_2$ | Et | Et | $RP(O)(OH)^+$ | 273 |
| $CH=CHPh$ | Et | Et | $P(O)(OX)^+$ $(X = R'-H)$ | 273 |
| $CH=C(Cl)Ph$ | Et | Et | | 273 |
| $C\equiv CBu^n$ | Et | Et | | 273 |
| $C=CC_8H_{17}{}^n$ | Et | Et | | 273 |
| $CH_2Ph$ | Et | Et | $R^+$ and $P(O)(OX)^+$ $(X = R'-H)$ | 273 |
| $CH_2Ph$ | $Pr^i$ | $Pr^i$ | $R^+$ | 273 |
| $4'\text{-}NO_2C_6H_4CH_2$ | Et | Et | | 273 |
| $4'\text{-}MeOC_6H_4CH_2$ | Et | Et | $R^+$ | 273 |
| $CH_2Cl$ | Ph | Ph | | 273 |
| $CH_2=CHPh$ | Ph | Ph | | 273 |
| $CH_2=CHPh$ | $4'\text{-}MeC_6H_4$ | $4'\text{-}MeC_6H_4$ | | 273 |
| $CH_2Ph$ | Ph | Ph | $CH_2Ph^+$ | 273 |
| $CH_2Ph$ | $4'\text{-}MeC_6H_4$ | $4'\text{-}MeC_6H_4$ | $4\text{-}MeC_6H_4^+$ | 273 |
| $CH_2CH_2OEt$ | Et | Et | | 274 |
| $CH_2CH_2OPr^n$ | Et | Et | | 274 |
| $CH_2CH_2OBu^n$ | Et | Et | | 274 |
| $CH_2CH_2OBu^i$ | Et | Et | | 274 |

Table 8.39 (continued)

| R(OR')(OR'')PO | | | Base peak (parent molecular ion observed unless stated otherwise) | Ref. |
|---|---|---|---|---|
| R | R' | R'' | | |
| $CH_2CH_2OBu^n$ | $Pr^i$ | $Pr^i$ | | 274 |
| $CH_2CH_2OBu^n$ | $Bu^n$ | $Bu^n$ | | 274 |
| $(CH_2)_2OCH_2CH(Et)$<br>│<br>$H_3C(CH_2)_3$ | Et | Et | | 274 |
| $(CH_2)_2OCH_2CH(Et)$<br>│<br>$H_3C(CH_2)_3$ | $Pr^n$ | $Pr^n$ | | 274 |
| $Bu^n$ | Et | Et | | 274 |
| $n\text{-}C_8H_{12}$ | Et | Et | | 274 |
| $CH_2CH(Et)$<br>│<br>$(CH_2)_3CH_3$ | Et | Et | | 274 |
| $CH(Et)CO_2Et$ | Et | Et | | 275 |
| $CH(Et)CO_2Bu^n$ | Et | Et | $RP(O)OH^+?$ | 275 |
| $CH(Me)CO_2Et$ | Et | Et | $(R'O)P(O)OH^+?$ | 275 |
| $(CH_2)_3CO_2Et$ | Et | Et | $(R'O)_2P(X)^+,(X=R-HCO_2Et)?$ | 275 |
| $(CH_2)_5CO_2Et$ | Et | Et | | 275 |
| $CH_2CO_2Me$ | Et | Et | $(R'O)P(OH)_2R'^+?$ | 275 |
| $CH_2CO_2Et$ | Et | Et | $(R'O)P(OH)(O)CH_2^+?$ | 275 |

$RP(OR')(OH)_2^+$, $RP(O)(OR')(OH)^+$, $RP(O)(OH)(OX)^+$, $RP(OH)_3^+$, $RPO\text{-}(OR')^+$, and $RP(O)(OH)^+$.

Some of the more common decomposition paths are given in reaction schemes (8.127) and (8.128). The decompositions were frequently associated with a metastable peak.

(a) Reactions with retention of the R group (OR' = OR'') in all cases in Table 8.38.

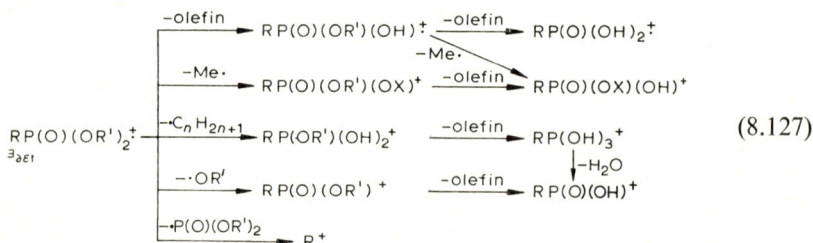

$$RP(O)(OR')_2^+ \quad (8.127)$$

($b$) With fragmentation of the R group,

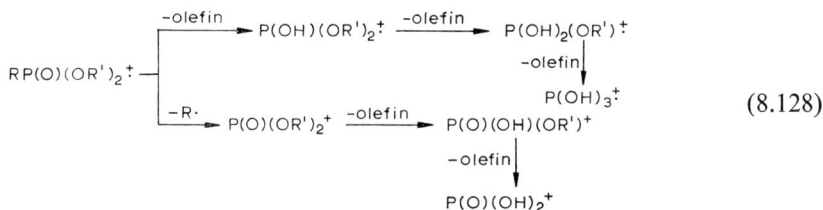

$$
\begin{array}{l}
RP(O)(OR')_2{}^{+} \\
\quad \xrightarrow{-\text{olefin}} P(OH)(OR')_2{}^{+} \xrightarrow{-\text{olefin}} P(OH)_2(OR'){}^{+} \\
\qquad\qquad\qquad\qquad\qquad\qquad \downarrow -\text{olefin} \\
\qquad\qquad\qquad\qquad\qquad\qquad P(OH)_3{}^{+} \\
\quad \xrightarrow{-R\cdot} P(O)(OR')_2{}^{+} \xrightarrow{-\text{olefin}} P(O)(OH)(OR'){}^{+} \\
\qquad\qquad\qquad\qquad\qquad\qquad \downarrow -\text{olefin} \\
\qquad\qquad\qquad\qquad\qquad\qquad P(O)(OH)_2{}^{+}
\end{array}
\tag{8.128}
$$

Features in the spectra of some compounds of the type (OR')(OR'')(R)PS where R' and R'' are organic radicals and R is an organic derivative bonded through nitrogen have been published[264]. They showed several similarities to compounds of the type $(RO)_3P{=}S$ which were discussed above. The compounds studied were

| R | R' | R'' |
|---|----|-----|
| $C_6H_{11}NH$ | Me | Me |
| $C_5H_{12}N$ | Me | Me |
| $C_6H_{11}NH$ | Me | $4\text{-}NO_2C_6H_4$ |
| $C_4H_9NO$ | Me | $4\text{-}NO_2C_6H_4$ |
| $C_6H_{11}NH$ | Me | $Pr^i$ |
| $C_4H_9NO$ | Ph | Ph |

All compounds showed parent molecular ions. Fragmentation by loss of an SH radical was important; all compounds showed an ion due to (parent molecule $-\cdot SH$)$^+$ and it was the base peak in several spectra, including (8.129).

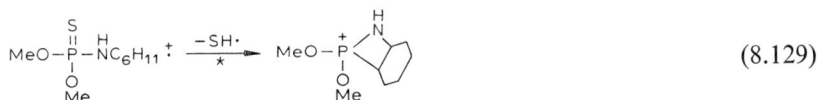

$$
\underset{\substack{\text{Me}}}{\underset{\text{O}}{\overset{\text{S}}{\text{MeO}-\text{P}-\overset{\text{H}}{\text{N}}\text{C}_6\text{H}_{11}}}}{}^{+} \xrightarrow[\ast]{-\text{SH}\cdot} \underset{\substack{\text{Me}}}{\underset{\text{O}}{\text{MeO}-\overset{+}{\text{P}}}}\!\!\overset{\text{H}}{\underset{}{\text{N}}}\!\! \tag{8.129}
$$

The formation of bonds between substituents (particularly aryl groups) on N, S, or O atoms and the central phosphorus atom was suggested to be a quite common feature. The spectra of deuterated derivatives, $D_3CO-$

instead of $H_3CO-$ and $N(D)C_6H_{11}$ instead of $N(H)C_6H_{11}$, in the above case, showed that the H atom in the eliminated $\cdot SH$ came from an alkyl (or aryl) group and not the N–H entity. Rearrangements of groups from –OR to –SR occurred apparently quite readily, *e.g.*

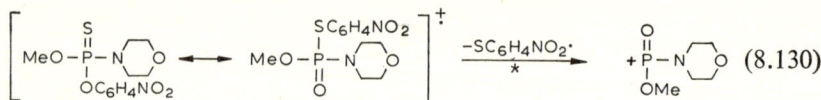

(8.130)

McMurray *et al.*[276] have reported some points in the spectra of two *N*-phenylphosphoramidate esters (R = Et or Ph).

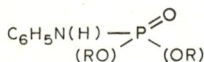

They observed parent molecular ions which were the base peaks in both spectra and the metastable supported fragmentations.

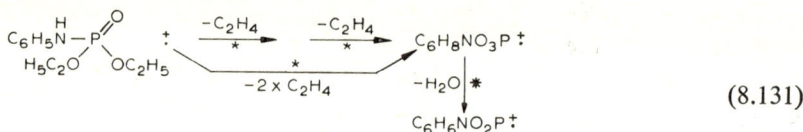

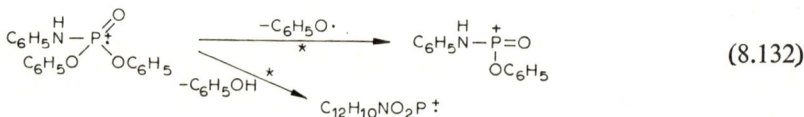

(8.131)

(8.132)

Compounds with R = H, R′, R″ = organic species, have been studied by several workers. Harless[277] and McLafferty[263] have reported some features for compounds where R′ = R″ = Me, Et, $Pr^n$, $Pr^i$, $Bu^n$, and allyl. Harless found that all compounds gave molecular ions and the base peaks were due to the $(HO)_3\overset{+}{P}H$ ion in all cases except for R = Me or allyl where they were $H\overset{+\cdot}{P}O_3$ and $C_3H_5^+$ respectively. McLafferty suggested reaction (8.133) to produce the $(HO)_3\overset{+}{P}H$ ion from $(EtO)_2HP(O)$,

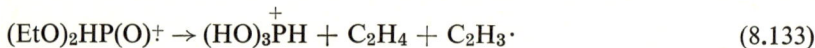

$$(EtO)_2HP(O)^{\ddagger} \rightarrow (HO)_3\overset{+}{P}H + C_2H_4 + C_2H_3\cdot \qquad (8.133)$$

Pritchard[278] has reported some detailed observations of the fragmentation of (MeO)$_2$HPO and (EtO)$_2$HPO. They are shown in reaction schemes (8.134) and (8.135)

$$(H_3CO)_2\underset{H}{P}=O\; \dagger\; \xrightarrow[*]{-CH_2O}\; (H_3CO)-\underset{H\quad H}{\overset{+\cdot}{P}}=O\; \xrightarrow[*]{-CH_3\cdot}\; P(OH)_2^{+}$$

$$(H_3CO)_2\overset{+}{P}=O\; \xrightarrow[*]{-CH_2O}\; \xrightarrow[*]{-CH_2O}\; H_2PO^{+} \tag{8.134}$$

$$\underset{HO}{\overset{H_3CO}{\diagdown}}\overset{+}{P}=O\; \xrightarrow[*]{-CH_2O}\; H(OH)\overset{+}{P}=O\quad (or\ (OH)_2P^{+})$$

The ethyl compound behaved similarly and also eliminated ethylene and the C$_2$H$_3$ radical, but as a two-stage process, not as a one-stage process as in reaction (8.135).

$$(C_2H_5O)_2\underset{H}{P}=O\; \dagger\; \xrightarrow[*]{-C_2H_3\cdot}\; (C_2H_5O)-\underset{H\quad OH}{\overset{+}{P}}-OH\; \xrightarrow[*]{-C_2H_4}\; (HO)_3PH^{+} \tag{8.135}$$

The cyclic compound LXXXV is reported to give a parent molecular ion[279].

$$\underset{CH_2-O}{\overset{CH_2-O}{|}}\diagdown\underset{\diagup}{P}\underset{H}{\overset{O}{\diagdown}}$$

LXXXV

The fragmentations of two allenephosphinic acids (R = organic, R' = R'' = H) have been reported[280]. Both compounds gave parent molecular ions and (parent molecule + 1)$^+$ ions. The reaction schemes suggested are given in (8.136) and (8.137).

$$\underset{HO}{\overset{HO}{\diagdown}}\underset{\diagup O}{\overset{}{}}P-CH=C=C\underset{Me}{\overset{Me}{\diagup}}\; \xrightarrow[*]{+H\cdot}\; (parent\ +1)^{+}$$

$$\underset{?}{\overset{}{}}\quad \overset{-H\cdot}{\diagdown *}\;\overset{-Me\cdot}{\diagdown *}$$

$$\underset{HO}{\overset{HO}{\diagdown}}P-\overset{+}{C}H=C=\overset{}{C}Me\; \xrightarrow[*]{-H_2O}\; \underset{O}{\overset{O}{\diagdown}}P-CH=C=\overset{+}{C}Me \tag{8.136}$$

(base peak)

$$
\begin{array}{c}
\mathrm{HO} \\
\phantom{xx}\mathrm{P-CH=C=C} \langle\!\!\!\bigcirc\!\!\!\rangle \overset{+}{\phantom{.}} \;\xrightarrow[\ast]{+\,\mathrm{H\cdot}}\; (\text{parent}+1)^{+} \\
\mathrm{HO}\;\; \mathrm{O}
\end{array}
$$

$$m/e = 187 \;\xrightarrow[\ast]{-\mathrm{H_2O}}\; ?$$

$$-\mathrm{C_2H_4}/\ast \qquad -\mathrm{H\cdot}\qquad \ast\backslash -\mathrm{H_2O}$$

$$
\mathrm{O=P-CH=C=C}\langle\!\!\!\bigcirc\!\!\!\rangle\overset{+}{\phantom{.}} \\
\;\;\;\overset{\|}{\mathrm{O}}
$$

$$m/e = 160 \qquad\qquad -\mathrm{PO_3H}\,|\ast$$

$$-\mathrm{C_3H_6}|\ast \qquad (-\mathrm{H\cdot})$$

$$m/e = 146 \quad \overset{+}{\mathrm{C}}\mathrm{H=C=C}\langle\!\!\!\bigcirc\!\!\!\rangle$$

$$-\mathrm{H_2O}\,\overset{\ast}{\searrow}$$

$$
\begin{array}{c}
\mathrm{H_2C} \\
\phantom{x}| \phantom{xx}\mathrm{C=C=\overset{\cdot\cdot}{C}H} \\
\mathrm{H_2C}\phantom{xx}\mathrm{O=P=O}
\end{array}
$$

$$(8.137)$$

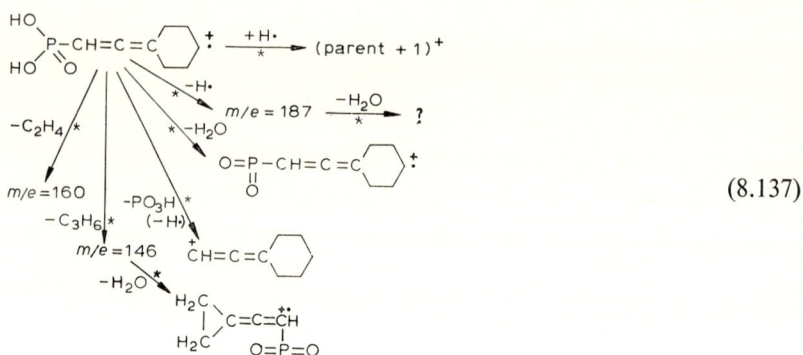

The base peak in the spectrum of this compound was the

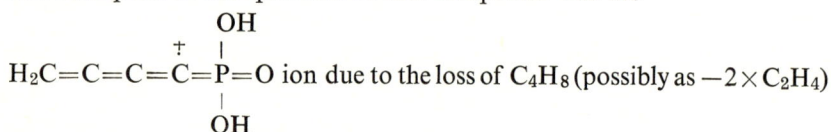

$$
\overset{\phantom{x}\mathrm{OH}}{\underset{\underset{\mathrm{OH}}{|}}{\mathrm{H_2C=C=C=C=\overset{+}{P}=O}}}
$$

ion due to the loss of $C_4H_8$ (possibly as $-2 \times C_2H_4$)

from the parent molecular ion.

Compounds with $R$ = halogen, $R'$, $R''$ = organic species have been reported in some detail by Cooks and Gerrard[264] and by Pritchard[278]. They show several similar fragmentation paths to those compounds of the types $(RO)_3P=L$, $(RO)_2HPO$, and $(RO)_2XP=L$ already discussed. Cooks and Gerrard studied the compounds $(RO)_2ClP=S$ ($R$ = Me or Ph) and were able to suggest several decomposition paths to account for the observed fragment ions. These paths included simple elimination reactions, (8.138), hydrogen rearrangements to give phenolate type ions, (8.139), and bond formation between substituents on an O atom and the central phosphorus atom, (8.140).

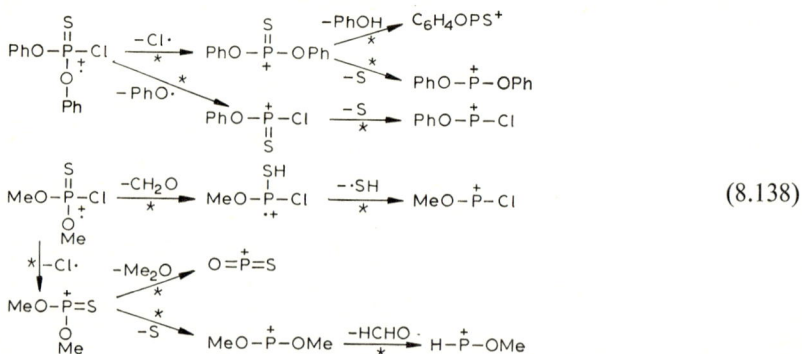

$$
\begin{array}{l}
\mathrm{PhO-\overset{\overset{\displaystyle S}{\|}}{\underset{\underset{\displaystyle Ph}{|}}{\underset{\displaystyle O}{P}}}-Cl}\;\overset{+}{\phantom{.}} \;\xrightarrow[\ast]{-Cl\cdot}\; \mathrm{PhO-\overset{\overset{\displaystyle S}{\|}}{P}-OPh}\;\overset{+}{\phantom{.}}\;\xrightarrow[\ast]{-PhOH}\; \mathrm{C_6H_4OPS^{+}} \\[2mm]
\qquad\qquad\qquad\qquad\qquad \xrightarrow[\ast]{-S}\; \mathrm{PhO-\overset{+}{P}-OPh} \\[2mm]
\qquad \overset{-PhO\cdot}{\searrow}\;\mathrm{PhO-\overset{+}{\underset{\underset{\displaystyle S}{\|}}{P}}-Cl}\;\xrightarrow[\ast]{-S}\; \mathrm{PhO-\overset{+}{P}-Cl}
\end{array}
$$

$$
\mathrm{MeO-\overset{\overset{\displaystyle S}{\|}}{\underset{\underset{\displaystyle Me}{|}}{\underset{\displaystyle O}{P}}}-Cl}\;\overset{+}{\phantom{.}}\;\xrightarrow[\ast]{-CH_2O}\; \mathrm{MeO-\overset{\overset{\displaystyle SH}{|}}{P}-Cl}\;\overset{\cdot+}{\phantom{.}}\;\xrightarrow[\ast]{-\cdot SH}\; \mathrm{MeO-\overset{+}{P}-Cl}
$$

$$(8.138)$$

$$
\begin{array}{l}
\ast\!\mid\!-Cl\cdot \qquad \overset{-Me_2O}{\searrow}\; \mathrm{O=\overset{+}{P}=S} \\[2mm]
\mathrm{MeO-\overset{+}{\underset{\underset{\displaystyle Me}{|}}{\underset{\displaystyle O}{P}}}=S}\;\overset{\ast}{\nearrow} \\[2mm]
\qquad\qquad \overset{-S}{\searrow}\; \mathrm{MeO-\overset{+}{P}-OMe}\;\xrightarrow[\ast]{-HCHO\cdot}\; \mathrm{H-\overset{+}{P}-OMe}
\end{array}
$$

$$
\begin{array}{c}
\overset{S}{\underset{\underset{Cl}{|}}{PhO-\overset{\|}{P}-OPh}} \longrightarrow PhO-\overset{SPh}{\underset{\underset{Cl}{|}}{P}=O}
\end{array}
$$

$$
PhOSPh \xleftarrow{\ -POCl\ }_{*} PhO-\overset{Cl}{\underset{\underset{\cdot SPh}{+|}}{P}}=O \longrightarrow \overset{SPh}{\underset{\underset{Cl}{|}}{P}=O} \ (via\ PhSPh^{+})
$$

(8.139)

$$
\overset{S}{\underset{\underset{Cl\ +}{|}}{PhO-\overset{\|}{P}-OPh}} \xrightarrow[*]{-\cdot SH} PhO-\overset{+}{\underset{\underset{Cl}{|}}{P}}-O
$$

(8.140)

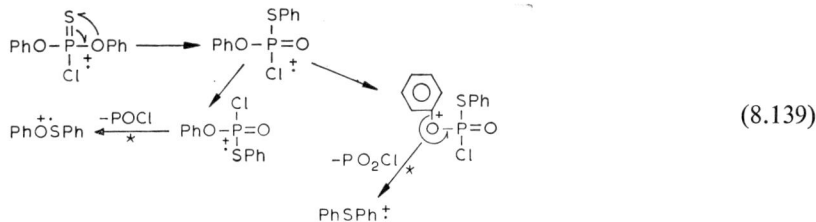

As with the similar compounds mentioned above, the loss of the SH radical from the parent molecular ion was an important process giving rise to a large peak due to the corresponding (parent molecule $-\cdot SH)^+$ ion. Pritchard studied the compounds $(RO)_2ClP=O$ (R = Me or Et) and $(RO)_2ClP=S$ (R = Me or Et). He observed some similar reactions to those reported by Cooks and Gerrard for $(MeO)_2ClP=S$ and the reactions (8.141) for R = Me and (8.142) for R = Et.

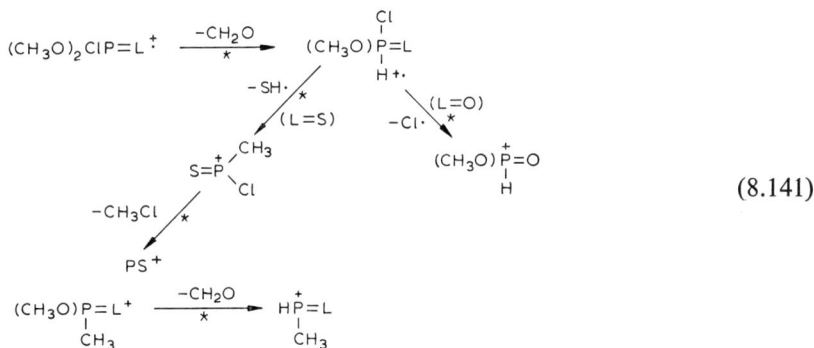

$$
(CH_3O)_2ClP=L^+ \xrightarrow[*]{-CH_2O} (CH_3O)\overset{Cl}{\underset{\underset{H}{|}}{P}}=L^{+\cdot}
$$

(8.141)

$$
(CH_3O)P=L^+ \xrightarrow[*]{-CH_2O} H\overset{+}{P}=L
$$

It is interesting to note that the ions which Pritchard writes as

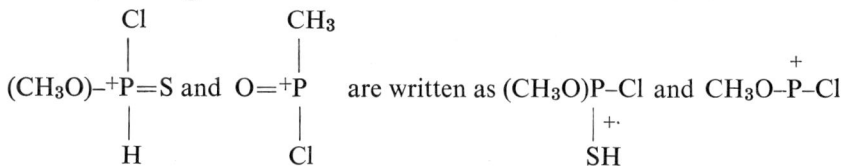

$$
(CH_3O)-\overset{Cl}{\underset{\underset{H}{|}}{^+P}}=S \quad and \quad O=\overset{CH_3}{\underset{\underset{Cl}{|}}{^+P}} \quad are\ written\ as\ (CH_3O)P-Cl\ and\ CH_3O-P-Cl
$$

by Cooks and Gerrard.

References pp. 439–447

$$(C_2H_5O)_2\overset{+}{P}=O \xrightarrow[\ast]{-C_2H_3\cdot} \xrightarrow[\ast]{-C_2H_4} (HO)_3\overset{+}{P}$$
$$\overset{|}{Cl} \qquad\qquad\qquad\qquad\qquad\qquad \overset{\ast}{\diagup}\overset{|}{Cl}$$
$$\qquad\qquad\qquad\qquad\qquad\qquad\qquad -HCl$$
$$(HO)_2\overset{+}{P}=O$$

$$(C_2H_5O)_2\overset{+}{P}=S \xrightarrow[\ast]{-CH_3CHO} \quad C_2H_5O\diagdown\;\diagup S$$
$$\overset{|}{Cl} \qquad\qquad\qquad\qquad\qquad\qquad\; P \qquad +\cdot$$
$$\qquad\qquad\qquad\qquad\qquad\qquad H\diagup\;\diagdown Cl$$
$$\qquad\qquad\quad H \qquad \overset{\ast}{\diagup}-C_2H_4$$
$$\qquad\qquad (OH)\overset{+}{P}S$$
$$\qquad\qquad\qquad\overset{|}{Cl}$$

(8.142)

$$(C_2H_5O)\overset{\diagup S}{P}\; + \xrightarrow[\ast]{-C_2H_4} (HO)_2\overset{S}{\overset{\|}{P}}\overset{+}{Cl} \xrightarrow[\ast]{-Cl\cdot} (OH)_2\overset{+}{P}=S$$
$$HO\diagdown Cl$$

$$(C_2H_5O)_2\overset{+}{P}=S \xrightarrow[\ast]{-C_2H_4} \xrightarrow[\ast]{-C_2H_4} (HO)_2\overset{+}{P}=S$$

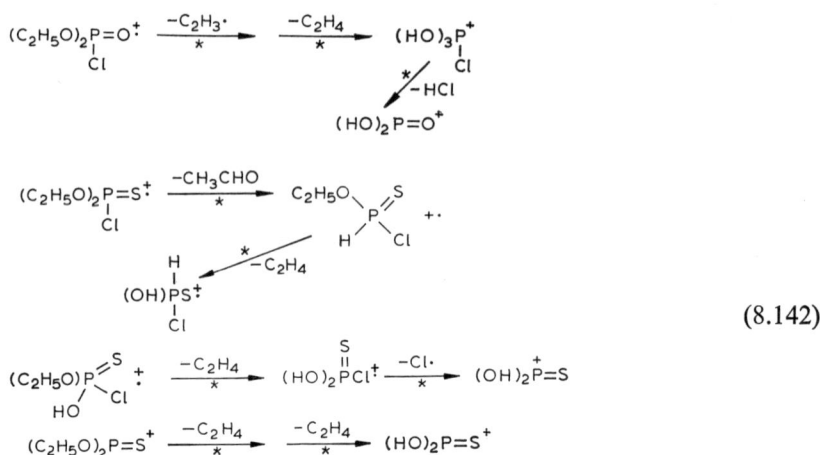

Other compounds of this type for which mass spectral data have been reported are

| R | OR′ | LR″ | L | Ref. |
|---|-----|-----|---|------|
| F | Ph | OPh | O | 241 |
| F | Et | SPh | S | 281 |
| F | Me | SEt | S | 282 |
| F | Et | SMe | S | 282 |
| F | Et | OEt | S | 282 |

Two compounds with R, R′, and R″ different have been studied by Cooks and Gerrard[264]. They were $(MeO)(HO)(HNC_6H_{11})P=S$ and $(MeO)(Cl)$-$(HNC_6H_{11})P=S$. The comments about fragmentation which were made concerning the other compounds reported by Cooks and Gerrard above apply here as well and these compounds will not be discussed further.

### 7.13.7 Compounds of the formula RR′P(L)(L′R″)

A large number of phosphinic esters has been studied mass spectrometrically. Most investigations have been concerned with decomposition paths, but Haake and Ossip[183] have reported the spectra of some alkyl esters of dialkylphosphinic acids (Table 8.40).

As was found for other alkoxy–phosphorus compounds, the majority of ions were produced by simple bond cleavages or elimination of olefins, water or $-CH_2O$ groups. All compounds except $Et_2P(O)(OPr^i)$ gave parent molecular ions. The base peaks in the spectra of higher alkyl methyl esters were

Table 8.40

Ion Abundances from $R_2(OR')P=O$ Compounds

| Composition of ion | R = Me R' = Me | Me Et | Et Me | $Pr^n$ Me | Et $Pr^i$ |
|---|---|---|---|---|---|
| $R_2(OR')PO$ | 6.4 | 0.9 | 2.3 | 3.1 | |
| $R_2P(OH)_2$ | | 15.8 | | | 24.1 |
| $R_2P(O)OH$ | 1.2[a] | 1.5[a] | | | 0.7 |
| $RHPO(OR')$ | 1.2[a] | 0.6 | 20.4 | 19.9 | 1.0 |
| $RPO(OR')$ | 40.2 | 2.2[b] | 26.9 | 9.9 | 9.5 |
| $R_2P(O)H$ | 18.1 | 11.4 | | | 2.7 |
| $R_2P(O)$ | 5.6 | 17.4 | 0.8 | | 10.4 |
| $(HO)HP(OR')$ | | 1.5 | 0.8 | 3.1 | |
| $HOP(OR')$ | | | 38.4 | 62.1 | |
| $RP(OH)_2$ | | 0.6 | | | 6.5 |
| $RP(O)OH$ | | 30.9 | | | 22.8 |
| $R(H)POH$ | | | | | 1.5 |
| $R(H)PO$ | 19.3 | 4.3 | 1.2 | | 3.6 |
| $RPO$ | 1.2 | 1.2 | | | 0.5 |
| $(HO)_2P$ | 0.8 | 1.2 | 0.8 | | 9.2 |
| $H_2PO$ | 1.2 | 2.5 | 2.3 | | 1.2 |
| $HPO$ | 0.8 | 1.2 | 1.5 | | 0.5 |
| $PO$ | 4.0 | 6.8 | 4.6 | 1.9 | 1.0 |
| $(p.m.i. - \cdot CH_3)$ | | | | | 4.8 |

[a] Sum of intensities of indistinguishable isomeric ions.
[b] Could also be the $(p.m.i. - \cdot CH_3)^+$ ion.

due to the $H_3C\,P(O)\,OH^+$ ion but for the methyl dimethylphosphinate it was due to the $H_3CP(O)(OCH_3)^+$ ion. This illustrates the favourable olefin elimination pathways in the former cases. Some metastable supported reactions observed in the methyl dimethylphosphinate and the ethyl ester are given in schemes (8.143) and (8.144).

$(CH_3)_2P(O)(OCH_3)^{\cdot+} \xrightarrow[*]{-CH_2O} (CH_3)_2P(O)H^{\cdot+}$

$(CH_3)P(O)(OCH_3)^+ \xrightarrow[*]{-CH_2O} CH_3P(O)^+ \;|\; H$

$(CH_3)_2P(O)(OC_2H_5)^{\cdot+} \xrightarrow[*]{-OCHCH_3} (CH_3)_2P(O)H^{\cdot+} \xrightarrow{-CH_3\cdot} CH_3P(O)H^+$

(8.143)

$(CH_3)_2P(O)OH^{\cdot+} \xrightarrow[*]{-CH_3\cdot} CH_3P(O)^+OH$

$CH_3^+P(O)OC_2H_5 \xrightarrow[*]{-C_2H_4} CH_3P(O)^+OH$

$(CH_3)_2P(O)^+OCH_2 \xleftarrow[*]{-CH_2O} (CH_3)_2\overset{+}{P}=O \xrightarrow[*]{-H_2O} (CH_3)_2\overset{+}{P}(OH)_2$

with $\xrightarrow[*]{-CH_2O}$

(8.144)

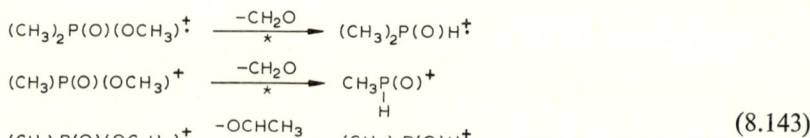

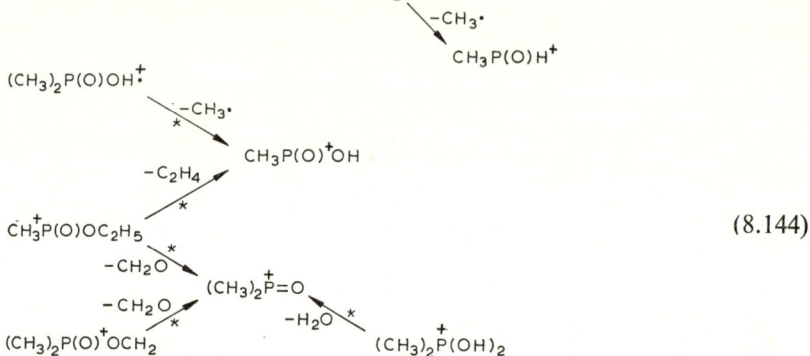

Haake *et al.*[284] have also studied compounds containing aryl groups and cyclic compounds, *i.e.* (R)(R′) P(O) (OR″) where R, R′, and R″ were

| R | R′ | R″ |
|---|---|---|
| Ph | Ph | Me |
| Ph | Ph | Et |
| Ph | Ph | Ph |
| Ph | Ph | CH₂Ph |
| Ph | 2-MeC₆H₄ | Ph |
| 4-Me₂NC₆H₄ | 4-Me₂NC₆H₄ | Me |
| 3-NO₂C₆H₄ | 3-NO₂C₆H₄ | Me |

and

LXXXVI          LXXXVII

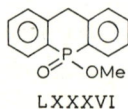

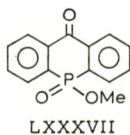

In general, the non-cyclic compounds behaved similarly to the alkyl compounds discussed previously, *e.g.*

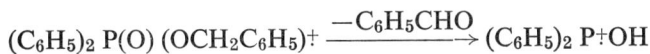

$$(C_6H_5)_2 \, P(O) \, (OC_2H_5)^+ \xrightarrow{-C_2H_4} (C_6H_5)_2P(O)^+OH \qquad (8.145)$$

$$(C_6H_5)_2 \, P(O) \, (OCH_2C_6H_5)^+ \xrightarrow{-C_6H_5CHO} (C_6H_5)_2 \, P^+OH$$

But fragmentation by cyclisation as found for other compounds containing aryl–phosphorus groups was important as shown by strong peaks due to (parent molecule $-1)^+$ ions and metastable supported reactions such as

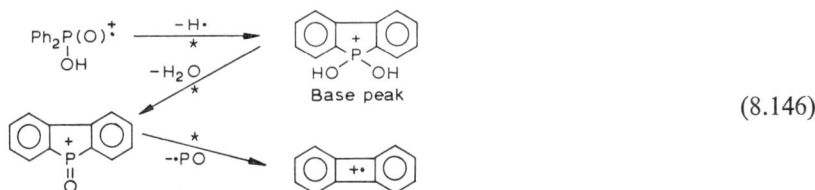

(8.146)

The compounds which contained nitro groups lost O, $\cdot$NO, and $\cdot$NO$_2$ species as well as fragmenting by cyclisation and showed noncyclic fragment ions of the type $(NO_2C_6H_4) \, (OCH_3)P(O)^+$. The cyclic compounds showed (parent molecule $-1)^+$ peaks but these were not as important as in the other compounds. The base peak in the spectrum of LXXXVI was produced by the reaction

(8.147)

This compound also lost an Me radical, an unusual path for such a compound, and the ion formed then lost a PO$_2$H group, *viz.*

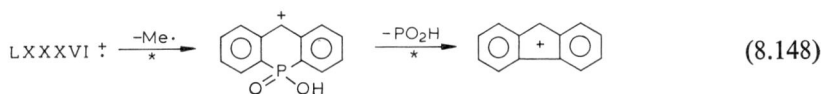

(8.148)

The cyclic compound LXXXVII produced a base peak due to loss of CH$_2$O. Other ions were formed by loss of CO, $\cdot$OH, and $\cdot$PO.

*References pp. 439–447*

Budzikiewicz and Pelah[285] have studied the compounds RR′P(O)(OR″)

| R | R′ | R″ |
|---|---|---|
| Ph | Me | Me |
| Ph | Et | Me |
| Ph | CH(OH)CCl$_3$ | Me |
| Ph | CH(OH)CCl$_3$ | Et |
| | CH(OH)CCl$_3$ | Me |
| | CH(OH)CCl$_3$ | Bu$^n$ |

Base peaks corresponding to the (parent molecule — · R)$^+$ ions were found for all compounds except Ph(CH(OH)CCl$_3$)P(O)OEt and

for which the base peaks were due to RP(O)OH$^+$ ions. The first two compounds listed exhibited ions due to loss of R· and H atoms. The later compounds showed loss of the CH(OH)CCl$_3$ group when R = Me but a more complicated fragmentation when R = Et or Bu$^n$ involving a McLafferty-type rearrangement, viz.

where X = H for R″ = Et, X = CH$_2$–CH$_3$ for R″ = Bu$^n$.

Other compounds in this class for which mass spectral data have been reported are given in Table 8.41.

Table 8.41

Other Compounds of the Type $R(R')(OR'')P=L$ Reported

| Compound | Comments | Ref. |
|---|---|---|
| $\begin{array}{ccc} C_6X_5 & H & C_6X_5 \\ \mid & \mid & \mid \\ (O)P - O - C - P(O) \\ \mid & \mid & \mid \\ C_6X_5 & CF_3 & C_6X_5 \end{array}$ | P.m.i. for X = H or D  (O)P(C_6X_5)_2^+ base peaks | 286 |
| $Ph_2P(S)SCH_2CN$ | P.m.i., $Ph_2P^+=S$ base peak | 254 |
| $Ph_2P(S)SCH_2CONH_2$ | P.m.i., $Ph_2P^+=S$ base peak | 254 |
| $Ph_2P(S)S-\langle\bigcirc\rangle-NO_2$ | P.m.i., $Ph_2P^+=S$ base peak | 254 |
| $Ph_2P(S)-O-\overset{}{\underset{Cl}{\langle\bigcirc\rangle}}-NO_2$ | P.m.i., $Ph_2P^+=S$ base peak | 254 |
| $Ph_2P(S)-S-\langle\bigcirc\rangle\begin{smallmatrix}N-Me\\N-Me\end{smallmatrix}$ | P.m.i.,$Ph_2P^+=S$ base peak | 254 |
| $\begin{array}{c} Ph_2P(O)-O-CPh \\ Ph-C \diagup \diagdown P(O)Ph_2 \\ \overset{\parallel}{O} \end{array}$ | P.m.i. | 247 |
| $\begin{array}{c} Ph_2P(O)-O-CHR \\ \mid \\ P(O)Ph_2 \\ (R = Ph \; or \; (CH_2)_2Ph) \end{array}$ | P.m.i. | 247 |
| $Bu^t-\langle\bigcirc\rangle\overset{Bu^t}{\underset{Bu^t}{\overset{}{\underset{H}{}}}}P\diagup\overset{O}{\diagdown}_{OMe}$ | P.m.i. | 288 |
| $\left(\begin{array}{c} 4\,MeO-C_6H_4 \\ 4\,MeO-C_6H_4-\langle\bigcirc\rangle-P\diagdown^O_O \\ 4\,MeO-C_6H_4 \quad H \end{array}\right)_2$ | P.m.i. | 288 |

Compounds with the general formula $(R)(R')P(L)OH$ have been reported
in detail by several workers. The dialkylphosphinic acids with $R = R' =$
Me, Et, $Pr^i$, and $Bu^n$ were studied by Haake and Ossip[283]. Their results for
the first three compounds mentioned are summarised in Table 8.42. Some of

Table 8.42

Ion Abundances of $R_2P(O)OH$ Compounds

| Composition of ion | R | | |
|---|---|---|---|
| | Me | Et | $Pr^i$ |
| $R_2P(O)OH$ | 17.7 | 2.9 | 2.8 |
| $R_2P(OH)_2$ | 7.5 | 1.8 | 9.9 |
| $R(H)P(OH)_2$ | | 1.1 | 2.5 |
| $RP(OH)_2$ | | 25.7 | 29.7 |
| $RP(O)OH$ | 44.2 | 13.2 | 1.9 |
| $R_2PO$ | 9.7 | 2.2 | |
| $R(H)PO$ | 1.8 | 1.8 | 0.6 |
| $RPO$ | 0.9 | 0.7 | 0.6 |
| $(HO)_2PH$ | | 2.2 | 9.6 |
| $(HO)_2P$ | | 36.1 | 30.9 |
| $(HO)PO$ | 4.0 | 2.2 | 0.6 |
| $HPO$ | 1.8 | 1.4 | |
| $PO$ | 12.4 | 4.7 | 1.9 |
| $m/e = 121$ | | 1.1 | |
| $m/e = 108$ | | 1.8 | |
| $m/e = 103$ | | 1.1 | |
| $m/e = 93$ | | | 1.9 |
| $m/e = 80$ | | | 4.3 |
| $m/e = 79$ | | | 2.8 |

the ions found were produced by processes of the same type as those involved
in the spectra of the phosphinic esters described above. Other ions were
formed by processes not observed in the previous spectra such as the production of (parent molecule $+1)^+$ ions for all the phosphinic acids except the
butyl compound. Compounds with alkyl groups higher than methyl were
seen to undergo several interesting rearrangement reactions. From the ethyl
compound, $m/e = 121$ and $m/e = 103$ were observed and a metastable
supported reaction (8.151) was found.

$$(C_2H_5)_2PO_2^+ \xrightarrow[\substack{\ast}]{-H_2O} \underset{\substack{m/e = 103}}{\overset{CH_2=CH}{\underset{C_2H_5}{\overset{O}{+P}\!\!\diagup\!\!\diagdown}}}$$

$$m/e = 121 \qquad\qquad m/e = 103$$

(8.151)

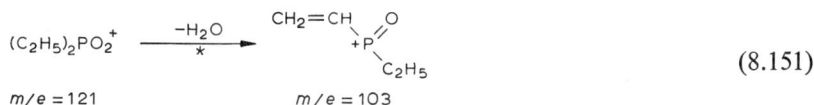

The appearance of an ion at $m/e = 108$ may be assigned to the formation of a rearrangement ion from the parent molecular ion by loss of $:CH_2$ i.e.

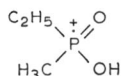

$$\underset{H_3C}{\overset{C_2H_5}{\diagdown}}\overset{+}{P}\overset{O}{\diagup\!\!\diagdown OH}$$

The spectrum of the isopropyl compound showed similar ions, LXXXVIII–XC.

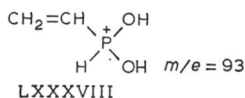

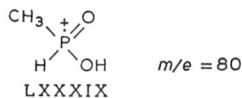

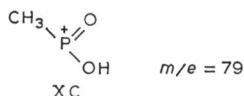

$$\underset{H}{\overset{CH_2=CH}{\diagdown}}\overset{+}{P}\overset{OH}{\diagup\diagdown OH} \quad m/e = 93$$

LXXXVIII

$$\underset{H}{\overset{CH_3}{\diagdown}}\overset{+}{P}\overset{O}{\diagup\!\!\diagdown OH} \qquad m/e = 80$$

LXXXIX

$$\overset{CH_3}{\diagdown}\overset{+}{P}\overset{O}{\diagup\!\!\diagdown OH} \qquad m/e = 79$$

XC

Similar processes to those described above were found for the butyl compound. They are summarised in the reaction scheme

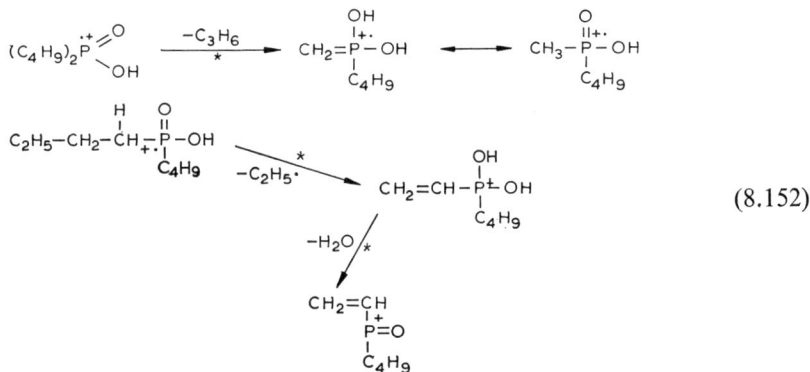

$$(C_4H_9)_2\overset{\cdot+}{P}\overset{O}{\diagup\!\!\diagdown OH} \xrightarrow[\substack{\ast}]{-C_3H_6} \underset{C_4H_9}{\overset{OH}{\underset{|}{CH_2=\overset{|+\cdot}{P}-OH}}} \longleftarrow \underset{C_4H_9}{\overset{O}{\underset{|}{CH_3-\overset{||+\cdot}{P}-OH}}}$$

$$\underset{C_4H_9}{\overset{H\quad O}{C_2H_5-CH_2-\overset{|}{C}\underset{+\cdot}{H}-\overset{||}{P}-OH}} \xrightarrow[-C_2H_5\cdot]{\ast} \underset{C_4H_9}{\overset{OH}{\underset{|}{CH_2=CH-\overset{|+}{P}-OH}}}$$

$$-H_2O\diagup_\ast$$

$$\underset{C_4H_9}{\overset{CH_2=\underset{|+}{C}H}{\underset{|}{P=O}}}$$

(8.152)

A series of diarylphosphinic acids (RR'P(O)OH) has been studied by Haake *et al.* who also studied the cyclic compounds XCI and XCII

XCI

XCII

In the compounds (R)(R')P(O)(OH), R = R' = Ph; R = Ph, R' = 2-Me–C₆H₄; R = 2-Me–C₆H₄, R' = 2-Me–C₆H₄; R = 3-NO₂C₆H₄, R' = 3-NO₂C₆H₄; and (Ph)₂PO(OX) with X = H, 80%; D, 20%. Fragmentation by loss of an H atom, then $H_2O$ and PO was an important pathway for the noncyclic compounds, *e.g.*

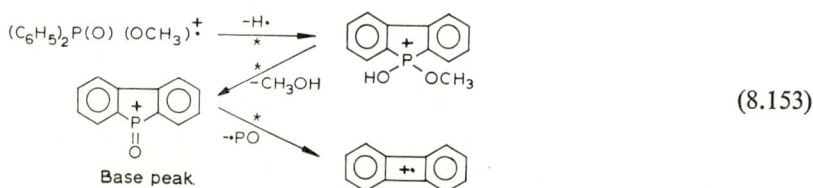

(8.153)

Evidence that the H atom came from a phenyl ring was provided by the spectrum of the partially deuterated compound. The nitro compound lost O, NO·, and ·NO₂ species as did the corresponding methyl ester discussed above. The (parent molecule −1)⁺ ion was not so important in the spectra of the cyclic compounds. Their preferred fragmentation reactions are given in schemes (8.154) and (8.155).

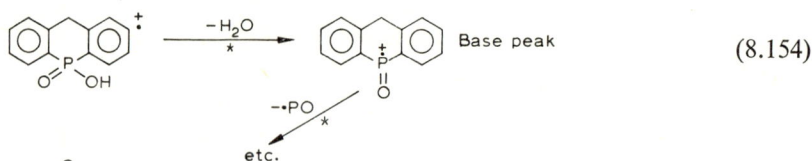

(8.154)

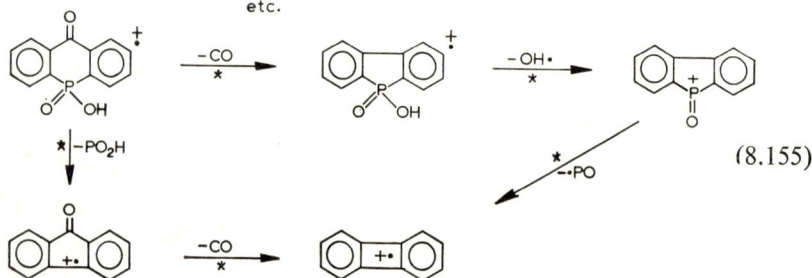

(8.155)

The spectra of two derivatives of phenylallenylphosphinic acid have been reported[280]. Both compounds showed parent molecular ions and (parent molecule $+ 1)^+$ ions. The fragmentation reactions observed for the parent compound are given in the scheme

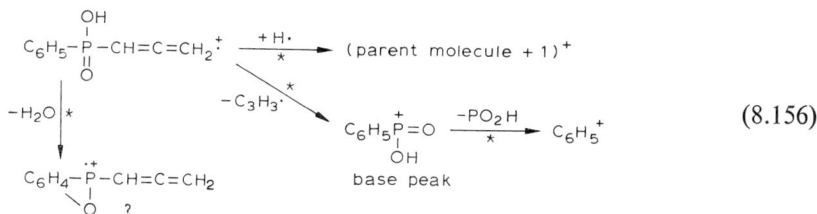

$$(8.156)$$

The other compound decomposed by loss of an H atom, $CH_3$ radical, and elimination of $C_2H_4$ and $C_3H_6$ from the parent molecular ion which was the base peak. The compound XCIII has been shown to be dimeric in the mass spectrometer[288].

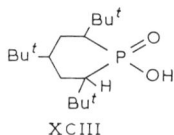

XCIII

Parent molecular ions were observed[254] for the compounds $Ph_2P(S)LH$ (L = O or S), and $Ph_2P(O)OH$. The base peaks were reported to be due to the parent molecular ion, $Ph_2P^+S$, and (parent molecule $-1)^+$ respectively. The last observation is in agreement with the work of Haake et al.[284].

Relatively few compounds of the type $R(H)P(O)(OR')$ have been studied. Budzikiewicz and Pelah[285] have briefly discussed some points in the spectra of $Ph(H)P(O)(OR')$, (R' = Me, Et, and $Pr^n$). Major ions are formed by loss of H atoms or elimination of olefins from R'.

Compounds for which parent molecular ions have been reported, including a number of compounds of the type $(R)(R')P(O)(OR'')$ with R, R', and R'' different species are, given in Table 8.43.

Table 8.43

Other Compounds of the Type $(RO)X_2P=L$ Reported

| Compound | Comments | Ref. |
|---|---|---|
| $(MeO)\overset{\displaystyle O}{\underset{\displaystyle F}{\overset{\|}{P}}}\text{-}(NH_2)$ | P.m.i.  $P(O)FNH_2^+$ base peak | 151 |
| $(EtO)P(O)F(NH_2)$ | P.m.i., $P(O)F(OH)_2^+$ base peak | 151 |
| $(PhO)P(O)F(NH_2)$ | P.m.i. | 151 |
| $(PhS)P(O)F(NH_2)$ | P.m.i. | 151 |
| $(MeS)P(S)F(NMe_2)$ | P.m.i. | 282 |
| $(EtS)P(S)F(NMe_2)$ | P.m.i., $PF(NMe_2)^+$ base peak | 282 |
| $(MeS)P(S)F(NEt_2)$ | P.m.i., $P(S)F(NEt_2)^+$ base peak | 282 |
| $(EtS)P(S)F(NEt_2)$ | P.m.i., $(HS)P(S)FNEt_2^+$ and $C_2H_6NPF^+$ both base peaks | 282 |
| $(MeO)P(S)F(NH_2)$ | P.m.i., $OPS^+$ base peak | 256 |
| $(EtS)P(S)F(NH_2)$ | P.m.i., $SPF^+$ base peak | 256 |
| $(EtO)P(S)F(NMe_2)$ | P.m.i., base peak | 256 |
| $(MeO)P(O)F(Me)$ | P.m.i. | 241 |
| $(MeO)P(O)F_2$ | P.m.i. | 241 |
| $(MeO)P(S)F_2$ | P.m.i. | 241 |
| $(PhS)P(S)FCl$ | P.m.i. | 281 |
| $MeP(S)(SH)F$ | P.m.i., $H_2CP^+$ base peak | 265 |
| $\underset{\displaystyle S}{\overset{\displaystyle F}{Me\overset{\|}{\underset{\|}{P}}}}\text{-S-}\underset{\displaystyle S}{\overset{\displaystyle F}{\overset{\|}{\underset{\|}{P}}}}\text{-Me}$ | P.m.i. | 265 |
| $\underset{\displaystyle S}{\overset{\displaystyle F}{Et\overset{\|}{\underset{\|}{P}}}}\text{-S-}\underset{\displaystyle S}{\overset{\displaystyle F}{\overset{\|}{\underset{\|}{P}}}}\text{-Et}$ | P.m.i. | 265 |
| $(MeS)P(S)FCl$ | P.m.i., base peak | 282 |
| $(MeS)P(S)F(Me)$ | P.m.i., base peak | 282 |
| $(EtS)P(S)F(Et)$ | P.m.i. | 282 |
| $\underset{\displaystyle }{\overset{\displaystyle S \quad S}{Ph_2\overset{\|}{P}\text{-S-}\overset{\|}{P}Ph_2}}$ | P.m.i., $Ph_2P=S^+$ base peak | 254 |
| $\underset{\displaystyle }{\overset{\displaystyle S \quad S}{Ph_2\overset{\|}{P}\text{-O-}\overset{\|}{P}Ph_2}}$ | P.m.i., base peak | 254 |

*7.14 Compounds containing fluorocarbon or chlorocarbon groups bonded to a Group V element*

*7.14.1 Fluoroalkyls*

Several fluoromethyl-containing compounds of P and As have been studied. The differences in the main features of these spectra and those of the corresponding methyl compounds are largely due to the higher ionisation potentials of $\cdot CF_3$ (10.1 eV) and $F\cdot$ (17.4 eV) compared with $\cdot CH_3$ (9.9 eV) and $H\cdot$ (13.6 eV)[315]. Rearrangements leading to M–F-containing ions are also important.

Cavell and Dobbie have studied compounds of the types $(CF_3)_2PX$ (X = $CF_3$, H, F, Cl, Br, I) and $(CF_3)PX_2$ (X = H, F, Cl, Br, I)[289] and $(CF_3)_2AsX$ (X = $CF_3$, H) and $(CF_3)AsX_2$ (X = H, Cl)[290]. Similar trends were noted in the spectra of analogous P and As compounds. The abundances of ions from the P compounds only were reported (Table 8.44). The symmetrical compound $(CF_3)_3P$ has also been reported by Hawthorne et al.[182] (Table 8.44). The agreement between Hawthorne et al. and Dobbie and Cavell is not very good. In particular, the ions $P^+$ and $CF_3PF^+$ are not reported by the latter

Table 8.44

Ion Abundances from Some Trifluoromethyl Phosphines

| Composition of ion | X | | | | | |
|---|---|---|---|---|---|---|
| | I* | Br* | Cl* | F* | H* | CF₃* |
| $(CF_3)_2PX$ | 48.0[b] | 23.5 | 19.5 | 15.5[c] | 19.0[d] | 6.0 | 7.7[a] |
| $C_2F_5PX$ | | | | 2.5 | 8.0 | 11.5 | 11.3 |
| $CF_3PX$ | 9.5 | 5.0 | 2.0 | 9.5 | 7.5 | | |
| $CF_2PX$ | 3.5 | 17.0 | 13.0 | | 18.0 | 7.5 | 5.1 |
| $CFPX$ | | | 3.0 | | 3.5 | 21.0 | 14.7 |
| $PX$ | 4.5 | 4.0 | 4.0 | 6.5 | 1.5 | 11.0 | 8.0 |
| $P$ | | 1.5 | 1.5 | | 1.5 | | 8.8 |
| $CF_3P$ | 2.5 | 3.0 | 2.0 | | 3.0 | | |
| $CF_2P$ | 1.5 | 2.0 | 2.0 | | 3.0 | 4.5 | 4.0 |
| $CFP$ | | 1.0 | | | 2.0 | 3.5 | 2.6 |
| $CF_4P$ | 1.0 | | 1.0 | | | | |
| $PF_2$ | 10.5 | 9.5 | 14.5 | 63.0 | 23.5 | 31.5 | 15.5 |
| $PFX$ | 17.0 | 31.0 | 38.0 | | 6.0 | | 1.4 |
| $PF$ | 2.0 | 2.5 | 2.5 | | 3.5 | 3.5 | 7.0 |

Table 8.44 (continued)

| Composition of ion | X | | | | |
|---|---|---|---|---|---|
| | I | Br | Cl | F* | H* |
| $(CF_3)PX_2$ | 41.5[b] | 17.5[b] | 16.0 | 11.0 | 48.5[e] |
| $CF_3PX$ | | | | 9.0 | 1.0 |
| $CF_2PX$ | | | 2.0 | | 5.5 |
| $PX_2$ | 27.5 | 55.5 | 58.5 | 75.0 | 4.0 |
| $PX$ | 10.0 | 7.5 | 6.0 | 5.0 | 5.5 |
| $P$ | 1.5 | 1.5 | 1.0 | | 2.5 |
| $PF_2$ | 3.5 | 3.0 | 3.5 | | 24.0 |
| $PF$ | 2.5 | 1.5 | 1.0 | | 5.0 |
| $PFX$ | 13.5 | 13.5 | 12.0 | | 4.0 |

* Spectrum had a large $CF_3^+$ peak and, in some cases, $CF_2^+$ and $CF^+$ also.
[a] Other ions were found at $m/e = 133$ (7.4%) and $m/e = 114$ (6.5%). They were unassigned. Trace amounts of $(CF_3)_2PF^+$ and $(CF_3)_2P^+$ were also observed.
[b] A peak due to $X^+$ was observed.
[c] $C_2F_4^+$ observed.
[d] $CF_2H^+$ observed.
[e] Other ions were $CF_2PH^+$, $CF_2P^+$, $CFPH^+$, $PFH_2^+$, $CF_2H^+$, $CFH_2^+$ all in low abundances (i.e. 2–5%).

workers, neither were ions at $m/e = 133$ and 114 found by Hawthorne et al. but not assigned a formula. Parent ions were observed, usually in quite high abundance, for all compounds. For the series I, Br, Cl, F, the parent molecular ion abundances decreased along the series, possibly reflecting the ease of ionisation of the X atom[289].

Fragmentation occurred with simple bond cleavages of $P–CF_3$ and $P–X$ bonds and elimination of species such as F atoms, $CF_4$, $C_2F_6$, $PF_3$, and $\cdot HF_2$. Several metastable supported decompositions were observed by Dobbie and Cavell, viz.

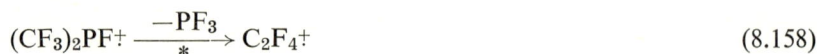

$$(CF_3)_2PH^+ \xrightarrow[*]{-\cdot CF_3} CF_3PH^+ \tag{8.157}$$
$$\downarrow * -CF_4$$
$$CF_2PH^+$$

$$(CF_3)_2PF^+ \xrightarrow[*]{-PF_3} C_2F_4^+ \tag{8.158}$$

References pp. 439–447

$$CF_3PH_2^+ \xrightarrow[*]{-\cdot HF_2} CFPH^+$$

$m/e = 51\downarrow*$

$(CF_2H^+ + PFH) \ and \ (PFH^+ + CF_2H)$ \hspace{2cm} (8.159)

$m/e = 51$ (doublet observed)

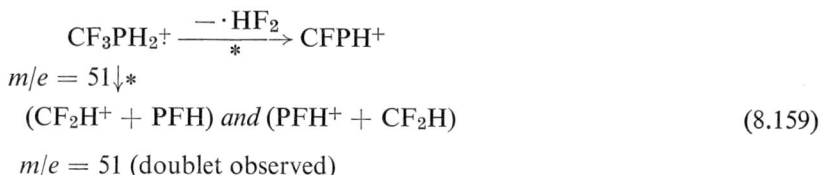

Dobbie and Cavell observed a metastable peak for the elimination of $CF_4$ from the $(CF_3)_2PCF_2^+$ ion. This was also found by Hawthorne *et al.* who observed the reactions in the scheme

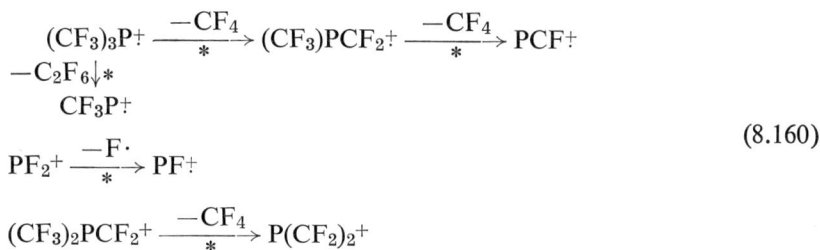

$$(CF_3)_3P^+ \xrightarrow[*]{-CF_4} (CF_3)PCF_2^+ \xrightarrow[*]{-CF_4} PCF^+$$
$$-C_2F_6\downarrow*$$
$$CF_3P^+$$

$$PF_2^+ \xrightarrow[*]{-F\cdot} PF^+$$

$$(CF_3)_2PCF_2^+ \xrightarrow[*]{-CF_4} P(CF_2)_2^+$$

(8.160)

This last reaction in scheme (8.160) is in marked contrast to the behaviour of the analogous As compound where another route to $(CF_2)_2As^+$ is favoured

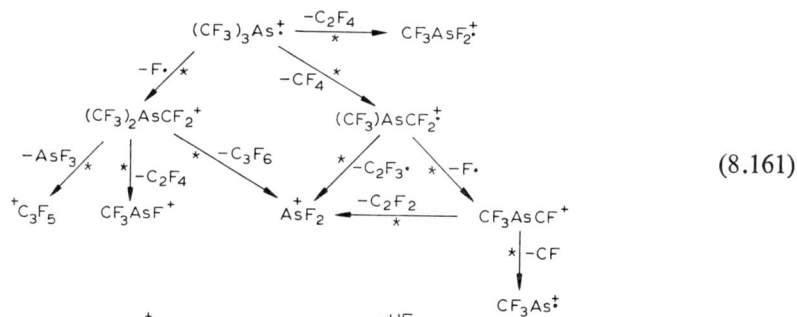

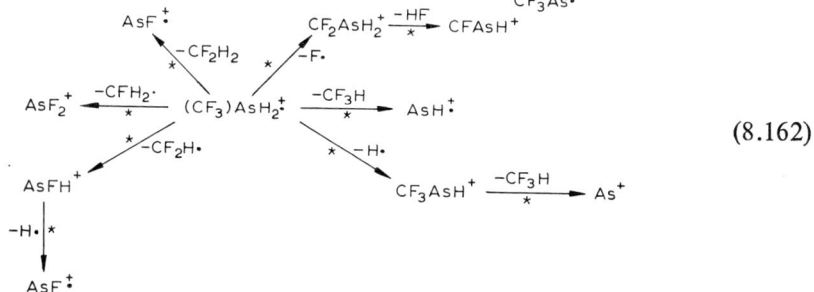

(8.161)

(8.162)

and $(CF_3)_2AsCF_2{}^+$ prefers to lose $AsF_3$ instead. The decompositions of $(CF_3)_3As$ and $(CF_3)AsH_2$ are given in schemes (8.161) and (8.162)[290].
No metastable transitions were found with the corresponding $CF_3AsCl_2$ compound.

Cowley and Pinnell[164] have reported the spectra of the cyclophosphines $(CF_3P)_4$ and $(CF_3P)_5$. Major peaks in the spectrum of the tetracyclic compound were due to the ions $(CF_3P)_4{}^+$, $(CF_3P)_nP^+$ ($n = 3$ or $2$), and $CF_3CP_2{}^+$ (base peak). From the pentamer, $(CF_3P)_5{}^+$, $(CF_3P)_4{}^+$, $(CF_3)_2P^+$, and $CF_3$-$CP_2{}^+$ (base peak) ions were observed. Later work by Dobbie and Cavell reported the base peak from $[(CF_3)P]_4$ to be due to the $PF_2{}^+$ ion.

Dobbie and Cavell[290] reported some metastable transitions observed in the spectra of $(CF_3As)_5$ and $(CF_3As)_4$ cyclic compounds.

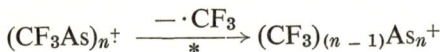

and

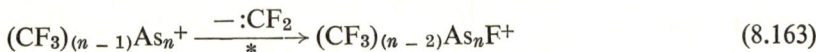
                                                                                                          (8.163)

Several rearrangement ions were observed including some of the types $As_nF_{(n-1)}{}^+$, $As_nF^+$ ($n = 2$–$4$), and $AsF_2{}^+$.

Some other compounds containing bonds between Group V elements which were studied by Dobbie and Cavell were $(CF_3)_4P_2$, $(CF_3P)_4$ (ref. 292), $(CF_3)_2P$–$P(CH_3)_2$, $(CH_3)_2P$–$As(CF_3)_2$, $(CF_3)_2P$–$As(CH_3)_2$, $(CH_3)_2As$–$As$-$(CF_3)_2$ (ref. 291), and $(CF_3)_4As_2$. Like the mononuclear species, typical fragmentation was by loss of $\cdot CF_3$, $:CF_2$, $CF_4$, $C_2F_4$, and $MF_3$ species. Rearranged ions containing M–F bonds were common (*e.g.* $PF_2{}^+$ being $16.5\%$ and $15.2\%$ for the $(CF_3)_4P_2$ and $(CF_3)_4P_4$ compounds respectively). Parent ions were reasonably abundant ($10$–$15\%$). In the case of $(CF_3)_4P_4$, the parent ion ($14.2\%$) was only slightly less abundant than the reported base peak ion, $PF_2{}^+$, ($15.2\%$). From the compounds which contained both $\cdot CH_3$ and $\cdot CF_3$ groups Dobbie and Cavell found base peaks $(CH_3)_2P$–$P(CF_3)^+$ ($16.6\%$), $(CH_3)_2P$–$As(CF_3)^+$ ($25\%$), and $As(CH_3)_2{}^+$ from both $(CH_3)_2As$–$As(CF_3)_2$ and $(CH_3)_2As$–$P(CF_3)_2$ in $38.2\%$ and $31.3\%$ abundance, respectively. They attempted to rationalise these results in terms of the comparative M–$CF_3$ and M–$CH_3$ bond strengths which they suggested were in the order $H_3C$–$P > F_3C$–$P \sim H_3C$–$As \sim F_3C$–$As$.

Several metastable transitions were observed from the spectra of the

$(CH_3)_2M-M'(CF_3)_2$ compounds. A generalised fragmentation scheme is given in (8.164).

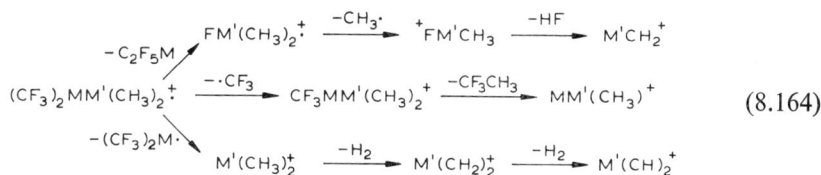

$$(8.164)$$

Compounds containing bonds between phosphorus and Group VI elements such as $(CF_3)_2P(S)X$, $[(CF_3)_2P]_2L$ have been studied by several workers. Dobbie et al.[293] have reported mass spectra for a series of $(CF_3)_2P(S)X$ compounds (Table 8.45). Parent molecular ions were established by accurate mass measurement in each case and were the base peaks for X = $NH_2$, F, Cl, Br, and $CF_3$. The base peaks in the spectra of X = $NMe_2$ and SH were $(CF_3)$-

Table 8.45

Ion Abundances from $(CF_3)_2P(S)X$ Compounds

| Composition of ion | % Abundance for X = | | | | | | |
|---|---|---|---|---|---|---|---|
| | NH₂[a] | NMe₂[b] | F | Cl[c] | Br | SH[d] | CF₃ |
| $(CF_3)_2P(S)X$ | 25.4 | 4.3 | 40.4 | 36.0 | 29.3 | 9.1 | 36.8 |
| $(CF_3)P(S)X$ | 19.4 | 25.7 | 7.1 | 8.3 | 20.2 | 5.1 | 2.8 |
| $(CF_3)PX$ | 2.5 | 3.0 | 14.1 | 4.7 | 3.6 | 11.0 | 2.9 |
| $(CF_3)PF$ | | | 5.9 | 2.7 | 1.9 | | |
| $FP(S)X$ | 19.8 | 16.2 | | 2.4 | 2.2 | 2.2 | |
| $FPX$ | 16.8 | 9.5 | | 16.8 | 10.8 | 13.9 | 14.2 |
| $P(S)X$ | 2.3 | 0.9 | 4.4 | 1.9 | 1.6 | | |
| $PF_2$ | 1.0 | 1.1 | 20.1 | 1.6 | 11.0 | 9.4 | 21.8 |
| $PS$ | 5.8 | 5.7 | 13.9 | 18.9 | 18.6 | 20.0 | 18.7 |

Other ions observed were:
[a] $PNH^+$ 7.0%.
[b] $P(S)NC_2H_5^+$ 1.8%, $CF_2NC_2H_6^+$ 3.2%, $PNCH_3^+$ 2.4%, $NC_2H_6^+$ 7.4%, $NC_2H_5^+$ 7.2%, $NC_2H_4^+$ 11.6%.
[c] $(CF_3)_2PS^+$ 3.5%.
[d] $(CF_3)_2PSH^+$ 12.2%, $F_2PS_2H^+$ 3.9%, $CF_3PH^+$ 1.4%, $CF_3P^+$ 1.9%, $CF_2PS^+$ 1.5%, $PS_2^+$ 1.7%, $CF_2SH^+$ 1.4%, $PSH^+$ 3.4%.

PS(NMe₂)⁺ and PS⁺, respectively. Fragmentation patterns were remarkably similar. Loss of $CF_3$ radicals, followed by loss of $CF_2S$ or $:CF_2$ were important processes.

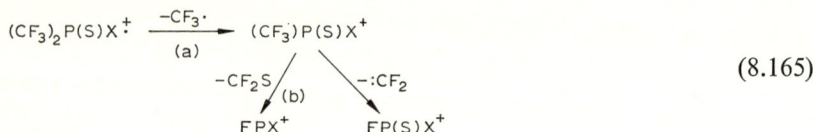

$$(CF_3)_2 P(S)X^{+\bullet} \xrightarrow[\text{(a)}]{-CF_3\bullet} (CF_3)P(S)X^{+}$$

$$-CF_2S \diagdown_{\text{(b)}} \qquad \diagdown -:CF_2$$

$$FPX^+ \qquad\qquad FP(S)X^+$$

(8.165)

Route (a) was metastable supported for X = NH₂, NMe₂, F, Cl, Br, SH, and CF₃; (b) was for X = Cl, Br, and CF₃. The metastable supported elimination of $:CF_2$ was observed from $(CF_3)P(NMe_2)^+$.

$$(CF_3)P-N(CH_3)_2^{+} \xrightarrow[*]{-:CF_2} F\overset{+}{P}N(CH_3)_2$$

(8.166)

Rearrangement ions with a P–F bond were quite abundant, (see Table 8.45 for abundances of FP(S)X and FPX species).

The mass spectra of the compounds [(CF₃)₂P]₂L (L = O or S) and [CF₃P]₄S were reported by Dobbie and Cavell[292]. As well as fragmentation by the loss of $CF_n$ and $PF_n$ fragments, these compounds also lost species with the P–L bond intact, viz.

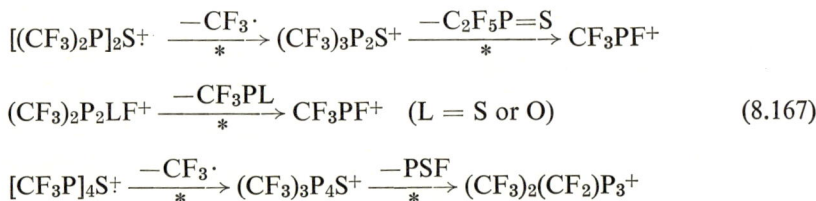

$$[(CF_3)_2P]_2S^{+\bullet} \xrightarrow[*]{-CF_3\bullet} (CF_3)_3P_2S^+ \xrightarrow[*]{-C_2F_5P=S} CF_3PF^+$$

$$(CF_3)_2P_2LF^+ \xrightarrow[*]{-CF_3PL} CF_3PF^+ \quad (L = S \text{ or } O)$$

(8.167)

$$[CF_3P]_4S^{+\bullet} \xrightarrow[*]{-CF_3\bullet} (CF_3)_3P_4S^+ \xrightarrow[*]{-PSF} (CF_3)_2(CF_2)P_3^+$$

Parent ions were observed for all compounds and the base peaks were $P_2F_3O^+$ (21.2%), $C_3F_9P_2S^+$ (16.5%), and $P=S^+$ (13.6%), respectively.

Other compounds containing fluorocarbon alkyl groups are listed in Table 8.46.

### 7.14.2 Compounds containing aromatic groups

The spectra of the compounds $(C_6F_5)_3M$ (M = P (refs. 174, 177), As (ref. 177) and Sb (ref. 177)) and $(C_6Cl_5)_3P$ (ref. 174) have been reported and are given in Table (8.47). All compounds formed parent molecular ions,

Table 8.46

Other Halocarbon Derivatives Reported

| Compound | Comments | Ref. |
|---|---|---|
| $F_2NCFCl_2$ | No p.m.i., but $F_2NCFCl^+$ and | 239 |
| | $FNCCl_2^+$, $CFCl_3^+$ base peak | 297 |
| $FN=CF_2$ | P.m.i., $CNF_2^+$ base peak | 294 |
| $CNF_2Cl$ | P.m.i., base peak | 294 |
| $CNF_2Br$ | P.m.i., $CNF_2^+$ base peak | 294 |
| $F_2N-CFBr_2$ | No p.m.i., expected fragment ions | 294 |
| $FN=C(CN)_2$ | P.m.i., $(CN)_2^+$ base peak | 295 |
| $FN=C(CN)F$ | P.m.i., $CNF^+$ base peak | 295 |
| $FN=C(CN)SF_5$ | P.m.i., base peak | 295 |
| $F_2C=N-NCF_2$ | P.m.i., $CNF^+$ base peak | 296 |
| $F_2C=N-NC(F)Cl$ | P.m.i., $CF^+$ base peak | 296 |
| $F_2C=N-NC(F)Br$ | P.m.i., $C_2N_2F_3^+$ base peak | 296 |
| $F_3CNH-NHCF_3$ | P.m.i., $CF_3^+$ base peak | 296 |
| $F_2NCF_2Cl$ | No p.m.i. but $CF_4N^+$ and $CClF_2^+$ | 297 |
| $CF_3NClF$ | No p.m.i., $CF_3^+$ base peak | 297 |
| $FC(O)NCO$ | P.m.i., base peak | 257 |
| $FC(O)N=SF_2$ | P.m.i. | 116 |
| $FC(O)N=SF_2(O)$ | P.m.i. | 129 |
| $(CF_3)_2NCF=CF_2$ | P.m.i. | 298 |
| $(CF_3)_2NCH=CF_2$ | P.m.i. | 298 |
| $(CF_3)_2NCF=C(F)H$ | P.m.i. | 298 |
| $(CF_3)_2N(CFBr)CF_2Br$ | No p.m.i., $C_4F_9N^+$ highest $m/e$ | 298 |
| $(CF_3)_2NCF=C(F)Br$ | P.m.i. | 298 |
| $(CF_3)_2NCF=C(F)Cl$ | P.m.i. | 298 |
| $(CF_3)_2NCBr=CF_2$ | P.m.i. | 298 |
| $(CF_3)_2NCHBrCF_2Br$ | No p.m.i., $C_4HBrF_8N^+$ highest $m/e$ | 298 |
| $CF_3N=CFC_2F_5$ | P.m.i. | 298 |
| $CF_3N=CFCF(Cl)CF_3$ | | 298 |
| $(CF_3)_2N-CH=CH_2$ | P.m.i. | 299 |
| $(CF_3)_2NCH=CHN(CF_3)_2$ | P.m.i. | 299 |
| $[(CF_3)_2N]_2C=CH_2$ | P.m.i. | 299 |
| $(CF_3)_2N-CH_2-CH_2X$ | | 299 |
| $(X = Br, I)$ | | |
| $(CF_3)_2NCH(Br)CH_2N(CF_3)_2$ | | 299 |
| $(CF_3)_2NCH(Br)CH_3$ | | 299 |
| $(CF_3)_2NCF-CF_2$ | P.m.i. | 300 |
| $\quad\quad\mid\quad\mid$ | | |
| $F_3COCF-CF_2$ | | |

Table 8.46 (continued)

| Compounds | Comments | Ref. |
|---|---|---|
| $(CF_3)_2NCF–CF_2$<br>$\quad\ \ \ \|\quad\ \|$<br>$(CF_3)_2NCF–CF_2$ | P.m.i. | 300 |
| $(CF_3)_2NCF_2N(CF_3)OCF_3$ | P.m.i. | 300 |
| $(CF_3)_2NN(CF_3)CF_2OCF_3$ | No p.m.i. but $C_4F_{11}N_2O^+$ at highest $m/e$ | 300 |
| $(CF_3)_2NN(CF_3)CF_2N(CF_3)COF$ | No p.m.i. but $C_5F_{14}N_3^+$ at highest $m/e$ | |
| $CF_3OCF\ –CF\ –\ \ CF–CF–N(CF_3)_2$<br>$\quad\ \ \|\quad\ \|\quad\ \ \|\quad\ \|$<br>$\quad CF_2–CF_2\ \ F_2C\ –CF_2$ | P.m.i. | 300 |
| $(CF_3)_2NN(CF_3)NO$ | No p.m.i. but $C_3F_8N_2^+$ at highest $m/e$ | 301 |
| $(CF_3)_2NOCF_3$ | P.m.i. | 301 |
| $(CF_3)_2N–N(CF_3)_2$ | P.m.i. and metastable supported reaction; | 301 |
| $(CF_3)_2N–N(CF_3)_2^+ \xrightarrow[*]{-CF_4N\cdot} (CF_3)_2NCF_2^+$ | | |
| $[(CF_3)_2N–N(CF_3)]_2$ | No p.m.i. but $C_4F_{12}N_3^+$ at highest $m/e$ | 301 |
| $(CF_3)_2N–N(CF_3)NO_2$ | No p.m.i. but $C_3F_8N_2^+$ at highest $m/e$ | 301 |
| $(CF_3)_2N(O)COF$ | P.m.i. | 301 |
| $(CF_3)_2N–O–N(CF_3)_2$ | P.m.i. | 301 |
| $F_2N–C–N(CF_2)NF_2$<br>$\quad\quad\ \|\|$<br>$\quad\quad NF$ | $C_2N_3F_5^+$ base peak | 302 |
| $F_2C=N–CF_2–N=CF_2$ | P.m.i., $CF_2N^+$ base peak | 303 |
| $(F_2C=N–CF_2)_2$ | No p.m.i. but $C_2F_3N^+$ base peak | 303 |
| $F_2C=N–CF_2–CF–N=CF_2$<br>$\quad\quad\quad\quad\ \ \|$<br>$\quad\quad\quad\quad\ CF_3$ | No p.m.i. observed | 303 |
| $[F_2C=N–CF(CF_3)_2]$ | No p.m.i. observed | 303 |
| $F_2C=NCF_2–CF–N=CF_2$<br>$\quad\quad\quad\quad\ \|$<br>$\quad\quad\quad\quad Cl$ | No p.m.i. observed | 303 |
| $CF_3NOCF(COF)CF_2$<br>$\quad\underline{\qquad\qquad\qquad}$ | P.m.i. | 304 |
| $CF_3N=CF(COF)$ | P.m.i. | 304 |
| $(CF_3)_2NOCOX$ | | |
| $\quad X = F$ | P.m.i. | 305 |
| $\quad X = Cl$ | No p.m.i. observed | 305 |
| $\quad X = CF_3$ | No p.m.i. observed | 305 |
| $\quad X = C_3F_7$ | No p.m.i. observed | 305 |

Table 8.46 (continued)

| Compounds | Comments | Ref. |
|---|---|---|
| X = $(CF_3)_2NO$ | No p.m.i. observed | 305 |
| $(CF_3)_2NPOF_2$ | P.m.i. | 306 |
| $(CF_3)_2NONMe_2$ | No p.m.i., $C_2F_5NON(CH_3)_2^+$ at highest $m/e$ | 307 |
| $[(CF_3)_2N]_2PCF_3$ | P.m.i. | 308 |
| $(CF_3)_2NAs(CF_3)_2$ | P.m.i. | 308 |
| $CH_3CF_2NFCl$ | P.m.i. | 309 |
| $(CF_3)_2C\text{–}C(O)F$ <br> | <br> $NF_2$ | No p.m.i. observed | 310 |
| $(CF_3)_2C(NF_2)C(O)OSO_2F$ | No p.m.i. observed | 310 |
| $F_2NOC(O)F$ | P.m.i. | 310 |
| $(C_2F_5P)_3$ | P.m.i. | 131 |
| $CF_3(Et)P(CH_2)_2CF_3$ | P.m.i. | 158 |
| $CF_3P[N(CH_3)_2]_2$ | P.m.i. | 312 |
| $C_3F_7P[N(CH_3)_2]_2$ | No p.m.i. but $C_3F_7PN(CH_3)_2^+$ base peak | 312 |
| | P.m.i. | 313 |
| $(CF_3)_2MH$ | P.m.i. ⎫ M = P, As | 314 |
| $(CF_3)MH_2$ | P.m.i. ⎭ | 314 |
| $F_2PCO_2R$ |  | 315 |
| R = $CF_3$ |  | 315 |
| R = $C_2F_5$ |  | 315 |
| R = $C_3F_7$ |  | 315 |
| $(CF_3)_2C(CN)OPF_2$ | P.m.i. | 309 |
| $(CF_3)_2C(CCN)X$ |  | 309 |
| X = $OP(O)F_2$ | No p.m.i. observed | 309 |
| X = $OP(S)F_2$ | No p.m.i. observed | 309 |
| $(CF_3)_2C=NC(OPF_2)(CF_3)_2$ |  | 309 |
| $(CF_3)_2C(N_3)OP(O)F_2$ |  | 309 |
| $(CF_3)_2(NCS)OP(O)F_2$ | P.m.i. observed in a mixture of | 309 |
| $(CF_3)_2(SCN)OP(O)F_2$ | compounds | 309 |
| | P.m.i., high resolution | 316 |
| $C_6F_5P[N(CH_3)_2]_2$ | P.m.i. | 317 |
| $C_6F_5P(Cl)N(CH_3)_2$ | P.m.i. | 317 |

Table 8.46 (continued)

| Compounds | Comments | Ref. |
|---|---|---|
| $C_6F_5PCl_2$ | P.m.i. | 317 |
| $C_6F_5P[N(Bu^t)_2]_2$ | P.m.i. | 317 |
| $(C_6F_5)_2PCl$ | P.m.i. | 317 |
| $(C_6F_5)_2PN(CH_3)_2$ | P.m.i. | 317 |
| $(C_6F_5)_2PN(H)CH_3$ | P.m.i. | 317 |
| $(C_6F_5)_2P[N(H)CH_2Ph]$ | P.m.i. | 317 |
| $C_6F_5PPh_2$ | P.m.i. | 318 |
| $C_6Cl_5PPh_2$ | P.m.i. | 318 |
| $(C_6Cl_5)_2PPh$ | P.m.i. | 318 |
| $C_6Cl_5P(O)Ph_2$ | P.m.i. | 318 |
| $(C_6Cl_5)_2P(O)Ph$ | P.m.i. | 318 |
| $(C_6Cl_5)_3P$ | P.m.i. | 318 |
| $(C_5Cl_4N)PPh_2$[a] | P.m.i. | 318 |
| $(C_5Cl_4N)_2PPh$[a] | P.m.i. | 318 |
| $(C_5Cl_4N)_3P$[a] | P.m.i. | 318 |
| $(C_5Cl_4N)_3P(O)$[a] | P.m.i. | 318 |
| $(C_5Cl_4N)_2P(O)Ph$[a] | P.m.i. | 318 |
| $(C_5Cl_4N)P(O)Ph_2$[a] | P.m.i. | 318 |

[a] $C_3Cl_4N$ is the 2,3,5,6-tetrachloro 4-pyridyl group.

Table 8.47

Ion Abundances from the Compounds $(C_6F_5)_3M$ (M = P, As, Sb) and $(C_6Cl_5)_3P$

| Composition of ion | % Abundance | | | |
|---|---|---|---|---|
| | $(C_6F_5)_3P$ | $(C_6F_5)_3As$ | $(C_6F_5)_3Sb$ | $(C_6Cl_5)_3P$ |
| $(C_6X_5)_3M$ | 61.9 | 16.3 | 26.9 | 29.2 |
| $(C_6X_5)_2MC_6X_4$ | 2.5 | 0.9 | 1.4 | 11.9 |
| $(C_6X_5)_2M$ | 24.8 | 15.7 | 14.3 | 6.0 |
| $(C_6X_4)_2M$ | | 1.4 | | 15.1 |
| $(C_6X_5)MX$ | 7.1 | 16.3 | 12.7 | |
| $(C_6X_5)M$ | 2.7 | 26.2 | 15.4 | 37.8 |
| $(C_6X_4)M$ | | 1.4 | | |
| $MX_2$ | 1.0 | 19.7 | 24.3 | |
| $MX$ | | 2.1 | 4.4 | |
| $M$ | | | 0.6 | |

which were the base peaks for $(C_6F_5)_3P$ and $(C_6F_5)_3Sb$. The other compounds had the ions $C_6F_5As^+$ and $C_6Cl_5P^+$ as base peaks. The spectra of the fluoro compounds showed ions due to rearrangement processes forming M–F bonds. However, the straightforward elimination of one or more fluorine atoms to give $(C_6F_5)_2MC_6F_4^+$ or $(C_6F_4)_2M^+$ ions was apparently not as favourable as the corresponding eliminations of Cl· (or H· from $(C_6H_5)_3M$). The metastable supported fragmentation reactions of $(C_6F_5)_3P$ observed by Rake and Miller are shown in scheme (8.168)[177].

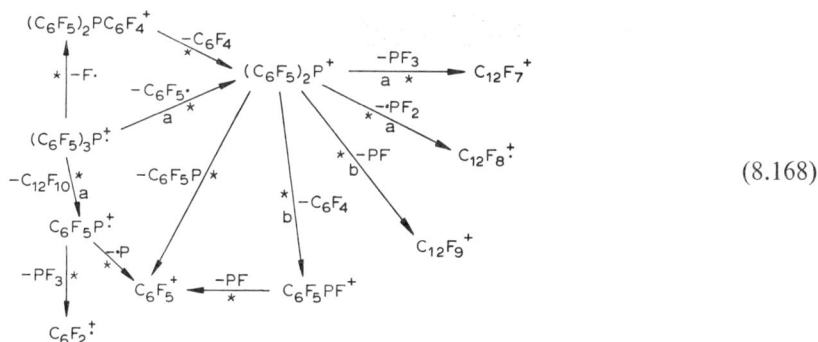

$$(8.168)$$

N.B. a indicates similar reactions for the As and Sb compounds;
   b indicates similar reactions for the As compound.

The fluorocarbon ions were found to decompose by loss of species such as $F_2$, $:CF_2$, $\cdot F$, $\cdot CF$, $C_3F_2$, $\cdot C_3F_3$, $\cdot C_5F$, $\cdot C_5F_3$, and $C_6F_4$. Other reactions observed for the arsenic compounds were

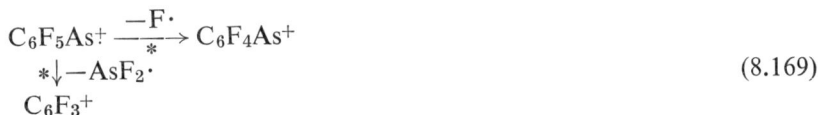

$$C_6F_5As^+ \xrightarrow[*]{-F\cdot} C_6F_4As^+$$
$$*\downarrow -AsF_2\cdot$$
$$C_6F_3^+$$

$$(8.169)$$

Miller[174] has reported some points in the spectra of the compounds $C_6F_5PX_2$ (X = F, Cl, Br), $(C_6F_5)_2PX$ (X = Cl, Br), $(C_6F_5)_3PCl_2$, $(C_6F_5)_2PCl_3$, $(C_6F_5)_3P=L$ (L = O or S) and $(C_6F_5)_2P-P(C_6F_5)_2$. The spectrum of $(C_6F_5)PF_2$ showed an abundant parent molecular ion and $C_6F_5PF^+$. A metastable supported reaction (8.170) was observed

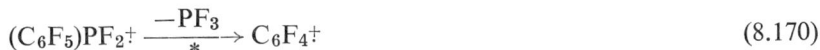

$$(C_6F_5)PF_2^+ \xrightarrow[*]{-PF_3} C_6F_4^+$$

$$(8.170)$$

Apparently the compounds $C_6F_5PBr_2$ and $(C_6F_5)_2PBr$ underwent redistributions reactions in the sample inlet since they gave rise to ions $(C_6F_5)_2PBr^+$ and $(C_6F_5)_3PBr^+$, respectively. (The compounds $PhPCl_2$, $Ph_2PCl$ reacted similarly.) A molecule of chlorine was lost from $(C_6F_5)PCl_2$ and $(C_6F_5)_3PCl_2$ to give the ions $(C_6F_5)_3P^+$ and $(C_6F_5)_2PCl^+$ which were the highest $m/e$ ions observed.

The P–P bonded compound $(C_6F_5)_4P_2$ produced an abundant parent molecular ion and $(C_6F_5)_2P^+$ as the base peak. Miller reported the following fragmentation scheme of metastable supported decompositions for $(C_6F_5)_4P_2$.

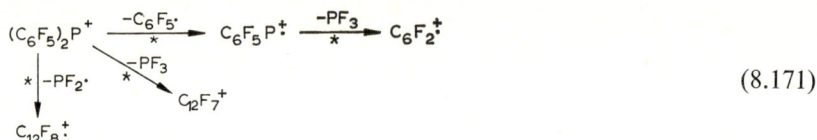

$$(C_6F_5)_2P^+ \xrightarrow[*]{-C_6F_5\cdot} C_6F_5P^+ \xrightarrow[*]{-PF_3} C_6F_2^+ \tag{8.171}$$

The $(C_6F_5)_3P=L$ compounds also had $(C_6F_5)_2P^+$ as the base peak.

Rake and Miller[177] have studied some mixed phenyl–fluorophenyl compounds of phosphorus, arsenic, and antimony in detail. Parent molecular ions were found for all the compounds $Ph_2P(C_6F_5)$, $PhP(C_6F_5)_2$, $Ph_2As-(C_6F_5)$, $PhAs(C_6F_5)_2$, and $PhSb(C_6F_5)_2$; they were the base peaks in the spectra of the phosphorus compounds. The base peaks in the monophenyl-arsenic and antimony compounds were due to the $C_6F_5M^+$ ion; that of $Ph_2AsC_6F_5$ was due to $C_6H_5As^+$. Fragmentation was by loss of H or F atoms, simple P–aromatic group bond cleavage, elimination of entities such as $C_6F_4$, $C_6F_5H$, $C_{12}F_{10}$ or HF and a number of M–F containing ions were formed by rearrangement processes. Rake and Miller observed several metastable supported decomposition reactions. The reactions of ions containing both hydrocarbon and fluorocarbon residues are shown in schemes (8.172)–(8.174).

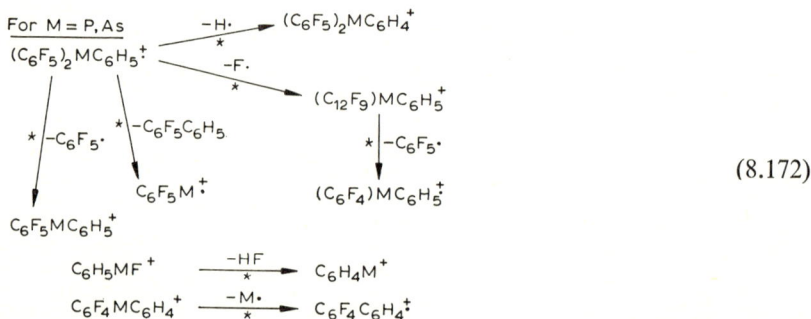

$$\tag{8.172}$$

From $C_6F_5P(C_6H_5)_2$,

$C_6F_5P(C_6H_5)_2^{+\cdot} \xrightarrow[\ast]{-F\cdot} C_6F_4P(C_6H_5)_2^{+} \xrightarrow[\ast]{-C_6H_6} C_6F_4PC_6H_4^{+}$

$-C_6F_5P\cdot \downarrow \ast \qquad \ast \diagdown -C_6F_5H \qquad \xrightarrow[\ast]{-C_6H_5\cdot} C_6F_5PC_6H_5^{+} \qquad \ast \diagup -HF$

$C_6H_5PC_6H_4^{+\cdot}$

$\ast \diagdown {-HF, \atop -P\cdot}$

$C_{12}H_{10}^{+\cdot}$

$C_6H_4C_6F_4^{+\cdot}$

(8.173)

From $C_6F_5As(C_6H_5)_2$,

$C_6F_5As(C_6H_5)_2^{+\cdot} \xrightarrow[\ast]{-HF, -C_6H_5\cdot \atop \ast} C_6F_4AsC_6H_4^{+}$

$\diagdown -C_6F_5As\cdot \atop \ast$

$C_{12}H_{10}^{+\cdot}$

(8.174)

and $C_6F_5AsC_6H_5^{+} \xrightarrow[\ast]{-HF} C_6F_4AsC_6H_4^{+}$

$C_6H_5As^{+}F \xrightarrow[\ast]{-HF} C_6H_4As^{+}$

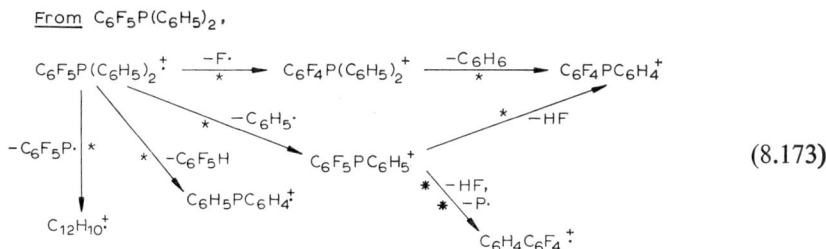

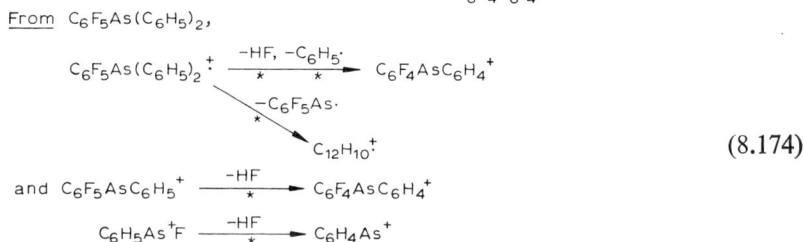

A large number of fluorocarbon derivatives of Group V elements have been characterised or identified by mass spectrometry. These compounds are listed in Table 8.46.

REFERENCES

1 J. H. Beynon, *Mass Spectrometry and Its Applications to Organic Chemistry*, Elsevier, Amsterdam, 1960.
2 H. Budzikiewicz, C. Djerassi and D. H. Williams, *Mass Spectrometry of Organic Compounds*, Holden-Day, New York, 1967.
3 J. H. Beynon, R. A. Saunders and A. E. Williams, *The Mass Spectra of Organic Molecules*, Elsevier, Amsterdam, 1968.
4 R. C. Weast (Ed.), *Handbook of Chemistry and Physics*, The Chemical Rubber Publishing Co., Cleveland, Ohio, U.S.A., 49th edn., 1968–69.
5 S. Trajmar, J. K. Rice and A. Kuppermann, *Advan. Chem. Phys.*, 18 (1970) 42.
6 E. N. Lassettre, A. Skerbele and V. D. Meyer, *J. Chem. Phys.*, 45 (1966) 3214.
7 W. C. Price, *Endeavour*, 26 (1967) 75.
8 D. W. Turner, C. Baker, A. D. Baker and C. R. Brundle, *Molecular Photoelectron Spectroscopy*, Wiley-Interscience, London, 1970, 34.
9 L. J. Kieffer and R. J. Van Brunt, *J. Chem. Phys.*, 46 (1967) 2728.
10 D. W. Vance, *J. Chem. Phys.*, 48 (1968) 1872.
11 J. J. Leventhal, T. F. Moran and L. Friedman, *J. Chem. Phys.*, 46 (1967) 4666.
12 W. B. Maier, *J. Chem. Phys.*, 47 (1967) 859.
13 M. V. Tikhomirov, V. N. Komarov and N. N. Tunitskii, *Russ. J. Phys. Chem.*, 38 (1964) 515.
14 M. S. B. Munson, F. H. Field and J. L. Franklin, *J. Chem. Phys.*, 37 (1962) 1790.

15 R. C. Lao, R. W. Rozett and W. S. Koski, *154th Meeting Am. Chem. Soc., Sept. 10–15th 1967, Abstr. V. 90*.

16 R. Johnson, H. L. Brown and M. A. Biondi, *J. Chem. Phys.*, 52 (1970) 5080.

17 T. M. Donahue, *Science*, 159 (1968) 489.

18 K. D. Carlson, F. J. Kohl and O. M. Uy, in *Advances in Chemistry Series*, J. L. Margrave (Ed.), Vol. 72, 1968, p. 245.

19 J. Carrette and L. Kerwin, *Can. J. Phys.*, 39 (1961) 1300.

20 L. Kerwin, *Can. J. Phys.*, 32 (1954) 757.

21 J. S. Kane and J. H. Reynolds, *J. Chem. Phys.*, 25 (1956) 342.

22 V. S. Ban and B. E. Knox, *Intern. J. Mass Spectrom. Ion Phys.*, 3 (1969) 131.

23 L. Brewer and J. S. Kane, *J. Phys. Chem.*, 59 (1955) 105.

24 J. Drowart and P. Goldfinger, *J. Chim. Phys.*, 55 (1958) 721.

25 K. A. Gingerich, *J. Phys. Chem.*, 68 (1964) 768.

26 A. J. H. Boerboom, H. W. Reyn, H. F. Vugt and J. Kistemaker, *Physica*, 30 (1964) 2137.

27 G. De Maria, J. Drowart and M. G. Ingham, *J. Chem. Phys.*, 31 (1959) 1076.

28 F. J. Kohl, O. M. Uy and K. D. Carlson, *J. Chem. Phys.*, 47 (1967) 2667.

29 G. F. Voronin, *Russ. J. Phys. Chem.*, 40 (1966) 744.

30 K. A. Gingerich, *J. Phys. Chem.*, 73 (1969) 2734.

31 F. J. Kohl, J. E. Prusaczyk and K. D. Carlson, *J. Am. Chem. Soc.*, 89 (1967) 5501.

32 L. Rovner, A. Drowart and J. Drowart, *Trans. Faraday Soc.*, 63 (1967) 2906.

33 G. De Maria, K. A. Gingerich and V. Piacente, *J. Chem. Phys.*, 49 (1968) 4705.

34 M. Hock and K. S. Hinge, *J. Chem. Phys.*, 35 (1961) 450.

35 D. R. Stull and G. C. Sinke, *Thermodynamic Properties of the Elements, Advances in Chemistry Series*, Vol. 18, Am. Chem. Soc., 1956.

36 J. B. Westmore, H. Fujisaki and A. W. Tickner, in *Advances in Chemistry Series*, J. L. Margrave (Ed.), Vol. 72, Am. Chem. Soc., 1968, p. 231.

37 J. B. Westmore, K. H. Mann and A. W. Tickner, *J. Phys. Chem.*, 68 (1964) 606.

38 D. W. Muenow, O. M. Uy and J. L. Margrave, *J. Inorg. Nucl. Chem.*, 32 (1970) 3459.

39 V. S. Ban and B. E. Knox, *J. Chem. Phys.*, 52 (1970) 248.

40 P. Goldfinger and M. Jeunehomme, in *Advan. Mass Spectrometry, Proc. Conf. Univ. London, 1958*, J. D. Waldron (Ed.), Pergamon Press, London, 1959, p. 534.

41 R. F. Porter and C. W. Spencer, *J. Chem. Phys.*, 32 (1960) 943.

42 C. L. Sullivan, J. E. Prusacyzyk and K. D. Carlson, *J. Chem. Phys.*, 53 (1970) 1289.

43 V. S. Ban and B. E. Knox, *J. Chem. Phys.*, 52 (1970) 243.

44 D. J. Cubicciotti, *J. Phys. Chem.*, 67 (1963) 1385.

45 O. M. Uy and J. Drowart, *Trans. Faraday Soc.*, 65 (1969) 3221.

46 F. E. Saalfeld and H. J. Svec, *Inorg. Chem.*, 2 (1963) 46.

47 N. V. Larin and I. L. Agafonov, *Tr. Khim. Tekhnol.*, 1 (1967) 90.

48 N. V. Larin, G. G. Devyatykh and I. L. Agafonov, *Russ. J. Inorg. Chem.*, 9 (1964) 110.

49 N. V. Larin, I. L. Agafonov and S. M. Vlasov, *Russ. J. Inorg. Chem.*, 13 (1968) 1.

50 H. Neuert and H. Clasen, *Z. Naturforsch.*, 7a (1952) 410.

51 T. P. Fehlner and R. B. Callen, in *Advances in Chemistry Series*, J. L. Margrave (Ed.), Vol. 72, Am. Chem. Soc., 1968, p. 181.

52 Y. Wada and R. W. Kiser, *Inorg. Chem.*, 3 (1964) 174.

53 M. Halmann, *J. Chem. Soc.*, (1962) 3270.

THE MAIN GROUP V ELEMENTS                                                441

54 F. E. Saalfeld and M. V. McDowell, *Inorg. Chem.*, 6 (1967) 96.
55 J. Fischler and M. Halmann, *J. Chem. Soc.*, (1964) 31.
56 W. C. Price and T. R. Passmore, *Discussions Faraday Soc.*, 35 (1963) 232.
57 G. R. Branton, D. C. Frost, C. A. McDowell and I. A. Stenhouse, *Chem. Phys. Letters*, 5 (1970) 1.
58 S. R. Gunn and L. G. Green, *J. Phys. Chem.*, 65 (1961) 779.
59 W. R. Cullen and D. C. Frost, *Can. J. Chem.*, 40 (1960) 390.
60 H. Ebinghaus, K. Kraus, W. Meuller-Duysing and H. Neuert, *Z. Naturforsch.*, 19a (1964) 732.
61 F. E. Saalfeld and H. J. Svec, *Inorg. Chem.*, 2 (1963) 50.
62 T. P. Fehlner, *J. Am. Chem. Soc.*, 90 (1968) 6062.
63 N. N. Grishin, G. M. Bogolyubov and A. A. Petrov, *J. Gen. Chem. USSR*, 38 (1968) 2595.
64 P. Royen and C. Rocktaschel, *Z. Anorg. Allgem. Chem.*, 346 (1966) 290.
65 T. P. Fehlner, *J. Am. Chem. Soc.*, 89 (1967) 6477.
66 J. R. Eyler, *Inorg. Chem.*, 9 (1970) 981.
67 M. Halmann and I. Platzner, *J. Phys. Chem.*, 71 (1967) 4522.
68 I. Platzner, *Israel J. Chem.*, 6 (1968) P34.
69 R. M. Reese and V. H. Dibeler, *J. Chem. Phys.*, 24 (1956) 1175.
70 R. C. Petry, *J. Am. Chem. Soc.*, 82 (1960) 2400.
71 F. D. Rossini et al., Selected Values of Chemical Thermodynamic Properties, *Natl. Bur. Std., Circ. 500* (1952).
72 J. C. Green, D. I. King and J. H. D. Eland, *J. Chem. Soc. D*, (1970) 1121.
73 A. A. Sandoval, H. C. Moser and R. W. Kiser, *J. Phys. Chem.*, 67 (1963) 124.
74 G. G. Devyatkh, V. G. Rachkov and I. L. Agafonov, *Russ. J. Inorg. Chem.*, 13 (1968) 1497.
75 M. Halmann and Y. Klein, *J. Chem. Soc.*, (1964) 4324.
76 P. Kusch, A. Hustrulid and J. T. Tate, *Phys. Rev.*, 52 (1937) 840.
77 R. W. Kiser, J. G. Dillard and D. L. Dugger, in Advances in Chemistry Series, J. L. Margrave (Ed.), Vol. 72, Am. Chem. Soc., 1968, p. 153.
78 A. Finch, A. Hameed, P. J. Gardner and N. Paul, *Chem. Commun.*, (1969) 391.
79 J. K. Ruff and G. Paulett, *Inorg. Chem.*, 3 (1964) 998.
80 T. A. Blazer, R. Schmutzler and I. K. Gregor, *Z. Naturforsch.*, 24b (1969) 1081.
81 T. Kennedy, D. S. Payne, R. I. Reed and W. Sneddon, *Proc. Chem. Soc.*, (1959) 133.
82 R. Rogowski and K. Cohn, *Inorg. Chem.*, 7 (1968) 2193.
83 E. D. Loughran and C. Mader, *J. Chem. Phys.*, 32 (1960) 1578.
84 C. B. Colburn and F. A. Johnson, *J. Chem. Phys.*, 32 (1960) 1869.
85 A. Kennedy and C. B. Colburn, *J. Chem. Phys.*, 35 (1961) 1892.
86 G. T. Armstrong, S. Marantz and C. F. Coyle, *Natl. Bur. Std., Rpt. No. 6584*, Oct., 1959.
87 L. H. Piette, F. A. Johnson, K. A. Booman and C. B. Colburn, *J. Chem. Phys.*, 35 (1961) 1481.
88 J. T. Herron and V. H. Dibeler, *J. Res. Natl. Bur. Std.*, 65A, (1961), 405; *J. Chem. Phys.*, 33 (1960) 1695.
89 F. A. Johnson, *Inorg. Chem.*, 5 (1966) 149.
90 D. Sloan and P. L. Timms, *Chem. Commun.*, (1968) 1540.

91  A. H. Cowley and S. T. Cohen, *Inorg. Chem.*, 4 (1965) 1221.
92  C. D. Schmulbach, A. G. Cook and V. R. Miller, *Inorg. Chem.*, 7 (1968) 2463.
93  C. E. Brion and N. L. Paddock, *J. Chem. Soc. A*, (1968) 388.
94  C. E. Brion and N. L. Paddock, *J. Chem. Soc. A*, (1968) 392.
95  G. E. Coxon, T. F. Palmer and D. B. Sowerby, *J. Chem. Soc. A*, (1967) 1568.
96  G. E. Coxon, T. F. Palmer and D. B. Sowerby, *J. Chem. Soc. A*, (1969) 358.
97  H. W. Roesky and W. Grosse-Bowring, *Z. Naturforsch.*, 24b (1969) 1250.
98  E. Niecke, O. Glemser and H. Thamm, *Chem. Ber.*, 103 (1970) 2865.
99  M. K. Feldt and T. Moeller, *J. Inorg. Nucl. Chem.*, 30 (1968) 2351.
100 M. Biddlestone and R. A. Shaw, *J. Chem. Soc. A*, (1970) 1750.
101 G. E. Brion, D. J. Oldfield and N. L. Paddock, *Chem. Commun.*, (1966) 226.
102 G. R. Branton, C. E. Brion, D. C. Frost, K. A. R. Mitchell and N. L. Paddock, *J. Chem. Soc. A*, (1970) 151.
103 J. P. Guertin, K. O. Christie and A. E. Pavlath, *Inorg. Chem.*, 5 (1966) 1921.
104 O. Glemser and A. Smalc, *Angew. Chem. Intern. Ed. Engl.*, 8 (1969) 517.
105 A. R. Young, T. Hirota and S. I. Morrow, *J. Am. Chem. Soc.*, 86 (1964) 20.
106 M. H. Cheng, M. Chiang, E. A. Gidason, B. H. Mahon, C. W. Tsao and A. S. Werner, *J. Chem. Phys.*, 52 (1970) 5518.
107 J. F. Paulson, *J. Chem. Phys.*, 52 (1970) 959.
108 H. D. Sharma, R. E. Jervis and K. Y. Wong, *J. Phys. Chem.*, 74 (1970) 923.
109 R. P. H. Gasser, M. F. King and P. R. Vaight, *Trans. Faraday Soc.*, 64 (1968) 2852.
110 D. R. Bates, *Contemp. Phys.*, 11 (1970) 105.
111 J. A. D. Stockdale, R. N. Compton, G. S. Hurst and P. W. Reinhardt, *J. Chem.Phys.*, 50 (1969) 2176.
112 R. F. Mathis, B. R. Turner and J. A. Rutherford, *J. Chem. Phys.*, 49 (1968) 2051.
113 E. Lindholm and G. Sahlstrom, *Intern. J. Mass Spectrom. Ion Phys.*, 4 (1970) 465.
114 D. W. Turner *et al.*, ref. 8 pp. 61, 37, and 86, respectively.
115 A. F. Clifford and G. R. Zeilenga, *Inorg. Chem.*, 8 (1969) 1789.
116 U. Biermann and O. Glemser, *Chem. Ber.*. 100 (1967) 3795.
117 H. W. Roesky, G. Holtschneider and H. H. Giere, *Z. Naturforsch.*, 25b (1970) 252.
118 A. Hass and P. Schott, *Chem. Ber.*, 101 (1968) 3407.
119 O. Glemser, H. W. Roesky and P. R. Heinge, *Angew. Chem. Intern. Ed. Engl.*, 6 (1967) 710.
120 R. J. Shozda and J. A. Vernon, *J. Org. Chem.*, 32 (1967) 2876.
121 R. W. Rudolph and R. W. Parry, *Inorg. Chem.*, 4 (1965) 1339.
122 P. M. Treichel, R. A. Goodrich and S. B. Pierce, *J. Am. Chem. Soc.*, 89 (1967) 2017.
123 M. Lustig and H. W. Roesky, *Inorg. Chem.*, 9 (1970) 1289.
124 P. W. Schenk and B. Leutner, *Angew. Chem. Intern. Ed. Engl.*, 5 (1966) 898.
125 T. L. Charlton and R. G. Cavell, *Inorg. Chem.*, 7 (1968) 2195.
126 T. L. Charlton and R. G. Cavell, *Inorg. Chem.*, 9 (1970) 379.
127 T. L. Charlton and R. G. Cavell, *Inorg. Chem.*, 8 (1969) 281.
128 K. Seppelt and W. Sundermeyer, *Angew. Chem. Intern. Ed. Engl.*, 8 (1969) 771.
129 J. K. Ruff, *Inorg. Chem.*, 5 (1966) 1787.
130 O. Glemser, R. Mews and H. W. Roesky, *Chem. Ber.*, 102 (1969) 1523.
131 W. Sundermeyer, *Angew. Chem. Intern. Ed. Engl.*, 6 (1967) 90.
132 M. Lustig and G. H. Cady, *Inorg. Chem.*, 2 (1963) 389.

133 H. W. Roesky, *Angew. Chem. Intern. Ed. Engl.*, 7 (1968) 630.
134 R. L. Patton and W. L. Jolly, *Inorg. Chem.*, 9 (1970) 1079.
135 H. W. Roesky and W. Grosse-Bowring, *Chem. Ber.*, 103 (1970) 2281.
136 R. L. McKenny and N. R. Fetter, *J. Inorg. Nucl. Chem.*, 30 (1968) 2927.
137 H. W. Roesky and W. Grosse-Bowring, *Inorg. Nucl. Chem. Letters*, 5 (1969) 597.
138 H. W. Roesky and L. F. Grimm, *Chem. Ber.*, 102 (1969) 2319.
139 H. W. Roesky and L. F. Grimm, *Chem. Ber.*, 103 (1970) 1665.
140 H. W. Roesky, *Chem. Ber.*, 101 (1968) 3679.
141 H. W. Roesky and L. F. Grimm, *Angew. Chem. Intern. Ed. Engl.*, 9 (1970) 244.
142 H. W. Roesky, *J. Inorg. Nucl. Chem.*, 32 (1970) 1845.
143 M. Lustig, *Inorg. Chem.*, 8 (1969) 443.
144 H. W. Roesky, *Inorg. Nucl. Chem. Letters*, 6 (1970) 129.
145 C. B. Colburn, W. E. Hill and D. W. A. Sharpe, *J. Chem. Soc. A*, (1970) 2221.
146 M. Lustig, *Angew. Chem. Intern. Ed. Engl.*, 6 (1967) 959.
147 T. L. Charlton and R. G. Cavell, *Inorg. Chem.*, 6 (1967) 2204.
148 L. F. Centofanti and R. W. Parry, *Inorg. Chem.*, 7 (1969) 1005.
149 L. F. Centofanti and R. W. Parry, *Inorg. Chem.*, 9 (1970) 744.
150 T. L. Charlton and R. G. Cavell, *Chem. Commun.*, (1966) 763.
151 H. W. Roesky and W. Kloker, *Z. Anorg. Allgem. Chem.*, 375 (1970) 140.
152 H. W. Roesky and L. F. Grimm, *Chem. Ber.*, 103 (1970) 3114.
153 R. G. Kostyanovsky and V. V. Yakshin, *Bull. Acad. Sci. USSR*, 10 (1967) 2261.
154 Y. Wada and R. W. Kiser, *J. Phys. Chem.*, 68 (1964) 2290.
155 R. E. Winters and R. W. Kiser, *J. Organometal. Chem.*, 10 (1967) 7.
156 R. G. Gillis and G. L. Long, *Org. Mass Spectrom.*, 2 (1969) 1315.
157 G. M. Boglyubov, N. N. Grishin and A. A. Petrov, *J. Gen. Chem. USSR*, 39 (1969) 2190.
158 R. Fields, R. N. Hazeldine and N. F. Wood, *J. Chem. Soc. C*, (1970) 1370.
159 R. G. Kostyanovsky, V. V. Yashin and Z. L. Zimont, *Tetrahedron*, 24 (1968) 2995.
160 S. Chan, H. Goldwhite, H. Keyser, D. G. Rowsell and R. Tang, *Tetrahedron*, 25 (1969) 1097.
161 L. D. Quin, J. G. Bryson and C. G. Moreland, *J. Am. Chem. Soc.*, 91 (1969) 3308.
162 F. Seel and K. Rudolph, *Z. Anorg. Allgem. Chem.*, 363 (1968) 233.
163 L. K. Krannick and H. H. Sisler, *Inorg. Chem.*, 8 (1969) 1032.
164 A. H. Cowley and R. P. Pinnell, *Inorg. Chem.*, 5 (1966) 1459.
165 H. G. Ang and B. O. West, *Australian J. Chem.*, 20 (1967) 1133.
166 U. Schmidt, I. Boie, C. Osterroht, R. Schroer and H.-F. Grutzmacher, *Chem. Ber.*, 101 (1968) 1381.
167 E. J. Wells, R. C. Ferguson, J. G. Hallet and L. K. Peterson, *Can. J. Chem.*, 46 (1968) 2733.
168 P. S. Elmes, S. Middleton and B. O. West, *Australian J. Chem.*, 23 (1970) 1559.
169 W. Z. Borer and K. Cohn, *Anal. Chim. Acta*, 47 (1969) 355.
170 P. S. Braterman, *J. Organometal. Chem.*, 11 (1968) 198.
171 R. G. Kostyanovskii, I. A. Nuretdinov, N. P. Grehkin and I. I. Chervin, *Bull. Acad. Sci., USSR (Chem. Sect.)*, 11 (1969) 2429.
172 D. B. Whigan, J. W. Gilje and A. E. Goya, *Inorg. Chem.*, 9 (1970) 1279.
173 D. E. Bublitz and A. W. Baker, *J. Organometal. Chem.*, 9 (1967) 383.

174  J. M. Miller, *J. Chem. Soc. A*, (1967) 828.

175  R. Colton and Q. N. Porter, *Australian J. Chem.*, 21 (1968) 2215.

176  D. H. Williams, R. S. Ward and R. G. Cooks, *J. Am. Chem. Soc.*, 90 (1968) 966.

177  A. T. Rake and J. M. Miller, *J. Chem. Soc. A*, (1970) 1881.

178  B. Zeeh and J. B. Thomson, *Tetrahedron Letters*, 2 (1969) 111.

179  A. T. Rake and J. M. Miller, *Org. Mass Spectrom.*, 3 (1970) 237.

180  J. H. Bowie and B. Nussey, *Org. Mass Spectrom.*, 3 (1970) 933.

181  J. H. Bowie and B. Nussey, *J. Chem. Soc. D*, (1970) 17.

182  J. D. Hawthorne, M. J. Mays and R. N. F. Simpson, *J. Organometal. Chem.*, 12 (1968) 407.

183  R. De Ketelaere, E. Muglle, W. Vanerman, E. Claeys and G. P. Van de Kelen, *Bull. Soc. Chim. Belges*, 78 (1969) 219.

184  D. Hellwinkel, B. Knabe and G. Kilthau, *J. Organometal. Chem.*, 24 (1970) 165.

185  E. Müller, B. Teisser, H. Eggenspreger, A. Pieker and K. Schaffer, *Ann. Chem.*, 705 (1967) 54.

186  D. Hellwinkel and M. Bach, *J. Organometal. Chem.*, 20 (1969) 273.

187  J. Müller, *Chem. Ber.*, 102 (1969) 152.

188  W. A. Henderson, M. Epstein and F. S. Seichter, *J. Am. Chem. Soc.*, 85 (1963) 2482.

189  U. Schmidt and I. M. Bone, *Angew. Chem. Intern. Ed. Engl.*, 5 (1966) 1038.

190  R. I. Wagner, L. D. Freeman, H. Goldwhite and D. G. Rowsell, *J. Am. Chem. Soc.*, 89 (1967) 1102.

191  P. Tavs, *Angew. Chem. Intern. Ed. Engl.*, 8 (1969) 751.

192  H. R. Hays and T. J. Logan, *J. Org. Chem.*, 31 (1966) 3391.

193  R. G. Cooks, R. S. Ward, D. H. Williams, M. A. Shaw and J. C. Tebby, *Tetrahedron*, 24 (1968) 3289.

194  A. P. Gara, R. A. Massy-Westrop and J. H. Bowie, *Australian J. Chem.*, 23 (1970) 307.

195  R. T. Alpin, A. R. Hands and A. J. H. Mercer, *Org. Mass Spectrom.*, 2 (1969) 1017.

196  G. H. Birum and C. N. Matthews, *J. Am. Chem. Soc.*, 90 (1968) 3842.

197  H. J. Bestmann and R. Kunstmann, *Chem. Ber.*, 102 (1969) 1816.

198  M. A. Shaw, J. C. Tebby, R. S. Ward and D. H. Williams, *J. Chem. Soc. C* (1967) 2442.

199  M. A. Shaw, J. C. Tebby, J. Ronayne and D. H. Williams, *J. Chem. Soc. C*, (1967) 944.

200  M. A. Shaw, J. C. Tebby, R. S. Ward and D. H. Williams, *J. Chem. Soc. C*, (1968) 1609.

201  H. Diefenbach, H. Ringsdorf and R. E. Wilhelms, *Chem. Ber.*, 103 (1970) 183.

202  E. D. Bergmann, M. Rabinovitz, C. Lifshitz, D. Shapiro and I. Agranat, *Org. Mass Spectrom.*, 4 (supplementary) (1970) 89.

203  L. Tokes and S. C. K. Wong, *Org. Mass Spectrom.*, 4 (supplementary) (1970) 59.

204  B. Zeeh and R. Beutler, *Org. Mass Spectrom.*, 1 (1968) 791.

205  R. A. Baldwin, C. O. Wilson and R. I. Wagner, *J. Org. Chem.*, 32 (1967) 2172.

206  D. Hellwinkel and W. Schenk, *Angew. Chem. Intern. Ed. Engl.*, 8 (1969) 987.

207  P. de Koe, R. van Veen and F. Bickelhaupt, *Angew. Chem. Intern. Ed. Engl.*, 7 (1968) 465.

208  P. de Koe and F. Bickelhaupt, *Angew. Chem. Intern. Ed. Engl.*, 7 (1968) 889.

209  M. A. Shaw, J. C. Tebby, R. S. Ward and D. H. Williams, *J. Chem. Soc. C*, (1970) 504.

210 N. E. Waite, J. C. Tebby, R. S. Ward and D. H. Williams, *J. Chem. Soc. C*, (1969) 1100.
211 N. E. Waite and J. C. Tebby, *J. Chem. Soc. C*, (1970) 386.
212 I. Granoth, A. Kalir and Z. Pelah, *Israel J. Chem.*, 6 (1968) 651.
213 I. Granoth, A. Kalir, Z. Pelah and E. D. Bergmann, *Org. Mass Spectrom.*, 3 (1970) 1359.
214 I. Granoth, A. Kalir, Z. Pelah and E. D. Bergmann, *Tetrahedron*, 25 (1969) 3919.
215 E. E. Schweizer, W. S. Creasey, J. G. Leihr, M. E. Jenkins and D. L. Dalrymple, *J. Org. Chem.*, 35 (1970) 601.
216 G. Guardiano, R. Mandelli, P. P. Ponti, C. Ticozzi and A. Umani-Rochi, *J. Org. Chem.*, 33 (1968) 4431.
217 G. Märkl and H. Schubert, *Tetrahedron Letters*, 15 (1970) 1273.
218 W. Hawes and S. Trippett, *J. Chem. Soc. C*, (1969) 1465.
219 E. J. Corey and D. E. Cane, *J. Org. Chem.*, 34 (1969) 3053.
220 M. Baudler, K. Kipler and H-W. Valpertz, *Naturwissenschaften*, 53 (1969) 612.
221 P. Jutzi and K. Deuchert, *Angew. Chem. Intern. Ed. Engl.*, 8 (1969) 991.
222 H. Vermeer and F. Bickelhaupt, *Tetrahedron Letters*, 12 (1970) 1007.
223 D. W. Allen, J. C. Coppola, O. Kenward, F. G. Mann, W. D. S. Motherwell and D. G. Watson, *J. Chem. Soc. C*, (1970) 810.
224 R. A. Earley and M. J. Gallagher, *Org. Mass Spectrom.*, 3 (1970) 1283.
225 R. A. Earley and M. J. Gallagher, *Org. Mass Spectrom.*, 3 (1970) 1287.
226 N. P. Buu-Hoi, M. Mangane and P. Jaquignon, *J. Heterocyclic Chem.*, 3 (1966) 149.
227 J. C. Tou and C. S. Wang, *Org. Mass Spectrom.*, 4 (Supplementary) (1970) 503.
228 J. C. Tou and C. S. Wang, *Org. Mass Spectrom.*, 3 (1970) 287.
229 J. C. Tou, C. S. Wang and E. G. Alley, *Org. Mass Spectrom.*, 3 (1970) 747.
230 R. Johnson, *J. Appl. Chem.*, 19 (1969) 204.
231 H. Goldwhite, *Chem. Commun.*, (1970) 651.
232 J. F. Seel and K. Rudolph, *Z. Anorg. Allgem. Chem.*, 359 (1968) 333.
233 R. A. Wiesbroek and J. K. Ruff, *Inorg. Chem.*, 5 (1966) 1629.
234 J. S. Harman and D. W. A. Sharpe, *J. Chem. Soc. A*, (1970) 1935.
235 J. E. Clune and K. Cohn, *Inorg. Chem.*, 7 (1968) 2067.
236 G. G. Flaskerud, K. E. Pullen and J. M. Shreeve, *Inorg. Chem.*, 8 (1969) 728.
237 R. Foester and K. Cohn, *Inorg. Chem.*, 9 (1970) 1571.
238 E. I. Babkina and I. V. Vereshchinskii, *J. Gen. Chem. USSR*, 38 (1968) 1727.
239 L. M. Zaboroski and J. M. Shreeve, *J. Am. Chem. Soc.*, 92 (1970) 3665.
240 M. J. McGlinchey, J. D. Odom, T. Reynoldson and F. G. A. Stone, *J. Chem. Soc. A*, (1970) 31.
241 G. S. Reddy and R. Schmutzler, *Z. Naturforsch.*, 25b (1970) 1199.
242 G. C. Demitras and A. G. MacDiarmid, *Inorg. Chem.*, 6 (1967) 1903.
243 R. Schmutzler, *Inorg. Chem.*, 3 (1964) 421.
244 M. A. Baldwin, A. G. Loudon, A. Maccoll, R. E. Dunmur, R. Schmutzler and I. K. Gregor, *Org. Mass Spectrom.*, 2 (1969) 765.
245 K. Utrary and M. Bermann, *Inorg. Chem.*, 8 (1969) 1038.
246 D. F. Clemens, M. L. Caspar, D. Rosenthal and R. Peluso, *Inorg. Chem.*, 9 (1970) 960.
247 R. S. Davidson, R. A. Sheldon and S. Trippett, *J. Chem. Soc. C*, (1967) 1547.
248 R. S. Davidson, R. A. Sheldon and S. Trippett, *J. Chem. Soc. C*, (1968) 1700.

249  J. A. Miller, *Tetrahedron Letters*, 50 (1969) 4335.

250  I. Ojima, K. Akiba and N. Inamoto, *Bull. Chem. Soc. Japan*, 42 (1969) 2975.

251  J. Ellerman, F. Poerech, R. Kunstmann and R. Kramolowsky, *Angew. Chem. Intern. Ed. Engl.*, 8 (1969) 203.

252  P. C. Crofts and D. M. Parker, *J. Chem. Soc. C*, (1970) 332.

253  H. P. Angstadt, *J. Am. Chem. Soc.*, 86 (1964) 5040.

254  R. A. Spence, J. M. Swan and S. H. B. Wright, *Australian J. Chem.*, 22 (1969) 2359.

255  R. G. Cavell, *Can. J. Chem.*, 45 (1967) 1309.

256  H. W. Roesky and H. Beyer, *Chem. Ber.*, 102 (1969) 2588.

257  O. Glemser, U. Biermann and M. Fild, *Chem. Ber.*, 100 (1967) 1082.

258  J. L. Occolowitz and G. L. White, *Anal. Chem.*, 35 (1963) 1179.

259  D. G. Hendricker, *J. Heterocyclic Chem.*, 4 (1967) 385.

260  D. B. Denney and S. L. Varga, *Tetrahedron Letters*, 40 (1966) 4935.

261  O. Vogl, B. C. Anderson and D. M. Simons, *J. Org. Chem.*, 34 (1970) 204.

262  F. W. McLafferty, *Anal. Chem.*, 28 (1956) 306.

263  F. W. McLafferty, *Appl. Spectry.*, 11 (1957) 148.

264  R. G. Cooks and A. F. Gerrard, *J. Chem. Soc. B*, (1968) 1327.

265  J. Jörg, R. Houriet and G. Spiteller, *Monatsh. Chem.*, 97 (1966) 1064.

266  J. N. Damico, *J. Assoc. Offic. Anal. Chem.*, 49 (1966) 1027.

267  J. N. Damico, R. P. Barron and J. A. Sphon, *Intern. J. Mass Spectrom. Ion Phys.*, 2 (1969) 161.

268  A. Y. Leung and A. G. Paul, *J. Pharm. Sci.*, 57 (1968) 1667.

269  J. Picker and R. W. Richards, *Australian J. Chem.*, 23 (1970) 853.

270  H. Okawa and M. Eto, *Agr. Biol. Chem. (Tokyo)*, 33 (1969) 443,

271  A. Tatematsu, H. Yoshijumi and T. Goto, *Japan Analyst*, 17 (1968) 774.

272  J. R. Pardue, E. A. Hansen, R. P. Berron and J.-Y. T. Chen, *J. Agr. Food Chem.*, 18 (1970) 405.

273  J. L. Occolowitz and J. M. Swan, *Australian J. Chem.*, 19 (1966) 1187.

274  T. Nishiwaki, *Tetrahedron*, 22 (1966) 1383.

275  T. Nishiwaki, *Tetrahedron*, 23 (1967) 2181.

276  W. J. McMurray, S. R. Lipsky, C. Ziondrou and G. L. Schmir, *Org. Mass Spectrom.*, 3 (1970) 1031.

277  H. R. Harless, *Anal. Chem.*, 33 (1961) 1387.

278  J. G. Pritchard, *Org. Mass Spectrom.*, 3 (1970) 163.

279  E. E. Nifant'ev and I. S. Nasonovskii, *J. Gen. Chem. USSR*, 39 (1969) 1911.

280  E. Cherbuliez, A. Buchs, S. Jaccard, D. Janjic and J. Rabinowitz, *Helv. Chim. Acta*, 49 (1966) 2395.

281  H. W. Roesky, *Chem. Ber.*, 101 (1968) 636.

282  H. W. Roesky, *Chem. Ber.*, 101 (1968) 2977.

283  P. Haake and P. S. Ossip, *Tetrahedron*, 24 (1968) 565.

284  P. Haake, M. J. Frearson and C. E. Diebert, *J. Org. Chem.*, 34 (1969) 788.

285  H. Budzikiewicz and Z. Pelah, *Monatsh. Chem.*, 96 (1965) 1739.

286  E. Lindner, H.-D. Ebert and A. Haag, *Chem. Ber.*, 103 (1970) 1872.

287  E. Linder, H.-D. Ebert and P. Junkes, *Chem. Ber.*, 103 (1970) 1364.

288  K. Dimroth, K. Vogel, W. Mach and U. Schoeler, *Angew. Chem. Intern. Ed. Engl.*, 7 (1968) 371.

289 R. G. Cavell and R. C. Dobbie, *Inorg. Chem.*, 7 (1968) 101.
290 R. C. Dobbie and R. G. Cavell, *Inorg. Chem.*, 6 (1967) 1450.
291 R. G. Cavell and R. C. Dobbie, *J. Chem. Soc. A*, (1968) 1406.
292 R. G. Cavell and R. C. Dobbie, *Inorg. Chem.*, 7 (1968) 690.
293 R. C. Dobbie, L. F. Doty and R. G. Cavell, *J. Am. Chem. Soc.*, 90 (1968) 2015.
294 D. H. Dybvig, *Inorg. Chem.*, 5 (1966) 1795.
295 A. L. Logothetis and G. N. Sausen, *J. Org. Chem.*, 31 (1966) 3689.
296 R. A. Mitsch and R. H. Ogden, *J. Org. Chem.*, 31 (1966) 3833.
297 J. B. Hynes, B. C. Bishop and L. A. Bigelow, *Inorg. Chem.*, 6 (1967) 417.
298 R. N. Haszeldine and A. E. Tipping, *J. Chem. Soc. C*, (1968) 398.
299 E. S. Alexander, R. N. Haszeldine, M. J. Newlands and A. E. Tipping, *J. Chem. Soc. C*, (1968) 796.
300 R. N. Haszeldine and A. E. Tipping, *J. Chem. Soc. C*, (1967) 1241.
301 R. N. Haszeldine and A. E. Tipping, *J. Chem. Soc. C*, (1966) 1236.
302 D. A. Rausch and J. J. Hoekstra, *J. Org. Chem.*, 33 (1968) 2522.
303 P. H. Ogden and R. A. Mitsch, *J. Am. Chem. Soc.*, 89 (1967) 3868.
304 R. E. Banks, R. N. Haszeldine, M. J. Stevenson and B. G. Willoughby, *J. Chem. Soc. C*, (1969) 2119.
305 D. P. Babb and J. M. Shreeve, *Inorg. Chem.*, 6 (1967) 351.
306 H. J. Emeléus and T. Onak, *J. Chem. Soc. A*, (1966) 1291.
307 Y. O. El Nigumi and H. J. Emeléus, *J. Inorg. Nucl. Chem.*, 32 (1970) 3213.
308 H. G. Ang and H. J. Emeléus, *J. Chem. Soc. A*, (1968) 1334.
309 M. Lustig, *Inorg. Chem.*, 7 (1968) 2054.
310 W. B. Fox, G. Franz and L. R. Anderson, *Inorg. Chem.*, 7 (1968) 383.
311 A. H. Cowley, T. A. Furtsch and D. S. Dierdorf, *J. Chem. Soc. D*, (1970) 523.
312 Y. O. El Nigumi and H. J. Emeléus, *J. Inorg. Nucl. Chem.*, 32 (1970) 3211.
313 R. C. Dobbie, M. Green and F. G. A. Stone, *J. Chem. Soc. A*, (1969) 1881.
314 R. G. Cavell and R. C. Dobbie, *J. Chem. Soc. A*, (1967) 1308.
315 V. I. Vedeneyev, L. V. Gurvich, V. N. Kondvat'yev, V. A. Medvedev and Ye. L. Frankevich, *Bond Energies, Ionisation Potentials and Electron Affinities*, Arnold, London, 1966.
316 D. M. Roundhill and G. Wilkinson, *J. Org. Chem.*, 35 (1970) 3561.
317 M. G. Barlow, M. Green, R. N. Hazeldine and H. G. Higson, *J. Chem. Soc. C*, (1966) 1592.
318 S. S. Dua, R. C. Edmundson and H. Gilman, *J. Organometal. Chem.*, 24 (1970) 703.

*Chapter 9*

# The Main Group VI Elements

M. R. LITZOW

## 1. INTRODUCTION

Vast numbers of organic compounds contain elements of main Group VI, especially oxygen and sulphur. Many of these have been examined in detail mass spectrometrically but are not of interest in the present survey. Considerable effort has been expended on energetics studies of the elements themselves and on other simple inorganic compounds containing these elements. Only recently have the mass spectra of more complex systems been investigated but unfortunately published details are sparse so that it is difficult, if not impossible, to discern trends and extract information which would be of value in future studies.

## 2. THE ELEMENTS

All of the elements of the group are polyisotopic, the complexity of the

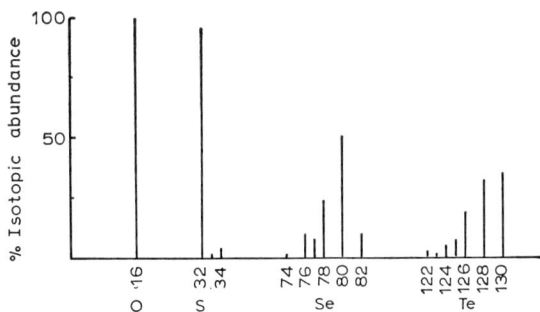

Fig. 9.1. Natural isotopic abundances of main Group VI elements (ref. 1).

Table 9.1

Natural Abundances of the Isotopes of Main Group VI Elements[1]

| Element | Mass no. | % Natural abundance | Element | Mass no. | % Natural abundance |
|---|---|---|---|---|---|
| O | 16 | 99.759 | S | 32 | 95.0 |
| | 17 | 0.037 | | 33 | 0.76 |
| | 18 | 0.024 | | 34 | 4.22 |
| | | | | 36 | 0.014 |
| Se | 74 | 0.87 | Te | 120 | 0.089 |
| | 76 | 9.02 | | 122 | 2.46 |
| | 77 | 7.58 | | 123 | 0.87 |
| | 78 | 23.52 | | 124 | 4.61 |
| | 80 | 49.82 | | 125 | 6.99 |
| | 82 | 9.19 | | 126 | 18.71 |
| | | | | 128 | 31.79 |
| | | | | 130 | 34.48 |

Table 9.2

Ionisation and Appearance Potential Values of Ions Derived from Molecular Oxygen (eV)
References are given in brackets.

| | Ionisation potential (eV) | | Appearance potential $O^+$ (eV) |
|---|---|---|---|
| | Electron impact | Photoionisation | |
| 1st I.P. $O_2$ | $12.5 \pm 0.1$ (4) | $12.075 \pm 0.01$ (8) | A.P. $O^+$ $17.267 \pm 0.024$ (16) |
| | $12.1 \pm 0.2$ (5) | $12.072 \pm 0.008$ (9) | $17.30 \pm 0.10$ (2) |
| | $12.21 \pm 0.04$ (2) | $12.063 \pm 0.001$ (10) | 17.25 (9) |
| | $12.20 \pm 0.02$ (6) | $12.08 \pm 0.01$ (11) | |
| | $12.07 \pm 0.02$ (7) | $12.065 \pm 0.003$ (12) | |
| | | $12.04 \pm 0.01$ (13) | |
| | | 12.06 (14) | |
| | | 12.075 (15) | |
| 2nd I.P. $O_2$ | $16.30 \pm 0.03$ (2) | 16.3 (9) | |
| | | 16.1 (17) | |
| | | 16.11 (15) | |
| 3rd I.P. $O_2$ | $17.18 \pm 0.02$ (2) | 16.8 (17) | |
| 4th I.P. $O_2$ | $18.42 \pm 0.02$ (2) | 18.4 (9) | |
| | | 18.2 (17) | |
| | | 18.19 (15) | |

Table 9.3

Ionisation and Appearance Potential Values and Some Bond Dissociation Energies Obtained in Studies of Group VI Compounds

| Compound | Ion | A.P. (eV) | Bond dissociation energy (kcal.mole$^{-1}$); radical I.P. (eV) | | Method employed | Ref. |
|---|---|---|---|---|---|---|
| CO | CO$^+$ | 14.00$\pm$0.05 | | | RPD electron impact | 18 |
| | | 13.98$\pm$0.02 | | | RPD electron impact | 19 |
| | | 14.05$\pm$0.02 | | | Electron impact | 20 |
| | | 14.1 $\pm$0.2 | | | Electron impact | 5 |
| | | 14.0 | | | Photoionisation | 21 |
| | | 14.01$\pm$0.01 | | | Photoionisation | 8 |
| | | 14.009 | | | Spectroscopic | 22 |
| | | 14.01 | | | Photoelectron spectroscopy | 23 |
| | | 14.01 | | | Photoelectron spectroscopy | 15 |
| CO$_2$ | CO$_2^+$ | 13.75$\pm$0.05 | | | Electron impact | 24 |
| | | 13.85 | | | RPD electron impact | 25 |
| | | 13.79$\pm$0.01 | | | Photoionisation | 8 |
| | | 13.77$_5$$\pm$0.01 | | | Photoionisation | 9 |
| | | 13.73 | | | Spectroscopic | 26 |
| | | 13.68 | | | Photoelectron spectroscopy | 23 |
| | | 13.78 | | | Photoelectron spectroscopy | 27 |
| | O$^+$ | 19.10$\pm$0.01 | $D$(OC–O) 126.5 | | Photoionisation | 9 |
| COS | COS$^+$ | 11.18$\pm$0.01 | | | Photoionisation | 9 |
| | | 11.17$\pm$0.02 | | | Photoionisation | 17 |
| | | 11.23 | | | Spectroscopic | 28 |
| | | 11.06 | | | Photoelectron spectroscopy | 18 |
| | | 11.18, 11.23 | | | Photoelectron spectroscopy | 27 |
| | S$^+$ | 13.65$\pm$0.03 | $D$(OC–S) 75.9$\pm$0.7 | | Photoionisation | 9 |
| CS$_2$ | CS$_2^+$ | 10.15 | | | RPD electron impact | 25 |
| | | 10.08$\pm$0.01 | | | Photoionisation | 8 |
| | | 10.059$\pm$0.008 | | | Photoionisation | 9 |
| | | 10.07 | | | Photoelectron spectroscopy | 23 |
| | | 10.06, 10.12 | | | Photoelectron spectroscopy | 27 |
| | S$^+$ | 14.81$\pm$0.03 | $D$(SC–S) | 102.7 | Photoionisation | 9 |
| | CS$^+$ | 16.16$\pm$0.01 | $I$(CS) | 11.71 | Photoionisation | 9 |
| | | | | 11.8$\pm$0.2 | Electron impact | 29 |
| NO | NO$^+$ | 9.28$\pm$0.03 | | | RPD electron impact | 19 |
| | | 9.4 $\pm$0.2 | | | Electron impact | 5 |
| | | 9.34 | | | Photoelectron spectroscopy | 23 |
| | | 9.25$\pm$0.02 | | | Photoionisation | 8 |
| N$_2$O | N$_2$O$^+$ | 12.8 $\pm$0.05 | | | Electron impact | 24 |
| | | 12.82 | | | Photoelectron spectroscopy | 23 |
| | | 12.89 | | | Photoelectron spectroscopy | 27 |
| | | 12.90$\pm$0.01 | | | Photoionisation | 8 |

natural abundance distribution increasing as the group is descended[1] (Table 9.1, Fig. 9.1).

## 3. MASS SPECTRAL STUDIES

Because of the complexity of the absorption spectrum of oxygen in the region close to threshold, it is extremely difficult to determine the first ionisation potential from this spectrum. The $O_2$ molecule has been the subject of a large number of fairly detailed electron-impact and photoionisation studies, however; the first ionisation potential has been determined and all the spectroscopically known excited states of molecular $O_2$ have been observed. In addition, processes involved in the formation of the atomic oxygen ions $O^+$ and $O^-$ have been proposed[2]. Several reported ionisation and appearance potential values are summarised in Table 9.2. Detailed discussions may be found in publications referred to in this table. Similar studies have been carried out with compounds such as $CO_2$ and $CS_2$ (Table 9.3). Ion–molecule reactions between excited $O^+$ ions and $O_2$ have been reported[3] and the formation of the $O^-$ ion from molecules such as $O_2$, CO, NO, $SO_2$, and $H_2O$ has been studied extensively; the results have provided values for the electron affinity of oxygen.

The degree of association of saturated sulphur vapour has been the subject of physicochemical investigations since the early vapour density measurements of Preuner and Schupp[30]. Recent mass spectral studies have proved to be most successful, although the first attempts using this technique were somewhat uncertain. Goldfinger et al.[31] concluded that a considerable part of the ionic intensities of $S_8^+$, $S_7^+$, $S_6^+$, and $S_2^+$ was probably due to ionisation of parent molecules and they also reported the presence, in very low abundance at 60 °C, of $S_9^+$ as a primary ion. Zeitz[32], however, concluded that sulphur vapour in equilibrium with the condensed phase in the temperature range of 120–210 °C contained appreciable quantities of $S_8$, $S_7$, $S_6$, and $S_5$ molecules only. Berkowitz and Marquart[33] pointed out that, from a consideration of the heats of formation of $S_2$ and $S_8$, it could be shown that $S_2$ represents less than one part in $10^4$ of the saturated vapour under the conditions of the previous mass spectral experiments. These workers carried out a detailed investigation of the problem in an effort to unravel many of the reported anomalies. Their results indicated the presence of all possible $S_n$ molecules between $S_2$ and $S_8$ in measurable concentration between room temperature and the boiling point of sulphur, plus insignificant but detectable amounts of $S_9$ and $S_{10}$. Equilibrium constants relating these species were deduced and enthalpy changes for the reactions $n/8\ S_8 \rightarrow S_n$ ($n = 2$–$7$) reported. The dissociation energy of $S_2$ was found to be $\sim 101$ kcal.mole$^{-1}$ (4.4 eV). Similarly, the enthalpies and entropies of the reactions $S_x \rightleftarrows S_y$ were measured by Colin et al.[34] who determined a value of $97 \pm 5$ kcal.mole$^{-1}$ for $D_0^0$ ($S_2$). Berkowitz and Chupka[35] turned to a study of the vapour above several of the allotropic forms of condensed sulphur. For example, free evaporation of rhombic sulphur ($S_8$) was shown to give rise solely to $S_8$ vapour molecules, whereas the sublimation of Engel's sulphur ($S_6$) produced only $S_6$ vapour molecules. However, the previously discussed vapour composition characteristic of equilibrium[33] could be produced by mixing any allotropic form of sulphur with commercial alumina which evidently catalyses the transformation among sulphur species. The modifications cyclohepta- ($S_7$), cyclodeca- ($S_{10}$), and cyclododeca-($S_{12}$) sulphur have recently been prepared and their mass spectra reported[36,37]. Hagemann[38] reported the ionisation potentials of the molecules $S_2$, $S_4$, $S_6$, and $S_8$.

In the mass spectrum of the vapour from grey selenium subliming in the temperature range 102–187 °C, all possible $Se_n^+$ ions between $n = 1$ and $n = 8$ were detected in measurable quantities while small amounts of $Se_9^+$ and $Se_{10}^+$ were also detected[39,40]. Fujisaki et al.[39] concluded that the ions $Se_5^+$, $Se_6^+$, $Se_7^+$, and $Se_8^+$ arise mainly from ionisation of the corresponding parent molecules; $Se^+$, $Se_2^+$, $Se_3^+$, and $Se_4^+$, on the other hand, were apparently fragment ions although it was suggested that small concentrations of Se and $Se_2$ may be present in the vapour. Goldfinger and Jeunehomme[40] suggested, however, that most or perhaps all of the species from $Se_2$ to $Se_8$ were present as parent molecules. This was substantiated by Berkowitz and Chupka[41] who determined the heats of formation of the species $Se_n$ ($n = 2, 3, 5$–$8$). Appearance potentials and thermodynamic data were also reported by Fujisaki et al.[39] but agreement between the two sets of values was poor.

The dissociation energies of the diatomic species $S_2$, $Se_2$, and $Te_2$ have been the subject of many investigations and considerable uncertainty[42]. It appears now that the value for $S_2$ is very close to 101 kcal.mole$^{-1}$ but the uncertainty with respect to the corresponding values for selenium and tellurium has not been resolved. Some of the reported values are shown in Table 9.4.

Both positive and negative ion mass spectra of sulphur dioxide[48–50], thionyl chloride[50], $SOCl_2$, sulphuryl chloride[50], $SO_2Cl_2$, and sulphuryl

Table 9.4

Some Reported Values of the Dissociation Energies of $S_2$, $Se_2$, and $Te_2$ Obtained by Various Methods

References are given in brackets.
Method                 Dissociation energy (kcal.mole$^{-1}$)

| | $S_2$ | $Se_2$ | $Te_2$ |
|---|---|---|---|
| Spectroscopic | $\leq 101.7$ (43) | $72.94 \pm 0.03$ (45) | |
| Thermochemical | $101.7 \pm 2.9$ (44) | $75.7 \pm 2.5$ (44) | $61.3 \pm 1.1$ (44) |
| | $96.4 \pm 5$ (42) | $74.5 \pm 2$ (42) | $55 \pm 2$ (42) |
| | 98.4 (42) | | |
| | $100.9 \pm 2.5$ (42) | | |
| | $103.6 \pm 2.5$ (42) | | |
| Mass Spectrometric | | | |
| Electron impact | 101 (33) | $75.3 \pm 2$ (41[a]) | |
| | $101.5 \pm 5$ (46) | | |
| | $97 \pm 5$ (34) | | |
| Photoionisation[a] | $101.0 \pm 0.2$ (47) | $78.6 - 83.2$ (47) | $62.3 \pm 0.2$ (47) |

[a] The value quoted in reference 41 was $71.2 \pm 2$ but a corrected value of $75.3 \pm 2$ was reported in ref. 47.

fluoride[48], $SO_2F_2$, have been reported and appearance potentials of the principal ions measured, leading to values for $D(S_2)$ (ref. 48), D(SO) (ref. 48), and the electron affinity of SO (ref. 49). The formation of $S_2O$ from sulphur and oxygen was studied mass spectrometrically by Hagemann[38,51]; its heat of formation was determined.

Cooper and Culka[52] employed mass spectrometry to analyse crystalline products obtained by solvent extraction of the high-temperature reaction mixture of elemental sulphur and selenium. Berkowitz and Chupka[41] found that S–Se mixed molecules readily formed at temperatures of about 300 °C and upon cooling they remained unchanged, so that their vapour could be readily studied mass spectrometrically. All members of the octatomic series $S_nSe_{(8-n)}$ were identified[52]. Significant amounts of hexatomic ions were also observed but it was not known if these were the result of fragmentation of the eight-membered ring systems or not. However, Umilin et al.[53] reported that, in the mass spectrum of sublimed sulphur–selenium mixtures, they observed a considerable number of ions of composition $S_nSe_k{}^+$, where $n = 1$–7, $k = 1$–4. Similar experiments[54] with the sulphur–tellurium system resulted in the observation of only one species containing tellurium, $S_7Te$. Effusion cell studies resulted in the identification of the molecule SeTe and its bond dissociation energy was determined[55] to be 2.5 eV or 58.7 kcal.mole$^{-1}$. Dillard and Franklin[124] studied the ion–molecule reactions between S$^-$ ions, formed from both carbonyl sulphide and carbon disulphide, and the parent molecules. The dominant secondary ion observed at low electron energies was $S_2^-$. At electron energies above 15 eV, the ions $S_n^-$ ($n = 1$–6) were observed and their formation involved a consecutive mechanism of the type

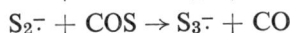

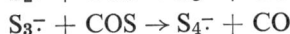

$$S^- + COS \rightarrow S_2^- + CO$$
$$S_2^- + COS \rightarrow S_3^- + CO$$
$$S_3^- + COS \rightarrow S_4^- + CO$$

Studies of ion–molecule reactions in $O_2$ (refs. 57, 58), $H_2O$ (refs. 58–60), $D_2O$ (ref. 60), $H_2S$ (refs. 61, 62), $CO_2$ (ref. 63), and CO (ref. 57) have also

Table 9.5

Results of Energetics Studies on Hydride Species of Group VI Elements

(i) *Onset of various ionisation processes* (eV)

| | | | | | |
|---|---|---|---|---|---|
| $H_2O$ | $12.60 \pm 0.01$ | $14.35 \pm 0.03$ | $16.34 \pm 0.06$[a] | | 122[b] |
| | $12.6 \pm 0.1$ | $14.5 \pm 0.3$ | $16.2 \pm 0.3$ | $18.0 \pm 0.5$ | 66 |
| | $12.69 \pm 0.08$ | | | | 56 |
| | $12.59 \pm 0.01$ | | | | 8 |
| | $12.593$[c] | | | | 67 |
| | | $14.6 \pm 0.3$ | | | 68 |
| | $12.61$[d] | $14.23$[d] | | $18.02$[d] | 69 |
| | $12.65 \pm 0.25$ | | | | 70 |
| | $12.56 \pm 0.02$ | | | | 71 |
| $H_2S$ | $10.45 \pm 0.03$ | $12.46 \pm 0.03$ | $14.18 \pm 0.04$ | | 122 |
| | $10.5 \pm 0.1$ | $12.2 \pm 0.2$ | $14.0 \pm 0.2$ | | 66 |
| | $10.42$[d] | $12.62$[d] | $14.82$[d] | | 18 |
| | $10.42 \pm 0.01$[e] | | | | 71 |
| | $10.46 \pm 0.01$[c] | | | | 8 |

[a] It was suggested[23] that this value corresponds to an autoionisation process.
[b] Reference number.
[c] Photoionisation technique.
[d] Photoelectron spectroscopy technique.
[e] Spectroscopic technique.

Table 9.5 (continued)

(ii) *Ionisation potentials of other molecules and radicals* (eV)
References are given in brackets.

| $H_2O_2$ | $\cdot HO_2$ | $\cdot H_3O$ | $\cdot OH$ | $H_2S_2$ | $HS\cdot$ |
|---|---|---|---|---|---|
| $11.26\pm0.05$ (56) | 11.53 (74) | 9–10.9 (75) | $13.53\pm0.08$ (56) | 10.09 (77) | $10.50\pm1$ (78) |
| $12.1\ \pm0.3$ (72) | | | $13.18\pm0.1$ (76) | 9.90 (77) | |
| $10.92\pm0.05$ (73) | | | $13.49\pm0.08$ (65) | | |
| $11.30\pm0.20$ (70) | | | $13.25\pm0.30$ (70) | | |
| | | | $12.94\pm0.02$ (67) | | |

(iii) *Appearance potentials of various ions*
References are given in brackets.

| Molecule | Ion | A.P. (eV) | Molecule | Ion | A.P. (eV) |
|---|---|---|---|---|---|
| $H_2O$ | $OH^+$ | $18.59\pm0.08$ (56) | $H_2O_2$ | $OH^+$ | $15.60\pm0.08$ (56) |
| | | $18.19\pm0.1$ (76) | | | $16.0\ \pm0.3$ (72) |
| | | $18.7\ \pm0.2$ (79) | | | $15.35\pm0.10$ (73) |
| | | | | $HO_2^+$ | $15.36\pm0.05$ (73,74) |
| | | | | | $16.1\ \pm0.4$ (72) |
| $H_2S$ | $HS^-$ | 2.2 (50) | | $H_2O^+$ | $14.09\pm0.10$ (73) |
| | | | | $O_2^+$ | $15.8\ \pm0.5$ (73) |
| $H_2Se$ | $HSe^-$ | 1.8 (50) | | $O^+$ | $17.0\ \pm1.0$ (73) |
| | | | | | $18.9\ \pm0.4$ (72) |

(iv) *Bond dissociation energies* (kcal·mole$^{-1}$)
References are given in brackets.

| | | | |
|---|---|---|---|
| $D(H–OH)$ | 117.6 (65) | $D(H–SH)$ | 95.3 (80) |
| $D(H–OOH)$ | 88.4 (74) | | $92.2\pm2$ (81) |
| $D(H–O_2)$ | 45.9 (74) | | $\leq90.6$ (78) |
| $D(HO–OH)$ | $47.8\pm3$ (56) | $D(H–S)$ | 67.0 (80) |
| | $48.7\pm2.5$ (65) | | 84 (78) |
| | | $D(CH_3S–H)$ | 88.8 (80) |
| | | $D(C_2H_5S–H)$ | 86.8 (80) |
| | | $D(HS–SH)$ | $59\pm3$ (77) |
| | | | 80.4 (80) |

been reported. The unusual ion $H_4O^+$ was observed[64] after exposing water vapour to a palladium catalyst.

A number of mass spectrometric studies resulting in values for the ionisation potentials of various main Group VI hydride compounds (including the radicals $HO_2$ and HS) as well as the energy levels of excited states of molecular ions, have been reported. Many of these values are shown in Table 9.5. Excellent agreement is evident between results from photoionisation and electron-impact methods. Values of various appearance potentials and bond dissociation energies have also been collected in Table 9.5. The negative ions $HS^-$ and $HSe^-$ are formed on electron bombardment of $H_2S$ and $H_2Se$, respectively[50].

Dibeler et al.[82] carried out a mass spectral investigation of oxygen difluoride, $OF_2$, and reported abundances and, where possible, appearance potentials of the ions which they observed (Table 9.6). Special techniques were

Table 9.6

Results of Energetics Studies on $OF_2$ and $O_2F_2$

| Molecule | Ion | A.P. (eV) | Bond dissociation energies | |
|---|---|---|---|---|
| $OF_2$[a] | $OF_2^+$ | $13.7 \pm 0.2$ | $D(FO-F)$ | 2.8 eV, 64.6 kcal.mole$^{-1}$ |
| | $OF^+$ | $15.8 \pm 0.2$ | $D(O-F)$ | 1.1 eV, 25.4 kcal.mole$^{-1}$ |
| | $F^-$ | $1.2 \pm 0.2$ | | |
| $O_2F_2$[b] | $O_2F^+$ | $14.0 \pm 0.1$ | $D(FO-OF)$ | $4.5 \pm 0.2$ eV, $103.5 \pm 5$ kcal.mole$^{-1}$ |
| | $OF^+$ | $17.5 \pm 0.2$ | $D(F-O_2F)$ | $0.8 \pm 0.3$ eV, $18.4 \pm 7$ kcal.mole$^{-1}$ |

[a] Results for $OF_2$ taken from ref. 82.
[b] Results for $O_2F_2$, taken from refs. 83, 84.

required to study dioxydifluoride, $O_2F_2$, and ozone difluoride, $O_3F_2$, since these compounds are only stable at very low temperatures. Malone and McGee[83,84] have reported detailed energetics studies resulting in values for the bond energies of the $O_2F_2$ molecule as well as of the $O_2F$ free radical (Table 9.6). These workers suggested[85] that the $O_3F_2$ molecule possesses the basic features of an $O_2F$ and an OF radical loosely bonded together and that the larger positive and negative ions were not stable for the required 50 $\mu$sec ion processing time of the instrument employed. The actual existence of the $O_3F_2$ molecule is still a matter of controversy, however; from $^{19}F$ and $^{17}O$ NMR evidence, for example, Solomon et al.[86] contend that the "compound" is actually a mixture of $O_2F_2$ and $(OOF)_n$.

Perchlorylfluoride, $ClO_3F$, has achieved some renown because of its unusual negative ion mass spectrum. In general, the abundance of negative ions is very small compared with that of positive ions; for this compound, however, $ClO_3^-$ is the most abundant ion, positive or negative[87]. Appearance potentials of the negative ions were measured and many have several resonance capture potentials. The mass spectra of the bis-(perfluoroalkyl) trioxides, $CF_3OOOCF_3$ (ref. 89) and $CF_3OOOC_2F_5$ (ref. 123), have been reported to show many similarities to those of the corresponding peroxides and ethers, although the intensities of the higher mass ions certainly decrease as the number of oxygens increases.

Two isomers of sulphur monofluoride, $S_2F_2$, have been isolated and mass spectrometry was employed, together with other physical methods, in their characterisation and structure determination. The predominant difference[90-93] between the spectra is the presence of the ion $SF_2^+$ in that of $S=SF_2$ and its absence in that of FSSF. Seel and Budenz[93] found that an equimolar mixture of the two compounds had a lower vapour pressure than either but the mass spectrum showed no ions at $m/e$ values higher than that corresponding to $S_2F_2$. Negative ions formed from sulphur monochloride, $S_2Cl_2$, have been observed and their appearance potentials measured[50].

A parent molecular ion was not originally observed in the mass spectrum of sulphur hexafluoride[88] and it was suggested that the most probable process in the production of positive ions was removal of $F^-$ rather than an electron.

$$SF_6 + e \rightarrow SF_5^+ + F^- + e$$

The doubly charged ions $SF_4^{2+}$ and $SF_2^{2+}$ were more abundant than the corresponding singly charged species at 100 eV ionising voltage, and it was further suggested that the former was formed by removal of two $F^-$ ions from $SF_6$. Appearance potentials of all ions were measured, that of $SF_5^+$ being $15.9 \pm 0.2$ eV. A later study[94] employing an RPD source provided a value of $15.85 \pm 0.15$ eV, in excellent agreement. The value obtained by photoionisation mass spectrometry was 15.29 eV, and Dibeler and Walker[95] suggested that, in the threshold region, the observed curve represented the formation of $SF_6^+$ which very quickly dissociated to $SF_5^+$ and a neutral F atom. The photoelectron spectrum[96] also yielded a low value, 15.35 eV.

The formation of negative ions from $SF_6$ has been studied by Ahearn and Hannay[97] who used normal mass spectroscopic techniques and reported that the maximum of the ionisation efficiency curve for $SF_6^-$ formation occurred at about 2 eV. Hickam and Fox[98] later employed an RPD source similar to that used in their positive-ion studies; the electron energy spread was only about 0.1 eV and a value of 0.08 eV was obtained for the ionisation efficiency curve maximum. Because of the narrow energy range over which the formation of $SF_6^{\bar{}}$ by the electron-capture process can occur, the compound is useful as a calibrant for low values of the electron energy in appearance potential measurements of negative ions.

The monoenergetic beam of electrons also appears to be essential in determining the true intensities of negative ions formed by electron-capture processes. Hickam and Fox, for example, found that the abundance of the $SF_6^{\bar{}}$ ion was 25 times that of $SF_5^-$ (and believed that it could be increased further with a narrow energy spread) while in Ahearn and Hannay's experiments, the abundances of both ions were the same.

Buchel'nikova[99] plotted negative ion current against the velocity of slow moving electrons (0.3 eV) and found that $SF_6$ yielded $SF_6^{\bar{}}$, $SF_5^-$, and $F^-$, the current ratios being 2500:100:1. $SF_6$ was also studied mass spectrometrically by Curran[100] who determined the appearance potential of $F^-$ to be 0–0.5 eV and $D(SF_5-F)$ to be $\leq 3.39$ eV.

The results of a further study of negative ions formed by electron impact on $SF_6$, and in addition $SeF_6$ and $TeF_6$, have been published recently[101]. The parent molecular ions $SeF_6^{\bar{}}$ and $TeF_6^{\bar{}}$ were not detected but it was observed that when $SF_6$ was mixed with these compounds, $SeF_6^{\bar{}}$ and $TeF_6^{\bar{}}$ were formed in large quantities, the $SF_6$ effecting collison stabilisation.

Paulett and Lustig[121] recorded the mass spectra of the compounds $FSO_2NF_2$, $(FSO_2)_2NF$, $SF_5NF_2$, and $FSO_2ONF_2$ and measured the appearance potentials of the principal ions. From these measurements they determined the values for $D(S-N)$ to be 39, 48, and 32 kcal.mole$^{-1}$, respectively, and $D(O-N)$ for the latter compound was found to be 35 kcal.mole$^{-1}$.

APPENDIX

The mass spectra of the following newly prepared main Group VI compounds have been recorded during their characterisation but full details of the spectra have not been published in many cases.

| Compound | Ref. | Compound | Ref. |
|---|---|---|---|
| OC(NCO)$_2$ | 102 | FCO(NCS) | 102 |
| CF$_3$SFO | 103 | CF$_3$OCFO | 114 |

| Compound | Ref. | Compound | Ref. |
|---|---|---|---|
| $CF_3SO_2OSO_2F$ | 104 | $CF_3SO_2OCF_3$ | 115 |
| $FCON=SF_2$ | 105 | $CF_3CFBrCF(OSO_2F)CF_3$ | 116 |
| $FSO_2N=SF_2$ | 105 | $ClSO_2N=SF_2$ | 105 |
| $FSO_2NHF$ | 106 | $FSO_2-NF-SF_5$ | 106 |
| $ClSO_2-N=CCl_2$ | 107 | $FSO_2-N=CCl_2$ | 107 |
| $CCl_3SSCCl_3$ | 108 | $CCl_3SSSCCl_3$ | 108 |
| $CCl_3SO_2SCCl_3$ | 108 | $CCl_3SO_2SSCCl_3$ | 108 |
| $NC-N=SF_2$ | 109 | $FCl(SCN)C-SCl$ | 117 |
| $CH_3N=SF_2$ | 110 | $(CH_3)_2N-SF_3$ | 118 |
| $CH_3N=S=NCH_3$ | 110 | $CF_3SN=S=NSCF_3$ | 111 |
| $CF_3SN=SF_2$ | 111 | $S=CF=NCS$ | 117 |
| $S_3O_8F_2$ | 112 | $(SF_5N=)_2SF_2$ | 119 |
| $(CH_3)_2NSF_3O$ | 113 | | |
| $(SPF_2)_2X$ | 120 | | |
| $(X = O, S, NCH_3 NH(CH_3)_3N)$ | | | |
| $(OPF_2)_2X$ | 120 | | |
| $(X = NCH_3, NH(CH_3)_3N)$ | | | |

REFERENCES

1  R. C. Weast (Ed.), *Handbook of Chemistry and Physics*, The Chemical Rubber Publishing Co., Cleveland, Ohio, 49th edn., 1968–69.
2  D. C. Frost and C. A. McDowell, *J. Am. Chem. Soc.*, 80 (1958) 6183.
3  H. Sjogren and E. Lindholm, *Arkiv Fysik*, 32 (1966) 275.
4  J. T. Tate and P. S. Smith, *Phys. Rev.*, 39 (1932) 270.
5  H. D. Hagstrum, *Rev. Mod. Phys.*, 23 (1951) 185.
6  C. E. Brion, *J. Chem. Phys.*, 40 (1964) 2995.
7  R. E. Winters, J. H. Collins and W. L. Courchene, *J. Chem. Phys.*, 45 (1966) 1931.
8  K. Watanabe, *J. Chem. Phys.*, 26 (1957) 542.
9  V. H. Dibeler and J. A. Walker, *J. Opt. Soc. Am.*, 57 (1967) 1007.
10  J. A. R. Samson and R. B. Cairns, *J. Opt. Soc. Am.*, 56 (1966) 769.
11  K. Watanabe and F. F. Marm, *J. Chem. Phys.*, 25 (1956) 965.
12  A. J. C. Nicholson, *J. Chem. Phys.*, 39 (1963) 954.
13  E. C. Y. Inn, *Phys. Rev.*, 91 (1953) 1194.
14  T. Thorburn, *Report on Conference of Applied Mass Spectrometry*, Institute of Petroleum, London, 1954, p. 129.
15  J. E. Collin and P. Natalis, *Intern. J. Mass Spectrom. Ion Phys.*, 2 (1969) 231.
16  F. E. Elder, D. Villarejo and M. G. Inghram, *J. Chem. Phys.*, 43 (1966) 758.

17 K. Watanabe, T. Nakayama and J. Mottl, *J. Quant. Spectry. Radiative Transfer.*, 2 (1962) 369.

18 M. I. Al-Joboury, D. P. May and D. W. Turner, *J. Chem. Soc.*, (1965) 6350.

19 P. M. Hierl and J. L. Franklin, *J. Chem. Phys.*, 47 (1967) 3154.

20 R. E. Fox and W. M. Hickam, *J. Chem. Phys.*, 22 (1954) 2059.

21 D. C. Frost, D. Mak, C. A. McDowell and D. A. Vroom, *Advan. Mass Spectrometry, Proc. Conf. Montreal, Canada, June 1964.*

22 G. Herzberg, *Molecular Spectra and Molecular Structure I. Spectra of Diatomic Molecules*, Van Nostrand, New York, 1950.

23 M. I. Al-Joboury and D. W. Turner, *J. Chem. Soc.*, (1964) 4434.

24 J. D. Carette, *Can. J. Phys.*, 45 (1967) 2931.

25 J. Collin, *J. Chim. Phys.*, 57 (1960) 424.

26 W. C. Price and D. M. Simpson, *Proc. Roy. Soc. (London), Ser. A*, 169 (1939) 501.

27 C. R. Brundle and D. W. Turner, *Intern. J. Mass Spectrom. Ion Phys.*, 2 (1969) 195.

28 Y. Tanaka, A. S. Jursa and F. J. Le Blanc, *J. Chem. Phys.*, 32 (1960) 1199.

29 L. P. Blanchard and P. Le Goff, *Can. J. Chem.*, 35 (1957) 89.

30 G. Preuner and W. Schupp, *Z. Physik. Chem.*, 68 (1909) 129.

31 P. Goldfinger, M. Ackermann and M. Jeunehomme, *ASTIA, AFOSR-5932*, under Contract AF 61(052)-19.

32 M. C. Zietz, *Thesis*, University of California, 1960.

33 J. Berkowitz and J. M. Marquart, *J. Chem. Phys.*, 39 (1963) 275.

34 R. Colin, P. Goldfinger and M. Jeunehomme, *Trans. Faraday Soc.*, 60 (1964) 306.

35 J. Berkowitz and W. A. Chupka, *J. Chem. Phys.*, 40 (1964) 287.

36 J. Buchler, *Angew. Chem. Intern. Ed. Engl.*, 5 (1966) 965.

37 V-I. Zahorszky, *Angew. Chem. Intern. Ed. Engl.*, 7 (1968) 633.

38 R. Hagemann, *Compt. Rend.*, 255 (1962) 1102.

39 H. Fujisaki, J. B. Westmore and A. W. Tickner, *Can. J. Chem.*, 44 (1966) 3063.

40 P. Goldfinger and M. Jeunehomme, *Advan. Mass Spectrometry, Proc. Conf. Univ. London, 1958*, J. D. Waldron (Ed.), Vol. 1, Pergamon Press, London, 1959, p. 534.

41 J. Berkowitz and W. A. Chupka, *J. Chem. Phys.*, 45 (1966) 4289.

42 J. Drowart and P. Goldfinger, *Quart. Rev.*, 20 (1966) 545.

43 R. F. Barrow and R. F. du Parcq, in *Elemental Sulphur*, B. Meyer (Ed.), Interscience, New York, 1965, Chap. 13.

44 P. Budininkas, R. K. Edwards and P. G. Wahlbeck, *J. Chem. Phys.*, 48 (1968) 2859; 48 (1968) 2867; 48 (1968) 2870.

45 R. F. Barrow, G. G. Chandler and C. B. Meyer, *Phil. Trans. Roy. Soc. London, Ser. A*, 260 (1966) 395.

46 R. Colin, P. Goldfinger and M. Jeunehomme, *Nature*, 187 (1960) 408.

47 J. Berkowitz and W. A. Chupka, *J. Chem. Phys.*, 50 (1969) 4245.

48 R. M. Reese, V. H. Dibeler and J. L. Franklin, *J. Chem. Phys.*, 29 (1958) 880.

49 H. Neuert and O. Rosenbaum, *Naturwissenschaften*, 41 (1954) 85.

50 O. Rosenbaum and H. Neuert, *Z. Naturforsch.*, 9a (1954) 990.

51 R. Hagemann, *Compt. Rend.*, 255 (1962) 899.

52 R. Cooper and J. V. Culka, *J. Inorg. Nucl. Chem.*, 29 (1967) 1217.

53 V. A. Umilin, I. L. Agafonov, L. N. Kornev and G. G. Devyatykh, *Zh. Neorgan. Khim.*, 9 (1964) 2492; *Russ. J. Inorg. Chem.*, 9 (1964) 1345.

54 R. Cooper and J. V. Culka, *J. Inorg. Nucl. Chem.*, 29 (1967) 1877.
55 R. F. Porter and C. W. Spencer, *J. Chem. Phys.*, 32 (1960) 943.
56 L. P. Lindeman and J. C. Guffy, *J. Chem. Phys.*, 29 (1958) 247.
57 J. J. Leventhal and L. Friedman, *J. Chem. Phys.*, 46 (1967) 997.
58 P. Dong and M. Cottin, *J. Chim. Phys.*, 57 (1960) 557.
59 A. M. Peers and M. Cottin, *J. Chim. Phys.*, 63 (1966) 1346.
60 J. C. J. Thynne and A. G. Harrison, *Trans. Faraday Soc.*, 62 (1966) 2468.
61 A. G. Harrison and J. C. Thynne, *Trans. Faraday Soc.*, 62 (1966) 3345.
62 W. E. W. Ruska and J. L. Franklin, *Intern. J. Mass Spectrom. Ion Phys.*, 3 (1969) 221.
63 J. F. Paulson, R. L. Mosher and F. Dale, *J. Chem. Phys.*, 44 (1966) 3025.
64 T. W. Martin, *J. Chem. Phys.*, 43 (1965) 1422.
65 L. P. Lindeman and J. C. Guffy, *J. Chem. Phys.*, 30 (1959) 322.
66 W. C. Price and T. M. Sugden, *Trans. Faraday Soc.*, 44 (1948) 108.
67 V. H. Dibeler, J. A. Walker and H. M. Rosenstock, *J. Res. Natl. Bur. Std.*, 70A (1966) 459.
68 G. J. Schulz, *J. Chem. Phys.*, 33 (1960) 1661.
69 M. I. Al-Joboury and D. W. Turner, *J .Chem. Soc. B*, (1967) 373.
70 P. Dong and M. Cottin, *J. Chim. Phys.*, 58 (1961) 803.
71 W. C. Price, *J. Chem. Phys.*, 4 (1936) 147.
72 A. J. B. Robertson, *Trans. Faraday Soc.*, 48 (1952) 228.
73 S. N. Foner and R. L. Hudson, *J. Chem. Phys.*, 36 (1962) 2676.
74 S. N. Foner and R. L. Hudson, *J. Chem. Phys.*, 36 (1962) 2681.
75 C. E. Melton and H. W. Joy, *J. Chem. Phys.*, 46 (1967) 4275.
76 S. N. Foner and R. L. Hudson, *J. Chem. Phys.*, 25 (1956) 602.
77 R. W. Kiser and B. G. Hobrock, *J. Phys. Chem.*, 66 (1962) 1214.
78 T. F. Palmer and F. P. Lossing, *J. Am. Chem. Soc.*, 84 (1962) 4661.
79 M. M. Mann, A. Hustrulid and J. T. Tate, *Phys. Rev.*, 58 (1940) 340.
80 J. L. Franklin and H. E. Lumpkin, *J. Am. Chem. Soc.*, 74 (1952) 1023.
81 D. P. Stevenson, *Trans. Faraday Soc.*, 49 (1953) 867.
82 V. H. Dibeler, R. M. Reese and J. L. Franklin, *J. Chem. Phys.*, 27 (1957) 1296.
83 T. L. Malone and H. A. McGee, Jr., *J. Phys. Chem.*, 69 (1965) 4338.
84 T. J. Malone and H. A. McGee, Jr., *J. Phys. Chem.*, 70 (1966) 316.
85 T. J. Malone and H. A. McGee, Jr., *J. Phys. Chem.*, 71 (1967) 3060.
86 I. J. Solomon, J. N. Keith, A. J. Kacmarek and J. K. Rainey, *J. Am. Chem. Soc.*, 90 (1968) 5408.
87 R. M. Reese, V. H. Dibeler and F. L. Mohler, *J. Res. Natl. Bur. Std.*, 57 (1956) 367.
88 V. H. Dibeler and F. L. Mohler, *J. Res. Natl. Bur. Std.*, 40 (1948) 25.
89 L. R. Anderson and W. B. Fox, *J. Am. Chem. Soc.*, 89 (1967) 4313.
90 R. Kuczkowski and E. B. Wilson, Jr., *J. Am. Chem. Soc.*, 85 (1963) 2028.
91 R. Kuczkowski, *J. Am. Chem. Soc.*, 85 (1963) 3047; 86 (1964) 3617.
92 F. Seel and R. Budenz, *Chimia (Aarau)*, 17 (1963) 355.
93 F. Seel and R. Budenz, *Chem. Ber.*, 98 (1965) 251.
94 R. E. Fox and R. K. Curran, *J. Chem. Phys.*, 34 (1961) 1595.
95 V. H. Dibeler and J. A. Walker, *J. Chem. Phys.*, 44 (1966) 4405.
96 D. C. Frost, C. A. McDowell, J. S. Sandhu and D. A. Vroom, *J. Chem. Phys.*, 46 (1967) 2008.

97 A. J. Ahearn and N. B. Hannay, *J. Chem. Phys.*, 21 (1953) 119.
98 W. M. Hickam and R. E. Fox, *J. Chem. Phys.*, 25 (1956) 642.
99 N. S. Buchel'nikova, *Zh. Eksperim. i Teor. Fiz.*, 35 (1958) 1119.
100 R. K. Curran, *J. Chem. Phys.*, 34 (1961) 1069.
101 C. E. Brion, *Intern. J. Mass Spectrom. Ion Phys.*, 3 (1969) 197.
102 W. Verbeek and W. Sundermeyer, *Angew. Chem. Intern. Ed. Engl.*, 6 (1967) 871.
103 E. W. Lawless and L. D. Harman, *Inorg. Chem.*, 7 (1968) 391.
104 R. E. Noftle, *Inorg. Chem.*, 7 (1968) 2167.
105 U. Biermann and O. Glemsen, *Chem. Ber.*, 100 (1967) 3795.
106 H. W. Roesky, *Angew. Chem. Intern. Ed. Engl.*, 7 (1968) 630.
107 H. W. Roesky and U. Biermann, *Angew. Chem. Intern. Ed. Engl.*, 6 (1967) 882.
108 S. Kaal and A. Senning, *Acta Chem. Scand.*, 22 (1968) 159.
109 W. Sundermeyer, *Angew. Chem. Intern. Ed. Engl.*, 6 (1967) 90.
110 B. Cohen and A. G. MacDiarmid, *J. Chem. Soc. A*, (1966) 1780.
111 A. Haas and P. Schott, *Angew. Chem. Intern. Ed. Engl.*, 6 (1967) 370.
112 J. E. Roberts and G. H. Cady, *J. Am. Chem. Soc.*, 81 (1959) 4166.
113 O. Glemsen, S. P. v. Halasz and U. Biermann, *Z. Naturforsch.*, 23b (1968) 1381.
114 T. Johnston, J. Heicklen and W. Stuckey, *Can. J. Chem.*, 46 (1968) 332.
115 R. E. Noftle and G. H. Cady, *Inorg. Chem.*, 4 (1965) 1010.
116 B. L. Earl, B. K. Hill and J. M. Shreeve, *Inorg. Chem.*, 5 (1966) 2184.
117 A. Haas and W. Klug, *Angew. Chem. Intern. Ed. Engl.*, 6 (1967) 940.
118 G. C. Demitras and A. G. MacDiarmid, *Inorg. Chem.*, 6 (1967) 1903.
119 A. F. Clifford and G. R. Zeilenga, *Inorg. Chem.*, 8 (1969) 1789.
120 T. L. Charlton and R. G. Cavell, *Inorg. Chem.*, 9 (1970) 379.
121 G. S. Paulett and M. Lustig, *J. Am. Chem. Soc.*, 87 (1965) 1020.
122 D. C. Frost and C. A. McDowell, *Can. J. Chem.*, 36 (1958) 39.
123 P. G. Thompson ,*J. Am. Chem. Soc.*, 89 (1967) 4316.
124 J. G. Dillard and J. L. Franklin, *J. Chem. Phys.*, 48 (1968) 2349, 2353.

*Chapter 10*

# The Main Group VII Elements

M. R. LITZOW

## 1. INTRODUCTION

Mass spectra of large numbers of halide compounds of various elements have been studied and, in general, these are discussed in the appropriate sections. Electron-impact studies of the diatomic halogen molecules, interhalogen compounds, and other simple halogen-containing molecules which have not been mentioned in these sections are discussed briefly here.

## 2. THE ELEMENTS

The natural abundances[1] of the various halogen isotopes are shown in Table 10.1 and Figure 10.1.

Table 10.1

Natural Abundances of the Halogen Isotopes[1]

| Element | Mass no. | % Abundance |
|---|---|---|
| F | 19 | 100 |
| Cl | 35 | 75.53 |
| | 37 | 24.47 |
| Br | 79 | 50.54 |
| | 81 | 49.46 |
| I | 127 | 100 |

*References pp. 468–469*

% Isotopic abundance

100

50

| 19 | 35 37 | 79 81 | 127 |
| F | Cl | Br | I |

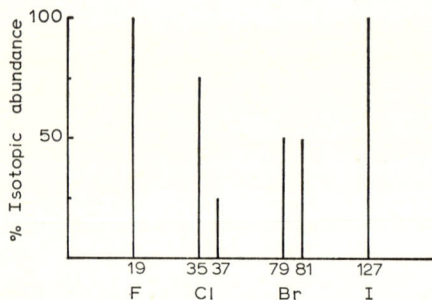

Fig. 10.1. Natural isotopic abundances of the halogens (ref. 1).

Table 10.2

Ionisation Potentials of Diatomic Halogen Molecules

References are given in brackets.

| Molecule | Ionisation potential (eV) | | | |
|---|---|---|---|---|
| | Electron impact | | Photoionisation (ref. 8) | Photoelectron[a] (ref. 9) |
| $F_2$ | $15.83 \pm 0.05$ (5) | $16.6 \pm 0.2$ (6) | | 15.63 |
| | | | | 17.35 |
| | | | | 18.46 |
| $Cl_2$ | $11.64 \pm 0.05$ (5) | $11.63 \pm 0.04$ (7) | $11.48 \pm 0.01$ | 11.50 |
| | | $14.09 \pm 0.03$ | | 14.11 |
| | | $20.61 \pm 0.06$ | | 15.94 |
| $Br_2$ | $10.58 \pm 0.08$ (5) | $10.69 \pm 0.03$ (7) | $10.55 \pm 0.02$ | 10.51 |
| | | $11.05 \pm 0.05$ | | 10.90 |
| | | $11.97 \pm 0.03$ | | |
| | | $12.36 \pm 0.04$ | | 12.52 |
| | | $13.72 \pm 0.04$ | | 14.44 |
| $I_2$ | | $9.35 \pm 0.03$ (7) | $9.28 \pm 0.02$ | 9.33 |
| | | $9.97 \pm 0.02$ | | 9.96 |
| | | $10.91 \pm 0.04$ | | 10.87 |
| | | $11.72 \pm 0.04$ | | 11.68 |
| | | $13.64 \pm 0.06$ | | 12.79 |
| ICl | | $10.31 \pm 0.02$ (7) | | |
| | | $10.79 \pm 0.03$ | | |
| | | $12.13 \pm 0.04$ | | |
| | | $\sim 12.2$ | | |
| IBr | | $9.98 \pm 0.03$ (7) | | |
| | | $10.49 \pm 0.03$ | | |
| | | $11.59 \pm 0.05$ | | |
| | | $\sim 11.7$ | | |

[a] Listed as successive energy levels.

3. MASS SPECTRAL STUDIES

The ionisation and dissociation of the diatomic molecules $F_2$, $Cl_2$, $Br_2$, $I_2$, $ICl$, and $IBr$ by electron impact have been studied and results are compared with those of other techniques in Table 10.2. A number of the very reactive polyatomic halogen fluorides has also been studied mass spectrometrically. As with the boron trihalides, conditioning of the instrument was found to be necessary before spectra of the compounds could be obtained. Ion abundances were reported for $IF_5$, $BrF_5$, $ClF_3$, $BrF_3$ (ref. 2) and $IF_7$ (ref. 3), while the astatine compounds AtCl, AtBr, AtI, as well as HAt and $CH_3At$, were detected and identified in a mass spectrometer[4]. Parent molecular ions were not observed for $BrF_5$ or $IF_7$. Irsa and Friedman[2] also determined appearance potentials of the various ions formed from the first four compounds but insufficient data were available to enable decomposition mechanisms to be established.

Frost and McDowell[10,11] studied the ionisation of the hydrogen halide molecules using an essentially monoenergetic electron beam, observed excited states of the parent molecular ions as with the diatomic halogen

Table 10.3

Ionisation Potentials of the Hydrogen Halides

| Molecule | Ionisation potential (eV) | | | |
|---|---|---|---|---|
| | Spectroscopic | Electron impact | Photoionisation | Photoelectron[b] (ref. 9) |
| HF | | 15.77 (10,11) | 16.04 (14) | 16.06 |
| | 16.88 (13) | 16.91 (10,11) | | 16.48 |
| HCl | | 12.72 (11) | 12.74 (15) | 12.75 |
| | | | | 12.85 |
| | 16.29, 16.27[a] | 15.92 (11) | | 16.28 |
| HBr | | 11.82 (11) | 11.62 (15) | 11.71 |
| | | 12.60 (11) | | 12.03 |
| | 14.91, 15.24[a] | 14.32 (11) | | 15.31 |
| HI | | 10.44 (10,11) | 10.38 (15) | 10.42 |
| | 11.094 (16) | 11.14 | | 11.08 |
| | | 13.27 | | 14.03 |

[a] Values calculated by Frost et al.[9] from previously reported results.
[b] Listed as successive energy levels.

molecules, and calculated the various ionisation potentials. The results obtained are compared with the results of spectroscopic, photoionisation and photoelectron studies in Table 10.3. A mass spectral determination of rate constants for H atom reactions with $Cl_2$ and $F_2$ has been reported[12].

The reaction of oxygen and $BrF_5$ was briefly studied[2] by mixing the two compounds within a mass spectrometer. Halogen oxyfluorides were also detected in $BrF_3$ and $ClF_3$ samples[2]. Mass spectra of $HOI_4$, $HIO_3$, and $I_2O_5$ were reported as photographs of oscilloscope displays[17]. Schack et al.[3] recorded the mass spectrum of $IOF_5$ and found that by far the most abundant ion was $IF_4^+$, while thermal degradation was also said to involve rupture of the I–O bond. Sloth et al.[18] studied the products formed on hydrolysis of bromine fluorides by mass spectrometry and observed a number of oxyhalogen ions. The fragmentation pattern of cyanogen fluoride, FCN, was determined[19] as an aid in the characterisation of the compound; the parent molecular ion was by far the most abundant.

Irsa and Friedman[2] observed a number of negative ions to be formed from $BrF_3$, while Stuckey and Kiser[20] reported both singly and doubly charged negative ions of fluorine, chlorine, and bromine arising from $CCl_4$, $CF_4$, $CF_3Cl$, $CF_3Br$, and $CF_2Cl_2$. Resonance capture processes leading to formation of $Cl^-$ from $Cl_2$ and $I^-$ from $I_2$ were investigated by Frost and McDowell[21]. Taylor and Grimsrud[22] determined chlorine isotope ratios from negative ion mass spectra; values obtained agreed very well with those previously accepted[1].

REFERENCES

1 R. C. Weast (Ed.), *Handbook of Chemistry and Physics*, The Chemical Rubber Publishing Co., Cleveland, Ohio, 49th edn., 1968–69.
2 A. P. Irsa and L. Friedman, *J. Inorg. Nucl. Chem.*, 6 (1958) 77.
3 C. J. Schack, D. Pilipovich, S. N. Cohz and D. F. Sheehan, *J. Phys. Chem.*, 72 (1968) 4697.
4 E. H. Appelman, E. N. Sloth and M. H. Studier, *Inorg. Chem.*, 5 (1966) 766.
5 J. T. Herron and V. H. Dibeler, *J. Chem. Phys.*, 32 (1960) 1884.
6 R. Thorburn, *Proc. Phys. Soc. (London)*, 73 (1959) 122.
7 D. C. Frost and C. A. McDowell, *Can. J. Chem.*, 38 (1960) 407.
8 K. Watanabe, *J. Chem. Phys.*, 26 (1957) 542.
9 D. C. Frost, C. A. McDowell and D. A. Vroom, *J. Chem. Phys.*, 46 (1967) 4255.
10 D. C. Frost and C. A. McDowell, *Can. J. Chem.*, 36 (1958) 39.
11 D. C. Frost, *Ph.D. Thesis*, Liverpool University, 1958.
12 R. G. Albright, A. F. Dodonov, G. K. Lavrovskaya, I. I. Morosov and V. L. Tal'roze, *J. Chem. Phys.*, 50 (1969) 3632.

13 J. W. C. Johns and R. F. Barrow, *Nature*, 179 (1957) 374.

14 D. C. Frost, D. Mak, C. A. McDowell and D. A. Vroom, *Twelfth Annual Conference on Mass Spectrometry and Allied Topics, Montreal, 1964*, ASTM Committee E-14.

15 K. Watanabe, T. Nakayama and J. R. Mottl, *J. Quant. Spectry. Radiative Transfer*, 2 (1962) 369.

16 W. C. Price, *Proc. Roy. Soc. (London), Ser. A*, 167 (1938) 216.

17 M. H. Studier and J. L. Huston, *J. Phys. Chem.*, 71 (1967) 457.

18 E. N. Sloth, L. Stein and C. W. Williams, *J. Phys. Chem.*, 73 (1969) 278.

19 F. S. Fawcett and R. D. Lipscomb, *J. Am. Chem. Soc.*, 86 (1964) 2576.

20 W. K. Stuckey and R. W. Kiser, *Nature*, 211 (1966) 963.

21 D. C. Frost and C. A. McDowell, *Advan. Mass Spectrometry, Proc. Conf. Univ. London, 1958*, J. D. Waldron (Ed.), Pergamon Press, London, 1959, p. 413.

22 J. W. Taylor and E. P. Grimsrud, *Anal. Chem.*, 41 (1969) 805.

*Chapter 11*

# Compounds of the Transition Metals

M. R. LITZOW

## 1. INTRODUCTION

Low-resolution mass spectra of a large number of transition metal compounds have been reported in the last few years and a considerable volume of information concerning the behaviour of them under electron impact has been accumulated. General fragmentation pathways of many ligand systems such as carbonyl, $\pi$-cyclopentadienyl, $\pi$-arene, fluorocarbon, etc. have been established. The mass spectra of metal carbonyl compounds are characterised by sequential loss of carbon monoxide from the parent molecular ion and only processes with very low energy requirements can compete with this decarbonylation. Occasionally, simultaneous loss of more than one CO group has been observed (Sections 13 and 15). Fragmentation of acetylene from hydrocarbon ligands is common. Olefin ligands tend to lose neutral species such as $H_2$, $\cdot CH_3$, $C_2H_4$, etc. Extrusion of neutral metal fluorides is a very frequent decomposition step in the fragmentation of fluorocarbon metal complexes. Many fragment ions appear to incorporate the benzene ligand as a result of rearrangement processes.

Although a large amount of information has thus been accumulated, many questions remain unanswered. Structures are often postulated for various ions but only in two cases have detailed studies been undertaken to determine these structures. Rarely have the energetics of these systems been investigated. It is often tempting to assign orders of bond dissociation energies on the basis of relative intensities of ions, but this can result in erroneous conclusions (see, for example, Section 4.2). Bond dissociation energies in a neutral molecule and in the corresponding positively charged ion may be significantly different. Metastable decompositions occurring in the field-free region between the electrostatic and magnetic analyser of a double-focusing mass

*References pp. 600–610*

spectrometer are observed in low-resolution spectra as diffuse, low-intensity peaks. Those that have been so observed have provided valuable information concerning fragmentation pathways. To establish complete pathways in most cases, however, more metastable decompositions need to be detected. Detailed study of the more numerous ones occurring in the region between the ion source and the electrostatic analyser (Chapter 2, Section 6) would provide this information.

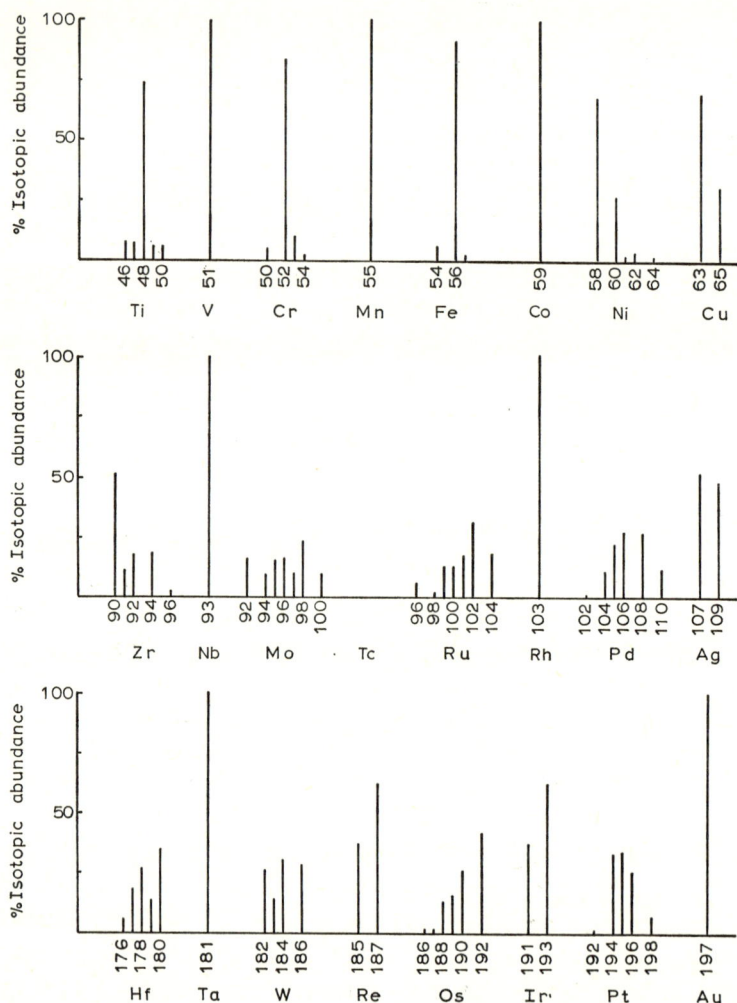

Fig. 11.1. Natural isotopic abundances of the transition metals (ref. 1).

Table 11.1

Naturally Occurring Isotopes of the Transition Elements[1]

| First transition series | | | Second transition series | | | Third transition series | | |
|---|---|---|---|---|---|---|---|---|
| Element | Mass no. | % Abundance | Element | Mass no. | % Abundance | Element | Mass no. | % Abundance |
| Ti | 46 | 7.93 | Zr | 90 | 51.46 | Hf | 174 | 0.18 |
| | 47 | 7.28 | | 91 | 11.23 | | 176 | 5.20 |
| | 48 | 73.94 | | 92 | 17.11 | | 177 | 18.50 |
| | 49 | 5.51 | | 94 | 17.40 | | 178 | 27.14 |
| | 50 | 5.34 | | 96 | 2.80 | | 179 | 13.75 |
| | | | | | | | 180 | 35.24 |
| V | 50 | 0.24 | Nb | 93 | 100 | Ta | 180 | 0.01 |
| | 51 | 99.76 | | | | | 181 | 99.99 |
| Cr | 50 | 4.31 | Mo | 92 | 15.84 | W | 180 | 0.14 |
| | 52 | 83.76 | | 94 | 9.04 | | 182 | 26.41 |
| | 53 | 9.55 | | 95 | 15.72 | | 183 | 14.40 |
| | 54 | 2.38 | | 96 | 16.53 | | 184 | 30.64 |
| | | | | 97 | 9.46 | | 186 | 28.41 |
| | | | | 98 | 23.78 | | | |
| | | | | 100 | 9.13 | | | |
| Mn | 55 | 100 | Tc | | | Re | 185 | 37.07 |
| | | | | | | | 187 | 62.93 |
| Fe | 54 | 5.82 | Ru | 96 | 5.51 | Os | 184 | 0.02 |
| | 56 | 91.66 | | 98 | 1.87 | | 186 | 1.59 |
| | 57 | 2.19 | | 99 | 12.72 | | 187 | 1.64 |
| | 58 | 0.33 | | 100 | 12.62 | | 188 | 13.3 |
| | | | | 101 | 17.07 | | 189 | 16.1 |
| | | | | 102 | 31.61 | | 190 | 26.4 |
| | | | | 104 | 18.58 | | 192 | 41.0 |
| Co | 59 | 100 | Rh | 103 | 100 | Ir | 191 | 37.3 |
| | | | | | | | 193 | 62.7 |
| Ni | 58 | 67.88 | Pd | 102 | 0.96 | Pt | 190 | 0.01 |
| | 60 | 26.23 | | 104 | 10.97 | | 192 | 0.78 |
| | 61 | 1.19 | | 105 | 22.23 | | 194 | 32.9 |
| | 62 | 3.66 | | 106 | 27.33 | | 195 | 33.8 |
| | 64 | 1.08 | | 108 | 26.71 | | 196 | 25.3 |
| | | | | 110 | 11.81 | | 198 | 7.21 |
| Cu | 63 | 69.09 | Ag | 107 | 51.82 | Au | 197 | 100 |
| | 65 | 30.91 | | 109 | 48.18 | | | |

Thermal decomposition has been largely overlooked with these compounds but many organometallic compounds are extremely susceptible. Pyrolysis should always be suspected whenever the ligand ions have by far the largest peak intensity in the spectrum. Some elementary precautions can be taken to at least minimise, if not eliminate, these effects (Chapter 3, Section 1).

## 2. THE ELEMENTS

All isotopes of technetium are unstable towards $\beta$ decay or electron capture and traces exist in nature only as fragments from the spontaneous emission of uranium. Naturally occurring isotopes of other transition elements, together with their abundances, are shown in Table 11.1 and Fig. 11.1.

## 3. HIGH-TEMPERATURE STUDIES

A number of diatomic species has been observed by a combination of Knudsen cell effusion techniques and mass spectrometry. The values of their

Table 11.2

References are given in brackets.

(a) Dissociation Energies $D$ of Homonuclear Diatomic Molecules of the Transition Elements, Determined Mass Spectomertrically (kcal . mole$^{-1}$)

| | | | | | | | |
|---|---|---|---|---|---|---|---|
| $Sc_2$ | $38.0\pm5$ (7) | $Mn_2$ | 18 (7) | $Cu_2$ | 46.1–48.4 (12) | $Ag_2$ | 36.8–43.8 (12) |
| | | | <21 (8) | | 46.1 (13) | | 37.6 (13) |
| $Ti_2$ | <58 (8) | | 10 (10) | | $45.5\pm2.2$ (14) | | $37.6\pm2.2$ (14) |
| $Cr_2$ | <44 (8) | $Co_2$ | $39\pm6$ (8) | $Y_2$ | $37.3\pm5$ (7) | $Au_2$ | 50.7–62.3 (12) |
| | $36.0\pm7$ (9) | | | | | | 48.4 (13) |
| | $41\pm8$ (9) | $Ni_2$ | $54.5\pm5$ (11) | $Pd_2$ | $\leq33$ (7) | | $51.5\pm2.2$ (14) |

(b) Dissociation Energies $D$ of Heteronuclear Diatomic Molecules Determined Mass Spectrometrically (kcal . mole$^{-1}$)

| | | | | | | | |
|---|---|---|---|---|---|---|---|
| CuSn | $41.4\pm4$ (15) | AgMn | $23\pm11$ (10) | AuBe | 73 (7) | AuAl | 70 (7) |
| AgSn | $31.6\pm5$ (15) | AuCr | $50.4\pm3.5$ (16) | AuMg | 61 (7) | AuGa | 46 (7) |
| AuSn | $57.5\pm4$ (15) | AuPd | $33.3\pm5$ (16) | AuCa | 57 (7) | AuSi | 81 (7) |
| CuAg | $40.7\pm2.2$ (14) | NiGe | $66.4\pm4$ (18) | AuSr | 88 (7) | AuPb | 30 (7) |
| CuAu | $54.5\pm2.2$ (14) | | | AuBa | 66 (7) | AuCe | $71\pm5$ (17) |
| AgAu | $47.6\pm2.2$ (14) | | | | | | |

dissociation energies have, in most cases, been determined (Table 11.2), as well as other thermodynamic values in some instances. No homonuclear diatomic molecules were observed by Zandberg et al.[2] in the vapours of Re, W, Ta, or Mo, nor by Panish and Reif[3] in the vapour over Hf but the heats of sublimation of the metals were determined. Similarly, Trulson and Schissel[4] were not able to detect any $Pd_2$ under the conditions of their experiments but the diatomic molecule was later shown to exist[5]. A large number of gaseous heteronuclear diatomic molecules containing gold have been studied and have proved to be remarkably stable (Table 11.2). A study of the activities in the Fe–Co and Fe–Ni systems has been reported[6].

Various transition metal oxide systems have been studied by mass spectrometry. The vapour above titanium[19] and hafnium[3] dioxides was investigated; so, too, were the oxides of chromium[20] and a mass spectral investigation of the oxidation of molybdenum and tungsten was reported[21]. One of two investigations of the tungsten–oxygen system reported[22] that $W_4O_{12}$, $W_3O_9$, $W_3O_8$, and $W_2O_6$ were the thermodynamically important vapour species, while the second concluded[23] that the main products of the reaction were $W_3O_9$, $W_2O_6$, $WO_3$, and $WO_2$. Both the positive and negative ion mass spectra of ruthenium and osmium tetroxides have been published[24]; appearance potentials and bond dissociation energies were determined. Gaseous oxides of iridium[25], palladium[26], and platinum[27] were also studied using this technique, as was manganous sulphide[28] and molybdenum monophosphide[29]. Values which have been determined for the dissociation energies of gaseous monoxides, atomisation energies of gaseous dioxides and trioxides, and polymerisation energies of gaseous oxides have been summarised by Drowart and Goldfinger[7].

The volatility of $TiCl_4$, $NbCl_4$, $NbCl_5$, and $TaCl_5$ enabled them to be studied at normal operating temperatures[30]. The spectra were recorded, appearance potentials of the ions measured, enabling their heats of formation to be calculated, and clastograms plotted.

## 4. METAL CARBONYLS

### 4.1 Mononuclear carbonyls

A very early attempt was made by J. J. Thomson to study nickel tetracarbonyl in his mass spectrograph, but decomposition within the instrument rendered the attempt a failure[31]. Subsequently, Aston[32] was successful in utilising the carbonyls of chromium, molybdenum, tungsten, iron, and nickel in determinations of isotopic abundances of the metals. As with alkyl derivatives of main Group elements utilised in the same way, few details of the spectra were published. Other investigations were made by de Gier and Zeeman[33] and Dempster[34]. In 1951 Baldock et al.[35,36] reported investigations into the productions of large quantities of metal ions from tungsten, iron, and nickel carbonyls, and the electromagnetic separation of the isotopes. Since 1960 a spate of publications on mass spectral investigations of these compounds has appeared, paralleling the intense general interest in them.

The first of these was by Vilesov and Kurbatov[37] who determined the ionisation potentials of $Ni(CO)_4$, $Fe(CO)_5$, $Cr(CO)_6$, $Mo(CO)_6$, and $W(CO)_6$ by means of a photoionisation technique. This was followed by a comprehensive study of the nickel and iron compounds by Winters and Kiser[38] who determined isotopic abundances of the metals, ion abundances, and appearance potentials of these ions. The most prolific ions were the metal ion itself and the monocarbonylmetal ion but all possible ions of type $M(CO)_x^+$ were present together with some similar doubly charged species. Clastograms obtained for the singly charged metal-containing ions and appearance potential values which consistently increased in magnitude as the number of fragmented carbonyl groups increased, led to the postulate that the respective ions were formed by successive losses of CO groups (reaction 11.1). This has since been shown to be a characteristic of the metal carbonyls in general, and also of the many substituted metal carbonyls.

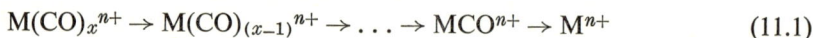

$$M(CO)_x^{n+} \rightarrow M(CO)_{(x-1)}^{n+} \rightarrow \ldots \rightarrow MCO^{n+} \rightarrow M^{n+} \tag{11.1}$$

Conclusive evidence for this proposed fragmentation scheme was soon forthcoming. Employing experimental techniques and conditions which enabled the detection of ions of very low abundance, Winters and Collins[39] not only observed a number of ions such as $FeC_3O_2^{n+}$, $FeC_2O^{n+}$, $FeO^{n+}$, $FeC^{n+}$, ($n = 1$ or 2) and $Fe(CO)_x^{2+}$ ($x = 0, 2, 3, 4$ or 5)* which had not previously been detected, but also ten metastable transitions corresponding to the loss of neutral CO (reactions 11.2–11.4). The previously proposed series of unimolecular reactions was therefore substantiated.

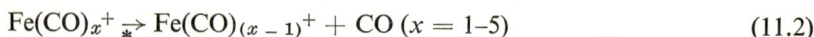

$$Fe(CO)_x^+ \xrightarrow{*} Fe(CO)_{(x-1)}^+ + CO \ (x = 1\text{-}5) \tag{11.2}$$

---

* $Fe(CO)^{2+}$ had been observed previously.

$$Fe(CO)_x^{2+} \underset{*}{\rightarrow} Fe(CO)_{(x-1)}^{2+} + CO \ (x = 2\text{–}4) \tag{11.3}$$

$$Fe(CO)_xC^+ \underset{*}{\rightarrow} Fe(CO)_{(x-1)}C^+ + CO \ (x = 1 \text{ or } 2) \tag{11.4}$$

No evidence for the ion reaction

$$Fe(CO)_x^{2+} \rightarrow Fe(CO)_{(x-1)}^+ + CO^+ \tag{11.5}$$

was found[39] although a specific search was made. Winters and Kiser[40] showed that a definite relationship existed between the intensities of singly and doubly charged metal carbonyl ions.

A detailed study of nickel tetracarbonyl was carried out by Schildcrout et al.[41] on a molecular beam or zero source contact mass spectrometer. A wide range of singly and doubly charged ions was observed analogous to those reported for $Fe(CO)_5$ by Winters and Collins[39] and, in addition, the successive unimolecular decompositions involving loss of neutral CO molecules were metastable supported for this compound also.

Results of two independent studies[42,43] of the Transition Metal Group VI carbonyls appeared in 1965. In direct analogy with the nickel and iron carbonyls, Winters and Kiser[42] observed all possible ions of type $M(CO)_x^+$ $(x = 0\text{–}6)$. All three $M(CO)_5^+$ ions were present in low abundance, while the intensity of $M(CO)_4^+$ and $M(CO)_3^+$ increased considerably on descending the Group. Clastograms and appearance potentials again suggested successive loss of CO groups rather than simultaneous loss of more than one such group.

Instead of plotting full clastograms as done by Winters and Kiser, Cantone et al.[44] concerned themselves with the initial steeply rising section of the ionisation efficiency curves of the various ions formed from $W(CO)_6$. Their results indicated that, in addition to the stepwise decomposition process, the ion $W(CO)_3^+$ was also formed from an excited state of the parent molecular ion, "$W(CO)_6^+$".

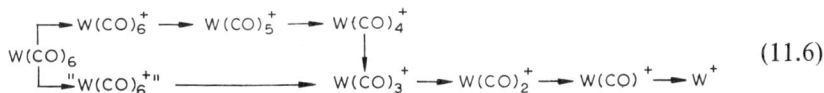

$$\tag{11.6}$$

More recent studies of the Group VI metal carbonyls have also been reported[45,46].

All of the compounds previously mentioned form ions of the type

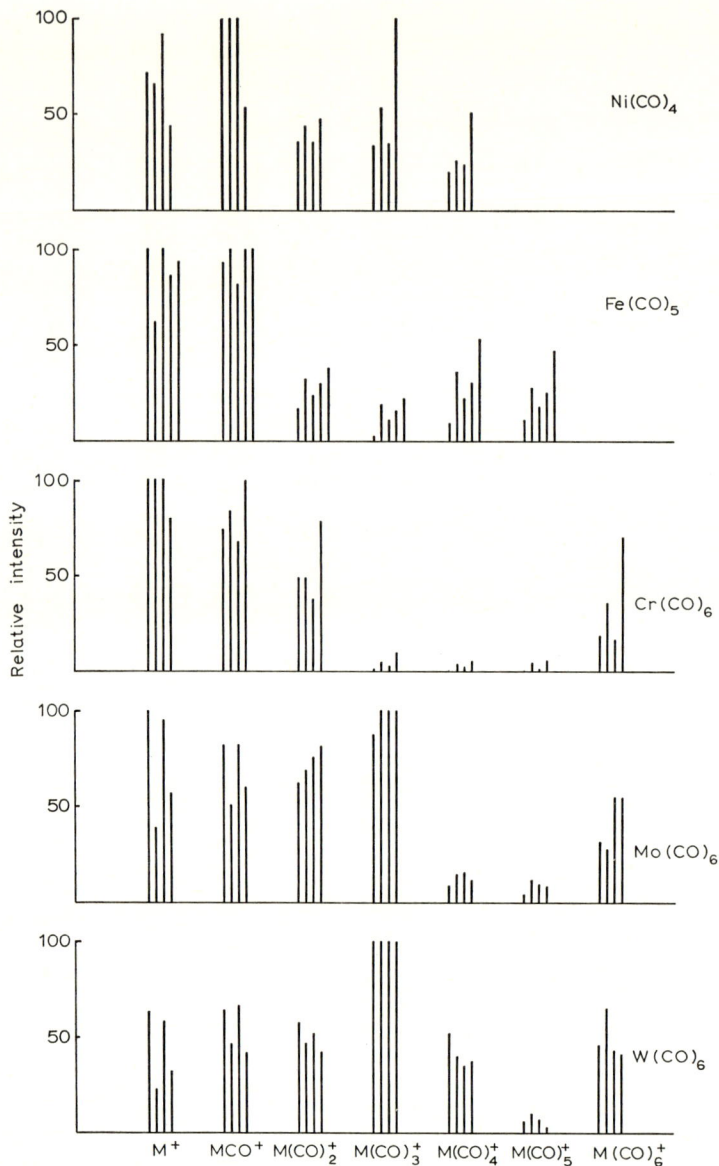

Fig. 11.2. Comparison of reported intensities of ions of type $M(CO)_x{}^+$ formed from mononuclear metal carbonyls (refs. 38, 39, 41–43, 45, 46).

Table 11.3

Reported Appearance Potentials of Ions Formed from Mononuclear Metal Carbonyls

(1) $Ni(CO)_4$ I.P. Ni = 7.633 eV (ref. 54)

| Ion | A.P. (eV) | Method[a] | $\Delta H_f$ (kcal·mole⁻¹) | Ref. |
|---|---|---|---|---|
| $Ni(CO)_4^+$ | 8.28 ±0.03 | Phot. | | 37 |
| | 8.22 ±0.02 | Phot. | | 55 |
| | 8.64 ±0.15 | EVD | 54 | 38 |
| | 8.35 ±0.15 | VC | | 41 |
| | 8.57 ±0.10 | EVD | 53 | 45 |
| | 8.75 ±0.07 | VC | 56 | 46 |
| | 8.28 ±0.01[b] | Phot. | | 47 |
| | 8.93 ±0.05[c] | PES | | 47 |
| $Ni(CO)_3^+$ | 9.36 ±0.15 | EVD | 97 | 38 |
| | 8.89 ±0.15 | VC | | 41 |
| | 9.22 ±0.10 | EVD | 94 | 45 |
| | 9.34[d] | VC | 96 | 46 |
| $Ni(CO)_2^+$ | 10.7 ±0.2 | EVD | 155 | 38 |
| | 10.21 ±0.15 | VC | | 41 |
| | 10.48 ±0.05 | EVD | 149 | 45 |
| | 10.63[d] | VC | 152 | 46 |
| $NiCO^+$ | 13.5 ±0.2 | EVD | 246 | 38 |
| | 12.17 ±0.15 | VC | | 41 |
| | 12.96 ±0.10 | EVD | 233 | 45 |
| | 12.84[d] | VC | 229 | 46 |
| $Ni^+$ | 16.0 ±0.3 | EVD | 330 | 38 |
| | 14.45 ±0.15 | VC | | 41 |
| | 15.1 ±0.3 | EVD | 309 | 45 |
| | 15.51[d] | VC | 317 | 46 |
| $Ni(CO)_4^{2+}$ | 25.1 ±0.5 | VC | | 41 |
| $Ni(CO)_3^{2+}$ | 25.2 ±0.5 | VC | | 41 |
| $Ni(CO)_2^{2+}$ | 28.3 ±1 | EVD | 560 | 38 |
| | 27.2 ±0.5 | VC | | 41 |
| $NiCO^{2+}$ | 30.2 ±0.5 | VC | | 41 |
| $NiC_3O_2^+$ | 20.1 ±0.5 | VC | | 41 |
| $NiC_2O^+$ | 20.6 ±0.3 | VC | | 41 |
| $NiC_2O_2^{2+}$ | 38.5 ±0.5 | VC | | 41 |
| $NiC_2^+$ | 30.1 ±1 | VC | | 41 |
| $NiC^+$ | 24.2 ±0.2 | EVD | | 45 |
| | 22.1 ±0.3 | VC | | 41 |
| $NiO^+$ | 26.4 ±1 | VC | | 41 |
| $C_2O_2^+$ | 21.6 ±0.5 | VC | | 41 |
| $C_2O^+$ | 31.7 ±1 | VC | | 41 |
| $C_2^+$ | 39.9 ±2 | VC | | 41 |

References pp. 600–610

Table 11.3 (continued)

| Metastable transition | A.P. (eV) | Ref. |
|---|---|---|
| $Ni(CO)_4^+ \rightarrow Ni(CO)_3^+ + CO$ | $8.90 \pm 0.15$ | 41 |
| $Ni(CO)_3^+ \rightarrow Ni(CO)_2^+ + CO$ | $10.6 \pm 1$ | 41 |
| $Ni(CO)_2^+ \rightarrow NiCO^+ + CO$ | $12.7 \pm 1$ | 41 |
| $NiCO^+ \rightarrow Ni^+ + CO$ | $15.1 \pm 2$ | 41 |
| $Ni(CO)_4^{2+} \rightarrow Ni(CO)_3^{2+} + CO$ | $26.9 \pm 2$ | 41 |
| $Ni(CO)_3^{2+} \rightarrow Ni(CO)_2^{2+} + CO$ | $27.6 \pm 0.5$ | 41 |

(2) $Fe(CO)_5$ I.P. $Fe = 7.87$ eV (ref. 54)

| Ion | A.P. (eV) | Method[a] | $\Delta H_f$ (kcal.mole$^{-1}$) | Ref. |
|---|---|---|---|---|
| $Fe(CO)_5^+$ | $7.95 \pm 0.03$ | Phot. | | 37 |
| | $7.85$ | Phot. | | 56 |
| | $8.53 \pm 0.2$ | EVD | 23 | 38 |
| | $8.14 \pm 0.06$ | CS | | 43 |
| | $8.16 \pm 0.05$ | EVD | 14.2 | 45 |
| | $8.40 \pm 0.03$ | VC | 20 | 46 |
| | $7.96 \pm 0.02^b$ | Phot. | | 47 |
| | $8.60 \pm 0.04^c$ | PES | | 47 |

| Ion | A.P. (eV) | Method[a] | $\Delta H_f$ (kcal.mole$^{-1}$) | Ref. |
|---|---|---|---|---|
| $FeCO^+$ | $14.0 \pm 0.2$ | EVD | 255 | 38 |
| | $12.9 \pm 0.1$ | CS | | 43 |
| | $13.39 \pm 0.07$ | EVD | 235 | 45 |
| | $13.76^d$ | VC | 248 | 46 |
| $Fe^+$ | $16.1 \pm 0.2$ | EVD | 330 | 38 |
| | $14.7 \pm 0.1$ | CS | | 43 |
| | $15.31 \pm 0.1$ | EVD | 304 | 45 |
| | $15.99^d$ | VC | 326 | 46 |

Table 11.3 (continued)

| Ion | A.P. (eV) | Method[a] | $\Delta H_f$ (kcal.mole⁻¹) | Ref. |
|---|---|---|---|---|
| Fe(CO)₄⁺ | 10.0 ±0.2 | EVD | 83 | 38 |
| | 8.34±0.12 | CS | | 43 |
| | 8.73±0.08 | EVD | 49 | 45 |
| | 9.17ᵈ | VC | 64 | 46 |
| Fe(CO)₃⁺ | 10.3 ±0.3 | EVD | 117 | 38 |
| | 9.89±0.05 | CS | | 43 |
| | 10.01±0.04 | EVD | 109 | 45 |
| | 10.04ᵈ | VC | 110 | 46 |
| Fe(CO)₂⁺ | 11.8 ±0.2 | EVD | 178 | 38 |
| | 10.92±0.04 | CS | | 43 |
| | 11.27±0.05 | EVD | 162 | 45 |
| | 11.12ᵈ | VC | 161 | 46 |

(3) Cr(CO)₆ I.P. Cr = 6.749 eV (ref. 54)

| Ion | A.P. (eV) | Method[a] | $\Delta H_f$ (kcal.mole⁻¹) | Ref. |
|---|---|---|---|---|
| Cr(CO)₆⁺ | 8.03±0.03 | Phot. | | 37 |
| | 8.15±0.17 | EVD, SL, EC | −52 | 42 |
| | 8.18±0.07 | CS | | 43 |
| | 8.48±0.08 | EVD | −52 | 45 |
| | 8.44±0.05 | VC | −47 | 46 |
| | 8.142±0.017ᵇ | Phot. | | 47 |
| | 8.38±0.02ᶜ | Phot. | | 47 |
| | 8.24 | Calc. | | 57 |

| Ion | A.P. (eV) | Method[a] | $\Delta H_f$ (kcal.mole⁻¹) | Ref. |
|---|---|---|---|---|
| FeCO²⁺ | 30.2 ±2 | EVD | 629 | 38 |
| FeC⁺ | 23.6 ±0.3 | EVD | | 45 |
| CO⁺ | 14.15±0.02 | CS | | 43 |

| Ion | A.P. (eV) | Method[a] | $\Delta H_f$ (kcal.mole⁻¹) | Ref. |
|---|---|---|---|---|
| Cr(CO)₂⁺ | 13.1 ±0.2 | EVD | 167 | 42 |
| | 11.56±0.2 | CS | | 43 |
| | 11.94±0.1 | EVD | 136 | 45 |
| | 12.56ᵈ | VC | 154 | 45 |
| CrCO⁺ | 14.9 ±0.2 | EVD, SL | 235 | 42 |
| | 13.3 ±0.2 | CS | | 43 |
| | 13.63±0.2 | EVD | 202 | 45 |
| | 14.12ᵈ | VC | 216 | 46 |
| Cr⁺ | 17.7 ±0.3 | EVD, SL | 326 | 42 |

Table 11.3 (continued)

| Ion | A.P. (eV) | Method[a] | $\Delta H_f$ (kcal.mole$^{-1}$) | Ref. |
|---|---|---|---|---|
| Cr(CO)$_5^+$ | (9.5)$^e$ | | (5) | 42 |
| | 9.17±0.04 | CS | | 43 |
| | 8.95±0.1 | EVD | −5 | 45 |
| | 9.32$^d$ | VC | 0 | 46 |
| | 9.26 | Phot. | | 47 |
| Cr(CO)$_4^+$ | (10.7) | | (60) | 42 |
| | 9.97±0.04 | CS | | 43 |
| | 9.64±0.1 | EVD | 39 | 45 |
| | 9.52$^d$ | VC | 31 | 46 |
| Cr(CO)$_3^+$ | (12.0) | | (115) | 42 |
| | 10.62±0.15 | CS | | 43 |
| | 11.2 ±0.2 | EVD | 89 | 45 |
| | 10.42$^d$ | VC | 79 | 46 |

| Ion | A.P. (eV) | Method[a] | $\Delta H_f$ (kcal.mole$^{-1}$) | Ref. |
|---|---|---|---|---|
| | 14.7 ±0.1 | CS | | 43 |
| | 15.1 ±0.2 | EVD | 261 | 45 |
| | 17.07$^d$ | VC | 311 | 46 |
| CrCO$^{2+}$ | 30.8±1.0 | EVD | | 45 |
| CrC$^+$ | (24.7) | | (530) | 42 |
| CrO$^+$ | 23.15±0.3 | EVD | | 45 |
| | 23.45±0.3 | EVD | | 45 |
| CO$^+$ | 14.15±0.03 | CS | | 43 |

(4) Mo(CO)$_6$ I.P. Mo = 7.10 eV (ref. 54)

| Ion | A.P. (eV) | Method[a] | $\Delta H_f$ (kcal.mole$^{-1}$) | Ref. |
|---|---|---|---|---|
| Mo(CO)$_6^+$ | 8.12±0.03 | Phot. | | 37 |
| | 8.23±0.12 | EVD, SL, EC | | 42 |
| | 8.30±0.3 | CS | −29 | 43 |
| | 8.46±0.08 | EVD | −28 | 45 |
| | 8.43±0.05 | VC | −27 | 46 |
| | 8.227±0.011$^b$ | Phot. | | 47 |

| Ion | A.P. (eV) | Method[a] | $\Delta H_f$ (kcal.mole$^{-1}$) | Ref. |
|---|---|---|---|---|
| Mo(CO)$_2^+$ | 15.6 ±0.3 | EVD, SL | 247 | 42 |
| | 13.90±0.3 | CS | | 43 |
| | 14.5 ±0.1 | EVD | 215 | 45 |
| | 14.76$^d$ | VC | 226 | 46 |
| MoCO$^+$ | 18.1 ±0.3 | EVD, SL | 331 | 42 |
| | 15.8 ±0.06 | CS | | 43 |
| | 15.7 ±0.2 | EVD | 278 | 45 |

Table 11.3 (continued)

| Ion | A.P. (eV) | Method[a] | $\Delta H_f$ (kcal·mole⁻¹) | Ref. |
|---|---|---|---|---|
| | 8.35±0.07[c] | Phot. | | 47 |
| | 8.42 | Calc. | | 57 |
| Mo(CO)₅ | 9.80±0.15 | EVD, SL | 34 | 42 |
| | 9.64±0.05 | CS | | 43 |
| | 9.43±0.1 | EVD | 28 | 45 |
| | 9.14[d] | VC | 18 | 46 |
| | 9.21±0.03 | Phot. | | 47 |
| Mo(CO)₄⁺ | 11.9±0.2 | EVD, SL | 109 | 42 |
| | 11.28±0.14 | CS | | 43 |
| | 10.63±0.15 | EVD | 87 | 45 |
| | 10.72[d] | VC | 81 | 46 |
| Mo(CO)₃⁺ | 13.7±0.3 | EVD, SL | 177 | 42 |
| | 12.36±0.12 | CS | | 43 |
| | 12.82±0.1 | EVD | 151 | 45 |
| | 13.18[d] | VC | 164 | 46 |

(5) $W(CO)_6$ I.P. $W = 7.10$ eV (ref. 54)

| Ion | A.P. (eV) | Method[a] | $\Delta H_f$ (kcal·mole⁻¹) | Ref. |
|---|---|---|---|---|
| W(CO)₆⁺ | 8.18±0.03 | Phot. | | 37 |
| | 8.56±0.13 | EVD, SL, | | |
| | | EC | −12 | 42 |
| | 8.46±0.02 | CS | | 43 |
| | 8.47±0.1 | EVD | −15.1 | 45 |

| Ion | A.P. (eV) | Method[a] | $\Delta H_f$ (kcal·mole⁻¹) | Ref. |
|---|---|---|---|---|
| Mo⁺ | 15.61[d] | VC | 273 | 46 |
| | 20.7 ±0.5 | EVD, SL | 417 | 42 |
| | 18.3 ±0.3 | CS | 367 | 43 |
| | 18.6 ±0.2 | EVD | 391 | 45 |
| | 19.63[d] | VC | | 46 |
| Mo(CO)₃²⁺ | 29.1 ±1.2 | EVD, SL | 532 | 42 |
| Mo(CO)₂²⁺ | 30.8 ±0.5 | EVD, SL | 597 | 42 |
| MoCO²⁺ | 31.6 ±1 | EVD | | 45 |
| | 34.5 ±0.5 | EVD, SL | 709 | 42 |
| | 35.7 ±1 | EVD | | 45 |
| MoC⁺ | 27.2 ±0.4 | EVD, SL | 608 | 42 |
| | 24.3 ±1 | EVD | | 45 |
| MoO⁺ | 24.3 ±1 | EVD | | 45 |
| CO⁺ | 14.15±0.04 | CS | | 43 |

| Ion | A.P. (eV) | Method[a] | $\Delta H_f$ (kcal·mole⁻¹) | Ref. |
|---|---|---|---|---|
| W(CO)₂⁺ | 17.6 ±0.2 | EVD, SL | 302 | 42 |
| | 16.07±0.04 | CS | | 43 |
| | 15.8 ±0.3 | EVD | 264 | 45 |
| WCO⁺ | 18.51[d] | VC | 266 | 46 |
| | 20.2 ±0.3 | EVD, SL | 388 | 42 |

Table 11.3. (continued)

| Ion | A.P. (eV) | Method[a] | $\Delta H_f$ (kcal.mole⁻¹) | Ref. | Ion | A.P. (eV) | Method[a] | $\Delta H_f$ (kcal.mole⁻¹) | Ref. |
|---|---|---|---|---|---|---|---|---|---|
| | 8.48±0.05 | VC | −16 | 46 | | 18.5 ±0.16 | CS | | 43 |
| | 8.242±0.006[b] | Phot. | | 47 | | 18.7 ±0.3 | EVD | 351 | 45 |
| | 8.60±0.03[c] | Phot. | | 47 | | 18.51[d] | VC | 348 | 46 |
| | 8.29 | Calc. | | 57 | W+ | 22.9 ±0.6 | EVD, SL | 477 | 42 |
| W(CO)5+ | 9.80±0.17 | EVD, SL | 43 | 43 | | 20.6 ±0.2 | CS | | 43 |
| | 9.97±0.04 | CS | | 43 | | 21.7 ±0.3 | EVD | 435 | 45 |
| | 9.86±0.01 | EVD | 46 | 45 | | 22.25[d] | VC | 461 | 46 |
| | 9.21[d] | VC | 26 | 46 | W(CO)2²+ | 35.0 ±1 | EVD | | 45 |
| | 9.43 | Phot. | | 47 | WCO²+ | 31.7 ±1 | EVD | | 45 |
| W(CO)4+ | 12.7 ±0.2 | EVD, SL | 136 | 42 | WC2O+ | 25.9 ±0.6 | EVD, SL | | 42 |
| | 11.82±0.02 | CS | | 43 | WC+ | 28.8 ±0.5 | EVD, SL | | 42 |
| | 11.93±0.15 | EVD | 117 | 45 | CO+ | 14.15±0.02 | CS | | 43 |
| | 12.05[d] | VC | 120 | 46 | | | | | |
| W(CO)3+ | 14.9 ±0.2 | EVD, SL | 213 | 42 | | | | | |
| | 13.6 ±0.02 | ES | | 43 | | | | | |
| | 13.70±0.15 | EVD | 184 | 45 | | | | | |
| | 13.87[d] | VC | 189 | 46 | | | | | |

(6) V(CO)6 I.P. V = 6.74 eV (ref. 54)

| Ion | A.P. (eV) | Method[a] | $\Delta H_f$ (kcal.mole⁻¹) | Ref. | Ion | A.P. (eV) | Method[a] | $\Delta H_f$ (kcal.mole⁻¹) | Ref. |
|---|---|---|---|---|---|---|---|---|---|
| V(CO)6+ | 7.53±0.15 | EVD | −31 | 45 | V(CO)2+ | 12.3±0.2 | EVD | 186 | 45 |
| V(CO)5+ | 8.24±0.15 | | 12 | | VCO+ | 13.8±0.2 | | 246 | |
| V(CO)4+ | 9.70±0.2 | | 72 | | V+ | 15.5±0.2 | | 311 | |
| V(CO)3+ | 10.98±0.15 | | 128 | | VC+ | 23.8±0.8 | | | |

a Phot. A.P. value obtained by a photoionisation technique.
EVD A.P. value obtained by the extrapolated voltage difference method (Warren[58]) of analysing the electron-impact data.
VC Vanishing current method.
PES Photoelectron spectroscopy.
CS Critical slope method (Honig[59]).
SL Semi-logarithmic plot (Lossing et al.[60]).
EC Energy compensation method (Kiser and Gallegos[61]).
Calc. Calculated energy of the highest occupied molecular orbital[57].
b Adiabatic ionisation potential.
c Vertical ionisation potential.
d Reproducibility of appearance potential values reported to lie between $\pm 0.05$ and $\pm 0.2$.
e Values estimated by interpolation and extrapolation are shown in parenthesis.

$MC_xO_{(x-1)}{}^+$ ($x = 1$–$3$) under electron impact; presumably the C–O bond is only cleaved after several CO groups are lost from the molecule.

Although the mode of decomposition of metal carbonyls under electron impact has been firmly established, the lack of agreement among the reported ion intensities (Fig. 11.2) and, more especially, the appearance potential values (Table 11.3) is disturbing. Lloyd and Schlag[47] have very recently published the results of a detailed photoionisation study of the mononuclear metal carbonyls. They concluded from the variety of results which have been reported that the electron-impact values represent complex functions of the instruments on which the values were determined and that even within such a closely related set of compounds as the five carbonyls, the errors on any one instrument may vary from one compound to another by at least 0.3 eV. They therefore suggested that, in general, electron-impact measurements cannot be expected to distinguish between adiabatic and vertical ionisation potentials. This appears to confirm the now generally accepted belief that electron-impact values are upper limits to adiabatic ionisation potentials.

Foffani et al.[43] found that the appearance potentials of the $CO^+$ ions were, in all cases, equal to the ionisation potential of the CO molecule within experimental error. This immediately suggests that the reaction

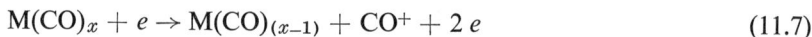

$$M(CO)_x + e \rightarrow M(CO)_{(x-1)} + CO^+ + 2\,e \qquad (11.7)$$

did not occur, but instead the $CO^+$ ions arise from ionisation of CO molecules formed by thermal decomposition of the metal carbonyl. Evidence for such decomposition in the inlet system was not obtained by these workers, however, although they considered that thermal decomposition on the rhenium filament to give free CO which was then ionised was a possibility. While maintaining the temperature of the inlet system as low as possible will undoubtedly reduce any thermal decomposition, the temperature of the walls of the ion source itself must be considered. Bidinosti and McIntyre[45] pointed out that $Ni(CO)_4$, $Fe(CO)_5$, and $M(CO)_6$ (M = Cr, Mo, W, or V) all decompose readily at conventional ion source temperatures, and therefore a significant fraction of the metal carbonyl molecules would be expected to be thermally excited to higher vibrational and possibly electronic states before ionisation and fragmentation. They claimed that ionisation from these levels could lead to different fragmentation patterns than those obtained from unexcited molecules. More important, the population of these higher molecular energy states may lead to anomalously low values for appearance potentials.

The lowest value obtained for an appearance potential by electron-impact techniques is usually considered the most reliable but this would not necessarily be so in such a situation.

Pignataro and Lossing[48] found that $Fe(CO)_5$ decomposed in the mass spectrometer at low temperatures giving mainly $Fe(CO)_4$ and $CO$. As the temperature was raised, the ion abundance ratio $Fe(CO_4)^+/Fe(CO)_5^+$ increased and the appearance potential of $Fe(CO)_4^+$, which was found to be 9.10 eV at 60–70 °C, decreased to 8.48 eV at 250–300 °C; the value measured at the higher temperature probably corresponded to the ionisation potential of the $Fe(CO)_4$ species. In the zero source contact mass spectral studies of Schildcrout et al.[41] on $Ni(CO)_4$, it was found that the absolute ion intensities decreased as the temperature was raised from 19° to 200°, presumably due to decomposition, but the relative ion intensities changed only slightly, in contrast to the behaviour of $Fe(CO)_5$. No evidence was found for the presence of gaseous intermediates such as $Ni(CO)_3$ in the pyrolysis of nickel tetracarbonyl although the existence of such intermediates could not definitely be precluded. Junk and Svec[46] also recognised that the discrepancies existing between the various studies on these compounds could be due to varying amounts of thermal decomposition. They suggested that the appearance potential values, especially of the lighter fragment ions, would vary with the temperature of the ion source and, as mentioned previously, work published by Pignataro and Lossing[48] at about the same time showed this to be so for $Fe(CO)_4^+$ formed from $Fe(CO)_5$. Results on substituted metal carbonyls showed[49] that thermal decompositions in the ion source were appreciable even at temperatures well below the expected thermal stability of these samples.

Even studies on pyrolysis effects are not entirely in harmony. Pignataro and Lossing[48] reported that the Group VI hexacarbonyls decomposed very little below ion source temperatures of 300° so that pyrolysis had little effect on appearance potentials. Above 300°, however, they suggested that the $M(CO)_5$ species were apparently formed. Svec and Junk[49], however, reported that decomposition of these hexacarbonyls began at lower temperatures. An important observation of theirs was that $Cr(CO)_6$ and $Mo(CO)_6$ began to decompose at an appreciably lower temperature during a second heating cycle of the pyrolysis furnace. These results suggested that the compounds can generate catalysts by depositing active metals and/or free radicals on heated wall surfaces inside the mass spectrometer.

The ionisation process of these molecules has been discussed in some detail. Vilesov and Kurbatov[37] postulated that the electron removed from the metal carbonyls during this process originated either from a $p$ orbital of the oxygen or else from a metal–carbon bond, thus indicating some metal–carbon double bond character. Winters and Kiser[42] pointed out (*i*) the ionisation potentials of the Group VI metal carbonyls increased slightly with increasing atomic number of the metal; (*ii*) this trend followed the ionisation potentials of the metals themselves; and (*iii*) the ionisation potentials of the carbonyls were not very different from those of the metals. They therefore suggested that the electron removed from the carbonyls on ionisation was one largely associated with the metal atom rather than with the carbonyl groups. It should be pointed out, however, that the two most recent energetics studies[45,46] on the Group VI hexacarbonyls produced ionisation potential values which did not follow the same order as those of the metals. Foffani *et al.*[43,50] compared trends in the ionisation potentials of the Group VI metals and their hexacarbonyls with trends in the force constants of the M–C and C–O bonds and with the M–C bond lengths; they postulated that the degree of metal–ligand $\pi$-bonding is not a relevant factor in the ionisation potentials. Thus, they too, argued that the electron removed by electron impact must belong to a molecular orbital very largely associated with the metal atom. In addition, the metal carbonyl ionisation potentials were considerably lower than that of carbon monoxide making it unlikely that the electron removed was from the CO groups. Kiser has recently pointed out[51] that molecular orbital calculations indicate that the highest occupied molecular orbital in $Ni(CO)_4$ arises from the $3d$ nickel orbitals with relatively little contribution from the ligand orbitals. This means that, in $Ni(CO)_4$ at least, the electron removed on ionisation may be regarded as having been localised on the metal atom and similar situations probably exist in other metal carbonyls. Calculations by Gray and Beach[52] indicate that the highest occupied molecular orbitals in the Group VI hexacarbonyls should be the $t_{2g}(\pi)$. These arise from a combination of the $d_\varepsilon$ metal atomic orbitals with $\pi$ and $\pi^*$ orbitals of the CO ligand and are mainly localised on the metal; the authors suggest a virtually non-bonding character for them. Kiser has also shown[51] that the second ionisation potential for several of the metal carbonyl ions suggest that the second electron withdrawn also is removed from a molecular orbital principally constituted from the metal $d$ orbitals.

Winters and Kiser[42,51] have advanced an excited-state hypothesis suggesting that the measured appearance potential of the metal ion may well refer to its formation in an excited state. The heats of formation which they determined for the Group VI metal ions, $M^+$, formed from the corresponding

Table 11.4

Mean Bond Dissociation Energies in Metal Carbonyls (kcal.mole$^{-1}$)
References are given in brackets.

**(a) $\bar{D}$(M–CO)**

| | | | | | | | | |
|---|---|---|---|---|---|---|---|---|
| Ni(CO)$_4$ | | 48[a] | 39(41) | | | 43(45) | 46(46) | 35.2(53) |
| Fe(CO)$_5$ | | 38 | | | 32(43) | 33 | 37 | 27.7 |
| Cr(CO)$_6$ | 39.2(47) | | | 42[c] | 31 | 31 | 39 | 27.1 |
| Mo(CO)$_6$ | 30.9 | | | 52 | 43 | 42 | 48 | 35.9 |
| W(CO)$_6$ | 40.4 | | | 57 | 49 | 50 | 55 | 42.1 |
| V(CO)$_6$ | | | | | | 34 | | |

**(b) $\bar{D}$(M$^+$–CO)**

| | | | | | | | | |
|---|---|---|---|---|---|---|---|---|
| Ni(CO)$_4$ | | 42[a] | 35[b] | | | 37(45) | 39[d] | |
| Fe(CO)$_5$ | | 35 | | | 30(43) | 32 | 35 | |
| Cr(CO)$_6$ | | | | 37[c] | 25 | 31 | 33 | |
| Mo(CO)$_6$ | | | | 48 | 38 | 39 | 43 | |
| W(CO)$_6$ | | | | 55 | 47 | 48 | 53 | |
| V(CO)$_6$ | | | | | | 31 | | |

[a],[b],[c],[d] Values calculated from data in refs. 38, 41, 42, and 46 respectively.

hexacarbonyls, appeared to support this hypothesis. Similar results were obtained by Junk and Svec.[46].

Values for bond dissociation energies may be calculated from the appearance potentials, neglecting any excess kinetic and excitation energy associated with the various ions. In each case the average metal–CO bond dissociation energy was larger than the more accurate values calculated from heats of combustion[53] although the same order of increasing bond strength is apparent (Table 11.4). Junk and Svec[46] assumed that the metal ion was formed in an excited state, viz.

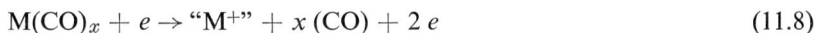

$$M(CO)_x + e \rightarrow \text{“}M^+\text{”} + x\,(CO) + 2\,e \tag{11.8}$$

They therefore used eqn. 11.9 to calculate the dissociation energies and found that these values agreed remarkably well with those obtained from the calorimetric measurements[53] for Fe(CO)$_5$, Cr(CO)$_6$, Mo(CO)$_6$, and W(CO)$_6$.

$$D[M(CO)_x] = \text{A.P. }[\text{“}M^+\text{”}] - \text{I.P. }[M \rightarrow \text{“}M^+\text{”}] \tag{11.9}$$

The mean dissociation energies in both the neutral and ionised states of the Group VI metal carbonyls increase in the same order as the corresponding ionisation potentials. Junk and Svec[46] calculated values of the energy required for dissociative loss of a single CO group from the parent molecular ion and from all the fragment ions. They found that the most constant values were those required for the loss of the first CO group from the parent molecular ion but this constancy disappeared rapidly as successive CO groups were dissociated. In most cases, but not all, the dissociation energy increased with each successive rupture of a metal–CO bond. Whereas Junk and Svec[46] suggested that some mass spectral measurements are not consistent with extensive $\pi$-bonding in metal carbonyls, Lloyd and Schlag[47] interpreted their results in terms of strong $\pi$ "back-donation" from metal to ligand, which for the hexacarbonyls increases in the sequence $Mo(CO)_6 < Cr(CO)_6 < W(CO)_6$.

*4.2 Binuclear carbonyls*

Dimanganese decacarbonyl was first studied mass spectrometrically by Hurd et al.[62] in 1949. The parent molecular ion was not detected although observation of the ions $Mn_2(CO)_x^+$ ($x = 5$, 6 or 7) indicated a dimetallic compound; relative intensities of the ions were not reported. Winters and Kiser[63] published the results of a much more detailed study in 1965. They were successful in observing the parent molecular ion, $Mn_2(CO)_{10}^+$, suggesting that pyrolytic decomposition occurred in the original study, but not the ions $Mn_2(CO)_x^+$, where $x = 8$ or 9. The ion $Mn_2(CO)_7^+$ was only observed when the electron energy was lowered from the normal 70 eV. Later workers[64] were successful in identifying each of these ions which were present in small quantities. The spectra of $Tc_2(CO)_{10}$, $Re_2(CO)_{10}$, and $MnRe(CO)_{10}$ have likewise been recorded[64-67] and the relative intensities of the dimetallic ions are represented in Fig. 11.3.

Winters and Kiser[63] plotted clastograms for the principal ions observed in the mass spectrum of $Mn_2(CO)_{10}$. A close similarity with corresponding studies on mononuclear carbonyls was noted and it was postulated that the unimolecular decomposition of $Mn_2(CO)_{10}^+$ involved successive losses of CO groups. Svec and Junk[66] arrived at the same conclusion from appearance potential measurements. Metastable transitions supporting this fragmentation mechanism were originally reported[64] for $Re_2(CO)_{10}$ and have recently been observed[68] for $Mn_2(CO)_{10}$ but details have not been published in the

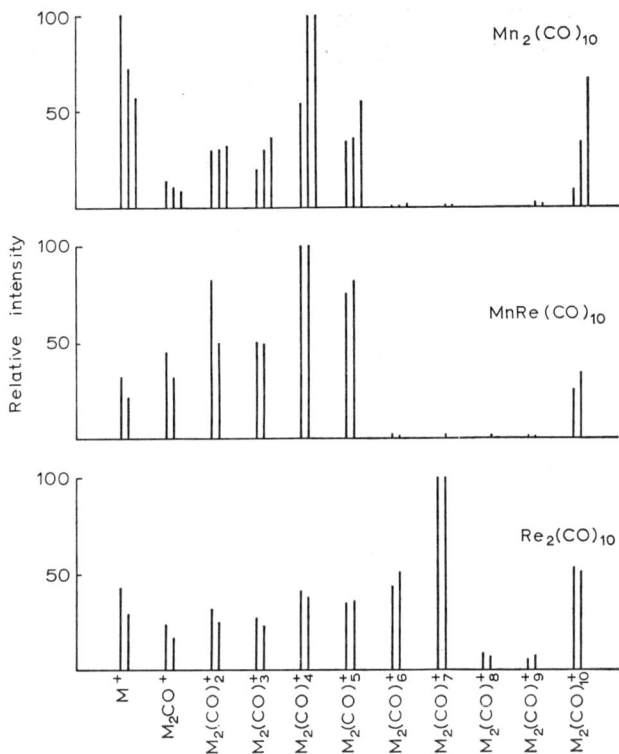

Fig. 11.3. Comparison of reported intensities of ions of type $M_2(CO)_x^+$ formed from the compounds $M_2(CO)_{10}$ (refs. 63–67).

latter case. A similar mechanism has been shown to operate for the compound $CoRe(CO)_9$; metastable transitions have been observed[69] corresponding to the decompositions

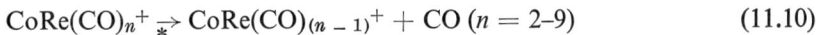

$$CoRe(CO)_n^+ \xrightarrow{*} CoRe(CO)_{(n-1)}^+ + CO \ (n = 2–9) \tag{11.10}$$

Rupture of the metal–metal bond occurs under electron impact for all five compounds $Tc_2(CO)_{10}$, $Mn_2(CO)_{10}$, $MnRe(CO)_{10}$, $Re_2(CO)_{10}$, and $CoRe(CO)_9$ so that the mononuclear ions $M(CO)_x^+$ (M = Mn or Re, $x = 0$–5; M = Co, $x = 0$–4) have all been observed. Winters and Kiser[63] believed that the manganese pentacarbonyl radical was not formed during fragmentation of $Mn_2(CO)_{10}$ but that decomposition (11.11) was taking place. However, the small difference between the subsequently reported value for the ionisation potential of $\cdot Mn(CO)_5$ and the appearance potential of $Mn(CO_5^+$ formed from $Mn_2(CO)_{10}$ suggested that reaction (11.12) rather than (11.11) was accounting for mononuclear ions in the spectrum. The participation of (11.12) in the fragmentation mechanism has since been confirmed by the observation[66,68] of the appropriate metastable transition at $m/e$ 97.5.

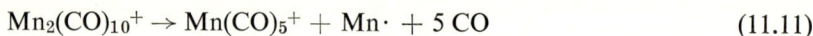

$$Mn_2(CO)_{10}^+ \rightarrow Mn(CO)_5^+ + Mn\cdot + 5\,CO \tag{11.11}$$

$$Mn_2(CO)_{10}^+ \xrightarrow{*} Mn(CO)_5^+ + \cdot Mn(CO)_5 \tag{11.12}$$

No other reaction resulting in the formation of monometallic ions from di-metallic species has been substantiated by metastable transitions, but those corresponding to the decompositions

$$Mn(CO)_x^+ \xrightarrow{*} Mn(CO)_{(x-1)}^+ + CO\ (x = 3\text{--}5) \tag{11.13}$$

were noted[66,68].

Bidinosti and McIntyre[70] were successful in generating the $\cdot Mn(CO)_5$ radical by pyrolysing $Mn_2(CO)_{10}$ in an effusion cell adjacent to the ion source of a mass spectrometer. Direct measurement of its ionisation potential, coupled with the appearance potential for the $Mn(CO)_5^+$ ion from the dimetallic compound, enabled the Mn–Mn bond dissociation energy to be calculated. Svec and Junk[66] also used this radical ionisation potential and an estimated ionisation potential of the $\cdot Re(CO)_5$ radical to calculate metal–metal bond dissociation energies for the compounds $Mn_2(CO)_{10}$, I, MnRe $(CO)_{10}$, II, and $Re_2(CO)_{10}$, III. Their results indicated that the Re–Re bond was stronger than Mn–Mn, but suggested that the Mn–Re bond was stronger still. More recent estimates of the $\cdot M(CO)_5$ radical ionisation potentials have been employed to calculate the metal–metal bond dissociation energies[67] (Table 11.5).

Lewis et al.[64,65] had previously compared the spectra of the three compounds I, II, and III; they found that in the spectrum of I 59% of the total ionisation was carried by ions containing two metal atoms but this rose to 71% in II and 96% in III. They pointed out that caution should be exercised in relating ionic abundances with bond stability but concluded that the metal–metal bond in the $Re_2(CO)_x^+$ ions was much stronger than in the

Table 11.5

Metal–Metal Bond Dissociation Energies for $M_2(CO)_{10}$ (ions and neutral molecules)[67]

| Process | Bond dissociation energy | |
|---|---|---|
| | eV | kcal.mole$^{-1}$ |
| $(CO)_5Mn–Mn(CO)_5 \rightarrow \cdot Mn(CO)_5 + \cdot Mn(CO)_5$ | $1.08 \pm 0.03$ | 24.9 |
| $[(CO)_5Mn–Mn(CO)_5]^+ \rightarrow Mn(CO)_5{}^+ + \cdot Mn(CO)_5$ | $0.94 \pm 0.05$ | 21.7 |
| $(CO)_5Tc–Tc(CO)_5 \rightarrow \cdot Tc(CO)_5 + \cdot Tc(CO)_5$ | $1.84 \pm 0.02$ | 42.5 |
| $[(CO)_5Tc–Tc(CO)_5]^+ \rightarrow Tc(CO)_5{}^+ + \cdot Tc(CO)_5$ | $1.83 \pm 0.05$ | 42.2 |
| $(CO)_5Re–Re(CO)_5 \rightarrow \cdot Re(CO)_5 + \cdot Re(CO)_5$ | $1.94 \pm 0.05$ | 44.8 |
| $[(CO)_5Re–Re(CO)_5]^+ \rightarrow Re(CO)_5{}^+ + \cdot Re(CO)_5$ | $1.98 \pm 0.05$ | 45.7 |
| $(CO)_5Re–Mn(CO)_5 \rightarrow \cdot Re(CO)_5 + \cdot Mn(CO)_5$ | $2.18 \pm 0.10$ | 50.4 |
| $[(CO)_5Re–Mn(CO)_5]^+ \rightarrow Re(CO)_5{}^+ + \cdot Mn(CO)_5$ | $2.36 \pm 0.05$ | 54.5 |

corresponding Mn species, while the Mn–Re bond in ions from II was intermediate in strength as might be expected; no energetics studies were made. Svec and Junk[66] later carried out a comprehensive mass spectral investigation of the three compounds. As with studies of the mononuclear carbonyls, disagreement in fragmentation patterns is evident here also (Fig. 11.3) and for the proportion of the total ion current carried by dimetallic ions these workers obtained corresponding values of 83%, 89%, and 95% respectively for I, II and III, the same order of increasing magnitude as found by Lewis et al.*. However, their appearance potential measurements indicated that the bond dissociation energies of the metal–metal bonds were in the order Mn–Mn < Re–Re < Mn–Re, illustrating the possible danger involved in assigning orders of bond stabilities in different molecules from ion intensity measurements.

The various electron-impact values obtained for $D(Mn–Mn)$, $18.9 \pm 1.4$ (ref. 70), $24.9 \pm 0.7$ (ref. 67), 24.6 (ref. 68), 25.3 (ref. 68), and $21 \pm 3$ (ref. 71) kcal.mole$^{-1}$, are much lower than the thermochemical value[72] of $34 \pm 13$ kcal.mole$^{-1}$ but Bidinosti and McIntyre[70] suggested that they were in keeping with the very long Mn–Mn bond length[73] of 2.93 Å. It does indicate, however, that the process involved in the formation of the $Mn(CO)_5{}^+$ ion was

---

* The spectra were recorded at 70 and 55 eV but the clastograms of Winters and Kiser[63] suggest that little change in ion abundances could be expected in such a variation of the electron energy.

reaction (11.12) and not (11.11) since, in the latter, a total of five metal–carbon bonds must also be broken; this would result in a negative value for $D$(Mn–Mn).

Winters and Kiser[63] suggested that the electron removed on ionisation of binuclear carbonyls was, as with the mononuclear compounds, associated with the metal atom. Svec and Junk[66] believed that the site of ionisation of these molecules can be considered to be "isolated" on one of the metal atoms and the M–CO bonds of the uncharged metal atom were the first to be ruptured. Johnson et al.[65] had also previously suggested, solely on the basis of ion intensities, that five CO groups were preferentially lost from the one metal.

Carbide ions $M(CO)_nC^+$ (M = Mn, $n$ = 0–2 (refs. 63, 64); M = Re, $n$ = 0 (ref. 64), $M_2(CO)_nC^+$ (M = Mn or Re, $n$ = 0–5)[64] and $M_2(CO)_nC^{2+}$ (M = Mn or Re, $n$ = 0–6)[64] were observed, as were the ions $M_2(CO)_n{}^{2+}$ (M = Mn or Re, $n$ = 0–6)[64].

A significant observation of Svec and Junk[66] was that of the ion $Re(CO)_6{}^+$ in the mass spectrum of $MnRe(CO)_{10}$. This ion, which would have the electron configuration of the stable compound $W(CO)_6$, must be formed by rearrangement of a CO group but was present in high abundance ($\sim$31 % of the intensity of the base peak; 7 % of the total ion intensity) although the corresponding ion was not detected by them in the spectra of either $Mn_2(CO)_{10}$ or $Re_2(CO)_{10}$. $Re(CO)_6{}^+$ was, however, reported by Smith et al.[74] to be formed from dirhenium decacarbonyl (5 % of base peak). The same ion was noted by Bruce[69] to be formed in low yield from $CoRe(CO)_9$; again, CO migration must have occurred. Such carbonyl group migration appears to occur especially in iron compounds. One must remember that pyrolysis of the compound under investigation, with formation of $Fe(CO)_5$, may account for what appears to be rearrangement of these groups but, in some cases, for example $HFeCo(CO)_{12}$ (ref. 75), this possibility is ruled out by the authors.

The mass spectrum[65,76] of $Fe_2(CO)_9$ provides a striking contrast to those of the previously discussed $M_2(CO)_{10}$ compounds. The dimetallic ions were of very low abundance compared with mononuclear ions of type $Fe(CO)_x{}^+$ ($x$ = 0–5) and Johnson et al.[65] suggested that, on initial fragmentation, cleavage of the bridging carbonyl bonds occurs in preference to cleavage of terminal CO bonds and the stability of the $Fe(CO)_3Fe$ nucleus is low. The ions $Fe(CO)_x{}^+$ ($x$ = 4 or 5) would appear to incorporate CO groups which were originally in bridging positions but $Fe(CO)_6{}^+$ was not observed.

King[77] has therefore suggested that the $Fe(CO)_5^+$ ion may arise from iron pentacarbonyl formed by the pyrolysis process

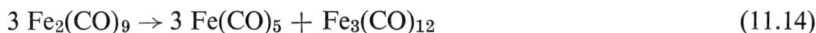

$$3\ Fe_2(CO)_9 \rightarrow 3\ Fe(CO)_5 + Fe_3(CO)_{12} \tag{11.14}$$

If such a pyrolysis reaction did occur, one would also expect ions formed from $Fe_3(CO)_{12}$ to be observed; trinuclear ions were very prominent in its spectrum[64]. However, no mention of any such ions was made. Energetics measurements should serve to resolve this uncertainty. Kiser[51] has also suggested that the low intensity of the dimetallic ions may be due to thermal decomposition and noted the failure to observe any metastable transitions of the type

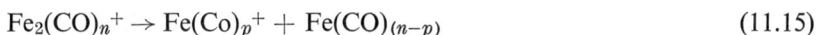

$$Fe_2(CO)_n^+ \rightarrow Fe(Co)_p^+ + Fe(CO)_{(n-p)} \tag{11.15}$$

as tentative support for this conclusion. The intensities of $Co_2$-containing ions formed from dicobalt octacarbonyl were higher[63,65]. Again, the successive loss of CO groups from the parent molecular ion was postulated[65] from plotted clastograms and subsequently metastable supported[68], as were reactions analogous to (11.13) where $x = 3$ and 4. The decomposition

$$Co_2(CO)_8^+ \xrightarrow{*} Co(CO)_4^+ + \cdot CO(CO)_4 \tag{11.16}$$

was also postulated by Winters and Kiser[63] and later supported by the observation[68] of the appropriate metastable transition. The $\cdot Co(CO)_4$ radical was identified and its ionisation potential determined[78] by pyrolysing $Co_2(CO)_8$ using a similar technique to that applied to the production of $\cdot Mn(CO)_5$ from $Mn_2(CO)_{10}$. This led to a value of $11.5 \pm 4.6$ kcal.mole$^{-1}$ being obtained for $D[(CO)_4Co-Co(CO)_4]$. A more recent value of $13 \pm 3$ kcal.mole$^{-1}$ was reported[71]. Bidinosti and McIntyre suggested that the $\cdot Co(CO)_4$ radical arose from the high energy form with no bridging carbonyl groups* so that the value obtained corresponds to the Co–Co bond dissociation energy. A subsequent mass spectral temperature study is reported[68] to yield values of 13.4, 14.0, 14.9, and 17.7 kcal.mole$^{-1}$.

---

* The crystalline form of $Co_2(CO)_8$ contains two bridging carbonyls; however, in solution this structural form is reported[79] also to be accompanied by a second isomer (to the extent of 40%) devoid of such bridges and having two $Co(CO)_4$ units joined by a Co–Co bond.

*4.3 Tri- and polynuclear carbonyls*

All of the possible trinuclear ions formed by losses of CO groups, including the bare metal cluster ions $M_3^+$, were observed in the mass spectra of the compounds $M_3(CO)_{12}$, where $M = Fe$ (refs. 64, 80), Ru (refs. 64, 80), or Os (refs. 65, 81) and metastable transitions reported[68] for the iron compound indicated that the previously discussed mechanism involving stepwise loss of CO groups is in operation here also. As the atomic number of the metal increases, fragmentation of the metal cluster becomes more difficult[64,65] and the only ions observed in the spectrum of the osmium compound containing the metal are trinuclear species. The doubly charged ions $Os_3(CO)_n^{2+}$ were present in unusually high abundance[65,81]. Observation of the ion $Fe(CO)_5^+$ formed from $Fe_3(CO)_{12}$, coupled with the absence of the corresponding ion in the spectrum of $Re_3(CO)_{12}$, was interpreted[64] as evidence for a structure of the former compound in which five carbonyl groups are attached to an iron atom. The correct structure was later shown[82] to have this arrangement. However, it has also been shown that, contrary to the assumptions of Lewis *et al.*[64], CO migration between metal atoms does occur, as discussed previously.

King[80] postulated that fragmentation of $Fe_3(CO)_{12}$ occurred by stepwise loss of CO groups from the parent molecular ion down to $Fe_3(CO)_3^+$ when rupture of the Fe–Fe bonds took place to produce mono- and binuclear species, indicating weaker metal–metal bonds, relative to metal–CO bonds, than in $Ru_3(CO)_{12}$. However, no energetics study was made.

The spectrum of $Os_4O_4(CO)_{12}$ was similar to that of $Os_3(CO)_{12}$ in that fragmentation of the $Os_4$ nucleus did not occur until all of the CO groups had been removed[65,81]. $RuFe_2(CO)_{12}$ and $Ru_2Fe(CO)_{12}$ also show parent molecular ions followed by loss of twelve CO groups[83,84]. Fragmentation of $Co_4(CO)_{12}$ (refs. 65, 80), $Rh_4(CO)_{12}$ (ref. 65), and $Rh_6(CO)_{16}$ (ref. 85) involved, predominantly, loss of carbonyl groups from the metal cluster which appears to be very stable under electron-impact conditions. Few additional details have been reported.

Johnson *et al.*[65,81] have proposed an unusual disproportionation (11.17) to account for the absence of $Os_3(CO)_n^+$ ions and the relatively high abundance of $Os_3(CO)_n^{2+}$ in the spectrum of $Os_4O_4(CO)_{12}$ but Kiser[51] has suggested that the doubly charged positive ions arise instead from reactions (11.18) and (11.19).

$$Os_4O_4(CO)_{12}^+ \rightarrow OsO_4^- + Os_3(CO)_{12}^{2+} \tag{11.17}$$

$$Os_4O_4(CO)_{12} + e \rightarrow Os_4O_4(CO)_{12}^{2+} + 3\ e \tag{11.18}$$

$$Os_4O_4(CO)_{12}^{2+} \rightarrow Os_3(CO)_{12}^{2+} + OsO_4 \tag{11.19}$$

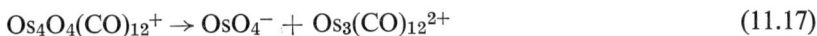

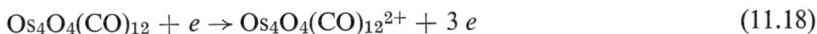

*4.4 Negative ion mass spectra*

The negative ion mass spectra of $M(CO)_6$ (M = Cr, Mo or W), $Fe(CO)_5$, and $Ni(CO)_4$ have been reported by Pignataro et al.[86] and by Winters and Kiser[87]. Parent molecular ions were not observed, the ion of highest mass in each case being (parent molecule $-$ CO)$^-$ and these were also the most abundant. These ions were formed by processes indicative of electron capture and ion-pair production. The peaks were very sharp and well-resolved and it has been suggested that they could be used as "indicators" in the study of secondary capture processes[86] and also for calibration of the electron energy scale[51]. All possible ions formed by further loss of CO groups were observed but apparently rupture of C–O bonds did not occur to give $M(CO)_nC^-$-type ions in contrast to the positive ion mass spectra. A series of consecutive unimolecular reactions involving elimination of CO groups was proposed by Winters and Kiser[87] similar to the decomposition processes of the corresponding positive ions. This mechanism has been supported by the negative ion clastogram for $Cr(CO)_6$ and by metastable transition studies[51].

Kiser[51] has suggested that the parent molecular ions would not be expected to be observed in these compounds since the addition of an electron to $Ni(CO)_4$ or the Group VI hexacarbonyls would require it to enter an antibonding molecular orbital. He pointed out, however, that the $V(CO)_6$ molecule would differ in that the added electron could enter a non-bonding orbital. He and his coworkers have since detected the ion $V(CO)_6^-$ in the mass spectrum of this compound[88]; its ease of formation in solution is, of course, well known. Metastable transitions corresponding to the following decompositions were also observed.

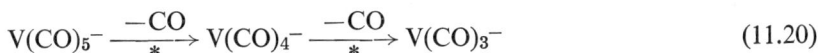

$$V(CO)_5^- \xrightarrow[*]{-CO} V(CO)_4^- \xrightarrow[*]{-CO} V(CO)_3^- \tag{11.20}$$

Negative ion mass spectra of $Mn_2(CO)_{10}$ and $Co_2(Co)_8$ were also briefly investigated[63].

## 5. CARBONYL HALIDES

Mass spectra of the pentacarbonylmanganese and -rhenium halides, $M(CO)_5X$ (X = Cl, Br, or I), have been reported[64,69,89,90]. Those of the three rhenium compounds are quite similar with the halogen exerting little influence on the fragmentation pattern. The similarity between the spectra of the three manganese compounds, however, is not as pronounced. In general, carbonyl groups and halogen atoms appear to break away from the parent molecular ion at about the same rate (Fig. 11.4). Junk et al.[90] commented on the absence of all three $Mn(CO)_4X^+$ ions, while the corresponding rhenium ions were all present in high abundance, and suggested that such a difference may be indicative of a significant difference in bonding and structure of these

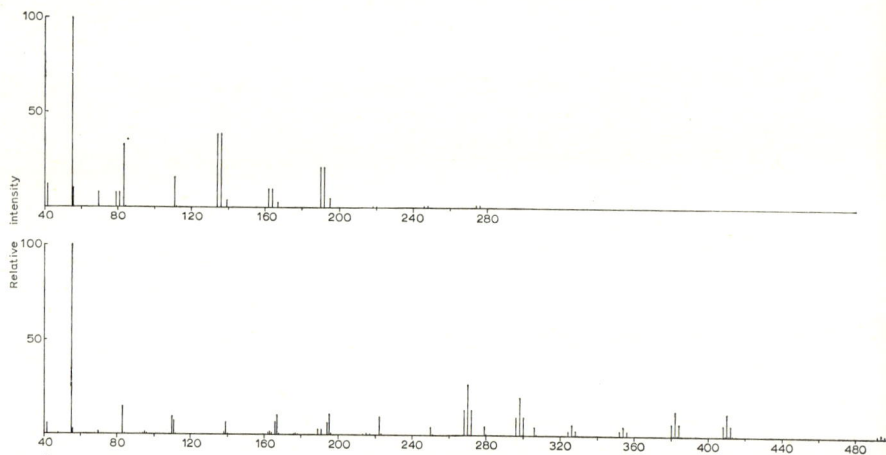

Fig. 11.4. Mass spectra of $Mn(CO)_5Br$ (top) and $Mn_2(CO)_8Br_2$ (bottom).

compounds. From an examination of the fragmentation patterns, Bruce[69] has claimed that ions of the type $M(CO)_nX^+$ are more stable than the similar ions $M(CO)_n^+$ and that the general trend of stability of the M–X bond is I > Br > Cl. Unfortunately, the appearance potentials of the $M(CO)_5^+$ ions have not been determined; in general the intensities of these ions are quite low.

Junk et al.[90] demonstrated that the ionisation potentials of these compounds were related not only to the metal atom but also to the halide ligand. Average bond energies calculated from appearance potential measurements indicate that (*i*) the Re–CO bonds are much more stable than the Mn–CO bonds; (*ii*) the first two M–CO bonds are weaker than the last three; (*iii*) the energetics for loss of the first and second CO groups follow the trend $M(CO)_5I > M(CO)_5Br > Mn(CO)_5Cl$; and (*iv*) no consistent halogen effect is observed for the average M–CO bond energy for rupture of three, four, or five CO groups. These results were correlated with the relative kinetic reactivities of $M(CO)_5X$ compounds in reactions where M–CO bond rupture is rate-determining.

Competitive loss of carbonyl groups and halogen atoms again occurs in the compounds $M(CO)_4I_2$ where M = Fe (ref. 89) or Ru (ref. 69). In the latter case, the ions devoid of halogen are much less abundant. Ions observed in the spectra of the corresponding chloro and bromo complexes of ruthenium have been reported[91] but no further details given. The compound $[Os(CO)_4I]_2$, assumed to contain terminal iodines but none acting in a bridging capacity, likewise appears to fragment by competitive loss of CO groups and iodine atoms but comprehensive details of the spectrum were not published[92]. The trinuclear compounds $Ru_3(CO)_{12}X_6$ (X = Cl, Br, I) appear also to undergo competitive loss of carbonyl groups and halogen atoms[91], indicating that at least some of the halogens occupy terminal positions although the structure is still uncertain. Loss of both CO groups and halogen atoms from the parent molecular ions $[Os_3(CO)_{12}X_2]^+$ and the observation of fragment ions $[Os_3(CO)_nX_2]^+$ and $[Os_3(CO)_nX]^+$ ($n$=0–11), $[Os_2(CO)_nX]^+$ ($n$ = 0–8) and $[Os(CO)_nX]^+$ ($n$ = 0–4) were consistent with the proposed linear structure (IV)[93].

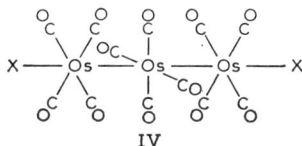

IV

A number of binuclear carbonyl halides containing bridging halogens has been examined mass spectrometrically. A notable contrast with the compounds containing only terminal halogens is apparent in that fragmentation of the former, for example $Mn_2(CO)_8X_2$ (X = Cl (ref. 64), Br (ref. 64) or I (ref. 84)), $Re_2(CO)_8X_2$ (X = Cl or I)[89], $Ru_2(CO)_6X_4$ (X = Cl or Br)[76,91], $Os_2(CO)_6I_2$ (ref. 92), and $Rh_2(CO)_4Cl_2$ (ref. 89), does not appear to involve competitive loss of carbonyl groups and halogen atoms (*e.g.* $Mn_2(CO)_8Br_2$,

Fig. 11.4) but rather the metal–halogen bridge system tends to remain intact until loss of all CO groups occurs. The two isomers of each compound $Ru_2(CO)_6X_4$ (X = Cl, Br, or I) were responsible for mass spectra with no detectable differences[91]. The iodo compounds give rise to the ions $[Ru_2(CO)_nI_4]^+$ ($n$ = 0–6), $[Ru_2(CO)_nI_3]^+$ ($n$ = 0–6), $[Ru_2(CO)_nI_2]^+$ ($n$ = 0–3), and $[Ru_2I_n]^+$ ($n$ = 0 or 1), and Johnson et al.[91] point out that a competitive loss of CO and I from the parent molecular ion occurs. All of these ions except $[Ru_2I_n]^+$ are consistent with retention of the intact $[Ru_2I_2]$ bridged nucleus, however, while presumably the terminal iodines are the ones which are lost; the non-appearance of the ion $[Ru_2I_2]^+$ is somewhat surprising. A similar competitive loss of CO groups and terminal halogens appears to be involved in the decomposition of $Os_2(CO)_6Cl_4$, although full details of the spectrum were not published[92].

Johnson et al.[91] again attempt to correlate differences in the mass spectra (of the $Ru_2(CO)_6X_4$ compounds) with metal–halogen bond strength Ru–I < Ru–Br < Ru–Cl.

## 6. CARBONYL HYDRIDES

It appears that fragmentation of metal carbonyl hydrides under electron impact involves consecutive loss of CO groups as in the metal carbonyls themselves, although observation of metastable transitions to support such a mode of decomposition has not been made in most cases. As with the metal carbonyl halides, fragmentation of hydrides containing terminal metal–hydrogen bonds appears to involve competitive loss of CO and ·H. In most cases, mass spectrometry has proved extremely useful in the difficult problem of determining the number of hydrogens present in the molecule but care must be exercised as the following examples illustrate. The original report[94] of the mass spectrum of $H_3Mn_3(CO)_{12}$ noted that the parent molecular ion was not observed but it was possible to establish that the molecule contained three hydrogen atoms from the presence of lower mass fragments of the type $H_3Mn_3(CO)_n^+$ ($n$ < 12); its parent molecular ion has since been detected[74,95,96]. The ion of highest mass observed in the spectrum of $H_4Ru_4(CO)_{12}$ was $H_2Ru_4(CO)_{12}^+$, but no ion containing four hydrogens was present in the spectrum[97,98] so that in a case such as this an erroneous conclusion could be reached in deducing the number of hydrogens present in the molecule. Kiser[51] has pointed out that the tendency of some polynuclear carbonyl hydrides not to exhibit parent molecular ions is somewhat surprising when one considers that metal hydride species are often formed by rearrangement processes, for example from various metal alkyls; this suggests that the metal hydride ions should be rather stable.

### 6.1 Mononuclear hydride complexes

Mass spectra of the acidic mononuclear hydride complexes $HMn(CO)_5$ (refs. 99, 100), $HRe(CO)_5$ (ref. 100), and $HCo(CO)_4$ (ref. 100) have been reported. Metastable decompositions observed[93] for the manganese compound partially confirmed the previously suggested[99] fragmentation mechanism

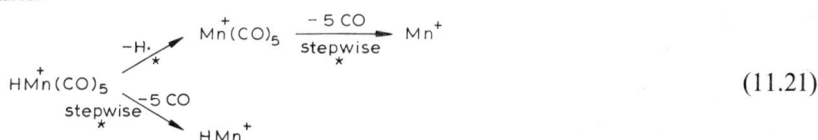

$$
\begin{array}{c}
& \overset{+}{Mn}(CO)_5 \xrightarrow[\underset{*}{\text{stepwise}}]{-5\ CO} Mn^+ \\
\overset{-H\cdot}{\underset{*}{\nearrow}} & \\
\overset{+}{HMn}(CO)_5 & \\
\underset{\underset{*}{\text{stepwise}}}{\searrow -5\ CO} & \\
& HMn^+
\end{array}
\qquad (11.21)
$$

Competitive loss of CO and H radicals was again observed[100] in the fragmentation of $HM(CO)_3(\pi\text{-}C_5H_5)$ (M = Mo or W) to give, ultimately, $MC_5H_5^+$; however, loss of H from the parent molecular ion of the tungsten compound was not detected by King[101].

Parent molecular ions were also found in high abundance in the spectra of the more basic compounds $HRe(\pi\text{-}C_5H_5)_2$, $H_2W(\pi\text{-}C_5H_5)_2$ (ref. 100) and $HTc(\pi\text{-}C_5H_5)_2$ (ref. 102) and it was assumed that stepwise loss of hydrogen occurred before fission of the cyclopenadienyl ring system. Similarly, parent molecular ions of each of the square planar platinum compounds *trans*-$HPtX(PEt_3)_2$ (X = Cl, Br, CN, or CNO) were observed[100] but initial fragmentation was dependent on the group X. The ions (parent molecule $- \cdot X)^+$ and (parent molecule $-$ HX)$^+$ were present in all spectra but (parent molecule $- \cdot H)^+$ ions were formed only when X was CN or Br. The complexes $HZr(\pi\text{-}C_5H_5)_2BH_4$ (ref. 103), $HCo(CO)_n(PF_3)_{(4-n)}$ ($n = 0$–4) (ref. 104), and $HCo(PF_3)_3PH_3$ (ref. 105) also gave rise to parent molecular ions.

### 6.2 Polynuclear hydride complexes

A number of polynuclear metal carbonyl hydrides has been subjected to mass spectral analysis. Smith *et al.*[74] found that the spectrum of $H_7B_2Mn_3$-$(CO)_{12}$, which contains only bridging hydrogens, showed a parent molecular ion followed by the ions $H_7B_2Mn_3(CO)_n^+$ ($n = 7$–$9$). The first loss of hydrogen occurred from the $H_7B_2Mn_3(CO)_6^+$ ion. Similarly, the spectrum of $HRe_3(CO)_{14}$ indicated all possible ions due to the loss of the fourteen CO groups, but no hydrogen loss until $HRe_3(CO)_2^+$ was reached. Structure V was postulated[74,106] for the compound.

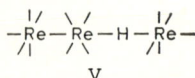

V

Fragmentation of the ions $HRe_2(CO)_9^+$ and $HRe(CO)_5^+$ followed a competitive decomposition process. The ions $Re_2(CO)_{10}^+$ and $Re(CO)_6^+$ surprisingly were observed, the intensity of the former being significantly higher than that of any of the ions $HRe_3(CO)_n^+$. More importantly, it was significantly above the level of detection (by infrared spectroscopy) of any $Re_2(CO)_{10}$ impurity. Its formation was assumed to result from CO transfer during fragmentation but pyrolysis appears to be a more likely explanation. No operating temperatures were reported. The parent molecular ion of the mixed compound $HMnRe_2(CO)_{14}$ showed CO loss but no fragmentation of H occurred down to $HMnRe_2(CO)_7^+$; below this the spectrum was obscured by ions formed from an impurity[74].

The mass spectra of the compounds $H_3M_3(CO)_{12}$ (M = Mn or Re) have not been of much assistance in the elucidation of their structure with respect to the function of the hydrogens in the molecules[94–96]. Failure to detect any monomeric ions with more than four carbonyl groups per metal atom is consistent with a trinuclear ($M_3$) structure involving four terminal CO groups per metal atom. Infrared[107] and Raman[108] studies indicate the presence of bridging hydrogen. However, loss of two hydrogens under electron impact competed effectively with CO loss in all multiplets but closer examination[95] reveals that no ions completely devoid of hydrogen are encountered prior to $M_3(CO)_3^+$. If one were to consider the hydrides to be analogous to the metal carbonyl halides, it would be tempting to postulate that one H is bridging while the other two are in terminal positions but this appears unlikely. Structure VI has been postulated[74] for the compound.

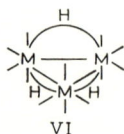

VI

The compound $H_2Os_3(CO)_{10}$ exhibited loss of hydrogen from the parent molecular ion to give the ions $HOs_3(CO)_{10}^+$ and $Os_3(CO)_{10}^+$ but since these ions occurred in fairly low abundances Johnson et al.[93] did not believe that terminally bonded H atoms were present.

The postulated fragmentation scheme[84,98] for $H_2Ru_4(CO)_{13}$ may be represented by

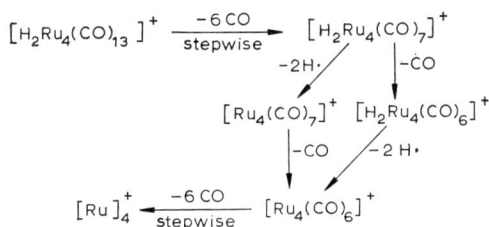

$$[H_2Ru_4(CO)_{13}]^+ \xrightarrow[\text{stepwise}]{-6\,CO} [H_2Ru_4(CO)_7]^+$$

$$[H_2Ru_4(CO)_7]^+ \xrightarrow{-2H\cdot} [Ru_4(CO)_7]^+ \quad [H_2Ru_4(CO)_6]^+ \xleftarrow{-CO} $$

$$[Ru]_4^+ \xleftarrow[\text{stepwise}]{-6\,CO} [Ru_4(CO)_6]^+$$

and is consistent with the presence of bridging H ligands. Similarly, the spectrum of the mixed metal hydride $H_2FeRu_3(CO)_{13}$ indicates loss of seven CO groups before loss of H· occurs, again suggesting a H-bridged structure[83,84]. The compounds $HFeCo_3(CO)_{12}$ and $HRuCo_3(CO)_{12}$ provide an interesting and unusual situation[75,109]. The mass spectra reveal that no loss of hydrogen occurs from ions of the type $HMCo_3(CO)_n^+$ ($n = 6$–$12$) while only a small percentage of H· loss occurs from the ions where $n = 0$–$5$. Once the metal cluster is broken, however, loss of H· becomes an important process. Ions containing H and Co only ($HCo_3^+$, $HCo_2^+$) were observed in both spectra. On the basis of this and infrared evidence, a structure was suggested in which the H atom was located inside a tetrahedral metal cage, VII. While the H is still formally bonded to the Fe (or Ru), it can also interact with the three cobalt atoms.

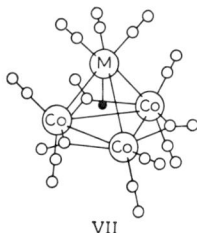

VII

Parent molecular ions were not observed[94,98] in the mass spectra of either

$\alpha$- or $\beta$-$H_4Ru_4(CO)_{12}$, the ions of highest mass being attributed in each case to $H_2Ru_4(CO)_{12}^{+}$*. Assignment of lower mass peaks was reported[98] to be difficult due to complicated loss of both CO and $\cdot$H, and the large number of Ru isotopes. A parent molecular ion was present in the spectrum of the osmium complex, $H_4Os_4(CO)_{12}$. Structures incorporating bridging hydrogens have been proposed for these compounds.

Competitive loss of H$\cdot$ and CO occurred in the fragmentation of the compounds $H_4Os_4(CO)_{12}$, $HOs_3(CO)_{10}X$ (X = H, OH, or OMe), and $HRu_3$-$(CO)_{10}SR$ (R = Et, $Bu^n$ or Ph) only after one or more CO groups had been lost from the parent molecular ion.

In summary, parent molecular ions have been recorded in the mass spectra of all mononuclear metal hydride complexes investigated. Loss of hydrogen atoms from this parent was common to all except $HCo(CO)_n(PF_3)_{(4-n)}$, while most polynuclear carbonyl complexes with bridging hydrogens but none in terminal positions undergo fission of CO groups before loss of H$\cdot$ occurs. The previously noted exceptions, however, prevent the designation of hydrogen in new metal hydride complexes as occupying either bridging or terminal positions solely on the basis of mass spectral data. Similarly, the detection of MH$^+$ ions can be taken as indicative of metal hydride species only where hydrogen-containing ligands are absent, since such ions formed by rearrangement processes are possible in the latter class of compounds. The $\pi$-cyclopentadienylchromium-tricarbonyl dimer is an excellent example; ions such as $HCr(CO)_3(C_5H_5)^+$ were observed[80] in its spectrum.

## 7. METAL NITROSYL COMPOUNDS

Foffani et al.[110] reported parallel cascade processes in the decomposition of $Fe(CO)_2(NO)_2$ and $Co(CO)_3NO$ due to loss of both CO and NO$\cdot$from the parent molecular ions but in each case the (parent molecule $-$ CO)$^+$ ion had a very much higher abundance than the (parent molecule $-$ NO$\cdot$)$^+$ ion. Winters and Kiser[111] had similarly found that fragment ions were formed by loss of CO in preference to cleavage of nitrosyl or cyclopentadienyl groups in

---

* The existence of a compound described as $\beta$-$H_4Ru_4(CO)_{12}$ has been questioned in a later paper by Kaesz et al.[385]. However, these workers were able to prepare and characterize both $D_2H_2Ru_4(CO)_{12}$ and $H_4FeRu_3(CO)_{12}$ mass spectrometrically. The latter compound was observed to undergo rearrangement in the spectrometer to give $H_4Ru_4(CO)_{12}$.

the compound $C_5H_5Mo(CO)_2NO$ in agreement with the known order of bond stabilities. The ions $C_3H_3Mo^+$ and $C_5H_5Mo^{2+}$ were highly abundant (95.0% and 34.1% of the base peak respectively). The fragmentation pattern of $C_5H_5NiNO$ indicated that the NO group was lost more readily than the cyclopentadienyl ligand[110]. The ion $FeN^+$, analogous to the carbide ions observed in the mass spectra of metal carbonyl compounds, was reported in the spectrum of $Fe(CO)_2(NO)_2$. The compound $Co(CO)_3NO$ was shown[48] to undergo thermal decomposition to $Co(CO)_2NO$ in the mass spectrometer; as the temperature was raised, the ratio $Co(CO)_2NO^+/Co(CO)_3NO^+$ increased and at the same time the appearance potential of $Co(CO)_2NO^+$ decreased.

Ionisation potentials of only a few metal nitrosyl complexes have been determined (Table 11.6) but these values indicate that, as with the metal

Table 11.6

Ionisation Potentials of Metal Nitrosyl Complexes

| Compound | Ionisation potential (eV) | Ref. |
|---|---|---|
| $Co(CO)_3NO$ | $8.75 \pm 0.1$ | 110 |
|  | $8.11 \pm 0.03^a$ | 47 |
|  | $9.05 \pm 0.17^b$ | 47 |
| $Fe(CO)_2(NO)_2$ | $8.45 \pm 0.1$ | 110 |
|  | $8.25 \pm 0.12^a$ | 47 |
|  | $9.01 \pm 0.15^b$ | 47 |
| $Co(CO)_2NOPCl_3$ | $8.40 \pm 0.1$ | 110 |
| $FeCO(NO)_2P(OC_2H_5)_3$ | $7.50 \pm 0.1$ | 110 |
| $C_5H_5NiNO$ | $8.50 \pm 0.1$ | 110 |
| $C_5H_5Mo(CO)_2NO$ | $8.1 \pm 0.2$ | 111 |

a Adiabatic I.P.
b Vertical I.P. Value is probably a lower limit.
Both of the above values were determined by a photoionisation technique.

carbonyls, the electron removed was associated largely with the metal atom[111].

The mass spectra of the binuclear compounds $[C_5H_5CoNO]_2$ (ref. 112) and $[C_5H_5FeNO]_2$ (ref. 113) (postulated to contain an iron–iron double bond) showed the base peak in each case to be due to $(C_5H_5)_2M^+$. The rearrangement of a cyclopentadienyl group, common in the mass spectra of binuclear cyclopentadienylmetal carbonyl compounds (Section 8.2) occurs here also and a metastable transition corresponding to the formation of $(C_5H_5)_2Co^+$ from the parent molecular ion of $[C_5H_5CoNO]_2$ was observed[112], viz.

$$[C_5H_5CoNO]_2{}^+ \xrightarrow[\ast]{-(Co\,+\,2\,NO\cdot)} (C_5H_5)_2Co^+$$
$$\qquad\qquad \ast\downarrow -C_5H_5\cdot \qquad (11.23)$$
$$C_5H_5CoNO^+ \xrightarrow[\ast]{-NO\cdot} C_5H_5Co^+$$

Pyrolytic decomposition of the iron compound to ferrocene cannot be dismissed since no metastable transition corresponding to the formation of the $(C_5H_5)_2Fe^+$ ion was detected[113]. A very intense peak in its spectrum was assigned to the unusual rearrangement ion $C_5H_5Fe_2O^+$.

$$[C_5H_5FeNO]_2{}^+ \xrightarrow[\ast]{-NO\cdot} (C_5H_5)_2Fe_2NO^+ \xrightarrow[\ast]{-(C_5H_5\cdot\,+\,N\cdot)} C_5H_5Fe_2O^+$$
$$(11.24)$$

A similar decomposition occurred in the fragmentation of $C_5H_5V(NO)_2CO$ (ref. 114).

$$C_5H_5V(NO)_2CO^+ \xrightarrow[\ast]{-CO} C_5H_5V(NO)_2{}^+ \xrightarrow[\ast]{-NO} C_5H_5VNO\cdot^+$$
$$\qquad\qquad\qquad \ast\downarrow -(\cdot C_5H_5 + N\cdot)$$
$$\qquad\qquad\qquad VO^+ \qquad (11.25)$$

Fragmentation of the halogen-bridged compounds $M_2(NO)_4X_2$ (M = Fe, X = Cl or Br; M = Co, X = Cl, Br, or I) involved successive removal of the nitrosyl ligands with, presumably, maintenance of the $M_2X_2$ cluster but, in addition, cleavage of the dimeric molecular ion to the monomeric ion $M(NO)_2X^+$ occurred[115]. Subsequent fragmentation of this ion involved loss of both halide and NO ligands. The much lower abundance of the mononuclear ions in the spectrum of the iron compound than in that of the cobalt complex was related to metal–metal interaction in the former. The behaviour of these compounds under electron impact contrasted with that of halogen-bridged metal carbonyls; the latter tend to lose CO groups with retention of the metal–halogen cluster and little fission to mononuclear species occurred[115]. The principal decomposition mechanisms are represented by

$$Fe_2(NO)_4X_2{}^+ \rightarrow Fe_2(NO)_nX_2{}^+ + (4 - n)NO\cdot \qquad (n = 0\text{–}4) \qquad (11.26)$$

$$Co_2(NO)_4X_2^+ \rightarrow Co(NO)_2X^+ + Co(NO)_2X \qquad (11.27)$$

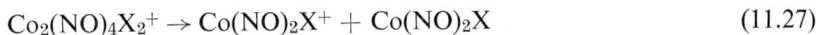

The previously discussed studies on carbonyl–nitrosyl complexes appear to indicate that CO groups are more readily lost than NO· groups in the decomposition of the ions. The fragmentation pattern observed for the cobalt nitrosyl halide complexes, with preferential fission of the metal–halogen bridge to give high-intensity mononuclear ions, was related to the stability of the NO–metal unit. The higher stability of the iron–halogen cluster, on the other hand, was correlated with the presence of further interaction between the iron atoms, which stabilises the cluster relative to loss of · NO.

8. $\pi$-CYCLOPENTADIENYL METAL COMPOUNDS

*8.1 Bis($\pi$-cyclopentadienyl) metal compounds*

In 1955, Friedman *et al.*[116] reported the results of a mass spectral study of the new and interesting class of compounds of formula $M(C_5H_5)_2$. This study served to divide the compounds into two types, in agreement with their other known physical and chemical properties. The "sandwich"-bonded compounds of V, Cr, Fe, Co, Ni, and Ru gave rise to spectra in which the parent molecular ions were the most abundant whereas the corresponding ions of Mg and Mn were relatively unstable and underwent a greater degree of fragmentation resulting in higher yields of $C_5H_5M^+$ and $M^+$. These results were correlated with a higher degree of ionic bonding in the metal–ring bonds of the Mg and Mn compounds. The principal ions formed from all compounds were $(C_5H_5)_nM^+$ ($n = 0$, 1, or 2) with very little further fragmentation occurring. Similar fragmentation patterns have been obtained by Müller and D'Or[117] in a much more recent study, except that many more ions formed by elimination of various neutral hydrocarbon fragments have been detected in small quantities. These ions were much more abundant, however, in the spectra of complexes of second and third row elements. Metastable transitions observed by Denning and Wentworth[118] and by Schumacher and Taubenest[119] established a stepwise loss of cyclopentadienyl rings from ferrocene.

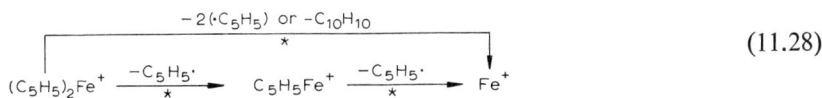

$$(C_5H_5)_2Fe^+ \xrightarrow[\ast]{-C_5H_5\cdot} C_5H_5Fe^+ \xrightarrow[\ast]{-C_5H_5\cdot} Fe^+ \qquad (11.28)$$

with the top reaction $\xrightarrow[\ast]{-2(\cdot C_5H_5)\ \text{or}\ -C_{10}H_{10}}$

The intensity of the parent molecular ion from $(C_5H_5)_2Ru$ was especially high in comparison with the intensity of $C_5H_5Ru^+$ and $Ru^+$, indicating a higher stability than shown by ferrocene. With very few exceptions, the parent molecular ion is the most abundant in the mass spectra of bis($\pi$-cyclopentadienyl) compounds and in substituted derivatives of these. However, the spectra of tri- and tetra-cyclopentadienyl complexes (where the metal was either titanium or a member of the lanthanide or actinide series) which have been reported[120,121] indicate that, in each case, the most abundant ion in the spectrum was $(C_5H_5)_{(n-1)}M^+$. Stepwise loss of the cyclopentadienyl rings was metastable supported for many of these compounds and metal–ring bond dissociation energies calculated.

Some fragmentation of a cyclopentadienyl ring was indicated by the observation of the ions $FeC_3H_3^+$, $FeC_3H_2^+$, and $FeC_2H^+$ from ferrocene[118,119] (Fig. 11.5) and the corresponding ions from nickelocene[119] in low yield.

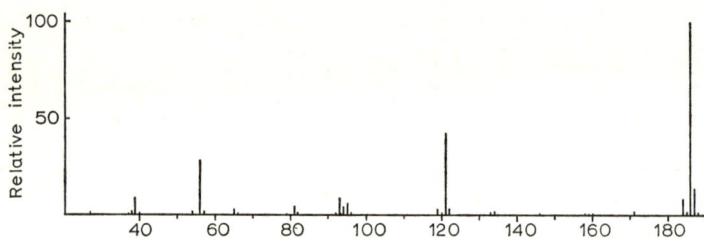

Fig. 11.5. Mass spectrum of ferrocene.

Metastable transitions observed by Schumacher and Taubenest[119] support the loss of acetylene from $C_5H_5Ni^+$ so that the fragmentation process may be represented by

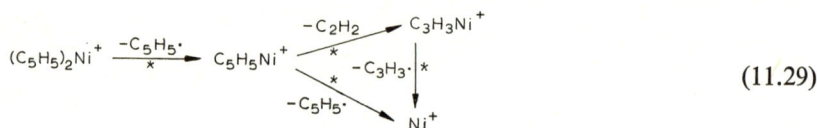

$$(C_5H_5)_2Ni^+ \xrightarrow[*]{-C_5H_5\cdot} C_5H_5Ni^+ \begin{array}{c} \xrightarrow[*]{-C_2H_2} C_3H_3Ni^+ \\ \xrightarrow[*]{-C_3H_3\cdot} \Big|* \\ \xrightarrow[*]{-C_5H_5\cdot} Ni^+ \end{array} \qquad (11.29)$$

Azaferrocene, $C_5H_5FeC_4H_4N$, underwent elimination of HCN to form[122,123] the ion $C_8H_8Fe^+$. Observed metastable transitions[123] corresponded to the decompositions

$$\text{(11.30)}$$

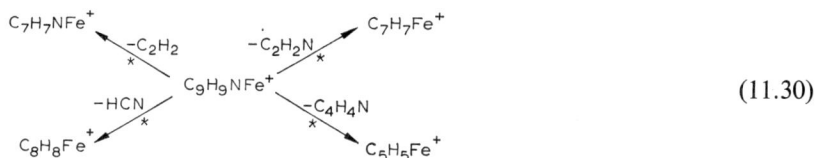

The peak at $m/e = 131$ was assigned to the ion $C_9H_9N^+$ by King[123] but to the ion $FeC_6H_3^+$ by Seel and Sperber[122]. The intensity of the ion $C_5H_5Fe^+$ was much greater than that of $FeC_4H_4N^+$ in the spectrum of this compound suggesting that the ability of the $C_4H_4N$ ring to act as a polydentate ligand was lower than that of the cyclopentadienyl ring. A fairly detailed fragmentation path can be constructed for tris($\pi$-cyclopentadienyl)titanium from the metastable supported decompositions reported by Müller[120].

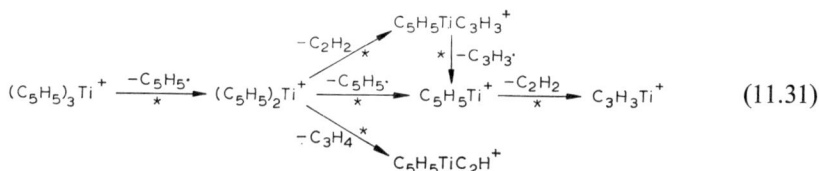

$$\text{(11.31)}$$

Of considerable interest were the metastable transitions corresponding to the decompositions

$$(C_5H_5)_3M^{2+} \underset{*}{\rightarrow} C_5H_5MC_5H_4^{2+} + C_5H_6 \ (M = Pr, Ho, Lu) \qquad \text{(11.32)}$$

The parent molecular ion of cobaltocene undergoes a unique fragmentation process involving loss of a methyl group[117], viz.

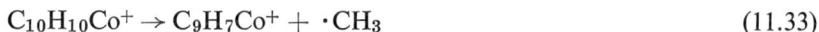

$$C_{10}H_{10}Co^+ \rightarrow C_9H_7Co^+ + \cdot CH_3 \qquad \text{(11.33)}$$

A metastable transition corresponding to the decomposition process

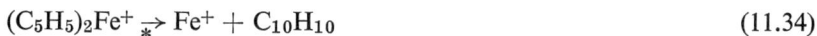

$$(C_5H_5)_2Fe^+ \underset{*}{\rightarrow} Fe^+ + C_{10}H_{10} \qquad \text{(11.34)}$$

(which could be caused by the simultaneous expulsion of two discrete $C_5H_5$ units rather than a $C_{10}H_{10}$ neutral species) was reported by Schumacher and Taubenest[119] while Mendelbaum and Cais[124] observed that the process

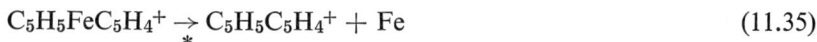

$$C_5H_5FeC_5H_4^+ \underset{*}{\rightarrow} C_5H_5C_5H_4^+ + Fe \qquad \text{(11.35)}$$

was metastable supported in the mass spectra of a number of monosubstituted ferrocenes. The ions $C_{10}H_{10}{}^+$ (ref. 122) and $C_{10}H_9{}^+$ (refs. 122, 124) have both been observed in the spectrum of ferrocene. It appears, therefore, that unique processes in which the iron is eliminated and the cyclopentadienyl ring systems joined take place on electron impact of ferrocenes. A similar metastable supported reaction was observed[125] in the spectrum of the compound $C_5H_5Fe(CO)_2C\equiv CPh$, *viz.*

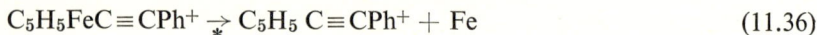

$$C_5H_5FeC\equiv CPh^+ \underset{*}{\rightarrow} C_5H_5\, C\equiv CPh^+ + Fe \qquad (11.36)$$

The elimination of neutral metal fluorides with the formation of complex organic ions is common in fluorocarbon complexes of the transition metals (Section 13).

A number of ions, namely $(C_5H_5)_2Fe_2{}^+$, $(C_5H_5)_2Fe_2C_3H_3{}^+$, $C_5H_5Fe_2C_3H_3{}^+$, and $(C_5H_5)_3Fe_2{}^+$, were observed in low yields in the spectrum of ferrocene, while a mixture of ferrocene and nickelocene resulted in the ions $(C_5H_5)_3Fe_2{}^+$, $(C_5H_5)_3Ni_2{}^+$, and $(C_5H_5)_3FeNi^+$. "Triple-decker" sandwich-type structures, *e.g.* VIII and IX, have been postulated for these ions, assumed to have been formed by ion–molecule reactions[119].

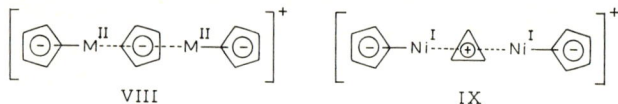

VIII                                IX

Kiser[51] prefers the explanation that molecular dimers exist to a limited extent after sublimation of the material into the ion source of the mass spectrometer and that ionisation of these dimers is responsible for the ions observed. However, parent molecular ions corresponding to dimers were not detected. The occurrence of ion–molecule reactions other than those resulting in (parent molecule + ·H)⁺ ions is somewhat unusual at normal operating pressures. However, Coutts and Wailes[333] have reported the ions $(C_5H_5)_2Ti(NCO)C_3H_3{}^+$ and $(C_5H_5)_2Ti(NCO)C_3H^+$ in the mass spectrum of the monomer bis($\pi$-cyclopentadienyl)titanium isocyanate. Since no other ions of mass greater than the monomer were observed, it is highly unlikely that they originated from dimeric species since fragmentation of the $C_5H_5$ ring is a high-energy process. In this case, at least, ion–molecule reactions appear responsible for their formation. Ions at *m/e* values higher than the parent molecular ion were also observed in tris(acetylacetonato)metal(III) complexes[292].

Appearance potential measurements carried out by Friedman *et al.*[116] allowed the calculation of values for the metal–ring bond energies; weaker bonding in $(C_5H_5)_2Mg$, $(C_5H_5)_2Mn$, and $(C_5H_5)_2Ni$ as compared with the corresponding compounds of Fe, Co, Cr, and V was indicated. Significantly, the results predicted that about three times as much energy was required for reaction (11.37) as for (11.38) in the highly delocalised systems where M = Cr, Fe, Co, or Ni

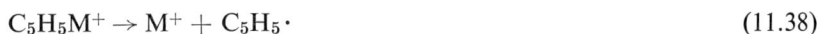

$$(C_5H_5)_2M^+ \rightarrow C_5H_5M^+ + C_5H_5 \cdot \qquad (11.37)$$

$$C_5H_5M^+ \rightarrow M^+ + C_5H_5 \cdot \qquad (11.38)$$

(V appeared anomalous.) The values for Mg and Mn were similar for both

Table 11.7

Ionisation Potentials of Metal Cyclopentadienyls

| Molecule | Ionisation potential (eV) | Ref. | I.P. of metal atom (eV) |
|---|---|---|---|
| $(C_5H_5)_2Mg$ | $7.76 \pm 0.1$ | 116 | 7.64 |
| $(C_5H_5)_2V$ | $7.56 \pm 0.1$ | 116 | 6.74 |
| | $7.33 \pm 0.1$ | 117 | |
| $(C_5H_5)_2Cr$ | $6.91 \pm 0.2$ | 116 | 6.75 |
| | $6.26 \pm 0.1$ | 117 | |
| $(C_5H_5)_2Mn$ | $7.25 \pm 0.1$ | 116 | 7.43 |
| | $7.32 \pm 0.1$ | 117 | |
| $(C_5H_5)_2Fe$ | $7.05 \pm 0.1$ | 116 | 7.87 |
| | $6.99 \pm 0.1$ | 110 | |
| | $7.96 \pm 0.28$ | 127 | |
| | $7.15 \pm 0.1$ | 117 | |
| $(C_5H_5)_2Co$ | $6.2 \pm 0.3$ | 116 | 7.86 |
| | $5.95 \pm 0.1$ | 126 | |
| | $6.21 \pm 0.1$ | 117 | |
| $(C_5H_5)_2Ni$ | $7.06 \pm 0.1$ | 116 | 7.63 |
| | $6.75 \pm 0.1$ | 110 | |
| | $8.38 \pm 0.12$ | 127 | |
| | $7.16 \pm 0.1$ | 117 | |
| $(C_5H_5)_2Ru$ | $7.82 \pm 0.1$ | 117 | 7.36 |
| $(C_5H_5)_2Os$ | $7.59 \pm 0.1$ | 117 | 8.5 |
| $(C_5H_5)_3Ti$ | $6.47 \pm 0.1$ | 120 | 6.82 |

steps. Although the more recent measurements of Müller and D'Or[117] differ slightly, the same overall pattern is observed. The values obtained for the molecular ionisation potentials were less than the ionisation potentials of the metals (which in turn were less than the ionisation potentials of the $C_5H_5$ ligand) for the Mn, Fe, Co, and Ni compounds and only the value obtained for $(C_5H_5)_2V$ was significantly greater than that of the metal (Table 11.7). Their figures were later substantiated for ferrocene[110], cobaltocene[126], and nickelocene[110]. Winters[127] has recently remeasured the ionisation potentials of ferrocene and nickelocene and obtained much higher values intermediate between those of the metal and the $C_5H_5$ ligand. A very recent publication[120] reports the ionisation potential of $(C_5H_5)_3Ti$ to be less than that of Ti itself; the compounds of a number of lanthanide and actinide elements, however, were all reported to have ionisation potentials significantly greater than those of the elements. It is therefore impossible to make assumptions regarding the origin of the electron removed on ionisation of metal cyclopentadienyl compounds. The very low value obtained for $(C_5H_5)_2Co$ may be significant since the ion $(C_5H_5)_2Co^+$ has a closed shell configuration[116].

### 8.1.1 Substituted bis($\pi$-cyclopentadienyl)metal compounds

Bis($\pi$-indenyl)iron behaves under electron impact[128] in a manner which contrasts sharply with the behaviour of ferrocene. The parent molecular ion readily loses one of the indenyl ligands but the resultant ion then decomposes so that the charge remains on the second ligand and an iron atom is eliminated. The only other metal-containing ion present in significant abundance was the doubly charged parent molecular ion. Any fragmentation of the ligands analogous to that in ferrocene which produced cyclopentadienyl–metal ions was insignificant. However, the ion $C_9H_7Mo^+$, formed from $\pi$-$C_3H_5Mo(CO)_2C_9H_7$, underwent successive eliminations of acetylene to form $C_7H_5Mo^+$ and $C_5H_3Mo^+$. The complete decomposition[128] of $(\pi$-$C_9H_7)_2Fe$ is

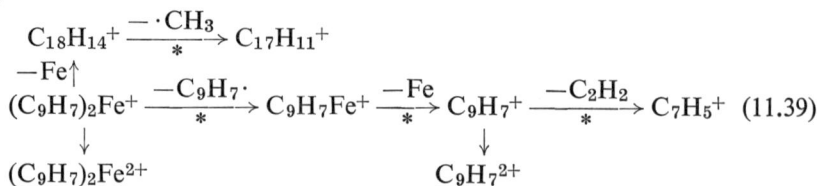

$$C_{18}H_{14}^+ \xrightarrow[*]{-\cdot CH_3} C_{17}H_{11}^+$$
$$\uparrow -Fe$$
$$(C_9H_7)_2Fe^+ \xrightarrow[*]{-C_9H_7\cdot} C_9H_7Fe^+ \xrightarrow[*]{-Fe} C_9H_7^+ \xrightarrow[*]{-C_2H_2} C_7H_5^+ \quad (11.39)$$
$$\downarrow \qquad\qquad\qquad\qquad\qquad \downarrow$$
$$(C_9H_7)_2Fe^{2+} \qquad\qquad\qquad C_9H_7^{2+}$$

The compound $(\pi$-$C_5H_5)Fe(\pi$-$C_9H_7)$ fragments[128] by loss of either its cyclopentadienyl ligand to form $C_9H_7Fe^+$ or its indenyl ligand to give

$C_5H_5Fe^+$, but the latter is heavily favoured; this is in harmony with the compound's chemical properties which indicate lower stability of the metal–indenyl bonds compared with the analogous metal–cyclopentadienyl bonds. Whereas two $C_5H_5$ ligands couple in the decomposition of ferrocene giving rise to the ion $C_{10}H_{10}^+$ and a similar reaction produces $C_{18}H_{14}^+$ from $(C_9H_7)_2Fe$, in the mixed compound coupling of the indenyl and cyclopentadienyl ligands occurs but this is followed by dehydrogenation so that $C_{14}H_{11}^+$ and $C_{14}H_{10}^+$ are observed but not $C_{14}H_{12}^+$. The hydrogenated compounds $C_5H_5FeC_9H_9$ and $C_5H_5FeC_9H_{11}$ were also studied[128].

The mass spectrum of the novel bis(as-indacenyliron), X, was similar to that of other ferrocenes in its simplicity, and the parent molecular ion was again the most abundant[129]. Peaks were observed at $m/e$ values corresponding to loss of both Fe atoms and also three and four hydrogens; presumably the two ring residues were linked as in the previously discussed substituted ferrocenes. After the parent molecular ion, the next most intense metal-containing ion in the spectrum of bis($\pi$-hydropentalenyl)iron, $C_{16}H_{14}Fe$ (XI), was $C_8H_6Fe^+$; the abundance of the $C_8H_7Fe^+$ ion was much lower so that loss of a H atom appears particularly facile in the fragmentation of this compound. Other transition metal complexes containing this ligand were also investigated[130].

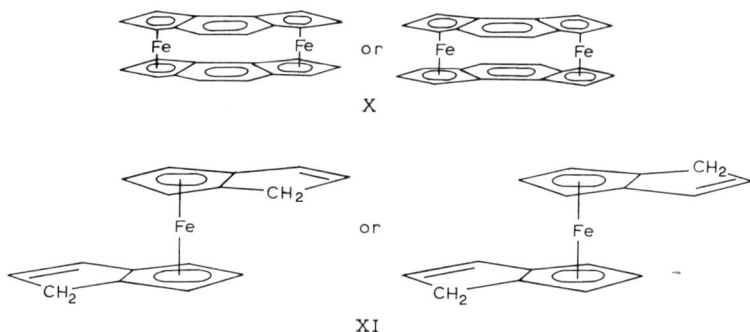

X

XI

From an examination of the mass spectra of a series of substituted ferrocenes, Reed and Tabrizi[131] concluded that there was no similarity in the modes of fission of the ferrocenes and corresponding aromatic compounds. However, ions of formula $C_6H_6Fe^+$ and $C_7H_7Fe^+$ have since been observed in the spectra of a large number of substituted ferrocenes[132–134]. Roberts et al.[135] noted the latter as a fairly intense ion in the spectra of ferrocenes containing at least a two-carbon side chain but found that the hydrogens need not necessarily be attached to the first two carbons. In compound XII where R = –CH=CH–CO–CH$_3$ and R' = H, the seventh hydrogen was shown, in fact, to come from the methyl group. A detailed study of ion structures was carried out on the simple derivative 1,1'-divinylferrocene (XII: R = R' = –CH=CH$_2$)[135]. Labelling experiments indicated that formation of the $C_7H_7^+$ ion from this compound involved rearrangement so that it was, in fact, the tropylium ion XIII, a situation reminiscent of the

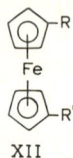

XII             XIII

classical studies on the formation of this ion from toluene and ethylbenzene (ref. 136). Decompositions employed in studying the structures of the ligands were[135]

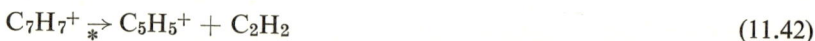

$$C_7H_7FeC_7H_7^+ \rightarrow C_7H_7FeC_5H_5^+ + C_2H_2 \qquad (11.40)$$

$$C_7H_7Fe^+ \underset{*}{\rightarrow} C_5H_5Fe^+ + C_2H_2 \qquad (11.41)$$

$$C_7H_7^+ \underset{*}{\rightarrow} C_5H_5^+ + C_2H_2 \qquad (11.42)$$

Competition was found to take place between the loss of acetylene (the carbons and hydrogens of the vinyl group) and equilibration of all carbons and hydrogens of the ligand. The parent molecular ion lost acetylene faster than it could rearrange but apparently sufficient time was available for practically all of the $C_7H_7$ ligand of $C_7H_7Fe^+$ to be reorganised to a symmetrical structure. This would explain its stability and prominence in the mass spectra of substituted ferrocenes.

Numerous ring-substituted $\pi$-cyclopentadienylmetal complexes have been studied. Bis($\pi$-pentamethylcyclopentadienyl)iron, $(C_5Me_5)_2Fe$, loses methyl groups to form the ions $C_{20}H_{30}Fe^+$, $C_{19}H_{27}Fe^+$, and $C_{18}H_{24}Fe^+$, while a doubly charged parent molecular ion of relatively high abundance was noted[137].

Fragmentation of the ligand $C_5Me_5$ was not observed in the electron-impact spectrum of $(C_5Me_5RhCl_2)_2$ except for the loss of a single hydrogen[138]. However, metastable transitions corresponding to the loss of $H_2$, $\cdot CH_3$, and $C_2H_6$ (or simultaneous loss of two $\cdot CH_3$ groups) from the parent molecular ion $C_5Me_5IrC_8H_{12}^+$ were observed[139]: the latter two groups may have originated from $C_5Me_5$ although the $C_8H_{12}$ ligand has been observed[140] to fragment by loss of $H_2$ and $C_2H_6$. The parent molecular ions were the most abundant in the spectra of a large number of alkylferrocenes and bi-ferrocenyls[141]. The predominant fragment ions were formed by $\beta$ cleavage of the alkyl group. Gamma and alpha cleavages were the only other significant modes of alkyl side-chain fragmentation. The cleavage of the alkyl group was often accompanied or followed by cleavage of either one or both of the cyclopentadienyl–iron bonds. The $m/e$ 121 ($C_5H_5Fe^+$) peak was predominant in the spectra of ferrocenes with an unsubstituted cyclopentadienyl group but weak peaks were also observed at this $m/e$ value in the spectra of 1,1'-dialkyl- and 1,1'-disilylferrocenes. This peak can therefore not be used as a criterion to determine the presence of unsubstituted cyclopentadienyl rings, although Spilners and Larson[141] suggest that its relative intensity may serve as an indicator. These workers observed a metastable decomposition in the spectrum of 1,1'-diethylferrocene corresponding to its formation.

$$FeC_7H_7^+ \xrightarrow[\ast]{-C_2H_2} FeC_5H_5^+ \xrightarrow[\ast]{-C_5H_5\cdot} Fe^+ \qquad (11.43)$$

The ion $FeC_6H_6^+$ was observed in the spectra of a number of these compounds and it was assumed that the hydrocarbon portion was a six-carbon ring.

The monosubstituted ferrocenes $C_5H_5FeC_5H_4COR$, where $R = CH_3$, $C_6H_5$, $p\text{-}CH_3OC_6H_4$, OH, OD, $OCH_3$, and $NHCH_3$, followed a number of decomposition paths[124]. One of particular interest (which was not observed when $R = CH_3$) involved rearrangement of the R group to form the ion $C_5H_5FeR^+$ which in turn decomposed to $(FeR)^+$.

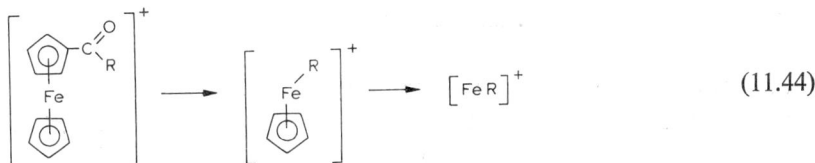

$$\qquad (11.44)$$

The parent molecular ion for all of the compounds except where R = OH was the most intense in the spectrum. A similar rearrangement was found to occur in the series of ferrocene derivatives XIV with an alkylhydroxy side chain and, in fact, the ion $C_5H_5FeOH^+$ was responsible for the base peak in

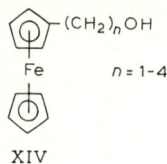

XIV

the spectrum of hydroxymethylferrocene[132]. Although the migrating group in each case was –OH yielding the ion $C_5H_5FeOH^+$, its abundance decreased rapidly with increasing chain length. Cleavage of the side chain, however, to form the ion $C_5H_5FeC_5H_4CH_2^+$ (which may or may not rearrange to incorporate a $\pi$-benzene ligand) increased in importance at the same time. This work was later extended and showed that such rearrangements occurred in a wide range of compounds including substituted arene and olefin complexes[142,143], e.g.

$$\longrightarrow \quad [MnR]^+ \qquad\qquad (11.45)$$

In other compounds where R was not attached to the $\alpha$-carbon atom, Cais and co-workers[142,143] observed that a similar rearrangement took place. Examples of such compounds were $(CO)_3MnC_5H_4-CH_2-CH_2-C(O)OH$ and $(CO)_3MnC_5H_4-CH=CH-C(O)OH$; both gave rise to the ion $[Mn-OH]^+$. The authors suggest that homo- and hetero-disubstituted ferrocenes may be readily distinguished by noting the presence or absence of the $m/e$ 121 peak. However, a low-intensity peak has been observed at $m/e$ 121 in the spectra of certain compounds in the latter class also[141].

In contrast to the behaviour of these substituted ferrocenes, none of the more abundant ions in the spectra of ferrocenylboronic acid, $[(\pi\text{-}C_5H_5)Fe(\pi\text{-}C_5H_4)]B(OH)_2$, or diferrocenylborinic acid, $[(\pi\text{-}C_5H_5)Fe(\pi\text{-}C_5H_4)]_2BOH$, arose by processes involving Fe–O bond formation[144]. The main decomposition reaction observed in the fragmentation of the latter was a rearrangement, presumably four-centred, which gave either the ferrocene parent molecular ion or $[(C_5H_5)_2Fe]_2B=O^+$ depending on which fragment retained the charge.

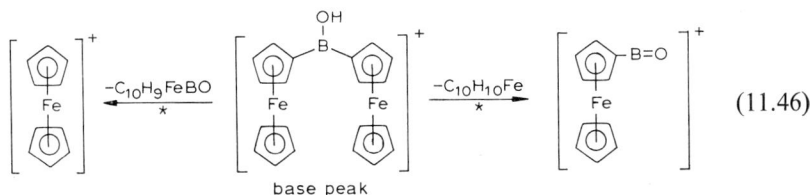

$$(11.46)$$

A significant feature observed in the spectra of the compounds investigated was the difficulty of B–O cleavage.

The mass spectra of configurational isomers of bis-($\alpha$-hydroxytetramethylene)ferrocene were investigated by Egger and Falk[145] and characteristic differences noted.

A number of carboxylato derivatives of bis(cyclopentadienyl)titanium(III) of general formula $(C_5H_5)_2TiOOCR$ were investigated[146] and the mass spectra of all were found to contain low-intensity ions of the type (parent molecule $+ \cdot C_3H_3)^+$ and (parent molecule $+ xCH_2)^+$ ($x = 1$, 2, or 3), apparently formed by ion–molecule reactions.

Thermal decomposition of ferrocene, cobaltocene, nickelocene, and acetylferrocene within the mass spectrometer has been reported[48] to be insignificant.

### 8.2 $\pi$-Cyclopentadienylmetal carbonyls

The mass spectra of a number of cyclopentadienylmetal carbonyl compounds have been reported. Winters and Kiser[147] noted that the relative intensities of the various ions formed indicated that the CO groups were fragmented more readily than the cyclopentadienyl ring system. Cleavage of the C–O bonds, which is fairly widespread in the metal carbonyls, was only rarely observed as in[69], for example, $[C_5H_5Ru(CO)_2]_2$, but cyclopropenyl-metal ions formed by rupture of the cyclopentadienyl ring were apparently more prevalent than in ferrocene and related compounds. If, as previously suggested, such rupture of C–O bonds in the metal carbonyls only occurs as the number of carbonyl groups becomes low and the remaining metal–carbon bonds as a result become stronger, absence of fragment ions of the type $C_5H_5MC_nO_{(n-1)}^+$ is understandable if the CO groups are lost before the $C_5H_5$ rings. This certainly appears to be the case since, in general,

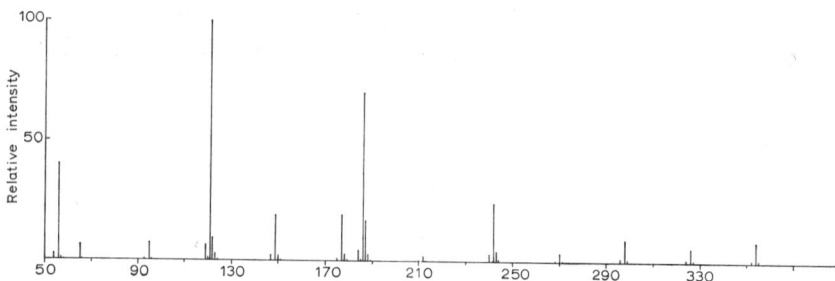

Fig. 11.6. Mass spectrum of $[(\pi\text{-}C_5H_5)Fe(CO)_2]_2$.

the only ions of type $M(CO)_n{}^+$ reported were those where $n = 1$ and these were in low abundance[143,147].

Fragmentation of $[C_5H_5FeCO]_4$, however, appears exceptional in this respect. The ions $(C_5H_5)_4Fe_4(CO)_nC^+$ ($n = 0$ or 2) had relative abundances similar to those of $(C_5H_5)_4Fe_4(CO)_n{}^+$ ($n = 1$, 2, or 3)[148]. King[148] suggested that this was indicative of strong M–CO bonds while the C–O bonds were relatively weak, consistent with the unusually low $\nu_{co}$ frequency observed in the infrared spectrum. The $Ru_4$ cluster in the analogous ruthenium compound appeared to be more stable[149]. A number of relatively intense doubly charged ions containing the $Ru_4$ cluster were observed.

Clastograms[147,150] indicated successive loss of carbonyl groups similar to the metal carbonyls and this mode of decomposition was subsequently substantiated for the compounds $[(\pi\text{-}C_5H_5)Fe(CO)_2]_2$ (ref. 64), $[(\pi\text{-}C_5H_5)Ru(CO)_2]_2$ (ref. 69), $(\pi\text{-}C_5H_5)V(CO)_4$ (ref. 101), and $(\pi\text{-}C_5H_5)RuRe(CO)_7$ (ref. 69) by the observation of the appropriate metastable transitions. Appearance potential measurements and calculation of heats of formation of the gaseous cyclopentadienyl- and cyclopropenyl-metal ions allowed the following estimation of the bond strengths of the metal–ring bonds of the gaseous ions[147].

$$D(C_5H_5\text{-}Co^+) \sim D(C_5H_5\text{-}V^+) > D(C_5H_5\text{-}Mn^+) \tag{11.47}$$

and

$$D(C_5H_5\text{-}M^+) > D(C_3H_3\text{-}M^+) \tag{11.48}$$

The compound fulvalenehexacarbonyldimanganese, XV, also loses carbonyl groups before the hydrocarbon ligand but fragmentation of the ligand

XV                                 XVI

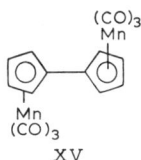

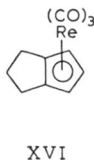

itself was not observed[151]. Another substituted ($\pi$-cyclopentadienyl)metal carbonyl, XVI, behaved somewhat differently[152]. Initial fragmentation of the parent molecular ion involved loss of a CO group but then a molecule of hydrogen was expelled before further significant loss of carbonyl groups occurred. Cais et al.[143] noted that, not only were the CO groups lost before the substituted cyclopentadienyl ring in a large number of mono-substituted cymantrenes*, but they were also generally lost before any fragmentation of the side chain took place. The ion (parent molecule — CO)$^+$ corresponding to loss of a single carbonyl group from the parent molecular ion was observed in only a few cases, and was then present in low abundance.

The parent molecular ion in the mass spectrum of $CH_3Mo(CO)_3(\pi\text{-}C_9H_7)$ exhibited stepwise loss of its carbonyl groups but, in addition, fragmentation of the $CH_3$ group competed with CO loss after the first of these groups had been eliminated[128]. Other $\pi$-indenylmolybdenum carbonyl compounds[128] appeared to behave in an analogous manner to that of $\pi$-cyclopentadienyl-metal carbonyls. Neither of the compounds $C_6H_5COFe(CO)_2(\pi\text{-}C_5H_5)$ or $CH_2{=}CH{-}CO{-}W(CO)_3(\pi\text{-}C_5H_5)$ displayed a parent molecular ion but rather the ion of highest $m/e$ was 28 mass units lower[101]. Decarbonylation could have occurred in the instrument before ionisation. The observed mass spectrum of $[(\pi\text{-}C_5H_5)CoCO]_3$ corresponded to that expected for the unknown tetranuclear compound $(C_5H_5)_4Co_4(CO)_2$; it appears that pyrolysis occurred within the instrument[153].

Ions of type $(C_5H_5)_2M^+$ have been observed in the spectra of many complexes which contain only one cyclopentadienyl ring bonded to a metal atom. King[101], for example, has observed the ion $(C_5H_5)_2Fe^+$ in the spectra of a number of mononuclear compounds. He recorded the spectrum of $CH_3COFe(CO)_2(\pi\text{-}C_5H_5)$ on two instruments having source temperatures of 70° and 200 °C, respectively, and differences observed indicated that ferrocene was produced by pyrolysis within the instruments. Pignataro and Lossing[48] have shown that cobaltocene was produced from $(\pi\text{-}C_5H_5)Co(CO)_2$

---

* Cymantrene is a trivial name for $\pi$-cyclopentadienylmanganese tricarbonyl.

by thermal decomposition in the mass spectrometer. King[101] has, in addition, observed that compounds of the type $RFe(CO)_2(\pi\text{-}C_5H_5)$ produced substituted ferrocenes $C_{10}H_9RFe$ and $C_{10}H_8R_2Fe$ presumably by similar pyrolytic processes. Certain acyl derivatives such as $PhCOFe(CO)_2C_5H_5$ undergo decarbonylation in the mass spectrometer to form, for example, the corresponding phenyl derivatives.

It is also well known that $\pi$-cyclopentadienyl-iron compounds containing bridging carbonyl groups thermally decompose to ferrocenes[154] and it was therefore assumed that the ion $(C_5H_5)_2Fe^+$ present in the spectra of many such compounds arose from ferrocene produced by pyrolytic decomposition within the mass spectrometer. However, studies of $[(\pi\text{-}C_5H_5)Fe(CO)_2]_2$ showed that, while pyrolysis may have been partially responsible for the highly abundant ion $(C_5H_5)_2Fe^+$ (Fig. 11.6), a metastable transition was observed corresponding to its formation from $(C_5H_5)_2Fe_2^+$ by migration of a cyclopentadienyl ring from one iron atom to the other[64,150].

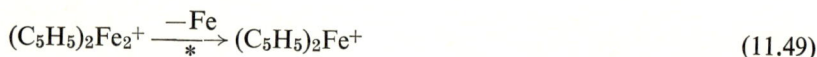

$$(C_5H_5)_2Fe_2^+ \xrightarrow[\;*\;]{-Fe} (C_5H_5)_2Fe^+ \tag{11.49}$$

Similar rearrangement processes have since been observed in a large number of compounds, *e.g.*

(*i*) $[(\pi\text{-}C_5H_5)FeCO]_4$ (ref. 148)

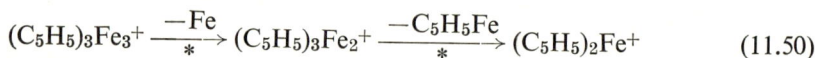

$$(C_5H_5)_3Fe_3^+ \xrightarrow[\;*\;]{-Fe} (C_5H_5)_3Fe_2^+ \xrightarrow[\;*\;]{-C_5H_5Fe} (C_5H_5)_2Fe^+ \tag{11.50}$$

The "triple-decker" sandwich ion XVII was postulated as the structure of the ion $(C_5H_5)_3Fe_2^+$. King[153] had previously shown that the tetramer decomposed to ferrocene at the approximate temperature of the ion source employed, so it appeared reasonable to assume that the $(C_5H_5)_2Fe^+$ ion arose from ferrocene produced by thermal decomposition within the instrument.

(*ii*) $[(\pi\text{-}C_5H_5)CoNO]_2$ (ref. 112)

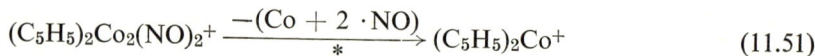

$$(C_5H_5)_2Co_2(NO)_2^+ \xrightarrow[\;*\;]{-(Co\,+\,2\,\cdot NO)} (C_5H_5)_2Co^+ \tag{11.51}$$

(*iii*) $[(\pi\text{-}C_5H_5)NiCO]_2$ (ref. 150)

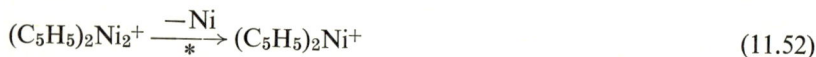

$$(C_5H_5)_2Ni_2^+ \xrightarrow[\;*\;]{-Ni} (C_5H_5)_2Ni^+ \tag{11.52}$$

(*iv*) [CF$_3$C$_2$HNi($\pi$-C$_5$H$_5$)]$_2$  XVIII  (ref. 155)

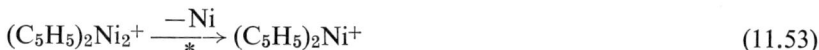

$$(C_5H_5)_2Ni_2^+ \xrightarrow[*]{-Ni} (C_5H_5)_2Ni^+ \qquad (11.53)$$

(*v*) ($\pi$-C$_5$H$_5$)$_2$Cr$_2$(NO)$_2$(SCH$_3$)$_2$  XIX  (ref. 156)

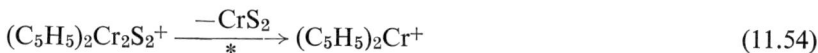

$$(C_5H_5)_2Cr_2S_2^+ \xrightarrow[*]{-CrS_2} (C_5H_5)_2Cr^+ \qquad (11.54)$$

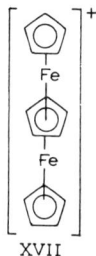
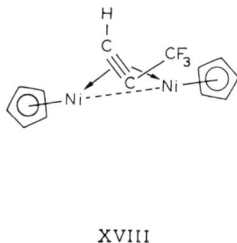
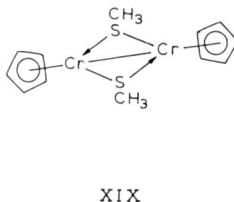

XVII                    XVIII                    XIX

These examples cover a wide range of complexes. The ion of type $(C_5H_5)_2M^+$ is observed in the mass spectra of a large number of bi- and polynuclear complexes but it can no longer be assumed that this indicates thermal decomposition in the mass spectrometer even though such decomposition is most likely partially responsible in a great many cases.

Loss of carbonyl groups from ($\pi$-C$_5$H$_5$)M(CO)$_2$X (M = Fe or Ru; X = Cl, Br, or I) is reported[69] to be favoured over loss of halogen; metastable transitions again support the successive loss of CO groups. This mode of decomposition appears to be general for cyclopentadienylmetal carbonyl halide compounds. Extensive decomposition of the compounds ($\pi$-C$_5$H$_5$)M-(CO)$_3$Cl (M = Mo or W) and [($\pi$-C$_5$H$_5$)MoNOI$_2$]$_2$ occurred under the conditions employed in recording their spectra[157].

King[80] has reported that the mass spectrum of [($\pi$-C$_5$H$_5$)Cr(CO)$_3$]$_2$ did not exhibit any ions containing two chromium atoms, the ions C$_5$H$_5$Cr(CO)$_3^+$ and C$_5$H$_5$Cr(CO)$_3$H$^+$ being those of highest mass in the spectrum. The corresponding ions were observed in the spectra of the analogous molybdenum and tungsten compounds but, in addition, the parent molecular ions were present in high abundance and ions formed by loss of CO groups from them were also observed[80,100]. Since bridging carbonyl groups are absent, a much stronger metal–metal bond in the molybdenum and tungsten compounds is indicated.

The compounds $HM(CO)_3(\pi\text{-}C_5H_5)$ (M = Mo or W) have been discussed in Section 6.

An unusual decomposition observed in the fragmentation of the $\pi$-cyclopentadienylmetal nitrosyl compounds $(\pi\text{-}C_5H_5)V(NO)_2CO$ (ref. 114) and $[(\pi\text{-}C_5H_5)FeNO]_2$ (ref. 113) was the simultaneous expulsion of $C_5H_5\cdot$ and $N\cdot$, resulting in M–O rearrangement ions. Reactions for which metastable transitions were observed are

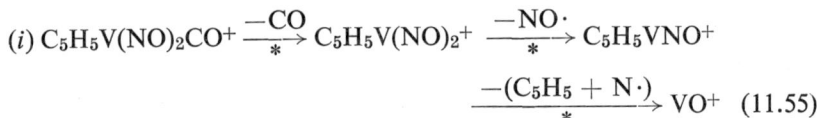

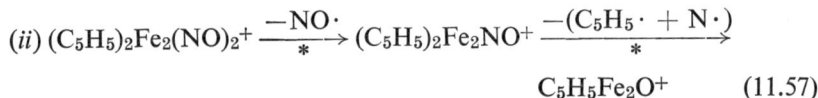

$$(i)\ C_5H_5V(NO)_2CO^+ \xrightarrow[*]{-CO} C_5H_5V(NO)_2^+ \xrightarrow[*]{-NO\cdot} C_5H_5VNO^+$$

$$\xrightarrow[*]{-(C_5H_5\,+\,N\cdot)} VO^+ \quad (11.55)$$

$$C_5H_5V^+ \xrightarrow[*]{-C_2H_2} C_3H_3V^+ \qquad\qquad\qquad (11.56)$$

$$(ii)\ (C_5H_5)_2Fe_2(NO)_2^+ \xrightarrow[*]{-NO\cdot} (C_5H_5)_2Fe_2NO^+ \xrightarrow[*]{-(C_5H_5\cdot\,+\,N\cdot)}$$

$$C_5H_5Fe_2O^+ \qquad (11.57)$$

This decomposition reaction was not observed in the mass spectra of other cyclopentadienylmetal nitrosyl compounds[158] such as $(\pi\text{-}C_5H_5)FeNOI_2$, $[(\pi\text{-}C_5H_5)FeNOCH_3]_2$, or $C_5H_5CoNOI$.

### 8.3 Other cyclopentadienylmetal derivatives

The parent molecular ion was the most abundant in the mass spectrum of bis($\pi$-cyclopentadienyl)rhenium hydride[102,116], while that corresponding to loss of a hydrogen was also present in high abundance ($\sim$65–70% of the parent molecular ion) as might be expected. Fischer and Schmidt[102] found that the ions $(\pi\text{-}C_5H_5)_2TcH^+$ and $(\pi\text{-}C_5H_5)_2Tc^+$ in the spectrum of bis($\pi$-cyclopentadienyl)technetium hydride had approximately equal relative intensities, a finding compatible with other data which suggested that the Tc–H bond was weaker than the corresponding Re–H bond. In addition, fragmentation of a $C_5H_5$ ring apparently occurred to a greater extent in the rhenium compound but, rather surprisingly, no $C_5H_5Re^+$ ion was observed.

Considerable use has been made of isotopic mass distributions in identifying ions in the mass spectra of $(\pi\text{-}C_5H_5)_2ZrCl_2$ and $[(\pi\text{-}C_5H_5)_2ZrCl]_2O$ (ref. 159). The ions derived from the former compound were readily explained by successive losses of cyclopentadienyl radicals or chlorine atoms from the parent molecular ion. Considerable fragmentation of the $C_5H_5$ rings was also evident. Fragmentation of the oxygen-bridged compound was much more complex; a detailed fragmentation mechanism incorporating some novel structures was proposed.

Results of a second mass spectral study of bis($\pi$-cyclopentadienyl)zirconium dichloride were essentially in agreement, and a similar investigation of the corresponding titanium compound was reported at the same time[160]. The most abundant ion in the spectrum of both the titanium and zirconium compounds was $C_5H_5MCl_2{}^+$, formed by loss of $\cdot C_5H_5$ from the parent molecular ion. A detailed series of unimolecular decompositions was established from both clastogram curves and a number of observed metastable transitions. Consecutive and competitive reactions were shown to occur involving loss of either $C_5H_5\cdot$ or $\cdot Cl$ from the parent molecular ion and from subsequent fragment ions. This decomposition is therefore in distinct contrast to that of cyclopentadienylmetal carbonyls[147] and also ($\pi$-$C_5H_5$)-$TiCl_3$ (ref. 161) where loss of CO or $\cdot Cl$ occurred much more readily than loss of $\cdot C_5H_5$ indicating the probable greater bond strength and ionic character of the metal–chlorine bond in the ($\pi$-$C_5H_5$)$_2MCl_2$ compounds (M = Ti or Zr). In addition, the ions $C_3H_3ZrCl_n{}^+$ and $C_2H_2ZrCl_m{}^+$ ($n =$ 0–2; $m = 1$ or 2) were observed[159,160]; to date this has been the only cyclopentadienylmetal compound which has been shown to give rise to ions containing a $C_2H_2$ ligand attached to the metal. Appearance potentials of the principal ions were determined and their heats of formation calculated by Dillard and Kiser[160] who postulated that the electron removed on ionisation of the molecules originated from the molecular orbital composed predominantly of contributions from the metal and chlorine.

Mass spectra of the complete series $(C_5H_5)_2MX_2$ (M = Ti, Zr, or Hf; X = F, Cl, Br, or I) have recently been reported[162], as have the results of mass spectral studies[163] of the pseudohalide complexes $(C_5H_5)_2M(NCS)_2$ (M = Ti, Zr, Hf or V), $(C_5H_5)_2Ti(NCSe)_2$, $(C_5H_5)_2M(OCN)_2$ (M = Ti, Zr or Hf), $(C_5H_5)_2V(NCO)_2$, and $(C_5H_5)_2TiNCO$. These results were discussed in connection with the mode of bonding of the pseudohalide group.

The parent molecular ion derived from bis($\pi$-cyclopentadienyl)zirconium hydride tetrahydroborate, $(C_5H_5)_2Zr(H)BH_4$, under electron impact was present in low abundance but $(C_5H_5)_2ZrBH_4{}^+$ was prominent[103]. The ion $(C_5H_5)_2Zr^+$ appeared to possess considerable stability. It was the most abundant ion in the spectrum and the corresponding doubly charged ion was also prominent. Fragment ions such as $C_5H_5Zr^+$, or others formed by extrusion of neutral hydrocarbon species from the cyclopentadienyl rings, were not observed. A fragmentation scheme has been proposed[164] for the compound $\pi$-C$_5$H$_5$Pt(CH$_3$)$_3$.

### 8.3.1 Negative ion mass spectra

The negative ions formed by fragmentation of the compounds ($\pi$-C$_5$H$_5$)-Co(CO)$_2$, ($\pi$-C$_5$H$_5$)Mn(CO)$_3$, and ($\pi$-C$_5$H$_5$)V(CO)$_4$ were investigated by Winters and Kiser[147]. Ions of type C$_5$H$_y$M(CO)$_x^-$ were observed but none of type M(CO)$_x^-$, again indicating the ease of removal of CO groups as compared with removal of the C$_5$H$_5$ ligand.

Negative parent molecular ions were produced from the compounds (C$_5$H$_5$)$_2$MCl$_2$ (M = Ti, Zr, Hf) at low electron energies by direct capture of thermal electrons, while fragment negative ions were detected in dissociative resonance capture reactions[165].

### 9. $\pi$-ARENE COMPLEXES

A number of workers have reported the mass spectrum of dibenzene chromium[118,126,166,167]. The intensity of the C$_6$H$_6^+$ ion varied greatly, no doubt due to varying thermal decomposition, but when the metal-containing ions only are considered, the spectra in refs. 126, 166, and 167 are in fairly good agreement. The spectrum of ($\pi$-C$_6$H$_6$) has also been reported[166]. Decomposition to benzene was shown[48] to increase as the temperature was raised; a metallic mirror was deposited on the walls of the ion source and this catalytic surface enhanced further decomposition. A number of ions formed by fragmentation of the C$_6$H$_6$ ligand have been observed in very low abundances but decomposition of the parent molecular ions proceeded predominantly by successive elimination of the ligands[166]. This was also true[168] for the dibenzenechromium derivatives Cr[$\pi$-1,3,5-(CH$_3$)$_3$C$_6$H$_3$]$_2$, Cr[$\pi$-(CH$_3$)$_6$-C$_6$]$_2$, Cr[($\pi$-C$_6$H$_6$)($\pi$-C$_6$H$_5$-C$_6$H$_5$)], and Cr($\pi$-C$_6$H$_5$-C$_6$H$_5$)$_2$. Metastable transitions corresponding to the unusual decompositions

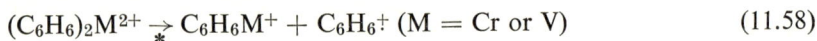

$$(C_6H_6)_2M^{2+} \underset{*}{\rightarrow} C_6H_6M^+ + C_6H_6^{+} \quad (M = Cr \text{ or } V) \qquad (11.58)$$

involving the decay of doubly charged to singly charged ions have been observed[166].

The ionisation potential of ($\pi$-C$_6$H$_6$)$_2$Cr was reported[116] to be significantly lower than that of the metal atom. Appearance potential measurements were used[168] to calculate dissociation energies and total bonding energies for the gaseous ions as well as the neutral complexes. The stability sequence $C_6H_6 < C_6H_3(CH_3)_3 \leq C_6H_5\text{-}C_6H_5 \ll C_6(CH_3)_6$ was established for the metal–ligand bond strength in the neutral complexes. The mass spectra of the mixed $\pi$-complexes $(\pi\text{-}C_5H_5)M(\pi\text{-}C_6H_6)$ (M = Cr or Mn) indicated preferential elimination of the $C_6H_6$ ring from the parent molecular ion and a number of low-intensity ions of the type $C_xH_yM^+$ were again identified[166]. Decay of the doubly charged ions into two singly charged species was observed in these compounds also.

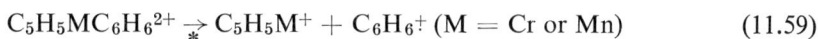

$$C_5H_5MC_6H_6{}^{2+} \xrightarrow{*} C_5H_5M^+ + C_6H_6{}^+ \text{ (M = Cr or Mn)} \qquad (11.59)$$

$$C_5H_5MnC_6H_6{}^{2+} \xrightarrow{*} C_5H_5{}^+ + MnC_6H_6{}^+ \qquad (11.60)$$

Denning and Wentworth[118] compared the mass spectrum of $\pi$-benzene-($\pi$-cyclopentadienyl)manganese(I) with those of ferrocene and dibenzenechromium. The most abundant ions were $C_5H_5Mn^+$, $(C_5H_5)_2Fe^+$, and $Cr^+$ respectively. The abundance of the ion $C_6H_6Mn^+$ was very low but that of the hydride species $C_6H_6MnH^+$ was four times greater. This ratio increased further as the ionising energy was decreased and at 20 eV $C_6H_6Mn^+$ could not be detected but $C_6H_6MnH^+$ and $Mn^+$ were of similar intensity. The ions $C_5H_5Mn^+$ and $C_6H_6MnH^+$ had similar appearance potentials and a concerted mechanism was invoked to explain the formation of the latter.

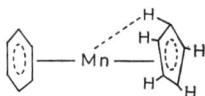

The ion $MnH^+$ was also observed in appreciable quantities. Metastable transitions supported the decay process

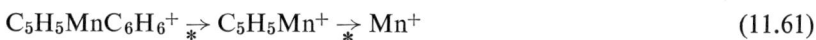

$$C_5H_5MnC_6H_6{}^+ \xrightarrow{*} C_5H_5Mn^+ \xrightarrow{*} Mn^+ \qquad (11.61)$$

The spectrum of $\pi$-benzenechromium tricarbonyl was simple and straightforward[126,169]. No metal-containing ions formed by elimination of hydrocarbon fragments from the benzene ring were observed and no C–O bond rupture was apparent. The ions $C_6H_6Cr(CO)_n{}^+$ ($n = 0\text{–}3$) and $Cr(CO)_n{}^+$

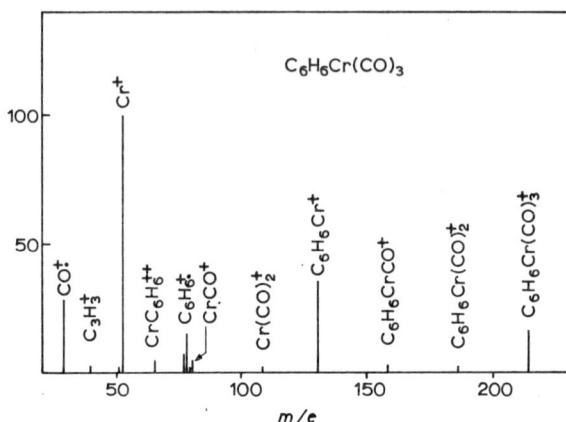

Fig. 11.7. Mass spectrum of $\pi$-benzenechromium tricarbonyl, $\pi$-$C_6H_6Cr(CO)_3$. (Reproduced with permission from ref. 126.)

($n = 0$–2) were formed, $Cr^+$ being by far the most abundant in the spectrum. As mentioned previously, ions of type $M(CO)_n^+$ are very unusual in the spectra of $\pi$-cyclopentadienylmetal carbonyl compounds; it is generally accepted that the stability of $C_5H_5$–M bonds is greater than that of $C_6H_6$–M bonds.

Bursey et al.[170] have recently found that the benzene molecular ions produced by fragmentation of $\pi$-benzenechromium tricarbonyl on electron impact (Fig. 11.7) reproducibly showed a fragmentation pattern which distinctly differed from that of benzene ionised directly under similar conditions. Similar results were obtained with substituted $\pi$-arenechromium tricarbonyls and the parent arenes. The metastable spectrum of $C_6H_6^+$ generated from $\pi$-benzenetungsten tricarbonyl was similar to that of $C_6H_6^+$ from benzene itself but this could have arisen from thermal decomposition of the complex to produce benzene under the conditions necessary to obtain a spectrum. A great deal of further study is required before the implications are fully understood. Under the conditions employed by Pignataro and Lossing[48], thermal decomposition of the chromium compound occurred within the instrument to produce benzene and CO but ($\pi$-$C_6H_6)_2Cr$ was not formed.

Mass spectra of the complexes ($\pi$-arene)$W(CO)_3$ (arene = toluene, p-xylene, mesitylene) were appreciably more complex due to extensive fragmentation of the arene ligand[169].

Metastable transitions indicated that a number of ($\pi$-benzobarrelene)iron tricarbonyl complexes lose the three CO groups in a stepwise manner, while decomposition of the resultant ion LFe$^+$ depended quite markedly[171] on the number and position of methyl groups in the ligand L. Rearrangements of

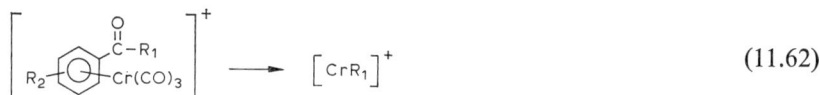

$$\tag{11.62}$$

type (11.62) were shown to occur in a series of substituted arenemetal complexes[142]. Even bimetallic ions such as [Cr–Cr]$^+$ were observed in the spectra of a number of complexes containing two metal atoms, one example being

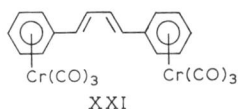

XXI

Hexamethylborazinechromium tricarbonyl is reported[172] to form the ions $B_3N_3(CH_3)_6Cr(CO)_n^+$ ($n = 0, 1, 3$) only.

The ionisation potentials of several ring-substituted benzenechromium tricarbonyls were measured by Müller[173]. As a first approximation, a linear correlation was found to exist between the $\nu(CO)$ force constants of the complexes and their ionisation potentials but no simple connection with the ionisation potentials of the free aromatic ligands could be deduced.

## 10. $\pi$-CYCLOHEPTATRIENYL COMPLEXES

Observed metastable decompositions have indicated four fragmentation pathways leading from the parent molecular ions of the $\pi$-cyclopentadienyl-$\pi$-cycloheptatrienyl compounds ($\pi$-C$_5$H$_5$)M($\pi$-C$_7$H$_7$) (M = Cr or V) to the "bare" metal ions[123,166].

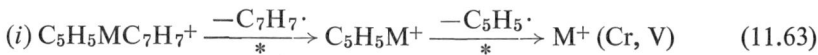

$$\tag{11.63}$$

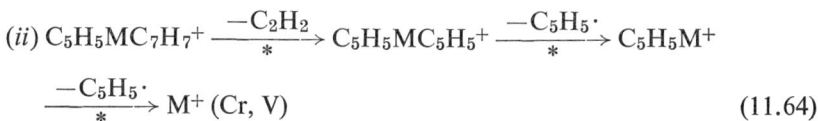

$$\tag{11.64}$$

$$(iii)\ C_5H_5MC_7H_7{}^+ \xrightarrow[*]{-C_6H_6} C_6H_6M^+ \xrightarrow[*]{-C_6H_6} M^+\ (Cr,\ V) \qquad (11.65)$$

$$(iv)\ C_5H_5MC_7H_7{}^+ \xrightarrow[*]{-(\cdot C_5H_5\ +\ C_7H_7\cdot)} M^+\ (Cr) \qquad (11.66)$$

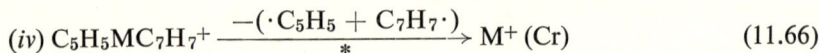

Additional metastable supported decompositions were

$$(v)\ C_7H_7M^+ \xrightarrow[*]{-M} C_7H_7{}^+\ (Cr,\ V) \qquad (11.67)$$

producing the highly stable tropylium cation, and

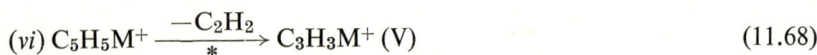

$$(vi)\ C_5H_5M^+ \xrightarrow[*]{-C_2H_2} C_3H_3M^+\ (V) \qquad (11.68)$$

Decomposition (iii) is particularly unusual. Müller and Göser[166] suggested that the $(C_6H_6)V^+$ fragment ion was not $C_5H_5VCH^+$ but rather an ion containing a benzene ligand. They concluded, mainly on the basis of appearance potential measurements, that bis(benzene)vanadium was not involved in the fragmentation. From a study of the compound incorporating a partially deuterated cyclopentadienyl ligand, however, Rettig et al.[174] showed that rearrangement of the parent molecular ion to $C_6H_6VC_6H_6{}^+$ did indeed take place prior to fragmentation. $VC_6{}^+$ fragments containing deuterium were produced, even though the deuterium was introduced only into the five-membered ring in the $C_5H_5VC_7H_7$, while $(C_6H_6)V^+$ containing no deuterium was also observed. Metastable transitions corresponding to these decompositions were detected.

$$ (11.69) $$

The following mechanism was suggested for the rearrangement.

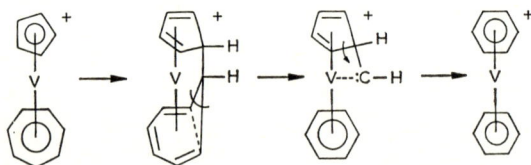

The ions $C_5H_5MoC_5H_5^+$, $C_6H_6Mo^+$, and $C_5H_5Mo^+$ were also observed[175] in the mass spectrum of $C_5H_5MoC_7H_7$, as well as $C_8H_8Mo^+$ and $C_4H_4Mo^+$. The expulsion of acetylene from the $C_7H_7$ ligand also appears to occur in the decay of the compounds $C_5H_5M(CO)_2C_7H_7$ (M = Mo or W); the ion $(C_5H_5)_2M^+$ was observed in each case[101]. The $\pi$-benzyl ligand might be expected to rearrange to the cycloheptatrienyl ligand which in turn would fragment by loss of acetylene similar to the decomposition mechanism shown[135] to occur for 1,1'-divinylferrocene, $(C_5H_4CH{=}CH_2)_2Fe$, discussed previously. Such a rearrangement also occurs in toluene and ethylbenzene under electron impact[136]. However, a low concentration of the ion $(C_5H_5)_2$-$Mo^+$ in the spectrum of $C_5H_5Mo(CO)_2CH_2C_6H_5$ has prompted King[101] to suggest that this was indicative of more difficult loss of acetylene from the $\pi$-benzyl ligand which instead tended to lose ethylene, since the ion $C_5H_5MoC_5H_3^+$ was in much higher abundance. Nevertheless, inspection reveals that the abundance of the ion $(C_5H_5)_2Mo^+$ was not much lower than it was in the spectrum of $C_5H_5Mo(CO)_2C_7H_7$ and the ion "$C_5H_5MoC_5H_3^+$" could have been of the form $C_3HMoCH_2C_6H_5^+$; fragmentation of the cyclopentadienyl ligand to give $C_3HM^+$ ions has been observed *. Thus rearrangement of the $\pi$-benzyl ligand to $\pi$-cycloheptatrienyl followed by loss of acetylene to form $\pi$-cyclopentadienyl cannot be ruled out, although loss of the entire $C_7H_7$ ligand is likely to be a much more highly favoured process, therefore explaining the low abundance of the $(C_5H_5)_2Mo^+$ ion.

Appearance potentials of the principal ions in the spectra of the $C_5H_5MC_7H_7$ compounds (M = Cr, V) have been determined and dissociation energies of the metal–ligand bonds calculated[166].

## 11. $\pi$-OLEFIN COMPLEXES

The mass spectra of a number of cyclopentadienyl metal olefin complexes are similar in that the parent molecular ion and the fragment ions $C_5H_5M^+$ and $M^+$ are all abundant. The parent molecular ion of the compound $\pi$-cyclopentadienyldiethylenerhodium, $C_5H_5Rh(C_2H_4)_2$, decomposed by successive loss of ethylene groups similar to the well-known loss of carbonyl groups in metal carbonyls, cyclopentadienylmetal carbonyls, etc., followed by loss of the $C_5H_5$ ring[123,176], viz.

---

* For example, $C_3HW^+$, relative intensity 23% in the spectrum of $C_5H_5W(CO)_3H$ (ref. 101); $C_3HRe^+$, relative intensity 20% in the spectrum of $C_5H_5Re(CO)_3$ (ref. 101).

$$C_5H_5Rh(C_2H_4)_2^+ \xrightarrow[*]{-C_2H_4} C_5H_5RhC_2H_4^+ \xrightarrow[*]{-C_2H_4} C_5H_5Rh^+$$

$$\xrightarrow[*]{-C_5H_5 \cdot} Rh^+ \qquad (11.71)$$

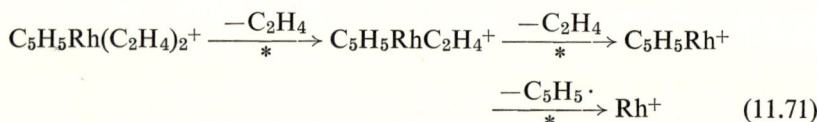

The olefinic ligand was also lost readily[177] from the parent molecular ion of $\pi$-cyclopentadienylbutadienecobalt(I), $C_5H_5CoC_4H_6$. Fragmentation of the cyclopentadienyl-1,5-cyclooctadienemetal complexes, $C_5H_5CoC_8H_{12}$ and $C_5H_5RhC_8H_{12}$, on the other hand, was much more complex[123,176,177]. The postulated decay schemes, based on several observed metastable transitions, are reproduced in Fig. 11.8. Of particular interest are the ions $C_5H_5MC_6H_6^+$, presumably containing the $\pi$-benzene ligand.

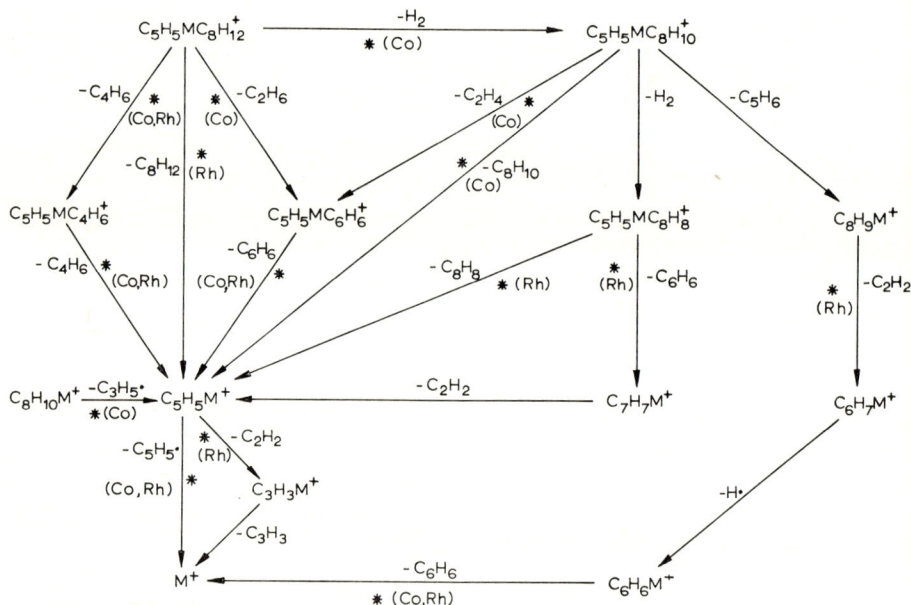

Fig. 11.8. Fragmentation scheme for $\pi$-cyclopentadienyl-1,5-cyclooctadiene cobalt and rhodium complexes (refs. 176, 177).

King[169,178] has investigated a number of olefinmetal carbonyl derivatives and found that stepwise loss of CO groups occurred. The metal–hydrocarbon ions then generally underwent further fragmentations by dehydrogenation and elimination of two-carbon fragments, particularly $C_2H_2$. Metastable transitions indicated that the $C_8H_{10}Fe^+$ ion formed in the decomposition of the 1,3,5-cyclooctatriene complex $C_8H_{10}Fe_2(CO)_6$ fragmented by the steps[179]

$$C_8H_{10}Fe^+ \xrightarrow[*]{-C_2H_4} C_6H_6Fe^+ \xrightarrow[*]{-C_6H_6} Fe^+ \qquad (11.72)$$

Thus the stable $\pi$-benzene ligand is formed by loss of ethane from $C_8H_{12}$ (ref. 176) and by the loss of ethylene from $C_8H_{10}$ (ref. 179). The $C_5H_5Fe^+$ ion was also observed in the spectrum of this complex. The ions $C_2H_2W-(CO)_n^+$ ($n = 0$–2) were present in the spectrum of bicyclo[2,2,1]-heptadiene-tetracarbonyltungsten, XXII (ref. 178), while the ion $C_7H_8Fe_2S_2^+$, formed in the decomposition of $C_7H_8Fe_2(CO)_4(SCH_3)_2$, apparently lost acetylene

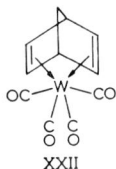

XXII

to give $C_5H_6Fe_2S_2^+$ which then underwent loss of $\cdot$H followed by the complete$\cdot C_5H_5$ unit[180]. The mass spectrum of the cyclohexa-1,3-diene compound $\pi$-$C_5H_5CoC_6H_8$ has been reported[177] while Haas and Wilson[181] examined the spectra of a number of cyclohexa-1,3-dieneiron tricarbonyl derivatives specifically to determine the effect of the presence of the metal atom on the fragmentation processes of the organic ligand. They found that elimination of $H_2$ from the complexed ligand was an important process. On the other hand, elimination of a single hydrogen atom, or of a radical, was much more probable in the decomposition of the uncomplexed ligands which were examined. This dehydrogenation was reported by King[77] also to be particularly facile for other olefinic ligands containing two adjacent $sp^3$ carbon atoms such as 1,3,5-cyclooctatriene and 1,5-cyclooctadiene. Haas and Wilson observed two series of ions formed from each complex: $RC_6H_7Fe(CO)^+$ and $RC_6H_5Fe(CO)_n^+$. The ion $RC_6H_5Fe(CO)_3^+$ was, in all cases, either absent or of very low abundance. They concluded that (*i*) in general the complexes decomposed to odd-electron species more readily than do non-metal-containing compounds, and (*ii*) elimination of one or more CO groups resulted in the metal atom being electron deficient; subsequent loss of $H_2$ converted the organic moiety from a $4\pi$ to a $6\pi$ electron system which could better stabilise the electron configuration of the iron atom. Loss of neutral species such as $CH_4$, $C_2H_4$, and $C_3H_6$ in the fragmentation of 1,3-dieneiron tricarbonyl complexes resulted in the formation[182] of highly unsaturated odd-electron ions such as $FeC_5H_6^+$, $FeC_4H_4^+$, $FeC_3H_4^+$, and $FeC_2H_2^+$.

Winters and Kiser[111] had previously reported the mass spectrum of the parent compound $C_6H_8Fe(CO)_3$ and compared it with that of the perfluoro derivative $C_6F_8Fe(CO)_3$ (ref. 183). Stepwise loss of CO groups from the parent molecular ion was suggested in both cases. The ions $C_6H_6FeCO^+$ and $C_6H_6Fe^+$ were formed (the latter in large abundance) but corresponding ions were not detected in the spectrum of the perfluoro compound. Ions of type $Fe(CO)_n^+$ were of much lower abundance in the spectrum of $C_6H_8Fe(CO)_3$ than in the spectrum of $C_6F_8Fe(CO)_3$. It was suggested[111] that the above results may have been due to the greater strength of the C–F bonds as compared with the C–H bonds in the ionic species and to the $C_6H_8$–$Fe^+$ bond being stronger than the $C_6F_8$–$Fe^+$ bond, respectively.

Müller and Herberhold[184] have studied a series of complexes of formula $\pi$-$C_5H_5Mn(CO)_2L$ where L was a cycloolefin ($C_5H_8$, $C_7H_{12}$, $C_8H_{14}$, nor-$C_7H_8$, nor-$C_7H_{10}$, and maleic anhydride) and also $C_5H_5Mn(CO)$butadiene. They found that initial decomposition of the parent molecular ions $C_5H_5Mn(CO)_2L^+$ involved either simultaneous loss of both CO groups or loss of the olefin. The most abundant ion in each case was $C_5H_5Mn^+$. Metastable transitions enabled the following decomposition pathways to be established.

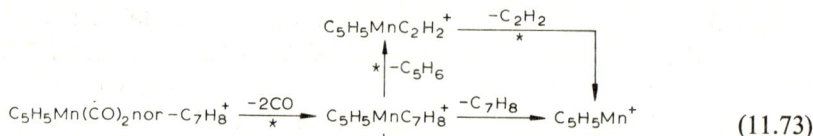

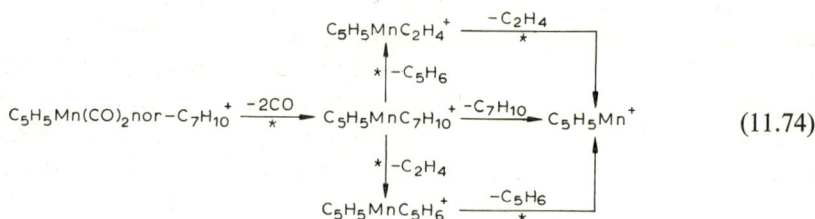

$$
\begin{array}{c}
C_5H_5MnC_2H_2^+ \xrightarrow[*]{-C_2H_2} \\
\uparrow * \big| -C_5H_6 \\
C_5H_5Mn(CO)_2nor\text{-}C_7H_8^+ \xrightarrow[*]{-2CO} C_5H_5MnC_7H_8^+ \xrightarrow{-C_7H_8} C_5H_5Mn^+ \\
* \big| -C_2H_2 \\
C_5H_5MnC_5H_6^+ \xrightarrow[*]{-C_5H_6}
\end{array}
\qquad (11.73)
$$

$$
\begin{array}{c}
C_5H_5MnC_2H_4^+ \xrightarrow[*]{-C_2H_4} \\
\uparrow * \big| -C_5H_6 \\
C_5H_5Mn(CO)_2nor\text{-}C_7H_{10}^+ \xrightarrow[*]{-2CO} C_5H_5MnC_7H_{10}^+ \xrightarrow{-C_7H_{10}} C_5H_5Mn^+ \\
* \big| -C_2H_4 \\
C_5H_5MnC_5H_6^+ \xrightarrow[*]{-C_5H_6}
\end{array}
\qquad (11.74)
$$

Energetics measurements were also carried out.

The halogen bridged 1,5-cyclooctadiene complexes $(C_8H_{12}RhX)_2$ (X = Cl, Br, or I) all underwent some thermal decomposition in the mass spectrometer to form[140] free $C_8H_{12}$; the same was true[185] for $(C_8H_{12})_2Ni$ and $(C_8H_{12})_2Pt$. All three halogen compounds decayed via the very unusual fragmentation involving successive loss of two $H_3X$ units from the parent molecular ion. These metastable supported decompositions presumably consisted of simultaneous loss of HX and $H_2$. Loss of $H_2$ from the $C_8H_{12}$ ligand had previously been observed[176] but this ready loss of HX contrasts with the stability of the metal–halogen unit in halogen-bridged metal carbonyls. The compounds do, however, display a close analogy with di-$\mu$-chlorotetraallylrhodium(III); the concept of metal–metal bond formation was invoked by Lupin and Cais[186] in certain stable fragment ions formed from this compound by loss of the two bridging chlorines. Loss of neutral $C_2H_6$, and possibly $C_3H_8$, from the parent molecular ion of $(C_8H_{12}RhX)_2$ was also observed. A fairly detailed fragmentation scheme was postulated but only a few of the proposed steps were supported by metastable transitions. Several groups of low intensity peaks arising from successive loss of $C_2$ and $C_3$ units from the parent molecular ion were also observed[185] in the spectrum of $(C_8H_{12})_2Pt$ and metastable transitions corresponding to loss of $H_2$, $\cdot CH_3$, and $C_2H_6$ from the parent molecular ion were observed[139] in the spectrum of $C_5(CH_3)_5IrC_8H_{12}$. In this latter case, however, the $\cdot CH_3$ especially and also the $C_2H_6$ neutral fragments probably originated from the pentamethylcyclopentadienyl ligand; $[C_5(CH_3)_5]_2Fe$ has been reported[137] to lose methyl groups.

The mass spectrum of hexamethylbenzenemanganese(I)-cyclohexadienyl, XXIII, was distinguished by the high intensity of rearrangement ions, presumably formed by transfer of a hydrogen from the $CH_2$ group in the $C_6H_7$ ligand to the metal[187]. The base peak was due to the ion $C_6(CH_3)_6MnH^+$. The unusual ion $(CH_3)_6C_6H^+$ was also present in high abundance.

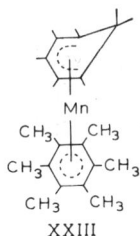

XXIII

Cycloheptadiene was readily lost from the parent molecular ion cycloheptatrienylchromium 1,3-cycloheptadiene, $C_7H_7CrC_7H_{10}$, so that $C_7H_7Cr^+$ was the base peak, whereas $CrC_7H_{10}^+$ was not detected[188]. However, $CrC_6H_7^+$, whose formation presumably involved loss of $\cdot CH_3$ from the $C_7H_{10}$ ligand, was observed in low abundance.

The mass spectrum of cyclobutadieneiron tricarbonyl showed the ions $C_4H_4Fe(CO)_n^+$ ($n = 0$–3) but none due to loss of $C_2H_2$ fragments were observed[189]. That of tetramethylcyclobutadieneiron tricarbonyl also showed the parent molecular ion $(CH_3)_4C_4Fe(CO)_3^+$ and ions formed by successive loss of CO groups down to $(CH_3)_4C_4Fe^+$, the most abundant in the spectrum[190]. Further fragmentation of this ion resulted in the observation of $C_8H_8Fe^+$, $C_7H_{10}Fe^+$, and $C_4H_6Fe^+$. The ions $Ph_4C_4Ru(CO)_n^+$ ($n = 0$–3) were all observed in the spectrum of (tetraphenylcyclobutadiene)ruthenium tricarbonyl[191].

Many dienonemetal carbonyl complexes not only lose the metal-bonded CO groups readily but also the CO group from the organic ligand. Thus the spectra of complexes XXIV, XXV, XXVI, and XXVII, for example, all exhibit peaks corresponding to the parent molecular ion and fragments resulting from loss of one, two, three, and four CO groups[192–194]. However, the ion corresponding to loss of four carbonyl groups was not reported[92] for the compound tentatively given structure XXVIII nor for the compound[193] $C_5H_2(CF_3)_2OFe(CO)_3$. $C_6H_6^+$ was also present[194] in the spectrum of XXVII.

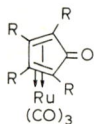 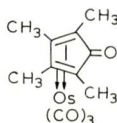

XXIV  R = C$_6$H$_5$
XXV   R = CF$_3$
XXVI  R = C$_6$F$_5$

XXVII

XXVIII

Rearrangements of the type shown in equation (11.69) were observed in the fragmentation of a number of dieneiron tricarbonyl complexes[142]. These are analogous to those reported for a series of substituted $\pi$-cyclopentadienyl complexes (Section 8.1).

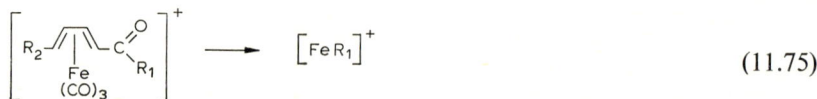

$$\left[ R_2 \overset{O}{\underset{\substack{Fe \\ (CO)_3}}{\diagup\!\!\!\bigg\backslash}}\!\!C\!\!\diagdown_{R_1} \right]^+ \longrightarrow \left[ FeR_1 \right]^+ \qquad (11.75)$$

Decarbonylation of the ligand also occurred in a number of complexes; metastable transitions were observed supporting the decompositions

$$[(L-\overset{\overset{O}{\|}}{C}-R_1)Fe(CO)_3]^+ \xrightarrow[*]{-nCO} [(L-\overset{\overset{O}{\|}}{C}-R_1)Fe]^+ \xrightarrow[*]{-CO} [LR_1Fe]^+ \qquad (11.76)$$

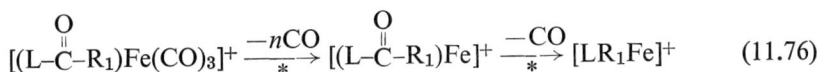

The following mechanism was postulated.

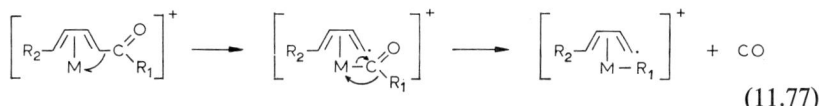

$$(11.77)$$

The mass spectra of a number of complexes of the type $LFe(CO)_4$, where L was a simple olefin, have been reported[195]. Ions of type $LFe(CO)_n^+$ and $Fe(CO)_n^+$ ($n = 0-4$) were all observed. Rearrangement ions $FeX_2^+$ and $FeX^+$ were very prominent when L was a halogenated olefin such as, for example, $CHX{=}CHX$. The butatriene complexes $(CHR{=}C{=}C{=}CHR)$-$Fe_2(CO)_6$ (R = H, $CH_3$) have also been the subject of a mass spectral study[196].

The characteristic stepwise loss of carbonyl groups is shown by cyclooctatetraenemetal carbonyl complexes, resulting in the $C_8H_8M^+$ ion[197,198]. Bruce[197] has suggested that in the complexes $C_8H_8M(CO)_3$ (M = Fe or Ru) loss of acetylene from this $C_8H_8M^+$ ion is the predominant process since he observed the ions $C_nH_nM^+$ ($n = 0, 2, 4$ or 6). He further suggested a structure derived from Dewar benzene for the $C_6H_6M^+$ ion based on the fact that normal benzene metal complexes do not usually fragment by loss of acetylene but rather by loss of benzene as one unit. However, no evidence was presented that the $C_6H_6M^+$ ion did lose acetylene but a metastable transition was reported[197] corresponding to the loss of benzene as one unit.

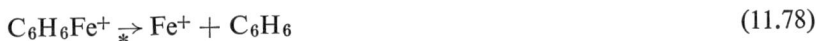

$$C_6H_6Fe^+ \xrightarrow{*} Fe^+ + C_6H_6 \qquad (11.78)$$

Alternatively, the $C_4H_4M^+$ ion may have been formed by the reaction

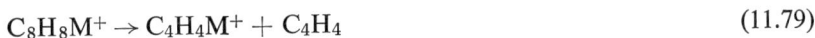

$$C_8H_8M^+ \to C_4H_4M^+ + C_4H_4 \qquad (11.79)$$

Again, observation of the ion $C_2H_2M^+$ was cited[197] as evidence against the $\pi$-cyclobutadiene structure for the ion $C_4H_4M^+$ since the compound $C_4H_4$-$Fe(CO)_3$ was reported not to fragment by loss of acetylene[189]. However, the $C_2H_2M^+$ ion could have been formed by expulsion of a molecule of benzene from $C_8H_8M^+$. A metastable transition corresponding to loss of $C_4H_4$ as one unit from $C_4H_4Fe^+$ was observed[197]. The observation of more metastable transitions would certainly be of interest in deciding between these possibilities but it is felt that on the evidence presently available, a considerable degree of speculation is involved in suggesting structures and decay mechanisms. Loss of acetylene from the parent molecular ion of $\pi$-cyclopentadienylcyclooctatetraenetitanium, $C_5H_5TiC_8H_8$, resulted[268] in the ion $C_5H_5TiC_6H_6^+$.

Binuclear complexes investigated showed a pronounced tendency for the formation of doubly charged ions.

### 12. $\pi$-ALLYL COMPLEXES

Both field ionisation and electron-impact techniques were employed[199] to record the mass spectra of several isoleptic* transition metal $\pi$-allyl complexes; the former proved to be particularly useful in identifying the compounds since the parent molecular ions were present as the most abundant ions in the spectra and fragmentation was low. No metal-containing fragment ions were detected in the spectrum of either $Ni(C_3H_5)_2$ or $Pt(C_3H_5)_2$ for example, while the intensity of $PdC_3H_5^+$ in the spectrum of $Pd(C_3H_5)_2$ was only $3\%$ of that of the parent molecular ion. Elimination of an allyl radical apparently occurred much more readily in $Hf(\pi\text{-}C_3H_5)_4$ and $Zr(\pi\text{-}C_3H_5)_4$, the intensity of the $M(C_3H_5)_3^+$ ions being similar to that of the parent molecular ions.

The electron-impact mass spectra, on the other hand, indicated extensive fragmentation of the parent molecular ions. Loss of allyl radicals was reported to occur readily from $M(C_3H_5)_2^+$ (M = Ni, Pd, or Pt) while loss of hydrogen atoms from these various ions was also noted. Becconsall et al.[199], however, suggested that loss of propylene from $Pd(C_3H_5)_2^+$ was the favoured decomposition process of this parent molecular ion while a second fragmentation path of $Ni(C_3H_5)_2$ involved loss of ethylene from the parent molecular ion. The conditions of relatively high resolution under which the electron impact spectra were recorded were not conducive to the observation of metastable transitions and energetics studies were not made. Thus decomposition paths involved in the formation of any of the ions observed in the

---

* The term isoleptic was used by the authors to indicate that all the ligands attached to the central metal atom were identical in constitution.

spectra of these three compounds could not be established with certainty; this is especially so with the lower fragment ions.

The one metastable transition detected in this study corresponded to loss of ethylene from $Zr(C_3H_5)_3^+$ (11.80), the most intense ion in the spectrum of $Zr(\pi\text{-}C_3H_5)_4$.

$$^{90}ZrC_9H_{15}^+ \underset{*}{\rightleftharpoons} {}^{90}ZrC_7H_{11}^+ + C_2H_4 \qquad (11.80)$$

Structure XXIX was suggested as a possible intermediate in this type of decomposition; such an ion may subsequently lose ethylene, leaving two vinyl groups attached to the metal atom. The authors concluded that $Zr(C_3H_5)_3^+$ was a particularly stable ion while $Pt(C_3H_5)_2^+$ was more stable than $Pd(C_3H_5)_2^+$. The compound $Rh(\pi\text{-allyl})_3$ was reported[200] to give rise to a strong parent molecular ion, while peaks due to $Rh(C_3H_5)_2^+$, $RhC_3H_5^+$ and $Rh^+$ were also observed. No further details were given.

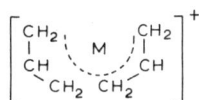

XXIX

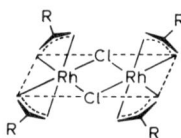

XXX   R = H
XXXI   R = CH₃

A very recent mass spectral investigation of rhodium and palladium π-allyl and π-methylallyl complexes containing halogen bridges was successful in establishing in detail the various fragmentation paths involved[186]. Although the complexes di-μ-chlorotetra-allyldirhodium(III), XXX, and di-μ-chloro-tetra(2-methylallyl)dirhodium(III), XXXI, have been shown to have similar structures incorporating asymmetrically bonded allylic groups, the mass spectra of these two compounds were significantly different, even though some common features were apparent (Fig. 11.9). Both gave rise to parent molecular ions of low abundance. The concept of metal–metal bond formation was invoked in certain stable fragment ions, especially where such ions contain two rhodium atoms but no chlorines. The metastable-supported elimination of propene in the fragmentation of the π-allyl complex required the transfer of · H between allyl groups on the same rhodium; the authors suggested that such a process resulted in ring closure involving the allyl group which lost the hydrogen, forming a cyclopropene ring π-bonded to the metal atom. A similar decomposition step was postulated for the methyl-allyl compound but no metastable was observed to support it. The spectrum of the 2-methylallyl complex differed in two main respects: fragmentation to give half the dimeric molecule was observed and the major peaks contained only one Rh atom. An interesting feature in the fragmentation of this compound was the possibility of ring expansion occurring to form the cyclobuta-dienylcyclopropenylrhodium fragment. Transfer of a methylallyl group from one rhodium atom to another also occurred. Lupin and Cais[186]

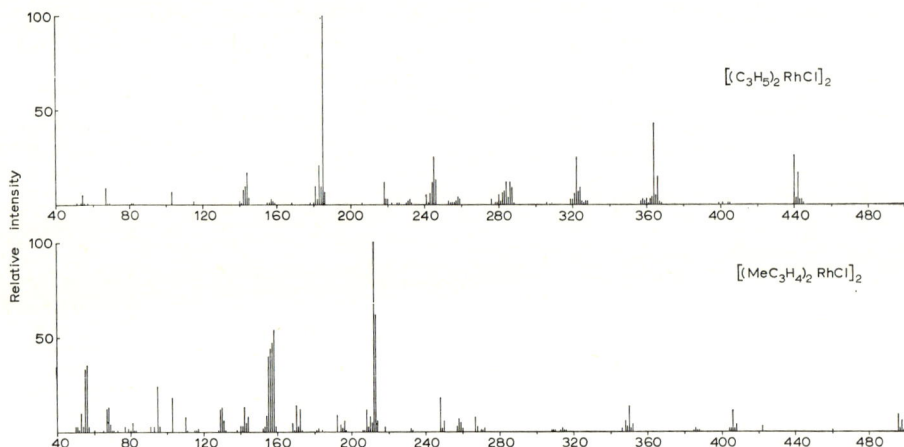

Fig. 11.9. Mass spectra of di-$\mu$-chlorotetra-allyldirhodium(III), [($C_3H_5$)$_2$RhCl]$_2$, and di-$\mu$-chlorotetra(2-methylallyl)dirhodium(III), [(MeC$_3$H$_4$)$_2$RhCl]$_2$ (ref. 186).

suggested that the oxidation state change from octahedral Rh(III) to square planar Rh(I) was an important factor in the fragmentation processes of these $\pi$-allyl complexes.

The mass spectrum of di-$\mu$-chlorodiallyldipalladium(II), [$C_3H_5$PdCl]$_2$, was also recorded; a decomposition scheme was suggested, but no metastable transitions were observed to confirm it. As with the allylrhodium complex, no fragment corresponding to half the dimeric molecule was observed. Because several fragment ions containing the Pd$_2$Cl unit were observed, Lupin and Cais suggested that one of the bridging bonds in the parent molecular ion was broken while a new and perhaps stronger bond was formed between the two Pd atoms (11.81). They also suggested that this process could operate for the rhodium complexes.

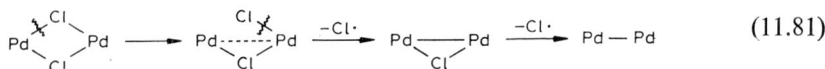

$$\text{(11.81)}$$

King[157] also studied the palladium compound, did not observe the $Pd_2^+$ ion, and suggested that this was due to the absence of a strong metal–metal bond so that a chlorine bridge was necessary to hold the two Pd atoms together. The mass spectrum of $C_3H_5PdC_5H_5$ has also been reported[123].

The principal ions and their intensities observed in the mass spectrum of the bis($\pi$-allyl) complex of suggested structure XXXII have been reported[201]. All ions of composition $(PhCH=C=CH_2)_2Fe_2(CO)_n^+$ ($n = 0$–6) were observed suggesting a stepwise loss of CO groups but no metastables were mentioned. Formation of ions of type $(PhCH=C=CH_2)Fe(CO)_n^+$ ($n = 0$ or 2), present in fairly high abundance, must have involved rupture of both the Fe–Fe bond and the C–C bond joining the two halves of the ligand.

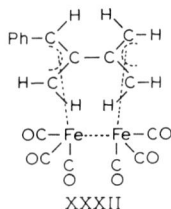

XXXII

It does not appear possible to assign a general fragmentation scheme to $\pi$-allyl metal carbonyl compounds. The parent molecular ions of the compounds $C_3H_5M(CO)_2C_5H_5$ ($M = Mo$ or W) were reported to lose one of the CO groups to give the ion $C_3H_5M(CO)C_5H_5^+$. However, fragmentation of this ion then appeared to involve a nearly simultaneous loss of CO and $H_2$ resulting in formation of the ion $C_3H_3MC_5H_5^+$, apparently containing the $\pi$-bonded cyclopropenyl ligand[101]. The compound $C_3H_5Fe(CO)_3I$ followed a number of different fragmentation paths to form the ions $C_3H_5Fe(CO)_nI^+$ ($n = 0$–3), $Fe(CO)_nI^+$ ($n = 0$–2), $C_3H_5Fe(CO)_n^+$ ($n = 0$–3), $C_2H_2Fe(CO)_n^+$ ($n = 0$–3), and $Fe(CO)_n^+$ ($n = 0$–2) (ref. 157). The spectrum of the simple $\pi$-allylmetal carbonyl $C_3H_5Rh(CO)_2$ was reported[202] to show the ions $C_3H_5Rh(CO)_n^+$ ($n = 0$–2) and $Rh^+$. No ions indicative of a dimeric species were observed although cryoscopic molecular weight data were in accord with a monomer–dimer equilibrium.

## 13. FLUOROCARBON COMPLEXES

The mass spectra of fluorocarbon derivatives of the transition metals display some unique features. While any carbonyl groups present tend to be eliminated in the familiar stepwise fashion, fragmentation paths involving the fluorocarbon entities provide sharp contrasts with corresponding hydrocarbon derivatives. Decompositions of metal-containing ions include the elimination of (*i*) neutral metal fluorides, (*ii*) HF, (*iii*) neutral hydrocarbon species such as $\cdot CF$, $CF_2$, $\cdot CF_3$, and (*iv*) F atoms. Loss of (*i*), (*ii*), and (*iii*) usually occurred after any carbonyl groups originally present had been expelled while loss of (*iv*) occurred both before and after. The rearrangement of one or more F atoms from the ligand to the metal and subsequent elimination of the neutral metal fluoride species occurred with a wide range of fluorocarbon ligands. It appears to be especially favoured with iron cpmplexes. The fluorocarbon entity need not be directly attached to the metal. The compound

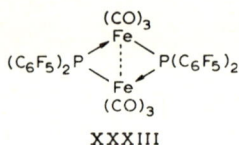

$$
(C_6F_5)_2P \underset{\underset{(CO)_3}{Fe}}{\overset{\overset{(CO)_3}{Fe}}{\diagup\diagdown}} P(C_6F_5)_2
$$

XXXIII

was prepared[203] by the reaction of tetrakis(pentafluorophenyl)diphosphine with iron pentacarbonyl and during decomposition within the mass spectrometer the P–P bond was reformed, if not during the first step, certainly during the second[204].

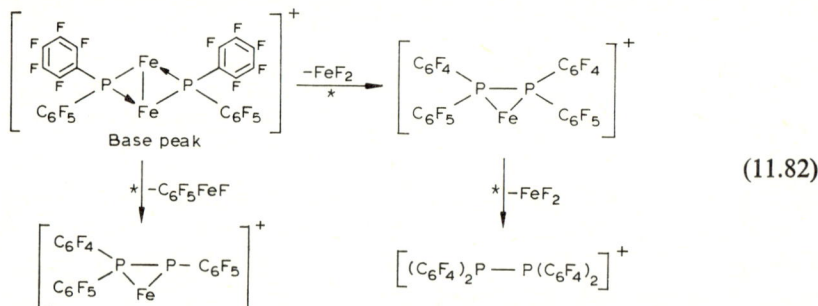

(11.82)

Metastable transitions corresponding to many other metal fluoride elimina-

Table 11.8

Metastable Supported Decompositions Observed in the Mass Spectra of Fluorocarbon Derivatives of Transition Metals and Involving Elimination of Neutral Metal Fluoride Species

| Compound | Process | Neutral fragment lost | Ref. |
|---|---|---|---|
| $(CO)_5CrC(CH_3)NHC_6H_4CF_3$ (XXXIV) | $CrC(CH_3)NHC_6H_4CF_3^+ \xrightarrow{*} C(CH_3)NHC_6H_4CF_2^+$ | $CrF$ | 207 |
| $C_6F_9Mn(CO)_5$ (XXXV) | $C_6F_9Mn^+ \xrightarrow{*} C_6F_7^+$ | $MnF_2$ | 205 |
| $C_6F_9Fe(CO)_2C_5H_5$ (XXXVI) | $C_6F_9FeC_5H_5^+ \xrightarrow{*} C_6F_7C_5H_5^+$ | $FeF_2$ | 205 |
| $C_5F_6ClFe(CO)_2C_5H_5$ (XXXVII) | $C_5F_6ClFeC_5H_5^+ \xrightarrow{*} C_5F_5C_5H_5^+$ | $FeFCl$ | 205 |
| | $C_5C_6ClFeC_5H_5^+ \xrightarrow{*} C_5F_4ClC_5H_5^{+\cdot}$ | $FeF_2$ | 205 |
| $C_2F_4S_2Fe_2(CO)_6$ (XXXVIII) | $C_2F_4S_2Fe_2(CO)_3^+ \xrightarrow{*} C_2F_2S_2Fe(CO)_3^+$ | $FeF_2$ | 208, 209 |
| $C_4F_6S_2Fe_2(CO)_6$ (XXXIX) | $C_4F_6S_2Fe_2(CO)_3^+ \xrightarrow{*} C_4F_4S_2Fe(CO)_3^+$ | $FeF_2$ | 208 |
| $C_6F_5Fe(CO)_2C_5H_5$ (XL) | $C_6F_5FeC_5H_5^+ \xrightarrow{*} C_6F_3C_5H_5^+$ | $FeF_2$ | 210 |
| | $C_6F_4FeC_5H_4^+ \xrightarrow{*} C_{11}H_4F_3^+$ | $FeF$ | 211 |
| | $C_6F_4FeC_5H_4^+ \xrightarrow{*} C_{11}H_4F_2^{+\cdot}$ | $FeF_2$ | 208, 210, 211 |
| | $C_6F_3FeC_5H_5^+ \xrightarrow{*} C_{11}H_5F^{+\cdot}$ | $FeF_2$ | 211 |
| p-$HC_6F_4Fe(CO)_2C_5H_5$ (XLI) | $HC_6F_4FeC_5H_5^+ \xrightarrow{*} HC_6F_3C_5H_5^+$ | $FeF$ | 211 |
| | $HC_6F_4FeC_5H_5^+ \xrightarrow{*} HC_6F_2C_5H_5^{+\cdot}$ | $FeF_2$ | 210, 211 |
| | $HC_6F_3FeC_5H_4^+ \xrightarrow{*} HC_6FC_5H_4^{+\cdot}$ | $FeF_2$ | 210, 211 |
| 2,5- and 3,4-$H_2C_6F_3Fe(CO)_2C_5H_5$ (XLII) | $H_2C_6F_3FeC_5H_5^+ \xrightarrow{*} H_2C_6F_2C_5H_5^+$ | $FeF$ | 211 |
| | $H_2C_6F_3FeC_5H_5^+ \xrightarrow{*} H_2C_6FC_5H_5^{+\cdot}$ | $FeF_2$ | 208, 211 |

| Compound | Reaction | Product |
|---|---|---|
| $p$-$CF_3C_6F_4Fe(CO)_2C_5H_5$ (XLIII) | $H_2C_6F_2FeC_5H_4^+ \rightarrow H_2C_6FC_5H_4^+$ * | FeF   211 |
| | $H_2C_6F_2FeC_5H_4^+ \rightarrow H_2C_6C_5H_4^{+\cdot}$ * | FeF$_2$   208, 211 |
| | $H_2C_5F_3FeC_5H_5^+ \rightarrow H_2C_6FC_5H_5^{+\cdot}$ * | FeF$_2$   210 |
| | $H_2C_5F_2FeC_5H_5^+ \rightarrow H_2C_5C_5H_5^+$ * | FeF$_2$   210 |
| | $C_7F_7FeC_5H_5^+ \rightarrow C_{12}H_5F_5^{+\cdot}$ * | FeF$_2$   208, 210 |
| | $C_7F_6FeC_5H_4^+ \rightarrow C_{12}H_4F_4^{+\cdot}$ * | FeF$_2$   208, 210 |
| $C_{14}H_{14}F_6Fe(CO)_3$ (XLIV) | $C_{14}H_{14}F_6Fe^+ \rightarrow C_{14}H_{14}F_4^{+\cdot}$ * | FeF$_2$   208, 210 |
| $CF_3CH{=}CHFe(CO)_2C_5H_5$ (XLV) | $CF_3C_2H_2FeC_5H_5^+ \rightarrow C_3FH_2C_5H_5^{+\cdot}$ * | FeF$_2$   155 |
| $C_3F_7Fe(CO)_4I$ (XLVI) | $C_3F_7Fe^+ \rightarrow C_3F_5^+$ * | FeF$_2$   212 |
| $CF_3CF{=}CFFe(CO)_2C_5H_5$ (XLVII) | $C_3F_5FeC_5H_5^+ \rightarrow C_3F_3C_5H_5^{+\cdot}$ * | FeF$_2$   212 |
| | $C_3F_4FeC_5H_5^+ \rightarrow C_3F_2C_5H_5^+$ * | FeF$_2$   212 |
| $(CF_3)_2H_2C_5OFe(CO)_3$ (XLVIII) | $C_7F_5H_2OFe^+ \rightarrow C_7F_3H_2O^+$ | FeF$_2$   193 |
| | $C_7F_6H_2OFe^+ \rightarrow C_7F_4H_2O^{+\cdot}$ | FeF$_2$   193 |
| $C_6F_5Ru(CO)_2C_5H_5$ (XLIX) | $C_6F_5RuC_5H_5^+ \rightarrow C_6F_3C_5H_5^{+\cdot}$ * | RuF$_2$   211 |
| $(CF_3)_4C_5ORu(CO)_3$ (L) | $C_9F_{12}ORu^+ \rightarrow C_9F_{10}O^{+\cdot}$ | RuF$_2$   193 |
| | $C_9F_{12}ORu^+ \rightarrow C_9F_9O^+$ | RuF$_3$   193 |
| | $C_8F_{12}Ru^+ \rightarrow C_8F_{10}^{+\cdot}$ | RuF$_2$   193 |
| | $C_8F_{12}Ru^+ \rightarrow C_8F_9^+$ | RuF$_3$   193 |
| $C_{14}H_{14}F_6CoC_5H_5$ (LI) | $C_{14}H_{14}F_6CoC_5H_5^+ \rightarrow C_{14}H_{14}F_5^+$ * | C$_5$H$_5$CoF   208 |
| | $C_{10}H_8F_6CoC_5H_5^+ \rightarrow C_{10}H_8F_5^+$ * | C$_5$H$_5$CoF   208, 210 |
| $(CF_3C{\equiv}CH)Co_2(CO)_6$ (LII) | $C_3F_3HCo_2^+ \rightarrow C_3FHCo^+$ | CoF$_2$   155 |
| $(CF_3)_2C_2S_2Co(CO)_3$ (LIII) | $C_{12}F_{18}S_6Co_3^+ \rightarrow C_{12}F_{16}S_6Co_2^+$ * | CoF$_2$   213 |

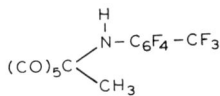

XXXIV

XXXV  X = Mn(CO)₅
XXXVI  X = Fe(CO)₂C₅H₅

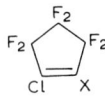

XXXVII  X = Fe(CO)₂C₅H₅
LIV  X = Mn(CO)₅
LV  X = Re(CO)₅

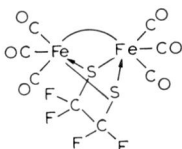

XXXVIII

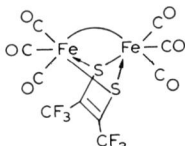

XXXIX

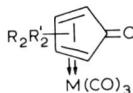

XLVIII  M = Fe; R=H, R'=CF₃
L  M = Ru; R=R'=CF₃

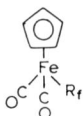

XL  R_f = C₆F₅
XLI  R_f = p-HC₆F₄
XLII  R_f = 2,5-& 3,4-H₂C₆F₃
XLIII  R_f = p-CF₃C₆F₄

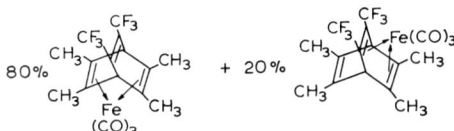

80%  + 20%

XLIV

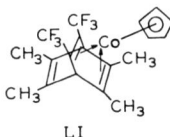

LI

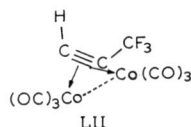

LII

tions have been reported, most of which are listed in Table 11.8. Several other similar decomposition steps were postulated in fragmentation mechanisms but were not metastable supported; these included the extrusion of ReF₃ (ref. 205), C₅H₅FeF (refs. 155, 205), C₅H₅RuF (ref. 205), CoF (refs. 155, 205), and C₅H₅NiF (ref. 155). Fluorine migration from carbon to metal has also been reported to occur[206] during fragmentation of the compound [π-C₅H₅NiP(CF₃)₂]₂.

In the fragmentation of the compounds C₂F₄S₂Fe₂(CO)₆, XXXVIII, and C₄F₆S₂Fe₂(CO)₆, XXXIX, an unusual mode of decomposition was noted, namely extrusion of FeF₂ after only three CO groups had been lost. These reactions suggested that in each case the first three CO groups lost were all from the same iron atom; the remaining carbonyls in the resultant ions R_fS₂Fe(CO)₃⁺ were then lost in the familiar stepwise manner.

The effect of the metal on the breakdown pattern of the fluorocarbon was well demonstrated[205] in the spectrum of perfluorocyclobutenemanganese pentacarbonyl, C₄F₆Mn(CO)₅. Elimination of the neutral MnF₂ molecule from $C_4F_5Mn^+$ gave the ion $C_4F_3^+$ in high abundance but, in comparison, this same ion was only present to the extent of $<1\%$ of the base peak in the spectrum of the free fluorocarbon $C_4F_6$. In general, the ready elimination of $MnF_2$ resulted in metal-free ions which were much more abundant than corresponding ions in the spectra of rhenium complexes; in the latter, elimination of neutral fluorocarbon species resulted in a charged metal fluoride[155,212]. However, this tendency to lose neutral fluorocarbon species was also reported[205,212,214] in certain manganese compounds, resulting in a high abundance of $MnF^+$. The elimination of $MnF_2$ was certainly not as facile as the elimination of $FeF_2$ from iron complexes.

Müller and Fenderl[215] have recently shown that such an elimination of neutral metal species is not restricted to compounds containing fluorinated ligands. If X is highly electronegative (*e.g.* Cl, Br, or $OC_6H_5$) in complexes of the type $(\pi\text{-}C_5H_5)Mn(CO)_2PX_3$, a rearrangement consisting of migration of the X group to the central metal atom is preferred. Heterocyclic ions of general formula $C_5H_5PX^+$ among others are formed.

Loss of HF was another common decomposition step and was observed in the spectra of most fluorocarbon metal complexes, *e.g.*[208]

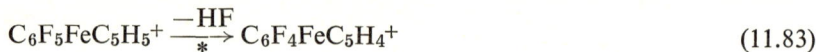

$$C_6F_5FeC_5H_5^+ \xrightarrow[*]{-HF} C_6F_4FeC_5H_4^+ \tag{11.83}$$

The metal carbonyl derivatives were usually responsible for two series of ions due to the successive loss of CO groups from (*i*) the parent molecular ion and (*ii*) the (parent molecule $- \cdot$ F)$^+$ ion[155,193,212,214,216,217]. This occurred also[218] in bis(pentafluorophenylphosphine) complexes such as $(C_6F_5)_2$-$PHCr(CO)_5$. In some cases metastable transitions were observed corresponding to the simultaneous loss of two, and even three, CO groups[155,205]. An unusual decomposition was reported[211] in the breakdown of $H_2C_6F_3Fe$-$(CO)_2C_5H_5$.

$$H_2C_6F_3Fe(CO)C_5H_5^+ \xrightarrow[*]{-FeCO} H_2C_6F_3C_5H_5^{+} \tag{11.84}$$

The base peak in the spectrum of the benzyl complex $(\pi\text{-}C_5H_5)Fe(CO)_2$-$CH_2C_6F_5$ was $C_7H_2F_5^+$ (refs. 216, 217). Bruce suggested that the low relative abundance of $FeC_7H_2F_5^+$ was due to the $C_5H_5$–Fe bond being much stronger than the Fe–$C_7H_2F_5$ bond. Also the ion $C_5H_5FeC_7H_7^+$ in the spectrum of $C_5H_5Fe(CO)_2CH_2C_6H_5$ was much more intense than the ion $C_5H_5FeC_7H_2F_5^+$ in the spectrum of $C_5H_5Fe(CO)_2CH_2C_6F_5$. Thus the former ion may have contained the $\pi$-tropylium ligand while the corresponding 7-membered ring system was not present in the latter; this is in keeping with the idea that fluorocarbons do not form $\pi$-complexes of the same type as do the corresponding hydrocarbons.

The mass spectra of the complexes $C_5F_6ClM(CO)_5$, $M = Mn$ (LIV) or Re (LV), provided no evidence for elimination of metal fluoride entitities in the decay process but the losses of $\cdot F$, $\cdot Cl$, and Mn atoms and the neutral fragments $\cdot CF$, $CF_2$, and $\cdot CF_3$ from carbonyl-free ions were metastable supported[219].

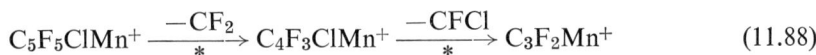

$$C_5F_6ClMn^+ \xrightarrow[\ast]{-Mn} C_5F_6Cl^+ \xrightarrow[\ast]{-CF_2} C_4F_4Cl^+ \qquad (11.85)$$

$$C_5F_6ClMn^+ \xrightarrow[\ast]{-F\cdot} C_5F_5ClMn^+ \qquad (11.86)$$

$$C_5F_6ClMn^+ \xrightarrow[\ast]{-Cl\cdot} C_5F_6Mn^+ \qquad (11.87)^\ast$$

$$C_5F_5ClMn^+ \xrightarrow[\ast]{-CF_2} C_4F_3ClMn^+ \xrightarrow[\ast]{-CFCl} C_3F_2Mn^+ \qquad (11.88)$$

$$C_5F_6ClRe^+ \xrightarrow[\ast]{-\cdot CF_3} C_4F_3ClRe^+ \qquad (11.89)$$

Similar decompositions were also observed in complexes which do fragment by loss of metal fluoride species.

### 14. METAL CARBONYL DERIVATIVES WITH NITROGEN DONOR LIGANDS

The compound $[H_2NFe(CO)_3]_2$ was initially postulated[220], on the basis of elemental analysis and infrared data, to contain two NH bridging ligands although the authors acknowledged that it could, instead, contain $NH_2$ bridging units. Two subsequent and independent mass spectral studies[221,222]

---

$^\ast$ Table 2 of ref. 219 assigns the metastable transitions at $m/e$ 198.1 to the decomposition

$$C_5F_6ClMn^+ \rightarrow Mn^+ + C_5F_6Cl\cdot$$

This $m/e$ value, however, corresponds to the decomposition

$$C_5F_6ClMn^+ \rightarrow C_5F_6Mn^+ + Cl\cdot$$

(calc. 198.6) as in Chart 2 of ref. 219.

*References pp. 600–610*

showed conclusively that the compound possessed $\mu$-amido groups, LVI, by observation of the parent molecular ion $Fe_2(CO)_6N_2H_4^+$. This ion underwent the usual stepwise loss of carbonyl groups; the ions $Fe_2(CO)_nN_2H_4^+$ ($n = 0$–6) were all observed, as were the corresponding doubly charged ions ($n = 1$–3, 6). The mass spectrum of the conjugated olefin complex $Fe(CO)_3$-$(CH_3N=N-C(CH_3)=CH_2)$ has also been reported[223].

LVI                                    LVII

The mass spectral cracking pattern for the compound $(C_2H_5N)_2Fe_2(CO)_7$ was reported to show that elimination of $C_2H_5NCO$ from the (parent molecule $-$ $6CO)^+$ ion was the preferential mode of degradation, in accord with the proposed structure LVII[224]. At about the same time, this was shown by X-ray diffraction to be the structure of the phenyl derivative[225] and its mass spectrum has since been reported in some detail by King[226]. All seven CO groups were lost in a stepwise fashion but the intensities of the ions were interpreted as indicating that the last CO was lost with much greater difficulty than were the first six, as one might expect from the structure of the compound. The ion $(C_6H_5N)_2COFe_2^+$ was also shown to decompose by three additional pathways: ($i$) elimination of an FeNCO fragment, ($ii$) elimination of an HNCO fragment, and ($iii$) elimination of a phenyl isocyanide fragment,

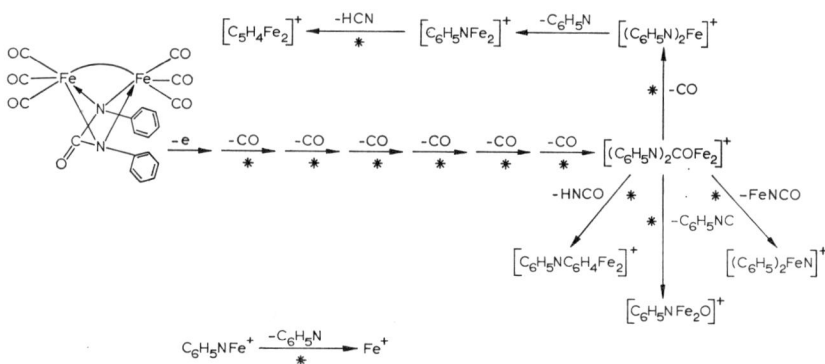

Fig. 11.10. Fragmentation scheme for the compound $(C_6H_5N)_2COFe_2(CO)_6$.

$C_6H_5NC$. From the reported metastable transitions, the fragmentation scheme shown in Fig. 11.10 may be constructed.

The spectra of several other complexes containing nitrogen donor ligands were also reported[226]. Some novel fragmentation steps were noted for the unusual compound $CF_3C(NH)Fe(CO)(NCCF_3)(C_5H_5)$, LVIII, containing the trifluoroacetonitrile ligand, and the following breakdown postulated[227].

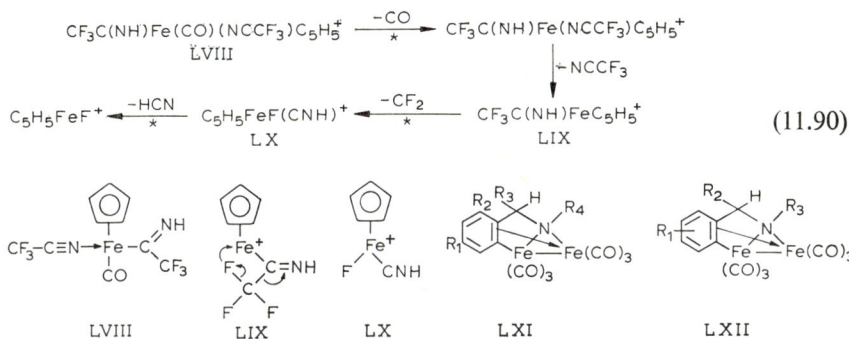

(11.90)

Bagga *et al.*[228] have prepared a number of iron carbonyl complexes of type LXI by means of the reaction between aromatic Schiff's bases and $Fe_2(CO)_9$. A detailed mass spectral study of the products, utilising labelled compounds, provided a means of determining the primary structural arrangements of these complexes. Decomposition paths of the N-benzylideneaniline

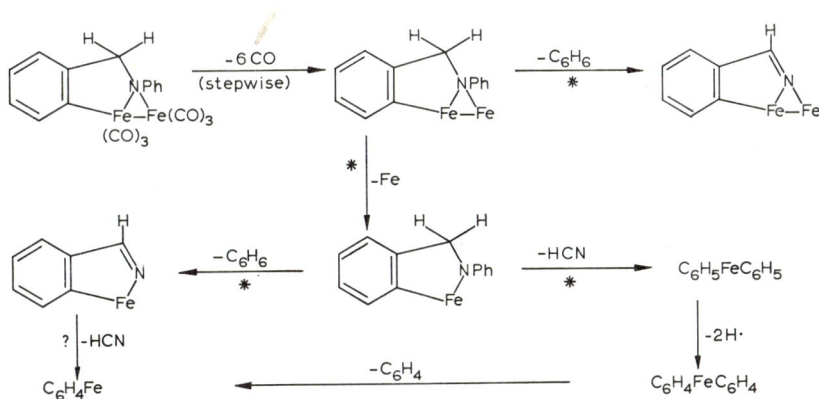

Fig. 11.11. Postulated fragmentation pathway of a benzylidene–aniline iron complex. (Reproduced with permission from ref. 228.)

*References pp. 600–610*

compound are shown in Fig. 11.11. A further study of complexes of this type, LXII, showed[229] that fragmentation depended upon the nature of the substituents $R_1$, $R_2$, and $R_3$. $R_1$ substituents were found to affect the decomposition process primarily by electronic effects whereas effects of $R_2$ were primarily steric in character. Variations in the mass spectra caused by $R_3$ were very small and could not be attributed to either of these.

Preston[230] has reported preliminary results of an investigation into a possible relationship between ion abundance and structure. He examined complexes of type $XX'[Fe_2(CO)_6]$ where X and X' were 3-electron donor systems containing nitrogen, phosphorus, or sulphur and showed that the nature of the ligand system had a recognisable effect on the abundances of fragment ions formed by successive elimination of CO from the parent molecular ion.

### 15. METAL CARBONYL COMPLEXES WITH PHOSPHORUS DONOR LIGANDS

In these complexes, the familiar stepwise loss of carbonyl groups is again evident. This was confirmed by the observation of a large number of metastable transitions[64]. In addition, however, metastables corresponding to the simultaneous loss of two, three, and in one case four, CO groups were also reported. In contrast to the behaviour of metal carbonyl complexes containing halogen atoms in terminal positions, competitive loss of the CO and phosphorus ligands is more the exception than the rule. However, with certain ligands such as trialkyl phosphites, $P(OR)_3$, and trichlorophosphine, cleavage of P–O and P–Cl bonds, respectively, was responsible for the occurrence of secondary fragmentation pathways.

Two preliminary reports of the preparation and characterisation of the phosphine complexes $M(CO)_4(PH_3)_2$ (M = Cr, Mo, or W) have recently appeared[231,232]. Mass spectra were recorded and the parent molecular ions observed; decomposition of these ions appeared to involve competitive loss of both CO and $PH_3$ groups. Thus the ions $Cr(CO)_n(PH_3)_2^+$ ($n = 0$–4), $Co(CO)_nPH_3^+$ ($n = 0$–4), $Cr(CO)_n^+$ ($n = 0$–2), and $Cr(PH_3)_n^+$ ($n = 0$–2) were formed from the chromium complex, as well as several others resulting from P–H bond fission[232]. The compound $Mn(CO)_5PH_3$ was reported to show a "parent molecular ion and a fragmentation pattern consistent with the loss of four carbonyl groups"[231]. The mass spectral behaviour of the compound $C_5H_5Mn(CO)_2AsH_3$ was similar to that of the phosphine complexes[233]. The arsine and CO groups were considered to be equally strongly bonded to the Mn atom. Rupture of As–H bonds was prevalent. Fragmentation of hydridophosphine(trifluorophosphine)cobalt(I), $HCo(PF_3)_3PH_3$, was characterised by cleavage of both P–F and P–H bonds[105].

The complexes $M(CO)_5L$, where M = Cr, Mo, or W and L = $P(OCH_3)_3$ or $P(OC_2H_5)_3$, underwent stepwise loss of CO groups but an additional fragmentation path involving loss of an OR group followed then by loss of carbonyls from the resultant ion was observed[234]. The very low intensity of the ions formed by loss of the complete phosphorus ligand was interpreted as indicative of considerable strengthening of the M–P bonds relative to the M–C bonds in the ion as compared with the situation in the neutral molecule. Braterman[235] has observed loss of a methyl group from the $P(OCH_3)_3$ ligand after all CO groups had been lost and also metastable transitions corresponding to expulsion of formaldehyde, $CH_2O$, from the complex trans-$Cr(CO)_4$-$[P(OCH_3)_3]_2$. These decompositions also occurred in the free ligand.

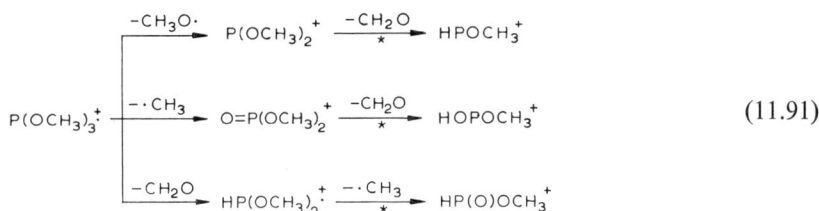

$$\begin{array}{c}
P(OCH_3)_3^{+\cdot} \left\{ \begin{array}{l}
\xrightarrow{-CH_3O\cdot} P(OCH_3)_2^+ \xrightarrow[*]{-CH_2O} HPOCH_3^+ \\[2mm]
\xrightarrow{-\cdot CH_3} O=P(OCH_3)_2^+ \xrightarrow[*]{-CH_2O} HOPOCH_3^+ \\[2mm]
\xrightarrow{-CH_2O} HP(OCH_3)_2^{+\cdot} \xrightarrow[*]{-\cdot CH_3} HP(O)OCH_3^+
\end{array} \right.
\end{array} \qquad (11.91)$$

The favoured decomposition path in the complex $Ru(CO)_2[P(OCH_3)_3]_2$-$(C_2F_4)$, however, was loss of methyl rather than methoxy radicals; the ions (parent molecule $- n\cdot CH_3)^+$ ($n = 0–3, 5$) were observed but no ions formed by loss of $\cdot OCH_3$ from the parent molecular ion were reported[236].

Cleavage of P–C bonds in trialkyl- and triarylphosphine metal complexes apparently occurred only to a very limited extent[64]. The rearrangement ions $R_2PHMn^+$ observed in the spectra of the complexes $[LMn(CO)_4]_2$ (L = $PEt_3$, $PPr^n_3$, $PBu^n_3$, $PEt_2Ph$, $PEtPh_2$, or $PPh_3$) may incorporate a secondary phosphine ligand rather than a metal–hydrogen bond since triethylphosphine has been shown to undergo expulsion of ethylene

$$P(C_2H_5)_3^{+\cdot} \xrightarrow{-C_2H_4} PH(C_2H_5)_2^{+\cdot} \qquad (11.92)$$

and $PH(CH_3)_2^+$ was observed in the mass spectrum of trimethylphosphine[237]. However, the ion $PH(C_6H_5)_2^+$ was apparently not formed during fragmentation of $P(C_6H_5)_3^+$ (ref. 238) and $(C_6H_5)_2PHMn^+$ was of very low intensity in the spectrum of $[(C_6H_5)_3PMn(CO)_4]_2$. Phosphorus–carbon bonds[184] were readily ruptured, however, in the fragmentation of $(\pi\text{-}C_5H_5)Mn(CO)_2P(\text{iso-}C_3H_7)_3$; metastable transitions enabled the following decomposition scheme for the parent molecular ion to be established.

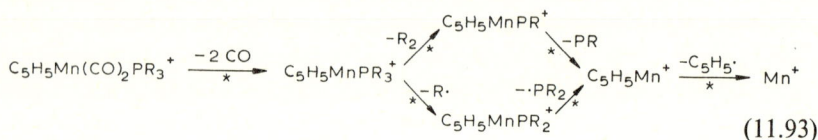

$$
C_5H_5Mn(CO)_2PR_3^+ \xrightarrow[\quad*\quad]{-2\,CO} C_5H_5MnPR_3^+
\begin{array}{c}
\nearrow^{-R_2}\!/_* \;\; {}^{C_5H_5MnPR^+} \searrow^{-PR}\!_* \\[4pt]
\searrow_{-R\cdot}^{*} \qquad\qquad \nearrow^{-\cdot PR_2}_{*}{}^{} \!C_5H_5Mn^+ \xrightarrow[\quad*\quad]{-C_5H_5\cdot} Mn^+ \\[4pt]
\qquad\qquad {}^{C_5H_5MnPR_2^{+}}
\end{array}
$$

$$(11.93)$$

Breakdown of the ligand bis(pentafluorophenyl)phosphine while still attached to the metal was observed[239] with the complexes $[(C_6F_5)_2PH]M(CO)_5$ (M = Mo or W). The novel rearrangement ions $[(C_6F_5)_2PCFM(CO)_n]^+$ ($n = 0-5$) were noted[239] in the spectra of the complexes $(C_6F_5)_4P_2M(CO)_5$ (M = Cr or Mo) in which the diphosphine is acting as a monodentate ligand.

Complexes of $(PR)_5$ ring compounds $(PR)_5M(CO)_n$ (M = Cr, Mo or W; $n = 3$, 4, or 5) fragment in the mass spectrometer to form ions of type $(PR)_5^+$. The iron compound $Fe_2(CO)_6P_4R_4$ did not give rise to $(PR)_4^+$ ions, however, leading Ang and West[240] to postulate that these latter complexes may not contain rings of phosphorus atoms but be bridged "diphosphido" derivatives instead.

Dimetallic fragments were much less abundant when carbonyl groups in $Mn_2(CO)_{10}$ were substituted by phosphine ligands[64]. This was especially so in arylphosphine complexes. The spectrum of the 1,2-bis(diphenylphosphino)-ethane (diphos) complex, (diphos)$[W(CO)_5]_2$, indicated that no ions containing only one tungsten atom were present. The complexes (diphos)$M(CO)_4$ (M = Mo or W) underwent expulsion of the ethylene bridge to form the ions $(Ph_2P)_2M^{n+}$ ($n = 1$ or 2).

The mass spectra of several tris(dimethylamino)phosphine complexes of metal carbonyls have been reported[235,241,242]. The main features of the fragmentations may be summarised as

(i) usual stepwise loss of CO groups,

(ii) loss of a dimethylamino group, $(CH_3)_2N\cdot$, from $[(CH_3)_2N]_3PM(CO)_n^+$ ions (this process competed effectively with (i)),

(iii) elimination of a $CH_3NCH_2$ (azapropene) fragment (cf. $PR_3$ complexes). This occurred in ions of first-row transition metals which were devoid of carbonyl groups, e.g.

$$[(CH_3)_2N]_3PM^+ \underset{*}{\rightleftharpoons} [(CH_3)_2N]_2PHM^+ + CH_3NCH_2 \quad (M = Cr \text{ or } Fe)$$
$$(11.94)$$

$$\left[[(CH_3)_2N]_3P\right]_2Fe^+ \xrightarrow[\;\;*\;\;]{-CH_3NCH_2} [(CH_3)_2N]_2PHFeP[N(CH_3)_2]_3^+$$

$$\xrightarrow[\;\;*\;\;]{-CH_3NCH_2} \left[[(CH_3)_2N]_2PH\right]_2Fe^+ \qquad\qquad\qquad (11.95)$$

$$C_5H_5VP[N(CH_3)_2]_3^+ \underset{*}{\rightleftharpoons} C_5H_5VPH[N(CH_3)_2]_2^+ + CH_2NCH_2 \quad (11.96)$$

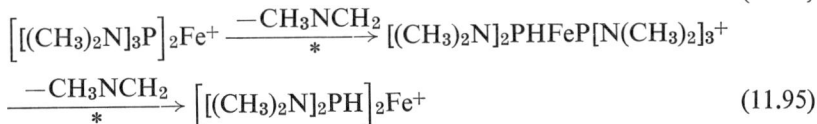

This rearrangement was not observed during fragmentation of the ligand itself[235].

(*iv*) elimination of neutral phosphorus hydride species, *e.g.*

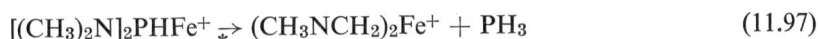

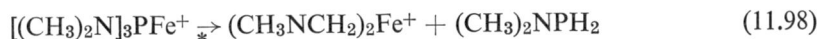

$$[(CH_3)_2N]_2PHFe^+ \underset{*}{\rightleftharpoons} (CH_3NCH_2)_2Fe^+ + PH_3 \qquad (11.97)$$

$$[(CH_3)_2N]_3PFe^+ \underset{*}{\rightleftharpoons} (CH_3NCH_2)_2Fe^+ + (CH_3)_2NPH_2 \qquad (11.98)$$

As a result of these and other minor decomposition steps, such as the loss of methylamine, the fragmentation of these compounds is rather complex.

Kiser *et al.*[243] have studied tetrakis(trifluorophosphine)nickel(0) mass spectrometrically and their results indicated several close similarities to the metal carbonyl systems. Clastograms indicated (*i*) stepwise loss of $PF_3$ groups from $Ni(PF_3)_4$; this was supported by an energetic study and confirmed by the observation of the appropriate metastable transitions, and (*ii*) the occurrence of the processes $NiP_3F_9^+ \rightarrow NiP_3F_8^+ + \cdot F$ and $NiP_2F_6^+ \rightarrow NiP_2F_5^+ + \cdot F$. This type of decomposition, however, was much more extensive in $PH_3$ complexes[232]. The mass spectrum of the compound $Mo(CO)_5PCl_3$ was reported to show the series of ions $Mo(CO)_nPCl_2^+$ in addition to the "normal" series formed by CO losses from the parent molecular ion[234]. Müller and Fenderl[215] reported that the fragmentation of the compounds $C_5H_5Mn(CO)_2PX_3$ ($X = H, Cl, Br, C_6H_5, iso\text{-}C_3H_7, OC_6F_5, \text{ or } iso\text{-}OC_3H_7$) was initiated by the elimination of CO groups but in some cases X radicals were also primarily removed. The nature of the substituent X was found to play a decisive role in the various dissociation processes of the ion $C_5H_5MnPX_3^+$, migration of the X group to the central metal being prefered when $\cdot X$ was highly electronegative. Elimination of neutral $MnX_2$ species to form ions of type $C_5H_5PX^+$ is similar to the elimination of neutral metal fluorides from fluorocarbon derivatives of transition metals.

One significant difference between the spectrum of $Ni(PF_3)_4$ and those of the metal carbonyls was the absence of doubly charged positive ions in that of the former compound. This is in sharp contrast to the rather intense $Ni(CO)_n^{2+}$ ($n = 1$ or 2) ions formed from $Ni(CO)_4$ and was interpreted as indicating that the nickel–ligand bonding differed to some extent in the two molecules. Similarly, doubly charged ions were not observed in the spectra of the complexes $HCo(CO)_n(PF_3)_{(4-n)}$ ($n = 0$–3) although they were present (ref. 104) in that of $HCo(CO)_4$. This study showed, however, that the Co–CO and Co–PF$_3$ bond dissociation energies were essentially identical ($56 \pm 15$ kcal.mole$^{-1}$). In addition, the heats of formation of these compounds became more negative by approximately 200 kcal.mole$^{-1}$ for each CO moiety that was replaced by PF$_3$ and since this variation was the difference between the heats of formation of PF$_3$ and CO, it was postulated that the nature of the Co–CO and Co–PF$_3$ bonds must be quite similar.

Appearance potential measurements on a series of iron carbonyl nitrosyl phosphite complexes enabled differences in the Fe–CO and Fe–P(OR)$_3$ bond energies to be calculated[244]. For the complexes $FeCO(NO)_2P(OR)_3$ and $Fe(NO)_2[P(OR)_3]_2$ ($R = CH_3$ or $C_2H_5$), a value of $33 \pm 10$ kcal.mole$^{-1}$ was obtained. On going from these neutral molecules to the parent molecular ions, the difference increased to about 45 kcal.mole$^{-1}$. This increase was said to indicate poor contribution of back-donation from Fe to the Fe–P bond in comparison with that of the Fe–CO; ionisation of the complex lowers the back-donation capabilities of the central metal[245] and this would cause a greater weakening of the Fe–CO bond than of the Fe–P bond.

The low ionisation potential of $Ni(PF_3)_4$, similar to that of $Ni(CO)_4$, suggested that the electron removed had originated from a molecular orbital associated largely with the metal atom[243]. This has been assumed to be the case with the metal carbonyls (Section 4). Foffani et al.[110] carried out ionisation potential measurements on metal carbonyl nitrosyl complexes and found that the various values were decreased by substitution of a PX$_3$ ligand ($X = Cl$ or $OC_2H_5$) for a carbonyl group and, in fact, $P(OC_2H_5)_3$, with an ionisation potential lower than that of PCl$_3$, decreased the ionisation potential of the complexes by a greater amount. The change in I.P. of the complexes by this ligand substitution has also been shown to parallel the change in C–O and N–O stretching frequencies in the infrared spectra[246,247]. The investigation was extended to a large number of monosubstituted metal carbonyl complexes of the type $M(CO)_5L$, $Fe(CO)_4L$, and $Co(CO)_2NOL$ ($M = Cr$, Mo or W; $L = PX_3$, CNR)[234,245]. Linear relationships were obtained between the ionisation potentials of the complexes and ligands for all of the series examined (Fig. 11.12). They argued that the ligand ionisation

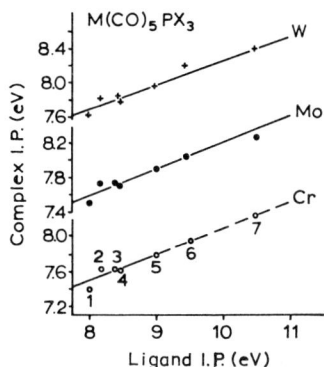

Fig. 11.12. Plots of complex I.P. *vs.* ligand I.P. for hexacoordinated complexes. Numbering of the ligands is as follows: 1, P(C$_4$H$_9$)$_3$; 2, P(C$_2$H$_5$)$_3$; 3, P(OC$_2$H$_5$)$_3$; 4, P(OC$_4$H$_9$)$_3$; 5, P(OCH$_3$)$_3$; 6, PCl$_2$C$_6$H$_5$; 7, PCl$_3$. (Reproduced with permission from ref. 245.)

potentials could be correlated with their $\sigma$-donor ability and polarisability but the $\pi$-acceptor ability was much less important. As the ligand donor ability was increased (I.P. decreased), the ionisation potential of the complex was found to be lowered (Table 11.9). The increased donor capacity of the ligands was not balanced by an increased $\pi$-acceptor capacity and as a consequence the charge density at the central metal atom was increased; this in turn would result in a lowering of the energy required to remove an electron from that point in the molecule. The results indicated, therefore, that

Table 11.9

Ionisation Potentials of Phosphine Substituted Metal Carbonyl Complexes[245]

| Ligand | Ligand I.P. (eV) | I.P. metal complex (eV) | | | | |
|---|---|---|---|---|---|---|
| | | Cr(CO)$_5$L | Mo(CO)$_5$L | W(CO)$_5$L | Fe(CO)$_4$L | Co(CO)$_2$NOL |
| P(C$_4$H$_9$)$_3$ | 8.00 | 7.37 | 7.51 | 7.63 | 7.29 | 7.51 |
| P(C$_2$H$_5$)$_3$ | 8.11 | 7.63 | 7.73 | 7.82 | | 7.62 |
| P(OC$_2$H$_5$)$_3$ | 8.40 | 7.62 | 7.72 | 7.80 | 7.43 | 7.82 |
| P(OC$_4$H$_9$)$_3$ | 8.44 | 7.63 | 7.71 | 7.85 | | |
| P(iso-OC$_3$H$_7$)$_3$ | 8.46 | | | | | 7.64 |
| P(OCH$_3$)$_3$ | 9.00 | 7.80 | 7.89 | 7.96 | 7.65 | 7.92 |
| PCl$_2$C$_6$H$_5$ | 9.45 | | 8.03 | 8.20 | | |
| PCl$_3$ | 10.50 | | 8.25 | 8.50 | 8.05 | 8.40 |
| CO | 14.01 | 8.18 | 8.30 | 8.46 | 8.14 | 8.75 |

the electron removed on ionisation originated from an orbital with high metal character. From Table 11.9 one can also see that the ionisation potentials of the complexes $M(CO)_5L$ (M = Cr, Mo, or W) increased in the order Cr < Mo < W; a similar trend had been previously noted in the metal hexacarbonyls although its authenticity has been questioned (Section 4.1).

The slopes of all the plots incorporating mono-substituted complexes were very similar. However, different kinds of ligands, including CO, provide values which lie well removed from the phosphine plots. The plot of complex I.P. against ligand I.P. for the isonitrile complexes $Mo(CO)_5CNR$ gave a line whose slope was lower and reversed in sign with respect to that for the corresponding phosphine complexes[245]. However, the ionisation potentials of the isonitrile ligands cannot be related to their donor capacity since the electron removed on ionisation was most likely from the CN $\pi$-system and not from the $\sigma$-donor orbital on the isonitrilic carbon. This illustrates the important point that the correlations previously discussed are only valid when the ligand ionisation potential is a measure of the energy required to remove an electron from the ligand lone pair involved in the $\sigma$-bond with the central metal. As a result of this work, it appears that, provided that the nature of the orbital involved in the ionisation process of the complex is unchanged along a given series as in a series containing structurally related ligands with the same donor atom such as the $PX_3$ compounds, the ionisation potentials of the complexes are linearly related to the ionisation potentials of the corresponding ligands; other ligands such as CO, CNR, etc. provide values which do not fit this relationship.

Innorta et al.[248] applied the semi-empirical equivalent orbital method to the calculation of the first ionisation potentials of a number of metal complexes such as $M(CO)_{(6-x)}L_x$ and $M(CO)_x(NO)_yL_{(n-x)}$ (M = Cr, Mo, W, Fe, or Co; L = phosphine or isonitrilic ligands). Good agreement was obtained with experimental values, illustrating the value of the method in predicting these ionisation potentials.

Mass spectra of two classes of metal carbonyl derivatives containing bridging phosphorus units have been studied[249-252], LXIII and LXIV; a metal–metal bond may or may not be present in LXIV.

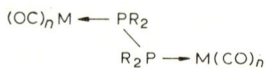

LXIII

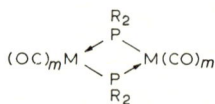

LXIV

In all cases, the parent molecular ion was observed; progressive loss of the CO groups then occurred before appreciable fragmentation of the $M_2P_2R_4$ nucleus began. A number of metastable transitions confirmed this breakdown pattern. The spectra of compounds of type LXIII were characterised by highly abundant dimeric ions of the type $[M_2(CO)_{2n}P_2R_4]^+$, LXV, (M = Cr, Mo, or W, $n = 0$–5; M = Fe, $n = 0$–4) and also the corresponding doubly charged ions but, in addition, the monomeric ions $[M(CO)_nP_2R_4]^+$, LXVI, and $[M(CO)_nPR_2]^+$, LXVII, were also present. As with similar compounds, the abundance of the doubly charged ions increased in the order Cr $\ll$ Mo $<$ W.

The spectra of compounds of structure LXIV were again dominated by the appearance of dimeric ions $[M_2(CO)_{2m}P_2R_4]^+$ (M = Cr, Mo, W, or Mn, $m = 0$–4; M = Fe, $m = 0$–3). Significantly, however, monomeric ions of type LXVI or LXVII were not observed. The behaviour of the pentafluorophenylphosphine compound LXIV (M = Fe, $m = 3$, R = $C_6F_5$) was similar except that neutral iron fluoride species were expelled; this is discussed in more detail in Section 13.

In the mass spectra of the halogen derivatives believed to have structure LXVIII, the parent molecular ion was not observed for any of the compounds studied and, in fact, only in the spectrum of the iodo derivative were any appreciable amounts of halogen-containing fragments observed.

A brief report of the mass spectrum of the compound LXIX has been published[253].

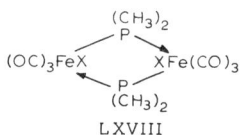

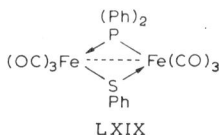

16. METAL CARBONYL THIOLS

A number of metal carbonyl derivatives containing bridging organo-sulphur ligands has been studied[89,180,254–259] and successive loss of CO groups appeared to be the common fragmentation path resulting in ions of general formula $M_m(SR)_n^+$. Rupture of R–S bonds occurred next, resulting in ions of type $M_mS_n^+$, even when dithiols were involved and two sulphur atoms were joined by a single organic group such as $-CH_2-CH_2-$, $-CH=CH$,

CH$_3$ , etc. The ion $M_2S_2^+$, for example, then decomposed in one of two ways:

$$M_2S_2^+ \rightarrow MS_2^+ + M \tag{11.99}$$

or

$$M_2S_2^+ \rightarrow M^+ + MS_2 \tag{11.100}$$

In some cases, for example when R was an ethyl, iso-propyl or $n$-butyl group, an olefin fragment $C_nH_{2n}$ was lost from the $M_2(SR)_2^+$ ion, presumably forming species with either metal–hydrogen or sulphur–hydrogen bonds (refs. 89, 115, 260).

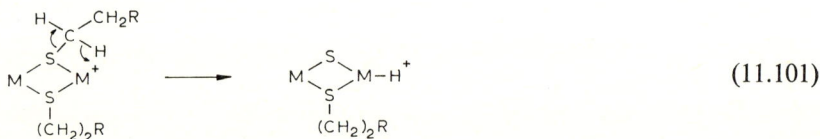

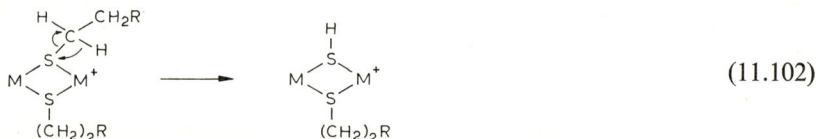

$$\tag{11.101}$$

$$\tag{11.102}$$

Since no ions of the type $MH^+$ were reported, reaction (11.102) appears to represent the more likely mechanism. This decomposition step could not occur when R was a methyl or phenyl group. Thus fragmentation of the iron compounds $[Fe(CO)_3SR]_2$ may be represented by the scheme[89]

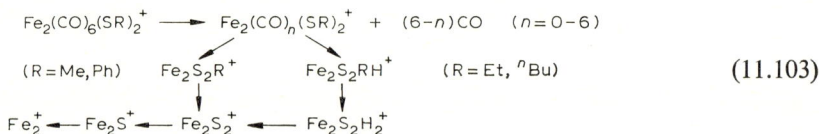

$$\tag{11.103}$$

Metastable transitions observed[258] in the mass spectrum of $Fe_2(CO)_6$-$(SC_6F_5)_2$ corresponded to the decompositions

$$Fe_2S_2(C_6F_5)_2^+ \xrightarrow[*]{-FeF_2} FeS_2C_{12}F_8^+ \xrightarrow[*]{-FeF_2} S_2C_{12}F_6^+ \qquad (11.104)$$

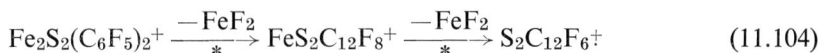

King[180] has assumed that the $(C_5H_5)_2Fe^+$ ion in the mass spectrum of $[(CH_3SFe(CO)C_5H_5]_2$, LXX, arose from ferrocene formed by pyrolysis within the instrument. However, rearrangements of the cyclopentadienyl group have been shown to occur in a number of similar binuclear complexes to give ions of this type (Section 8.2), so this possibility must be considered here also.

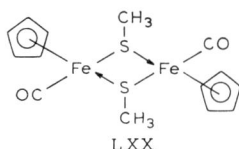

LXX

Stepwise loss of carbonyl groups also occurs in compounds containing terminal sulphur-bonded ligands[180] and in the $Fe_3S_2$ cluster compound $Fe_3(CO)_9S_2$ where all possible ions of the type $Fe_3(CO)_nS_2^+$ ($n = 0$–9)were readily observed[261]. Cleavage of the sulphur–carbon bond does not appear to occur readily where the metal is bonded to a terminal sulphur ligand. Instead, the ion $C_5H_5FeSCH_3^+$ arising from the compound $C_5H_5Fe(CO)_2$-$SCH_3$ was postulated[180] to undergo dehydrogenation to the ion $C_6H_6FeS^+$ and $H_2S$ elimination to form $C_6H_6Fe^+$ while the concentration of $C_5H_5FeS^+$ was negligible. Competitive loss of CO and $CF_3S$ ligands occurred during fragmentation of compounds containing terminal S ligands[180,262].

A number of rearrangements have been noted in the electron-impact induced fragmentation of the complexes

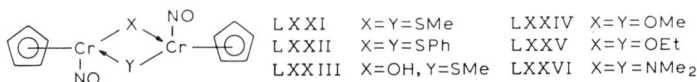

| | |
|---|---|
| LXXI  X=Y=SMe | LXXIV  X=Y=OMe |
| LXXII  X=Y=SPh | LXXV  X=Y=OEt |
| LXXIII  X=OH, Y=SMe | LXXVI  X=Y=NMe₂ |

and some novel structures postulated to account for the resultant ions[263]. The effect of the bridging atom was related to the relative strengths of the ligand–metal and metal–nitrosyl bonds.

Rearrangement has been observed[264] to occur during fragmentation of thenyl derivatives so that ions containing metal–sulphur bonds are formed.

## 17. MISCELLANEOUS MIXED LIGAND COMPLEXES

An extremely stable metal cluster was indicated[265] in the carbido-complex $Ru_6C(CO)_{17}$, LXXVII. All possible singly and doubly charged ions formed by loss of carbonyl groups were observed: $Ru_6C(CO)_x^{n+}$ ($x = 0$–17; $n = 1$, 2). The ion $Ru_6C^+$ was observed in low abundance but no ions corresponding to fragmentation of the $Ru_6C$ cluster were present. Fragmentation of the compounds $Ru_6C(CO)_{14}$(arene) was very similar except for additional complicated degradations associated with the arene ligands.

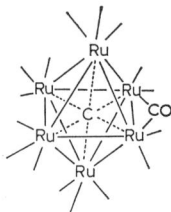

LXXVII

Some unusual features were provided by a series of methinyltricobalt enneacarbonyls, $Co_3(CO)_9CY$ (Y = H, F, Cl, Br, $CH_3$, $CF_3$, or $C_6H_5$) in two independent but almost identical studies[266]. Rupture of the $Co_3C$ nucleus did not occur until all the carbonyl groups had been lost in a stepwise fashion and then preferential rupture of Co–Co rather than Co–C bonds occurred. Singly charged trinuclear ions accounted for 73% of the metal-containing fragments. The C–Y bond was not fragmented until all the carbonyls had been lost, so that ions of the type $Co_3(CO)_nC^+$ were not observed. The principal decay scheme, established by metastable transitions, was

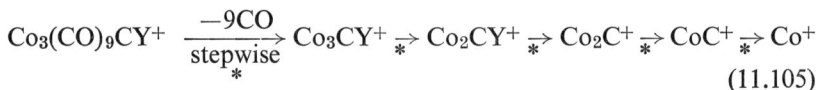

$$Co_3(CO)_9CY^+ \xrightarrow[\text{stepwise}]{-9CO} Co_3CY^+ \xrightarrow{*} Co_2CY^+ \xrightarrow{*} Co_2C^+ \xrightarrow{*} CoC^+ \xrightarrow{*} Co^+ \tag{11.105}$$

Thus an analogy may be drawn with the most stable carbonyl cluster compounds such as $Os_3(CO)_{12}$. As noted by Robinson and Tham[266], the corresponding compounds of rhodium and iridium should be of interest because of the apparent increase in metal–metal bond stability with increasing atomic number.

Fragmentation patterns of a number of pentacarbonylchromium carbene complexes $(CO)_5CrC(X)Y$ (X = OR, NHR, $NR_2$, SR; Y = $CH_3$, $C_6H_5$)

were discussed[207]. Ionisation potentials were determined; they (*i*) increased linearly with increasing values of the Hammett constant of R in

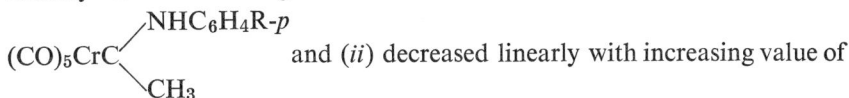

$(CO)_5CrC\begin{smallmatrix} \nearrow NHC_6H_4R\text{-}p \\ \searrow CH_3 \end{smallmatrix}$ and (*ii*) decreased linearly with increasing value of

the ratio of the intensity of the $CrC(X)CH_3^+$ ion to the sum of the intensities of the $[(CO)_nCrC(X)CH_3]^+$ ions ($n = 1\text{–}5$) in the compounds $(CO)_5CrC(X)\text{-}CH_3$.

The main features of the mass spectra of the isonitrile complexes $M(CO)_5\text{-}CNR$ (M = Mo or W, R = $C_6H_5$; M = Mo, R = $C_4H_9$)[234] and $C_5H_5Mn\text{-}(CO)_2L$ (L = CNH or $CNC_6H_{11}$)[184] have been reported. Loss of both carbonyl groups from the parent molecular ion of the latter compound must have occurred readily since the ions $C_5H_5MnL^+$ were prominent while $C_5H_5Mn(CO)L^+$ were not observed. Of particular interest was the rearrangement ion $C_5H_5MnCNH^+$ which was responsible for the base peak in the spectrum of $C_5H_5Mn(CO)_2CNC_6H_{11}$. The ion $C_5H_5MnCNC_6H_{11}^+$ was postulated to decompose by the following paths:

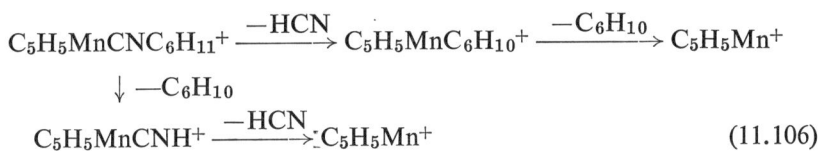

$$C_5H_5MnCNC_6H_{11}^+ \xrightarrow{-HCN} C_5H_5MnC_6H_{10}^+ \xrightarrow{-C_6H_{10}} C_5H_5Mn^+$$

$$\downarrow -C_6H_{10}$$

$$C_5H_5MnCNH^+ \xrightarrow{-HCN} C_5H_5Mn^+ \qquad (11.106)$$

The complex LXXVIII did not exhibit a parent molecular ion, that of highest $m/e$ being $NiI_2C_6H_4$ (ref. 267). A large number of rearrangement ions were prominent, namely $C_6H_5I_2^+$, $C_6H_5CO^+$, $C_6H_5^+$, and $I_2^+$.

$[C_6H_4Ni(CO)I]_2$

LXXVIII

## 18. COORDINATION COMPLEXES

The compounds discussed in this section are those with $\sigma$-donor bonds between ligands containing elements from either main Group V or main Group VI, the normal coordination complexes. In general, their volatility is low but with recent advances in instrumentation it has been possible to record the mass spectra of many of them. However, precautions in interpretation of the spectra are often necessary. High source temperatures must usually be employed because of the low volatility so that thermal decomposition of the sample is always a possibility. Frazer et al.[269] studied the influences of source temperature on the mass spectra of a number of metal chelate compounds and found that significantly different spectra were obtained for the same compound at different temperatures, and ions at $m/e$ values above that of the parent molecular ion often appeared at the higher temperatures. Other examples of ions with high mass numbers have also been reported (refs. 270, 271). Frazer et al. suggested, therefore, that when mass spectra are obtained with a direct insertion probe, studies should be made over a range of source temperatures and that the results, if used for molecular weight measurements, should be interpreted with caution. The presence of an uncomplexed potential donor atom in a ligand may lead to rearrangement reactions occurring during fragmentation which could result in an incorrect conclusion concerning the structure of the compound.

Mass spectrometry has recently been shown[271–274] to be a convenient technique for the detection and estimation of metals in the form of chelates in amounts down to $10^{-12}$ g. The procedure employed has been extended to enable the detection of isomers in a sample mixture[319].

### 18.1 Complexes having ligands with main Group V donor atoms

The outstanding feature of the mass spectral behaviour of macrocyclic ligands with nitrogen donor atoms is the stability of the macrocyclic ring and its resistance to fragmentation. Hill and Reed[275] studied copper phthalocyanine, LXXIX, and found that at normal operating temperatures only ions of $m/e$ 128 and below were obtained. As the temperature was raised, the parent molecular ion was observed but on further increasing the temperature only ions of $m/e$ 128 and below were again recorded. The singly charged parent molecular ion was by far the most abundant in the spectrum, while the corresponding doubly charged ion was next in intensity, demonstrating the unusual stability of the molecule. Mass spectra of copper complexes of a number of chlorophthalocyanines have also been reported and again illustrate this stability. The spectrum of tetrachlorocopper phthalocyanine[276] shows the following major ions to be present: $P^{n+}$, $(P-\cdot Cl)^{n+}$, $(P-HCl)^{n+}$, $(P-\cdot H_2Cl)^{n+}$, $(P-Cl_2)^{n+}$, $(P-\cdot HCl_2)^{n+}$, $(P-\cdot Cl_3)^{n+}$, and $(P-Cl_4)^{n+}$ where $n = 1$ or 2 and P represents the parent molecular ion. In addition, peaks were observed due to quarters of the outer skeleton of the molecule plus the copper atom, LXXX. The spectra of other chloro derivatives have also been reported; this includes an electron attachment study[277].

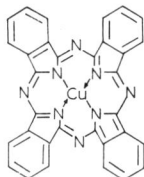

LXXIX

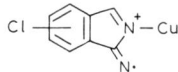

LXXX

Cobalt monomethylphthalocyanine was decomposed homolytically in the mass spectrometer with the formation of methyl radicals which then attacked the bridging nitrogen atoms of the phthalocyanine ring, so that ions of type $[CoPc + n \cdot CH_3]^+$ ($n = 2$–5, Pc = phthalocyanine ligand system) were observed in the mass spectrum in addition[278] to the parent molecular ion $[CoCH_3Pc]^+$ and the ions $(CoPc)^+$ and $[CoPc]^{2+}$. Phthalocyanine-oxo-molybdenum(IV), on the other hand, merely gave rise[279] to the ions $(MoOPc)^{n+}$, where $n = 1$ or 2.

The mass spectrum of ferric mesoporphyrin IX chloride dimethyl ester[280], LXXXI, and nickel and copper etioporphyrins[281] were similar. The relatively high stability of the singly and doubly charged parent molecular ions gave rise to little fragmentation of the porphyrin nucleus except for cleavage of $\cdot Cl$ and the esters at the $\beta$ position in the former compound. A cationic palladium(II) complex, LXXXII, of the similar corphin (corrinoid–porphinoid) ligand system has recently been reported[282]. Its mass spectrum showed a prominent parent molecular ion but again the only fragment ions reported were (parent molecule $- H\cdot$)$^+$ and (parent molecule $- H\cdot - n \cdot CH_3$)$^+$ where $n = 1$–4.

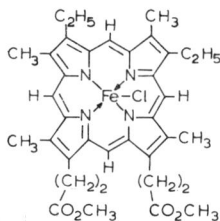

LXXXI

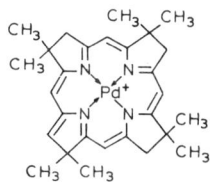

LXXXII

References pp. 600–610

Octahedral cobalt(III) complexes of type LXXXIII underwent thermal reactions in the inlet system of the mass spectrometer[283] yielding information concerning the identity of the ligands R and R'. Similarly, thermal reactions in complexes of type LXXXIV produced HX so that the anion X⁻ was easily identified. Thermal cyclisation of open ligands also took place. The fragmentation of a number of complexes with other macrocyclic ligand systems was also investigated and the mechanisms involved discussed. The stability of the ligand system was again evident; compound LXXXV, for example, fragmented by the loss of methyl groups but the macrocyclic ring

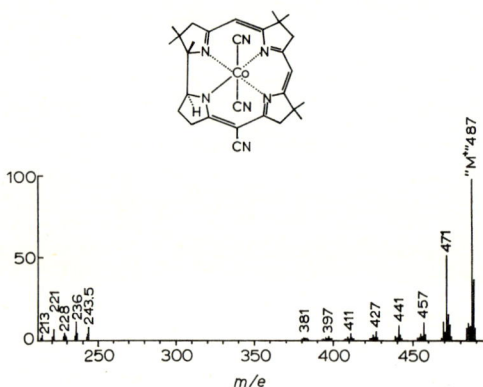

Fig. 11.13. Mass spectrum of an octahedral cobalt(III) macrocyclic ligand complex. (Reproduced with permission from ref. 283.)

remained intact (Fig. 11.13). A number of doubly charged ions were also prominent.

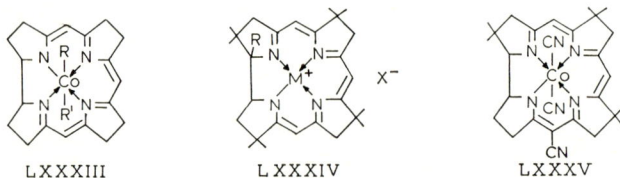

LXXXIII                LXXXIV                LXXXV

Taylor et al.[284] found that the alkoxide groups in the dialkoxide derivatives of tetraazacyclohexadecinenickel(II), LXXXVI, were readily cleaved so that the parent molecular ion and (parent molecule − ·OR)⁺ ions were of low abundance whereas (parent molecule $-$ $2 \cdot OR)^+$ was the most intense in the spectrum. However, the authors suggested that the conditions necessary to obtain the mass spectra were responsible for the cleavage of the alkoxide groups.

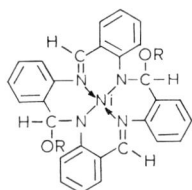

LXXXVI

The mass spectrum of a reaction product of bis(acetylacetonato)nickel(II) indicated that the compound had a different structure than had been previously assigned to it[285]. Mass spectrometry was also utilised in the determination[286] of the structure of the complexes $\pi$-$C_5H_5CoN_4R_2$ (R = $CH_3$, $C_6H_5$). The phenyl derivative showed weak peaks at $m/e$ values higher than that of the parent molecular ion, and these were assigned to (parent molecule $+ C_6H_4)^+$ and (parent molecule $+ C_6H_4 - N_2)^+$.

Numerous metal carbonyl complexes with phosphorus donor ligands have been investigated mass spectrometrically and have been discussed in Section 15. However, the mass spectra of very few other phosphorus–ligand complexes have been reported. The behaviour of $Ni(PF_3)_4$ under electron impact has also been discussed (Section 14). The fragmentation patterns of $Fe(PF_3)_5$ (ref. 287), $Fe(PF_3)_4PF_2OC_2H_5$ (ref. 288), $Ni(PF_3)_3PF_2OC_2H_5$ (ref. 288), and $HCo(PF_3)_3PH_3$ (ref. 289) indicate that P–X bond rupture occurred readily in addition to the elimination of the complete $PX_3$ ligand molecules. $Pd(PPh_3)_4$ gave only the spectrum of triphenylphosphine[290].

*18.2 Complexes having ligands with main Group VI donor atoms*

*18.2.1 β-diketone complexes*

The mass spectrum of tris(acetylacetonato)chromium(III) was first reported by McLafferty[291] in 1957. Several years elapsed before further mass spectra of compounds of this type were reported but the results of several independent and detailed studies have been published in the last few years. Complexes of acetylacetone and its derivatives have received the most attention.

Macdonald and Shannon[292] studied the mass spectra of acetylacetonato complexes of a large number of di- and trivalent elements The spectrum of bis(acetylacetonato)nickel(II) is shown in Fig. 11.14. They found that the most intense ions in the spectra were usually derived from the monomeric forms but in a number of cases, including tris(acetylacetonato) complexes of trivalent metals, they observed ions of low intensity derived from dimers

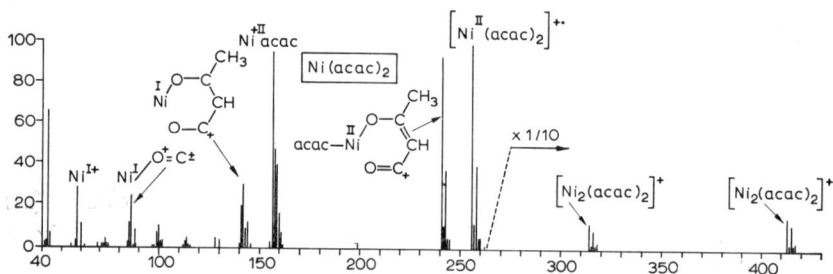

Fig. 11.14. Mass spectrum of bis(acetylacetonato)nickel(II). (Reproduced with permission from ref. 292.)

and trimers. Peaks due to the loss of even-electron neutral fragments from the parent molecular ion were small in comparison with those arising from loss of odd-electron neutral fragments. In addition, odd-electron ions of general structure $M(acac)_n^+$ typically lost the odd-electron fragments $\cdot CH_3$ when $n = 1$ or 2 or $acac\cdot$ when $n = 3$ or 4, while even-electron ions of this structure typically lost an even-electron fragment of mass 82, viz.

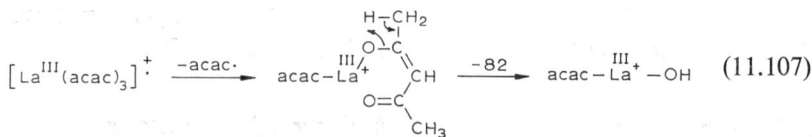

Shannon and Swan[293] suggested that odd-electron ions would be changed to even-electron ions, and vice versa, by a change of oxidation state of the metal atom in the ions. Thus in the spectrum of $Co(acac)_3$, for example, ions of structure $Co(acac)_n^+$ were assumed mainly to dissociate as odd-electron ions even when $n = 1$ or 2 because of changes in metal oxidation state.

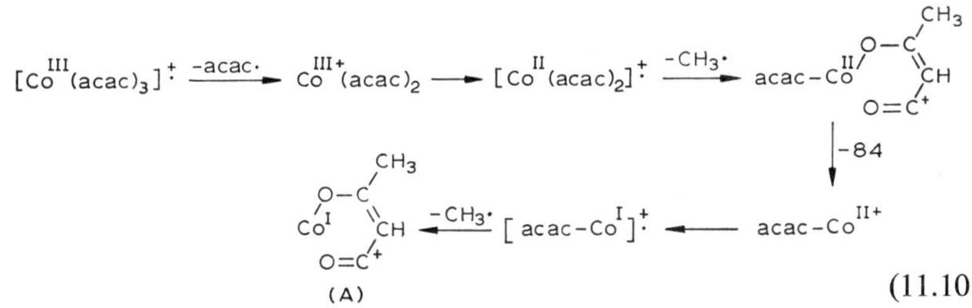

$$(11.108)$$

Two fragmentation schemes were proposed and these accounted for most peaks in the spectra of the complexes examined. Figure 11.15 shows the sequence of reactions which occurred when there was no change of metal oxidation state while the scheme in Fig. 11.16 applied to complexes in which a decrease of metal oxidation state occurred. Once a ligand group started to fragment it usually continued to do so until all of it, or all except –OH or =O, was lost before fragmentation of another ligand group commenced. Rearrangement ions such as $(M–CH_3)^+$ were also noted. Bancroft *et al.*[294]

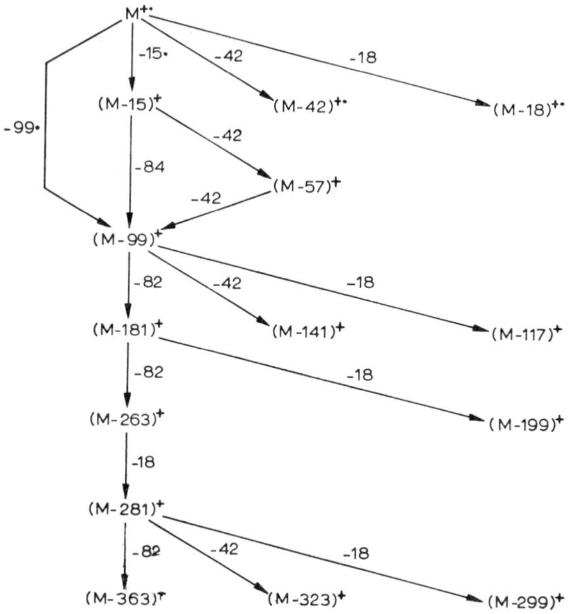

Fig. 11.15. Metal acetylacetonato reaction paths in which not more than one odd-electron neutral fragment is lost. (Reproduced with permission from ref. 292).

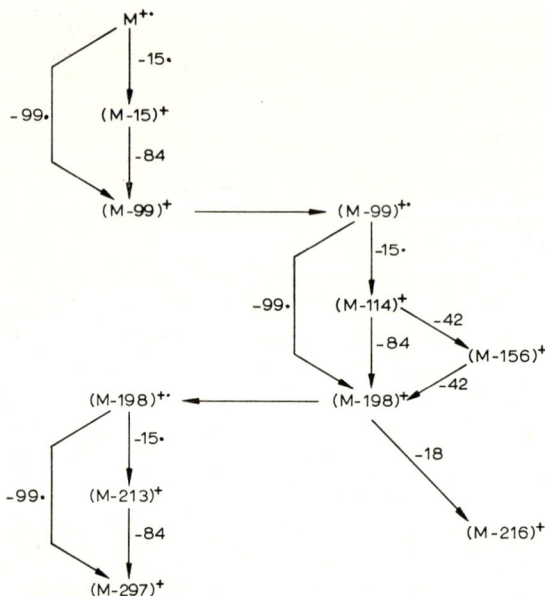

Fig. 11.16. Metal acetylacetonato reaction paths on which more than one odd-electron neutral fragments are lost. (Reproduced with permission from ref. 292.)

agreed that the most important factor in the mass spectra is the tendency towards formation of even-electron ions accompaned, when necessary, by oxidation state changes in the metal. Mechanisms they proposed to account for the important peaks in most of the spectra are shown in Fig. 11.17. They also considered that this electron transfer from one of the ligand orbitals to a metal orbital was due to changes in electron distribution caused by differences in electron-withdrawing power between the metals and the ligands.

From a study of metastable decompositions occurring in both of the field-free regions of a double-focusing mass spectrometer, Koob et al.[295] have suggested that the general decomposition scheme for metal acetylacetonates involves delayed consecutive dissociations. Thus fragmentation of tris(hexafluoroacetylacetonato)aluminium(III) was observed to proceed via the decompositions

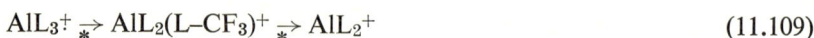

$$AlL_3^+ \underset{*}{\rightarrow} AlL_2(L-CF_3)^+ \underset{*}{\rightarrow} AlL_2^+ \qquad (11.109)$$

Macdonald and Shannon[292] further postulated, on the basis of metastable

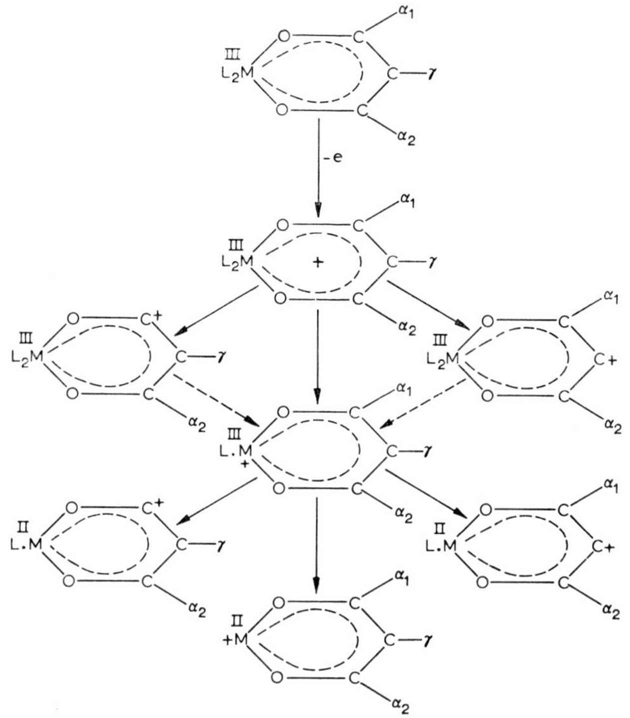

Fig. 11.17. Postulated partial fragmentation scheme for acetylacetonatometal complexes. (Reproduced with permission from ref. 294.)

transition data, that decomposition of the $[Co + 99]^+$ ion in the mass spectra of Co(acac)$_3$ and Co(acac)$_2$ occurred through the initial loss of a methyl radical followed by sequential losses of carbon monoxide, ethylene, and carbon monoxide, ultimately yielding the $M^+$ ion, *viz.*

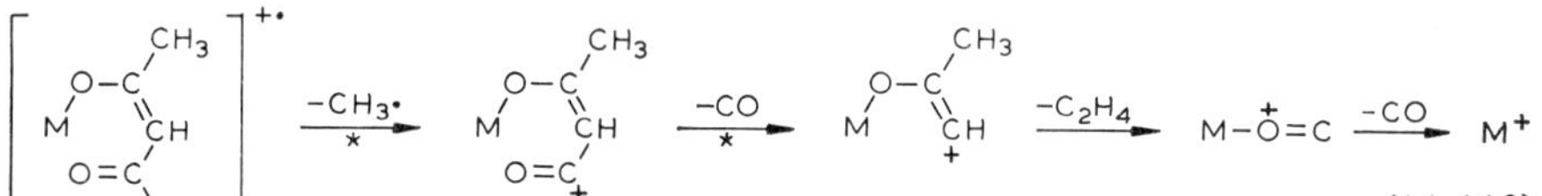

$$(11.110)$$

The same reactions may occur during decomposition of Ni(acac)$_2$. Using precise mass measurements and high-resolution mass spectrometry, Fraas *et al.*[296] revealed that the $[M + 28]^+$ ion consisted of both $[MCO]^+$ and

$[MC_2H_4]^+$ ions. Thus the mechanism postulated by Macdonald and Shannon should be extended to

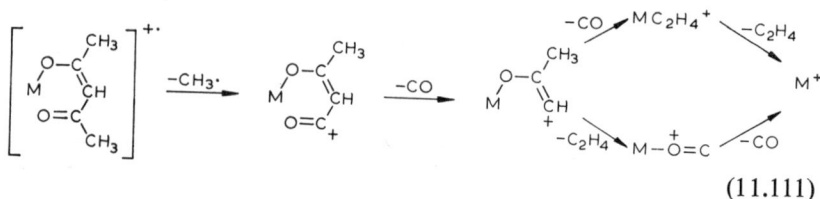

$$(11.111)$$

Bancroft et al.[297] obtained spectra which were essentially in agreement with those of Macdonald and Shannon. Whereas the latter workers were unable to obtain a spectrum of Mn(acac)₃, however, Bancroft et al. were successful. They also reported the spectra of the tris(acetylacetonato) complexes of titanium and vanadium which had not appeared previously. The method of introduction of the samples into the ion source proved to be of considerable importance for the compounds M(acac)₃ where M = Mn, Fe, or Co. Vaporisation via the heated inlet system led to a much less abundant parent molecular ion and a lower appearance potential for M(acac)₂⁺ than that obtained when the compound was vapourised directly into the ionisation chamber. Thus decomposition to the bis(acetylacetonato) complexes apparently occurred. Variation of the chamber temperature between 150° and 250 °C, however, had very little effect on the spectra of complexes introduced by direct insertion. This behaviour probably explains the failure of Sasaki et al.[298] to observe the parent molecular ion from Co(acac)₃ and from Fe(acac)₃. However, the reason for this non-detection of Co(acac)₃⁺ is uncertain since Bancroft et al.[297] found that ion intensities and appearance potential values did not vary significantly with either mode of introduction or temperature variation of the ionisation chamber between 150° and 250°C.

As mentioned previously, Macdonald and Shannon[292] postulated oxidation state changes in the metal to account for the observed mass spectra and correlated the occurrence or non-occurrence of oxidation state changes with the established chemistry of the metal. Although Bancroft et al.[297] considered this an oversimplification, they noted that peaks were present in the mass spectra of Ti(acac)₃ and V(acac)₃ which were believed to be characteristic of the +4 oxidation state of the metal. In addition, fragmentation proceeded with the formation of oxy or hydroxy species which were absent in the spectra of other chelates, in accord with the established chemistry of the elements. Peaks believed to be characteristic of the +2 oxidation state of Mn,

Fe, and Co were observed in the spectra of the corresponding tris(acetyl-acetonato) complexes of the $+3$ oxidation state metals, while peaks charac-teristic of the Cr(IV) oxidation state were absent and those characteristic of Cr(II) were small, apparently because $Cr^{3+}$ is much more stable than $Cr^{2+}$ or $Cr^{4+}$. From a comparison of mass spectra of acetylacetonates of first row transition elements, Bancroft et al.[297] suggested that the intensity of the [parent molecule $- (\cdot acac) - (\cdot CH_3)]^+$ and [parent molecule $- 2(\cdot acac)]^+$ ions paralleled the ease with which the $+3$ state of the metals was reduced.

A number of mass spectral studies of these and similar complexes (see, for example, ref. 299) provided additional indirect evidence of metal oxida-tion state changes in the fragmentation processes but Clobes et al.[300] have recently presented direct evidence that such a reduction in oxidation state was accompanying the elimination of an odd-electron fragment from an even-electron ion. They recorded the mass spectra of tris(1,1,1,5,5,5-hexafluoro-2,4-pentanedionato) or hexafluoroacetylacetonato (hfac) complexes of Cr(III), Fe(III), and Co(III) and were able to detect and assign a large num-ber of metastable transitions which enabled the principal fragmentation

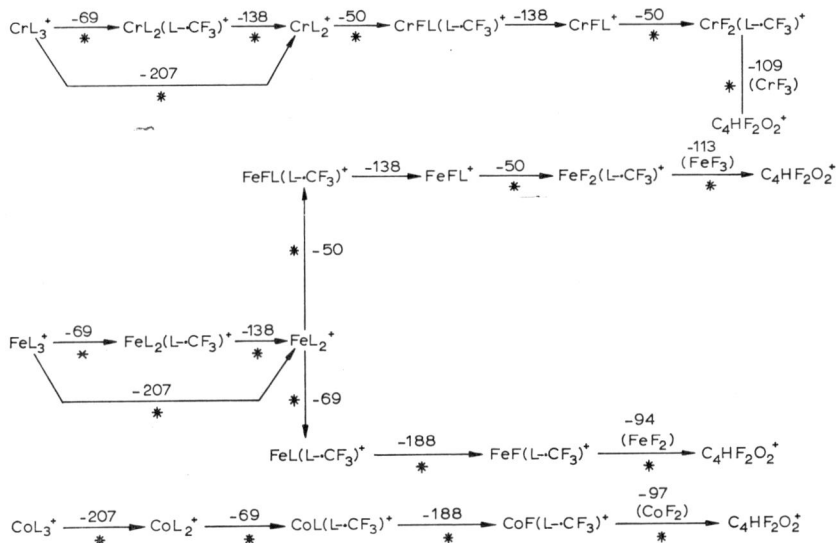

Fig. 11.18. Decomposition paths of tris(hexafluoroacetylacetonato)metal complexes leading to the elimination of neutral metal fluorides. (Reproduced with permission from ref. 300.)

paths to be determined (Fig. 11.18). Of particular interest are the metastable supported eliminations of neutral metal fluoride species. Thus the iron complex was shown to decompose *via* two fragmentation paths, one of which involves the elimination of $FeF_3$ while in the other $FeF_2$ is lost. The reaction sequence of the chromium complex incorporates loss of the trivalent fluoride $CrF_3$ only, while that of cobalt illustrates reduction of the metal since $CoF_2$ is eliminated. Clobes *et al.* point out the correlation between the relative stability of the (III)/(II) oxidation states of the metals involved and the stoichiometry of the eliminated neutral metal fluorides.

In an independent study, Sasaki *et al.*[298] examined the behaviour of fifteen metal acetylacetonates under electron impact, as a result of which they divided the compounds into three groups according to their stability under the conditions of the experiment. They decided that the smaller the electronegativity of the metal the more stable the complex, while the introduction of electron-attracting groups into the ligand also enhanced the stability. However, the members of one group comprise the tris(acetylacetonates) of chromium, iron, and cobalt since parent molecular ions were not observed for these compounds. Bancroft *et al.*[297] later showed, however, that this was due to thermal decomposition in the inlet system and was not connected with stability of the complexes under electron impact.

Holtzclaw *et al.*[301] attempted to establish a relationship between the electronic structure and the mass spectra of the copper(II) chelates of systematically substituted 1,3-diones through consideration of ion intensities and electron-impact ionisation potentials. Stepwise elimination of the alkyl groups around the periphery of the chelate ring structure was noted and as the stability of the radical formed from the alkyl group increased [$\cdot CH_3 < \cdot CH_2CH_3 < \cdot CH(CH_3)_2 < \cdot C(CH_3)_3$] the population of the copper species containing the alkyl group decreased. The postulated fragmentation scheme for the copper-containing species is shown in Fig. 11.19. Ligand cleavage is also possible, leading to fragments associated with the ligand where the ligand carried the charge after fragmentation. However, this pathway is apparently only favoured if the organic species acquires an extra hydrogen from some source giving fragments characteristic of the dione itself.

Shannon and co-workers extended their studies to complexes of dibenzoylmethane[302] and found that most reactions were completely analogous to those of the metal acetylacetonates. However, reactions attributed to unit decrease in the oxidation state of the metal atom in the ion were less common

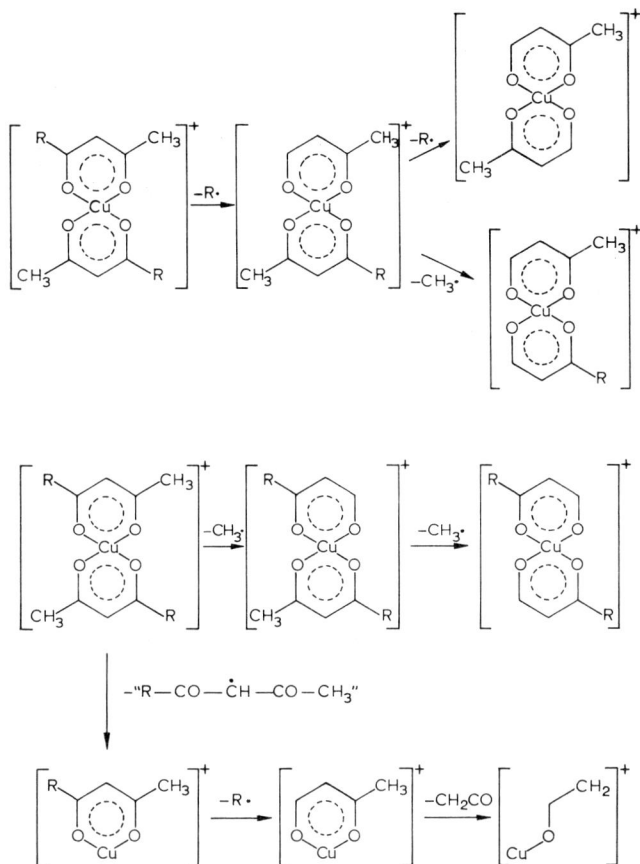

Fig. 11.19. Proposed fragmentation scheme of substituted acetylacetonato complexes of copper(II) (ref. 301).

in the spectra of the dibenzoylmethane complexes. Some even-electron ions such as, for example, Al(III)$^+$(dbm)$_2$, Cr(III)$^+$(dbm)$_2$, and C$_6$H$_5$Al(III)$^+$dbm underwent a phenyl migration reaction in which the whole of an intact dbm ligand, except for one phenyl group, was lost. Reactions of the parent molecular ions involving either hydrogen loss or transfer, or both, were common.

Complexes of the analogous ligand 2,2,6,6-tetramethylheptanedione with all of the rare earths except Ce and Pm and with Al, Co, Sc, Y, and Zr have also been examined mass spectrometrically[303].

Johnson *et al.*[304] concluded that mass spectrometry served to distinguish between the O-bonded $\beta$-diketone complexes, LXXXVII, and the C-bonded $\beta$-diketone derivatives, LXXXVIII, since the former fragmented by loss of the radical R whereas the latter lost acyl fragments, RCO$\cdot$. However, a large number of the O-bonded complexes do not exhibit ions formed by loss of R$\cdot$ in their mass spectra, so the absence of such an ion cannot be taken as diagnostic of a structure of type LXXXVIII. On the other hand, insufficient $\beta$-diketonyl complexes LXXXVIII have been investigated to determine if the (parent molecule–$\cdot$RCO)$^+$ ion is a general characteristic of their spectra. Only in favourable cases, therefore, is it possible for mass spectra to differentiate between LXXVII and LXXXVIII.

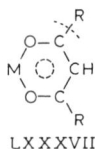

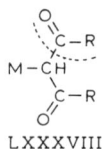

LXXXVII                 LXXXVIII

The mass spectra of metal complexes of both trifluoroacetylacetone, tfac (1,1,1-trifluoro-2,4-pentanedione, $CF_3$–CO–$CH_2$–CO–$CH_3$) and hexafluoroacetylacetone, hfac (1,1,1,5,5,5-hexafluoro-2,4-pentanedione, $CF_3$–CO–$CH_2$–CO–$CF_3$) are, in general, similar to those of acetylacetone although more extensive fragmentation of the fluorinated complexes occurred. One notable difference, however, was the loss of $CF_2$ fragments with rearrangement of the remaining fluorine to the metal during fragmentation of both tfac[299] and hfac[299,300] complexes. The greater fragmentation of the fluorinated chelates has been attributed[305] to (*a*) the relative instability of the parent molecular ion due, for example, to weak C–$CF_3$ bonds in it, (*b*) the relative stability of fragment ions in which rearrangement allows formation of strong metal–fluorine bonds, and (*c*) the destabilisation of a C–$CH_3$ or C–$CF_3$ bond owing to the increase of positive charge on the carbon atom *via* the electron-withdrawing power of the $CF_3$ substituents. Loss of one fluorine from the parent molecular ions of Zr(hfac)$_4$ and Hf(hfac)$_4$ has been reported and the observation of the ion Hf(hfac)$_3$F$^+$ in the mass spectrum of the latter compound noted[306]. Surprisingly, it was found possible to obtain[307] the mass spectrum of the salt Cs[Y(hfac)$_4$]. In addition to the parent molecular ion, the ion Cs[Y(hfac)$_3$]$^+$ was observed and it was postulated that the caesium ion is held so strongly to the $\beta$-diketonate complex anion by means of a strong ion pair between $Cs^+$ and the sheath of electronegative fluorine atoms of the chelating ligands surrounding the yttrium atom that loss of a ligand molecule can readily occur without the prior departure of $Cs^+$. A number of metal chelates of 1,1,1,2,2,3,3-heptafluoro-7,7-dimethyl-4,6-octanedione have been studied in the development of a mass spectrometric metal analysis technique[274]. Quantitative analysis of chromium as chromium(III) tris(hexafluoro-acetylacetonate) has also been described[308].

Majer and Perry[309] have noted extensive rearrangements during the evaporation of metal chelates into the ion source of a mass spectrometer. These reactions were demonstrated most simply by the resultant spectrum when simultaneous evaporation of two different metal chelates into the ion source was carried out. This technique was used by Holtzclaw et al.[301] to obtain the spectrum of (acac)Cu(hfac) mentioned previously.

The mass spectra of the acetylacetone complexes of Sr(II), Ca(II), Mg(II), and Na(I) contained intense peaks corresponding to polymeric species[310]. Elimination of a neutral acetone molecule ($CH_3COCH_3$, mass 58) was noted but has apparently not been observed in the fragmentation of other complexes. Polymeric ions containing at least three alkali metal atoms were also noted in other $\beta$-diketone complexes[311].

*Appearance potentials*

Appearance potential measurements of Bancroft et al.[297] were not in accord with Barnum's simple theoretical calculations[312] suggesting that either these calculations are not even qualitatively correct or else the electron is not removed from the highest occupied molecular orbital in all cases. The authors suggested that the nature of the ligand has a far greater effect than the metal on the appearance potentials of the metal chelates. To investigate this further, Bancroft et al.[294] studied a series of acetylacetonato complexes of chromium(III). The results tended to support this suggestion and appeared to be consistent with the ionisation of an electron from the $\pi$ system of the chelate ring. Similar results were obtained[301] on a series of copper(II) complexes (Table 11.10). Comparison of the ionisation potentials of the isomeric compounds $Cu(tfac)_2$ and (acac)Cu(hfac) suggested that the orbital from which the electron was removed was affected equally by both ligands. Further experiments with nitro-substituted acetylacetone complexes of chromium(III) agreed with this conclusion[313]. Schildcrout et al.[314] arrived at a similar conclusion concerning the origin of the electron removed. They found that the ionisation potentials of tris complexes of acetylacetone, tri-

Table 11.10

Ionisation and Appearance Potentials of Ions Derived from Acetylacetonato–Metal Chelates

References are given in brackets.

(a) *Complexes of trivalent metals*

| Complex | I.P. of neutral atom M (eV) | I.P. of complex (eV) | A.P. $ML_2^+$ ions (eV) |
|---------|---------|---------|---------|
| Al(acac)$_3$[a] | 5.98 | 8.27 ± 0.13 (314) | |
| | | 7.95 ± 0.05 (297) | 9.1 ± 0.2 (297) |
| Al(tfac)$_3$ | | 9.05 ± 0.1  (305) | 10.2 ± 0.1 (305) |
| Al(hfac)$_3$ | | 10.30 ± 0.11 (314) | |
| | | 9.80 ± 0.1  (305) | 11.2 ± 0.1 (305) |
| Ti(acac)$_3$ | 6.82 | 7.1  ± 0.1  (297) | 11.8 ± 0.1 (297) |
| V(acac)$_3$ | 6.76 | 7.72 ± 0.10 (314) | |
| | | 7.9  ± 0.1  (297) | 11.8 ± 0.1 (297) |
| Cr(acac)$_3$ | 6.76 | 7.87 ± 0.12 (314) | |
| | | 8.10 ± 0.05 (297) | 11.3 ± 0.1 (297) |
| Cr(tfac)$_3$ | | 9.09 ± 0.05 (305) | 11.9 ± 0.1 (305) |
| Cr(hfac)$_3$ | | 9.97 ± 0.08 (314) | |
| | | 10.13 ± 0.05 (294) | 14.3 ± 0.1 (305) |
| Mn(acac)$_3$ | 7.43 | 7.95 ± 0.10 (314) | |
| | | 7.85 ± 0.05 (297) | ≥8.7 ± 0.1 (297) |
| Fe(acac)$_3$ | 7.87 | 8.64 ± 0.11 (314) | |
| | | 8.45 ± 0.05 (297) | ≥9.4 ± 0.1 (297) |
| Fe(tfac)$_3$ | | 9.38 ± 0.11 (314) | |
| | | 9.10 ± 0.05 (305) | 9.2 ± 0.1[b] (305) |
| Fe(hfac)$_3$ | | 10.34 ± 0.10 (314) | |
| | | 10.2  ± 0.1  (305) | 10.2 ± 0.1[b] (305) |
| Co(acac)$_3$ | 7.86 | 7.81 ± 0.10 (314) | |
| | | 7.80 ± 0.05 (297) | 10.7 ± 0.3 (297) |
| Co(hfac)$_3$ | | 10.12 ± 0.15 (314) | |
| Rh(hfac)$_3$ | 7.46 | 10.15 ± 0.12 (314) | |
| H(acac) | 13.60 | 9.2  ± 0.1 (297) | |
| H(tfac) | | 9.96 ± 0.10 (314) | |
| H(hfac) | | 10.68 ± 0.09 (314) | |
| | | 10.55 ± 0.05 (305) | |

Table 11.10 (continued)

*(b) Complexes of divalent metals*

| Complex | I.P. (eV) | A.P. $(P-\cdot CH_3)^+$ $(eV)^d$ | A.P. $(P-\cdot CF_3)^+$ $(eV)^d$ | A.P. $(P-\cdot L)^+$ $(eV)^d$ | Ref. |
|---|---|---|---|---|---|
| Mn(acac)$_2$ | 8.34 ± 0.05 | 11.7 ± 0.1 | | 13.7 ± 0.1 | 315 |
| Fe(acac)$_2$ | 8.10 ± 0.05 | 11.7 ± 0.1 | | 13.9 ± 0.1 | 315 |
| | 7.50 ± 0.04 | | | | 301 |
| Fe(tfac)$_2$ | 8.75 ± 0.1 | | 12.6 ± 0.1 | 14.5 ± 0.1 | 305 |
| Fe(hfac)$_2$ | 9.7 ± 0.1 | | 13.2 ± 0.2 | | 305 |
| Co(acac)$_2$ | 8.54 ± 0.05 | 11.5 ± 0.1 | | 13.9 ± 0.2 | 315 |
| Ni(acac)$_2$ | 8.23 ± 0.05 | 11.6 ± 0.1 | | 13.5 ± 0.2 | 315 |
| Cu(acac)$_2$ | 8.31 ± 0.05 | 10.9 ± 0.1 | | 13.1 ± 0.2 | 315 |
| | 7.75 ± 0.05 | | | | 301 |
| Cu(tfac)$_2$ | 9.05 ± 0.1 | | 11.5 ± 0.1 | 13.1 ± 0.1 | 305 |
| Cu(hfac)$_2$ | 9.86 ± 0.05 | | 11.6 ± 0.1 | | 305 |
| Zn(acac)$_2$ | 8.62 ± 0.05 | 10.9 ± 0.1 | | 14.1 ± 0.2 | 315 |
| Zn(tfac)$_2$ | 9.40 ± 0.1 | 11.7 ± 0.1 | 11.3 ± 0.1 | 14.6 ± 0.1 | 305 |
| Zn(hfac)$_2$ | 10.07 ± 0.05 | | 11.35 ± 0.1 | 15.3 ± 0.2 | 305 |
| H(acac) | 9.2 ± 0.05 | 10.7 ± 0.1 | | | 315 |
| H(tfac) | 9.8 ± 0.1 | 11.7 ± 0.1 | 10.6 ± 0.2 | | 305 |
| H(hfac) | 10.55 ± 0.05 | | 11.2 ± 0.1 | | 305 |

*(c) Substituted acetylacetonato complexes of chromium(III)[294]*

| Compound | I.P. (eV) | A.P. $(P - \cdot L)^+$ $(eV)^d$ |
|---|---|---|
| Cr(acac)$_3$ | 8.11 ± 0.05 | 11.3 ± 0.10 |
| Cr(CH$_3$acac)$_3$ | 7.81 | 10.7 |
| Cr(Clacac)$_3$ | 8.16 | 11.1 |
| Cr(Bracac)$_3$ | 8.05 | 11.0 |
| Cr(Iacac)$_3$ | 8.03 | 10.8 |
| Cr(NO$_2$acac)$_3$ | 8.63 | 11.6 |
| Cr(tfac)$_3$ | 9.09 | 11.9 |
| Cr(hfac)$_3$ | 10.13 | 14.3 |

Table 11.10 (continued)

(d) Substituted acetylacetonato complexes of copper(II), $(R-CO-CH-CO-R')_2Cu$[301]

| Ring substituents | | I.P. (eV) |
|---|---|---|
| R | R' | |
| CH₃ | CH₃ | 7.75 ± 0.05 |
| CH₃ | CH₂CH₃ | 7.68 ± 0.03 |
| CH₃ | CH(CH₃)₂ | 7.61 ± 0.06 |
| CH₃ | C(CH₃)₃ | 7.59 ± 0.05 |
| CH₃ | CF₃ | 8.61 ± 0.05 |
| CF₃ | CF₃ | 9.68 ± 0.01 |
| { CH₃   CF₃ | CH₃ } c   CF₃ | 8.65 ± 0.01 |

[a] Acac, tfac, and hfac represent the anions of acetylacetone, trifluoroacetylacetone, and hexafluoroacetylacetone, respectively.

[b] May be low because of the possibility of thermal decomposition to the bischelate in the ion source.

[c] Mixed ligand complex (CH₃–CO–CH–CO–CH₃)Cu(CF₃–CO–CH–CO–CF₃), (acac)Cu(hfac).

[d] P represents the parent molecule.

fluoroacetylacetone, and hexafluoroacetylacetone with various metals changed very little as the metal atom was varied but were strongly dependent upon the nature of the ligand. They pointed out that this is contrary to predictions based on Koopman's theorem and to LCAO–MO calculations although these calculations are qualitatively consistent with the generally accepted view of the electronic structure of these complexes. Appearance potential measurements enabled Reichert and Westmore[315] to calculate M–O coordinate bond energies.

### 18.2.2 Complexes of other ligands with two oxygen donor atoms

Mass spectra of some metal complexes of 2-carbamoyldimedone (2-carbamoyl-5,5-dimethylcyclohexane-1,3-dione) have been investigated by Dudek and Barber[317]. Those of the bis-ligand complexes, in which considerable intermolecular bonding is present, reveal peaks corresponding to the parent molecular ion and to the ligand; no metal-containing ions other than the parent were observed. However, the N-phenyl-2-carbamoyldimedone, LXXXIX, and tris-2-carbamoyldimedone complexes underwent significant fragmentation. It was deduced from the mass spectra that the metals were coordinated to the two oxygens in agreement with the conclusion reached from other physical data. However, during fragmentation, rearrangement reactions occurred to form fragment ions in which the nitrogen was bonded to the metal. The same was found to occur with bis(acetoacetanilidato)-copper(II), XC, provoking caution in the assignment of structure of certain metal complexes from their mass spectra. Distinction between O–O and O–N chelation of O-nitroso-phenol ligands was not possible from mass spectra of the complexes[318].

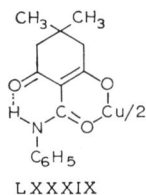

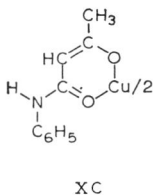

LXXXIX                    XC

The mass spectra of both bis(dimethylglyoximato)- and bis(benzildioxi-mato)nickel(II) have been reported[272] and a method developed for the detection and estimation of the metal in the form of the chelates down to $10^{-12}$ g. The integrated ion current at a significant mass number was directly proportional to the concentration of metal chelate.

A distinctive feature of the mass spectra of basic beryllium[320] and zinc[320,321] carboxylates is the intensity of doubly charged ions. The ion $[(RCO_2)_4Zn_4O]^{2+}$ was one of the major peaks in the spectrum of each zinc compound and gave rise to metastable ions at mass numbers greater than the parent molecular ion[321].

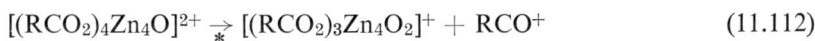

$$[(RCO_2)_4Zn_4O]^{2+} \underset{*}{\rightarrow} [(RCO_2)_3Zn_4O_2]^+ + RCO^+ \tag{11.112}$$

The mass spectrum of copper(I) benzoate[322] contained several ions arising from migration of the phenyl group such as $Cu_2Ph_2^+$, $CuPh_2^+$, $Cu_2Ph^+$, and $CuPh^+$. The spectrum of copper(I) acetate was straightforward.

*18.2.3 Complexes of ligands with sulphur donor atoms*

A number of complexes incorporating ligands with sulphur donor atoms in addition to other ligands such as CO and fluorocarbons have already been discussed in Sections 13 and 16.

Metal chelates derived from substituted *cis*-1,2-ethylene-dithiols, XCI, are of considerable theoretical interest and have been extensively studied in recent years. These complexes have been found to give well-defined mass spectra[330] with parent molecular ions of moderate intensity when R = methyl, phenyl, or *p*-anisyl, and M = Ni, Pd, or Pt. A common feature of the spectra was the elimination of neutral substituted acetylene fragments R–C=C–R (Fig. 11.20), the reverse of the original method of preparation

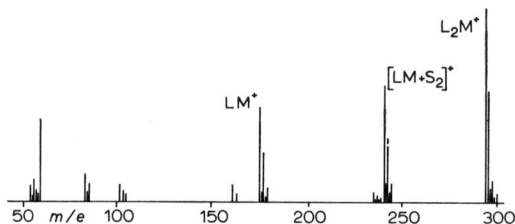

Fig. 11.20. Mass spectrum of bis-(2,3-butanedithione)nickel. (Reproduced with permission from ref. 330.)

of such complexes. In addition, the acetylenic ions R--C=C–R$^+$ were also observed. Other ions present in the various spectra were found to be dependent on both R and M. The differences in both the nature and degree of fragmentation were reported to be related to differences in the metal–ligand bonding.

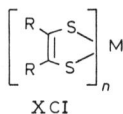

XCI

Mass spectra of other sulphur chelate complexes have also been reported[331,332]. The tris(dithioacetylacetonato) complexes of rhodium(III) and iridium(III) were reported[316] to give well-defined peaks corresponding to the parent molecular ions, whereas for the corresponding iron(III) and cobalt(III) complexes, the peak of highest mass corresponded to the fragment ions [M(ligand)$_2$]$^+$; no further details were given.

*18.3 Complexes of ligands with both nitrogen and oxygen donor atoms*

The mass spectra of complexes of iron(III)[323] and manganese(III)[324] with

$N$-$n$-propyl-salicylaldimine have been reported. Whereas the peaks of highest $m/e$ values arising from the iron complexes and also Mn(sal-$N$-$n$-propyl)$_3$ correspond to the parent molecular ions, the complexes Mn(sal-$N$-$n$-propyl)$_2$(acetate) and Mn(sal-$N$-$n$-propyl)$_2$Br give rise to peaks at higher mass numbers; these were assigned to species such as Mn$_2$(sal-$N$-$n$-propyl)$_4$$^+$, Mn(sal-$N$-$n$-propyl)$_3$(acetate)$_2$$^+$ and Mn(sal-$N$-$n$-propyl)$_3$(acetate)$^+$. This could be indicative, at least in the vapour phase, of dimeric compounds though the possibility of fragment ion recombination could not be ruled out[324]. The same was true for the complex Mn(salen)I, where salen is the Schiff's base formed from salicylaldehyde and ethylenediamine.

Similar complexes of nickel(II) also gave rise to high mass species under electron impact[270]. The Schiff's base complex XCII, for example, is known to be a four-coordinate square planar nickel(II) compound with a ligand: nickel (L:Ni) ratio of 2:1. However, in addition to the parent molecular ion, a multiplet of peaks of much higher intensity was observed at higher mass numbers in its mass spectrum; these could be assigned to the ion [(L–H)$_2$-Ni$_2$]$^+$ for which Dudek et al.[270] postulated structure XCIII.

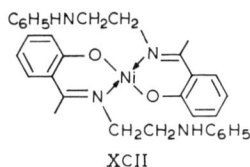

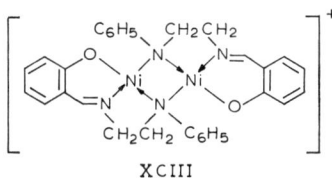

XCII           XCIII

Deuteration studies showed that the amine hydrogen was indeed the one lost on formation of this dinickel species. The $N$-methylamine complex behaved similarly but, in addition, low-intensity ions at still higher mass containing two metal ions and several ligands were observed. A molecule–molecule reaction at the sample surface yielding the dimer (L–·H)$_2$Ni$_2$ and neutral ligand LH was postulated rather than an ion–molecule reaction or thermal decomposition on the direct insertion probe; the LH$^+$ ion was also observed. A second group of nickel complexes in which the side chain of the salicylaldimine did not contain a readily removable hydrogen gave rise to the binuclear species L$_2$Ni$_2$$^+$ but only in very low abundance ($<1\%$ of the parent molecular ion).

Martell and co-workers[325] reported the Schiff's base "bisbenzoylacetone-ethylenediimine", prepared its metal chelates with Cu(II), Ni(II), Co(II), Pd(II), and on the basis of infrared and UV–visible spectroscopy and other

evidence concluded that the complexes had the structure XCIII. Chaston
et al.[326] later studied both the ligand and its complexes mass spectrometrically and showed that, instead, the Schiff's base was NN'-ethylenebis-
(benzoylisopropylideneimine) with the imino nitrogens attached to the same
carbon atoms as the methyl groups; the structure of the complexes is therefore represented by XCV. Further examples of metal oxidation state changes
and iron-rearrangement reactions were noted.

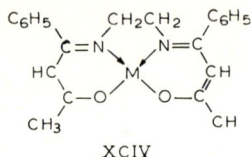

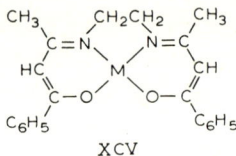

XCIV                          XCV

The fragmentation patterns of a number of metal complexes of 2-butyl-8-
hydroxyquinoline have been reported[327]. When the divalent metal ion possessed the ability to absorb an electron and exist in a stable monovalent form,
the main fragment was formed by loss of a molecule of 2-$n$-butyl-8-hydroxy-
quinoline. In other cases, elimination of an ethyl radical from the parent
molecular ion dominated the fragmentation, viz.

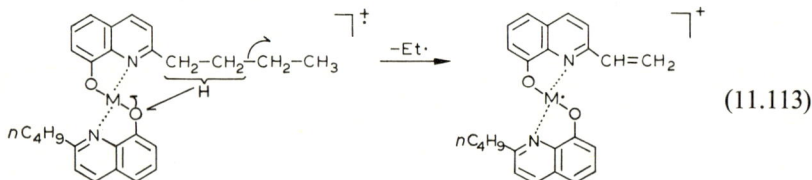

(11.113)

The mass spectra of Al(III) complexes of 2-methyl-8-hydroxyquinoline
have been reported[328] while those of 8-hydroxyquinoline complexes, or
oxinates of Mg(II), Al(III), Mn(II), Fe(III), Co(II), Ni(II), Cu(II), and Zn(II)
have been recorded in an investigation into the detection and estimation of
the metals[273]. Mass spectra of 2- and 4-(2-pyridylazo) phenol, 2-(2-pyridyl-
azo)-1- and 1-(2-pyridylazo)-2-naphthol complexes with Mn(II), Co(II),
Ni(II), Cu(II), and Zn(II) have been reported and discussed[329].

19. COMPOUNDS WHICH CONTAIN A TRANSITION METAL BONDED TO A
MAIN GROUP METAL (T. R. SPALDING)

In this section we will discuss compounds which contain transition metal to main Group metal bonds. Compounds of this type which have been studied mass spectrometrically have been almost exclusively those with transition metal–main Group IV, transition metal–Group II (Zn, Cd and Hg), or transition metal–main Group V bonds. Most of the discussion here will concern the two former situations since compounds of the main Group V elements usually have the Group V elements acting as "donor" sites and these complexes have already been discussed to some extent in Sections 15 and 13 on phosphines and related ligands. We will neither be primarily concerned with compounds containing metal–metal bonds solely between transition metals nor compounds which do not initially contain metal–metal bonds but fragment with the formation of such bonds in the mass spectrometer. We will discuss transition metal–main Group IV compounds first since these constitute the largest number, then we will deal with the Group II derivatives. Finally, compounds with the Group V elements arsenic and antimony, and then those containing other main group metals bonded to transition metals will be mentioned.

The reader will appreciate that in some cases, say for a compound with molybdenum–tin bonds where there are peaks due to combinations of seven isotopes of Mo and ten isotopes of Sn, extensive overlap of peaks can make interpretation of spectra very difficult. Computer programmes have proved useful in helping to calculate monoisotopic spectra from polyisotopic species[334].

### 19.1 Compounds containing main Group IV metals

#### 19.1.1 Hydrides with $H_3M-$ and $H_5M_2-$ groups

Mass spectrometric studies of $H_3SiCo(CO)_4$ (ref. 335), $H_3GeCo(CO)_4$ and $D_3GeCo(CO)_4$ (ref. 336), $H_3SiMn(CO)_5$ (ref. 337), $H_3GeMn(CO)_5$ (ref. 338), and $H_3GeRe(CO)_5$ (ref. 339) have been reported. In the case of the Si–Co compound, the full mass spectrum was reported but it was difficult to interpret with the low-resolution conditions used because of overlapping peaks due to ions such as $SiCo(CO)_n^+$ and $Co(CO)_{(n+1)}^+$ ($n = 0–3$) and the appearance of ions attributed to $Si_2OH_x^+$ species produced by hydrolysis in the spectrometer. Only partial spectra were reported for the other compounds.

Several features were common in these spectra. The parent ions were observed in low abundance, loss of carbonyl groups provided important fragmentation pathways, and ions of the types $H_3M–M'(CO)_n^+$ ($n = 0–4$ for Co or 0–5 for Mn, Re) were found in reasonable abundance. Apparently, in the silicon compounds, loss of H· was not observed from the parent molecular ions and in the case of the Si–Mn compound, ions of the types $HSiMn(CO)_n^+$ and $SiMn(CO)_n^+$ were found only with $n = 0–2$. The germanium compounds showed a tendency to undergo rearrangements producing M–H-containing ions, e.g. $HMn(CO)_5^+$ (ref. 338), $HCo(CO)_4^+$ and $DCo(CO)_4^+$ (ref. 336), and ions of the type $Re(CO)_xC^+$ were observed in the spectrum of the Ge–Re compound.

Table 11.11

Ion Abundances in the Mass Spectra of Me₃Si–Transition Metal Compounds

| Composition of ion | % Abundance | Composition of ion | % Abundance |
|---|---|---|---|
| $(CH_3)_3SiCo(CO)_4$ | 1.1[a] | $C_5H_5Fe(CO)_2Si(CH_3)_3$ | 5.7 |
| $(CH_3)_3SiCo(CO)_3$ | 3.1 | $C_5H_5Fe(CO)Si(CH_3)_3$ | 4.5 |
| $(CH_3)_3SiCo(CO)_2$ | 2.7 | $C_5H_5FeSi(CH_3)_3$ | 12.0 |
| $(CH_3)_3SiCo(CO)$ | 5.2 | $C_5H_5Fe(CO)_2Si(CH_3)_2$ | 1.3 |
| $(CH_3)_3SiCo$ | 7.3 | $C_5H_5Fe(CO)Si(CH_3)_2$ | 1.0 |
| $Co(CO)_2$ | 12.3 | $C_5H_5Fe(CO)SiCH_3$ | 2.9 |
| $Co(CO)$ | 16.0 | $C_5H_5FeSi(CH_3)_2$ | 4.5 |
| $(CH_3)_3Si$ | 21.6 | $C_5H_5FeSi(CH_3)CH_2$ | 7.9 |
| $Co$ | 30.7 | $C_5H_5FeSi(CH_3)$ | 2.2 |
| | | $C_5H_5FeSi$ | 0.6 |
| | | $C_5H_4FeSi$ | 0.6 |
| | | $C_6H_6Fe$ | 1.3 |
| | | $FeSi(CH_3)_2CH_2$ | 5.4 |
| | | $C_5H_5Fe$ | 7.2 |
| | | $C_6H_7Si$ | 3.7 |
| | | $FeSi(CH_3)H$ | 3.5 |
| | | $FeSiCH_2$ | 2.2 |
| | | $C_3H_3Fe$ | 1.9 |
| | | $C_3HFe$ | 6.3 |
| | | $FeSi$ | 1.9 |
| | | $Si(CH_3)_3$ | 11.4 |
| | | $Fe$ | 9.1 |
| | | $SiCH_3$ | 2.9 |

[a] The spectrum of this compound was contaminated by the presence of $(CH_3)_6Si_2O$. A previous reference[347] failed to report the parent molecular ion or ions $(CH_3)_3SiCo(CO)_n^+$ ($n = 0–3$). It was suggested that this was due to thermal decomposition in the heated inlet. Unfortunately, no details of temperature were given.

The germanium–manganese compound $H_5Ge_2Mn(CO)_4$ has been charac-terised by its mass spectrum[340]. The parent molecular ion and the ions $H_nGe_2Mn(CO)_5^+$ ($n = 1-4$) have all been identified by accurate mass measurement.

### 19.1.2 Derivatives with $R_nM-$ groups ($R = alkyl$ or $aryl$; $n = 1-3$) and $Me_{(2n + 1)}Si_n$ ($n = 1-4$)

The mass spectra of $Me_3SiCo(CO)_4$ (ref. 341), $\pi$-$C_5H_5Fe(CO)_2SiMe_3$ (ref. 342), and $\pi$-$C_5H_5M'(CO)_3MMe_3$ ($M' = $ Cr, Mo, or W; $M = $ Ge or Sn)[343], have been reported fully and features in the mass spectra of a large number of $Me_3M-$ containing compounds have received comment by various workers.

The metal-containing ions in the spectra of the trimethylsilyl derivatives are given in Table 11.11. Parent ions and ions formed by the loss of CO groups were found as expected. The spectrum of the Si–Co compound was very simple; apart from the ions $(CH_3)_3SiCo(CO)_n^+$ ($n = 0-4$), only $Co(CO)_2^+$, $Co(CO)^+$, $(CH_3)_3Si^+$, and $Co^+$ (30.7%, base peak) were reported. The spectrum of the Si–Fe compound was much more complicated. As well as the ions $C_5H_5Fe(CO)_nSi(CH_3)_3^+$ ($n = 0-2$), ions formed by the loss of $\cdot CH_3$, $:CH_2$, $\cdot H$, or $\cdot C_5H_5$ were observed. Two ions associated with the cleavage of the $C_5H_5$ ring, $C_3H_3Fe^+$, and $C_3HFe^+$ were found but no re-arrangement ion $C_5H_5Si^+$ was observed.

The spectra of the Group IV–Group VI compounds obtained by Lappert et al. are given in Table 11.12. The spectrum of $\pi$-$C_5H_5Mo(CO)_3SnMe_3$ has also been reported by King[344] and features of the mass spectra of germanium derivatives ($M' = $ Mo, $R = $ Me, Et, or $Pr^n$; $M' = $ W, $R = $ Me or Et) have been reported by Carrick and Glockling[345]. Overall, the reported spectra of $\pi$-$C_5H_5Mo(CO)_3SnMe_3$ are in good agreement except that the ion at $m/e = $ 185 suggested to be $(C_5H_5{}^{120}Sn)^+$ by Lappert et al. is reported to be $(CH_3)_2SnCl^+$ by King, who uses $m/e$ 184; $(CH_3)_2{}^{119}Sn^{35}Cl^+$. However, King does not report ions at $m/e$ 169 $(CH_3)^{119}Sn^{35}Cl^+$ or $m/e$ 154 $(^{119}Sn^{35}Cl)^+$ which would be expected if $(CH_3)_2SnCl^+$ was present[346].

Interpretation of the spectra is complicated by the number of isotopes of the metals present. In the "ions" marked with an asterisk (containing both metals), there may have been two or three types of ion present, e.g. $C_5H_5$-$(CO)_2M'M(CH_3)_2^+$ would overlap with $C_5H_5(CO)_3M'M^+$ and $C_5H_5(CO)$-$CM'M(CH_3)_3^+$ and have the same $m/e$ values to within two amu. Under the low-resolution conditions used, it was very difficult to evaluate the rela-tive amounts of the various ions present in such a mixture. All the compounds

Table 11.12

Major Metal-Containing Ions (>1% Total Abundance)

| | Total abundance (%) from $(\pi-C_5H_5)(CO)_3M'-MMe_3$ for M'-M = | | | | | |
|---|---|---|---|---|---|---|
| | Cr–Ge | Mo–Ge | W–Ge | Cr–Sn | Mo–Sn | W–Sn |
| $(C_5H_5)(CO)_3M'MMe_3^+$ | 3 | 7 | 9 | 5 | 6 | 11 |
| $(C_5H_5)(CO)_3M'MMe_2^+$ | 1 | 4 | 13 | 2 | 15 | 27 |
| $(C_5H_5)(CO)_2M'MMe_2^{+a}$ $[(C_5H_5)(CO)_3M'M]^+$ | 1 | 2 | 6 | | 4 | 6 |
| $(C_5H_5)(CO)_2M'MMe^{+a}$ $(C_5H_5)(CO)M'MMe^+$ | 2 | 3 | 3 | 1 | | 3 |
| $[(C_5H_5)(CO)_2M'M]^{+a}$ $(C_5H_5)(CO)M'MMe_2^+$ | 1 | 2 | 5 | 2 | 1 | 4 |
| $[(C_5H_5)(CO)M'MMe_2]^{+a}$ $[(C_5H_5)M'M]^+$ | 2 | 8 | 8 | 4 | 3 | 7 |
| $(C_5H_5)(CO)M'MMe^{+a}$ $(C_5H_5)M'MMe_3^+$ | 5 | 7 | 4 | 4 | 4 | 8 |
| $(C_5H_5)M'MMe^+$ | 1 | | 7 | 2 | 11 | 11 |
| $[(C_5H_5)M'M]^+$ | | 1 | 2 | 2 | 9 | 3 |
| $[(C_5H_5)(CO)_3M']^+$ | 2 | 3 | | 1 | 1 | 1 |
| $[(C_5H_5)(CO)_2M']^+$ | 2 | 2 | 5 | | | |
| $[(C_5H_5)(CO)M']^+$ | 5 | 2 | 4 | 4 | 3 | 2 |
| $[(C_5H_5)M']^+$ | 38 | 6 | 5 | 24 | 3 | 2 |
| $[M']^+$ | 25 | 1 | | 22 | 1 | |
| $MMe_3^+$ | 11 | 45 | 19 | 10 | 14 | 8 |
| $MMe_2^+$ | | | | 1 | 3 | 1 |
| $MMe^+$ | | 1 | 3 | 4 | 8 | 2 |
| $[(C_5H_5)M]^+$ | 1 | 6 | 7 | 12 | 14 | 4 |

[a] Other ions which may be present include $(C_5H_5)(CO)_xM'MMe_y$ ($x = 0$ or 1 and $y = 0$–3).

showed parent ions in reasonable abundance (3–11%). Fragmentation was by the loss of one or more CO or $\cdot CH_3$ groups initially giving ions of the type $C_5H_5(CO)_xM'M(CH_3)_y^+$. The fractional ion current carried by ions with the M'–M bond apparently intact were[343]

Cr–Ge (16%), Mo–Ge (34%), W–Ge (57%),
Cr–Sn (22%), Mo–Sn (53%), W–Sn (80%).

Generally, there are more similarities between the spectra of the molybdenum and tungsten compounds than between these and the chromium analogue. Thus, in the tin series, whereas the ion $C_5H_5(CO)_2CrSn(CH_3)_3^+$ was found in low abundance, under the same conditions the Mo- and W-containing species were not detected. Similarly, ions of the type $C_5H_5(CO)_2M'-Sn(CH_3)_2^+$ were much more abundant for $M' =$ Mo or W than for $M' =$ Cr. In the germanium series, however, the ion $C_5H_5(CO)_2M'-Ge(CH_3)_3^+$ occurred in the spectra of the compounds of $M' =$ Cr or Mo but not when $M' =$ W, whereas the ion $C_5H_5(CO)_2M'-Ge(CH_3)_2^+$ was abundant only when $M' =$ Mo or W.

The ions of the type $C_nH_xMoSn^+$, due to the fragmentation of the cyclopentadienyl ring, which were observed by King[344] in 2–3% abundance from the Mo–Sn compound and by Carrick and Glockling[345] from the Mo–Ge and W–Ge compounds, were also observed by Lappert et al. but in less than 1% abundance. Carrick and Glockling also reported ions formed by the loss of $m/e = 16$, 29, 30, and 44 which they attributed to O (or $CH_4$), HCO, $H_2CO$, and $CO_2$ species. Lappert et al. observed similar ions but in very low abundance.

Rearrangement ions of the type $C_5H_5M^+$ (M = Sn or Ge) were fairly abundant (4–14%) for all compounds except Cr–Ge (1%)[343]. Such rearrangement ions typify the mass spectra of $C_5H_5M'M$ compounds. Doubly charged ions were present in very low abundance from Mo and W compounds but not from the Cr analogues. Those observed included $[C_5H_5(CO)MoGe]^{2+}$ from Mo–Ge and $[C_5H_5MoSn(CH_3))]^{2+}$ from Mo–Sn compounds; $[C_5H_5WGeCH_3]^{2+}$, $[C_5H_5W]^{2+}$, and $[C_5H_5W(CO)]^{2+}$ from W–Ge; and $[C_5H_5WSnCH_3]^{2+}$, $[C_5H_5(CO)_3WSn(CH_3)_2]^{2+}$, $[C_5H_5(CO)_2WSnCH_3]^{2+}$, and $[C_5H_5(CO)WSnCH_3]^{2+}$ from the W–Sn compound. Carrick and Glockling found all the ions of the type $[C_5H_5(CO)_nWGe(CH_3)_m]^{2+}$ where $n = 0$–2 but fewer such ions in the spectrum of the Mo–Ge compound.

Lappert and coworkers[343] measured the appearance potentials of the $(CH_3)_3M^+$ ions; from these values and the ionisation potentials of the corresponding radicals they calculated the bond dissociation energies $D(\pi\text{-}C_5H_5\text{-}(CO)_3M'\text{-}MMe_3)$ given in Table 11.13.

Glockling and Hooton reported several features from the spectra of platinum–$GeMe_3$ compounds[348]. A parent molecular ion was observed from trans-$(Et_3P)_2Pt(GeMe_3)Cl$ only at high source pressures although the ion (parent molecule $- \cdot CH_3)^+$ was present even when the parent molecular ion was not seen. The eliminations of three ethylene molecules from the base peak

Table 11.13

Bond Dissociation Energies $D(\pi C_5H_5(CO)_3M'-MMe_3)$

| Compound | A.P. of $(CH_3)_3M^+$ (eV) | $D°(C_5H_5(CO)_3M'-MMe_3)$ (kcal . mole$^{-1}$) |
|---|---|---|
| $\pi$-$C_5H_5(CO)_3CrGeMe_3$ | $9.06 \pm 0.1$ | $46.7 \pm 2.3$ |
| $\pi$-$C_5H_5(CO)_3MoGeMe_3$ | $9.63 \pm 0.14$ | $59.9 \pm 3.2$ |
| $\pi$-$C_5H_5(CO)_3WGeMe_3$ | $9.84 \pm 0.1$ | $64.6 \pm 2.3$ |
| $\pi$-$C_5H_5(CO)_3CrSnMe_3$ | $9.09 \pm 0.1$ | $53.5 \pm 2.3$ |
| $\pi$-$C_5H_5(CO)_3MoSnMe_3$ | $9.85 \pm 0.1$ | $71.0 \pm 2.3$ |
| $\pi$-$C_5H_5(CO)_3WSnMe_3$ | $10.05 \pm 0.15$ | $75.6 \pm 3.5$ |

$Pt(P(C_2H_5)_3)_2^+$ were metastable supported processes. Ions produced by the loss from the parent molecule of $\cdot CH_3$, $CH_4$, $CH_3Cl$, $\cdot Ge(CH_3)_3$, and $(\cdot CH_3 + P(C_2H_5)_3)$ groups were identified by accurate mass measurement.

The parent molecular ion from the compound $Ph_3Ge-AuPEt_3$ was reported by Glockling and Wilbey[349]. Other ions formed by loss of $Ph\cdot$, $Et\cdot$, and elimination of $C_6H_6$ and $C_2H_4$ were observed. The base peak was due to the $(C_2H_5)_3PAu^+$ ion and the $(C_6H_5)_3Ge^+$ ion was also reported.

Mass spectrometry was used[350,351] to characterise compounds of the type $(\pi$-$C_5H_5)_2M'(Cl)MPh_3$ where $M' = Zr$ or $Hf$; $M = Si$ or $Ge$. Parent molecular ions were observed for all compounds. Fragmentation by loss of $\cdot C_5H_5$, $Cl\cdot$, or $\cdot MPh_3$ groups provided the major modes of decomposition. The base peaks were $(C_6H_5)_3M^+$ ions and the rearrangement ions $C_5H_5M^+$ were found in all the spectra. Interestingly, in comparison with the alkyl–Group IV compounds discussed earlier, very little trace of ions of the type $(C_5H_5)_xM'(Cl)_yM(C_6H_5)_z^+$ ($x = 0–2$; $y = 0$, or $1$; $z = 0–2$) were found, indicating that loss of a Ph group from the Group IV metal atom was not a favourable reaction. Ions produced by the cleavage of Sn–Ph bonds have been found in the spectra of $Ph_3SnMn(CO)_5$ (refs. 359, 353) and $\pi$-$C_5H_5Fe(CO)_2SnPh_3$ (ref. 352). Usually, stepwise loss of all CO groups precedes the loss of Ph radicals giving the ions $MnSn(C_6H_5)_n^+$ ($n = 0–2$) and $C_5H_5FeSn(C_6H_5)_n^+$ ($n = 0–2$) but in the case of the iron–tin compound an ion containing both CO groups, $C_5H_5Fe(CO)_2Sn(C_6H_5)_2^+$, was reported. Apparently the loss of simple $\cdot SnPh_3$ groups was not found from the manganese and iron compounds; this is in marked contrast to some alkyl–tin compounds already discussed (see Table 11.12). In both $Ph_3Sn$ compounds above, the

*References pp. 600–610*

base peak was due to the $^+Sn(C_6H_5)$ ion. The rearrangement ions $^+Mn$-$(C_6H_5)$, $^+Fe(C_6H_5)$ and $C_5H_5Sn^+$ were observed. The partial mass spectrum of $Ph_3SnMn(CO)_5$ reported by Mays and Simpson[353] showed a broad general agreement with that of Lewis *et al.*[352]. However, Mays and Simpson do

Table 11.14

Other Compounds Containing an $Me_3M$ Group

| Compound | Comments[a] | Ref. |
|---|---|---|
| $(Me_3M)_2Ru(CO)_4$<br>M = Si or Sn | P.m.i. | 355, 354 |
| $[R_3SiRu(CO)_4]_2$<br>R = Me, Et, $Pr^n$ | P.m.i. | 354, 355 |
| $Me_3SiRu(CO)_4SnR_3$<br>R = Me, Ph | P.m.i. | 355 |
| $Et_3SiRu(CO)_4SnMe_3$ | P.m.i. | 355 |
| $Me_3SnRu(CO)_4SnBu^n_3$ | P.m.i. | 355 |
| $Me_3SiRu(CO)_4AuPPh_3$ | P.m.i. | 355 |
| $Me_3MOs(CO)_4H$<br>M = Si, Sn | P.m.i. | 356 |
| $Et_3MOs(CO)_4H$<br>M = Si, Ge | P.m.i. | 356 |
| $[Me_3SiOs(CO)_4]_2$ | P.m.i. | 356 |
| $Me_3MOs(CO)_4M'Me_3$<br>M = M' = Si or Sn,<br>M = Si, M' = Sn | P.m.i. | 356 |
| $Me_3SiOs(CO)_4Ru(CO)_4SiMe_3$ | P.m.i. | 356 |
| $Me_3SiOs(CO)_4R$<br>R = Me or $C_2F_4H$ | P.m.i. | 356 |
| *trans*-$(PhMe_2P)_2Pt(GeMe_3)I$ | P.m.i. | 348 |
| *trans*-$(Et_3)P_2Pt(GeMe_3)X$<br>X = CN or NCS | P.m.i. | 348 |
| $(Et_3P)_2(Me_3Si)ClPh_2C_4H_2Pt$ | P.m.i.? | 348 |
| *trans*-$(Et_3P)_2Pt(SiMe_3)Cl$ | No p.m.i., highest $m/e$ due to<br>(parent molecule $- \cdot CH_3)^+$ | 348 |
| $(diphos)Pt(GePh_3)X$<br>X = H or Me,<br>X = Ph | P.m.i. only for X = H or Me<br>not for Ph | 357<br>358 |
| $(C_6F_5)_3SiM(CO)_5$<br>M = Mn or Re | P.m.i. | 358 |

[a] P.m.i. = parent molecular ion(s).

*References pp. 600–610*

report the formation of the ion $HMn^+$ in trace amounts which Lewis *et al.* apparently did not observe.

Table 11.14 lists some of the other compounds containing $R_3M-$ groups for which mass spectral data have been reported.

King *et al.*[342] report mass spectra for a series of compounds $Me_{(2n+1)}Si_n$-$Fe(CO)_2\pi$-$C_5H_5$ ($n = 1$–4). All spectra showed parent ions and ions due to the loss of CO groups. In the higher homologues, cleavage of Si–Me and Si–Si bonds was also an important process of fragmentation giving rise to ions of the type $C_5H_5Fe(CO)_xSi(CH_3)_2^+$ ($x = 0$–2). The base peaks were $Si_2(CH_3)_5^+$ for $n = 2$, and $(CH_3)_3Si^+$ for $n = 3$ or 4.

Parent ions have been observed for two compounds containing the $Me_2Sn$ group; $[Me_2SnRu(CO)_4]_2$ (ref. 355) and $Me_2Sn[Mn(CO)_5]_2$ (ref. 359).

The partial mass spectra of some compounds containing the RSn group have been reported. They were $RSn[Co(CO)_4]_3$ where $R = Me$ (refs. 360, 361), $Bu^n$ (ref. 360), vinyl (ref. 360), allyl (ref. 361), and Ph (ref. 361). Parent ions were observed for $R = Me$, vinyl, Ph, and all compounds showed $RSnCo_3(CO)_x^+$ ions in moderately high abundance, particularly for $x = 5$ and 11. Other ions found were of the types $RSnCo_2(CO)_x^+$, $RSnCo(CO)_x^+$, $SnCo(CO)_x^+$, $Co(CO)_x^+$, $Co^+$, and $Co_2^+$. The butyl compound showed (parent molecule $- 1)^+$ ion at highest $m/e$. Ball *et al.*[362] have observed the parent molecular ion from $PhGeCo_3(CO)_{11}$ and all $C_6H_5Ge(CO)_n^+$ ions ($n = 0$–10).

The tin–cobalt compound $Sn[Co(CO)_4]_4$ has been reported to show a (parent molecule $-$ CO)$^+$ ion but no parent molecular ion[363]. The complete series of ions $SnCo_4(CO)_n^+$ ($n = 0$–15) and $SnCo_3(CO)_n^+$ ($n = 0$–12) were observed and in this respect this compound was similar to the $RMCo_3(CO)_x$ compounds described above.

### 19.1.3 $X_nM$ groups ($X$ = halogen)

The mass spectra of the compounds $X_3Si$–$Co(CO)_4$ (X = F (ref. 364), Cl (ref. 341)) have been reported (Table 11.15). Parent ions were observed and ions due to CO loss, $X_3SiCo(CO)_n^+$ ($n = 0$–3 for X = F or Cl) and loss of $\cdot X$, *i.e.* $X_2SiCo(CO)_n^+$ ($n = 0$–4 for X = F and $n = 0$ for X = Cl) were found. Rearrangement ions $ClCo(CO)_2^+$ and $XCo^+$ were reported. In the X = Cl case, metastable transitions were observed which showed that the Cl–Co-containing ions were produced by elimination of $SiCl_2$, *viz.*

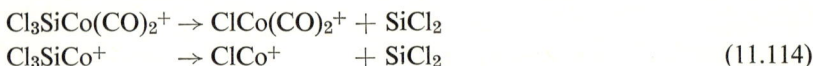

$$Cl_3SiCo(CO)_2^+ \rightarrow ClCo(CO)_2^+ + SiCl_2$$
$$Cl_3SiCo^+ \quad\;\; \rightarrow ClCo^+ \quad\;\;\; + SiCl_2 \tag{11.114}$$

Table 11.15

Ion Abundances from Compounds Containing $X_3M$ Groups

| Composition of ion | % Abundance for X = | | Composition of ion | % Abundance for M = | |
| --- | --- | --- | --- | --- | --- |
| | $F^a$ | Cl | | Si | Sn |
| $X_3SiCo(CO)_4$ | 3.1 | 1.1 | $C_5H_5Ru(CO)_2MCl_3$ | 4.8 | 1.8 |
| $X_3SiCo(CO)_3$ | 4.5 | 4.0 | $C_5H_5Ru(CO)MCl_3$ | 17.4 | |
| $X_3SiCo(CO)_2$ | 3.8 | 7.7 | $C_5H_5RuMCl_3$ | 18.4 | 1.3 |
| $X_3SiCo(CO)$ | 2.9 | 17.6 | $C_5H_5Ru(CO)_2MCl_2$ | | 3.1 |
| $X_3SiCo$ | 14.0 | 11.8 | $C_5H_5RuMCl_2$ | 0.9 | 0.8 |
| $X_2SiCo(CO)_4$ | 0.6 | | $C_5H_5RuMCl$ | 0.9 | 0.2 |
| $X_2SiCo(CO)_3$ | 0.3 | | $C_5H_5RuM$ | 2.6 | |
| $X_2SiCo(CO)_2$ | 0.1 | | $C_5H_5Ru(CO)_2Cl$ | | 13.3 |
| $X_2SiCo$ | 0.4 | 3.6 | $C_5H_5Ru(CO)Cl$ | | 11.0 |
| $XCo(CO)_2$ | | 0.2 | $C_5H_5RuCl$ | 14.5 | 18.6 |
| $Co(CO)_2$ | 13.5 | 0.6 | $C_5H_5Ru(CO)_2$ | 0.7 | 2.5 |
| $Co(CO)$ | 19.8 | 10.1 | $C_5H_5Ru(CO)$ | 2.6 | 2.9 |
| $XCo$ | 0.9 | 2.3 | $C_5H_5Ru$ | 15.6 | 22.5 |
| $CCo$ | 2.5 | | $MC_5H_5$ | 1.3 | 0.8 |
| $X_3Si$ | 3.9 | | $RuCl_2$ | 0.7 | 0.2 |
| $X_2Si$ | 1.5 | | $RuCl$ | 7.3 | 5.2 |
| $XSi$ | 2.0 | 24.5 | $RuC$ | 4.4 | 4.3 |
| $Co$ | 26.2 | 16.5 | $Ru$ | 2.2 | 2.0 |
| | | | $MCl$ | 5.7 | 5.2 |
| | | | $M$ | | 4.3 |

[a] (i) Two doubly charged ions were observed in low abundance: $Co(CO)_2^{2+}$ 2%, and $Co(CO)^{2+}$ 1%. (ii) A partial negative ion mass spectrum was also recorded. The ions $F_3SiCo(CO)_3^-$, $F_3SiCo(CO)_2^-$, and $F_3SiCo(CO)^-$ were found at 70 eV in decreasing abundance.

The spectra of the tri-halo compounds were remarkably different from that of the trimethyl compound (Table 11.11) especially in the % abundances of the $X_nSi^+$ species found. The ion $(CH_3)_3Si^+$ was the second most abundant in the spectrum of $Me_3SiCo(CO)_4$ (21.6%); in the halo compounds $F_3Si^+$ is only 3.9% and the corresponding $Cl_3Si^+$ ion was not observed but $ClSi^+$ was the base peak (24%). Evidently, elimination of the $Cl_2Si$ species is very facile.

The studies of the halo compounds report bond dissociation energies for the $X_3Si$–$Co(CO)_4$ bond. The values obtained from a combination of mass spectrometric and thermochemical data were

$$D(F_3Si–Co(CO)_4) \ = 105 \pm 12 \text{ or } 74 \pm 25 \text{ kcal.mole}^{-1}$$
$$D(Cl_3Si–Co(CO)_4) = 126 \pm 25 \text{ or } 95 \pm 25 \text{ kcal.mole}^{-1}$$

depending on what value of $\varDelta H_f Co(CO)_4$ was taken[341]. In the case of $F_3Si$–$Co(CO)_4$, one may compare these values with the bond dissociation energies

$$D(F_3Si–SiF_3) \sim 93\text{–}95 \text{ kcal.mole}^{-1} \text{ (ref. 365)}$$
$$D((CO)_4Co–Co(CO)_4) = 11.5 \pm 4.6 \text{ kcal.mole}^{-1} \text{ (ref. 366)}$$

It appears, then, that the values of MacDiarmid and coworkers[341] are rather high.

The mass spectra of $F_3SiM(CO)_5$ (M = Mn or Re) have briefly been reported[367]. Both compounds exhibited parent molecular ions, fragments due to CO losses, loss of $F_3Si\cdot$ (or $F_2Si$ from $F_2SiM(CO)_n{}^+$), loss of $\cdot F$, and rearrangement of $\cdot F$ on to the metal M. However, although the mass spectra were shown by "stick" diagrams, no actual % abundances were reported and thus interpretation of the spectra is more difficult.

The ruthenium compounds $(\pi\text{-}C_5H_5)Ru(CO)_2MCl_3$ (M = Si or Sn) have been studied by Blackmore et al.[149]. The fragmentation behaviour was fairly simple in that most ions were formed by loss of CO, $\cdot Cl$, $\cdot C_5H_5$, or $\cdot MCl_3$ groups (Table 11.15). Rearrangement ions containing Ru–Cl species were prominent, particularly the ions $C_5H_5RuCl^+$ and $RuCl^+$. It was suggested that the ions $C_5H_5(CO)_nRuCl^+$ ($n = 1$ or 2), which were only found in the spectrum of the tin compound, showed that the elimination of $SnCl_2$ was facile. It was pointed out that the insertion of $SnCl_2$ into $\pi\text{-}C_5H_5(CO)_2RuCl$ can be used as a method of preparing $\pi\text{-}C_5H_5(CO)_2RuSnCl_3$.

Mass spectrometry has been used to characterise $[Cl_3SiRu(CO)_4]_2$ (ref. 355), $\pi\text{-}C_5H_5Co(CO)(GeCl_3)_2$ (ref. 368), $Cl_3SnFe(Cl)(CO)_4$, $trans\text{-}(Cl_3Ge)_2$-$Fe(CO)_4$, and $trans\text{-}(I_3Ge)_2Fe(CO)_4$ (ref. 369).

The mass spectrum of $Cl_2Sn[Fe(CO)_2(\pi\text{-}C_5H_5)]_2$ has been reported by Lewis et al.[352]. Although a large number of ions was observed, interpretation of the spectrum was relatively simple. Fragment ions were formed mainly by the loss of CO, $\cdot C_5H_5$, and $\cdot Cl$ groups and the rearrangement ions $C_5H_5SnCl^+$ and $C_5H_5Sn^+$ (base peak 41.0%) were observed. No Fe–Cl-containing ions were reported in contrast to the Ru–Cl species found in the spectra of $\pi\text{-}C_5H_5(CO)_2RuMCl_3$ compounds[149]. Several series of ions were found including ions of the general types $(C_5H_5)_2Fe_2(CO)_nSnCl_m{}^+$ ($n =$

0–4; $m = 0$–2) and $C_5H_5Fe(CO)_nSnCl_m^+$ ($n. = 0$–2; $m = 0$–2), and $Fe_nSn$-$Cl_m^+$ ($n = 0$–2; $m = 0$–2).

The compound $[(Cl_2Ge)Fe(CO)_4]_2$ has been characterised by a molecular weight measurement[369]. A parent molecular ion was observed from $(\pi C_5H_5CoCO)_2(GeCl_2)_2Fe(CO)_4$, a compound which contains Co–Ge–Fe bonds[370].

Thompson and Graham[359] have used mass spectrometry to characterise the compounds $XSn[Mn(CO)_5]_3$ for X = Cl or I and Patmore and Graham[363] observed a parent molecular ion from $FSn[Co(CO)_4]_3$ as well as all the $FSnCo_3(CO)_n^+$ ions, ($n = 0$–11).

### 19.1.4 $R_nX_m$ groups ($R = $ alkyl or aryl; $X = $ halogen or hydrogen)

The mass spectra of both $\pi$-$C_5H_5Fe(CO)_2Si(H)Me_2$ (ref. 342) and MeSi-$(F_2)Co(CO)_4$ (ref. 371) have been reported. The spectrum of $\pi$-$C_5H_5Fe(CO)_2$-$Si(H)Me_2$ has many ions in common with that of $\pi$-$C_5H_5Fe(CO)_2SiMe_3$. However, the base peak in the case of the former compound is $C_5H_5FeSiCH_3^+$ whilst in the latter it is $C_5H_5FeSi(CH_3)_3^+$. Ions containing the SiH entity are prominent in the spectrum of $\pi C_5H_5Fe(CO)_2Si(H)Me_2$ but not in that of the trimethyl compound. There are many similar ions in the spectra of $MeSiF_2$-$Co(CO)_4$ and $F_3SiCo(CO)_4$, including $F_2SiCo(CO)_4^+$ and $F_2SiCo^+$. However, it is of interest to note the ion $CH_3Co^+$ is found in the spectrum of $MeSiF_2Co(CO)_4$ but not in that of $Me_3SiCo(CO)_4$. This is presumably due to the stability of the $SiF_2$ species which can eliminate easily from ions such as $CH_3SiF_2Co^+$ (base peak 18.0%). By using thermochemical and mass spectrometric data, MacDiarmid and coworkers[341,371] calculated the bond dissociation energy $D(MeSiF_2$–$Co(CO)_4)$ to be $127 \pm 15$ or $96 \pm 25$ kcal.mole$^{-1}$. These values are probably high (see Section 19.1.3).

Knox and Stone[355] have reported using mass spectrometry to characterise $[MeCl_2SiRu(CO)_4]_2$ and Mays and Simpson[353] have reported the partial mass spectrum of $PhBr_2SnMn(CO)_5$. From the latter compound the base peak was $C_6H_5Sn^+$ and the rearrangement ions $C_6H_5Mn^+$, $BrMn^+$, and $HMn^+$ were observed.

### 19.1.5 Other Group IV derivatives

The acetate derivatives $(CH_3CO_2)_2Sn[Co(CO)_4]_2$ and $CH_3CO_2Sn[Co$-$(CO)_4]_3$ have been studied[359,363]. The former compound does not exhibit a parent molecular ion but the ion $(CH_3CO_2)_2SnCo_2(CO)_6^+$ or $[(CH_3O)_2$-

$SnCo_2(CO)_8^+$] was found. Fragmentation by loss of CO groups was observed and the ions $(CH_3CO_2)_nSn_2^+$ ($n = 1$ or 2) were reported. Very little data were given for the monoacetate. Patmore and Graham[372] report that the acetylacetonato derivative $(C_5H_7O_2)_2SnCo_2(CO)_7$ does not exhibit a parent molecular ion although the (parent molecule $-$ CO)$^+$ ion was observed.

## 19.2 Compounds containing Group II metals (Zn, Cd and Hg)

### 19.2.1 With two transition metal–Group II metal bonds

Burlitch and Ferrari[373] have reported the mass spectra of $M[Co(CO)_4]_2$ and $M[Cr(CO)_3\pi\text{-}C_5H_5]_2$ (M = Zn, Cd, Hg) in some detail. The spectra were essentially very simple. For the cobalt compounds the following series of ions were observed: $MCo_2(CO)_n^+$ ($n = 0$–8 for M = Zn, Cd and 0–3, 7 and 8 for Hg) and $MCo(CO)_n^+$ ($n = 0$–3), $Co_2(CO)_n^+$ ($n = 0$–7), $Co(CO)_n^+$ ($n = 0$–3) for Zn, Cd, and Hg. The only other ions reported were M$^+$, CoC$^+$, and the rearrangement ion $Co_2C^+$. The production of ions containing the dicobalt species may have been due to an elimination reaction of M or some pyrolysis reaction even though the source temperature was quite low (70–80 °C). The base peaks in these spectra were due to the ions $Co_2(CO)_4^+$ for M = Zn or Cd but Hg$^+$ from the mercury–cobalt compound.

The chromium–Group II compounds showed fewer $MCr_n$-containing ions ($n = 1$ or 2). Parent molecular ions were observed and the base peaks were due to the ions $(C_5H_5)_2Cr^+$ from M = Zn and $C_5H_5Cr^+$ from M = Cd or Hg. Other ions found were of the general types $MCr(CO)_nC_5H_5^+$, M$^+$, $Cr_2(CO)_n(C_5H_5)_2^+$, $Cr(CO)_nC_5H_5^+$, and Cr$^+$.

Burlitch and Ferrari[373] identified by their parent ions the compounds $Hg[Mn(CO)_5]_2$, $M[Fe(CO)_2\pi C_5H_5]_2$, and $M[M'(CO)_3\pi C_5H_5]_2$ (M = Zn, Cd, or Hg and M' = Mo or W). The mercury–manganese and mercury–molybdenum compounds were also reported to give parent molecular ions in low abundance by King[344]. The mass spectrum of $Hg[Fe(CO)_2\pi\text{-}C_5H_5]_2$ has been reported by Lewis et al.[352]. A parent molecular ion was observed and the base peak was due to the Hg$^+$ ion. Fragmentation of this compound was very similar to the chromium compounds discussed above. Loss of CO groups was the predominant pathway giving the series of ions $HgFe_2(CO)_n$-$C_5H_5^+$ ($n = 0$–2), $Fe_2(CO)_n(C_5H_5)_2^+$ ($n = 0$–4) and $Fe(CO)_nC_5H_5^+$ ($n = 0$–2). The only other ions observed were due to Fe$^+$, Hg$^+$, and the rearrangement ion $(C_5H_5)_2Fe^+$. As with the chromium compounds, no rearrangement ions involving the transfer of a $\cdot C_5H_5$ group on to M were observed in contrast with many of the compounds containing transition metal–main Group IV bonds which have been discussed.

The spectrum of the analogous Ru compound[149,352] showed only the parent ion, $Hg^+$, and ions associated with the pyrolysis product $[\pi\text{-}C_5H_5(CO)_2Ru]_2$.

The mercury–iron compound $Hg(Fe(CO)_4C_3F_7)_2$ showed many Hg–Fe-containing ions including the parent molecular ion[374]. Fragmentation of $HgFe_2$- or HgFe-containing species often appeared to lose a fluorine-containing entity and one or more CO groups concurrently. For instance, the ion $HgFe(CO)_4C_3F_7^+$ appeared to lose $(F_2 + 2CO)$ or $(2 \cdot FCO)$ to give $HgFe(CO)_2C_3F_5^+$ as the ion at next lowest $m/e$. Ions which may have been due to a preceeding pyrolysis reaction or elimination of Hg were observed such as $Fe_2(CO)_5C_2F_5^+$. Other ions were produced by loss of $F\cdot$, $\cdot CF_3$, $\cdot C_2F_5$, and Fe and rearrangement ions containing Fe–F bonds were common, for example $HgFeF^+$ and $(CO)_nFeF^+$ ($n = 1$ or 2). The base peak was due to the $Fe(CO)_4C_3F_7^+$ ion.

Mays and Robb[375] have reported characterising compounds $\pi\text{-}C_5H_5Mo(CO)_3HgCo(CO)_4$, $\pi\text{-}C_5H_5W(CO)_3HgCo(CO)_4$, and $\pi\text{-}C_5H_5Fe(CO)_2HgCo(CO)_4$ by mass spectrometry.

### 19.2.2 With one Hg–transition metal bond

Compounds of the type $\pi\text{-}C_5H_5(CO)_2FeHgX$, $\pi\text{-}C_5H_5Mo(CO)_3HgX$, and $\pi\text{-}C_5H_5W(CO)_3HgX$ (X = Cl, Br, I, SCN) have been characterised by a mass spectrometric determination of their molecular weights[375]. No other ions were reported. King[344] reported the molecular ion from $\pi\text{-}C_5H_5(CO)_3Mo\text{-}HgCl$, but the spectrum also showed ions due to decomposition products and impurities from several sources.

## 19.3 Compounds containing main Group V metals

### 19.3.1 Hydrides with $H_3M$ groups

Fischer and co-workers[376] have reported several complexes containing $AsH_3$ or $SbH_3$ groups. A parent molecular ion was observed from $H_3AsMn(CO)_2\pi\text{-}C_5H_5$ and ions due to the loss of one or both CO groups. Other ions found were $C_5H_5MnAsH_n^+$ ($n = 0$–2), $C_5H_5Mn(CO)_n^+$ ($n = 0$–2), $C_5H_5Mn^+$ (base peak), and ions derived therefrom by fragmentation of the $C_5H_5$ ring, the hydride ions $MnAsH_n^+$ ($n = 0$–2), $MnH^+$, and $Mn^+$. The ionisation potential was reported to be $7.16 \pm 0.1$ eV. The compounds $H_3AsM(CO)_5$

(M = Cr, Mo, W) showed parent molecular ions in reasonable abundance[377] Most of the fragment ions observed were produced by CO and H· loss and corresponded to the general type $(CO)_nMAsH_m^+$ for $n = 0$–4 and $m = 0$–3 but the series $^+M(CO)_n$ ($n = 0$–5 for M = Mo or W and $n = 0$–4 for Cr) was also observed. The analogous chromium–antimony compound $(CO)_5$-

Table 11.16

Ion Abundances from Some Compounds Containing Manganese–$MPh_3$ Bonds (M = P, As, Sb)

| Composition of ion | % Ion abundance for M = | | |
|---|---|---|---|
| | P | As | Sb |
| *(a) π-Pyrrolyl derivatives* | | | |
| $C_4H_4NMn(CO)_2M(C_6H_5)_3$ | 1.3 | 0.6 | 0.3 |
| $C_4H_4NMnM(C_6H_5)_3$ | 5.0 | 19.0 | 6.4 |
| $MnM(C_6H_5)_3$ | 5.0 | 3.8 | 1.1 |
| $M(C_6H_5)_3$ | 7.5 | 3.4 | 2.3 |
| $C_{12}H_8MnM$ | 5.5 | 4.8 | 1.6 |
| $C_4H_4NMnMC_6H_6$ | | 0.5 | 0.5 |
| $M(C_6H_5)_2$ | 7.5 | 7.4 | 7.5 |
| $C_{12}H_8M$ | 7.5 | 17.3 | 5.7 |
| $MC_6H_5$ | 4.0 | 22.3 | 62.5 |
| $C_6H_5Mn$ | 3.5 | 3.6 | 3.7 |
| $C_4H_4NMn$ | 2.6 | 1.5 | 1.7 |
| $Mn$ | 50.0 | 15.4 | 6.7 |
| $C_4H_4NMnM(C_6H_5)_3$ (doubly charged) | 0.6 | 0.4 | |
| | | | |
| *(b) π-Indenyl derivatives* | | | |
| $C_9H_7Mn(CO)_2M(C_6H_5)_3$ | 0.2 | 1.3 | 1.3 |
| $C_9H_7MnM(C_6H_5)_3$ | 51.5 | 25.1 | 17.9 |
| $MnM(C_6H_5)_3$ | 0.1 | 1.3 | 1.3 |
| $M(C_6H_5)_3$ | 9.8 | 6.3 | 4.3 |
| $MnM(C_6H_5)_2$ | Trace | 2.8 | 1.6 |
| $C_{12}H_8MnM$ | 0.1 | | |
| $M(C_6H_5)_2$ | 4.1 | 4.0 | 4.6 |
| $C_{12}H_8M$ | 11.2 | 6.8 | 2.3 |
| $C_9H_7M$ | 5.6 | 2.5 | 0.9 |
| $MC_6H_5$ | 5.1 | 20.6 | 37.1 |
| $C_6H_5Mn$ | 5.6 | 7.7 | 8.2 |
| $Mn$ | 6.7 | 21.5 | 20.5 |

$CrSbH_3$ was reported to show a parent molecular ion and the ions $(CO)_nCrSbH_m{}^+$ for $n = 0$–4 and $m = 0$–3, $Cr^+$, and $SbH_3{}^+$. The compound $(CO)_4FeAsH_3$ behaved similarly, showing the ions $(CO)_nFeAsH_3{}^+$ ($n = 4$ or 3), $(CO)_nFeAsH_m{}^+$ ($n = 0$–2 and $m = 0$–3), $Fe^+$, and $^+AsH_3$.

### 19.3.2 Derivatives containing $R_3M$ groups

King and Efraty[378] have reported the mass spectra of the compounds ($\pi$-pyrrolyl)$Mn(CO)_2MPh_3$ and ($\pi$-indenyl)$Mn(CO)_2MPh_3$ for (M = P, As, Sb). Parent molecular ions were observed for all the compounds studied. Fragmentation by loss of CO groups and either the pyrrolyl or indenyl group was apparently favoured over the loss of $\cdot MPh_3$ or $\cdot Ph$ groups (Table 11.16). The rearrangement ion $C_6H_5Mn^+$ was observed in all spectra in reasonable abundance as also was the 9-heterofluorenyl ion, $C_{12}H_8M^+$ (M = P, As or Sb, see Chapter 8). The base peaks from the $\pi$-pyrrolyl compounds were the ions $MC_6H_5{}^+$ (for M = As or Sb) and $Mn^+$ (for M = P). Those of the $\pi$-indenyl compounds were $C_9H_7MnM(C_6H_5)_3{}^+$ (for M = P or As) and $MC_6H_5{}^+$ (for M = Sb). The following metastable supported decompositions were observed from $\pi$-$C_9H_7Mn(CO)_2AsPh_3$.

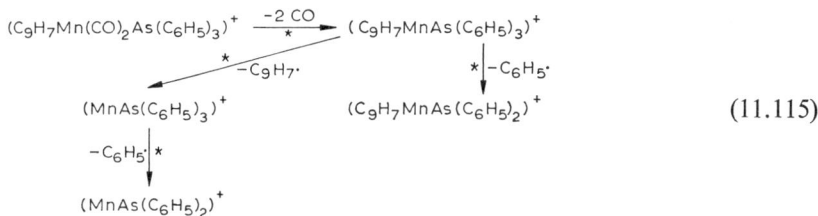

$$(C_9H_7Mn(CO)_2As(C_6H_5)_3)^+ \xrightarrow[\quad * \quad]{-2\,CO} (C_9H_7MnAs(C_6H_5)_3)^+$$

$$(MnAs(C_6H_5)_3)^+ \xleftarrow[\quad -C_9H_7\cdot \quad]{*} \qquad \qquad *\Big\downarrow -C_6H_5\cdot$$

$$-C_6H_5\Big\downarrow * \qquad \qquad (C_9H_7MnAs(C_6H_5)_2)^+$$

$$(MnAs(C_6H_5)_2)^+ \tag{11.115}$$

The compound $H_2Fe(CO)_4AsPh_3$ has been identified from its mass spectrum[379].

Several compounds of iron containing bidentate arsenic ligands have been characterised. Cullen et al. reported parent molecular ions from (diars)-$Fe_2(CO)_8$, (diars)$FeCO_3$ (ref. 380), (ffars)$Fe_3(CO)_{10}$, and (ffars)$Fe_3(CO)_9$ (ref. 381). The ffars ligand has the structure XCVI.

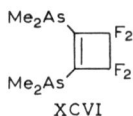

XCVI

Fragmentation of the two diars complexes was mainly by the loss of CO groups and $CH_3$ radicals, and elimination of $CH_4$. The ffars complexes lost CO, $\cdot F$, FeF, and $FeF_2$ entities. Two other arsenic–iron complexes, $C_6H_4(AsMe_2)CH_2Fe_2(CO)_8$ (ref. 380) and $As_2Me_2CH_2Fe_3(CO)_9$ (ref. 381) were reported to show parent molecular ions and some expected fragment ions.

### 19.4 Compounds containing other metals

Graham and coworkers[382] report that $Br_2GaMn(CO)_5$ gives a parent molecular ion in its mass spectrum. No further details are reported.

Several workers report the use of mass spectrometry in the characterisation of metal–selenium compounds. Abel *et al.*[383] studied compounds of the type $[M(CO)_4SeR]_2$ (for M = Mn or Re; R = Me, Et, or Ph) and $[M(CO)_3SeR]_n$ (for M = Mn or Re; R = Me or Ph). Fischer *et al.*[384] studied the compound

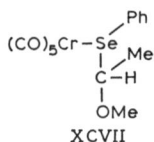

```
            Ph
           /
(CO)₅Cr−Se
          | /Me
          C−H
          |
          OMe
      XCVII
```

and found a parent molecular ion and fragment ions $(Cr–SeC_6H_5CH(CH_3)OCH_3)^+$, $(SeC_6H_5CH(CH_3)OCH_3)^+$, and $(CrSeC_6H_5)^+$.

### APPENDIX

*π-Cyclopentadienyl complexes*

| | |
|---|---|
| $(C_5H_5)_2Fe$ | F. W. McLafferty, *Anal. Chem.*, 28 (1956) 306. |
| $(C_5H_5)_4U$, $(C_5H_5)_3UCl$, $(C_5H_5)_3UBH_4$ | M. L. Anderson and L. R. Crisler, *J. Organometal. Chem.*, 17 (1969) 345. |
| $(C_5H_5)_4Rh_3H$ | O. S. Mills, E. O. Fischer, W. Wawersik and F. F. Paulus, *Chem. Commun.*, (1967) 643. |
| $(C_5H_5)_2MX_2$ (M = Mo, W) | M. L. H. Green and W. E. Lindsell, *J. Chem. Soc. A*, (1969) 2150. |

| | |
|---|---|
| $(C_5H_5)_2MCl_2$ (M = Ti, Zr, Hf) (negative ion spectra) | J. G. Dillard, *Inorg. Chem.*, 8 (1969) 1590. |
| $C_5H_5FeC_5H_4R$ | M. Cais, M. S. Lupin and J. Sharvit, *Israel J. Chem.*, 7 (1969) 73. |
| $C_5H_5FeC_5H_4R$ and $(C_5H_4R)_2Fe$ | M. I. Bruce, *Org. Mass. Spectrom.*, 2 (1969) 997. |
| $(C_5H_4-R-C_5H_4)Fe$ | T. H. Barr and W. E. Watts, *Tetrahedron*, 24 (1968) 3219. |
| $(C_9H_7)_2Ni$ | H. P. Fritz, F. H. Köhler and K. E. Schwarzhaus, *J. Organometal. Chem.*, 19 (1969) 449. |
| $[(C_5H_5)_2Ti(C\equiv CC_6H_5)]$ | J. H. Teuben and H. J. De Liefde Meijer, *J. Organometal. Chem.*, 17 (1969) 87. |
| $C_5H_5MoC_7H_7$ | E. O. Fischer and H. W. Wehner, *J. Organometal. Chem.*, 11 (1968) P29. |
| $(C_6F_5C_5H_4)(C_5H_5)MoH(C_6F_5)$ | M. L. H. Green and W. E. Lindsell, *J. Chem. Soc. A*, (1969) 2215. |
| $C_5H_5Fe(CO)C_7H_7$ | D. Ciappenelli and M. Rosenblum, *J. Am. Chem. Soc.*, 91 (1969) 6876. |
| $C_5H_5FeC_9H_9O_2$ | H. Egger and H. Falk, *Monatsh. Chem.*, 97 (1966) 1590. |
| $C_5H_5FeC_4H_3(CH_3)N$ 2-methylazaferrocene | K. Bauer, H. Falk and K. Schlögl, *Angew. Chem. Intern. Ed. Engl.*, 8 (1969) 135. |
| $(Azulene)Mn_2(CO)_6$ | P. S. Bird and M. R. Churchill, *Chem. Commun.*, (1968) 145. |
| $(6\text{-Methylfulvene})Fe_2(CO)_6$ | C. H. DePuy, V. M. Kobal and D. H. Gibson, *J. Organometal. Chem.*, 13 (1968) 266. |
| $C_5H_5Fe(CO)_2C_6F_4X$ | M. I. Bruce and C. H. Davies, *J. Chem. Soc. A*, (1969) 1077. |
| $C_5H_5Co(N_4R_2)$ | S. Otsuka and A. Nakamura, *Inorg. Chem.*, 7 (1968) 2542. |
| $C_5H_5(CO)_2Fe(CF_3)C=C(CF_3)\text{-}Sn(CH_3)_3$ | R. E. J. Bichler, M. R. Booth and H. C. Clark, *Inorg. Nucl. Chem. Letters*, 3 (1967) 71. |
| $C_5H_5V(CO_2CH_3)_2$ | R. B. King, *Inorg. Chem.*, 5 (1966) 2231. |
| $[C_5H_5Mo(NO)I_2]_2$ | R. B. King, *Inorg. Chem.*, 6 (1967) 30. |
| $C_5H_5Mo(CO)_3(CH_2)_3I$ | J. D. Hawthorne, M. J. Mays and R. N. |

$C_5H_5Fe(CO)_2CH_2C_6H_4CH_3$          F. Simpson, *J. Organometal. Chem.*, 12
                                        (1968) 407.
$C_5H_5Mn(CO)_2CNH$                     E. O. Fischer and R. J. J. Schneider, *J.*
                                        *Organometal. Chem.*, 12 (1968) P27.

*π-Arene complexes*

$(C_6H_6)_2Cr$                          C. Elschenbroich, *J. Organometal. Chem.*,
(partially deuterated)                  14 (1968) 157.
$C_6(CH_3)_6M(CO)_4$                    E. O. Fischer, W. Berngruber and C. G.
(M = Cr, Mo, W)                         Kreiter, *Chem. Ber.*, 101 (1968) 824.

*π-Olefin complexes*

$C_8H_{13}CoC_8H_{12}$                  S. Otsuka and M. Rossi, *J. Chem. Soc. A*,
                                        (1968) 2630.
$CF_3CCHCo_2(CO)_6$                     R. S. Dickson and D. B. W. Yawney,
                                        *Australian J. Chem.*, 20 (1967) 77.
$(C_3H_5)_2OPtCl_2$                     R. Jones, *J. Chem. Soc. A*, (1969) 2477.
$(C_8H_8)Ru_3(CO)_4$                    F. A. Cotton, A. Davison and A. Musco,
                                        *J. Am. Chem. Soc.*, 89 (1967) 6796.
$(C_8H_8XY)M(CO)_3$                     M. Green and D. C. Wood. *J. Chem. Soc.*
M = Ru, X = O, Y = C(CF_3)_2            *A*, (1969) 1172.
        X = C(CN)_2,
            Y = C(CF_3)_2
        X = Y = C(CN)CF_3
        X = Y = C(CN)_2
M = Fe, X = C(CN)_2,
            Y = C(CF_3)_2
$[(C_6H_5)_3P]_2Pt(C_8H_{12}acac)_2$    B. F. G. Johnson, T. Keating, J. Lewis,
$[(C_6H_5)_3P]_2Pt[C_8H_{12}(OCH_3)$-   M. S. Subramanian and D. A. White, *J.*
(acac)                                  *Chem. Soc. A*, (1969) 1793.
$(CH_3O_2C·CH{:}CH·CH{:}CH·CO_2$        B. L. Booth and R. G. Hargreaves, *J.*
$CH_3)Mn_2(CO)_6$                       *Chem. Soc. A*, (1969) 2766.

*Fluorocarbon complexes*

$R_fFe(CO)_3, R_fFe(CO)_4,$             W. R. Cullen, D. A. Harbourne, B. V.

| | |
|---|---|
| R$_f$Fe$_2$(CO)$_8$ | Liengme and J. R. Sams, *Inorg. Chem.*, 8 (1969) 1464. |
| *p*-FC$_6$H$_4$CH$_2$Mn(CO)$_5$ | J. D. Hawthorne, M. J. Mays and R. N. |
| C$_6$F$_5$CH$_2$Mn(CO)$_5$ | F. Simpson, *J. Organometal. Chem.*, 12 (1968) 407. |
| C$_2$F$_4$Ru(CO)$_2$(PR$_3$)$_2$ | M. Cooke and M. Green, *J. Chem. Soc. A*, (1969) 651. |
| C$_5$H$_5$Fe(CO)$_2$R$_f$ | M. I. Bruce, D. A. Harbourne, F. Waugh |
| R$_f$Re(CO)$_5$ | and F. G. A. Stone, *J. Chem. Soc. A*, (1968) 895. |

*Miscellaneous mixed ligand complexes*

| | |
|---|---|
| [(CH$_3$)$_2$N–CH$_2$–C(CH$_3$)=CH]-PtClP(C$_6$H$_5$)$_3$ | J. M. Kliegman and A. C. Cope, *J. Organometal. Chem.*, 16 (1969) 309. |
| [(C$_2$H$_5$)$_3$P]$_2$Fe(CO)$_2$SO$_2$ | R. Burt, M. Cooke and M. Green, *J.* |
| (CO)$_2$[(CH$_3$O)$_3$P]$_2$RuCl$_2$ | *Chem. Soc. A*, (1969) 2645. |
| Fe$_2$(CO)$_6$(RNS) | S. Otsuka, T. Yoshida and A. Nakamura, *Inorg. Chem.*, 7 (1968) 1833. |
| (C$_5$H$_8$N$_2$)Cr(CO)$_5$ | K. Öfele, *J. Organometal. Chem.*, 12 (1968) P 42. |
| (CH$_3$N)$_2$Fe$_3$(CO)$_9$ | M. Dekker and G. R. Knox, *Chem. Commun.*, (1967) 1243. |
| RMn(CO)$_5$ | M. J. Mays and R. N. F. Simpson, *J. Chem. Soc. A*, (1967) 1936. |
| (C$_5$H$_4$O$_2$)Fe(CO)$_3$ | M. Rosenblum and C. Gatsonis, *J. Am. Chem. Soc.*, 89 (1967) 5074. |
| (π-2-CH$_3$C$_3$B$_3$H$_5$)Mn(CO)$_3$ | J. H. Howard and R. N. Grimes, *J. Am. Chem. Soc.*, 91 (1969) 6499. |
| Co$_3$(CO)$_{10}$BH$_2$N(CH$_3$)$_3$ | F. Klanberg, W. B. Askew and L. J. Guggenberger, *Inorg. Chem.*, 7 (1968) 2265. |
| [C$_6$H$_5$(CH$_3$O)C=C(OCH$_3$)-C$_6$H$_5$]Cr(CO)$_3$ | E. O. Fischer, B. Heckl, K. H. Dötz, J. Müller and H. Werner, *J. Organometal. Chem.*, 16 (1969) P 29. |
| Pt$_2$Cl$_4$(C$_3$H$_6$)$_2$ | S. E. Binns, R. H. Cragg, R. D. Gillard, B. T. Heaton and M. F. Pilbrow, *J. Chem. Soc. A*, (1969) 1227. |

*Coordination compounds*

| | |
|---|---|
| Cu(acac)$_2$ | H. Junge, H. Musso and U. I. Zahorszky, *Chem. Ber.*, 101 (1968) 793. |
| M($\beta$-diketonate)$_3$ | J. R. Majer and R. Perry, *Chem. Commun.* (1969) 271. |
| (tfac)$_2$Cu($\gamma$-pic-$N$-oxide) | R. W. Kluiber and W. DeW. Horrocks, *Inorg. Chem.*, 6 (1967) 1427. |
| M[C$_6$H$_4$(NH)$_2$]$_2$ | A. L. Balch and R. H. Holm, *J. Am. Chem. Soc.*, 88 (1966) 5201. |

REFERENCES

1 R. C. Weast (Ed.), *Handbook of Chemistry and Physics*, The Chemical Rubber Publishing Co., Cleveland, Ohio, 49th edn., 1968–69.
2 E. Y. Zandberg, N. I. Inov and A. Y. Tontegode, *Zh. Tekhn. Fiz.*, 35 (1965) 1504.
3 M. B. Panish and L. Reif, *J. Chem. Phys.*, 38 (1963) 253.
4 O. C. Trulson and P. O. Schissel, *J. Less-Common Metals*, 8 (1965) 262.
5 K. A. Gingerich, *Naturwissenschaften*, 54 (1967) 43.
6 G. R. Belton and R. J. Fruehan, *J. Phys. Chem.*, 71 (1967) 1403.
7 J. Drowart and P. Goldfinger, *Angew. Chem. Intern. Ed. Engl.*, 6 (1967) 581.
8 A. Kant and B. Strauss, *J. Chem. Phys.*, 41 (1964) 3806.
9 A. Kant and B. Strauss, *J. Chem. Phys.*, 45 (1966) 3161.
10 A. Kant, *J. Chem. Phys.*, 48 (1968) 523.
11 A. Kant, *J. Chem. Phys.*, 41 (1964) 1872.
12 J. Drowart and R. E. Honig, *J. Chem. Phys.*, 25 (1956) 581.
13 J. Drowart and R. E. Honig, *J. Phys. Chem.*, 61 (1957) 980.
14 G. Verhaegen, F. E. Stafford and J. Drowart, *J. Chem. Phys.*, 33 (1960) 1784.
15 M. Ackerman, J. Drowart, F. E. Stafford and G. Verhaegen, *J. Chem. Phys.*, 36 (1962) 1557.
16 M. Ackerman, F. E. Stafford and G. Verhaegen, *J. Chem. Phys.*, 36 (1962) 1560.
17 K. A. Gingerich, *Chem. Commun.*, (1968) 1674.
18 A. Kant, *J. Chem. Phys.*, 44 (1966) 2450.
19 G. Mesnard, R. Uzan and B. Cabaud, *Rev. Phys. Appl.*, 1 (1966) 123.
20 H. Schäfer and K. Rinke, *Z. Naturforsch.*, 20b (1965) 702.
21 J. B. Berkowitz-Mattuck, A. Büchler, J. L. Engelke and S. N. Goldstein, *J. Chem. Phys.*, 39 (1963) 2722.
22 R. J. Ackerman and E. G. Rauk, *J. Phys. Chem.*, 67 (1963) 2596.
23 P. O. Schissel and O. C. Trulson, *J. Chem. Phys.*, 43 (1965) 737.
24 J. G. Dillard and R. W. Kiser, *J. Phys. Chem.*, 69 (1965) 3893.
25 J. H. Norman, H. G. Staley and W. E. Bell, *J. Chem. Phys.*, 42 (1965) 1123.
26 J. H. Norman, H. G. Staley and W. E. Bell, *J. Phys. Chem.*, 69 (1965) 1373.

27 J. H. Norman, H. G. Staley and W. E. Bell, *J. Phys. Chem.*, 71 (1967) 3686.

28 H. Wiedemeier and P. W. Gilles, *J. Chem. Phys.*, 42 (1965) 2765.

29 K. A. Gingerich, *J. Phys. Chem.*, 68 (1964) 2514.

30 R. W. Kiser, J. G. Dillard and D. L. Dugger, in *Advances in Chemistry Series*, R. G. Gould (Ed.), Vol. 72, Am. Chem. Soc., 1968, p. 153.

31 F. W. Aston, *Mass Spectra and Isotopes*, 2nd edn., Arnold, London, 1942, pp. 117ff.

32 F. W. Aston, *Phil. Mag.*, 45 (1923) 934; *Proc. Roy. Soc. (London), Ser. A*, 115 (1927) 487; 130 (1931) 302; 132 (1931) 487; 149 (1935) 396.

33 J. de Gier and P. Zeeman, *Proc. Roy. Acad. Amsterdam*, 38 (1935) 910; 38 (1935) 959; 39 (1935) 327.

34 A. J. Dempster, *Phys. Rev.*, 50 (1936) 98.

35 R. Baldock and J. R. Sites, *U.S. At. Energy Comm. Tech. Rept. Y-761*, Oak Ridge, Tenn., 1951; *Phys. Rev.*, 83 (1951) 488.

36 L. O. Gilpatrick and R. Baldock, *U.S. At. Energy Comm. Tech. Rept. Y-762*, Oak Ridge, Tenn., 1951.

37 F. I. Vilesov and B. L. Kurbatov, *Dokl. Akad. Nauk SSSR*, 140 (1961) 1364.

38 R. E. Winters and R. W. Kiser, *Inorg. Chem.*, 3 (1964) 699.

39 R. E. Winters and J. H. Collins, *J. Phys. Chem.*, 70 (1966) 2057.

40 R. E. Winters and R. W. Kiser, *J. Phys. Chem.*, 70 (1966) 1680.

41 S. M. Schildcrout, G. A. Pressley, Jr. and F. E. Stafford, *J. Am. Chem. Soc.*, 89 (1967) 1617.

42 R. E. Winters and R. W. Kiser, *Inorg. Chem.*, 4 (1965) 157.

43 A. Foffani, S. Pignataro, B. Cantone and F. Grasso, *Z. Physik. Chem. (Frankfurt)*, 45 (1965) 79.

44 B. Cantone, F. Grasso and S. Pignataro, *J. Chem. Phys.*, 44 (1966) 3115.

45 D. R. Bidinosti and N. S. McIntyre, *Can. J. Chem.*, 45 (1967) 641.

46 G. A. Junk and H. J. Svec, *Z. Naturforsch.*, 23b (1968) 1.

47 D. R. Lloyd and E. W. Schlag, *Inorg. Chem.*, 8 (1969) 2544.

48 S. Pignataro and F. P. Lossing, *J. Organometal. Chem.*, 11 (1968) 571.

49 H. J. Svec and G. A. Junk, *Inorg. Chem.*, 7 (1968) 1688.

50 A. Foffani, S. Pignataro, B. Cantone and F. Grasso, *Z. Physik. Chem. (Frankfurt)*, 42 (1964) 221.

51 R. W. Kiser, Mass Spectroscopy of Organometallic Compounds, in *Characterization of Organometallic Compounds*, M. Tsutsui (Ed.), Part I, Interscience, New York, 1969.

52 H. B. Gray and N. A. Beach, *J. Am. Chem. Soc.*, 85 (1963) 2922.

53 F. A. Cotton, A. K. Fischer and G. Wilkinson, *J. Am. Chem. Soc.*, 81 (1956) 800.

54 C. E. Moore, *Atomic Energy Levels*, Vol. 3, *Circ. 467*, Natl. Bur. Std., U.S. Govt. Printing Office, Washington, D.C., 1958.

55 D. R. Lloyd, private communication cited in refs. 41 and 51.

56 D. R. Lloyd, private communication cited in ref. 51.

57 D. A. Brown and R. M. Rawlinson, *J. Chem. Soc. A*, (1969) 1530.

58 J. W. Warren, *Nature*, 165 (1950) 810.

59 R. E. Honig, *J. Chem. Phys.*, 16 (1948) 105.

60 F. P. Lossing, A. W. Tickner and W. A. Bryce, *J. Chem. Phys.*, 19 (1951) 1254.

61 R. W. Kiser and E. J. Gallegos, *J. Phys. Chem.*, 66 (1962) 947.

62 D. T. Hurd, G. W. Sentell, Jr. and F. J. Norton, *J. Am. Chem. Soc.*, 71 (1949) 1899.

63 R. E. Winters and R. W. Kiser, *J. Phys. Chem.*, 69 (1965) 1618.

64 J. Lewis, A. R. Manning, J. R. Miller and J. M. Wilson, *J. Chem. Soc. A*, (1966) 1663.

65 B. F. G. Johnson, J. Lewis, I. G. Williams and J. M. Wilson, *J. Chem. Soc. A*, (1967) 341.

66 H. J. Svec and G. A. Junk, *J. Am. Chem. Soc.*, 89 (1967) 2836.

67 G. A. Junk and H. J. Svec, *J. Chem. Soc. A*, (1970) 2102.

68 D. R. Bidinosti, private communication cited in ref. 51.

69 M. I. Bruce, *Intern. J. Mass Spectrom. Ion Phys.*, 1 (1968) 141.

70 D. R. Bidinosti and N. S. McIntyre, *Chem. Commun.*, (1966) 555.

71 D. R. Bidinosti and N. S. McIntyre, *Can. J. Chem.*, 48 (1970) 593.

72 F. A. Cotton and R. R. Monchamp, *J. Chem. Soc.*, (1960) 533.

73 L. F. Dahl, E. Ishishi and R. E. Rundle, *J. Chem. Phys.*, 26 (1957) 1750.

74 J. M. Smith, K. Mehner and H. D. Kaesz, *J. Am. Chem. Soc.*, 89 (1967) 1759.

75 M. J. Mays and R. N. F. Simpson, *J. Chem. Soc. A*, (1968) 1444.

76 M. H. Chisholm, A. G. Massey and N. R. Thompson, *Nature*, 211 (1966) 67.

77 R. B. King, *Fortschr. Chem. Forsch.*, 14 (1970) 92.

78 D. R. Bidinosti and N. S. McIntyre, *Chem. Commun.*, (1967) 1.

79 K. Noack, *Spectrochim. Acta*, 19 (1963) 1925.

80 R. B. King, *J. Am. Chem. Soc.*, 88 (1966) 2075.

81 B. F. G. Johnson, J. Lewis, I. G. Williams and J. Wilson, *Chem. Commun.*, (1966) 391.

82 C. H. Wei and L. F. Dahl, *J. Am. Chem. Soc.*, 88 (1966) 1821.

83 D. B. W. Yawney and F. G. A. Stone, *Chem. Commun.*, (1968) 619.

84 D. B. W. Yawney and F. G. A. Stone, *J. Chem. Soc. A*, (1969) 502.

85 S. H. H. Chaston and F. G. A. Stone, *J. Chem. Soc. A*, (1969) 500.

86 S. Pignataro, A. Foffani, F. Grasso and B. Cantone, *Z. Physik. Chem. (Frankfurt)*, 47 (1965) 106.

87 R. E. Winters and R. W. Kiser, *J. Chem. Phys.*, 44 (1966) 1964.

88 R. E. Sullivan, M. S. Lupin and R. W. Kiser, *Chem. Commun.*, (1969) 655.

89 K. Edgar, B. F. G. Johnson, J. Lewis, I. G. Williams and J. M. Wilson, *J. Chem. Soc. A*, (1967) 379.

90 G. A. Junk, H. J. Svec and R. J. Angelici, *J. Am. Chem. Soc.*, 90 (1968) 5758.

91 B. F. G. Johnson, R. D. Johnston and J. Lewis, *J. Chem. Soc. A*, (1969) 792.

92 M. I. Bruce, M. Cooke, M. Green and D. J. Westlake, *J. Chem. Soc. A*, (1969) 987.

93 B. F. G. Johnson, J. Lewis and P. A. Kilty, *J. Chem. Soc. A*, (1968) 2859.

94 B. F. G. Johnson, R. D. Johnston, J. Lewis and B. H. Robinson, *Chem. Commun.*, (1966) 851.

95 B. F. G. Johnson, R. D. Johnston, J. Lewis and B. H. Robinson, *J. Organometal. Chem.*, 10 (1967) 105.

96 E. O. Fischer and R. Aumann, *J. Organometal. Chem.*, 8 (1967) P1.

97 J. W. S. Jamieson, J. V. Kingston and G. Wilkinson, *Chem. Commun.*, (1966) 569.

98 B. F. G. Johnson, R. D. Johnston, J. Lewis, B. H. Robinson and G. Wilkinson, *J. Chem. Soc. A*, (1968) 2856.

99 W. F. Edgell and W. M. Risen, *J. Am. Chem. Soc.*, 88 (1966) 5451.

100 B. F. G. Johnson, J. Lewis and P. W. Robinson, *J. Chem. Soc. A*, (1970) 1684.

101 R. B. King, *J. Am. Chem. Soc.*, 90 (1968) 1417.

102 E. O. Fischer and M. W. Schmidt, *Angew. Chem. Intern. Ed. Engl.*, 6 (1967) 93.

103 B. D. James, R. K. Nanda and M. G. H. Wallbridge, *Inorg. Chem.*, 6 (1967) 1979.
104 F. E. Saalfeld, M. V. McDowell, S. K. Gondal and A. G. Macdiarmid, *J. Am. Chem. Soc.*, 90 (1968) 3684.
105 J. M. Campbell and F. G. A. Stone, *Angew. Chem. Intern. Ed. Engl.*, 8 (1969) 140.
106 W. Fellmann and H. D. Kaesz, *Inorg. Nucl. Chem. Letters*, 2 (1966) 63.
107 D. K. Huggins, W. Fellmann, J. M. Smith and H. D. Kaesz, *J. Am. Chem. Soc.*, 86 (1964) 4841.
108 J. M. Smith, W. Fellmann and L. H. Jones, *Inorg. Chem.*, 4 (1965) 1361.
109 M. J. Mays and R. N. F. Simpson, *Chem. Commun.*, (1967) 1024.
110 A. Foffani, S. Pignataro, G. Distefano and G. Innorta, *J. Organometal. Chem.*, 7 (1967) 473.
111 R. E. Winters and R. W. Kiser, *J. Phys. Chem.*, 69 (1965) 3198.
112 H. Brunner, *J. Organometal. Chem.*, 12 (1968) 517.
113 H. Brunner, *J. Organometal. Chem.*, 14 (1968) 173.
114 E. O. Fischer, R. J. J. Schneider and J. Müller, *J. Organometal. Chem.*, 14 (1968)P4.
115 B. F. G. Johnson. J. Lewis, I. G. Williams and J. M. Wilson, *J. Chem. Soc. A*, (1967) 338.
116 H. L. Friedman, A. P. Irsa and G. Wilkinson, *J. Am. Chem. Soc.*, 77 (1955) 3689.
117 J. Müller and L. D'Or, *J. Organometal. Chem.*, 10 (1967) 313.
118 R. G. Denning and R. A. D. Wentworth, *J. Am. Chem. Soc.*, 88 (1966) 4619.
119 E. Schumacher and R. Taubenest, *Helv. Chim. Acta*, 47 (1964) 1525.
120 J. Müller, *Chem. Ber.*, 102 (1969) 152.
121 J. L. Thomas and R. G. Hayes, *J. Organometal. Chem.*, 23 (1970) 487.
122 F. Seel and V. Sperber, *J. Organometal. Chem.*, 14 (1968) 405.
123 R. B. King, *Appl. Spectry.*, 23 (1969) 148.
124 A. Mendelbaum and M. Cais, *Tetrahedron Letters*, (1964) 3847.
125 M. L. H. Green and T. Mole, *J. Organometal. Chem.*, 12 (1968) 404.
126 S. Pignataro and F. P. Lossing, *J. Organometal. Chem.*, 10 (1967) 531.
127 R. E. Winters, *Doctoral Dissertation*, Kansas State University, 1965.
128 R. B. King, *Can. J. Chem.*, 47 (1969) 559.
129 T. J. Katz, V. Balogh and J. Schulman, *J. Am. Chem. Soc.*, 90 (1968) 734.
130 T. J. Katz and J. J. Mrowca, *J. Am. Chem. Soc.*, 89 (1967) 1105.
131 R. I. Reed and F. M. Tabrizi, *Appl. Spectry.*, 17 (1963) 124.
132 H. Egger, *Monatsh. Chem.*, 97 (1966) 602.
133 D. T. Roberts, Jr., W. F. Little and M. M. Bursey, *J. Am. Chem. Soc.*, 89 (1967) 4917.
134 D. T. Roberts, Jr., W. F. Little and M. M. Bursey, *J. Am. Chem. Soc.*, 89 (1967) 6156.
135 D. T. Roberts, Jr., W. F. Little and M. M. Bursey, *J. Am. Chem. Soc.*, 90 (1968) 973.
136 H. Budzikiewicz, C. Djerassi and D. H. Williams, *Mass Spectrometry of Organic Compounds*, Holden-Day, San Francisco, 1967.
137 R. B. King and M. B. Bisnette, *J. Organometal. Chem.*, 8 (1967) 287.
138 B. L. Booth, R. N. Haszeldine and M. Hill, *J. Chem. Soc. A*, (1969) 1299.
139 B. L. Booth, R. N. Haszeldine and M. Hill, *J. Organometal. Chem.*, 16 (1969) 491.
140 I. M. T. Davidson, M. Jones and R. D. W. Kemmitt, *J. Organometal. Chem.*, 17 (1969) 169.
141 I. J. Spilners and J. G. Larson, *Org. Mass Spectrom.*, 3 (1970) 915.
142 N. Maoz, A. Mendelbaum and M. Cais, *Tetrahedron Letters*, (1965) 2087.

143 M. Cais, M. S. Lupin, N. Maoz and J. Sharvit, *J. Chem. Soc. A*, (1968) 3086.
144 E. W. Post, R. G. Cooks and J. C. Kotz, *Inorg. Chem.*, 9 (1970) 1670.
145 H. Egger and B. Falk, *Tetrahedron Letters*, (1966) 437.
146 R. S. P. Coutts and P. C. Wailes, *Australian J. Chem.*, 20 (1967) 1579.
147 R. E. Winters and R. W. Kiser, *J. Organometal. Chem.*, 4 (1965) 190.
148 R. B. King, *Chem. Commun.*, (1969) 436.
149 T. Blackmore, J. D. Cotton, M. I. Bruce and F. G. A. Stone, *J. Chem. Soc. A*,(1968) 2931.
150 E. Schumacher and R. Taubenest, *Helv. Chim. Acta*, 49 (1966) 1447.
151 M. Rausch, R. F. Kovar and C. S. Kraihanzel, *J. Am. Chem. Soc.*, 91 (1969) 1259.
152 K. K. Joshi, R. H. B. Mais, F. Nyman, P. G. Owston and A. M. Mood, *J. Chem. Soc. A*, (1968) 318.
153 R. B. King, *Inorg. Chem.*, 5 (1966) 2227.
154 B. F. Hallam and P. L. Pauson, *J. Chem. Soc.*, (1956) 3030.
155 M. I. Bruce, *Org. Mass Spectrom.*, 1 (1968) 687.
156 F. J. Preston and R. E. Reed, *Chem. Commun.*, (1966) 51.
157 R. B. King, *Org. Mass Spectrom.*, 2 (1969) 401.
158 H. Brunner and H. Wachsmann, *J. Organometal. Chem.*, 15 (1968) 409.
159 A. F. Reid, J. S. Shannon, J. M. Swan and P. C. Wailes, *Australian J. Chem.*, 18 (1965) 173.
160 J. G. Dillard and R. W. Kiser, *J. Organometal. Chem.*, 16 (1969) 265.
161 A. N. Nesmeyanov, V. A. Dubovitskii, O. V. Nogina and V. N. Bochkarev, *Dokl. Akad. Nauk SSSR*, 165 (1965) 125.
162 P. M. Druce, B. M. Kingston, M. F. Lappert, T. R. Spalding and R. C. Srivastava, *J. Chem. Soc. A*, (1969) 2106.
163 J. L. Burmeister, E. A. Deardorf, A. Jensen and V. Christiansen, *Inorg. Chem.*, 9 (1970) 58.
164 K. W. Egger, *J Organometal. Chem.*, 24 (1970) 501.
165 J. G. Dillard, *Inorg. Chem.*, 8 (1969) 2148.
166 J. Müller and P. Göser, *J. Organometal. Chem.*, 12 (1968) 163.
167 R. W. Kiser and M. A. Krassoi, results cited in ref. 51.
168 G. E. Herberich and J. Müller, *J. Organometal. Chem.*, 16 (1969) 111.
169 R. B. King, *Appl. Spectry.*, 23 (1969) 537.
170 M. M. Bursey, F. E. Tibbetts, III, W. F. Little, M. D. Rausch and G. A. Moser, *Tetrahedron Letters*, (1969), 3469.
171 D. M. Roe and A. G. Massey, *J. Organometal. Chem.*, 17 (1969) 429.
172 R. Prinz and H. Werner, *Angew. Chem. Intern. Ed. Engl.*, 6 (1967) 91.
173 J. Müller, *J. Organometal. Chem.*, 18 (1969) 321.
174 M. F. Rettig, C. D. Stout, A. Klug and P. Farnham, *J. Am. Chem. Soc.*, 92 (1970) 5100.
175 E. O. Fischer and H. W. Wehner, *J. Organometal. Chem.*, 11 (1968) P29.
176 R. B. King, *J. Organometal. Chem.*, 14 (1968) P19.
177 K. Yasufuku and H. Yamazaki, *Org. Mass Spectrom.*, 3 (1970) 23.
178 R. B. King, *Proc. First Intern. Symp. New Aspects Chem. Metal Carbonyls and Derivatives, Venice, Italy, September, 1968*, Paper E7; *Appl. Spectry.*, 23 (1969) 536.
179 R. B. King, *Org. Mass Spectrom.*, 2 (1969) 657.

180 R. B. King, *J. Am. Chem. Soc.*, 90 (1968) 1429.
181 M. A. Haas and J. M. Wilson, *J. Chem. Soc. B*, (1968) 104.
182 W. G. Dauben and M. E. Lorber, *Org. Mass Spectrom.*, 3 (1970) 211.
183 H. H. Hoehn, L. Pratt, K. F. Watterson and G. Wilkinson, *J. Chem. Soc.*, (1961) 2738.
184 J. Müller and M. Herberhold, *J. Organometal. Chem.*, 13 (1968) 399.
185 J. Müller and P. Göser, *Angew. Chem. Intern. Ed. Engl.*, 6 (1967) 364.
186 M. S. Lupin and M. Cais, *J. Chem. Soc. A*, (1968) 3095.
187 E. O. Fischer and M. W. Schmidt, *Chem. Ber.*, 100 (1967) 3782.
188 E. O. Fischer, A. Reckziegel, J. Müller and P. Göser, *J. Organometal. Chem.*, 11 (1968) P13.
189 G. F. Emerson, L. Watts and R. Pettit, *J. Am. Chem. Soc.*, 87 (1965) 131.
190 R. Bruce, K. Moseley and P. M. Maitlis, *Can. J. Chem.*, 45 (1967) 2011.
191 C. T. Sears and F. G. A. Stone, *J. Organometal. Chem.*, 11 (1968) 644.
192 M. I. Bruce and J. R. Knight, *J. Organometal. Chem.*, 12 (1968) 411.
193 M. I. Bruce, *Intern. J. Mass Spectrom. Ion Phys.*. 1 (1968) 335.
194 J. M. Landesberg and J. Sieczkowski, *J. Am. Chem. Soc.*, 90 (1968) 1655.
195 E. Koerner von Gustorf, M. C. Henry and D. J. McAdoo, *Ann. Chem.*, 707 (1967) 190.
196 S. Otsuka, A. Nakamura and T. Yoshida, *Bull. Chem. Soc. Japan*, 40 (1967) 1266.
197 M. I. Bruce, *Intern. J. Mass Spectrom. Ion Phys.*, 2 (1969) 349.
198 R. B. King, *J. Organometal. Chem.*, 8 (1967) 139.
199 J. K. Becconsall, B. E. Job and S. O'Brien, *J. Chem. Soc. A*, (1967) 423.
200 J. K. Becconsall and S. O'Brien, *Chem. Commun.*, (1966) 720.
201 S. Otsuka, A. Nakamura and K. Tani, *J. Chem. Soc. A*, (1968) 2248.
202 S. O'Brien, *Chem. Commun.*, (1968) 757.
203 H. G. Ang and J. M. Miller, *Chem. Ind. (London)*, (1966) 944.
204 J. M. Miller, *J. Chem. Soc. A*, (1967) 828.
205 M. I. Bruce, *Org. Mass Spectrom.*, 1 (1968) 503.
206 R. C. Dobbie, M. Green and F. G. A. Stone, *J. Chem. Soc. A*, (1969) 1881.
207 J. Müller and J. A. Connor, *Chem. Ber.*, 102 (1969) 1148.
208 R. B. King, *J. Am. Chem. Soc.*, 89 (1967) 6368.
209 R. B. King, *J. Am. Chem. Soc.*, 90 (1968) 1429.
210 R. B. King, *Appl. Spectry.*, 23 (1969) 137.
211 M. I. Bruce, *Org. Mass Spectrom.*, 1 (1968) 835.
212 M. I. Bruce, *Org. Mass Spectrom.*, 2 (1969) 63.
213 R. B. King and F. T. Korenowski, *Chem. Commun.*, (1966) 771.
214 M. J. Mays and R. N. F. Simpson, *J. Chem. Soc. A*, (1967) 1936.
215 J. Müller and K. Fenderl, *J. Organometal. Chem.*, 19 (1969) 123.
216 M. I. Bruce, *J. Organometal. Chem.*, 10 (1967) 495.
217 M. I. Bruce, *Inorg. Nucl. Chem. Letters*, 3 (1967) 157.
218 M. Green, A. Taunton-Rigby and F. G. A. Stone, *J. Chem. Soc. A*, (1969) 1875.
219 M. I. Bruce, *J. Organometal. Chem.*, 10 (1967) 95.
220 W. Hieber and H. Beutner, *Z. Anorg. Allgem. Chem.*, 317 (1962) 63.
221 V. Frey, W. Hieber and O. S. Mills, *Z. Naturforsch.*, 23b (1968) 105.
222 L. F. Dahl, W. R. Costello and R. B. King, *J. Am. Chem. Soc.*, 90 (1968) 5422.

223 S. Otsuka, T. Yoshida and A. Nakamura, *Inorg. Chem.*, 8 (1969) 2514.
224 W. T. Flannigan, G. R. Knox and P. L. Pauson, *Chem. Ind. (London)*, (1967) 1094.
225 J. A. Jarvis, B. E. Job, B. T. Kilburn, R. H. B. Mais, P. G. Owston and P. F. Todd, *Chem. Commun.*, (1967) 1149.
226 R. B. King, *Org. Mass Spectrom.*, 2 (1969) 387.
227 R. B. King and K. H. Pannell, *J. Am. Chem. Soc.*, 90 (1968) 3984.
228 M. M. Bagga, W. T. Flannigan, G. R. Knox, P. L. Pauson, F. J. Preston and R. I. Reed, *J. Chem. Soc. C*, (1968) 36.
229 F. J. Preston, W. T. Flannigan, G. R. Knox, P. L. Pauson and R. I. Reed, *Intern. J. Mass Spectrom. Ion Phys.*, 3 (1969) 63.
230 F. J. Preston, *Intern. J. Mass Spectrom. Ion Phys.*, 2 (1969) 391.
231 F. Klanberg and E. L. Muetterties, *J. Am. Chem. Soc.*, 90 (1968) 3296.
232 E. O. Fischer, E. Louis, W. Bathelt, E. Moser and J. Müller, *J. Organometal. Chem.*, 14 (1968) P9.
233 E. O. Fischer, W. Bathelt, M. Herberhold and J. Müller, *Angew. Chem. Intern. Ed. Engl.*, 7 (1968) 634.
234 S. Pignataro, A. Foffani, G. Innorta and G. Distefano, in *Mass Spectrometry, Proc. Conf. Mass Spectrometry, Berlin, 1967*, E. Kendrick (Ed.), Vol. 4, Inst. Petroleum, London, 1968, p. 323.
235 P. S. Braterman, *J. Organometal. Chem.*, 11 (1968) 198.
236 M. Cooke and M. Green, *J. Chem. Soc. A*, (1969) 651.
237 Y. Wada and R. W. Kiser, *J. Phys. Chem.*, 68 (1964) 2290.
238 D. H. Williams, R. S. Ward and R. G. Cooks, *J. Am. Chem. Soc.*, 90 (1968) 966.
239 M. Green, A. Taunton-Rigby and F. G. A. Stone, *J. Chem. Soc. A*, (1969) 1875.
240 H. G. Ang and B. O. West, *Australian J. Chem.*, 20 (1967) 1133.
241 R. B. King, *J. Am. Chem. Soc.*, 90 (1968) 1412.
242 R. B. King and T. F. Korenowski, *J. Organometal. Chem.*, 17 (1969) 95.
243 R. W. Kiser, M. A. Krassoi and R. J. Clark, *J. Am. Chem. Soc.*, 89 (1967) 3653.
244 G. Distefano and G. Innorta, *J. Organometal. Chem.*, 14 (1968) 465.
245 G. Distefano, G. Innorta, S. Pignataro and A. Foffani, *J. Organometal. Chem.*, 14 (1968) 165.
246 W. D. Horrocks, Jr. and R. C. Taylor, *Inorg. Chem.*, 2 (1963) 723.
247 W. Beck and K. Lottes, *Z. Naturforsch.*, 19b (1964) 987.
248 G. Innorta, G. Distefano and S. Pignataro, *Intern. J. Mass Spectrom. Ion Phys.*, 1 (1968) 435.
249 H. G. Ang and J. M. Miller, *Chem. Ind. (London)*, (1966) 944.
250 J. M. Miller, *J. Chem. Soc. A*, (1967) 828.
251 B. F. G. Johnson, J. Lewis, J. M. Wilson and D. T. Thompson, *J. Chem. Soc. A*, (1967) 1445.
252 N. W. Smallwood and L. W. Daasch, presented at ASTM Committee E-14, *15th Annual Conference on Mass Spectrometry and Allied Topics, Denver, Colorado, 1967*, Paper No. 154, p. 500.
253 B. E. Job, R. A. N. McLean and D. T. Thompson, *Chem. Commun.*, (1966) 895.
254 M. Ahmad, G. R. Knox, F. J. Preston and R. I. Reed, *Chem. Commun.*, (1967) 138.
255 R. B. King and M. B. Bisnette, *Inorg. Chem.*, 6 (1967) 469.

256 B. F. G. Johnson, P. J. Pollick, I. G. Williams and A. Wojcicki, *Inorg. Chem.*, 7 (1968) 831.

257 R. B. King and C. A. Eggers, *Inorg. Chem.*, 7 (1968) 1214.

258 E. Kostiner, M. L. N. Reddy, D. S. Urch and A. G. Massey, *J. Organometal. Chem.*, 15 (1968) 383.

259 C. R. Crooks, B. F. G. Johnson, J. Lewis and I. C. Williams, *J. Chem. Soc. A*, (1969) 797.

260 K. Farmery and M. Kilmer, *J. Organometal. Chem.*, 17 (1969) 127.

261 S. R. Smith, R. A. Krause and G. O. Dudek, *J. Inorg. Nucl. Chem.*, 29 (1967) 1533.

262 R. B. King and N. Welcman, *Inorg. Chem.*, 8 (1969) 2540.

263 F. J. Preston and R. I. Reed, *Org. Mass Spectrom.*, 1 (1968) 71.

264 R. B. King and R. N. Kapoor, *Inorg. Chem.*, 8 (1969) 2535.

265 B. F. G. Johnson, R. D. Johnston and J. Lewis, *J. Chem. Soc. A*, (1968) 2865.

266 B. H. Robinson and W. S. Tham, *J. Chem. Soc. A*, (1968) 1784.

267 F. W. Gowling, S. F. A. Kettle and G. N. Sharples, *Chem. Commun.*, (1968) 21.

268 H. O. Oven and H. J. de Liefde Meijer, *J. Organometal. Chem.*, 19 (1969) 373.

269 M. J. Frazer, W. E. Newton and B. Rimmer, *Chem. Commun.*, (1968) 1336.

270 E. P. Dudek, E. Chaffee and G. Dudek, *Inorg. Chem.*, 7 (1968) 1257.

271 J. R. Majer, M. J. A. Reade and W. I. Stephen, *Talanta*, 15 (1968) 373.

272 A. E. Jenkins and J. R. Majer, *Talanta*, 14 (1967) 777.

273 A. E. Jenkins, J. R. Majer, and M. J. A. Reade, *Talanta*, 14 (1967) 1213.

274 B. R. Kowalski, T. L. Isenhour and R. E. Sievers, *Anal. Chem.*, 41 (1969) 998.

275 H. C. Hill and R. I. Reed, *Tetrahedron*, 20 (1964) 1359.

276 J. H. Beynon, R. A. Saunders and A. E. Williams, *Appl. Spectry.*, 17 (1963) 63.

277 M. Starke and R. Tümmler, *Z. Chem.*, 7 (1967) 433.

278 P. Day, H. A. O. Hill and M. G. Price, *J. Chem. Soc. A*, (1968) 90.

279 H. A. O. Hill and M. M. Norgett, *J. Chem. Soc. A*, (1966) 1476.

280 D. G. Whitten, K. E. Bentley and D. Kuwada, *J. Org. Chem.*, 31 (1966) 322.

281 A. Hood, E. G. Carlson and M. J. O'Neal, in *Encyclopedia of Spectroscopy*, G. L. Clark (Ed.), Vol. I, Reinhold, New York, 1960, p. 616.

282 A. P. Johnson, P. Wehrli, R. Fletcher and A. Eschenmoser, *Angew. Chem. Intern. Ed. Engl.*, 7 (1968) 623.

283 J. Seibl, *Org. Mass Spectrom.*, 1 (1968) 215.

284 L. T. Taylor, F. L. Urbach and D. H. Busch, *J. Am. Chem. Soc.*, 91 (1969) 1072.

285 K. G. Das, D. N. Seu and N. Thankarajan, *Tetrahedron Letters*, (1968) 869.

286 S. Otsuka and A. Nakamura, *Inorg. Chem.*, 7 (1968) 2542.

287 T. Kruck and A. Prasch, *Z. Anorg. Allgem. Chem.*, 356 (1968) 118.

288 T. Kruck, M. Hofler, K. Baur, P. Junkes and K. Glinka, *Chem. Ber.*, 101 (1968) 3827.

289 J. M. Campbell and F. G. A. Stone, *Angew. Chem. Intern. Ed. Engl.*, 8 (1969) 140.

290 J. K. Becconsall, B. E. Job and S. O'Brien, *J. Chem. Soc. A*, (1967) 423.

291 F. W. McLafferty, *Appl. Spectry.*, 11 (1957) 148.

292 C. G. Macdonald and J. S. Shannon, *Australian J. Chem.*, 19 (1966) 1545.

293 J. S. Shannon and J. M. Swan, *Chem. Commun.*, (1965) 33.

294 G. M. Bancroft, C. Reichert, J. B. Westmore and H. D. Gesser, *Inorg. Chem.*, 8 (1969) 474.

295 R. D. Koob, M. L. Morris, A. L. Clobes, L. P. Hills and J. H. Futrell, *Chem. Commun.* (1969) 1177.
296 R. E. Fraas, R. W. Kiser and G. L. Chaney, *Org. Mass Spectrom.*, 2 (1969) 1171.
297 G. M. Bancroft, C. Reichert and J. B. Westmore, *Inorg. Chem.*, 7 (1968) 870.
298 S. Sasaki, Y. Itagaki, T. Kurokawa, K. Nakanishi and A. Kasahara, *Bull. Chem. Soc. Japan*, 40 (1967) 76.
299 C. Reichert, J. B. Westmore and H. D. Gesser, *Chem. Commun.*, (1967) 782.
300 A. L. Clobes, M. L. Morris and R. D. Koob, *J. Am. Chem. Soc.*, 91 (1969) 3087.
301 H. F. Holtzclaw, Jr., R. L. Lintvedt, H. E. Baumgarten, R. G. Parker, M. M. Bursey and P. F. Rogerson, *J. Am. Chem. Soc.*, 91 (1969) 3774.
302 M. J. Lacey, C. G. Macdonald and J. S. Shannon, *Org. Mass Spectrom.*, 1 (1968) 115.
303 J. D. McDonald and J. L. Margrave, *J. Less-Common Metals*, 14 (1968) 236.
304 B. F. G. Johnson, J. Lewis and M. S. Subramanian, *J. Chem. Soc. A*, (1968) 1993.
305 C. Reichert, C. M. Bancroft and J. B. Westmore, *Can. J. Chem.*, 48 (1970) 1362.
306 S. C. Chattoraj, C. T. Lynch and K. S. Mazdiyarni, *Inorg. Chem.*, 7 (1968) 2501.
307 S. J. Lippard, *J. Am. Chem. Soc.*, 88 (1966) 4300.
308 J. L. Booker, T. L. Isenhour and R. E. Sievers, *Anal. Chem.*, 41 (1969) 1705.
309 J. R. Majer and R. Perry, *Chem. Commun.*, (1969) 454.
310 J. Macklin and G. Dudek, *Inorg. Nucl. Chem. Letters*, 2 (1966) 403.
311 R. Belcher, J. R. Majer, R. Perry and W. I. Stephen, *Anal. Chim. Acta*, 45 (1969) 305.
312 D. W. Barnum, *J. Inorg. Nucl. Chem.*, 21 (1961) 221; 22 (1961) 183.
313 M. M. Bursey and P. F. Rogerson, *Inorg. Chem.*, 9 (1970) 676.
314 S. M. Schildcrout, R. G. Pearson and F. E. Stafford, *J. Am. Chem. Soc.*, 90 (1968) 4006.
315 C. Reichert and J. B. Westmore, *Inorg. Chem.*, 8 (1969) 1012.
316 G. A. Heath and R. L. Martin, *Chem. Commun.*, (1969) 951.
317 E. P. Dudek and M. Barber, *Inorg. Chem.*, 5 (1966) 375.
318 J. Charalambous, M. J. Frazer and F. B. Taylor, *J. Chem. Soc. A*, (1969) 2787.
319 J. R. Majer and R. Perry, *J. Chem. Soc. A*, (1970) 822.
320 J. G. Vogel and B. G. Hobrock, *153rd A.C.S. Meeting, Miami Beach, Florida, April, 1967.*
321 W. L. Mead, W. K. Reid and H. B. Silver, *Chem. Commun.*, (1968) 573.
322 C. Reichert, D. K. C. Fung, D. C. K. Lin and J. B. Westmore, *Chem. Commun.*, (1968) 1094.
323 A. van den Bergen, K. S. Murray, M. J. O'Connor, N. Rehak and B. O. West, *Australian J. Chem.*, 21 (1968) 1505.
324 A. van den Bergen, K. S. Murray, M. J. O'Connor and B. O. West, *Australian J. Chem.*, 22 (1969) 39.
325 P. J. McCarthy, R. J. Hovey, K. Ueno and A. E. Martell, *J. Am. Chem. Soc.*, 77 (1955) 5820; K. Ueno and A. E. Martell, *J. Phys. Chem.*, 59 (1955) 998; 61 (1957) 257; P. J. McCarthy and A. E. Martell, *J. Am. Chem. Soc.*, 78 (1956) 264.
326 S. H. H. Chaston, S. E. Livingstone, T. N. Lockyer and J. S. Shannon, *Australian J. Chem.*, 18 (1965) 1539.
327 H. Budzikiewicz and E. Plöger, *Org. Mass Spectrom.*, 3 (1970) 709.
328 P. R. Scherer and Q. Fernando, *Anal. Chem.*, 40 (1968) 1938.
329 D. Betteridge and D. John, *Talanta*, 15 (1968) 1227.

330 S. M. Bloom and G. O. Dudek, *Inorg. Nucl. Chem. Letters*, 2 (1966) 183.
331 A. L. Balch, I. G. Dance and L. H. Holm, *J. Am. Chem. Soc.*, 90 (1968) 1139.
332 J. P. Fackler, Jr., D. Coucouvanis, J. A. Fetchin and W. C. Seidel, *J. Am. Chem. Soc.*, 90 (1968) 2784.
333 R. S. P. Coutts and P. C. Wailes, *Inorg. Nucl. Chem. Letters*, 3 (1967) 1.
334 A. Garrick and F. Glockling, *J. Chem. Soc. A*, (1967) 40.
335 B. J. Aylett and J. M. Campbell, *J. Chem. Soc. A*, (1969) 1910.
336 K. M. Mackay and R. D. George, *Inorg. Nucl. Chem. Letters*, 6 (1970) 289.
337 B. J. Aylett and J. M. Campbell, *J. Chem. Soc. A*, (1969) 1916.
338 K. M. MacKay and R. D. George, *Inorg. Nucl. Chem. Letters*, 5 (1969) 797.
339 K. M. MacKay and S. R. Stobart, *Inorg. Nucl. Chem. Letters*, 6 (1970) 687.
340 S. R. Stobart, *J. Chem. Soc. A*, (1970) 999.
341 F. E. Saalfeld, M. V. McDowell and A. G. MacDiarmid, *J. Am. Chem. Soc.*, 92 (1970) 2324.
342 R. B. King, K. H. Pannell, C. R. Bennett and M. Ishaq, *J. Organometal. Chem.*, 19 (1969) 327.
343 D. J. Cardin, S. A. Keppie, M. F. Lappert, M. R. Litzow and T. R. Spalding, *J. Chem. Soc. A*, (1971) 2262.
344 R. B. King, *Org. Mass Spectrom.*, 2 (1969) 657.
345 A. Carrick and F. Glockling, *J. Chem. Soc. A*, (1968) 913.
346 M. Gielen and G. Mayence, *J. Organometal. Chem.*, 12 (1968) 363.
347 Y. L. Baay and A. G. MacDiarmid, *Inorg. Chem.*, 8 (1968) 986.
348 F. Glockling and K. A. Hooton, *J. Chem. Soc. A*, (1967) 1066.
349 F. Glockling and M. D. Wilbey, *J. Chem. Soc. A*, (1968) 2168.
350 B. M. Kingston and M. F. Lappert, *Inorg. Nucl. Chem. Letters*, 4 (1968) 371.
351 B. M. Kingston, M. F. Lappert, M. R. Litzow and T. R. Spalding, unpublished results, 1969.
352 J. Lewis, A. R. Manning, J. R. Miller and J. M. Wilson, *J. Chem. Soc. A*, (1968) 1663.
353 M. F. Mays and R. N. F. Simpson, *J. Chem. Soc. A*, (1967) 1936.
354 J. D. Cotton, S. A. R. Knox and F. G. A. Stone, *Chem. Commun.*, (1967) 965.
355 S. A. R. Knox and F. G. A. Stone, *J. Chem. Soc. A*, (1969) 2559.
356 S. A. R. Knox and F. G. A. Stone, *J. Chem. Soc. A*, (1970) 3147.
357 E. H. Brooks, R. J. Cross and F. Glockling, *Inorg. Chim. Acta*, 2 (1968) 17.
358 R. R. Schrieke and B. O. West, *Australian J. Chem.*, 22 (1969) 49.
359 J. A. J. Thompson and W. A. G. Graham, *Inorg. Chem.*, 6 (1967) 1365.
360 D. J. Patmore and W. A. G. Graham, *Inorg. Nucl. Chem. Letters*, 2 (1966) 179.
361 D. J. Patmore and W. A. G. Graham, *Inorg. Chem.*, 5 (1966) 2222.
362 R. Ball, M. J. Bennett, E. H. Brooks, W. A. G. Graham, J. Hoyano and S. M. Illingworth, *J. Chem. Soc. D*, (1970) 592.
363 D. J. Patmore and W. A. G. Graham, *Inorg. Chem.*, 7 (1968) 771.
364 F. E. Saalfeld, M. V. McDowell, A. P. Hagen and A. G. MacDiarmid, *Inorg. Chem.*, 7 (1968) 1665.
365 J. D. McDonald, C. H. Williams, J. C. Thompson and J. L. Margrave, in *Advances in Chemistry Series*, J. L. Margrave (Ed.), Vol. 72, Am. Chem. Soc., 1968, p. 261.
366 D. R. Bidinosti and N. S. McIntyre, *Chem. Commun.*, (1967) 1.
367 R. R. Schrieke and B. O. West, *Inorg. Nucl. Chem. Letters*, 5 (1969) 141.

368 R. Kummer and W. A. G. Graham, *Inorg. Chem.*, 7 (1968) 523.

369 R. Kummer and W. A. G. Graham, *Inorg. Chem.*, 7 (1968) 1208.

370 M. J. Bennett, W. Brooks, M. Elder, W. A. G. Graham, D. Hall and R. Kummer, *J. Am. Chem. Soc.*, 92 (1970) 208.

371 F. E. Saalfeld, M. V. McDowell, S. K. Gondal and A. G. MacDiarmid, *Inorg. Chem.*, 7 (1968) 1465.

372 D. J. Patmore and W. A. G. Graham, *Chem. Commun.*, (1967) 7.

373 J. M. Burlitch and A. Ferrari, *Inorg. Chem.*, 9 (1970) 563.

374 F. Seel and G.-V. Rosenthaler, *Z. Anorg. Allgem. Chem.*, 373 (1970) 182.

375 M. J. Mays and J. D. Robb, *J. Chem. Soc. A*, (1968) 329.

376 E. O. Fischer, W. Bathelt, M. Heberhold and J. Müller, *Angew. Chem. Intern. Ed. Engl.*, 7 (1968) 634.

377 E. O. Fischer, W. Bathelt and J. Müller, *Chem. Ber.*, 103 (1970) 1815.

378 R. B. King and A. Efraty, *Org. Mass Spectrom.*, 3 (1970) 1227.

379 K. Farmery and M. Kilner, *J. Chem. Soc. A*, (1970) 634.

380 W. R. Cullen and D. A. Harbourne, *Can. J. Chem.*, 47 (1969) 3371.

381 W. R. Cullen, D. A. Harbourne, B. V. Liengme and J. R. Sams, *Inorg. Chem.*, 9 (1970) 702.

382 J. Hoyano, D. J. Patmore and W. A. G. Graham, *Inorg. Nucl. Chem. Letters*, 4 (1968) 201.

383 E. W. Abel, B. C. Crosse and G. V. Hutson, *J. Chem. Soc. A*, (1967) 2014.

384 E. O. Fischer and V. Kiener, *Angew. Chem. Intern. Ed. Engl.*, 6 (1967) 961.

385 H. D. Kaesz, S. A. R. Knox, J. W. Koepke and R. B. Saillant, *Chem. Commun.*, (1971) 477.

*Chapter 12*

# The Rare Gases

M. R. LITZOW

## 1. INTRODUCTION

Members of the Group are frequently used to calibrate the energy scale during ionisation and appearance potential measurements since they are convenient and easy to handle and their ionisation potentials are accurately known. Although many compounds have been prepared and studied since 1962, they are still few in number compared with those of other Groups and, in general, they are of simple composition. The mass spectra of these compounds, therefore, are usually straight-forward.

## 2. THE ELEMENTS

The rare gases are all polyisotopic[1] (Table 12.1, Fig. 12.1).

Table 12.1

Natural Abundances of the Rare Gas Isotopes[1]

| Element | Mass no. | % Abundance | Element | Mass no. | % Abundance | Element | Mass no. | % Abundance |
|---------|----------|-------------|---------|----------|-------------|---------|----------|-------------|
| He | 3 | 0.00013 | Kr | 78 | 0.35 | Xe | 124 | 0.096 |
|    | 4 | 99.99987 |    | 80 | 2.27 |    | 126 | 0.090 |
|    |   |          |    | 82 | 11.56 |    | 128 | 1.92 |
| Ne | 20 | 90.92 |    | 83 | 11.55 |    | 129 | 26.44 |
|    | 21 | 0.26 |    | 84 | 56.90 |    | 130 | 4.08 |
|    | 22 | 8.82 |    | 86 | 17.37 |    | 131 | 21.18 |
| Ar | 36 | 0.337 |    |   |   |    | 132 | 26.89 |
|    | 38 | 0.063 |    |   |   |    | 134 | 10.44 |
|    | 40 | 99.60 |    |   |   |    | 136 | 8.87 |

*References pp. 615–616*

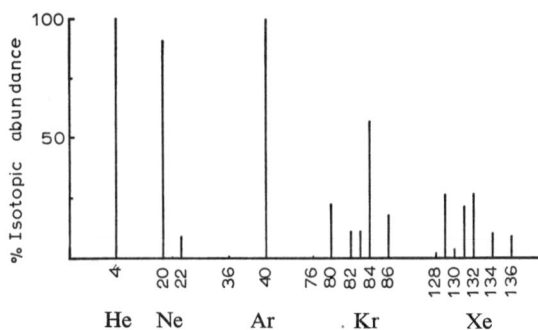

Fig. 12.1. Natural isotopic abundances of the rare gases (ref. 1).

## 3. MASS SPECTRAL STUDIES

Melton[2] studied the multiple ionisation of argon and krypton and detected the ions $Ar^{n+}$ and $Kr^{n+}$ ($n = 1$–8); the formation of triply and quadruply charged argon ions by electron impact has also been investigated by Fiquet-Fayard[3] and Fiquet-Fayard and Lahmani[4]. The homonuclear[5-7] and heteronuclear[6-10] diatomic ions of all the the rare gases except radon have been studied by mass spectrometry and appearance potentials of all except the $HeXe^+$ ion measured at pressures where three-body processes could be neglected.

A number of ion–molecule reactions involving inert gases has been studied. Kaul et al.[11] discussed the mechanisms and energetics of the reactions between rare gases and hydrogen; the appearance potentials, ionisation potentials, and dissociation energies of the hydrides HeH, NeH, ArH, KrH, and KrD were determined by these workers. The reactions of helium and hydrogen in the ion source of a mass spectrometer were also studied by Hertzberg et al.[12] and by Koch and Friedman[13]. The latter workers showed that practically all of the $HeH^+$ was formed by the reaction of $H_2^+$ and He rather than by the reaction of $H_2$ and $He^+$. Other ion–molecule investigations[14,15] have also included work on the $H_2$–He system. The ionic reaction of $CH_4^+$ and Xe was found to result in the products $Xe^+$, $XeH^+$, and $XeCH_3^+$, while that of $Xe^+$ and $CH_4$ produced $CH_3^+$, $CH_4^+$, and $XeCH_x^+$ ($x = $ 0–4)[16].

The mass spectra of both xenon trioxide[17] and tetroxide[17,18] have been reported. All possible ions $XeO_n^+$ were observed in each case but no evidence of formation of lower xenon oxides such as XeO or $XeO_2$ in the mass spectrometer was found. The xenon fluorides similarly form all possible ions

Table 12.2

Appearance Potentials of Ions Derived from $XeF_4$ and $XeF_2$

| Ion | Appearance potential (eV) | | |
|---|---|---|---|
| | $XeF_4$ (ref. 25) | $XeF_2$ (ref. 25) | $XeF_2$ (ref. 26) |
| $Xe_2F_5^+$ | $13.8 \pm 0.2$ | | |
| $XeF_4^+$ | $12.9 \pm 0.1$ | | |
| $XeF_3^+$ | $13.1 \pm 0.1$ | | |
| $XeF_2^+$ | $14.9 \pm 0.1$ | $12.6 \pm 0.1$ | 12.28 |
| $XeF^+$ | $13.3 \pm 0.1$ | $13.3 \pm 0.1$ | 12.78 |
| $Xe^+$ | $12.4 \pm 0.1$ | $12.0 \pm 0.1$ | |

$XeF_n^+$ in the mass spectrometer[19-23]. The negative ion spectrum of $XeF_6$ has been examined in some detail and the energetics investigated[24]. Ions observed were $F^-$, $F_2^-$, and $XeF_n^-$ ($n = 1$–4). The major negative ion peak in the spectrum of $XeF_4$ was $XeF_2^-$; $XeF_4^-$ was not observed[24].

Appearance potentials of positive ions derived from both $XeF_4$ and $XeF_2$ have been determined by Svec and Flesch[25] (Table 12.2). Standard heats of formation for the gaseous compounds and the average strength of the bonds in the molecules were then calculated. The experimental values were found to be compatible with theoretical estimates of these quantities and with figures obtained by calorimetric measurements (Table 12.3). Morrison et al.[26] recently reported a photoionisation study of xenon difluoride and observed that Xe was being formed in the ion source chamber by decomposition of the compound. This was undoubtedly happening during the measurements of Svec and Flesch also since the appearance pontential of the $Xe^+$ ion was the same as the value they obtained for the ionisation potential of free xenon. The appearance potential values of Morrison et al. for the ions $XeF_2^+$ and $XeF^+$ were considerably lower than those obtained by Svec and Flesch.

Mass spectra of the oxyfluorides $XeO_2F_2$ (refs. 22, 30), $XeO_3F_2$ (ref. 31), and $XeOF_4$ (ref. 22) have also been reported. The following fragmentation pattern was derived for xenon dioxide difluoride[30].

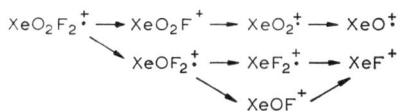

$$XeO_2F_2^+ \longrightarrow XeO_2F^+ \longrightarrow XeO_2^+ \longrightarrow XeO^+$$
$$\searrow XeOF_2^+ \longrightarrow XeF_2^+ \longrightarrow XeF^+$$
$$\searrow XeOF^+ \nearrow$$

Table 12.3

Heats of Formation for $XeF_4$ and $XeF_2$ and Mean Bond Dissociation Energies in These Compounds (kcal.mole$^{-1}$)

| $\Delta H_f(XeF_4)_{(g)}$ | $\Delta H_f(XeF_2)_{(g)}$ | $D(Xe-F)_{XeF_4}$ [a] | $D(Xe-F)_{1,2 XeF_4}$ [b] | $D(Xe-F)_{3,4 XeF_4}$ [c] | $D(Xe-F)_{XeF_2}$ [d] | Ref. |
|---|---|---|---|---|---|---|
| − 53 ± 5 | −37 ± 10 | 32 ± 2 | 26 ± 3 | 38 ± 7 | 39 ± 10 | 25 |
| − 50 | | 31 | | | | 27 |
| − 45 | | 31 | | | | 28 |
| − 54.8 | | 31.5 | | | | 29 |

[a] Average bond energy in $XeF_4$.
[b] Average bond energy for the first two Xe–F bonds to break in the dissociation of $XeF_4$.
[c] Average bond energy for the third and fourth Xe–F bonds to break in the dissociation of $XeF_4$.
[d] Average bond energy in $XeF_2$.

The negative ion mass spectra of $XeO_2F_2$ (ref. 30) and $XeOF_4$ (ref. 22) have been observed qualitatively. Ions observed were $XeF^-$, $XeF_2^-$, and $XeOF^-$ from the former compound and $XeF^-$, $XeF_2^-$, $XeF_3^-$, $XeF_4^-$, and $XeOF_3^-$ from the latter.

On the basis of mass spectral evidence, an unstable solid has been tentatively assigned the composition xenon tetrakistrifluoroacetate; peaks were ascribed[32] to $XeFCO_2^+$, $XeFCO^+$, and $XeCO^+$.

## REFERENCES

1  R. C. Weast (Ed.), *Handbook of Chemistry and Physics*, The Chemical Rubber Publishing Co., Cleveland, Ohio, 49th edn., 1968–69.

2  C. E. Melton, *J. Chem. Phys.*, 37 (1962) 562.

3  F. Fiquet-Fayard, *J. Chim. Phys.*, 59 (1962) 439.

4  F. Fiquet-Fayard and M. Lahmani, *J. Chim. Phys.*, 59 (1962) 1050.

5  J. A. Hornbeck and J. P. Molnar, *Phys. Rev.*, 84 (1951) 621.

6  W. Kaul and R. Taubert, *Z. Naturforsch.*, 17a (1962) 88.

7  M. S. B. Munson, J. L. Franklin and F. H. Field, *J. Phys. Chem.*, 67 (1963) 1542.

8  M. Pahl and V. Weimer, *Naturwissenschaften*, 44 (1957) 487; *Z. Naturforsch.*, 12a (1957) 926.

9  R. Fuchs and W. Kaul, *Z. Naturforsch.*, 15a (1960) 108.

10  W. Kaul, V. Lauterbach and R. Fuchs, *Naturwissenschaften*, 47 (1960) 353.

11  W. Kaul, V. Lauterbach and R. Taubert, *Z. Naturforsch.*, 16a (1961) 624.

12  M. Hertzberg, D. Rapp, I. B. Ortenburger and D. D. Briglia, *J. Chem. Phys.*, 34 (1961) 343.

13  H. von Koch and L. Friedman, *J. Chem. Phys.*, 38 (1963) 1115.

14  M. Pahl, *Ergeb. Exakt. Naturw.*, 35 (1962) 182.

15  C. F. Giese and W. B. Maier, III, *J. Chem. Phys.*, 35 (1962) 1913.

16  T. O. Tiernan and P. S. Gill, *J. Chem. Phys.*, 50 (1969) 5042.

17  M. H. Studier and J. L. Huston, *J. Phys. Chem.*, 71 (1967) 457.

18  J. L. Huston, M. H. Studier and E. N. Sloth, *Science*, 143 (1964) 1161.

19  C. L. Chernick, H. H. Claasen, P. R. Fields, H. H. Hyman, J. G. Malm, W. G. Manning, M. S. Matheson, L. A. Quarterman, F. Schreiner, H. H. Selig, I. Sheft, S. Siegel, E. N. Sloth, L. Stein, M. H. Studier, J. L. Weeks and M. H. Zirin, *Science*, 138 (1962) 136.

20  D. F. Smith, *J. Chem. Phys.*, 38 (1963) 270.

21  M. H. Studier and E. N. Sloth, *J. Phys. Chem.*, 67 (1963) 925.

22  M. H. Studier and E. N. Sloth, in *Noble-Gas Compounds*, H. H. Hyman (Ed.), University of Chicago Press, Chicago, 1963, p. 47.

23  J. Marsel and V. Vrscaj, *Croat. Chem. Acta*, 34 (1962) 191.

24  G. M. Begun and R. N. Compton, *J. Chem. Phys.*, 51 (1969) 2367.

25  H. J. Svec and G. D. Flesch, *Science*, 142 (1963) 954.

26  J. D. Morrison, A. J. C. Nicholson and T. A. O'Donnell, *J. Chem. Phys.*, 49 (1968) 959.

27 S. R. Gunn and S. M. Williamson, *Science*, 140 (1963) 177.
28 S. R. Gunn, *Conference on Noble Gas Compounds, Argonne National Laboratories, Argonne, Ill., April, 1963*.
29 L. Stein and P. Plurien, *Conference on Noble Gas Compounds, Argonne National Laboratories, Argonne, Ill., April, 1963*.
30 J. L. Huston, *J. Phys. Chem.*, 71 (1967) 3339.
31 J. L. Huston, *Inorg. Nucl. Chem. Letters*, 4 (1968) 29.
32 A. Iskraut, R. Taubenest and E. Schumacher, *Chimia (Aarau)*, 18 (1964) 188.

# Subject Index